Molecular Evolutionary Genetics

MAX K. HECHT
Queens College of the
City University of New York
Flushing, New York

BRUCE WALLACE
Virginia Polytechnic Institute
and State University
Blacksburg, Virginia

GHILLEAN T. PRANCE
New York Botanical Garden
Bronx, New York

MACROMOLECULAR SEQUENCES IN SYSTEMATIC AND EVOLUTIONARY BIOLOGY
Edited by Morris Goodman

EVOLUTIONARY GENETICS OF FISHES
Edited by Bruce J. Turner

[redacted]ICROORGANISMS AS MODEL SYSTEMS FOR STUDYING [redacted]UTION
E[redacted]obert P. Mortlock

MOLECU[redacted]UTIONARY GENETICS
Edited by Ro[redacted]tyre

Molecular Evolutionary Genetics

Edited by

Ross J. MacIntyre

Cornell University
Ithaca, New York

PLENUM PRESS • NEW YORK AND LONDON

Library of Congress Cataloging in Publication Data

Main entry under title:

Molecular evolutionary genetics.

(Monographs in evolutionary biology)
Includes bibliographies and index.
1. Chemical Evolution. 2. Molecular genetics. I. MacIntyre, Ross J. II. Series.
HQ371.M76 1985 574.87'3224 85-19312

ISBN-13: 978-1-4684-4990-7 e-ISBN-13: 978-1-4684-4988-4
DOI: 10.1007/978-1-4684-4988-4

Softcover reprint of the hardcover 1st edition 1985
A Division of Plenum Publishing Corporation
233 Spring Street, New York, N.Y. 10013

Contributors

Raoul E. Benveniste Laboratory of Viral Carcinogenesis, National Cancer Institute, Frederick, Maryland 21701

Wesley M. Brown Division of Biological Sciences and Museum of Zoology, University of Michigan, Ann Arbor, Michigan 48109

Susan A. Gerbi Division of Biology and Medicine, Brown University, Providence, Rhode Island 02912

Wen-Hsiung Li Center for Demographic and Population Genetics, University of Texas Health Science Center, Houston, Texas 77030

Chi-Cheng Luo Center for Demographic and Population Genetics, University of Texas Health Science Center, Houston, Texas 77030

George L. Gabor Miklos Department of Population Biology, and Center for Recombinant DNA Research, Research School of Biological Sciences, Australian National University, Canberra, A. C. T. 2600, Australia

Stephen J. O'Brien Section of Genetics, National Cancer Institute, Frederick, Maryland 21701

Jeffrey D. Palmer Division of Biological Sciences, University of Michigan, Ann Arbor, Michigan 48109

Carl W. Schmid Department of Chemistry, University of California, Davis, California 95616

Hector N. Seuanez Department of Genetics, Universidade Federal do Rio de Janeiro, Rio de Janeiro, Brazil

Che-Kun James Shen Department of Genetics, University of California, Davis, California 95616

James E. Womack Department of Veterinary Pathology, College of Veterinary Medicine, Texas A&M University, College Station, Texas 77843

Chung-I Wu Center for Demographic and Population Genetics, University of Texas Health Science Center, Houston, Texas 77030

Preface

This volume in the Monographs in Evolutionary Biology series addresses issues that are part of an emerging area of research loosely called "molecular evolution." Its practitioners include both molecular biologists curious about the evolutionary implications of their data and evolutionary biologists pushing their analyses to the molecular level. The union of these fields of molecular and organismal biology has been turbulent at times, and, as shall be seen, this dialectic has led to some very serious challenges to long-held notions about the role of natural selection in evolution and the economy of genome organization in eukaryotes. As an inevitable outgrowth of molecular biology, molecular evolution is necessarily a young discipline, but it can already point proudly to two major discoveries. The first, is the molecular clock, a concept that has emerged from the analysis of at least four data sets—amino acid sequences, immunologic data, DNA renaturation studies, and, recently, analyses of DNA sequences. The reality of a strong stochastic component in the evolution of nucleotide sequences can no longer be doubted, although the accuracy of the clock with regard to particular sequences and within particular groups of organisms should be independently measured each time it is used. Nevertheless, molecular clocks will assume increasingly important roles in phylogenetic reconstructions, especially since the fossil record is so fragmentary.

The second major discovery of molecular evolution has been the incredible complexity of the eukaryotic genome. A hint that the genome of higher organisms is not a tightly knit bundle of carefully regulated and essential genes came first from the so-called *C*-value paradox. The first measurements of genome sizes of eukaryotes indicated that there was simply too much DNA, and the large differences in the relative amounts failed to correlate with the morphological or behavioral complexities of the organisms. The Cot curves resulting from DNA renaturation studies defined the components of the genome that varied so paradoxically, but in a sense just deepened the mystery; e.g., why should an organism need

10^6–10^7 copies of a simple repeated sequence? Recombinant DNA technology has subsequently allowed us to analyze small, defined regions of the genome, but the paradoxes remain even when the interspersion patterns of unique and repeated sequences are analyzed and compared. In fact, new paradoxes continue to arise; e.g., pseudogenes and transposons. Structural genes themselves, which were thought to be understood fairly well in the early 1970s, turn out to be mosaics with introns, exons, and flanking regions each evolving at a different rate and presumably under different selective constraints.

The dialectic between the two main discoveries of molecular evolution demands a fundamental reanalysis of the traditional tenets of the neo-Darwinian view of evolution. No longer is it useful to think that every structure, sequence, or nucleotide must have a function, meshing it into a well-oiled, smoothly running biological machine. As many authors have recently pointed out, natural selection is a tinkerer, a discarder of outmoded parts. Molecular evolutionary studies are indicating that this is true to an extent that must surprise even the most dedicated neutralist. This volume presents abundant evidence that genomes are littered with discarded sequences that perhaps never had any functional importance and sequences that are invaders from other evolving lineages. Yet even against this bewildering background of genomic noise, natural selection is at work. Indeed, as is also shown in this volume, molecular evolutionary investigations are discovering some of the clearest and most elegant examples of the effectiveness of natural selection.

The primary focus of this volume is the second important discovery of molecular evolution, that eukaryotic genomes are surprisingly complex. Yet the question of sequence evolution is also treated in many of the chapters. In fact, one of the most powerful ways to understand the evolution of genomes is to examine the tempo and mode of the changes in their component nucleotide sequences. The authors were charged to write *interpretive reviews* of their topics. Thus, the chapters include relevant and often extensive data, which not only are essential for the subsequent interpretations, but also serve as important sources of information for others who may want to reexamine the raw data with different analytical procedures or extract some of the data for related investigations. The authors' interpretations involve two components: first, the conclusions that can be safely drawn from the information, and second, the questions now raised by those conclusions. Specifically, these questions point out how we have to rethink our current ideas about both gene and genome evolution. Below I provide an overview of the volume, briefly summarizing for each chapter the kind of information presented, the conclusions drawn, and the questions raised. In doing so, I hope it will be apparent that there is a certain logical order to these chapters despite the fact that there is an inevitable dilemma in this regard. The very complexity of the genes and genomes that these authors so admirably review and discuss makes a neat and sequential textbook presentation of the topics a rather unrealistic goal. However, by proceeding from simple organellar to complex nuclear genomes, I hope the reader will see not only just how much tinkering has gone on in genome evolution, but also an undercurrent of common themes with regard to evolutionary forces.

The initial chapter, by Wen-Hsiung Li, Chi-Cheng Luo, and Chung-I Wu, raises several of these common themes. It is also, in a sense, a methods section, stressing current and refined ways to estimate nucleotide sequence divergence. Data presented include evolutionary rates of change of the various bases within codons, of introns, and of 5′ and 3′ flanking sequences. Also reviewed are rates of pseudogene and duplicate gene evolution and the ever increasing information on codon usage in different genes and in different organisms. The data allow these authors to reevaluate the molecular clock and the effects of different generation times on rates of nucleotide sequence divergence. Furthermore, patterns of codon usage appear to be predominately influenced by tRNA abundance and codon–anticodon binding efficiencies. In this first chapter there is a thorough analysis of the concerted evolution of repeated sequences, including genetic mechanisms thought to be important and rates of sequence homogenization. This phenomenon, which is also documented and discussed by most of the subsequent authors in the volume, raises significant questions about the spread of advantageous mutations within the genome and especially about current estimates of the rates of duplicate gene divergence. Despite the wealth of information they present, Li *et al.* point out the paucity of information about nucleotide sequence divergence in most invertebrates and in the entire plant kingdom.

Wesley Brown next reviews the evolution of a relatively simple and compact genetic system, the animal mitochondrial genome. A rapid rate of sequence divergence and a transitional bias contrast dramatically with a virtual absence of rearrangement in the animal mitochondrial genome. Here also is the first of several examples to be found in the volume of coevolution at the molecular level, in this case the apparent coordinate evolution of the mitochondrial cytochrome oxidase II gene with the nuclear-encoded cytochrome *c* gene. Once again, however, the data raise as many questions as they answer; e.g., how can the rate of change of animal mitochondrial genes be so rapid if this rate is in part due to simultaneous selection pressures on nuclear genes?

Jeffrey Palmer expands the analysis to the substantially more complex chloroplast and plant mitochondrial genomes. An extensive review of gene content, gene arrangement, and sequence divergence in these larger organellar genomes reveals some striking similarities and differences between them and the animal mitochondrial genome. Compactness is clearly not a necessary condition for an organellar genome, as this chapter so convincingly demonstrates. Whereas animal mitochondrial and chloroplast genomes are structurally very stable, this is not true of the generally larger plant mitochondrial genome. Furthermore, there is no correlation between rates of gene evolution and rates of chromosome repatterning within organellar genomes; the evolutionary processes appear to be essentially uncoupled. Like Brown in the preceding chapter, Palmer raises many questions about the forces stabilizing the gene arrangements in these chromosomes. Finally, he points out how ignorant we are about the mechanisms behind the surprisingly frequent transfer of genes between nuclear and organellar genomes as well as between mitochondrial and chloroplast chromosomes.

Next George Miklos initiates a series of chapters devoted to the nuclear genome and its complexity. His analysis concerns the simplest of the repeated sequences, satellite DNA. He exhaustively reviews the organization and the variation in these sequences in a host of different organisms. Emerging from this review is an indisputable conclusion: selection at the level of the individual cannot explain the features of satellite DNA nor even its presence in the genome. Such DNA sequences are clearly not important in chromosome pairing, nor are they likely to be involved in chromosome rearrangements on an evolutionary time scale. Are they, then, selfish DNAs, which happen to have some replicative advantage over other genomic sequences, or is their role to suppress recombination in critical areas of the genome? Also, what are the processes involved in the extensive homogenization of these sequences; i.e., in their concerted, yet essentially independent evolution within different lineages?

The next look at the repeated sequences in the nuclear genome comes in the chapter by Carl Schmid and Che-Kung James Shen. Their comparisons focus on the highly repeated, interspersed, and so-called "retroinserted" *Alu* and *Kpn* sequences in primate and rodent genomes. These are reverse transcripts of processed mRNAs that randomly (?) find their way back into the genome. Superimposed on this ongoing process are the usual undefined processes of concerted evolution, and there are several dramatic examples of sequence homogenization in their review. These authors point out how the divergence between the two copies of the repeat flanking an insertion compared with the divergence of the insert itself (from a consensus sequence) allows one to infer the relative dates of insertion and homogenization. If the molecular clock for these sequences can be set with a high degree of accuracy, rates of sequence homogenization, at least in the mammalian genome, might eventually be determined. There is a larger issue here, however, for if these sequences serve no functional role, how can the genome tolerate such a load (up to 20%) of "selfish" or "ignorant" DNA? Perhaps their role as polymorphic elements may be to increase genetic stability in regions containing duplicate genes. The formal population genetics of potentially selfish DNA sequences that have bonus effects such as the stabilization of different regions of the genome needs to be explored.

Raoul Benveniste reviews still another class of repeated sequences in the next chapter. Endogenous retroviruses are not as highly repeated as the *Alu* sequences, and, unlike the latter, may well have a selective impact on their host, with concomitant selective constraints on their divergence. Furthermore, these are clearly intruders into the genome and as such are transmitted horizontally as well as vertically in evolution. Benveniste provides a thorough taxonomy of these sequences and documents an impressive number of instances of horizontal transmission of these virogenes. He also shows how they have helped us understand certain aspects of primate and felid evolution. Indeed, since their rate of change is generally more rapid than the unique sequences of their host genomes, they can serve as useful markers for establishing relationships between recently diverged species. It is clear that without molecular biology these sequences would never have been discovered; the question thus arises, is our ability to measure horizontal transmission limited only by the number of probes we currently have at our disposal? Another unanswered question is the extent to which endogenous retroviruses will serve as paradigms for understanding the evolutionary significance of transposons in general.

Susan Gerbi in her review considers the ribosomal DNA genes, a class of middle repetitive sequences with a profound functional role. Her extensive phylogenetic survey from bacteria to higher eukaryotes includes for each group the organization of the rDNA repeat unit, interspecific sequence divergence, mechanisms of transcript processing, and the presumed secondary structures of the processed transcripts. From these analyses comes excellent evidence for the coevolution of sequences within the rRNA transcripts, as well as impressive examples of structure and function at the molecular level. Here also we see the apparently random forces of concerted evolution at work in the context of extreme functional constraints. How is this interplay orchestrated; e.g., does it involve a reliance on one particular or several different homogenization mechanism(s)? Intriguing questions are also raised about the evolution of coordinate control in different lineages once the proto 5S genes were separated during evolution from the genes coding for the larger ribosomal RNAs.

The final chapter, by Stephen O'Brien, Hector Seuanez, and James Womack, contains an overview of genome evolution in the mammals as detected at the cytogenetic level. This larger picture of chromosomal repatterning is a fitting final verse for this volume on molecular evolution, since it gives one a feeling for the magnitude of the job of eventually understanding these events and processes at the molecular level. O'Brien *et al.* trace chromosome homologies primarily via banding patterns and syntenic linkage relationships. Chromosome reorganization in different mammalian orders runs the gamut from very little (primates, cats) to profound (canids, cervids). In the process of this analysis, some linkage relationships are discovered that have remained intact for at least 100 million years. They also provide an extensive catalogue of the chromosomal locations of satellite sequences, ribosomal RNA genes, endogenous retroviruses, and oncogenes and assess the relative merits of these different sequences as chromosomal markers for evolutionary analyses. As in organellar genomes, sequence and structural evolution are uncoupled, with chromosomal repatterning occurring in punctuated bursts during the descent of the various mammalian groups. Once again, a new generation of questions emerges from their data. Have these bursts of chromosomal change been associated with extensive speciation events, or have they been important in the morphological and/or physiological adaptations so evident in the mammalian orders? What is the role of natural selection in the preservation of those linkage relationships that have persisted in all the lineages, including those that have undergone such extensive chromosomal reshuffling?

In summary, these chapters provide data sets that have been instrumental in the development of molecular evolution as an important component of evolutionary biology. These data have been used to determine the existence and the accuracy of molecular clocks and have helped us to at least begin to understand the dynamic processes of genome evolution. The authors have also assessed the data as evolutionary biologists, for, as George Miklos points out, "such data in isolation from relevant biological contexts are as impotent as biological data without DNA sequences." Thus, they have drawn the legitimate conclusions and, when possible, posed the intriguing and important questions inherent in the data themselves.

Where do we, as molecular and evolutionary biologists interested in genome evolution, go now? I suggest that we will go where the genome itself, with all its complexity, leads us. In many ways we are like children in an enchanted forest, wandering almost aimlessly from discovery to discovery. For the moment, at least, that should be sufficient. At some point we will inevitably emerge into a clearing where principles and patterns in the organization and evolution of the genome are evident. Until then, let us be thankful that the pleasures of the forest are so numerous and diverse. In retrospect, the neo-Darwinian view of evolution was unpleasantly austere.

Ross J. MacIntyre

Ithaca, New York

Contents

Chapter 1

Evolution of DNA Sequences

Wen-Hsiung Li, Chi-Cheng Luo, and Chung-I Wu

Chapter 2

The Mitochondrial Genome of Animals

Wesley M. Brown

Chapter 3

Evolution of Chloroplast and Mitochondrial DNA in Plants and Algae

Jeffrey D. Palmer

Chapter 4

Localized Highly Repetitive DNA Sequences in Vertebrate and Invertebrate Genomes

George L. Gabor Miklos

Chapter 5

The Evolution of Interspersed Repetitive DNA Sequences in Mammals and Other Vertebrates

Carl W. Schmid and Che-Kun James Shen

Chapter 6

The Contributions of Retroviruses to the Study of Mammalian Evolution

Raoul E. Benveniste

Chapter 7

Evolution of Ribosomal DNA

Susan A. Gerbi

Chapter 8

On the Evolution of Genome Organization in Mammals

Stephen J. O'Brien, Hector N. Seuanez, and James E. Womack

CHAPTER 1

Evolution of DNA Sequences

WEN-HSIUNG LI, CHI-CHENG LUO, and CHUNG-I WU

1. Introduction

The advent of restriction enzyme analysis, gene cloning, and DNA sequencing techniques has brought an explosion of knowledge of molecular biology and also a new period of exciting development in the study of molecular evolution. Application of these techniques has led to the discovery of many novel features of eukaryotic genes and genomes and to a rapid accumulation of data on DNA sequences. It is now well known that a typical eukaryotic gene consists of amino acid-coding regions, intervening sequences (introns), 5′ and 3′ (transcribed) untranslated regions, and 5′ and 3′ flanking (untranscribed) regions (see, e.g., Minty and Newmark, 1980). It is also well known that the genome of higher organisms contains many pseudogenes, multigene families, repetitive DNA sequences, and transposons. These new discoveries and the availability of a large number of DNA sequences have greatly facilitated the study of molecular evolution and considerable progress has already been made in the comparative study of DNA sequences. In this chapter we review this progress.

Comparative study of DNA sequences requires mathematical models for estimating the number of nucleotide substitutions between homologous sequences. Until very recently the most frequently used model was the one by Jukes and Cantor (1969). This model assumes random substitution among the four types of nucleotide; however, studies using protein sequences and hemoglobin variants (Vogel and Kopun, 1977, and references

WEN-HSIUNG LI, CHI-CHENG LUO, AND CHUNG-I WU • Center for Demographic and Population Genetics, University of Texas Health Science Center, Houston, Texas 77030.

therein) have indicated that this assumption is not true. It is therefore desirable to replace this assumption by a more realistic one and a number of new models have been proposed (e.g., Holmquist and Pearl, 1980; Kimura, 1980, 1981a; Kaplan and Risko, 1982; Gojobori *et al.*, 1982a). Methods have also been proposed for evaluating the extent of sequence divergence in coding regions in terms of synonymous and nonsynonymous (amino acid-changing) substitutions (Kimura, 1980; Miyata and Yasunga, 1980; Perler *et al.*, 1980; Li *et al.*, 1985). Furthermore, simulation studies (Takahata and Kimura, 1981; Gojobori *et al.*, 1982a; Gojobori, 1983; Tajima and Nei, 1984) have been conducted to evaluate the performance of some of the methods proposed.

An advantage of DNA sequence data over protein sequence data is that the former allow a more detailed examination of the process of gene evolution. At the DNA level the extent of sequence divergence can be evaluated not only in coding regions, but also in noncoding regions, such as introns and untranslated regions. Moreover, in coding regions the extent of sequence divergence can be expressed in terms of two types of substitution: synonymous and nonsynonymous. Such detailed studies will lead to a better understanding of the relationship between functional constraint and rate of nucleotide substitution and help to identify DNA regions important for gene function. Comparison of DNA sequences from diverse organisms has shown that the rate of nucleotide substitution varies considerably with the DNA region examined and that the more important the function of the DNA region, the lower the rate of substitution (e.g., Kimura, 1977; Miyata *et al.*, 1980; Efstratiadis *et al.*, 1980). This inverse relationship between the stringency of functional constraint and the rate of substitution was further strengthened by the observation that pseudogenes, which are apparently subject to no functional constraint, evolve at the highest rate (Miyata and Ysaunaga, 1981; Li *et al.*, 1981). The availability of a large number of sequences allows a comparison of the substitution rates in many different kinds of genes (Li *et al.*, 1985).

The observation that the rate of amino acid substitution in hemoglobin was nearly constant among different evolutionary lineages led to the proposal of the molecular clock hypothesis (Zuckerkandl and Pauling, 1965). This proposal has generated a great deal of controversy (Goodman, 1976, 1981a; Wilson *et al.*, 1977; Kimura, 1979, 1981b), particularly after the rate constancy was taken as evidence supporting the neutral mutation hypothesis (Kimura, 1969). Although this hypothesis is now accepted by a large number of molecular evolutionists [see the reviews by Wilson *et al.* (1977); and Kimura (1983a)], it is also severely criticized by some authors (e.g., Goodman, 1981a; Czelusniak *et al.*, 1982). Since this is an important hypothesis in molecular evolution, it is desirable to see if DNA sequence data support it. We have recently conducted such a study (Wu and Li, 1985).

Molecular evolutionists (e.g., Beale and Lehmann, 1965; Fitch, 1967) have long been interested in knowing the pattern of spontaneous (point) mutation, for it can serve as a standard for inferring how far the observed frequencies of interchanges between amino acids have deviated from the values expected under selective neutrality [see the review by Doolittle (1979)]. In inferring this pattern, previous authors (Fowler *et al.*, 1974; Vogel and Kopun, 1977, and references therein) have relied on revertants of mutants with known base pair changes and hemoglobin variants. We have pointed out some drawbacks of using such data and have proposed to use pseudogene sequences, because pseudogenes are apparently not subject to functional constraints, so that the pattern of nucleotide substitution in them should reflect the pattern of spontaneous mutation (Gojobori *et al.*, 1982b; Li *et al.*, 1984). Our analyses have clearly indicated that mutation occurs nonrandomly. We have also inferred the pattern of nucleotide substitution in some functional genes and found some conspicuous differences between this pattern and that estimated from pseudogenes (Gojobori *et al.*, 1982b).

Comparative studies of DNA sequences have raised a number of new issues in the study of molecular evolution. One is why the usage of synonymous codons is distinctly nonrandom in both prokaryotic and eukaryotic genes (Grantham *et al.*, 1980a,b, 1981, and references therein). This question has become a hot issue for two reasons. First, synonymous mutations have been proposed as candidates for selectively neutral mutations (Kimura, 1968b; King and Jukes, 1969). If all synonymous mutations are indeed selectively neutral, the synonymous codons for an amino acid should be used more or less equally frequently. For this reason, nonrandom usage has often been taken as evidence against the neutral mutation hypothesis (see, e.g., Grantham, 1890; Modiano *et al.*, 1980). Second, it is difficult to know the causes of nonrandom usage, and numerous hypotheses have been proposed (Grosjean *et al.*, 1978; Grantham *et al.*, 1980b; Pieczenik, 1980; Ikemura, 1981a,b, 1982; Sharp *et al.*, 1985, and references therein). Fortunately, a clear answer to this question is emerging, since the studies by Ikemura (1981a,b, 1982), Grosjean and Fiers (1982), Bennetzen and Hall (1982), and others strongly suggest that tRNA availability and codon-anticodon interaction are the two major factors.

Another new issue is the concerted evolution of multigene families; i.e., the members of a gene family evolve together in a concerted manner, so that a high homogeneity is maintained among members within a species, despite the rapid divergence that may occur between members of different species. This concept was first proposed to explain the evolution of repetitive DNA, immunoglobulin genes, and ribosomal rRNA genes (Britten and Kohne, 1968; Edelman and Gally, 1970; Brown *et al.*, 1972), but the phenomenon was later found to be general [see the reviews by Hood *et al.* (1975), Ohta (1980), Long and Dawid (1980), and Arnheim (1983)]. Many models have been proposed to explain this phenomenon, including the master-slave hypothesis (Callan, 1967), the saltatory replication hypothesis (Britten and Kohne, 1968), unequal crossover (Edelman and Gally, 1970; Smith, 1974), and gene conversion (Edelman and Gally, 1970; Birky and Skavaril, 1976). Previously, unequal crossover was regarded as the most important mechanism of concerted evolution (Smith, 1974, 1976; Ohta, 1980), but gene conversion has recently been thought to be equally or even more important (Baltimore, 1981; Jackson and Fink, 1981; Nagylaki and Petes, 1982; Dover, 1982; Weiss *et al.*, 1983).

The main purpose of a comparative study of DNA sequences is, in our view, to understand the mechanisms of DNA evolution. As is well known, the question of what comprises the main cause of evolutionary change in gene sequences has been extremely controversial (see, e.g., Goodman, 1976, 1981a; Wilson *et al.*, 1977; Kimura, 1983a). After reviewing the topics mentioned in the preceding paragraphs, we shall express our view on this issue. At the end of the chapter, we shall point out some areas that call for further study.

2. Methods for Estimating the Number of Nucleotide Substitutions between Sequences

2.1. Methods

A basic quantity in the study of DNA evolution is the number of nucleotide substitutions that have occurred since the divergence of two sequences. Many methods have been proposed for estimating this number (e.g., Jukes and Cantor, 1969; Holmquist, 1972; Kimura, 1980, 1981a; Holmquist and Pearl, 1980; Kaplan and Risko, 1982; Lanvae *et al.*, 1984). Here we review some of the most frequently used methods and propose a new one.

All of the methods proposed to date can be divided into two categories. In the first the methods are intended to be applied to polypeptide-coding regions and nucleotide substitutions are classified into synonymous and nonsynonymous. The last three methods discussed below belong to this category. The methods in the other category do not make a distinction between synonymous and nonsynonymous substitutions. These methods

Table I
Schemes of Nucleotide Substitution

O/S[a]	A	T	C	G
a. One-parameter (1-p) scheme				
A	$1 - 3\alpha$	α	α	α
T	α	$1 - 3\alpha$	α	α
C	α	α	$1 - 3\alpha$	α
G	α	α	α	$1 - 3\alpha$
b. Two-parameter (2-p) scheme				
A	$1 - \alpha - 2\beta$	β	β	α
T	β	$1 - \alpha - 2\beta$	α	β
C	β	α	$1 - \alpha - 2\beta$	β
G	α	β	β	$1 - \alpha - 2\beta$
c. Four-parameter (4-p) scheme				
A	$1 - (\gamma + \theta\alpha + \alpha)$	γ	$\theta\alpha$	α
T	γ	$1 - (\gamma + \theta\alpha + \alpha)$	α	$\theta\alpha$
C	$\theta\beta$	β	$1 - (\gamma + \theta\beta + \beta)$	γ
G	β	$\theta\beta$	γ	$1 - (\gamma + \theta\beta + \beta)$
d. Six-parameter (6-p) scheme				
A	$1 - (2\alpha + \alpha_1)$	α_1	α	α
T	β_1	$1 - (2\alpha + \beta_1)$	α	α
C	β	β	$1 - (2\beta + \alpha_2)$	α_2
G	β	β	β_2	$1 - (2\beta + \beta_2)$

[a] O, Original nucleotide; S, substitute nucleotide.

can be applied to DNA sequences in general, but when applied to coding regions it is better to consider the three positions of codons separately, for they are known to evolve at different rates. The first five methods considered below belong to this category.

2.1.1. Jukes and Cantor's Method

This is the simplest method. It is represented by the first substitution scheme in Table I. It assumes that all nucleotides have the same substitution rate (3α per site per unit time) and that the direction of substitution is random (α in each direction). Since it involves only one parameter, it is also known as the one-parameter method.

Let L be the number of nucleotide sites studied and n the number of sites at which the nucleotides are different between the two sequences. The mean number of substitutions per nucleotide site is given by $K = 3\alpha(2T)$, where T is the divergence time between the two sequences. The number K can be estimated by the following formula:

$$K = -(3/4)\ln(1 - 4p/3) \tag{1}$$

where $p = n/L$ (Jukes and Cantor, 1969). The sampling variance is approximately given by

$$V(K) = \frac{p(1 - p)}{L(1 - 4p/3)^2} \tag{2}$$

(Kimura and Ohta, 1972).

2.1.2. Tajima and Nei's Method

Tajima and Nei (1984) noted that formula (1) can be written as

$$K = -b_1 \ln(1 - p/b_1) \tag{3}$$

where $b_1 = 1 - \Sigma\, q_i^2$. Here q_i is the equilibrium frequency of the ith nucleotide (i = 1–4 corresponding to A, T, C, G). In the Jukes-Cantor model, $q_i = 1/4$, so that $b_1 = 3/4$. Tajima and Nei showed that formula (3) also holds when the rate of substitution (a_{ij}) of the jth nucleotide for the ith nucleotide is independent of the ith nucleotide; i.e., $a_{ij} = a_j$ for all $i \neq j$. When $a_j = a$ for all j, this model becomes identical with the Jukes-Cantor model. If the condition $a_{ij} = a_j$ does not hold for all $i \neq j$, formula (3) tends to give an underestimate. To remove this deficiency, they proposed the following formula as a general estimator of K:

$$K = -b \ln(1 - p/b) \tag{4}$$

where $b = (b_1 + p^2/h)/2$, and

$$h = \sum_{i=1}^{3} \sum_{j=i+1}^{4} \frac{x^2_{ij}}{2q_iq_j}$$

in which x_{ij} is the proportion of pairs of nucleotides i and j between the two homologous DNA sequences. The sampling variance for this estimator is approximately given by

$$V(K) = \frac{b^2p(1-p)}{(b-p)^2L} \tag{5}$$

2.1.3. Two-Parameter Method

Kimura (1980) proposed a model that allows transitional and transversional substitutions to occur at different rates. This model is represented by the second substitution scheme in Table I. The four substitution types A to G, T to C, C to T, and G to A are transitional, while the other eight types are transversional. At each nucleotide site, the rate of transitional substitution and the rate for each of the two types of transversional substitution are α and β substitutions per unit time, respectively, with a total rate of $\alpha + 2\beta$ substitutions per unit time.

Let P and Q be, respectively, the proportions of transitional and transversional differences between the two sequences under comparison. The mean number of substitutions per site is then given by $K = (\alpha + 2\beta)(2T)$ and can be estimated by

$$K = \tfrac{1}{2} \ln a + \tfrac{1}{4} \ln b \tag{6}$$

and the sampling variance is approximately given by

$$V(K) = [a^2P + c^2Q - (aP + cQ)^2]/L \tag{7}$$

where $a = 1/(1 - 2P - Q)$, $b = 1/(1 - 2Q)$, and $c = (a + b)/2$ (Kimura, 1980).

2.1.4. Four-Parameter Method

Takahata and Kimura (1981) proposed a four-parameter method, which is represented by the third substitution scheme in Table I. This model allows the two types of transversional substitution at a nucleotide site to occur at different rates; e.g., the rates of A to T and A to C are α and $\theta\alpha$, respectively. Moreover, the four types of transition can occur at different rates. Let X_{ij} be the proportion of nucleotide sites where the first and second sequences have nucleotides i and j, respectively. (Note that the x_{ij} defined above is equal to $X_{ij} + X_{ji}$.) Then

$$K = -\tfrac{1}{4} \ln \left[\left(1 - \frac{S + R}{2q}\right)^{8q-1} \left((X_{11} + X_{22} + X_{12} + X_{21}) (X_{33} + X_{44} + X_{34} + X_{43}) - \tfrac{1}{4}(S-R)^2)/q \right) \right] \tag{8}$$

where $S = X_{23} + X_{32} + X_{14} + X_{41}$, $R = X_{24} + X_{42} + X_{13} + X_{31}$, $q = w(1 - w)$, and w is the proportion of A + T in the two sequences.

2.1.5. Six-Parameter Method

Kimura (1981a) proposed a six-parameter method, which is represented by the last substitution scheme in Table I. This is the most general method that has been solved (Gojobori *et al.*, 1982a). Let q_1, q_2, q_3, and q_4 be the average frequencies of A, T, C, and G in the two sequences. Then,

$$K = pq \ln\left(\frac{pq}{B_1}\right) - 2q_1q_2 \ln\left[\frac{p(F_{12} - B_1 + 3E_{12}/B_1)}{3q_1q_2}\right] - \frac{2q_3q_4}{q} \ln\left[\frac{q(F_{34} - B_1 + 3E_{34}/B_1)}{3q_3q_4}\right] \quad (9)$$

where

$$\begin{aligned}
p &= q_1 + q_2, \qquad q = q_3 + q_4 \\
B_1 &= pq - (X_{13} + X_{14} + X_{23} + X_{24}) \\
E_{12} &= (q_1q - X_{13} - X_{14})(q_2q - X_{23} - X_{24}) \\
E_{34} &= (q_3p - X_{13} - X_{23})(q_4p - X_{14} - X_{24}) \\
F_{12} &= X_{11} + X_{22} - X_{12} - p^2 + 3q_1q_2 \\
F_{34} &= X_{33} + X_{44} - X_{34} - q^2 + 3q_3q_4
\end{aligned}$$

in which X_{ij} is defined as above. Gojobori *et al.* (1982a) have also obtained an approximate formula for the large sampling variance of K, and have developed a computer program for computing the mean and standard error.

2.1.6. Percentage Corrected Divergence (PCD) Method

This method was proposed by Perler *et al.* (1980). In this method, potential synonymous sites are classified into three categories according to whether a site can afford one, two, or three synonymous changes. For each category, one counts the number of sites in each of the two sequences compared and computes the average. Potential nonsynonymous sites are treated in the same way.

The second step is to compare the two sequences codon by codon, score one point for each synonymous or nonsynonymous difference, and categorize it according to the type of site in which it has occurred. For example, if the codons compared are CCC (Pro)–CCT (Pro), there is one change of category 3 of synonymous substitution, because at the third position of these two codons all three possible changes are synonymous.

If the codons are ATC (Ile)–ACG (Thr), then there are two possible paths with two changes: ATC (Ile) ↔ ACC (Thr) ↔ (ACG (Thr) and ATC (Ile) ↔ ATG (Met) ↔ ACG (Thr). Both paths are assumed to be equally probable, so that the difference at the third codon position betwen ATC and ACG is half synonymous (category 3) and half nonsynonymous (category 1); in either path, the difference at the second position is nonsynonymous (category 3).

The third step is to divide the sum of synonymous or nonsynonymous differences in each category by the corresponding potential sites. Each percentage change (p') is then corrected for multiple events by one of the following formulas:

$$\text{category 1:} \quad 3[-\tfrac{1}{2}\ln(1 - 2p')] \tag{10}$$
$$\text{category 2:} \quad \tfrac{3}{2}[-\tfrac{2}{3}\ln(1 - \tfrac{3}{2}p')] \tag{11}$$
$$\text{category 3:} \quad \tfrac{3}{4}[-\ln(1 - \tfrac{4}{3}p')] \tag{12}$$

Finally, for each pair of sequences the overall divergence is the average of two weight averages of the corrected percentages, using separately as weighting factors the number of sites and the number of changes in each category.

2.1.7. Miyata and Yasunaga's Method

This method differs from the PCD method in several ways. First, in inferring the numbers of synonymous and nonsynonymous changes between two codons, it assigns different weights to different possible evolutionary paths [for details, see Miyata and Yasunga (1980)]. Second, the numbers of synonymous and nonsynonymous sites in a codon are defined as follows. Let j be the number of possible single nucleotide changes that will cause no change in amino acid. Then the above two numbers are $j/3$ and $3 - j/3$, respectively. For example, the two numbers are 1/3 and 8/3 for the codon TTT (Phe), because only one of the nine possible changes is synonymous. For the ith codon pair between the two sequences compared, the average numbers of synonymous (N_{Si}) and nonsynonymous (N_{Ai}) sites are obtained by first computing the averages for the codons in each possible path and then taking weighted averages over all possible paths. The average numbers of synonymous (N_S) and nonsynonymous (N_A) sites for the two sequences are then $N_S = \Sigma N_{Si}$ and $N_A = \Sigma N_{Ai}$, the summation being over codons.

The number K_S of (synonymous) substitutions per synonymous site and the number K_A of (nonsynonymous) substitutions per nonsynonymous site are given by $K_S = M_S/N_S$ and $K_A = M_A/N_A$, where M_S and

M_A are the inferred numbers of synonymous and nonsynonymous differences between the two sequences. These formulas obviously do not take into account the effect of multiple hits at a site. Miyata *et al.* (1982) have suggested that corrections be made by using Jukes and Cantor's formula; i.e., by respectively replacing p by K_S and K_A in formula (1).

2.1.8. A New Method

We discuss in some detail a new method we (Li *et al.*, 1985) developed recently; the results cited in Section 3 were mainly obtained by using this method. We consider the genetic code for nuclear genes, which are our primary interest; the method actually applies better to mammalian mitochondrial genes. We classify nucleotide sites into nondegenerate, twofold dengerate, and fourfold degenerate sites. A site is fourfold degenerate if all possible changes at the site are synonymous. The third positions of 32 of the 61 sense codons, e.g., GTT (Val), are of this type. A site is twofold degenerate if one of the three possible changes is synonymous. The third position of 24 of the 61 sense codons, e.g., CAT (His), and the first positions of four leucine codons (TTA, TTG, CTA, CTG) and four arginine codons (CGA, CGG, AGA, AGG) are of this type. [In mammalian mitochondrial (mt) genes, AGA and AGG are stop codons (Anderson *et al.*, 1981) and the first positions of CGA and CGG (Arg) are nondegenerate.] We also include the third positions of the three isoleucine codons in this class, though they are actually threefold degenerate sites. (In mammalian mt genes, there are only two codons for isoleucine and their third positions are indeed twofold degenerate sites.) A site is nondegenerate if all possible changes at this site are nonsynonymous or nonsense. The second position of all sense codons and the first position of most codons belong to this class, and so do the third positions of ATG (Met) and TGG (Trp). (In mammalian mt genes, there are two codons for methionine and their third positions are twofold degenerate sites.) We count the numbers of sites in these three classes in each of the two sequences compared and then compute the average numbers, denoting them by L_0 (nondegenerate), L_2 (twofold), and L_4 (fourfold), respectivley.

Next, we compare the two sequences codon by codon and infer the nucleotide differences between each pair of codons. We classify each difference according to the type of site at which it has occurred. This can be done by considering all possible paths between the two codons compared according to the weights obtained from the relative frequencies of codon changes in mammalian genes (Li *et al.*, 1985). The nucleotide differences in each class are further classified into transitional (P_i) and transversional (Q_i) differences ($i = 0, 2, 4$). We note that in the class of twofold degenerate sites transitions are synonymous and transversions are nonsynonymous. The only exceptions are the three positions of the third Ile codons and the first positions of the four Arg codons mentioned above (see above for mt genes). In all these cases, all synonymous changes are included in P_2 and all nonsynonymous changes are included in Q_2.

The means and approximate sampling variances of the numbers of transitional (A_i) and transversional (B_i) substitutions per ith type site are given by

$$A_i = \tfrac{1}{2} \ln a_i - \tfrac{1}{4} \ln b_i \tag{13}$$
$$V(A_i) = [a_i^2 P_i + c_i^2 Q_i - (a_i P_i + c_i Q_i)^2]/L_i \tag{14}$$
$$B_i = \tfrac{1}{2} \ln b_i \tag{15}$$
$$V(B_i) = b_i^2 Q_i(1 - Q_i) \tag{16}$$

where $a_i = 1/(1 - 2P_i - Q_i)$, $b_i = 1/(1 - 2Q_i)$, and $c_i = (a_i - b_i)/2$. The total rate K_i is given by

$$K_i = A_i + B_i \tag{17}$$

with an approximate sampling variance given by

$$V(K_i) = [a_i^2 P_i + d_i^2 Q_i - (a_i P_i + d_i Q_i)^2]/L_i \tag{18}$$

where $d_i = b_i + c_i$. We note that A_2 and B_2 denote the numbers of synonymous and nonsynonymous substitutions per twofold degenerate site, $K_4 = A_4 + B_4$ the number of synonymous substitutions per fourfold degenerate site, and $K_0 = A_0 + B_0$ the number of nonsynonymus substitutions per nondegenerate site.

Let K_S be the number of (synonymous) substitutions per synonymous site and K_A the number of (nonsynonymous) substitutions per nonsynonymous site. Following the convention, we count one-third of a twofold degenerate site as synonymous and two-thirds as nonsynonymous. We then obtain

$$K_S = \frac{L_2A_2 + L_4K_4}{L_2/3 + L_4} = 3\,\frac{L_2A_2 + L_4K_4}{L_2 + 3L_4} \tag{19}$$
$$V(K_S) = 9\,\frac{[L_2^2V(A_2) + L_4^2V(K_4)]}{(L_2 + 3L_4)^2} \tag{20}$$
$$K_A = \frac{L_2B_2 + L_0K_0}{2L_2/3 + L_0} = 3\,\frac{(L_2B_2 + L_0K_0)}{2L_2 + 3L_0} \tag{21}$$
$$V(K_A) = 9\,\frac{[L_2^2V(B_2) + L_0^2V(K_0)]}{(2L_2 + 3L_0)^2} \tag{22}$$

2.2. Comparison of Methods

We first compare the one-, two-, four-, and six-parameter (1-p, 2-p, 4-p, and 6-p) methods. The 1-p method assumes that substitution occurs randomly among the four types of nucleotide. There is evidence that this assumption is not realistic. Indeed, as will be seen below, mutation does not occur randomly among the four types of nucleotide. The 2-p method takes into account the possibility that transitional and transversional substitutions may occur at different rates, and the 4-p and 6-p methods take into account even more possibilities of unequal rates of substitution among the four types of nucleotide. It would seem that a model with a larger number of parameters is better than one with a fewer number of parameters. In practice, this is not necessarily the case, for two reasons. One is that in addition to the substitution scheme it is usually necessary to make further assumptions in the formulation of a model. For example, in the 6-p method, it is assumed that the common ancestral sequence had the equilibrium nucleotide frequencies and these frequencies are estimated from the two extant sequences (Gojobori *et al.*, 1982a). A similar assumption is required in the 4-p method, but no such assumption is explicitly required in the 1-p and 2-p methods because formulas (1), (2), (6), and (7) do not involve nucleotide frequencies. Obviously, additional assumptions can cause estimation errors. The other reason, which is probably more important, is that sampling errors, which arise because the number of nucleotides compared is finite, can render a method inapplicable. This is because all estimators of K involve logarithmic functions and the argument can become zero or negative. This chance seems to increase with the number of parameters involved.

Table II cites some simulation results of Gojobori *et al.* (1982a). It shows that when the number K of substitutions per nucleotide site is 1 or larger, the 4-p and 6-p methods are not applicable to a high proportion of cases. For example, in the case of $K = 2.0$ these two methods are not applicable in 108 and 129 comparisons out of a total of 160, respectively, when the 6-p substitution scheme is used. By contrast, the 1-p method is not applicable in only 22 cases. In Table II the number L of nucleotides compared is only 144. When L becomes larger, the proportion of inapplicable cases will decrease. However, if K is as large as 2, the proportion remains quite high, about one-fourth for the 4-p method and higher for the 6-p method, even if $L = 3000$ (Gojobori *et al.*, 1982a). Thus, when K is large, inapplicability is a serious problem for these two methods.

The mean and standard deviations shown in Table II were computed by excluding inapplicable cases. When the 1-p substitution scheme is used, all three methods give good estimates of expected values. When the 4-p

Table II

Mean $\overline{K}$ and Standard Deviation σ_K of K over All Replications and the Frequency of Inapplicable Cases[a]

Scheme	True K value	1-p method		4-p method		6-p method		Inapplicable cases		
		$\overline{K}$	σ_K	$\overline{K}$	σ_K	$\overline{K}$	σ_K	1-p	4-p	6-p
1-p	0.5	0.52	0.07	0.53	0.08	0.53	0.08	0/40[b]	0/40	0/40
	1.0	1.03	0.23	1.07	0.25	1.09	0.27	0/80[b]	2/80	4/80
	2.0	2.08	0.55	1.91	0.33	2.10	0.51	20/160[b]	100/160	115/160
4-p	0.5	0.48	0.07	0.51	0.08	0.51	0.08	0/40	0/40	0/40
	1.0	0.90	0.19	1.00	0.24	1.03	0.27	0/80	4/80	4/80
	2.0	1.84	0.44	1.71	0.36	1.78	0.41	2/160	116/160	130/160
6-p	0.5	0.47	0.08	0.56	0.18	0.54	0.11	0/40	3/40	6/40
	1.0	0.81	0.13	1.00	0.24	1.01	0.23	0/80	11/80	22/80
	2.0	1.24	0.18	1.36	0.30	1.70	0.52	0/160	108/160	129/160

[a] Taken from Gojobori *et al.* (1982a). One-parameter (1-p), four-parameter (4-p), and six-parameter (6-p) methods are used for three different substitution schemes. $L = 144$.

[b] The number of replications was ten, but since there were eight "contemporary sequences," the number of possible comparisons was larger than ten. However, all comparisons are not independent.

or 6-p substitution scheme is used, the 4-p and 6-p methods give good estimates for $K = 0.5$ and 1.0, but the 1-p method tends to give slight underestimates. If $K = 2.0$ and either the 4-p or 6-p substitution scheme is used, all methods give underestimates; however, among those cases where all three methods are applicable, the 6-p method tends to give the best estimate and the 1-p method the worst.

Tajima and Nei (1984) have compared their method with the above four methods, using computer simulation. Considering various schemes of nucleotide substitution, they drew the following conclusion. When K is around 0.5 or smaller, Eq. (4) gives the best estimate. When K is 1 or larger, the 6-p method is on the average superior to Eq. (4) for those cases where both methods are applicable, but the 6-p method generally has considerably more inapplicable cases than Eq. (4), particularly when the number of nucleotides compared is small. Equation (3) also gives good estimates when K is 0.5 or smaller.

From these simulation results and those by Takahata and Kimura (1981), we may conclude that if K is smaller than 0.5, any of the above methods can be applied to estimate K. If K is between 0.5 and 1, one can use either the 6-p method or Eq. (4). However, if K is 1 or larger, it is better to use the 6-p method; if this method is inapplicable, then one may have to use Eq. (4) or one of the other methods. Regardless of which method is used, the estimate should be taken with caution if K is substantially larger than 1.

Next, we consider the last three methods. In our view the PCD method has three drawbacks. First, it assigns equal weights to all possible paths between two codons, even if some paths require more nonsynonymous substitutions than the others. For instance, in the example of ACT (Ile)–ACG (Thr) the two possible paths are each assigned the probability 0.5, even though one path requires two nonsynonymous substitutions while the other path requires one synonymous and one nonsynonymous substitution (see above). This assumption is likely to lead to an underestimate of the number of synonymous substitutions because it is well known that synonymous substitution occurs much more often than nonsynonymous substitution. Second, it assumes that nucleotide substitution occurs randomly; i.e., the 1-p substitution scheme. This assusmption tends to underestimate the numbers of synonymous and of nonsynonymous substitutions (Kimura, 1980; Gojobori *et al.*, 1982a). Third, the correction formula for category 1 sites actually assumes that there are only two states at each site, or, in other words, a site of category 1 can be occupied by only two types of nucleotide. This assumption is obviously incorrect. A similar comment applies to the correction formula for sites in category

2. Gojobori's (1983) simulation study indicates that the PCD method tends to give a serious underestimate of K_S when it is 0.8 or larger.

One advantage of Miyata and Yasunaga's (1980) method over the PCD method is that it assigns different weights to different paths. However, their weights were inferred from amino acid sequences and tend to be more in favor of nonsynonymous than synonymous sustitutions (Li *et al.*, 1985). Moreover, in this method correction for multiple hits is made in an *ad hoc* manner. According to Miyata *et al.* (1982), this method tends to give an underestimate of K when it is larger than 1.

One advantage of our new method over that of Miyata and Yasunga is that we separate degenerate sites into twofold and fourfold degnerate sites, so that correction for multiple hits and the derivation of the error variances can be done in a more rigorous manner. Another advantage is that our weights were inferred from nucleotide sequences and should therefore be more reliable than those inferred from amino acid sequences. A third advantage is that we consider the possibility that transitional substitution may occur more often than transversional substitution. In addition, our method provides more information about the substitution rates at nucleotide site with different degrees of codon degeneracy. Because of these advantages, we believe that our method is better than the other two. Of course, if both K_S and K_A are 0.5 or smaller, the three methods are expected to give very similar estimates, a situation similar to that in the first five methods. Like other methods, our method will also tend to give underestimates when the degree of sequence divergence is large. This is because we do not consider the full extent of nonrandom substitution among the four types of nucleotide.

Finally, we consider the question of whether it is better to treat the three positions of codons separately or to treat synonymous and nonsynonymous substitutions separately when evaluating the extent of sequence divergence in coding regions. The advantage of the former approach is that the three positions of codons are well defined and generally do not change with time. The disadvantage is that first position sites can be twofold degenerate or nondegenerate and third position sites can be fourfold degenerate, twofold degenerate, or nondegenerate, so that the estimates of the numbers of substitutions at the first and third positions may not be reliable, for different types of sites generally evolve at different rates. The advantage of the latter approach is that the classification into synonymous and nonsynonymous substitutions is quite natural, because the former do not cause changes in amino acids, while the latter do. Molecular biologists tend to prefer this approach, since the results are easier to interpret. One disadvantge of this approach is that synonymous sites and nonsynonymous sites are not well defined, because a site may be partly synonymous and partly nonsynonymous. In our approach this difficulty is overcome to some extent, since we classify sites into nondegenerate, twofold degenerate, and fourfold degenerate rather than just synonymous and nonsynonymous.

3. Rates of Nucleotide Substitution in Various Regions of Genes

3.1. Coding Regions

A typical eukaryotic gene consists of a 5′ flanking (untranscribed) region, a 5′ untranslated region, polypeptide-coding regions, intervening sequences (introns), a 3′ untranslated region, and a 3′ flanking (untranscribed) region (e.g., Minty and Newmark, 1980). These regions serve different functions and are apparently subject to different stringencies of functional constraints. It has therefore been customary to treat them separately.

The coding regions are by far the most studied regions because of their functional importance and because of the availability of a large amount of sequence data for these regions. Many comparative studies of nucleotide sequences in these regions have been published (e.g., Salser and Issacson, 1976; Grunstein *et al.*, 1976; Kimura, 1977; Kafatos *et al.*, 1977; Jukes and King, 1979; Lomedico *et al.*, 1979; Miyata *et al.*, 1980; Hayashida and Miyata, 1983). These studies have shown that substitution at the third position of codons (or synonymous substitution) occurs much faster than substitution at the first two positions of codons (or nonsynonymous substitution). A most interesting observation is that in the gene for histone IV (histone H4), synonymous substitution occurs at a very high rate (Kimura, 1977) despite the fact that histone IV is known to be the most conservative protein sequence (Dayhoff, 1972; Wilson *et al.*, 1977). Miyata *et al.* (1980) have compared the rate of synonymous substitution with that of nonsynonymous substitution in 50 genes and found that the former is considerably higher than the latter in all but the mouse immunoglobulin λ-type variable region genes. The higher rate of synonymous substitution than the rate of nonsynonymous substitution is thought to be a consequence of less stringent functional constraints on synonymous changes. Another interesting finding by Miyata *et al.* (1980) and Hayashida and Miyata (1983) is that the rate of synonymous substitution is rather uniform among genes, about 5.5×10^{-9} substitutions per synonymous site per year.

Taking advantage of the availability of a large number of DNA sequences, we have used our method to conduct an extensive analysis (Li *et al.*, 1985). Our main purposes were to see if the rate of synonymous substitution is really uniform among genes and to compare the rates of nonsynonymous substitution in various kinds of genes. We have used mainly genes from mammals—human, rodent (mouse, rat, Chinese hamster), and artiodactyl (bovine, goat, porcine)—for several reasons. First, the fossil record for mammals is better than those for other organisms; we assumed that the above mammalian orders diverged 80 million years ago (mya). Second, many genes from mammals have been sequenced. Third, to get a reliable estimate of the rate of nucleotide substitution the degree of sequence divergence should not be too small or too large. For this reason, genes from different mammalian orders are generally very suitable for estimating both the synonymous and nonsynonymous rates. For extremely conservative genes, such as those coding for histones and glucagon, estimates of nonsynonymous rates from comparisons of mammalian species may not be reliable, because the number of substitutions is too small, and we have therefore also obtained estimates from comparisons between mammals and chicken or fish.

The results in Table III are cited from Li *et al.* (1985). In each gene group, the results are arranged in the order of the increasing rate of nonsynonymous substitution. In the following we discuss a number of interesting features.

First, the rate of nonsynonymous substitution is extremely variable among different kinds of genes; it ranges from 0.004×10^{-9} (histone H4) to 2.80×10^{-9} substitutions per nonsynonymous site per year (interferon α), a more than 100-fold difference (see the next to last column of Table III). As is well known, histone and actin genes are extremely conservative. Some hormone genes, e.g., those for somatostatin-28, glucagon, gastrin, and insulin, are also extremely conservative, but others (e.g., those for relaxin and prolactin) evolve rapidly. The insulin C-peptide has often been used as an example of rapid evolution, but it actually has evolved considerably more slowly than relaxin and the relaxin C-peptide. The nonsynonymous rate observed in the signal peptide region of the insulin genes (1.2×10^{-9}) is about the average for signal peptide regions (results not shown); the extremely high and low rates in the signal peptide regions of the relaxin and somatostatin genes, respectively, are probably extreme random deviates, for signal peptide regions are very short. Hemoglobin genes evolve at an intermediate rate, while immunoglobulin and interferon genes evolve rapidly. The average rate of nonsynonymous substitution for all the comparisons of genes from mammals is 0.88×10^{-9} substitutions per nonsynonymous site per year.

Table III
Rate of Nucleotide Substitution in Various Genes[a]

Gene	L	Nondegenerate sites TI	Nondegenerate sites TV	Nondegenerate sites Total	Twofold degenerate sites TI	Twofold degenerate sites TV	Twofold degenerate sites Total	Fourfold degenerate sites TI	Fourfold degenerate sites TV	Fourfold degenerate sites Total	Nonsynonymous rate	Synonymous rate
Histones												
H4	101	0.00	0.00	0.00 (0.00)	1.99	0.15	2.13 (0.78)	2.97	3.19	6.16 (1.54)	0.027 (0.03)	6.13 (1.32)
	101	0.00[b]	0.00	0.00 (0.00)	0.43	0.02	0.46 (0.18)	0.39	1.07	1.46 (0.34)	0.004 (0.01)	1.43 (0.29)
H3	135	0.00	0.00	0.00 (0.00)	2.33	0.00	2.33 (0.70)	3.14	3.02	6.16 (1.41)	0.000 (0.00)	6.38 (1.19)
	135	0.01[b]	0.03	0.04 (0.02)	1.24	0.00	1.24 (0.43)	1.07	1.57	2.64 (0.81)	0.038 (0.02)	2.89 (0.69)
H2A	126	0.02[b]	0.06	0.08 (0.03)	0.75	0.07	0.82 (0.24)	0.16	1.85	2.01 (0.52)	0.079 (0.02)	2.06 (0.44)
H2B	120	0.03	0.04	0.07 (0.04)	1.44	0.08	1.52 (0.50)	1.47	1.91	3.37 (0.78)	0.075 (0.04)	3.59 (0.69)
Contractile system protein												
Actin, α	376	0.00	0.01	0.01 (0.01)	1.81	0.00	1.81 (0.33)	2.11	0.94	3.05 (0.47)	0.014 (0.01)	3.67 (0.43)
Hormones and active peptides												
Somatostatin-28	28	0.00	0.00	0.00 (0.00)	0.52	0.00	0.52 (0.55)	4.75	0.48	5.06 (3.69)	0.000 (0.00)	4.23 (2.84)
Gastrin	82	0.00	0.09	0.09 (0.12)	0.78	0.35	1.13 (0.98)	2.52	1.84	4.35 (4.00)	0.15 (0.15)	3.52 (2.57)
Glucagon	29	0.00	0.00	0.00 (0.00)	3.42	0.00	3.42 (2.28)	0.46	1.64	2.02 (1.15)	0.18 (0.06)[c]	5.10 (2.61)
Insulin	51	0.02	0.09	0.11 (0.07)	2.34	0.27	2.60 (0.89)	3.08	1.46	4.54 (2.67)	0.16 (0.09)	5.41 (1.98)
ACTH	39	0.05	0.16	0.22 (0.11)	1.06	0.00	1.06 (0.59)	1.17	0.45	1.62 (0.75)	0.19 (0.09)	1.99 (0.71)
Parathyroid hormone	90	0.34	0.18	0.52 (0.15)	0.89	0.06	0.95 (0.37)	0.64	0.59	1.22 (0.52)	0.44 (0.12)	1.73 (0.51)
Corticotropin-releasing factor	41	0.32	0.34	0.66 (0.13)	0.75	0.37	1.12 (0.32)	1.66	1.31	2.98 (0.67)	0.64 (0.12)	2.78 (0.54)
Glycoprotein hormone, α	92	0.36	0.42	0.78 (0.14)	2.99	0.12	3.11 (0.98)	3.65	1.27	4.92 (1.83)	0.67 (0.28)	6.23 (1.57)

Growth hormone	189	0.67	0.35	1.02 (0.12)	2.28	0.42	2.70 (0.45)	1.48	1.71	3.19 (0.55)	0.95 (0.11)	4.37 (0.56)
Insulin C-peptide	35	0.29	0.75	1.04 (0.29)	1.51	0.50	2.01 (0.90)	2.84	2.18	5.02 (2.37)	0.99 (0.26)	4.93 (1.95)
Prolactin	197	0.77	0.57	1.34 (0.14)	1.84	0.72	2.56 (0.40)	3.82	1.83	5.64 (1.27)	1.29 (0.12)	5.59 (0.92)
Relaxin C-peptide	99	1.19	0.85	2.03 (0.27)	1.79	0.52	2.31 (0.52)	1.59	2.13	3.72 (0.90)	1.78 (0.22)	4.32 (0.77)
Relaxin	54	1.51	1.30	2.81 (0.44)	2.54	0.79	3.34 (1.00)	5.01	2.40	7.42 (8.90)	2.51 (0.37)	7.49 (6.10)
Signal peptides												
Somatostatin	23	0.00	0.31	0.31 (0.23)	0.70	0.00	0.70 (0.74)	1.50	0.81	2.31 (1.17)	0.27 (0.19)	2.28 (1.05)
Insulin	23	0.65	0.56	1.21 (0.39)	0.44	0.58	1.02 (0.81)	1.21	1.36	2.57 (0.98)	1.16 (0.36)	2.43 (0.89)
Relaxin	23	1.87	0.63	2.50 (0.71)	2.33	1.29	3.63 (2.00)	4.00	2.80	6.80 (5.86)	2.39 (0.62)	6.86 (7.59)
Hemoglobins												
α Chain	141	0.19	0.40	0.59 (0.10)	1.97	0.29	2.24 (0.56)	1.88	1.44	3.32 (0.61)	0.56 (0.09)	3.94 (0.60)
β Chain	144	0.35	0.53	0.89 (0.13)	1.48	0.56	2.03 (0.43)	1.42	0.96	2.38 (0.47)	0.87 (0.11)	2.96 (0.46)
Immunoglobulins and related proteins												
Ig V_H	100	0.51	0.62	1.12 (0.21)	1.97	0.51	2.48 (0.73)	3.48	2.10	5.58 (1.71)	1.07 (0.19)	5.67 (1.36)
Ig $\lambda 1$	105	0.41	0.77	1.18 (0.21)	2.58	0.52	3.13 (0.95)	3.58	2.31	5.89 (1.71)	1.13 (0.19)	6.36 (1.46)
$\beta 2$ Microglobulin	99	0.41	0.83	1.24 (0.22)	3.79	0.72	4.50 (1.36)	10.07	1.94	12.0 (15.5)	1.21 (0.20)	11.77 (9.91)
Ig $\gamma 1$	321	0.60	0.97	1.56 (0.14)	1.86	0.59	2.44 (0.41)	2.77	2.18	4.95 (0.76)	1.46 (0.13)	5.11 (0.64)
Ig κ	106	1.18	0.70	1.88 (0.31)	2.57	1.04	3.61 (0.98)	2.42	2.58	5.00 (1.41)	1.83 (0.29)	5.89 (1.29)
Interferons												
$\alpha 1$	166	0.70	0.83	1.54 (0.20)	1.54	0.60	2.14 (0.43)	0.97	1.90	2.87 (0.67)	1.41 (0.13)	3.53 (0.61)
$\alpha 2$	165	0.79	0.74	1.53 (0.20)	1.32	0.82	2.14 (0.42)	1.02	1.65	2.67 (0.64)	1.47 (0.18)	3.15 (0.56)
$\beta 1$	159	1.33	1.03	2.37 (0.28)	2.10	1.05	3.15 (0.58)	2.42	3.14	5.56 (1.48)	2.21 (0.24)	5.88 (1.08)
γ	136	1.85	1.13	2.97 (0.37)	3.18	1.45	4.63 (0.96)	6.22	1.51	7.73 (4.22)	2.80 (0.31)	8.59 (2.56)

(continued)

Table III
(continued)

Gene	L	Nondegenerate sites TI	TV	Total	Twofold degenerate sites TI	TV	Total	Fourfold degenerate sites TI	TV	Total	Nonsynonymous rate	Synonymous rate
Other proteins												
HPRT	217	0.13	0.03	0.16 (0.05)	0.83	0.02	0.85 (0.21)	0.97	0.76	1.73 (0.42)	0.13 (0.04)	2.13 (0.35)
Neurophysin II	93	0.18	0.29	0.48 (0.13)	0.47	0.12	0.59 (0.29)	1.61	0.91	2.52 (0.76)	0.43 (0.12)	2.23 (0.60)
Metallothionein II	60	0.34	0.16	0.50 (0.17)	0.66	0.06	0.72 (0.39)	1.15	2.13	3.28 (1.27)	0.43 (0.14)	2.83 (0.92)
Metallothionein I	60	0.34	0.28	0.62 (0.14)	1.89	0.35	2.24 (0.64)	1.11	1.17	2.28 (0.68)	0.61 (0.12)	3.38 (0.76)
Fibrinogen γ	411	0.33	0.27	0.58 (0.07)	2.99	0.28	3.27 (0.48)	2.80	1.21	4.01 (0.67)	0.55 (0.06)	5.82 (0.67)
Albumin	590	0.35	0.56	0.91 (0.08)	3.38	0.64	4.02 (0.47)	2.69	2.13	4.82 (0.61)	0.92 (0.07)	6.72 (0.62)
α-Lactalbumin	123	0.55	0.63	1.18 (0.20)	2.05	0.50	2.55 (0.59)	2.02	0.37	2.39 (0.92)	1.10 (0.17)	4.16 (0.92)
α-Fetoprotein	586	0.76	0.54	1.30 (0.10)	2.13	0.55	2.69 (0.29)	2.38	1.60	3.98 (0.52)	1.21 (0.08)	4.90 (0.45)
Average[d]		0.48	0.45	0.94 (0.81)	1.83	0.42	2.26 (1.10)	2.51	1.68	4.18 (2.10)	0.88 (0.75)	4.65 (2.06)

[a]Taken from Li *et al.* (1985). Unless stated otherwise, all rates are obtained from comparisons between different mammalian orders, which are assumed to have diverged 80 mya. All rates are in terms of 10^{-9} substitutions per site per year. L, Number of codons compared; TI, for transition; TV, transversion. Value in parentheses is the standard error.

[b]All rates in these cases are obtained from comparisons between mammals and birds or fish. The divergence time between mammals and birds is assumed to be 270 mya and that between mammals and fish 400 mya.

[c]In these cases, the mean and variance of the rate of nonsynonymous substitution are obtained from comparisons between mammals and birds or fish.

[d]The average is obtained only from comparisons between mammalian orders.

Second, the rate of synonymous substitution varies greatly from gene to gene (see the last column in Table III), although not so much as the rate of nonsynonymous substitution. In the $\beta 2$-microglobulin gene the rate is as high as 11.8×10^{-9}, while in the parathyroid hormone gene it is as low as 1.7×10^{-9}. Of course, these two values are probably extreme random deviates, because the numbers of synonymous sites in these two genes are small and because the values in other cases are quite far from the two extremes. However, it would not be unusual to observe a rate higher than 5×10^{-9} or lower than 3×10^{-9}. Thus, our results do not support Miyata *et al.*'s (1980) suggestion that the rate of synonymous substitution is similar among genes. The average rate of synonymous substitution for all the comparisons of genes from mammals is 4.7×10^{-9} substitutions per synonymous site per year. This is five times higher than the average rate of nonsynonymous substitution. Our estimate is not much different from Hayashida and Miyata's (1983) estimate, 5.5×10^{-9}. (Their estimate reduces to 5.1×10^{-9}. if 80 instead of 75 mya is used as the divergence time between mammalian orders. Note also that we used 36 genes, while they used only 11 genes.) It is interesting to note that in histones H4 and H3 the synonymous rates are very high, although the nonsynonymous rates are extremely low. (In histones H4, H3, and H2A, the synonymous rates estimated from comparisons between chicken and human are quite low. This is probably because our method tends to give a serious underestimate when the degree of divergence is large.) There is, however, a slight tendency for a gene with a high nonsynonymous rate to have a high synonymous rate; in the majority of genes with a nonsynonymous rate higher than 1.0×10^{-9}, the synonymous rate is higher than the average.

Third, at fourfold degenerate sites the rate of transitional substitution tends to be higher than the rate of transversional substitution, although at each site two types of transversional change and only one type of transitional change can occur. The averages for these two rates are 2.5×10^{-9} and 1.7×10^{-9}, respectively. This observation can be explained by the fact that transitional mutation occurs more frequently than transversional mutation (see section 5). At twofold degenerate sites, the rate of transitional substitution is on the average about the same as that at fourfold degenerate sites, but the rate of transversional substitution is usually considerably lower than the corresponding rate at fourfold degenerate sites; the averages for the two rates at twofold degenerate sites are 1.8×10^{-9} and 0.42×10^{-9}, respectively. This lower rate of transversional substitution occurs because all transversional changes at twofold degenerate sites are nonsynonymous. At nondegenerate sites, the rates of transitional and transversional substitution are on the average about the same (0.48×10^{-9} and 0.45×10^{-9}, respectively) and are usually considerably lower than the corresponding values at fourfold degenerate sites. These low rates occur because all changes at nondegenerate sites are nonsynonymous. Thus, the (total) rate of nucleotide substitution is lowest at nondegenerate sites, intermediate at twofold degenernate sites, and highest at fourfold degenreate sites; the averages are 0.94×10^{-9}, 2.3×10^{-9}, and 4.2×10^{-9}, respectively.

Fourth, the rate of nucleotide substitution at fourfold degenerate sites tends to be lower than the rate of synonymous substitution, although all changes at fourfold degenerate sites are synonymous. Indeed, the average value is only 4.2×10^{-9} for the former rate, but is 4.7×10^{-9} for the latter rate. This difference occurs because in computing the rate of synonymous substitution the transitional changes at twofold degenerate sites are also included. As mentioned above, only one-third of a twofold degenerate site is counted as synonymous, but transitional (synonymous) substitution at a twofold degenerate site occurs at a rate higher than one-third of the total rate at a four-fold degenerate site. For this reason, the synonymous rate obtained by the conventional definition tends to be higher than the rate of nucleotide substitution at fourfold degenerate sites. In inferring the stringency of functional constraint on synonymous changes, it has been customary to compare the rate of synonymous substitution with the substitution rate in pseudogenes. It now appears that for this purpose one should compare the substitution rate at fourfold degenerate sites instead of the synonymous rate with the substitution rate in pseudogenes. On the other hand, the rate of nucleotide substitution at nondegenerate sites is usually quite similar to the nonsynonymous rate. This is because the rate of transversional (nonsynonymous) substitution at twofold degenerate sites is on the average quite similar to the rates of transitional and transversional substitution at nondegenerate sites and also because in an average gene the number of twofold degenerate sites is about only one-fourth of the number of nondegenerate sites, so that a different rate at these sites would not much affect the (weighted) average. Therefore, there is no need to make a distinction between the substitution rate at nondegenerate sites and the nonsynonymous rate.

In computing the rates presented in Table III, we have assumed that the divergence time between two genes is the same as the divergence time between the two species from which the genes were obtained. In those genes that belong to multigene families there is a possibility that the genes compared were actually derived from a duplication before the species divergence, or, in other words, that we are comparing paralogous rather than orthologous genes. This possibility exists for the histone genes, immunoglobulin genes, and the interferon α and β genes. Another possibility that can inflate the estimated rates is gene conversion between nonallelic genes of a multigene family. This phenomenon will be discussed in section 7.

3.2. Noncoding Regions

We now consider the rate of nucleotide substitution in the 5′ flanking and untranslated regions, introns, and the 3′ untranslated and flanking regions. The 5′ flanking region refers to the region 5′ upstream from the transcription initiation site; i.e., the cap site of mRNA. This region is thought to contain signal sequences for controlling the initiation of transcription (e.g., Minty and Newmark, 1980; Grosschedl and Birnstiel, 1980a,b; Efstratiadis *et al.*, 1980). The 5′ untranslated region is the segment from the cap site to the initiation codon. This region might contain signal sequences for ribosome binding and initiation of protein synthesis (Lomedico *et al.*, 1979). Introns are DNA regions that are transcribed but deleted before mRNA maturation. An intron generally begins with the dinucleotide GT, ends with the dinucleotide AG, and also contains other unknown signals for RNA splicing (Breathnack *et al.*, 1978; Rogers and Wall, 1980). Except for these signals, other parts of introns apparently do not serve any function; introns, however, may also contain transposons and transcription enhancer elements (e.g., Schon *et al.*, 1981; Khoury and Gruss, 1983). The 3′ untranslated region is the segment from the end of the termination codon to the polyadenylation site. This region contains the AATAAA sequence that is thought to be required for mRNA 3′ end formation and is probably also a signal for endonucleolytic cleavage (Proudfoot, 1984). The 3′ flanking region is the region 3′ to the polyadenylation site. There is evidence that this region contains a signal for transcription termination (Proudfoot, 1984). However, DNA sequencing work usually ends far before this signal, so that we may assume that this region does not have any function.

Since the majority of published sequences are cDNAs or mRNAs, among the noncoding regions only the 5′ and 3′ untranslated regions can be studied in detail. The studies by Efstratiadis *et al.* (1980) and Miyata *et al.* (1980) have suggested that these two regions evolve at fairly high rates. In Table IV, we show the results of our analysis of 11 genes. Because these regions are usually very short, the standard errors for the estimates of the substitution rates are in most cases quite large. For example, among the three comparisons of parathyroid hormone genes the estimated substitution rate varies from 1.78×10^{-9} to 3.71×10^{-9} substitutions per nucleotide site per year. Thus, the large variation in rate among genes may be largely due to random fluctuations rather than due to differences

Table IV
Rate of Nucleotide Substitution in the 5′ Untranslated (5′UT) Region and 3′ Untranslated (3′UT) Region of Genes[a]

Gene	Species compared	5′UT L	5′UT Rate	3′UT L	3′UT Rate
α Globin	Human vs. goat	37	0.54 (0.32)	91	1.56 (0.39)
	Human vs. mouse	34	1.56 (0.65)	90	2.21 (0.49)
	Mouse vs. goat	34	1.06 (0.51)	86	2.82 (0.59)
β Globulin	Human vs. bovine	50	0.39 (0.23)	124	1.88 (0.38)
	Human vs. mouse	50	1.30 (0.46)	126	2.85 (0.48)
	Bovine vs. mouse	52	1.78 (0.57)	119	3.10 (0.54)
Insulin	Human vs. rat I &II	56	3.16 (0.85)	52	2.16 (0.61)
ACTH	Human vs. bovine	103	1.42 (0.34)	153	1.56 (0.30)
	Human vs. mouse	99	1.87 (0.42)	97	2.32 (0.49)
	Bovine vs. mouse	104	1.56 (0.35)	100	2.53 (0.51)
Parathyroid hormone	Human vs. bovine	74	1.78 (0.48)	212	1.52 (0.24)
	Human vs. rat	84	1.79 (0.43)	228	2.21 (0.30)
	Bovine vs. rat	69	3.71 (0.85)	160	2.93 (0.46)
Growth hormone	Human vs. bovine	25	1.50 (0.73)	90	0.92 (0.23)
	Human vs. rat	21	1.79 (0.85)	91	1.83 (0.42)
	Bovine vs. rat	18	2.76 (1.25)	85	1.99 (0.46)
Prolactin	Human vs. bovine	57	1.40 (0.44)	148	1.48 (0.29)
Glycoprotein hormone, α subunit	Human vs. bovine	100	1.24 (0.32)	251	1.32 (0.21)
Somatostatin	Human vs. rat	58	1.82 (0.52)	140	1.17 (0.25)
Corticotropin-releasing factor	Human vs. porcine	123	1.28 (0.29)	441	1.32 (0.16)
Arginine vasopressin neurophysin I	Bovine vs. rat	27	2.13 (0.91)	53	2.47 (0.70)
Average over genes[b]			1.74 (0.66)		1.88 (0.51)

[a] Rates are expressed in terms of number of substitutions per nucleotide site per 10^9 year, assuming that mammalian orders diverged 80 mya. The results are obtained by the two-parameter method. *L*, Number of nucleotides compared. Value in parentheses denotes the standard error.

[b] When the number of species compared is more than two, an average over comparisons is obtained and then used to compute the average over genes.

in intrinsic rates. There is, however, a clear indication that the substitution rates in both regions are somewhat higher in the rodent (mouse, rat) lineage than in the other lineages. It is also clear that both regions evolve at fairly high rates. The average rates for these two regions are 1.74×10^{-9} and 1.88×10^{-9}, respectively, which are almost the same and are about two times higher than the average nonsynonymous rate among genes (Table III).

Current data do not allow an extensive analysis of the substitution rate in the other noncoding regions. In particular, the intron and the 3′

flanking region are known to be subject to rapid nucleotide substitution and a high incidence of deletion and insertion (van Ooyen *et al.*, 1979; Efstratiadis *et al.*, 1980; Miyata *et al.*, 1980). Therefore, in order to obtain reliable estimates of the substitution rates in these regions, one must use fairly closely related sequences. Table V shows the results of our analysis of four gene pairs. It is seen that the number K of substitutions per nucleotide site tends to be slightly higher in the 5′ flanking region than in the 5′ untranslated region, suggesting a slighty higher rate in the 5′ flanking region. The intron regions apparently evolve substantially more rapidly than the 5′ flanking and untranslated regions but somewhat more slowly than fourfold degenerate sites. The 3′ flanking region appears to evolve more rapidly than fourfold degenerate sites, because in the last two comparisons in Table V no sensible alignment for this region can be obtained, although the K values at fourfold degenerate sites are still fairly small (Table V), and in the first two cases the average K value in the 3′ flanking region is about the same as that at fourfold degenerate sites. Rough estimates of the substitution rates in these noncoding regions can be obtained as follows. The sum of the K values for the four cases in Table V is $0.06 + 0.04 + 0.07 + 0.13 = 0.30$ for the 5′ flanking region and $0.04 + 0.12 + 0.13 + 0.24 = 0.53$ for the fourfold degenerate sites. Since the average rate for a fourfold degenerate site is 4.2×10^{-9} (Table III), the average substitution rate in the 5′ flanking region can be estimated as $(0.30/0.53) \times 4.2 \times 10^{-9} = 2.4 \times 10^{-9}$. In the same manner we obtain 1.5×10^{-9}, 3.7×10^{-9}, and 2.3×10^{-9} as the average substitution rates in the 5′ untranslated region, small introns, and the 3′ untranslated region, respectively. The substitution rates in large introns and the 3′ flanking region can be estimated from the K values in the first two comparisons and are given by 3.2×10^{-9} and 4.5×10^{-9}, respectively. The above rates for the 5′ and 3′ untranslated regions are not far from the average rates given in Table IV. Our estimate for the substitution rate in small introns is the same as that of Hayashida and Miyata (1983). Our estimate for the substitution rate in large introns is much smaller than Hayashida and Miyata's estimate, 5.3×10^{-9}. Since our estimate is based on very limited data and theirs was based on comparisons between mammalian orders, we feel that both estimates are not reliable. At any rate, the substitution rates in introns and in the 5′ and 3′ flanking regions need to be studied more carefully, when more sequence data becomes available.

3.3. Pseudogenes

A pseudogene is a DNA segment that shows high homology to a functional gene but contains defects such as nonsense and frameshift mutations that prevent it from producing a functional product. The ex-

Table V
Comparison of Sequence Divergence (number of substitutions per nucleotide site) in Various Noncoding Regions and at Fourfold Degenerate Sites[a]

Genes compared	5′FL	5′UT	Small introns	Large introns	3′UT	3′FL	Fourfold degenerate sites
Bovine vs. goat	0.06	0.02	0.11	0.06	0.11	0.09	0.04
β globin	(0.02)	(0.02)	(0.03)	(0.01)	(0.03)	(0.02)	(0.02)
Bovine vs. goat	0.04	0.04	0.05	0.06	0.06	0.08	0.12
γ globin	(0.01)	(0.03)	(0.02)	(0.01)	(0.02)	(0.02)	(0.04)
Human growth hormone	0.07	0.03	0.09	NA	0.02	—[b]	0.13
vs. lactogen	(0.01)	(0.02)	(0.01)		(0.01)		(0.04)
Rat proinsulin	0.13	0.10	0.22	NA	0.10	—[b]	0.24
I vs. II	(0.06)	(0.05)	(0.05)		(0.05)		(0.09)

[a] 5′FL and 5′UT, 5′ flanking and untranslated regions, respectively; 3′UT and 3′FL, 3′ untranslated and flanking regions, respectively. Small introns are introns with a length shorter than 300 base pairs. The value in parentheses denotes the standard error obtained from formula (4). NA, Not applicable.

[b] No reliable alignment has been obtained.

istence of pseudogenes appears to be a widespread phenomenon, for they are found in most multigene families that have been subject to extensive DNA sequencing [see the review by Li (1983)]. The major process of generating pseudogenes is probably by gene duplication and subsequent loss of function of redundant duplicates. However, many pseudogenes, called processed pseudogenes, have features characteristic of mature mRNAs; e.g., with 3′ poly A sequences but without introns. It has been proposed that these pseudogenes have arisen through reverse transcription of mRNAs carried by retroviruses and then been reinserted into the nuclear genome (e.g., Hollis *et al.*, 1982; Wilde *et al.*, 1982; Schmid and Shen, this volume, Chapter 5). Another way of generating pseudogenes is by duplication of existing pseudogenes (Li, 1983).

Pseudogenes are apparently subject to no functional constraint and thus would accumulate nucleotide substitution at a rate equal to the mutation rate. Under the neutral mutation hypothesis, this should be the maximum rate of nucleotide substitution. For this reason, there has been a great deal of interest in knowing the substitution rate in pseudogenes (Kimura, 1980; Miyata and Yasunaga, 1981; Li *et al.*, 1981).

We have proposed the model shown in Fig. 1 for estimating this rate b (Li *et al.*, 1981; see also Miyata and Yasunaga, 1981). The time T since divergence between mouse and rabbit is known to be about 80 mya, so that the rate a_i of nucleotide substitution at the ith position of codons can readily be estimated from sequence comparison. Considering the number of nucleotide substitutions between each sequence pair at each of the three positions of codons, we can set up equations for obtaining T_d, T_n, and b, where T_d is the time of divergence between a pseudogene and its functional counterpart and T_n is the time since the psuedogene became nonfunctional. (We assume that once a gene becomes nonfunctional the substitution rate b is the same for all positions of codons.)

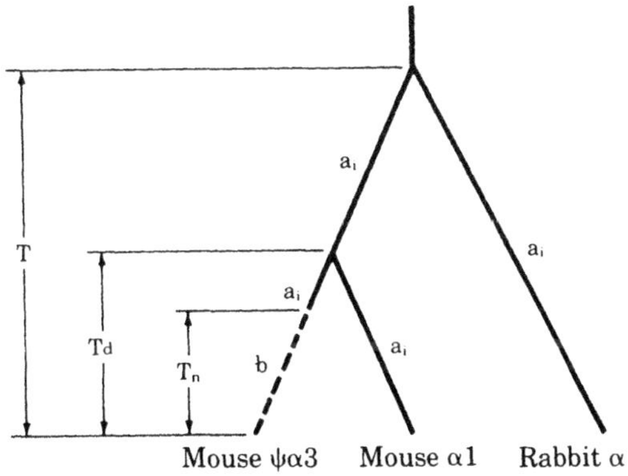

Figure 1. Plausible phylogenetic tree for mouse $\psi\alpha3$, mouse $\alpha1$, and rabbit α pseudogenes. Here T is the divergence time between mouse and rabbit, T_d the time since divergence of mouse $\psi\alpha3$ and $\alpha1$, and T_n the time sinze nonfunctionalization of mouse $\psi\alpha3$. a_i is the rate of nucleotide substitution per site per year at the ith position of codons in the normal globin genes, and b is the rate of substitution for mouse pseudogene $\psi\alpha3$. [From Li *et al.* (1981).]

The above method has been applied to a number of psuedogenes and the results are given in Table VI; the result for the goat globin pseudogenes was obtained by a different method (W. H. Li and T. Gojobori, unpublished results). The estimated rate varies from 4.1×10^{-9} to 6.4×10^{-9} and the average is 4.85×10^{-9}. This rate is indeed higher than any of the rates estimated above. It is also interesting to note that the difference between T_d and T_n is usually quite short. See Li (1983) for a more detailed discussion of the evolution of pseudogenes.

Table VII summarizes our results for the average substitution rates in various gene regions and in pseudogenes. The highest substitution rate is observed in pseudogenes. The 3′ flanking region of a gene evolves almost as rapidly as pseudogenes. Fourfold degenerate sites also evolve very rapidly, the rate being only 14% lower than the pseudogene rate. Introns evolve somewhat more slowly than fourfold degenerate sites. The 5′ flanking region and twofold degenerate sites evolve about half as rapidly as pseudogenes. The 5′ and 3′ untranslated regions evolve at a rate about one-third of the pseudogene rate. Nondegenerate sites are the most conservative, their rate of substitution being only one-fifth of the pseudogene rate. The implication of these observations for the mechanisms of DNA evolution are discussed in section 8.

4. Molecular Clock

4.1. The Controversy

In their comparative studies of amino acid sequences of hemoglobin and cytochrome *c*, Zuckerkandl and Pauling (1962, 1965) and Margoliash (1963) first noted that the rates of amino acid substitution in these proteins were approximately the same among different lineages. Zuckerkandl and Pauling therefore proposed that there exists a molecular evolutionary clock. This proposal was immediately challenged by some evolutionists (see Zuckerkandl and Pauling, 1965). The controversy became particularly heated following two subsequent developments. First, by studying the immunologic distance between the serum albumins of primates, Sarich and Wilson (1967) estimated that man and African apes diverged about 5 mya, much more recently than generally believed at that time. Second, the rate constancy was used to support the neutral mutation hypothesis (Kimura, 1969).

The strongest support for the molecular clock hypothesis is that, when one plots the estimated number of amino acid substitution between protein sequences against their divergence time, the points lie approxi-

Table VI

Rate[b] of Nucleotide Substitution per Year per Site in Pseudogenes, Time since Gene Duplication T_d, and Time since Nonfunctionalization T_n

Pseudogene	b ($\times 10^{-9}$)	T_d (million years)	T_n (million years)	Reference
Mouse $\psi\alpha 3$	5.0 ± 3.2	27 ± 6	23 ± 9	Li *et al.* (1981)
Human $\psi\alpha 1$	5.1 ± 3.3	49 ± 8	45 ± 37	Li *et al.* (1981)
Rabbit $\psi\beta 2$	4.1 ± 3.4	44 ± 10	43 ± 42	Li (1983)
Goat $\psi\beta^X$ and $\psi\beta^Z$	4.4	46	36	Li (1983)
Human Ig VHH16BR	4.1	29	13	Gojobori and Nei (1984)
Mouse Ig VH108B	6.4	25	17	Gojobori and Nei (1984)
Average	4.85			

Table VII

Average Substitution Rates v in Various Gene Regions and in Pseudogenes[a]

	5′FL	5′UT	Nondegenerate sites	Twofold degenerate sites	Fourfold degenerate sites	IVS	3′UT	3′FL	Pseudogenes
v	2.36	1.74	0.94	2.26	4.18	3.70	1.88	4.46	4.85
R	0.49	0.36	0.19	0.47	0.86	0.76	0.39	0.92	1.00

[a] Notations for gene regions: FL, flanking; UT, untranslated; IVS, intervening sequences (introns). All rates are in units of 10^{-9} substitutions per site per year. R, ratio of the average substitution rate in a gene region to the average substitution rate in pseudogenes.

mately along a straight line (Dickerson, 1971; Langley and Fitch, 1974; Wilson *et al.*, 1977; Kimura, 1983a). Further support comes from quantitative immunologic comparison of proteins and from DNA annealing studies [see the review by Wilson *et al.* (1977)]. It should be emphasized that proponents of the molecular clock do allow exceptional cases; a well-known case is that insulin evolved much more rapidly along the line leading to guinea pig than along other lines (King and Jukes, 1969). They also recognize that amino acid substitution usually does not follow a simple Poisson process; the observed variance in substitution rate among different lineages is about two times as large as the Poisson variance (Ohta and Kimura, 1971; Langely and Fitch, 1974; Kimura, 1983a). They propose only approximate constancy.

A number of evolutionists, particularly Goodman (1976, 1981a) and his associates (Czelusniak *et al.*, 1982), strongly deny even the existence of approximate constancy, on two grounds. First, they contend that accelerated evolution has often occurred following gene duplication. For example, Goodman (1976, 1981a) proposes that extremely high rates of amino acid substitution occurred following the gene duplication separating myoglobin and hemoglobin and the gene duplication separating α and β hemoglobins, and that the high rates were due to advantageous mutations improving the function of myoglobin and hemoglobin. We discuss this issue in a separate section (Section 4.3), because the possibility of accelerated evolution following gene duplication is also allowed by the neutralists (Kimura, 1981b), although they believe that acceleration is more likely to arise from relaxation of purifying selection than from advantageous mutations. Second, Goodman *et al.* contend that protein sequence evolution often proceeds much more rapidly at the time of adaptive radiation. For instance, Goodman (1981a) and Goodman *et al.* (1974, 1975) claim that amino acid substitution in cytochrome *c* and in α and β globins occurred rapidly in the period immediately after the teleost-tetrapod divergence, the early period of eutherian divergence, and the early period of primate divergence. There are, however, problems in the methods used by Goodman *et al.* (Nei and Tateno, 1978; Holmquist, 1979; Fitch, 1980; Kimura, 1981b, 1983a). First, sequence evolution may not follow the maximum parsimony principle (Kimura, 1981b). Second, their estimates are tree-dependent, so that if the inferred phylogenetic relationships are wrong, the estimates of substitution rates would be in error. Third, the maximum parsimony method tends to underestimate the number of substitutions in branch-poor lineages. In some studies Goodman *et al.* attempted to overcome this difficulty by using their augmentation procedure, but the simulation study by Tateno and Nei (1978) showed that this procedure can produce large overcompensation [however, see also the counter argument by Czelusniak *et al.* (1978)]. Thus, there are uncertainties in the high rates estimated by Goodman *et al.*

Although the majority of molecular evolutionists agree that sequence evolution is strongly dependent on elapsed time, there is a live controversy over whether the clock is geared to years or to generations (Wilson *et al.*, 1977). Based on DNA annealing data, Laird *et al.*, (1969), Kohne (1970), Kohne *et al.* (1972), Rice (1972), and Benveniste and Todaro (1975) found that the rate of nucleotide substitution between rat and mouse was much higher than that between human and chimpanzee. Their findings were challenged by Sarich and Wilson (1973), who argued that some of the studies were based on questionable estimates of divergence time. Some of the observed differences in rate were, however, just too large to be completely attributed to errors in estimates of divergence time. For example, the reported rate in the mouse–rat divergence was 16 times higher than in the human–chimpanzee divergence (Kohne, 1970), so that the difference is too large to be explained even by Wilson *et al.*'s (1977) preferred estimates of divergence time. Moreover, the study by Benveniste and Todaro (1975) actually employed the "relative rate test," which does not rely on estimates of divergence time. On the other hand, investigations with an immunologic approach (Sarich, 1972; Sarich and Wilson, 1973) have revealed no evident generation-time effect when the serum albumins of humans and tree shrews or of humans and mice were compared. Wilson *et al.* (1977) have also studied the rates of amino acid substitution in 11 proteins in a group of long-living species and a group of short-living species and concluded that there was no clear difference in rate between the two groups. Thus, the issue remains unsettled.

Since the molecular clock hypothesis is still controversial, it is desirable to reevaluate this hypothesis using DNA sequence data. We do so in the following section.

4.2. A Reevaluation

Few authors seem to have used DNA sequence data to examine the molecular clock hypothesis. This is probably largely because not many DNA sequences are available for this purpose. Our discussion will therefore be based mainly on our own study (Wu and Li, 1985, and unpublished results).

Miyata *et al.* (1980) and Hayashida and Miyata (1983) have found that the rate of synonymous substitution is relatively uniform among genes and have proposed to use synonymous substitution as a clock. As mentioned above, our study showed that the rate of synonymous substitution actually varies greatly from gene to gene (Table III), and therefore no single calibration of the synonymous rate is applicable to all genes. The large variation of synonymous rate among genes, however, should not be taken as evidence against the molecular clock hypothesis because what is at issue is not whether the rate is uniform among different kinds of genes, but whether the rate in a gene is constant over time or the same among different evolutionary lineages. We should also emphasize that we concur with Miyata *et al.* that it is better to estimate the divergence time between duplicate genes by using synonymous substitution than by using nonsynonymous substitution, because the rate of synonymous substitution tends to be more similar for duplicate genes than the rate of nonsynonymous substitution.

In their study of the evolution of globin and preproinsulin genes Perler *et al.* (1980) found that the rate of nonsynonymous substitution was approximately constant in the last 500 million years, but the rate of synonymous substitution was much faster in the last 80 million years (after the mammalian divergence) than in the 400 million years or so before the mammalian divergence. They therefore concluded that the accumulation of synonymous substitution becomes saturated after about 100 million years, so that synonymous substitution does not behave as an accurate evolutionary clock over a long period of time. However, the "saturation," phenomenon is probably an artifact, arising from the difficulty of detecting all synonymous substitutions between distantly related sequences (Kimura, 1983a; Gojobori, 1983). In Section 2 we pointed out that Perler *et al.*'s method tends to give a serious underestimate when the degree of sequence divergence is large.

The results shown in Fig. 2 were obtained by using our method. In the computation, however, we have used the weights given by Miyata and Yasunaga (1980) instead of our weights for different possible paths between codons. This is because at the time of computation we had not yet obtained our weights. We tried to redo the computation but found that for several of the distantly related genes, e.g., frog and human globins, no results could be obtained because the degrees of divergence at synonymous sites were too large. At any rate, using Miyata and Yasunaga's weights would tend to underestimate to some extent the synonymous rate, but would have only minor effects on the nonsynonymous rate.

Figure 2a shows the results of our analysis of the evolution of the preproinsulin gene. We used sequences from primates (human, macaque), rat, dog, chicken, and fish (anglerfish, salmon). Like Perler *et al.* (1980), we find that the rate of nonsynonymous substitution is approximately constant over time. We also find that the rate of synonymous substitution is faster after than before the mammalian radiation, which occurred about 80 mya, but the difference in rate is far smaller than that observed by

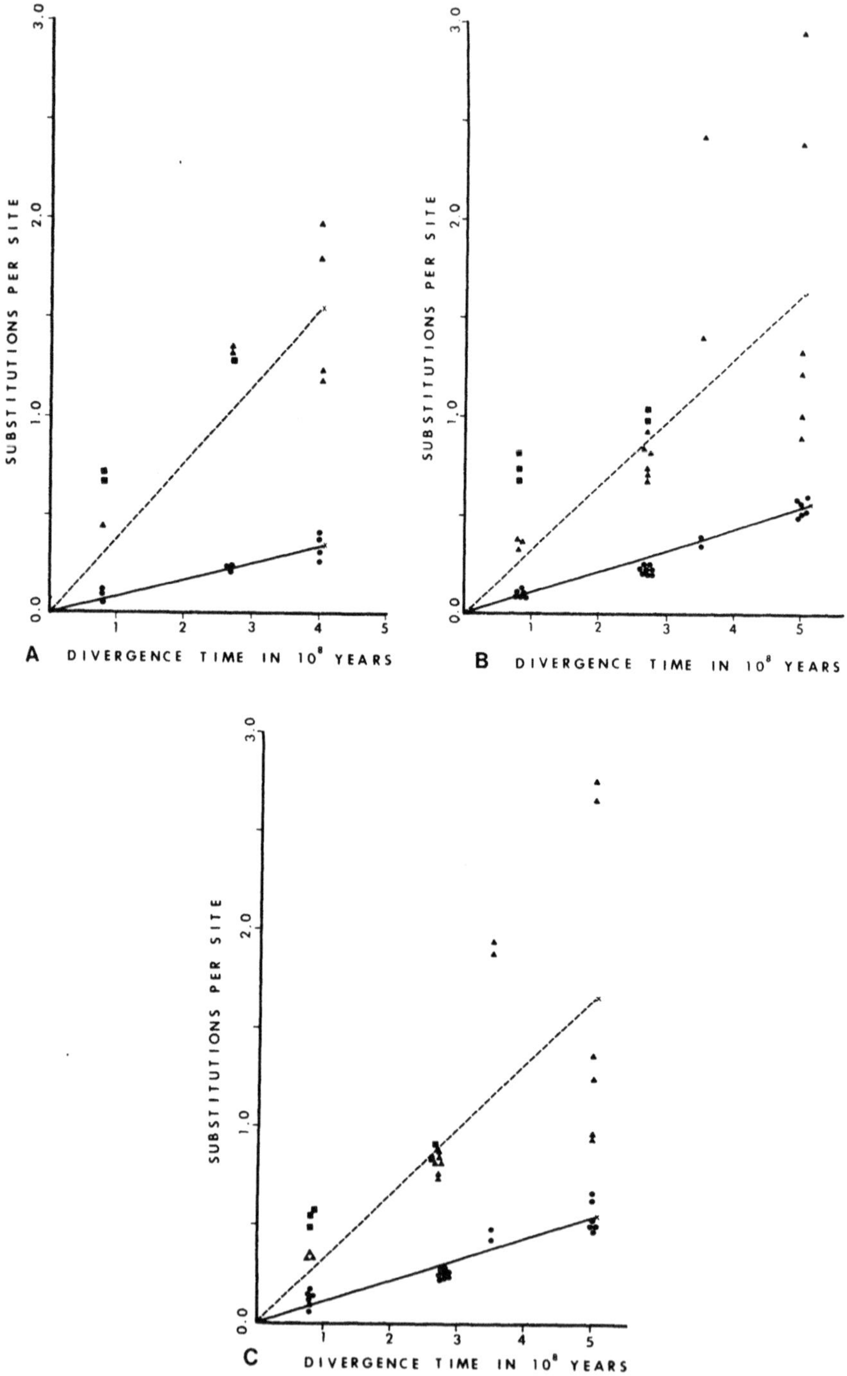
A
B
C
SUBSTITUTIONS PER SITE
DIVERGENCE TIME IN 10⁸ YEARS
0.0
1.0
2.0
3.0
1
2
3
4
5

Perler *et al.* (1980). We believe that the difference is largely due to a lower detectability of nucleotide substitutions between more distantly related sequences rather than due to an acceleration in substitution rate in the last 80 million years. Like other methods, our method also tends to underestimate the substitution rate when the degree of sequence divergence is large (see Section 2). Actually, except in the two comparisons between rat and other mammals, all points lie fairly close to the line connecting the origin and the mean of the comparisons between fish and mammals or chicken. As will be seen below, the above two exceptions are probably due to a higher rate of synonymous substitution in the rat.

Figures 2b and 2c shows the results of our analysis of the evolution of α- and β-globin genes. In the case of α globin the linear relationship between the number of nonsynonymous substitutions and divergence time holds surprising well—except for the comparisons between birds and mammals, all other points lie close to the line connecting the origin and the mean of the comparisons between α and β globins. In the case of β globin, the nonsynonymous rate tends to be somewhat higher in the mouse lineage and in the frog lineage and to be somewhat lower in the bird

---

Figure 2. (A) Divergence of preproinsulin genes. (●;—). Nonsynonymous substitutions [calculation from Eq. (21)]; (▲, ■;- - -) synonymous substitutions [from Eq. (19)]. Each of the two lines is simply drawn between the origin and the mean at the most ancient divergence (represented by ×). At 80 million years, the three pairwise comparisons are among primate (human and macaque monkey), rat (both genes I and II), and dog. At 270 million years the comparisons are between children and each of the three mammalian groups: primate, rat, and dog. The synonymous components involving rat are represented by squares, the rest by triangles. At 400 million years, the comparisons are between each of the two fishes (salmon and anglerfish) and the mammals (averaged over primate, rat, and dog) or chicken. (B) Divergence of α-globin genes. At 80 million years, the six pairwise comparisons are among human, mouse, rabbit, and goat. At 270 million years, the eight comparisons are between each of the four mammals and chicken or duck. The synonymous components involving mouse are represented by squares, the rest by triangles. At 350 million years, the two comparisons are between *Xenopus* and mammals (averaged over the four mammalian species) and between *Xenopus* and birds (averaged over chicken and duck). At 500 million years, the comparisons are between α and β globins. Both mammalian β globins (represented by human, mouse, and goat) and avian β globins (chicken and duck) are compared with mammalian α globins (averaged over the four species), avian α globins (averaged over chicken and duck), and *Xenopus* α globin. In all comparisons between two groups of genes, e.g., mammalian α vs. mammalian β, we take the average of the corrected values of divergence [Eq. (19) and (21)] between each species pair. The exception is the synonymous component between the *Xenopus* α-globin gene and the β-globin gene of duck, which is discarded because of inapplicability of Eq. (13). (C) Divergence of β-globin genes. Same as part B except that, at 500 million years, mammalian α globins (human, mouse, and goat) and avian α globins (chicken, duck, and pigeon) are compared with mammalian, avian, and *Xenopus* β globins.

lineage, but these deviations from linearity are not too large to be attributed to statistical fluctuations. In both genes, the estimated number of synonymous substitutions increases rather irregularly with divergence time, but there is no "saturation" phenomenon as observed by Perler *et al.* (1980). In fact, the synonymous rates obtained from the comparisons between frog and mammals and between frog and birds are rather high. If frog is removed from comparison, there is indeed a "slowdown" in synonymous rate for those distantly related sequences. This shows that in comparing distantly related sequences, adding or removing a sequence can completely change the picture. In other words, such sequences are not suitable for examining the molecular clock hypothesis, for it is difficult to know whether a large deviation is real or is a statistical artifact.

We now consider more closely related sequences. We shall use the so called "relative rate test" (Wilson *et al.*, 1977), which is depicted in Fig. 3. To compare the substitution rates in sequences 1 and 2, we use sequence 3 as a reference. The advantage of this approach is that no knowledge of divergence time is required. Let a, b, and c be the number of substitutions from O to sequences 1, 2, and 3, respectively. The null hypothesis to be tested is $a - b = 0$. Our estimator for $a - b$ is $d_{13} - d_{23}$ $[=(d_{10} + d_{03}) - (d_{20} + d_{03})]$, where d_{ij} is the number of substitutions between points i and j estimated from Eq. (13), (15), or (17). We are interested in testing the constancy of synonymous and nonsynonymous rates separately. For statistical reasons, it is better to consider nonde-

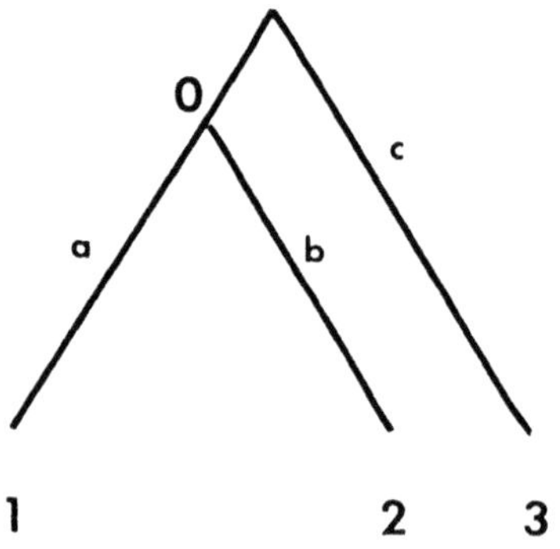

Figure 3. The phylogenetic relationship of species 1, 2, and 3. The symbols a, b, and c represent the numbers of nucleotide substitutions per site from 0 to 1, 2, and 3, respectively. Two cases are considered. In the first case, species 1 is mouse, species 2 is rat, and species 3 (the reference species) is human. In the second case, species 1 is human, species 2 is either rat or mouse, and species 3 is any of the following mammals: bovine, pig, goat, dog, or rabbit.

generate sites, twofold degenerate sites, and fourfold degenerate sites separately; substitutions at twofold degenerate sites are further classified into synonymous and nonsynonymous.

The variances of $d_{13} - d_{23}$ have been derived in Wu and Li (1985). Furthermore, numerical computations have indicated that the *difference* between two identical, independent Poisson variables approaches a normal distribution when the *sum* of their means exceeds 20. We may therefore determine the level of significance according to the procedure for the standardized normal test if the total number of substitutions between the two sequences compared is larger than 20 for the types of sites under consideration. The comparisons with too few substitutions to meet this criterion are generally uninformative and are not considered.

Table VIII shows the difference in substitution number (per site) between the mouse and rat lineages, using the human lineage as an external reference. (It would be better to use a lineage more closely related to mouse and rat as a reference, but this will greatly reduce the number of genes available for comparison.) We choose to compare the rates in mouse and rat because their generation times are quite similar, so that there would be little generation-time effect. We note from Table VIII that, with only one exception, the substitution rates are not significantly different between the two lineages. Although the average rates for all genes combined are higher (insignificantly) in the rat lineage, the sign of the difference varies from gene to gene. It therefore seems that both synonymous substitution and nonsynonymous substitution occur at similar rates in mouse and rat. This conclusion is, of course, quite tentative, because the data used are lmited and the reference species is not closely related to mouse and rat, so that the power of the statistical test is low.

Table IX shows the difference in substitution number between the rodent (mouse, rat) and human lineages, using dog, rabbit, bovine, porcine, or goat as a reference (Wu and Li, 1985). Although some of the reference species may not be equally related to man and rodents, this should introduce no serious error, because mammalian orders diverged at about the same time (Romer, 1966). Since the generation times in rodents are much shorter than that in man, it is interesting to see if the substitution rates are higher in rodents than in man. For synonymous substitution, the evidence is rather strong. In only three out of 24 comparisons is the synonymous rate higher in the human lineage and in all three cases the difference is not significant. In the remaining 21 comparisons, the synonymous rate is higher in the rodent lineage; the difference is significant in eight comparisons. When all 11 genes are combined, the synonymous rates at twofold degenerate sites and fourfold degenerate sites are very significantly higher in the rodent lineage than in the human

Table VIII
Difference in Number of Substitutions per Site between the Mouse (species 1) and Rat (species 2) Lineages in Various Genes[a]

	Nonsynonymous substitution				Synonymous substitution			
	Nondegenerate sites		Twofold dengerate sites		Twofold degenerate sites		Fourfold degenerate sites	
Gene	K_{12}	$K_{13}-K_{23}$	B_{12}	$B_{13}-B_{23}$	A_{12}	$A_{13}-A_{23}$	K_{12}	$K_{13}-K_{23}$
Actin	0.0	0.0	0.011	0.011[b]	0.116	0.005 (0.054)	0.164	0.053 (0.090)
AFP	0.083	−0.010 (0.011)	0.043	−0.031**(0.011)	0.163	−0.075 (0.045)	0.286	−0.120 (0.085)
ALB	0.047	0.009 (0.008)	0.036	0.002 (0.013)	0.205	−0.112 (0.072)	0.242	−0.152 (0.084)
GPH-A	0.006	0.007[b]	0.018	−0.008[b]	0.200	0.145 (0.151)	0.171	0.009[b]
All genes[c]	0.049	−0.002 (0.008)	0.032	−0.010 (0.006)	0.164	−0.036 (0.028)	0.229	−0.065 (0.044)
	$K_{13} = 0.150$		$B_{13} = 0.076$		$A_{13} = 0.368$		$K_{13} = 0.572$	
	$K_{23} = 0.152$		$B_{23} = 0.086$		$A_{23} = 0.404$		$K_{23} = 0.637$	
Number of sites	2534.7		863.3		863.3		547.0	

[a] A_{ij}, B_{ij}, and K_{ij} are, respectively, the number of transitional substitutions, the number of transversional substitutions, and the total number of substitutions between species i and species j. The numbers in the parentheses are the standard errors. **: Significant at the 1% level.

[b] Too little divergence is observed for a statistical test.

[c] The genes are actin, α-fetoprotein (AFP), albumin (ALB), glycoprotein hormone α subunit (GPH-A), the signal petides of GPH-A, metallothionein I, and ACTH. Each signal peptide is too small for comparison studies.

Table IX

Difference in Number of Substitutions per Site between the Human (species 1) and Rodent (species 2) Lineages in Various Genes[a]

	Nonsynonymous substitution				Synonymous substitution			
	Nondegenerate sites		Twofold degenerate sites		Twofold degenerate sites		Fourfold degenerate sites	
Gene	K_{12}	$K_{13}-K_{23}$	B_{12}	$B_{13}-B_{23}$	A_{12}	$A_{13}-A_{23}$	K_{12}	$K_{13}-K_{23}$
Proinsulin	0.074	−0.038 (0.023)	0.052	−0.061 (0.035)	0.341	0.035 (0.129)	0.767	−0.646* (0.328)
Actin	0.0	0.0	0.0	0.0	0.264	−0.096 (0.056)	0.424	−0.224* (0.096)
GH	0.211	0.136** (0.028)	0.089	0.062* (0.030)	0.389	−0.018 (0.096)	0.607	0.211 (0.122)
GPH-A	0.170	0.141** (0.034)	0.014	−0.004 (0.016)	0.403	−0.165 (0.127)	0.675	0.357 (0.538)
Prolactin	0.226	−0.118** (0.029)	0.130	−0.030 (0.038)	0.330	−0.009 (0.082)	1.11	−0.499* (0.252)
Relaxin	0.373	−0.074 (0.049)	0.093	−0.038 (0.033)	0.367	−0.163 (0.098)	0.896	−0.367 (0.229)
Four signal peptides	0.270	−0.080 (0.052)	0.110	−0.096 (0.059)	0.402	−0.122 (0.133)	0.712	−0.395* (0.182)
POMC	0.121	−0.056** (0.018)	0.082	−0.038 (0.026)	0.179	−0.073 (0.053)	0.646	−0.212* (0.104)
Ig κ	0.320	−0.009 (0.059)	0.150	0.012 (0.082)	0.459	−0.425 (0.241)	0.738	−0.243 (0.202)
Ig γ	0.226	−0.070* (0.031)	0.127	0.0 (0.041)	0.314	−0.205* (0.084)	0.829	−0.280* (0.143)
α Globin	0.093	−0.017 (0.020)	0.039	0.012 (0.025)	0.417	−0.351** (0.129)	0.696	−0.095 (0.130)
β Globin	0.141	−0.032 (0.025)	0.042	−0.087** (0.030)	0.274	−0.133 (0.091)	0.394	−0.153 (0.104)
All genes	0.168	−0.019* (0.008)	0.077	−0.018 (0.009)	0.322	−0.123** (0.026)	0.659	−0.194** (0.041)
		K_{13} = 0.146		B_{13} = 0.064		A_{13} = 0.183		K_{13} = 0.408
		K_{23} = 0.165		B_{23} = 0.083		A_{23} = 0.307		K_{23} = 0.603
Number of sites		3602.0		1094.3		1094.3		895.7

[a]Taken from Wu and Li (1985). A_{ij}, B_{ij}, and K_{ij} are, respectively, the number of transitional substitutions, the number of transversional substitutions, and the total number of substitutions between species i and species j. The numbers in the parentheses are the standard errors. GH, Growth hormone; GPH-A, glycoprotein hormone α subunit; POMC, pro-opiomelanocortin; Ig κ and Ig γ, immunoglobulin κ and γ contant regions; the four signal peptides are from prolactin, relaxin, GH, and GPH-A. *: Significant at 5% level. **: Significant at 1% level.

lineage ($p < 0.0001$ for both twofold and fourfold degenerate sites). From Table IX, it is straightforward to calculate the number of substitutions from the time of human-rodent divergence to the present time along each lineage. With all genes combined, there have been 0.099 and 0.223 synonymous substitutions per twofold degenerate site along the human and rodent lineages, respectively. The corresponding numbers are 0.232 and 0.427 for fourfold degenerate sites. In other words, the synonymous rate in the rodent lineage has been about twice as high as that in the human lineage since the time of their divergence.

For nonsynonymous substitution, the rates also appear to be faster in the rodent lineage, although the differences are not so dramatic as in the case of synonymous substitution. In 16 out of the 24 comparisons, the rodent lineage evolves more rapidly; among them, four (from four different genes) are significant. By contrast, the human lineage evolves more rapidly in only five comparisons; among them, three (from two different genes) are significant. With all 11 genes combined, the difference in rate between rodents and man is significant for nondegenerate sites ($p < 0.015$) and marginally significant for twofold degenerate sites ($p = 0.05$). A simple calculation will show that the rodent lineage has a nonsynonymous rate about 30% higher than that in the human lineage since the time of their divergence (0.090 versus 0.069 substitutions per nonsynonymous site).

We also found that rodents evolve 2–3 times more rapidly than man in the 5′ and 3′ untranslated regions of five different mRNAs and that the numbers of nucleotide substitutions between members of the β-globin gene family that were duplicated before the man-mouse split are also higher in mouse than in man (Wu and Li, 1985). Our results show that in general the differences in substitution rate between rodent and man are higher in DNA regions that are subject to weaker selective constraints.

We apparently have a dilemma, namely that the generation-time effect appears to be quite strong on synonymous substitutions but weak on nonsynonymous substitutions. This dilemma is, however, not difficult to explain. Since synonymous substitution is generally subject to only weak functional constraints, an increase in mutation rate would lead to a more or less proportional increase in synonymous rate. Thus, if the mutation rate is considerably higher in rodents than in man, so will be the synonymous rate. According to Wilson *et al.*'s (1977) estimate, the *average* number of cell divisions per unit time is about two times higher in the rodent lineage than in the human lineage. If the rate of mutation is directly proportional to the number of cell divisions per unit time, the synonymous rate would be about two times higher in rodents than in man. Interestingly, this is what we found above. On the other hand, nonsynonymous substitution is generally subject to strong selective constraints, so that the rate of nonsynonymous substitution is affected not only by the rate of mutation, but also by effective population size and the intensity of natural selection. Thus, it may not be necessarily much higher in organisms with short generations than in organisms with long generations [see Wu and Li (1985) for a more detailed discussion].

The above reasoning may also explain why DNA annealing data show generation-time effects whereas protein sequence data and immunologic data do not (see above). In the latter types of data, the changes detected are amino acid substitutions, which, like nonsynonymous substitutions, may not show conspicuous generation-time effects. On the other hand, most DNA sequences would contain parts (e.g., spacer regions and introns) that are subject to only weak functional constraints and would thus show strong generation-time effects. The same reasoning appears to apply to the 5′ and 3′ untranslated regions of genes. The functional constraints on these regions are apparently considerably weaker than those on nonsynonymous sites, for they evolve considerably more rapidly than nonsynonymous sites (Table VII). It is therefore not surprising to see that in these regions the rate of nucleotide substitution is higher in the rodent lineage than in the human lineage (Table IV); i.e., these regions also show strong generation-time effects.

In summary, the results presented in Fig. 2 and in Tables XIII and IX suggest that nonsynonymous substitution proceeds roughly in a clocklike manner, as long as there is no large change in generation time. Although in some genes large deviations may occur, the rate constancy holds approximately when many genes are considered together. Moreover, differences in generation time seem to have only relatively weak effects on the rate of nonsynonymous substitution. The results also suggest that the generation-time effect on synonymous substitution is fairly strong. However, synonymous substitution also proceeds roughly in a clocklike manner in a lineage if no large change in generation time occurs.

4.3. Rates of Nucleotide Substitution following Gene Duplication

After a gene is duplicated, the rate of sequence evolution may be accelerated for several reasons. First, some degree of relaxation in selective constraints may occur because two genes instead of one are now serving the same function. Second, the two resultant genes may become diversified or specialized in function, as in the case of myoglobin and hemoglobin, or may develop some interactions between their products, as in the case of α and β globins. Third, one of the two genes may acquire a new function and become a new gene. All these possibilities are allowed by both selectionists (Goodman, 1976, 1981a) and neutralists (Kimura, 1981b, 1983a). It has, however, been very controversial as to whether or not considerable acceleration in rate has really occurred in certain duplicate genes (Goodman, 1976,1981a; Wilson *et al.*, 1977; Kimura, 1981b). There is also a strong disagreement as to whether relaxation of selective constraints (purifying selection) or advantageous mutation (positive Darwinian selection) is a more likely cause of acceleration, if this has indeed occurred (Goodman, 1981a; Kimura, 1981b; Li and Gojobori, 1983).

The most frequently cited examples of accelerated evolution following gene duplication are the studies by Goodman (1976, 1981a) and his associates (Goodman *et al.*, 1974, 1975), who proposed that extremely high rates of amino acid substitution occurred following the gene duplication separating myoglobin and hemoglobin and the gene duplication separating α and β hemoglobins. These two examples have been challenged by Wilson *et al.* (1977) and Kimura (1981b), who contended that the alleged high rates are overestimates because the actual dates of gene duplication appear to be much older than those assumed by Goodman *et al.* For example, Goodman *et al.* assumed that the gene duplication leading to myoglobin versus hemoglobin divergence occurred about 470 mya, whereas Kimura (1981b) argued that the duplication time was probably about 800 mya or even earlier. Furthermore, as mentioned above, the methods used by Goodman *et al.* have been questioned. Some other cases of accelerated evolution — e.g.,the divergence between lysozyme C and lactalbumin (Dickerson and Geis, 1969; Dickerson, 1971) — have also been proposed, but, like the above two cases, they are all of ancient duplication and the dates of duplication are uncertain [see the review by Wilson *et al.* 1977)].

To resolve the above issues, Li and Gojobori (1983) proposed to use relatively closely related duplicate genes, for in such cases it is easier to obtain reliable estimates of the numbers of nucleotide substitution and to infer whether or not the rates of nonsynonymous substitution have been substantially elevated. In Table X, we show four examples of duplicate genes whose degree of divergence is still quite small. The comparison of rabbit β and mouse β major is given for reference. We note that in this comparison the nonsynonymous rate is four or five times lower than the synonymous rate and the substitution rate in introns. By contrast, in the comparison of goat adult (β^A) versus goat and sheep preadult (β^C) globin genes, the nonsynonymous rate is not too far from the synonymous rate and the substitution rate in introns. This is also true for the comparison of fetal (γ) globin versus the β^A and β^C in goat and sheep and that of human growth hormone versus placental lactogen, although in the case of rat proinsulin I versus II the nonsynonymous rate is much lower than

Table X
Comparison of Sequence Divergence at Nonsynonymous Sites, Synonymous Sites, and Nucleotide Sites in Introns in Duplicate Genes

Genes compared	Nonsynonymous sites	Synonymous sites	Introns
Rabbit β vs. mouse β major	0.14 ± 0.02	0.57 ± 0.11	0.69 ± 0.08
Goat β^A vs. goat and sheep β^C	0.07 ± 0.01	0.12 ± 0.03	0.11 ± 0.01
Goat and sheet γ vs. goat and sheep β^A and goat β^C	0.14 ± 0.02	0.17 ± 0.04	0.12 ± 0.01
Human growth hormone vs. placental lactogen	0.07 ± 0.01	0.15 ± 0.04	0.09 ± 0.01
Rat proinsulin I vs. II	0.02 ± 0.01	0.27 ± 0.08	0.22 ± 0.05

the synonymous rate and the substitution rate in introns. Thus, in three of the four cases considered the nonsynonymous rate appears to have been substantially elevated. This suggests that accelerated evolution can often occur following gene duplication.

In the case of goat and sheep globin genes Li and Gojobori (1983) examined further to see in which functional groups acceleration has occurred (see Table XI). In the β chain of hemoglobin, the positions with the most crucial functions are, first, the heme contacts (HC) and, second, those concerned with the interchain cooperativity (Coop) that facilitates oxygen delivery. These two regions in β^A, β^C, and γ are highly conserved because no nonsynonymous substitution, except the one in the HC region of goat γ, has occurred in these two regions (Table XI). The third most important fucntional group in the β chain is the positions for the $\alpha_1\beta_1$ contacts. In this region the nonsynonymous substitution rate has been accelerated in β^A and β^C but not in γ. The fourth most important functional group is the remaining interior positions (IP). The rate in this region has been accelerated in β^A and γ. The least important parts of the β chain are the remaining exterior positions (Other). In this region acceleration has occurred in β^A, β^C, and γ. Thus, acceleration has occurred only in less important functional groups of the β chain, particularly, the exterior positions. We believe that this pattern of acceleration can best be explained by relaxation of purifying selection, because some degree of relaxation would allow acceleration to occur in the less important parts but not in the most important parts of the gene.

5. Patterns of Nucleotide Substitution in Pseudogenes and Functional Genes

Since point mutation is one of the most important factors in the evolution of DNA sequences, molecular evolutionists have long been

Table XI
Rate of Nonsynonymous Substitution per Site per Year in Different Functional Groups of the β Chain[a]

Genes compared	Divergence time (mya)	Rate ($\times 10^9$) for given functional group[b]				
		HC	Coop	$\alpha_1\beta_2$	IP	Other
$H\beta$ vs. $M\beta^{maj}$	80	0.4 ± 0.2	0.1 ± 0.1	1.0 ± 0.5	1.0 ± 0.4	1.1 ± 0.2
$G\beta^A$ vs. $S\beta^A$	7	0.0	0.0	2.6 ± 2.6	2.1 ± 2.1	0.5 ± 0.5
$G\gamma$ vs. $S\gamma$	7	2.5 ± 2.5	0.0	0.0	0.0	1.9 ± 1.0
$G\beta^C$ vs. $G\beta^A$, $S\beta^A$	15	0.0	0.0	3.4 ± 1.4	1.3 ± 0.8	3.4 ± 0.6
$G\gamma$, $S\gamma$ vs. $G\beta^A$, $S\beta^A$, $G\beta^C$	30	0.2 ± 0.1	0.0	0.9 ± 0.3	5.4 ± 1.0	3.3 ± 0.3

[a]Taken from Li and Gojobori (1983). $H\beta$, Human β; $M\beta^{maj}$, mouse β major; $G\beta^A$ and $S\beta^A$, goat and sheep β^A; $G\beta^C$, goat β^C; $G\gamma$ and $S\gamma$, goat and sheep γ.

[b]The functional groups of codon positions follow the scheme used in Goodman (1981). HC, heme contacts; Coop, the cooperative sites, including sites for $\alpha_1\beta_2$ contacts, Bohr effect, and 2,3-diphosphoglycerate (2,3-DPG) binding; $\alpha_1\beta_1$, $\alpha_1\beta_1$ contacts; IP, interior positions; Other, remaining positions. Since the goat and sheep hemoglobins have lost the ability to bind 2,3-DPG, we have included the four codon positions for the 2,3-DPG binding under Other instead of Coop.

interested in knowing its pattern (Beale and Lehmann, 1965; Fitch, 1967; Zuckerkandl *et al.*, 1971; Vogel and Kopun, 1977, and references therein). In inferring this pattern three approaches have been used. One is to use the electrophoretic variants of hemoglobin (Fitch, 1967; Vogel and Kopun, 1977). The pattern thus obtained may be biased because such variants are unlikely to have the same fitness and because electrophoresis can detect only mutations that change the electrophoretic mobility of the protein. Another approach is to compare homologous protein sequences (Fitch, 1967; Vogel and Kopun, 1977). This approach is again likely to give a biased pattern because amino acid substitutions do not occur in a random manner (Dayhoff, 1972). The third approach is to study revertants of mutants with known base pair changes, using physiological and genetic tests (Fowler *et al.*, 1974; Sinha and Haimes, 1980). One drawback of this approach is that mutations can be studied only at particular sites and often only in certain directions. Recently we have proposed to infer the pattern by using pseudogene sequences (Gojobori *et al.*, 1982b; Li *et al.*, 1984). Since pseudogenes appear to have no function, all mutations in them would be selectively neutral and would beocme fixed in the population with equal probability. Thus, the pattern of nucleotide substitution in pseudogenes would reflect the pattern of spontaneous point mutation. The methodological aspects of our approach are given in Gojobori *et al.* (1982b) and Li *et al.* (1984). Roughly speaking, to infer the nucleotide substitutions in a pseudogene sequence, we need one or more reference sequences, which are usually functional genes closely related to the pseudogene.

The first matrix in Table XII represents the results of our analysis of 13 psueodgene sequences [Table II of Li *et al.* (1984)]; the average is weighted by the number of substitutions in each pseudogene. The elements f_{ij} in the matrix are called the relative substitution frequencies; f_{ij} represents the expected number of substitutions from the *i*th to the *j*th type nucleotide among every 100 substitutions in a sequence with equal frequencies of the four types of nucleotide (a random sequence); e.g., $f_{AT} = 4.7$ and $f_{AC} = 5.0$. The results are expressed in terms of the sense strand; i.e., the untranscribed strand. Thus, a G to A change actually means that a G:C pair is replaced by an A:T pair. This can occur as a result of either a G to A mutation in the sense strand or a C to T mutation in the other strand. Similarly, a C to T change can occur as a result of either a C to T mutation in the sense strand or a G to A mutation in the other strand. If there is no difference in the pattern of mutation between the two strands, we should have $f_{GA} = f_{CT}$. Under the same condition, we should also have $f_{AG} = f_{TC}$, $f_{AT} = f_{TA}$, $f_{AC} = f_{TG}$, $f_{CA} = f_{GT}$, and $f_{CG} = f_{GC}$. Since these equalities indeed hold approximately (Table XII), we assume that there is no difference in the pattern of mutation between

Table XII
Relative Substitution Frequency (%) in Pseudogenes and in Functional Genes[a]

	A	T	C	G	
			Pseudogenes		
	A	T	C	G	[59.2] (54.4)
A	—	4.7±1.3	5.0±0.7	9.4±1.3	19.1
	—	(5.3±1.4)	(5.6±0.8)	(10.3±1.4)	(21.2)
T	4.4±1.1	—	8.2±1.3	3.3±1.2	15.9
	(4.8±1.1)	—	(9.2±1.3)	(3.6±1.3)	(17.6)
C	6.5±1.1	21.0±2.1	—	4.2±0.5	31.7
	(7.1±1.3)	(18.2±2.3)	—	(4.2±0.6)	(29.5)
G	20.7±2.2	7.2±1.1	5.3±1.0	—	33.2
	(18.6±1.9)	(7.7±1.3)	(5.5±1.3)	—	(31.8)
	31.6	32.9	18.5	16.9	
	(30.5)	(31.2)	(20.3)	(18.1)	
			Estimated pattern (pseudogenes)		
	A	T	C	G	[59.2] (54.4)
A	—	4.6	4.2	8.8	17.6
	—	(5.0)	(4.6)	(9.7)	(19.3)
T	4.6	—	8.8	4.2	17.6
	(5.0)	—	(9.7)	(4.6)	(19.3)
C	6.8	20.8	—	4.8	32.4
	(7.4)	(18.4)	—	(4.9)	(30.7)
G	20.8	6.8	4.8	—	32.4
	(18.4)	(7.4)	(4.9)	—	(30.7)
	32.2	32.2	17.8	17.8	
	(30.8)	(30.8)	(19.2)	(19.2)	
			Functional genes		
	A	T	C	G	[45.5]
A	—	4.2±2.0	6.3±2.8	11.5±1.4	22.0
T	5.0±1.2	—	3.7±1.6	1.7±1.4	10.4
C	8.1±0.6	9.4±3.3	—	13.0±1.5	30.5
G	20.9±3.1	4.8±0.8	11.5±3.8	—	37.2
	34.0	18.4	21.5	26.2	

[a]Taken from Gojobori *et al.* (1982b) and Li *et al.* (1984).

the two strands. The second matrix in Table XII is obtained under this assumption, and is regarded as our estimate of the pattern.

In each of the matrices in Table XII, the four elements from the upper right corner to the lower left corner are the f_{ij} values for transitions, while the other eight elements are for transversions. It is seen from the first two matrices that all transitions, particularly C to T and G to A, occur more often than transversion. The sum of the relative frequencies of transitions is given in the brackets at the upper right corner and is 59.2%. We note that under random mutation the expected proportion of transitions is only 33%, for there are only four types of transitions but eight types of transversions. Therefore, the observed proportion is almost two times the value expected under random mutation.

In each matrix, we have computed the sum of the f_{ij} values in each column and presented the results in the row below the matrix. These sums represent the expected numbers of substitutions that result in A, T, C, and G, respectively, among every 100 substitutions in a random sequence. In the second matrix, these values are 32.2, 32.2, 17.3, and 17.3. We have also computed the relative frequencies that one type of nucleotide is replaced by any of the others. This quantity is given by the sum of the f_{ij} values in each row and is presented in the last column. The sums are 32.4 for both C and G but only 17.3 for A and T. From these observations, we may conclude that pseudogenes will become rich in A and T bases. The equilibrium frequencies of A, T, C, and G can be obtained by using the formula given by Wright (1969, p. 26) and Tajima and Nei (1982) and are 0.34, 0.34, 0.16, and 0.16, respectivley.

It is known that, in addition to base mispairing, the transition from C to T can also arise from conversion of methylated C residues to T residues upon deamination (Coulondre *et al.*, 1978; Razin and Riggs, 1980). The effect will elevate the frequencies of C:G → T:A and G:C → A:T; i.e., f_{CT} and f_{GA}. Since about 90% of methylated C residues in eukaryotic DNA occur at 5′-CG-3′ dinucleotides (Razin and Riggs, 1890), this effect should be expressed mainly as changes of the CG dinucleotide to TG or CA. After a gene becomes a pseudogene such changes would no longer be subject to any functional constraint and can therefore contribute significantly to C → T and G → A transitions if the frequency of CG is relatively high before silencing of the gene occurs. The substitution pattern obtained by excluding all nucleotide sites where the CG dinucleotides appear to have occurred in the ancestral sequences of these pseudogenes is given in the parentheses in the first two matrices in Table XII. This pattern is probably more suitable for predicting the pattern of mutations in a sequence that has not been subject to functional constraints for a long time (e.g., some parts of an intron) because in such a sequence there would exist few CG dinucleotides. The pattern with exclusion of CG dinucleotides is somewhat different from that obtained without excluding the CG dinucleotides. In particular, the differences among the relative frequencies of the four transitions become somewhat less conspicuous and the relative frequencies for the transversions become slightly higher except for G ↔ C. The sum of the relative frequencies of transitions is now reduced to 54% and the equilibrium frequencies of A, T, C, and G are 0.32, 0.32, 0.18, and 0.18, respectively.

The observation that transitional mutation occurs considerably more often than transversional mutation has the interesting consequence of in effect elevating the rate of synonymous mutation in a gene (Li and Gojobori, 1983; Li *et al.*, 1984). Indeed, for a gene consisting of only TTC (Phe) codons, the rate of synonymous mutation is expected to be 20.8/8.3 = 2.5 times the rate under random mutation, because f_{CT} is 20.8 under nonrandom mutation (Table XII) and 100/12 = 8.3 under random mutation. The increase, of course, would vary from one gene to another and would not be so large in general. Obviously, nonrandom mutation should have the tendency to reduce the rate of nonsynonymous mutation in a gene. Table XIII shows the ratios of the rate of synonymous or nonsynonymous mutation expected under the mutation pattern given in Table XII to that expected under random mutation in various genes. In all the sequences considered, the ratio of the synonymous rates is at least 1.1 and can be as high as 1.5, while the ratio of the nonsynonymous rates can be as low as 0.9. Thus, in some genes, such as the human α-globin gene, the rate of synonymous mutation per synonymous site is considerably higher than the rate of nonsynonymous mutation per nonsynonymous site. For this reason, the observation that the rate of synonymous substitution in a gene is generally much higher than the rate of nonsynonymous substitution might be partly the result of nonrandom mutation. Of course, the difference between the two rates is presumably mainly due to differences in functional constraints, because it is generally much larger than can be accounted for by nonrandom mutation.

Table XIII

Ratio of the Rate of Synonymous or Nonsynonymous Mutations Expected under Nonrandom Mutation to the Rate Expected under Random Mutation in Various Genes[a]

Genes	Synonymous	Nonsynonymous
Random sequences	1.15	0.96
Human α1 globin	1.47	0.97
Human β globin	1.33	0.94
Human albumin	1.15	0.91
Mouse α-fetoprotein	1.28	0.90
Mouse α-amylase	1.11	0.92
Mouse IgE	1.30	0.91
Drosophila ADH	1.42	0.92

[a]Taken from Li *et al.* (1984).

We have also considered some other evolutionary implications of nonrandom mutation (Li *et al.*, 1984). We showed that nonrandom mutation reduces (by about 10%) the proportion of polar changes among nonsynonymous mutations in a gene, but, as far as single nucleotide changes (in the codon table) are concerned, nonrandom mutation has only a slight bias favoring relatively conservative amino acid interchanges and has virtually no effect on the proportions of radical changes and nonsense mutations. The pattern of point mutation inferred from pseudogenes is partly in agreement with and partly contradictory to the predictions from Topal and Fresco's (1976) model for the molecular basis of point mutation (Gojobori *et al.*, 1982b; Li *et al.*, 1984).

We have also studied the substitution pattern in mammalian α-globin, β-globin and adrenocorticotropic hormone (ACTH) genes (Gojobori *et al.*, 1982b). Since our purpose was to know the effect of functional constraints on the pattern of nucleotide substitutions, we excluded the third position of codons because the functional constraints appear to be weak at this position in the majority of cases. There are some conspicuous differences between this pattern and the pattern based on pseudogenes (Table XII). The relative frequency of C $\rightarrow$ T has been drastically reduced, whereas those of C $\rightarrow$ G and G $\rightarrow$ C have been doubled (Table XII). Because of the reduction in the frequency of C $\rightarrow$ T, the proportion of transitions has been reduced to 46% (Table XII). The equilibrium frequencies of A, T, C, and G computed from this pattern are 0.30, 0.36, 0.17 and 0.17, respectively.

In order to understand what kinds of functional constraints were responsible for the differences between the two patterns, Gojobori et al. (1982b) studied the relationship between base exchangeability and amino acid exchangeability, because the majority of base changes at the first and second positions of codons cause changes in amino acids. A distance between nucleotides was defined using Grantham's (1974) distance between amino acids [for details, see Gojobori *et al.* (1982b)]. The average distance between bases at the first and second positions of codons is given at the bottom row in Table XIV. Above the chemical distances we show the relative substitution frequencies at the first and second positions of codons in functional genes and in six structural pseudogenes.

We note from Table XIV that if the average chemical distance between two bases is large, the relative substitution frequency in the functional genes is lower than that in the pseudogenes, but the situation is reversed if the average chemical distance is small. For example, the average chemical distance between C and T is large (120.5) and the relative substitution frequency between C and T in the functional genes is only half of that in the pseudogenes. On the other hand, the average chemical distance between C and G is small (84.5) and the relative substitution

Table XIV
Correlation between the Relative Frequency of Base Substitution and the Chemical Distance between Bases[a]

		Relative substitution frequency							
Sequence	N	C ↔ T	G ↔ A	C ↔ G	C ↔ A	T ↔ G	T ↔ A	Correlation coefficient	Rank correlation
Pseudogenes									
Human $\psi\alpha1$	49	27.6	33.5	10.1	17.2	4.6	7.0	−0.60	−0.77*
Mouse $\psi\alpha3$	25	23.8	35.1	4.1	11.4	13.0	12.7	−0.22	−0.03
Rabbit $\psi\beta2$	41	21.8	26.5	10.0	19.7	12.4	9.6	−0.51	−0.49
Goat $\psi\beta^{X}$	22	31.1	32.3	19.5	9.9	3.5	3.7	−0.61	−0.83*
Mouse $\psi\beta h3$	13	20.1	22.0	0.0	16.0	24.1	17.8	+0.47	+0.37
Human ψV_{K}	28	21.4	21.6	10.5	13.8	25.8	6.9	+0.15	+0.09
Weighted average	178	24.7	29.2	9.7	15.4	12.2	8.8	−0.43	−0.49
Functional genes									
α-globin genes	39	11.2	33.3	35.0	9.6	2.6	8.3	−0.81*	−0.89*
β-globin genes	44	10.4	22.6	19.6	20.1	11.3	16.2	−0.82*	−0.77[b]
ACTH genes	49	16.8	40.5	20.4	13.0	5.4	3.8	−0.80[b]	−0.89*
Weighted average	132	13.1	32.4	24.5	14.4	6.5	9.2	−0.88*	−1.00**
Chemical distance		120.5	76.5	84.5	89.0	153.0	142.5		

[a] Taken from Gojobori *et al.* (1982b). *N*, Number of substitutions. *: Significant at the 5% level. **: Significant at the 1% level.
[b] On the borderline of 5% significant level.

frequency between C and G in the functional genes is two times higher than that in the psuedogenes. Thus, the chemical distance between bases defined above appears to be a good indicator of the stringency of functional constraints.

We have also computed the moment-product correlation coefficient and the rank correlation coefficient between the average chemical distance and the relative substitution frequency (see the last two columns of Table XIV). We note that a significant negative correlation exists between the two quantities in the functional genes but not in the pseudogenes.

6. Nonrandom Usage of Synonymous Codons

6.1. Pattern of Codon Usage

When Adams *et al.* (1969) obtained the first nucleotide sequence, which was a fragment from the coat protein gene of R17 RNA, they were curious about which codons were actually used. They suggested that the sequence of a messenger RNA is determined not only by the need to specify an amino acid sequence, but also by its need to assume a particular secondary structure. Since then it has become almost a standard practice for molecular biologists to tabulate the codon usage when they obtain a sequence. As more sequences became available, it became evident that nonrandom usage of codons is the rule rather than an exception (e.g., Fiers *et al.*, 1975; Sanger *et al.*, 1977). Indeed, a recent extensive compilation by Grantham *et al.* (1980a,b, 1981) showed that usage is distinctly nonrandom in both prokaryotic and eukaryotic genes. Interestingly, genes in an organism or in related species generally show the same pattern of choices among synonymous codons. Thus, mammalian, bacterial, viral, mitochondrial, and yeast plus slime mold genes fall into distinct classes. Grantham *et al.* therefore proposed the genome hypothesis, postulating that genes in a genome use the same coding strategy with respect to choices among synonymous codons. Ikemura and Ozeki (1983) have recently compiled a table of codon usage in *Escherichia coli, Salmonella typhimurium,* and *Saccharomyces cerevisiae* (yeast). Like Grantham *et al.*, they found that genes in each of these species show clear similarities of choices among synonymous codons and that the usage pattern of *E. coli* genes is quite similar to that of *Salmonella* genes but very different from that of yeast genes.

Using Ikemura and Ozeki's compilation and data from T7 (Sharp *et al.*, 1985) and from man (28 genes), and following Grantham *et al.* (1980a) and Kimura (1983a), we have computed the frequencies of codon usage per thousand codons in *E. coli* (7898 codons in total), *S. typhimurium* (1374 codons), T7 (12045 codons), yeast (3136 codons), and man (5484 codons) (Table XV). We note that in all five species usage of the 61 sense codons is highly nonrandom; the expected usage frequency per codon is 1000/61, or approximately 16, if usage is random. We note further that strong biases also occur in the choice of synonymous codons. For example, among the six codons for Leu, CUG is predominantly used in *E. coli, S. typhimurium,* T7, and man, while UUG is predominantly used in yeast. The patterns of usage in the two bacterial species and in T7, which has the former two species as two of its hosts, are quite similar to each other, but are very different from the pattern in yeast, confirming the observations by Grantham *et al.* (1981), Ikemura and Ozeki (1983), and Sharp *et al.* (1984). Interestingly, synonymous codons are used more evenly in man than in bacteria and yeast. For instance, the six codons for Arg are fairly evenly used in man, but the two codons AGA and AGG are almost excluded from the *E. coli* genome.

Nonrandom codon usage has been found to be particularly conspicuous in highly expressed genes (e.g., Grantham *et al.*, 1981; Bennetzen and Hall, 1982; Grosjean and Fiers, 1982). For example, only 33 of the 61 sense codons are used in the two glyceraldehyde-3-phosphate dehydrogenase genes and the alcohol dehydrogenase isozyme I gene in yeast, although each of these genes consists of more than 300 codons; in particular, almost exclusive usage of a single synonymous codon is found among the six codons for Leu, the four codons for Pro, and the four codons for Gly (Bennetzen and Hall, 1982). By contrast, the two genes for iso-1 and iso-2 cytochrome *c*, which are weakly expressed, employ 41 and 43 different codons, respectively (Bennetzen and Hall, 1982). For further discussion about the difference in codon usage between highly and weakly expressed genes see section 6.2.2.

6.2. Factors Affecting Usage of Synonymous Codons

While usage is not expected to be random for all the 61 sense codons, because of the functional and structural requirements of proteins, it has been difficult to understand why there are strong biases in the choice of synonymous codons. If synonymous mutations are selectively neutral or almost neutral, as proposed by King and Jukes (1969) and others, and if mutation occurs randomly, the synonymous codons for an amino acid should be used more or less equally frequently. As mentioned above, mutation occurs nonrandomly, but in the majority of cases the direction of bias is contrary to that predicted by nonrandom mutation (see below) and this makes it even more difficult to explain the pattern observed. In

Table XV

Frequencies of Codon Usage per Thousand Codons in the Genomes of *Escherichia coli* (Eco), *Salmonella typhimurium* (Sal), Bacteriophage T7, Yeast (Yea), and Human (Hum)[a]

		Eco	Sal	T7	Yea	Hum			Eco	Sal	T7	Yea	Hum			Eco	Sal	T7	Yea	Hum			Eco	Sal	T7	Yea	Hum
Phe	UUU	11	20	11	8	18	Ser	UCU	13^{*5}	9	19	38^{*3}	11	Tyr	UAU	10	12	13	5	15	Cys	UGU	4	6	6	5	12
	UUC	21^{*4}	20	25	26^{*4}	27		UCC	15#	20#	13	31^{*3}	20		UAC	15^{*4}	18	21	23^{*4}	16		UGC	7	5	6	0	20
Leu	UUA	5	10	12	10	9		UCA	3	7	9	4	9	Ter	UAA	0	0	0	0	0	Ter	UGA	0	0	0	0	0
	UUG	8	7	10	60*	13		UCG	5	8	4	0	5		UAG	0	0	0	0	0	Trp	UGG	4	4	16	8	10
Leu	CUU	8	10	17	2	11	Pro	CCU	4	8	16	7	13	His	CAU	6	15	7	6	12	Arg	CGU	36^{*3}	16	24	3	4
	CUC	9	12	11	1	19		CCC	3	10	2	1	15		CAC	12	7	13	18	15		CGC	21^{*3}	23	14	0	12
	CUA	2	9	10	6	7		CCA	6*	3	10	30	10	Gln	CAA	10	12	17	26	14		CGA	2	2	8	0	5
	CUG	64*	53	21	2	43		CCG	28*	24	7	4	7		CAG	32*	31	20	3	36		CGG	2	2	3	0	4
Ile	AUU	20	31#	25#	29^{*3}	15	Thr	ACU	13*	4	20	30^{*3}	13	Asn	AAU	9	12	13	6	18	Ser	AGU	3	2	9	3	7
	AUC	41^{*4}	27	24	27^{*3}	21		ACC	26*	24	18	24^{*3}	23		AAC	27^{*4}	20	33	34^{*4}	22		AGC	11	9	7	1	18
	AUA	1	2	5	1	6		ACA	3	7	10	5	11	Lys	AAA	42^{*2}	44	23	20	33	Arg	AGA	0	0	5	36^{*2}	12
Met	AUG	26	23	27	17	21		ACG	6	14	8	1	4		AAG	16	16	45#	59*	41		AGG	0	1	3	1	12
Val	GUU	32^{*5}	14	20	41^{*3}	8	Ala	GCU	30^{*5}	17	44	79^{*3}	25	Asp	GAU	27	44	26	24	25	Gly	GGU	36*	25	39	76*	8
	GUC	11	21#	12	34^{*3}	13		GCC	23	39	14	25^{*3}	29		GAC	33	16	39	33	30		GGC	31*	43	14	2*	24
	GUA	18	4	17	2	6		GCA	23	12	17	4	14	Glu	GAA	58^{*2}	31	30	53^{*2}	36		GGA	3	7	12	1	13
	GUG	20	17	14	3	25		GCG	30	54#	14	1	6		GAG	19	26	43#	4	41		GGG	4	15	9	1	9

[a] The optimal codons predicted by the five rules of Ikemura and Ozeki (1983) are marked by asterisks and by the rule or rules according to which the optimal codons are determined. Rule 1 is given by an underline and the optimal codons in *Salmonella* and T7 are assumed to be the same as those in *E. coli*. The exceptional cases where the most frequently used codons are not the optimal codons are denoted by the symbol #. In man, the optimal codons are not determined, because tRNA availability in man has not been studied.

the past numerous explanations have been proposed for biased usage of synonymous codons. They can roughly be grouped into five categories as follows.

6.2.1. Nonrandom Mutation

It would be better to discuss the effect of nonrandom mutation on synonymous codon usage before considering other possible factors. The pattern of mutation estimated from pseudogenes predicts that nucleotide sites will tend to be occupied more often by A and T than by G and C if they are subject to no functional constraint (see above). Thus, in the absence of other evolutionary forces nonrandom mutation would, on the average, lead to higher frequencies of synonymous codons ending in A or U. In some cases in Table XV, the observed frequencies are consistent with this prediction. For example, GAA (Glu) is more frequent than GAG in the two bacterial species and yeast, although not in T7 and man. In the majority of cases, however, the observed frequencies are inconsistent with the above prediction. For instance, in the two codons for Phe, UUU is less frequent than UUC in all five organisms, and in the six codons for Leu, UUA, CUU, and CUA are much less frequent than CUG in four of the five organisms.

We mentioned above that the dinucleotide CG has a strong tendency to mutate to CA or TG. This tendency will have two consequences. First, a codon of the form NCG, N being any base, will tend to mutate to NCA (or NUG, but this change is nonsynonymous). We note, however, that in the majority of cases in Table XV, NCA is actually less frequent than NCG; e.g., CCA (Pro) is less frequent than CCG in the two bacterial species. Second, if the third position of a codon is C and the first position of the next codon is G, then C will tend to mutate to T (U in mRNA code); therefore, this effect will lead to a higher frequency for NNU than for NNC. We noted above that this prediction does not hold in the majority of cases in Table XV.

From the above discussion we can conclude that the nonrandom usage of synonymous codons shown in Table XV cannot be adequately explained by nonrandom mutation and that there must be some other factors that have exerted stronger effects on selection of synonymous codons than nonrandom mutation has. It should, however, be stressed that mutation pressure is generally very weak, so that it takes only a slight selective advantage to override the effect of nonrandom mutation.

6.2.2. tRNA Availability

The first substantial set of data supporting the possibility that tRNA availability can pose a constraint on codon usage was obtained by Nomura and his associates (Post *et al.*, 1979; Nomura *et al.*, 1980), although this possibility had been considered earlier by other authors (e.g., Fiers *et al.*, 1975; Grosjean *et al.*, 1978). Nomura and his associates found that *E. coli* ribosomal protein (r-protein) genes preferentially use synonymous codons that are recognized by the most abundant tRNA species. They suggested two possible factors that may have influenced the evolution of differential use of synonymous codons. One is the efficiency of translation of r-protein genes; the r-proteins are synthesized in large amounts and the use of abundant tRNA species would ensure efficient translation. The other factor is the fidelity of translation. It has been shown that when decreased levels of a correct aminoacyl tRNA limit the rate of protein synthesis, the frequency of translation errors increases, presumably because of more effective competition by incorrect aminoacyl tRNAs (Edelman and Gallant, 1977; O'Farrell, 1978; Parker *et al.*, 1978). Thus, the preferential use of codons recognized by major tRNA species may be expected to reduce translational errors in r-protein synthesis. Nomura *et al.* considered the second factor more important.

Recently more quantitative analyses of the relationship between codon usage and tRNA availability have been conducted by Ikemura (1981a,b, 1982) and Ikemura and Ozeki (1983). They studied the correlation between codon usage over the whole code (the 61 sense codons) and tRNA availability in *E. coli* and yeast. They defined the usage frequency of a tRNA responding to a single codon as the occurrence of the codon in a gene and that of a tRNA responding to multiple codons as the total of their occurrences. They found a strong positive correlation between the content of a tRNA and the frequency of its usage for most *E. coli* and yeast genes sequenced to date; the correlation was stronger for genes of high expressivity (e.g., ribosomal protein genes) and weaker for genes of low expressivity (e.g., repressor genes). These results indicate that tRNA availability is an important factor for the selection of codons. To further test this hypothesis, Sharp *et al.* (1985) have recently calculated the correlation of T7 codon usage (pooled over 50 genes) with tRNA abundance of *E. coli* reported by Ikemura (1981a); the propagation of T7 relies on the translation system of its hosts, of which *E. coli* is probably the best known. They obtained a correlation coefficient of 0.78. This is a high value, although it is substantially lower than the correlation, 0.95, calculated for the codon usage pooled over nine *E. coli* genes (Sharp *et al.*, 1985). This result further supports the importance of tRNA availability in the selection of codons.

Ikemura (1981a,b, 1982) and Ikemura and Ozeki (1983) also studied the relationship betwen isoaccepting tRNA content and synonymous codon usage (see also Bennetzen and Hall, 1982). Figure 4 gives some of their results. In this figure, both the content and usage frequency of the

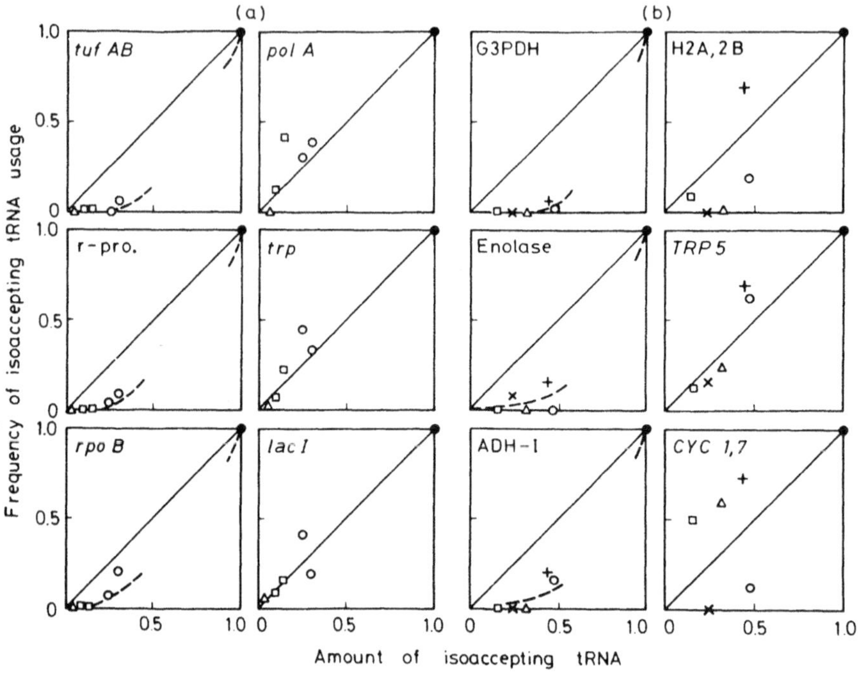

Figure 4. Relationship between the content of isoaccepting tRNA and the frequency of its usage for individual genes. The line is obtained by assuming that the use of isoaccepting tRNA is proportional to its abundance. (●) Most abundant isoacceptor of each amino acid. Other isoacceptors: (a) *E. coli* (○) Leu, (□) Gly, and (△) Ile; (b) yeast (×) Arg, (○) Leu, (△) Ser, (□) Val, and (+) Lys. [Taken from Ikemura and Ozeki (1983).]

most abundant isoacceptor of each amino acid are normalized to one. In each of these examples, the most abundant isoacceptor is used at the highest frequency. This is true for almost all *E. coli* and yeast genes studied, but often is not true for bacteriophage, transposon, and yeast plasmid genes (Ikemura, 1981a,b, 1982). It should be pointed out that the codon to which the most abundant yeast isoacceptor responds often differs from the respective codon of *E. coli* (Ikemura, 1981a,b, 1982; Ikemura and Ozeki, 1983). For example, the major Leu isoacceptor of yeast responds to UUG, but that of *E. coli* to CUG; these two codons are predominantly used in yeast and *E. coli*, respectively (Table XV). Thus, tRNA availability appears to be an important causal factor of the differential patterns of synonymous codon choices observed for the two organisms.

We note further from Fig. 4 that in highly expressed genes the data points are below the lines expected under the assumption that the frequency of isoacceptor use is directly proportional to its content. This means that in those genes the codons recognized by the most abundant isoacceptors are used more than proportionally to the isoacceptor contents, but other synonymous codons are almost completely avoided or strongly disfavored. By contrast, in modestly expressed genes such as amino acid-synthesizing genes or in weakly expressed genes, the data points seem to be represented roughly by linear functions. Grosjean and Fiers (1982) observed that in bacteriophage MS2 and *E. coli*, codons corresponding to minor tRNAs are clearly avoided in highly expressed genes. As mentioned above, other authors (e.g., Grantham *et al.*, 1981; Bennetzen and Hall, 1982) have also observed that nonrandom usage of synonymous codons is more pronounced in strongly expressed genes than in weakly expressed ones.

Now why does the usage pattern in highly expressed genes differ from that in modestly and weakly expressed genes? A simple explanation is that highly expressed genes require rapid and frequent translation of their mRNAs, so that use of codons recognized by abundant isoacceptors would not only increase the rate of translation but also reduce the chance of mistranslation (Post *et al.*, 1979; Ikemura, 1981a; Bennetzen and Hall, 1982). On the other hand, rapid translation is not required in modestly and weakly expressed genes and there is thus no need to use exclusively codons recognized by abundant isoacceptors. Also, it has been suggested that the presence of codons translated by minor tRNAs in weakly expressed genes serves to regulate the rate of translation (Grosjean and Fiers, 1982).

From his quantitative analyses, Ikemura (1981a,b, 1982) concluded that tRNA availability is the most important factor for the selection of synonymous codons; however, unlike Nomura *et al.* (1980), he apparently thought that translational efficiency is more important than translational fidelity for guiding the evolution of synonymous codon usage. Among his eight rules for synonymous codon selection (Ikemura and Ozeki, 1983), the first and the fifth are related to tRNA availabiity. Rule 1 says that among the synonymous codons for an amino acid the one that is recognized by the most abundant isoaccepting tRNA will be used most frequently, while Rule 5 says that if a codon (e.g., Gly GGA in *E. coli*) is recognized by only one tRNA species (Gly 1) but another codon (GGG) is recognized by an additional isoaccepting tRNA (Gly 2), then the latter will be used more frequently. Let us examine how well these two rules hold for *E. coli, Salmonella,* and T7, assuming that the pattern of tRNA availability estimated from *E. coli* also applies to the other two species. In the column for *E. coli* in Table XV, we have underlined the usage frequencies for those synonymous codons recognized by the most abundant isoaccepting tRNA species. There is only one exception to Rule 1: in *Salmonella,* GUC is the most frequently used codon for Val, although it is not recognized by a major tRNA species. Thus, Rule 1 holds well. Rule 5 together with Rule 1 predicts that GUU, UCU, and GCU should be the most frequently used codons for Val, Ser, and Ala, respectively. There are several exceptions to these predictions: GUC (Val) is the most frequent in *Salmonella;* UCC (Ser) in *E. coli* and *Salmonella;* and GCG (Ala) in *Salmonella*. This rule also predicts that GGG (Gly) should be more frequent than GGA. There is one exception to this prediction: in T7, GGA is more frequent than GGG. Therefore, although Rule 5 holds in the majority of cases, there are quite a few exceptions.

From the studies discussed above, we can conclude that tRNA availability provides a good explanation for why genes in a genome tend to have the same pattern of codon usage. It also provides a good explanation for why related species tend to have the same usage pattern, for they should have similar patterns of tRNA abundance; for example, *E. coli* and *S. typhimurium* have similar patterns of tRNA abundance (Ikemura and Ozeki, 1983) and similar patterns of codon usage (Table XV). Moreover, it can explain adequately why T7 has a usage pattern similar to that of *E. coli,* since T7 has *E. coli* as one of its hosts. In short, tRNA availability offers a good explanation for Grantham *et al.*'s (1980b) genome hypothesis. We mentioned above that the use of synonymous codons is more even in man than in *E. coli* and yeast. It will be interesting to see if the distribution of isoacceptor contents is more even in man than in the latter two species. If this is indeed the case, it will further support the importance of tRNA availability in codon selection.

6.2.3. Codon-Anticodon Interaction

Grosjean *et al.* (1978) and Grosjean and Fiers (1982) proposed that the efficiency of codon-anticodon interaction plays an important role in synonymous codon selection. Based on their study of the usage of synonymous codons in the MS2 RNA bacteriophage, they suggested that codons with an intermediate GC content and thus an intermediate codon-anticodon binding energy are optimal for translation. They claimed that their hypothesis is supported, at least in prokaryotic genes, by the selection of synonymous codons that contain a pyrimidine (U,C) as the third letter and that are recognized by the same tRNA. Specifically, they proposed the following two rules: First, codons in which the first two nucleotides NN are U and/or A should stabilize their interaction with their anticodons if a C is used in the wobble (third) position. They noted that in MS2 there is indeed a systematic preference for the NNC codons over NNU for Phe (UUU/UUC = 21/32), Ile (AUU/AUC = 12/28), Tyr (UAU/UAC = 10/33), and Asn (AAU/AAC = 19/28). Second, if the first two nucleotides of the codon are C and/or G, a preference would exist for U as a third letter to avoid too sticky a codon-anticodon interaction. Indeed, they noted that NNU is systematically preferred over NNC for Pro (CCU/CCC = 18/10), Ala (GCU/GCC = 26/21), and Gly (GGU/GGC = 37/16), albeit not for Arg (CGU/GGC = 21/22). They also found that these two rules applied to highly expressed genes in *E. coli*.

Ikemura (1981a,b, 1982) has also advocated the importance of codon-anticodon interaction. He included the above two Rules as Rules 4 and 6 of his eight rules for synonymous codon selection. In addition, he proposed the following rules: Rule 2: Introduction of thiolated uridine or of 5-carboxymethyl uridine at the anticodon wobble (first) position produces a tendency to restrict wobble G:U pairing and thus a preference for an A-ending codon over a G-ending codon (Nishimura, 1978; Weissenbach and Dirheimer, 1978); Rule 3: Introduction of inosine at the wobble position produces a preference for U- and C-ending codons over an A-ending codon, which leads to purine-purine pairing (Ikemura, 1981b; Bennetzen and Hall, 1982). His other two rules (Rules 7 and 8) are quite minor and not discussed here (Ikemura and Ozeki, 1983). Ikemura and Ozeki (1983) have shown that codon choices in both *E. coli* and yeast genes, especially choices in the highly expressed genes, conform well to the expectations deduced from Rules 2–4 (Table XVI). Interestingly, the tRNAs that have the modified anticodon nucleotides involved often differ between *E. coli* and yeast (e.g., see tRNAs in Table XVI that have inosine at the anticodon wobble position). Thus, this difference might also be a factor responsible for the organism-specific patterns of codon usage. Rule 4, which deals with codons for Asn, Phe, Tyr, and Ile (Table XVI), should produce the same codon preferences regardless of the organisms, as long as only one tRNA responds to the two codons involved. Actually, codon choices in the above amino acids conform well to Rule 4 in all the five organisms in Table XV. Similarly, Rule 6, which deals with the NNU and NNC codons for Pro, Ala, Gly, and Arg, should also produce the same codon preference if the two codons involved are recognized by the same tRNA. This rule does not hold as well as Rule 4; for example, in the case of Gly, GGC instead of GGU is used more often in *Salmonella* and man (Table XV).

Pieczenik (1980) found that anticodons in *E. coli* tRNA sequences are frequently bounded on the 3′ side by a purine (R) and on the 5′ side by a pyrimidine (Y) and suggested that codons of the form RNY, N being any base, are favored by natural selection because this would maximize tRNA-mRNA interaction. In fact, coding sequences from a wide variety

Table XVI
Preferential Usage Observed between Codons Corresponding to the Same tRNA[a]

			E. coli: Use of codons in individual genes					
tRNA	Rule	Codon	*tufAB*	r-Protein	*rpoBD*	*nuc*	*thrAB*	*trp*
Lys	2	AAA/AAG	35/11	90/24	77/37	52/19	24/21	53/11
Glu 2	2	GAA/GAG	60/13	61/16	147/46	60/22	52/20	74/31
Arg[b]	3	CGY/CGA	23/0	37/0	67.5/1	30/3	23.5/6	41.5/4
Phe	4	UUC/UUU	26/2	23/10	44/15	21/10	23/16	30/31
Ile 1	4	AUC/AUU	52/6	51/13	98/29	57/25	26/34	37/47
Tyr	4	UAC/UAU	17/3	13/3	38/18	22/7	11/17	18/31
Asn	4	AAC/AAU	14/0	42/3	66/4	31/9	24/28	33/25

			Yeast: Use of codons in individual genes					
tRNA	Rule	Codon	G3PDH	Enolase	ADH-I	H2A2B	TRP5	CYC1,7
Glu 3	2	GAA/GAG	29/0	53/0	20/0	25/1	33/9	8/5
Arg 3	2	AGA/AGG	22/0	26/0	8/0	30/2	22/2	5/1
Ile[b,c]	3	AUY/AUA	19.5/0	21.5/0	10.5/0	15/0	15.5/3	4/1
Val1[b]	3	GUY/GUA	36/0	34.5/0	18/0	11/0	24/6	1.5/1
Arg 2[b]	3	CGY/CGA	0/0	1/0	0/0	0/0	2/0	0/0
Thr 1[b]	3	ACY/ACA	23.5/0	20/0	7/0	14/5	17.5/9	4/5
Ala 1[b]	3	GCY/GCA	32.5/0	56.5/0	17.5/0	35.5/1	25.5/12	6/2
Phe	4	UUC/UUU	21/0	28/3	8/0	4/2	15/17	3/5
Tyr	4	UAC/UAU	20/0	18/1	13/0	9/7	10/9	5/5
Asn	4	AAC/AAU	26/0	38/2	11/0	17/5	15/11	11/3

[a] Taken from Ikemura and Ozeki (1983). Details and results of other *E. coli* and yeast genes were described by Ikemura (1981a,b, 1982).

[b] Half of the sum of U- and C-terminated codon use. Y, Pyrimidine.

[c] Rule 2 was not applied to yeast Ile codon use, since the respective tRNA has I at the anticodon wobble position (Ikemura, 1982).

of organisms have generally been found to have a surplus of RNY codons (Shepherd, 1981; Sharp *et al.*, 1985). Of course, this surplus may simply reflect the translational properties of the genetic code and the proportional amino acid requirements of typical proteins (Sharp *et al.*, 1985). To exclude this possibility, Sharp *et al.* (1985) examined the frequency of RNY and non-RNY codons within groups of synonymous codons in T7. They found a striking overall preference for RNY—only in the case of Ser codons is there an excess of non-RNY codons. However, we find that the surplus of RNY codons can also be explained by the rules discussed above. For example, the surplus of AUU and AUC (Ile) over AUA and that of GGU and GGC (Gly) over GGA and GGG (Table XV) can be explained by Rule 1; i.e., by tRNA availability. Since the above rules are more specific and appear to have more solid bases, we are more inclined to explain the surplus of RNY codon among synonymous codons by these rules than by Pieczenik's proposal.

Recently Andersson *et al.* (1984) questioned whether the nucleotide composition of codons can really influence the speed of accuracy of translation. They employed an *in vitro* system derived from *E. coli* that was previously shown to translate the codon UUU at a rate and with an accuracy very similar to those for polypeptide synthesis *in vivo*. Their data indicated that poly(U–G), which contains the alternating codons UGU (Cys) and GUG (Val), is translated at the same rate and with comparable missense errors as is poly U. They therefore suggested that the performance characteristics of the ribosome may not be sensitive to the presence or absence of G nucleotides in the mRNA. Their suggestion is, however, not supported by other *in vitro* studies, which have led to the general conclusion that poly U and poly C are uniquely error-prone messages and that purines (A, G) are generally read more accurately [see the review by Yarus (1979)]. Moreover, in an *in vitro* translation study of the MS2 coat protein message, Parker *et al.* (1983) have shown that the asparagine codon AAU at amino acid 12 is misread approximately eight times more frequently than the AAC at amino acid 3. This difference in translation fidelity may be taken as evidence that AAU is more error-prone than AAC, although it could instead be due to the "codon context effect" (see below).

So far we have not considered the so called "codon context effect," i.e., the translation of a codon during polypetide synthesis may be influenced in some way by the nucleotides preceding or following the translated codon (e.g., Bossi and Roth, 1980; Buckingham and Carrier, 1982; Murgola *et al.*, 1984). Since knowledge about this phenomenon is limited, it is not clear what effects codon context can have on codon usage. However, we speculate that the effect of codon context may be able to explain an interesting observation recently made by Blaisdell (1983a,b). He found that in the coding sequences of 30 eukaryotic genes the third base in a codon was often chosen to be unlike its neighbors in bonding energy on both sides up to two sites away; A and T (U) are weak (W), while G and C are strong (S), bonding bases. This preference produced an excess of runs of length one and two and a deficit of long runs. The neighbors in the codon 3′ to the third base were as important in determining the choices as were the neighbors 5′ in the same codon. In connection with this observation we note that *in vitro* data have shown that poly U and poly C are error-prone messages (Yarus, 1979) and that codon UGU (Cys) is misread about four times more frequently in a random U-rich polymer poly(U_5, G) than in a poly(U-G) message (Buckingham and Carrier, 1982). These data, although limited, suggest that messages with long runs of W or long runs of S may be more error-prone than messages with an excess of runs of length one or two. If this is true in general, then the effect of codon context may explain Blaisdell's observation.

6.2.4. Avoidance of Susceptibility to Mutational Damage

Clarke (1970) suggested that particular synonymous codons would be favored if they minimize mutational damage; e.g., mutations to stop codons. Modiano *et al.* (1981) observed that in human α and β genes "pretermination codons," which are codons that differ from termination codons by only one base, are never used when synonymous alternatives exist. They proposed that this avoidance of pretermination codons arose as an evolutionary strategy for reducing the rates of mutations with drastic effects. However, the selective advantage coming from such a strategy is too small (presumably the order of the mutation rate) to be effective in the actual course of evolution (Kimura, 1983a). In fact, Golding and Strobeck (1982) have shown that although mutation to termination codons would have an effect on the pattern of codon usage, the effect would be small and perhaps undetectable. Actually, our more extensive compilation of codon usage in man (Table XV) shows that there is really no avoidance of usage of pretermination codons. For instance, the two pretermination codons UUA and UUG are used as frequently as the other synonymous codons for Leu except for CUC and CUG, and the pretermination codon AGA is used more frequently than or as frequently as any of the other codons for Arg. We concur with Kimura (1983a) that the rarity of pretermination codons in the α- and α-globin genes observed by Fitch (1980) and Modiano *et al.* (1981) might be due to paucity of the tRNAs that recognize these codons. We note that these two genes are highly expressed. As mentioned above, codons recognized by minor tRNA species are rarely used in highly expressed genes.

6.2.5. Requirement and Avoidance of Certain RNA and DNA Structures

Molecular biologists have long been curious about what secondary and tertiary strucures mRNAs can form, for such structures may be important for mRNA stability, translation, and possibly other functions (e.g., Adams *et al.*, 1969; Fiers *et al.*, 1976; Heindell *et al.*, 1978). Unfortunately, little is known about the spatial organization of an mRNA molecule. However, from theoretical predictions as well as from physical measurements, it is clear that a large proportion of the bases in phage RNAs and presumably in most RNAs are involved in base pairing (see, e.g., Fitch, 1974; Steitz, 1979; Fiers, 1979; Muller and Fitch, 1982). Fiers and his associates (Min Jou *et al.*, 1972; Fiers *et al.*, 1975, 1976; Min Jou and Fiers, 1976) proposed secondary structure models for MS2 RNAs coding for proteins and found that double-stranded regions evolve at a rate two-fold lower than single-stranded regions, suggesting that the secondary structures serve certain functions. Like phage RNAs, globin mRNAs have been shown by physical studies to have considerable secondary structure (Holder and Lingrel, 1975; Favre *et al.*, 1975). So far the best known example that the spatial structure of an mRNA can affect translation is the attenuator in amino acid biosynthetic operons in bacteria [see the review by Kolter and Yanofsky (1982)]. Spatial structural requirements, for whatever purposes, obviously can affect codon selection (e.g., Fiers *et al.*, 1976).

We mentioned above that Blaisdell (1983a,b) has recently studied the distribution of weak and strong hydrogen-bonding bases in the coding sequences of 30 eukaryotic genes and found that the third position of codons tends to be unlike its neighbors in bonding energy on both sides up to two sites away. He suggested that this preference arose because it would produce an mRNA secondary structure of moderate stability, as a consequence of a deficit of long runs of strong-bonding bases and weak-bonding bases. He argued that too high a stability would impede the dissolution of the secondaray structure during decoding on the ribosome and too low a stability would produce an mRNA structure open and straggly, which might impede diffusion through or along the nuclear membrane and through the cytoplasm to the ribosome. Selection for a moderation of stability of the secondary structure of mRNAs may indeed produce nonrandom codon usage. However, whether there is really selection for moderation needs to be substantiated by data. As speculated above, the observed excess of W and S runs of length one and two might be due to the effect of codon context.

In eukaryotic genes, codon selection might also be affected by the structural requirements for RNA splicing. In fact, it has been found that the coding sequences adjacent to exon/intron jucntions are generally quite conservative (Miyata and Yasunaga, 1890; Efstratiadis *et al.*, 1980; Rogers and Wall, 1980; Lerner *et al.*, 1980).

Just as that there are structural requirements for mRNA stability, translation, etc., there should also be avoidance of certain mRNA structure. For example, there should be no splicing signal inside an exon. There might also be structural preferences and avoidances at the DNA level. For example, there should be no signal for transcriptional initiation inside a coding region. The deficit of CG dinucleotides mentioned above is another example. In an extensive compilation of dinucleotide frequencies in DNA sequences, Nussinov (1981) found that in the majority of cases a dinucleotide was either under- or overrepresented. For example, in prokaryotic genomes the AG dinucleotide occurred considerably less frequently than expected under random distribution of nucleotides. Since this occurred in both coding and noncoding regions, she concluded that the AG dinucleotide was disfavored in general and this could account partly for the extremely low frequencies of AGN codons in prokaryotic genes. However, the underrepresentation in the noncoding regions could be due to random effects, for it was not serious except in the insertion element in *E. coli*. Moreover, the AG dinucleotide was overrepresented in eukaryotic genes. We therefore have some reservation about Nussinov's conclusion. In general, we find it difficult to infer the cause of nonrandom codon usage from the distribution of dinucloetide frequencies, because, except in the case of CG, we do not know any evolutionary force operating on the dinucleotide as a unit of selection.

Rosenberg *et al.* (1979) and Sharp *et al.* (1985) surveyed the number of restriction recognition sites present in phage T7. They noted that in general such sites were underrepresented and, in particular, the recognition sequences for a number of restriction enzymes were completely absent. Sharp *et al.* (1985) suggested that the reduced frequency of certain potential four-base recognition sequences may explain another feature of the nonrandom codon usage in T7. There are ten pairs of synonymous codons (pairs in which both triplets end in either a pyrimidine or a purine) for which one triplet may form three-fourths of a potential four-base recognition sequence while the other triplet may not. Of these ten pairs, in seven cases (UCR, UGY, CUR, CCR, CAY, AGY, GAY) the potential recognition sites has been underrepresented. However, this avoidance of potential recognition sites might be peculiar to T7, for it has not been found in three other coliphages, ϕX174, G4, and fd (Adams and Rothman, 1982).

Requirements and avoidances of certain primary and spatial structures of DNA and mRNA, however, would be unlikely to produce a regular pattern of codon usage, for the requirements and avoidances would not be the same for all genes. Of course, if there is indeed selection for a moderate stability of mRNAs, as proposed by Blaisdell (1983a,b), then it can produce a pattern similar to that produced by selection for an intermediate strength of codon-anticodon binding.

In summary, tRNA availability and codon-anticodon interaction appear to be the two most important factors for synonymous codon selection; other factors are either quite minor or unlikely to produce systematic biases. We mentioned above that the first five rules of Ikemura and Ozeki (1983) are related either to tRNA availability or to codon-anticodon interaction and that they apparently give good predictions for preferred synonymous codons. From these five rules we have inferred the optimal synonymous codons in the first four organisms in Table XV; we have not done so for man, because tRNA availability has not been studied in man. The optimal codons are denoted by an asterisk and by the rule or rules according to which the optimal codons are determined; rule 1 is given by an underline and the optimal codons in *Salmonella* and T7 are assumed to be the same as those in *E. coli*. The exceptional cases where the most frequently used synonymous codons are not the optimal codons are denoted by the symbol #. We note that there are only a few exceptional cases (Table XV). We can therefore conclude that these five rules are very useful for predicting the pattern of synonymous codon usage.

7. Concerted Evolution of Multigene Families

7.1. Mechanisms of Concerted Evolution

From the mid 1960s to the mid 1970s a large number of DNA reannealing and hybridization studies were conducted to explore the structure and organization of eukaryotic genomes. These studies revealed that gneomes of higher organisms are composed of highly and moderately repeated sequences as well as single-copy sequences (Britten and Kohne, 1968). They also revealed an intriguing evolutionary phenomenon, namely that the members of a repeated sequence family are generally very similar within one species, although members of closely related species may differ greatly. For example, the nontranscribed spacer (NTS) regions within the ribosomal DNA of *Xenopus laevis* are highly homogeneous but are vastly different from those in *X. borealis* (previously misidentified as *X. mulleri*) (Brown *et al.*, 1972). The large interspecific difference indicates that the NTS regions accumulate mutations rapidly in evolutionary time, and thus the high intraspecific homogeneity is not maintained by selective constraints (purifying selection). Apparently, it is maintained by certain mechanisms by which mutations can spread horizontally to all members in the same family. This phenomenon has been called "horizontal evolution" (Brown *et al.*, 1972) or "coincidental evolution" (Hood *et al.*, 1975). Recently, it has also been called "concerted evolution" (Zimmer *et al.*, 1980), meaning that the members of a family evolve together in a concerted fashion. With the advent of restriction enzyme analysis and DNA sequencing techniques, a large body of data confirming the generality of concerted evolution in multigene families has been obtained [for reviews, see Brutlag (1980), Ohta (1980), Long and Dawid (1980), and Arnheim (1983)].

Several mechanisms have been proposed for concerted evolution, including the master-slave hypothesis (Callan, 1967), saltatory replication hypothesis (Britten and Kohne, 1968; Buongiorno-Nardelli *et al.*, 1972), unequal crossover (Edelman and Gally, 1970; Smith, 1974, 1976), gene conversion (Edelman and Gally, 1970; Birky and Shavaril, 1976; Baltimore, 1981; Jackson and Fink, 1981; Klein and Petes, 1981), and transposition (Dover and Coen, 1981; Dover, 1982). The master-slave hypothesis postulates that in each generation new slave copies are replicated from the master sequences; thus, variant repeats are removed in each generation. This hypothesis is incompatible with the observation that neighboring repeats in the rRNA family in *Xenopus* can be unequal in length (Wellauer *et al.*, 1976; see also Ohta, 1980). The saltatory replication hypothesis proposes that a recent amplification of a sequence is the basis for intraspecific homogeneity. Under this hypothesis, the homogeneity of a family would slowly disappear, due to the accumulation of mutations, and for two species that diverged after the sequence amplification the intraspecific variation should be approximately equal to the interspecific variation [for an explanation see Arnheim (1983)]. This prediction is not in agreement with observations (see above). Unequal crossover and gene conversion are currently considered the two most important mechanisms of concerted evolution (Ohta, 1980, 1983; Arnheim, 1983), for they are supported both by observations and by theoretical studies (see below). The importance of transposition as a mechanism of concerted evolution remains to be substantiated by data.

Unequal crossover is a reciprocal recombination process that creates a sequence duplication in one chromatid (or chromosome) and a corresponding deletion in the other. Figure 5A shows an example in which an unequal crossover event has led to duplication of three repeats in one daughter chromosome and deletion of three repeats in the other. Because of this unequal exchange, each of the daughter chromosomes has become more homogeneous than the parental chromosomes. If this process continues, the numbers of wild-type and mutant repeats on a chromosome will fluctuate with time and eventually one type of repeat will become dominant in the family. Indeed, both computer simulations and analytical treatments have shown that unequal crossover can produce homogeneous tandem arrays where all the members were derived from only one or a few of the original family members (Smith, 1974, 1976; Ohta, 1980, 1983). Unequal crossover has been suggested as a mechanism of concerted evolution in the 5S DNA family in *Xenopus* (Brown and Sugimoto, 1973), the rDNA family of several species (Brown *et al.*, 1972; Coen *et al.*,

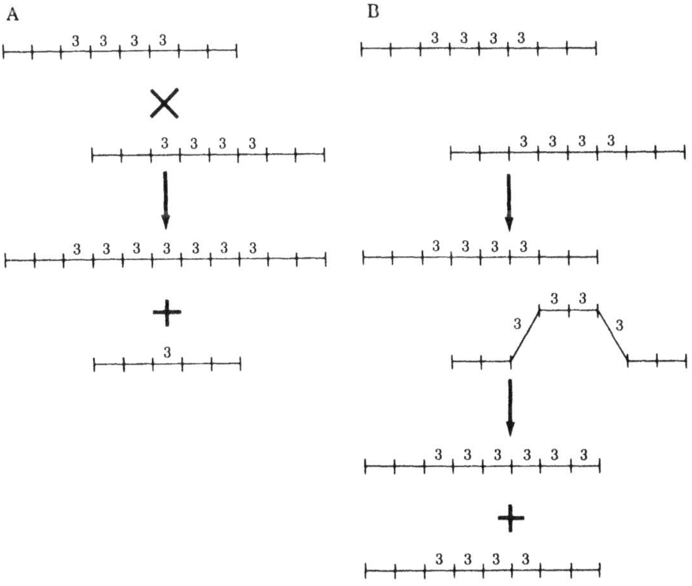

Figure 5. Models of (A) unequal crossing-over and (B) gene conversion. Both of these events are shown to occur between two homologous chromosomes, each possessing 50% wild-type repeats and 50% spontaneous mutation 3-containing repeats. (A) As a result of unequal crossing-over, both daughter chromosomes have an altered frequency of the two repeat types when compared to the parental frequencies. In addition, the number of copies of the repetitive element are also altered in the daughter chromosomes. (B) Gene conversion changes the frequency of the two types of elements in one of the daughter chromosomes, but does not alter the total number of repeats in either chromosomes. [Taken from Arnheim (1983).]

1982a,b; Arnheim, 1983), and in other families (Hood *et al.*, 1975; Zimmer *et al.*, 1980, Brutlag, 1980; Long and Dawid, 1980).

Gene conversion is a nonreciprocal recombination process in which two sequences interact in such a way that one is converted by the other [see the reviews by Radding (1978) and Szostak *et al.* (1983)]. Actually, what is more relevant to concerted evolution is nonallelic gene conversion, i.e., conversion between nonallelic genes. Figure 5B shows an example of nonallelic gene conversion in which two of the wild-type repeats are converted to the mutant type. As a consequence, the first daughter chromosome has become more homogeneous than the parental chromosomes, although no change has occurred in the second daughter chromosome. Theoretical studies have shown that, like unequal crossover, gene conversion can produce concerted evolution (Birky and Skavaril, 1976; Ohta, 1984; Nagylaki and Petes, 1982; Ohta and Dover, 1983; Nagylaki, 1984). Gene conversion has been suggested as a mechanism of homogenization in the γ-globin genes (Jeffreys, 1979; Slightom *et al.*, 1980; Scott *et al.*, 1984), the α-globin genes (Liebhaber *et al.*, 1981), and the heat shock genes of *Drosophila* (Brown and Ish-Horowicz, 1981). Moreover, it has been suggested as an important mechanism for the generation of polymorphism in histocompatibility antigens (Weiss *et. al.* 1983; Ohta, 1984).

Gene conversion appears to have several advantages over unequal crossover (Baltimore, 1981; Nagylaki and Petes; 1982, Dover, 1982). First, unequal crossover changes the number of repeated genes in a family and may sometimes cause a severe dosage imbalance. Gene conversion does not have this disadvantage, for it causes no change in gene number. Second, gene conversion can act as a correction mechanism not only on tandem repeats but also on dispersed repeats. In fact, gene conversion has been shown to occur between dispersed repeats within a chromosome (Jackson and Fink, 1981; Klein and Petes, 1981), between homologous chromosomes (Fogel *et al.*, 1978), and between nonhomologous chromosomes (Scherer and Davis, 1980; Ernst *et al.*, 1981). By contrast, unequal crossover has certain restrictions when repeats are dispersed. It probably can act effectively on repeats dispersed on nonhomologous chromosomes if they are located at the ends of chromosome arms as in the case of 5S RNA genes in *Xenopus* (Brown and Sugimoto, 1973) and that of rRNA genes in man and apes (Arnheim, 1983), but will be greatly restricted if the dispersed repeats are located inside chromosomes, as in the case of rRNA genes in mouse (Arnheim, 1983) and in *Drosophila melanogaster* (Coen and Dover, 1983). If repeats are dispersed on a chromosome, unequal crossover can cause deletion or duplication of the genes between repeats or an inversion, depending on the polarity of the repeats involved (Nagylaki and Petes, 1982). Third, gene conversion can have a preferred direction. Studies with fungi have shown that disparity (bias) in the direction of conversion is common and often strong (Lamb and Helmi, 1982). Nagylaki and Petes (1982) have shown that a small disparity can have a large effect on the probability of fixation of repeat variants. In addition to these advantages, gene conversion appears to occur more frequently than unequal crossover. Hurst *et al.* (1972) suggest that, although conversion can occur without reciprocal recombination, reciprocal recombination is always associated with formation of heteroduplexes, which are potential sites for gene conversion. Jackson and Fink (1981) show that between a linked pair of virtually identical genes, conversion occurs about 3–7 times more often than unequal crossover.

For the above reasons, some authors (Baltimore, 1981; Nagylaki and Petes, 1982; Dover, 1982) think that gene conversion plays a more important role in concerted evolution than unequal crossover. This is probably true for dispersed repeats, because in this case gene conversion can act more effectively than unequal crossover (see above). It is probably also true for small multigene families, because in such families unequal crossover may have a high chance of causing adverse effects. For example, deletion of one of the duplicated α-globin genes in man gives rise to the α-thalassemia-2 syndrome (Kan *et al.*, 1975). [Although it has been suggested that unequal crossover has played a major role in the concerted evolution of the α-globin genes in man (Zimmer *et al.*, 1980), one can argue equally well that the high homology between the two duplicates is mainly due to gene conversion.] In large, tandemly repeated sequence families, however, unequal crossover may not be less important than gene conversion, for two reasons. First, in such families, the number of repeats can apparently fluctuate greatly without causing significant adverse effects. This is suggested by the observations that the number of rRNA genes in *Drosophila* varies widely among individuals in the same species (Ritossa *et al.*, 1966) and that the number of 5S RNA genes is about 24,000 copies in *Xenopus laevis* but only 9000 copies or so in *X. borealis* (Brown and Sugimoto, 1973). Second, in a gene conversion event only the heteroduplex region is involved and thus the amount of information transferred is likely to be small. For example, in the case of the K^{bm-l} mutant gene in the major histocompatibility complex of mouse the number of nucleotides involved in the conversion event appears to be less than 60 (Weiss *et al.*, 1983). In the examples of gene conversion in the γ-globin genes proposed by Slightom *et al.* (1980) and Scott *et al.* (1984), the DNA segments involved are all relatively short. By contrast, the number of repeats exchanged in an unequal crossover event can be large. According to Szostak and Wu's (1980) study with yeast, a single unequal crossover event affects on the average seven repeats of rRNA genes. Obviously, the larger the number of repeats invovled in an unequal crossover event, the higher the rate of concerted evolution will be (Ohta 1976, 1980, 1983). This advantge of unequal crossover may be enough to offset the advantages of gene conversion. Actually, unequal crossover seems to have played a much more important role than gene conversion in the concerted evolution of immunoglobulin V_H gene family in mouse (Gojobori and Nei, 1984). Of course, the observed lower rate of gene conversion in this family might have been due to detection bias, for it is generally much easier to detect unequal crossover than gene conversion.

7.2. Factors Affecting the Rate of Concerted Evolution

7.2.1. Arrangement of Repeats

How cohesively the members of a repeated sequence family can evolve together depends on how they are arranged in the genome. There are, roughly speaking, two types of arrangement. In some families the members are highly dispersed all over the genome. For example, the human *Alu*I DNA family is composed of about 300,000 members, which are interspersed with single-copy sequences throughout the genome. This type of arrangement is least favorable for concerted evolution because it greatly reduces the chance of unequal crossover and gene conversion, and unequal crossover would often lead to disastrous cytological consequences. How concerted evolution has occurred in the *Alu*I DNA family is a puzzle. One suggestion is that some (but not all) *Alu*I segments are transcribed in the germ cells and are inserted back to the genome or are used as templates for gene conversion events (Arnheim, 1983).

In the other type of arrangement, all members of a family are clustered either in a single tandem array or in a number of tandem arrays located on different chromosomes. As mentioned above, the rDNA family in *Xenopus* is present in a tandem array. This arrangement is most favorable for unequal crossover and gene conversion to operate. If the repeats are located on more than one chromosome, the rate of unequal crossover would be greatly reduced, unless the clusters all occur at the ends of chromosome arms as in the case of the rDNA family in human. Moreover, the rate of gene conversion would also be reduced. However, Ohta and Dover (1983) have shown that such a reduction in gene conversion rate has only a minor effect on the extent of identity between genes, unless the conversion rate between genes on nonhomologous chromosomes becomes very low or unless the number of nonhomologous chromosomes on which gene members reside is large.

7.2.2. Functional Requirement

In terms of functional requirement, there are two extreme types of multigene families. In one the function has an extremely stringent structural requirement, and often requires a large amount of the same gene product. The rRNA genes and the histone genes are well-known examples. The other extreme is that the function requires a large amount of diversity. The immunoglobulin genes and the histocompatibility genes belong to this category. In general, the rate of concerted evolution is expected to be higher in the former type than in the latter type of families. Indeed, according to Gojobori and Nei's (1984) estimate, the rate is 100 times higher in the human rDNA family than in the mouse immunoglobulin V_H family. This large difference in rate seems to be mainly due to the difference in selection scheme between the two families (Gojobori and Nei, 1984). In rRNA genes, homogeneity appears to be advantageous, and since unequal crossover and gene conversion both have the effect of reducing the genetic variability among multigenes, they are apparently used effectively for maintaining the homogeneity of rRNAs. In the case of V_H genes, however, an individual who has many copies of the same DNA sequence due to unequal crossover or gene conversion seems to be at a disadvantage and is thus eliminated from the population. We note further that homogeneity, on the one hand, and unequal crossover and gene conversion, on the other hand, would go in a cycle. Unequal crossover and (nonallelic) gene conversion both come as a consequence of sequence misalignment between nonallelic repeats. Thus, the higher the degree of homogeneity among the repeats in a family, the higher the rates of unequal crossover and gene conversion would be (and *vice versa*). For this reason, the rates of unequal crossover and gene conversion in the rDNA family would be higher than those in the V_H family. This could be another reason for the large difference in the rate of concerted evolution between the two families. It goes without saying that, in a multigene family whose functional requirement is intermediate between the above two extremes, the rate of concerted evolution would also be intermediate.

7.2.3. Structure of Repeat Unit

By the structure of repeat unit we mean the numbers and sizes of noncoding regions (e.g., introns and spacer regions) and coding regions contained in a repeat unit. Since noncoding regions generally evolve rapidly, it would be difficult to maintain high homology among the repeats in a family if each repeat contains large or many noncoding regions. As just mentioned, the rates of unequal crossover and gene conversion are expected to decrease with decreasing intraspecific homogeneity. Thus, the structure of repeat unit should play a role in concerted evolution.

Zimmer *et al.* (1980) estimated that in higher primates the rate of concerted evolution in the α-globin gene region is 50 times higher than that in the β-globin gene region. They suggested that the rate in the β region has been greatly reduced because the introns and flanking sequences are highly divergent between the two β genes. It is interesting to note that the β genes have introns that are several times longer than those in the α genes and that the intergenic region between the two β genes is 2400 bases longer than that between the two α genes. Zimmer *et al.* (1980) suggested that the larger introns and intergenic region in β genes arose as a response to selection for reduction in the rate of unequal crossover, which produces a single gene state (the Lepore state) in which the major gene is under the control of the minor promoter. Because the noncoding sequences in the β region are highly divergent, they can act as barriers to unequal crossover. As the length of such sequences goes up, their effectiveness as barriers will rise. We note, however, that the selective advantage for a reduction in the rate of unequal crossover would be very small, at most as large as the amount of reduction in rate. Thus, the larger introns and intergenic region might have arisen by chance rather than by selection. It is also possible that the introns and the intergenic region were already large before the divergence of the higher primates, and this has promoted the divergence between the two β genes, including divergence in regulation. On the other hand, the introns and intergenic region in the α genes could have been small or have been reduced since the divergence of higher primates, and as a consequence, concerted evolution has occurred at a higher rate in the α genes than in the β genes.

7.2.4. Family Size and Population Size

It is quite easy to see that the rate of concerted evolution is dependent on the number of repeats in a family. For example, if there are only two repeats on a chromosome, a single intrachromosomal gene conversion will lead to homogeneity of the repeats on the chromosome, but this may not be the case if there are more than two repeats on the chromosome. It is also easy to see that the rate of concerted evolution is dependent on (effective) population size, because the time required for a variant to be eliminated from or to become fixed in a population is dependent on the population size.

Smith (1974) seems to have been the first author to conduct a quantitative study of the effect of family size on the rate of homogenization in a multigene family. His simulation study indicated that the number of unequal crossovers required for the fixation of a variant repeat in a single chromosomal lineage increases roughly with the number of repeats on the chromosome. In a series of papers, Ohta has investigated the effects of family size and population size on the evolution of multigene families under unequal crossover or gene conversion (Ohta, 1980, 1982, 1983, 1984). She proposed to study the probability f that two genes randomly chosen at the same locus are identical, the probability C_1 that two genes randomly chosen from the same chromosome are identical, and the probability C_2 that two genes randomly chosen from different loci on different chromosomes are identical. We note that $1 - f$ is a measure of intralocus variability, C_1 is an index of similarity between repeats within a chromosome, and C_2 incorporates both intralocus and interlocus variation. Nagylaki (1984) has also studied these quantities under the effect of intrachromosomal gene conversion.

The results shown in Table XVII were obtained by using Nagylaki's (1984) formulas (see also Ohta, 1983). We assume that gene conversion is asymmetric, i.e., nonreciprocal transfer of information from one gene to another. We assume further that each allele mutates to a new allele at the rate of 10^{-6} per generation. In the table, $\lambda = c/4n$, where n is the

Table XVII
Probability of Identity at Equilibrium[a]

		$\lambda = 0$	$\lambda = 10^{-5}$			$\lambda = 10^{-3}$		
N	n	f	f	C_1	C_2	f	C_1	C_2
10^3	10	0.996	0.978	0.516	0.516	0.973	0.969	0.968
	50	0.996	0.964	0.164	0.164	0.853	0.818	0.817
	100	0.996	0.961	0.088	0.088	0.743	0.680	0.679
	500	0.996	0.959	0.019	0.019	0.430	0.288	0.288
10^4	10	0.962	0.814	0.430	0.430	0.791	0.798	0.787
	50	0.962	0.729	0.124	0.124	0.368	0.354	0.353
	100	0.962	0.713	0.065	0.065	0.224	0.206	0.205
	500	0.962	0.698	0.014	0.014	0.070	0.047	0.047
10^5	10	0.714	0.304	0.161	0.161	0.275	0.315	0.274
	50	0.714	0.212	0.036	0.036	0.055	0.055	0.053
	100	0.714	0.199	0.018	0.018	0.028	0.026	0.026
	500	0.714	0.188	0.004	0.004	0.007	0.005	0.005

[a] N, Effective population size; n, number of repeats; $\lambda = c/4n$; c, probability of conversion per individual per generation; f, probability that two distinct genes at the same locus, chosen at random, are identical; C_1, probability that two distinct genes on the same chromosome, chosen at random, are identical; C_2, probability that two genes chosen randomly from different loci and on different chromosomes are identical.

family size and c is the probability of gene conversion per individual per generation. The probability c should increase at least linearly with n (Nagylaki and Petes, 1982), and thus λ depends more weakly on n than c does. In all cases, equal crossover between two adjacent loci occurs at the rate of 10^{-3} per generation. [For a more detailed description of the model, see Nagylaki (1984).] The case of no gene conversion, i.e., $\lambda = 0$, is given for comparison; in this case, C_1 and C_2 will decrease and eventually become zero because mutations at different loci are different. In the presence of gene conversion, C_1 and C_2 are not zero and can be high if N and n are small and λ is fairly large. We note that f, C_1, and C_2 all decrease as N and n increase and all become quite small when N and n are large. Thus, if the population size or the family size is large, it is difficult to maintain high sequence homogeneity in a family. Here, however, we have not considered natural selection; if the intensity of purifying selection is strong, high sequence homogeneity may be maintained in a large multigene family. In Table XVII we have considered the evolution of multigene families under intrachromosomal gene conversion. The reader may consult Ohta (1980, 1983) for the evolution of multigene families under unequal crossover.

7.3. Evolutionary Implications

7.3.1. Spread of Advantageous Mutations

Concerted evolution allows the spread of a variant repeat to all family members. This capability of horizontal spreading has profound evolutionary consequences, because an advantageous mutant repeat can replace all other repeats and becomes fixed in the family. We note that the advantage that a single variant can confer to an organism is usually very limited. The advantage would, however, be greatly amplified if the mutation spreads to many or all members. Thus, through concerted evolution a small selective advantage can become a great advantage. This is obviously a great evolutionary advantage of concerted evolution over independent evolution.

Arnheim (1983) has compared the evolution of RNA polymerase I transcriptional control signals with that of RNA polymerase II transcriptional control signals. The former RNA polymerase transcribes only the rRNA genes, whereas the latter transcribes all protein-coding genes. The former type of control signals appear to have evolved much more rapidly than the latter type. For example, in cell-free transcription systems, mouse rDNA clone does not work in HeLa cell extract, but clones of protein-coding genes from astonishingly diverse species are transcribed in heterologous systems; e.g., silk worm genes in HeLa cell extract. Arnheim (1983) argues that for RNA polymerase I transcription units, mutations that favorably affect transcription initiation could be propagated throughout the rDNA multigene family as a consequence of concerted evolution. On the other hand, for RNA polymerase II transcription units, advantageous mutations affecting transcription initiation that occur in any one gene would not be expected to be propagated throughout all genes, for they belong to many different families.

7.3.2. Retardation of Divergence of Duplicate Genes

The traditional view for the creation of a new gene is that a gene duplication occurs and one of the two resultant genes gradually diverges and becomes a new gene (e.g., Ohno, 1970). It is now clear that the process may not be as simple as assumed. As long as the degree of divergence between the two genes is not large, the diverged copy may be deleted by unequal crossover or converted to the conserved copy by gene conversion. In the former case it requires another duplication to create a redundant copy, while in the latter case divergence must start from scratch again. Thus, divergence of duplicate genes may proceed much more slowly than traditionally thought, and for this reason the chance of creation of a new gene from a redundant copy is reduced.

7.3.3. Evolutionary History of Duplicate Genes

It has been customary to assume that following a gene duplication the two resultant genes will diverge monotonically with time. Under this assumption it is rather simple to infer the origin of duplicate genes. For example, the protein sequences of human β and δ globins are more similar to each other than to rabbit $\beta 1$, mouse β major and minor sequences (Dayhoff, 1972). It has therefore been inferred that the two human genes were derived from a duplication event after the mammalian radiation, about 40 mya (Dayhoff, 1972). The evolutionary history of duplicate genes inferred in this traditional manner may be erroneous in view of the fact that duplicate genes can correct each other. In fact, it has recently been suggested that the β and δ genes originated from a duplication event before the mammalian radiation that occurred over 80 mya (Hardison and Margot, 1984). This suggestion is based on the observation that the large intron and the 3′ untranslated region of rabbit pseudogene $\psi\beta 2$ are more similar to human δ than to rabbit $\beta 1$ and that the pseudogene βh3 in mouse is similar to δ at its 3′ end (Edgell *et al.*, 1983). If this hypothesis, which is not supported by the observation that no obvious descendant of δ is found in the goat β-globin family so far analyzed (Cleary *et al.*, 1981; Shapiro *et al.*, 1983), turns out to be true, it will provide an excellent example of how gene correction events may obscure the evolutionary history of duplicate genes. In a large multigene family, gene correction events will occur more often and it will be even more difficult to trace the evolutionary relationships of the family members.

7.3.4. Generation of Genic Variation

From the evolutionary point of view there is a high degree of analogy between the evolution of multigene families and the evolution of subdivided populations. We may regard each repeat in a multigene family as a deme in a subdivided population. Then, transfer of information between repeats would be equivalent to migration of genes or individuals between demes. It is well known that migration will reduce the genetic difference between demes but will increase the amount of genic variation (e.g., the number of alleles) in a deme. Similarly, transfer of information between repeats will reduce the genetic difference between repeats but will increase the amount of genic variation at a locus. This can be seen from Table XVII. We note that the intralocus variation $1 - f$ is lower in the absence of gene conversion (i.e., $\lambda = 0$) than in the presence of gene conversion. We note that $1 - C_1$ and $1 - C_2$, the interlocus variations, are 1 in the absence of gene conversion but become lower in the presence of gene conversion.

It is well known that the *H-2*-antigen loci in the mouse major histocompatibility complex are highly polymorphic. Allelic gene products from different inbred strains differ extensively in amino acid sequences and hence in serologic properties. About 50 different alleles at each of the *H-2K* and *H-2D* loci have been observed (Klein, 1975). Weiss *et al.* (1983) suggest that the large numbers of alleles at the *H-2* loci are mainly generated by gene conversion-like events; donor sequences could be from the *H-2* loci and from nearby loci. This proposal can also explain the observation that a given allele of, say, the *H-2K* gene is commonly no more similar to another allele than it is to a nonallelic *H-2D* gene (Klein, 1975). Another interesting observation is that the differentiation antigen loci (*Qa* and *Tla*) are not as polymorphic as the histocompatibility antigen loci (*H-2*), although they all belong to the class I family of the major histocompatibility complex. In order to explain this observation, a model of preferential conversion has been proposed (Bregegere, 1983; Lalanne *et al.*, 1983; Ohta, 1984). It assumes that, in addition to the usual conversion process assumed in Ohta (1983), genes at *H-2* loci are preferentially converted by genes at *Qa* and *Tla* loci (Ohta, 1984). There are some data supporting this model (Weiss *et al.*, 1983).

8. Mechanisms of DNA Evolution

As is well known, there are two opposite views on the mechanisms of evolutionary change in macromolecules. The selectionists, who adhere to traditional neo-Darwinism, believe that the great majority of mutant substitutions in evolution are adaptive and caused by positive Darwinian selection. The neutralists, on the other hand, hold that chance events have been more important than natural selection in bringing about evolutionary changes in macromolecules. The latter hypothesis was proposed by Kimura (1968a) and King and Jukes (1969) and has been known as the neutral mutation–random drift hypothesis or simply the neutral theory. It should be emphasized that this theory does not assume that all mutations are selectively neutral. Rather, it assumes that a certain fraction of mutations in a gene, depending on the functional importance, are deleterious, and it does not deny that advantageous mutations play a role in protein evolution (Kimura, 1983a).

The major implication of the neutral theory is that the rate of evolution will be higher for molecules or parts of a molecule with weak or no selective constraints and low for those that are tightly constrained. This prediction can be represented by a simple formula (Kimura, 1983a). Suppose that advantageous mutations are negligibly rare, that a certain fraction f_0 of mutations are selectively neutral, and that the rest $1 - f_0$, are definitely deleterious and cannot become fixed in the population. Let v_T be the total mutation rate per nucleotide site per year. Then, the rate of neutral mutation per site is $v_T f_0$ and the rate of nucleotide substitutions per site per year is

$$k = v_T f_0 \tag{23}$$

(see Kimura, 1983a). Thus, the neutral theory predicts a negative correlation between rate and selective constraints, because the probability f_0 of a mutation being neutral should be higher if the selective constraints are fewer or weaker. According to this prediction, the substitution rate should be highest when there is no selective constraints; i.e., $f_0 = 1$.

This prediction is strongly supported by the results summarized in Section 3 for the rates of nucleotide substitution in various gene regions. Among the regions of a gene, nondegenerate sites are expected to be most tightly constrained, because all mutations at these sites are nonsynonymous, so that they are subject to constraints at both the RNA and protein levels. The constraints at the protein level would be particularly strong, for natural selection acts on phenotypes of the organism, for which the protein structure and function play a decisive role. Thus, the f_0 value is expected to be lowest at nondegenerate sites, and so is the substitution rate. This is indeed the case (Table VII). The rate should be higher at twofold degenerate sites, for a fraction of the mutations at these sites are synonymous, so that these sites should be less stringently constrained than nondegenerate sites. The rate should be even higher at fourfold degenerate sites, for all mutations at these sites are synonymous. These two predictions are seen to be true (Table VII). Among the noncoding regions, the 3′ flanking region is probably the least constrained, for it does not seem to serve any function or contain any important DNA signal. This region is seen to evolve very rapidly. Introns are also expected to evolve at a high rate, for they do not have any known function, although they must contain signals for RNA splicing. This prediction is again seen to be true. The 5′ and 3′ untranslated regions and the 5′ flanking region are seen to evolve at intermediate rates. These observations appear to be reasonable, for parts of these regions serve some important functions, such as transcriptional and translational control, but other parts do not seem to be important. Since pseudogenes appear to be free of selective constraints, all mutations in them would be neutral (i.e., $f_0 = 1$) and they would evolve at the highest rate. This is indeed true.

Under the neutralist view, we can also give a plausible explanation for why the rate of nonsynonymous substitution varies greatly among different kinds of genes. For those genes such as histones and actins that have extremely low nonsynonymous rates (Table III), we may assume that they are subject to tight selective constraints because their functions are important and require very specific protein structures. According to this view, nearly every nonsynonymous mutation in histone and actin genes is deleterious, and would not become fixed in the population, i.e., f_0 is almost zero. On the other hand, for genes such as those coding for relaxin, interferons, and immunoglobulins that have high nonsynonymous rates (Table III), we may assume that they are subject to fewer or weaker selective constraints because their functions do not require very specific protein structures or because their functions are not very important or can be carried out by other genes; i.e., these genes are "dispensable" (Wilson *et al.*, 1977). We also noted above that the variation in the synonymous rate among genes is large, although much smaller than that in the nonsynonymous rate. Under the neutralist view, this variation can be explained by assuming that it is largely due to chance effects and partly due to variation in selective constraints among genes. As will be discussed below, synonymous mutations are not completely free of selective constraints.

Although some of the above arguments were based on assumptions that remain to be substantiated by experimental data, the neutralist view can apparently give a consistent explanation for the various features of

DNA evolution. In contrast, the selectionist view cannot give a consistent explanation. To explain the negative correlation between rate and selective constraints by this view, one would have to assume that the probability of a mutation being selectively advantageous is higher in functionally less important DNA regions. It is, however, difficult to imagine how a mutation in a pseudogene can be advantageous to the organism and how the function of a gene can be constantly improved by synonymous mutations. One may concede that neutral or nearly neutral mutations are predominant in pseudogenes and at synonymous sites, but maintain that the majority of amino acid substitutions in protein evolution have been due to advantageous mutations (e.g., Goodman, 1981b). But it is again difficult to imagine why advantageous mutations should occur more often in less important coding regions, such as those coding for the signal peptides and the insulin C- and relaxin C-peptides, or, in other words, why these latter regions have constantly been improved by new mutations. Furthermore, it is difficult to imagine why the rate of improvement in function should be faster in a less important gene than in a more important one. In short, the selectionist view cannot give a satisfactory explanation for the negative correlation between rate and functional importance.

Under the neutralist view, one can use the substitution rate in pseudogenes as the neutrality standard to infer the stringency of selective constraints in various gene regions. The 3′ flanking region of a gene is almost free of selective constraint because it evolves almost as rapidly as pseudogenes (Table VII). The fourfold degenerate sites in a gene are on the average subject to only very weak constraints, because the substitution rate at these sites is only 14% lower than the pseudogene rate. Introns are apparently subject to somewhat stronger constraints, for they evolve somewhat more slowly than fourfold degenerate sites. The twofold degenerate sites and the 5′ flanking region are apparently subject to certain constraints, for they evolve only half as rapidly as pseudogenes. The 5′ and 3′ untranslated regions are apparently subject to even stronger selective constraints, for they evolve only one-third as rapidly as pseudogenes. The nondegenerate sites in a gene are by far the most tightly constrained region, for their average substitution rate is only one-fifth of the pseudogene rate. If we assume that the nucleotide substitutions at nondegenerate sites are totally due to neutral mutations, then the proportion of neutral mutations at nondegenerate sites is approximately 20%; i.e., $f_0 = 0.20$. This is quite similar to Kimura's (1983b) estimate of 0.14 ± 0.06. Of course, the proportion should vary from gene to gene, ranging from 62% in interferon γ to almost zero in histone H4 (Table III). Similarly, we can infer the stringency of selective constraints in various genes. For example, albumin has previously been used as an example of rapid evolution at the amino acid level and the argument for this is that albumin is a dispensable protein (Wilson *et al.*, 1977). Although the nonsynonymous rate in the gene for albumin is not low, 0.92×10^{-9} (Table VII), it is only 20% of the pseudogene rate. Thus, it appears that albumin is subject to fairly strong selective constraints. Among the genes listed in Table III, interferon α has the highest nonsynonymous rate, 2.80×10^{-9}. This is 59% of the pseudogene rate. Thus, it appears that, even in a rapidly evolving gene, a substantial fraction of nonsynonymous mutations are deleterious.

When the neutral mutation–random drift hypothesis was proposed, synonymous changes were considered as candidates for neutral mutations (Kimura, 1968b; King and Jukes, 1969). For this reason, the universal phenomenon of nonrandom usage of synonymous codons has often been taken as evidence against the neutral theory (e.g., Grantham, 1980). However, as mentioned in Section 6, there is now good evidence that the choice of synonymous codons may be constrained by tRNA availability and the interaction between codon and anticodon, as well as other factors. Thus, this phenomenon can also be explained within the framework of the neutral theory in terms of selective constraints. Actually, the selective constraints on synonymous changes appear to be generally quite weak, for the average substitution rate at fourfold degenerate sites is only 14% lower than the pseudogene rate. One may wonder if such weak selective constraints can produce the observed biases in codon usage. A definite answer to this question requires a quantitative treatment, but one should remember that the mutation pressure is generally exceedingly weak, so that even weak selective constraints can produce large biases in codon usage. Of course, in highly expressed genes, the codons recognized by only minor tRNA species are probably definitely disadvantageous, for they are rarely used. For a more quantitative discussion of nonrandom usage of synonymous codons, the reader is referred to Kimura (1983a).

The approximate constancy of the rate of amino acid substitution in protein evolution has been taken as strong evidence for the neutral theory (Kimura, 1969, 1983a). The argument is as follows. For strictly neutral mutations, the rate k of substitution is equal to the rate of neutral mutation $v_0 = v_T f_0$ [formula (23)]. Thus, the rate constancy follows readily from the neutral theory by assuming that v_0 is constant per year among different lineages. On the other hand, if mutant substitutions are due to definitely advantageous mutations, the substitution rate is

$$k = 4N_e s_A v_A \tag{24}$$

where N_e is the effective population size, s_A is the average advantage for advantageous mutations, and v_A is the rate of advantageous mutation

(Kimura, 1983a). Obviously, for k to be constant, the product $N_e s_A v_A$ must remain constant in the course of evolution among diverse lineages. This is very unlikely because, for example, N_e may vary widely among diverse groups such as human, bovine, and mouse.

In our reevaluation of the molecular clock or rate-constancy hypothesis, we also found that the rates of synonymous and nonsynonymous substitution are roughly constant over time, if no large change in generation time occurs. However, the generation-time effect is stronger on synonymous substitution than on nonsynonymous substitution, for the average nonsynonymous rate is only 30% higher in the rodent lineage than in the human lineage, but the average synonymous rate is about two times higher in the rodent lineage than in the human lineage. These latter observations are puzzling, but can be explained within the framework of the neutral theory. Let us assume that the total rate of mutation *per generation* is roughly proportional to the number of cell cycles per generation in the germ line. Under this assumption, the mutation rate per year in the rodent lineage would be higher than that in the human lineage; according to Wilson *et al.* (1977), the difference would be about twofold. Since the selective constraints on synonymous mutations appear to be generally very weak, we may assume that a large fraction of synonymous mutations are neutral or almost neutral, so that the rate of synonymous substitution would not be strongly affected by the effective population size. If these assumptions hold approximately, it is not surprising to see that the synonymous rate in the rodent lineage is about two times higher than in the human lineage, the difference being the order of the difference in mutation rate. On the other hand, nonsynonymous mutations are usually subject to strong selective constraints, so that the rate of mutation to "effectively neutral alleles" would decrease as the effective population size increases. It is reasonable to assume that the effective population size has been larger in the rodent lineage than in the human lineage; in fact, the fossil record suggests that rodents have been the most successful species in mammalian evolution (Romer, 1966). For this reason, the generation-time effect on the nonsynonymous rate is substantially weaker than on the synonymous rate. Under the selectionist view, it would be difficult to explain the above observations. According to formula (23), both the synonymous and the nonsynonymous rates should be considerably higher in the rodent lineage than in the human lineage, for the N_e value would have been considerably larger in the rodent lineage than in the human lineage and the v_A value would be about two times higher in the rodent lineage than in the human lineage. These predictions are not in agreement with the above observations.

We mentioned in Section 4 that many examples of accelerated evolution following gene duplication have been proposed and have been taken as evidence against the molecular clock hypothesis and the neutral theory. It should be pointed out that the neutral theory predicts a constant rate only under the condition that the selective constraints remain more or less the same over time. Thus, it does allow accelerated evolution to occur following gene duplication because the selective constraints are relaxed at least to some extent when a gene is duplicated. Moreover, it also allows a redundant duplicate to accumulate novel changes to improve its function or to become a new gene. However, it postulates that accelerated evolution is more likely to arise from relaxation of selective constraints than from advantageous mutations. Our study of the evolution of the goat and sheep globin genes supports this view (see Section 4).

Let us stress that we do believe that advantageous substitutions do occur in gene evolution. The considerably higher rates of nonsynonymous substitution in the genes coding for the growth hormone and for the α subunit of glycoprotein hormones in the human lineage than in the rodent lineage (Table IX) may have largely been due to advantageous mutations. Other exceptional cases to the molecular clock hypothesis, such as insulin in the goat lineage (King and Jukes, 1969), may also have been due to positive selection. We also believe that when a gene is duplicated, an opportunity is opened for novel changes in function. For example, gene duplication has allowed various interactions to develop between hemoglobin α and β chains. As another example, calmodulin is known to be extremely conservative, but a series of gene duplications has allowed the emergence of troponin-C, parvalbumin, and other members of the calmodulin superfamily, whose amino acid sequences are now considerably different from that of calmodulin (Baba *et al.*, 1984). Nevertheless, we believe that in the long history of protein evolution, advantageous substitutions can account for only a minor portion of amino acid substitutions—the majority are due to random fixation of neutral or nearly neutral mutations.

9. Concluding Remarks

In addition to the topics discussed above, there are several other important topics in molecular evolution. One is the so-called molecular phylogeny. The rapid accumulation of DNA sequences will be extremely useful for studying the phylogenetic relationships among organisms and for dating the divergence time between genes or organisms. The reader interested in the methodological aspects of phylogeny reconstruction or in the application of macromolecular sequences in evolutionary studies may refer to Dayhoff (1972), Sneath and Sokal (1973), Moore *et al.* (1973), Tateno *et al.* (1982), and Felsenstein (1983). Another topic is the evolutionary role of gene duplication. Sequencing work has revealed that the existence of duplicate genes in the genome of higher organisms is much more widespread than previously thought—it now appears that the majority of genes do not exist as single copies, but rather as clusters. Moreover, data on protein and DNA sequences strongly suggest that complex genes have been derived from primordial genes mainly by internal genes duplication, particularly duplication of functional or structural domains. We have discussed this topic in two recent reviews (Li, 1982, 1983); see also Ohno (1970), Dayhoff (1972), MacIntyre (1976), Wilson *et al.* (1977), and Doolittle (1979). The proposal of the concept of selfish DNA (Doolittle and Sapienza, 1980; Orgel and Crick, 1980) has provoked a great deal of controversy (Orgel *et al.*, 1980; Doolittle, 1982). Selfish DNAs are defined as pieces of DNA that have little or no phenotypic effect at the organismal level yet can spread in populations because of their rapid replication within the genome. Transposable elements are considered to be the most important class of DNAs that can be called "selfish" (Doolittle, 1982). The population genetics and evolution of transposable elements is currently a subject of considerable importance (e.g., Ohta and Kimura, 1981; Hickey, 1982; Campbell, 1983; Kidwell, 1983; Langley *et al.*, 1983).

Although considerable progress has been made in comparative studies of DNA sequences, there are many problems that call for further studies. First, as noted above, all current methods for estimating the number of nucleotide substitutions between sequences do not give reliable estimates when the degree of sequence divergence is large. Thus, how to overcome this difficulty is a challenging problem to theoretical molecular evolutionists. Second, the rates of nucleotide substitution in noncoding regions need to be studied more carefully, because current sequence data are not very suitable for this purpose. Moreover, current estimates of the substitution rates in various regions of genes have been obtained mainly from comparisons of vertebrate species, and it is not clear whether or not such estimates apply to invertebrates, plants, and prokaryotes. Obviously, much work remains to be done to obtain a general picture of the substitution rate in all sorts of organisms. Third, the molecular clock hypothesis needs to be examined much more carefully. Our reevaluation of this hypothesis has suggested the existence of a strong generation-time effect on synonymous substitution. If this is true, the generation-time effect should also occur in DNA regions that are subject to weak or no selective constraints. Introns and pseudogenes are therefore good materials for testing this effect. Furthermore, in lower organisms, the generation-time effect should be particularly strong, because their generation time is very short. These problems are worth pursuing. Fourth, so far the pattern of nucleotide substitution in functional genes has been studied in only a few genes. A detailed study of this pattern in each particular class of genes (i.e., genes coding for the same protein) may sharpen our understanding of the relationship between selective constraints and nucleotide exchangeability. Fifth, it is not clear whether gene conversion or unequal crossover has played a more important role in the concerted evolution of multigene families. To answer this question, we will need to know more about the rates of gene conversion and unequal crossover, the amount of information usually transferred in a gene conversion or unequal crossover event, and the factors that affect the rates of gene conversion and unequal crossover. We also need more detailed theoretical studies of the evolution of multigene families under gene conversion or unequal crossover. All current models of the evolution of multigene families involve some unrealistic assumptions (Ohta, 1980, 1983, 1984; Nagylaki, 1984). For example, in all models of gene conversion, it is assumed that conversion occurs randomly between members of a gene cluster, regardless of the position of the members.

In the past, our understanding of the mechanisms of molecular evolution has come mainly from comparative studies of protein and DNA sequences. Recent developments in directed mutagenesis and chemical synthesis of DNA sequences (e.g., Wallace *et al.*, 1981; Villafranca *et al.*, 1983; Nambiar *et al.*, 1984) have opened a new approach to this problem. With these techniques, the primary sequences of genes can be altered systematically, thus greatly facilitating structure–function studies of genes or proteins. In particular, it is now possible to prepare a series of DNA sequences that are possible immediate ancestral sequences of the genes in related species. By investigating the physicochemical properties of the proteins coded by these sequences, we may be able to infer whether the evolutionary changes have occurred mainly by improvement in protein function or by random fixation of neutral or nearly neutral mutations.

ACKNOWLEDGMENTS. We thank Dr. M. Nei for many suggestions. This work was supported by research grants from the National Institute of Health and the National Science Foundation.

References

Adams, J., and Rothman, E. D., 1982, Estimation of phylogenetic relationships from DNA restriction pattern and selection of endonuclease cleavage sites, *Proc. Natl. Acad. Sci. USA* **79:**3560–3564.

Adams, J. M., Jepperson, P. G. N., Sanger, F., and Barrell, B. G., 1969, Nucleotide sequence from the coat protein cistron of R17 bacteriophage RNA, *Nature* **223:**1009–1014.

Anderson, S., Bankier, A. T., Barrell, B. G., DeBruijn, M. H. L., Coulson, A. R., Drouin, J., Eperon, I. C., Nierlich, D. P., Roe, B. A., Sanger, F., Schreier, P. H., Smith, A. J. H., Staden, R., and Young, I. G., 1981, Sequence and organization of the human mitochondrial genome, *Nature* **290:**457–464.

Andersson, S. G. E., Buckingham, R. H., and Kurland, C. G., 1984, Does codon composition influence ribosome function? *EMBO J.* **3:**91–94.

Arnheim, N., 1983, Concerted evolution of multigene families, in: *Evolution of Genes and Proteins* (M. Nei and R. K. Koehn, eds.), Sinauer, Sunderland, Massachusetts, pp. 38–61.

Baba, M. L., Goodman, M., Berger-Cohn, J., Demaialle, J. G., and Matsuda, G., 1984, The early adaptive evolution of calmodulin, *Mol. Biol. Evol.* **1:**442–455.

Baltimore, D., 1981, Gene conversion: Some implications for immunoglobulin genes, *Cell* **24:**592–594.

Beale, D., and Lehmann, H., 1965, Abnormal hemoglobin and the genetic code, *Nature* **207:**259–261.

Beneveniste, R. E., and Todaro, G. J., 1975, Evolution of type C viral genes: Preservation of ancestral murine type C viral sequences in pig cellular DNA, *Proc. Natl. Acad. Sci. USA* **72:**4090–4094.

Bennetzen, J. L., and Hall, B. D., 1982, Codon selection in yeast, *J. Biol. Chem.* **257:**3026–3031.

Birky, C. W., Jr., and Skavaril, R. V., 1976, Maintenance of genetic homogeneity in systems with multiple genomes, *Genet. Res.* **27:**249–265.

Blaisdell, B. E., 1983a, A prevalent persistent global nonrandomness that distinguishes coding and noncoding eukaryotic nuclear DNA sequences, *J. Mol. Evol.* **19:**122–133.

Blaisdell, B. E., 1983b, Choice of base at silent codon site 3 is not selectively neutral in eucaryotic structural gene: It maintains excess short runs of weak and strong hydrogen bonding bases, *J. Mol. Evol.* **19:**226–236.

Bossi, L., and Roth, J. R., 1980, The influence of codon context on genetic code translation, *Nature* **286:**123–127.

Breathnach, R., Benoist, C., O'Hare, K., Gannon, F., and Chambon, P., 1978, Ovalbumin gene: Evidence for a leader sequence in mRNA and DNA sequencing at the exon–intron boundaries, *Proc. Natl. Acad. Sci. USA* **75:**4853–4857.

Bregegere, F., 1983, A directional process of gene conversion is expected to yield dynamic polymorphism associated with stability of alternate alleles in class I histocompatibility antigen gene family, *Biochimie* **65:**229–237.

Britten, R. J., and Kohne, D. E., 1968, Repeated sequences in DNA, *Science* **161:**529–540.

Brown, A. J. L., and Ish-Horowicz, D., 1981, Evolution of the 87A and 87C heat-shock loci in *Drosophila*, *Nature* **290:**677–682.

Brown, D. D., and Sugimoto, K., 1973, 5S DNAs of *Xenopus laevis* and *Xenopus mulleri:* Evolution of a gene family, *J. Mol. Biol.* **78:**397–415.

Brown, D. D., Wensink, P. C., and Jordan, E., 1972, A comparison of the ribosomal DNA's of *Xenopus laevis* and *Xenopus mulleri:* The evolution of tandem genes, *J. Mol. Biol.* **63:**57–73.

Brutlag, D. L., 1980, Molecular arrangement and evolution of heterochromatic DNA, *Annu. Rev. Genet.* **14:**121–144.

Buckingham, R. H., and Carrier, M. J., 1982, The effect of codon context on mistranslation of messenger RNA, in: *Interaction of Translational and Transcriptional Controls in the Regulation of Gene Expression* (M. Grunberg-Manago and B. Safer, eds.), Elsevier, New York, pp. 215–219.

Buongiorno-Nardelli, M., Amaldi, F., and Lava-Sanchez, P.A., 1972, Amplification as a rectification mechanism for the redundant rRNA genes, *Nature* **238**:134–137.

Callan, H. G., 1967, The organization of genetic units in chromosomes, *J. Cell Sci.* **2**:1–7.

Campbell, A., 1983, Transposons and their evolutionary significance, in: *Evolution of Genes and Proteins* (M. Nei and R. K. Koehn, eds.), Sinauer, Sunderland, Massachusetts, pp. 258–279.

Clarke, B. C., 1970, Darwinian evolution of proteins, *Science* **168**:1009–1011.

Cleary, M. L., Schon, E. A., and Lingrel, J. B., 1981, Two related pseudogenes are the result of a gene duplication in the goat β-globin locus, *Cell* **26**:181–190.

Coen, E. S., and Dover, G. A., 1983, Unequal exchanges and the coevolution of X and Y rDNA arrays in *Drosophila melanogaster, Cell* **33**:849–855.

Coen, E., Strachan, T., and Dover, G., 1982a, Dynamics of concerted evolution of ribosomal DNA and histone gene families in the *melanogaster* species subgroup of *Drosophila J. Mol. Biol.* **158**:17–35.

Coen, E. S., Thoday, J. M., and Dover, G., 1982b, Rate of turnover of structural variants in the rDNA gene family of *Drosophila melanogaster, Nature* **295**:564–568.

Coulondre, C., Miller, J. H., Farabaugh, P. J., and Gilbert, W., 1978, Molecular basis of base substitution hotspots in *Escherichia coli, Nature* **274**:775–780.

Czelusniak, J., Goodman, M., and Moore, G. W., 1978, On investigating the statistical properties of the populous path algorithm by computer simulation, *J. Mol. Evol.* **11**:75–85.

Czelusniak, J., Goodman, M., Hewett-Emmett, D., Weiss, M. L., Venta, P. J., and Tashian, R. E., 1982, Phylogenetic origins and adaptive evolution of avian and mammalian haemoglobin genes, *Nature* **298**:297–300.

Dayhoff, M. O., 1972, *Atlas of Protein Sequence and Structure,* Vol. 5, National Biomedical Research Foundation, Silver Spring, Maryland.

Dickerson, R. E., 1971, The structure of cytochrome c and the rates of molecular evolution, *J. Mol. Evol.* **1**:26–45.

Dickerson, R. E., and Geis, I., 1969, *The Structure and Action of Proteins,* Benjamin, London.

Doolittle, R. F., 1979, Protein evolution, in: *The Proteins* (H. Neurath and R. L. Hill, eds.), Academic Press, New York, pp. 1–119.

Doolittle, W. F., 1982, Selfish DNA after fourteen months, in: *Genome evolution* (G. A. Dover and R. B. Flavell, eds.), Academic Press, New York, pp. 3–28.

Doolittle, W. F., and Sapienza, C., 1980, Selfish genes, the phenotype paradigm and genome evolution, *Nature* **284**:601–603.

Dover, G., 1982, Molecular drive: A cohesive mode of species evolution, *Nature* **299**:111–117.

Dover, G., and Coen, E., 1981, Springcleaning ribosomal DNA: A model for multigene evolution? *Nature* **290**:731–732.

Edelman, G. M., and Gallant, J., 1977, Mistranslation in *E. coli, Cell* **10**:131–137.

Edelman, G. M., and Gally, J. A., 1970, Arrangement and evolution of eukaryotic genes, in: *The Neurosciences: Second Study Program* (F. O. Schmitt, ed.), Rockefeller University Press, New York, pp. 962–972.

Edgell, M. H., Hardies, S. C., Brown, B., Voliva, C., Hill, A., Philipps, S., Comer, M., Burton, F., Weaver, S., and Hutchison III, C. A., 1983, Evolution of the mouse β globin gene complex locus, in: *Evolution of Genes and Proteins* (M. Nei and R. K. Koehn, eds.), Sinauer, Sunderland, Massachusetts, pp. 1–13.

Efstratiadis, A., Posakony, J. W., Maniatis, T., Lawn, R. M., O'Connel, C., Spritz, R. A., DeRiel, J. K., Forget, B. G., Weissman, S. M., Slightom, J. L., Blechl, A. E., Smithies, O., Barralle, F. E., Shoulders, C. C., and Proudfoot, N. J., 1980, The structure and evolution of the human β-globin gene family, *Cell* **21**:653–668.

Ernst, J. F., Stewart, J. W., and Sherman, F., 1981, The *cyc* I–II mutation in yeast reverts by recombination with a nonallelic gene: Composite genes determining the isocytochromes c, *Proc. Natl. Acad. Sci. USA* **78:**6334–6338.

Favre, A., Morel, C., and Scherren, K., 1975, The secondary structure and poly(A) content of globin messenger RNA as a pure RNA and in polyribosome-derived ribonucleoprotein complexes, *Eur. J. Biol. Chem.* **57:**47–157.

Felsentein, J., 1983, Inferring evolutionary trees from DNA sequences, in: *Statistical Analysis of DNA Sequence Data* (B. S. Weir, ed.), Dekker, New York, pp. 133–150.

Fiers, W., 1979, Structure and function of RNA bacteriophages, in: *Comprehensive Virology,* Vol. 13 (H. Fraenkel-Conrat and R. R. Wagner, eds.), Plenum Press, New York, pp. 69–203.

Fiers, W., Contreras, R., Duerinck, F., Haegmean, G., Merregaert, J., Min Jou, W., Raeymakers, A., Volckaert, G., Ysebaert, M., Van de Kerckhove, J., Nolf, F., and Van Montagu, M., 1975, A-protein gene of bacteriophage MS2, *Nature* **256:**273–278.

Fiers, W., Contreras, R., Duerinck, F., Haegeman, G., Iserentant, D., Merregaert, J., Min Jou, W., Molemans, F., Raeymakers, A., Van Den Berghe, A., Volckaert, G., and Ysebaert, M., 1976, Complete nucleotide sequence of bacteriophage MS2 RNA: Primary and secondary structure of the replicase gene, *Nature* **260:**500–507.

Fitch, W. M., 1967, Evidence suggesting a non-random character to nucleotide replacements in naturally occurring mutations, *J. Mol. Biol.* **26:**499–507.

Fitch, W. M., 1974, The large extent of putative nucleic acid structure in random nucleotide sequences of amino acid derived mRNA, *J. Mol. Evol.* **3:**279–281.

Fitch, W. M., 1980, Estimating the total number of nucleotide substitutions since the common ancestor of a pair of homologous genes: Comparison of several methods and 3 β hemoglobin messenger RNA's, *J. Mol. Evol.* **16:**153–209.

Fogel, S., Mortimer, R., Lusnak, K., and Tavares, F., 1978, Meiotic gene conversion: A signal of the basic recombination event in yeast, *Cold Spring Harbor Symp. Quant. Biol.* **43:**1325–1341.

Fowler, R. G., Degnen, G. E., and Cox, E. C., 1974, Mutational specificity of a conditional *Escherichia coli* mutator, *mut* D5, *Mol. Gen. Genet.* **133:**179–191.

Gojobori, T., 1983, Codon substitution in evolution and the saturation of synonymous changes, *Genetics* **105:**1011–1027.

Gojobori, T., and Nei, M., 1984, Concerted evolution of the immunoglobulin V_H gene family, *Mol. Biol. Evol.* **1:**195–212.

Gojobori, T., Ishii, K., and Nei, M., 1982a, Estimation of average number of nucleotide substitution when the rate of substitution varies with nucleotide, *J. Mol. Evol.* **18:**414–423.

Gojobori, T., Li, W.-H., and Graur, D., 1982b, Patterns of nucleotide substitution in pseudogenes and functional genes, *J. Mol. Evol.* **18:**360–369.

Golding, G. B., and Strobeck, C., 1982, Expected frequencies of codon use as a function of mutation rates and codon fitnesses, *J. Mol. Evol.* **18:**379–386.

Goodman, M., 1976, Protein sequences in phylogeny, in: *Molecular Evolution* (F. J. Ayala, ed), Sinauer, Sunderland, Massachusetts, pp. 141–159.

Goodman, M., 1981a, Decoding the pattern of protein evolution, *Prog. Biophys. Mol. Biol.* **38:**105–164.

Goodman, M., 1981b, Positive selection causes purifying selection, *Nature* **295:**630.

Goodman, M., Moore, G. W., and Barnabas, J., 1974, The phylogeny of human hemoglobin genes investigated by the parsimony method, *J. Mol. Evol.* **3:**1–48.

Goodman, M., Moore, G. W., and Matsuda, G., 1975, Darwinian evolution in phylogeny of hemoglobin, *Nature* **253:**603–608.

Grantham, R., 1974, Amino acid difference formula to help explain protein evolution, *Science* **185:**862–864.

Grantham, R., 1980, Workings of the genetic code, *Trends Biochem. Sci.* **5**:327–31.

Grantham, R., Gautier, C., Gouy, M., Mercier, R., and Pave, A., 1980a, Codon catalog usage and the genome hypothesis, *Nucleic Acids Res.* **8**:r49–r62.

Grantham, R., Gautier, C., and Gouy, M., 1980b, Codon frequencies in 119 individual genes confirm consistent choices of degenerate bases according to genome type, *Nucleic Acids Res.* **8**:1883–1912.

Grantham, R., Gautier, C., Gouy, M., Jacobzone, M., and Mercier, R., 1981, Codon catalog usage is a genome strategy modulated for gene expressivity, *Nucleic Acids Res.* **9**:r43–r75.

Grosjean, H., and Fiers, W., 1982, Preferential codon usage in prokaryotic genes: The optimal codon–anticodon interaction energy and the selective codon usage in efficiently expressed genes, *Gene* **18**:199–209.

Grosjean, H., Sankoff, D., Min Jou, W., Fiers, W., and Cedergren, R. J., 1978, Bacteriophage MS2 RNA: A correlation between the stability of the codon–anticodon interaction and the choice of code words, *J. Mol. Evol.* **12**:113–119.

Grosschedl, R., and Birnstiel, M. L., 1980a, Identification of regulatory sequences of a H2A histone gene by the study of specific deletion *in vivo, Proc. Natl. Acad. Sci. USA* **77**:1432–1436.

Grosschedl, R., and Birnstiel, M. L., 1980b, Spacer DNA sequences upstream are essential for promotion of H2A histone gene transcription *in vivo, Proc. Natl. Acad. Sci. USA* **77**:7102–7106.

Grunstein, M., Schedl, P., and Kedes, L., 1976, Sequence analysis and evolution of sea urchin (*Lytechinus pictus* and *Strongylocentrotus pupuratus*) histone H4 messenger RNAs, *J. Mol. Biol.* **104**:351–369.

Hardison, R. C., and Margot, J. B., 1984, Rabbit globin pseudogene $\psi\beta 2$ is a hybrid of δ- and β-globin gene sequences, *Mol. Biol. Evol.* **1**:302–316.

Hayashida, H., and Miyata, T., 1983, Unusual evolutionary conservation and frequent DNA segment exchange in class I genes of the major histocompatability complex, *Proc. Natl. Acad. Sci. USA* **80**:2671–2675.

Heindell, H. C., Liu, A., Paddock, G. V., Studnicka, G. M., and Salser, W. A., 1978, The primary rabbit α-globin mRNA, *Cell* **15**:43–54.

Hickey, D. A., 1982, Selfish DNA: A sexually-transmitted nuclear parasite, *Genetics* **101**:519–531.

Holder, J. W., and Lingrel, J. B., 1975, Determination of secondary structure in rabbit globin messenger RNA by thermal denaturation, *Biochemistry* **14**:4209–4215.

Hollis, G. F., Heiter, P.A., McBride, O. W., Swan, D., and Leder, P., 1982, Processed genes: A dispersed human immunoglobulin gene bearing evidence of RNA-type processing, *Nature* **296**:321–325.

Holmquist, R., 1972, Theoretical foundations for a quantative approach to paleogenetics, Part 1: DNA, *J. Mol. Evol.* **1**:115–133.

Holmquist, R., 1979, The method of parsimony: An experimental test and theoretical analysis of the adequacy of molecular resolution studies, *J. Mol. Biol.* **135**:939–958.

Holmquist, R., and Pearl, D., 1980, Theoretical foundations for quantitative paleogenetics. III. The molecular divergence of nucleic acids and proteins for the case of genetic events of unequal probability, *J. Mol. Evol.* **16**:211–267.

Hood, L., Campbell, J. H., and Elgin, S. C. R., 1975, The organization expression, and evolution of antibody genes and other multigene families, *Annu. Rev. Genet.* **9**:305–353.

Hurst, D. D., Fogel, S., and Mortimer, R. K., 1972, Conversion-associated recombination in yeast, *Proc. Natl. Sci. USA* **69**:101–105.

Ikemura, T., 1981a, Correlation between the abundance of *Escherichia coli* transfer RNAs and the occurrence of the repetitive codons in its protein genes, *J. Mol. Biol.* **146**:1–21.

Ikemura, T., 1981b, Correlation between the abundance of *Escherichia coli* transfer RNAs and the occurrence of the repetitive codons in its protein genes: A proposal for a synonymous codon choice that is optimal for the *E. coli* translational system, *J. Mol. Biol.* **151**:389–409.

Ikemura, T., 1982, Correlation between the abundance of yeast transfer RNAs and the occurrence of the respective codons in protein genes: Differences in synonymous codon choice patterns of yeast and *Escherichia coli* with reference to the abundance of isoacceptor transfer RNAs, *J. Mol. Biol.* **158**:573–579.

Ikemura, T., and Ozeki, H., 1983, Codon usage and transfer RNA contents: Organism-specific codon-choice patterns in reference to isoacceptor contents, *Cold Spring Harbor Symp. Quant. Biol.* **42**:1087–1097.

Jackson, J. A., and Fink, G. R., 1981. Gene conversion between duplicated genetic elements in yeast, *Nature* **292**:306–311.

Jeffreys, A., 1979, DNA sequence variants in $^{G}\gamma$-, $^{A}\gamma$-, δ- and β-globin genes of man, *Cell*, **18**:1–10.

Jukes, T. H., and Cantor, C. R., 1969, Evolution of protein molecules, in: *Mammalian Protein Metabolism* (H. N. Munro, ed.), Academic Press, New York, pp. 21–123.

Jukes, T. H., and King, J. L., 1979, Evolutionary nucleotide replacements in DNA, *Nature* **281**:605–606.

Kafatos, F. C., Efstratiadis, A., Forget, B. G., and Weissman, S. M., 1977, Molecular evolution of human and rabbit β-globin mRNAs, *Proc. Natl. Acad. Sci. USA* **74**:5618–5622.

Kan, Y. W., Holland, J. P., Dozy, A. M., Charache, S., and Kazazian, H. H., 1975, Deletion of the β-globin structure gene in hereditary persistence of foetal haemoglobin, *Nature* **258**:162–163.

Kaplan, N., and Risko, K., 1982, A method for estimating rates of nucleotide substitution using DNA sequence data, *Theor. Popul. Biol.* **21**:318–328.

Khoury, G., and Gruss, P., 1983, Enhancer elements, *Cell* **33**:313–314.

Kidwell, M. G., 1983, Evolution of hybrid dysgenesis determinants in *Drosophila melanogaster, Proc. Natl. Acad. Sci. USA* **80**:1655–1659.

Kimura, M., 1968a, Evolutionary rate at the molecular level, *Nature* **217**:624–626.

Kimura, M., 1968b, Genetic variability maintained in a finite population due to mutational production of neutral and nearly neutral isoalleles, *Genet. Res.* **11**:247–269.

Kimura, M., 1969, The rate of molecular evolution considered from the standpoint of population genetics, *Proc. Natl. Acad. Sci. USA* **63**:1181–1188.

Kimura, M., 1977, Preponderence of synonymous changes as evidence for the neutral theory of molecular evolution, *Nature* **267**:275–276.

Kimura, M., 1979, The neutral theory of molecular evolution, *Sci. Am.* **241**:94–104.

Kimura, M., 1980, A simple method for estimating evolutionary rates of base substitutions through comparative studies of nucleotide sequences, *J. Mol. Evol.* **16**:111–120.

Kimura, M., 1981a, Estimation of evolutionary distances between homologous nucleotide sequences, *Proc. Natl. Acad. Sci. USA* **78**:454–458.

Kimura, M., 1981b, Was globin evolution very rapid in its early stages? A dubious case against the rate-constancy hypothesis, *J. Mol. Evol.* **17**:110–113.

Kimura, M., 1983a, *The Neutral Theory of Molecular Evolution*, Cambridge University Press, London.

Kimura, M., 1983b, Rare variant alleles in the light of the neutral theory, *Mol. Biol. Evol.* **1**:84–93.

Kimura, M., and Ohta, T., 1972, On the stochastic model for estimation of mutational distance between homologous proteins, *J. Mol. Evol.* **2**:87–90.

King, J. L., and Jukes, T. H., 1969, Non-Darwinian evolution, *Science* **164**:788–798.

Klein, H. L., and Petes, T. D., 1981, Intrachromosomal gene conversion in yeast, *Nature* **289:**144–148.

Klein, J., 1975, *Biology of the Mouse Histocompatibility-2-Complex*, Springer, New York.

Kohne, D. E., 1970, Evolution of high-organism DNA, *Q. Rev. Biophys.* **33:**1–48.

Kohne, D. E., Chiscon, J. A., and Hoyer, B. H., 1972, Evolution of primate DNA sequences, *J. Mol. Evol.* **1:**627–644.

Kolter, R., and Yanofsky, C., 1982, Attenuation in amino acid biosynthetic operons, *Annu. Rev. Genet.* **16:**113–134.

Laird, C. D., McConaughy, B. L., and McCarthy, B. J., 1969, Rate of fixation of nucleotide substitution in evolution, *Nature* **224:**149–154.

Lalanne, J. L., Bregegere, F., Delarbre, C., Abastado, J. P., Gachelin, G., and Kourilsky, P., 1983, Comparison of nucleotide sequences of mRNAs belonging to the mouse H-2 multigene family, *Nucleic Acids Res.* **10:**1039–1049.

Lamb, B. C., and Helmi, S., 1982, The extent to which gene conversion can change allele frequencies in populations, *Genet. Res.* **39:**199–217.

Lanave, C., Preparata, G., Saccone, C., and Serio, G., 1984, A new method for calculating evolutionary substitution rates, *J. Mol. Evol.* **20:**86–93.

Langley, C. H., and Fitch, W. M., 1974, An examination of the constancy of the rate of molecular evolution, *J. Mol. Evol.* **3:**161–177.

Langley, C. H., Brookfield, J. F. Y., and Kaplan, N., 1983, Transposable elements in Mendelian populations. I. A theory, *Genetics* **104:**457–471.

Lerner, M. R., Boyle, J. A., Mount, S. M., Wolin, S. L., and Steitz, J. A., 1980, Are snRNAs involved in splicing? *Nature* **283:**220–224.

Li, W.-H., 1982, Evolutionary change of duplicate genes, in: *Isozymes: Current Topics in Biological and Medical Research,* Vol. 6 (M. C. Rattazzi, J. G., Scandalios, and G. S. Whitt, eds.), Alan R. Liss, New York, pp. 55–92.

Li, W.-H., 1983, Evolution of duplicate genes and pseudogenes, in: *Evolution of Genes and Proteins* (M. Nei and R. K. Koehn, eds.), Sinauer, Sunderland, Massachusetts, pp. 14–37.

Li, W.-H., and Gojobori, T., 1983, Rapid evolution of goat and sheep globin genes following gene duplication, *Mol. Biol. Evol.* **1:**94–108.

Li, W.-H., Gojobori, T., and Nei, M., 1981, Pseudogenes as a paradigm of neutral evolution, *Nature* **292:**237–239.

Li, W.-H., Wu, C.-I., and Luo, C.-C., 1984, Nonrandom of point mutation as reflected in nucleotide substitutions and its evolutionary implications, *J. Mol. Evol.* **21:**58–71.

Li, W.-H., Wu, C.-I., and Luo, C.-C., 1985, A new method for estimating synonymous and nonsynonymous rates of nucleotide substitution considering the relative likelihood of nucleotide and codon changes, *Mol. Biol. Evol.,* **2:**150–174.

Liebhaber, S. A., Goossens, M., and Kan, Y. W., 1981, Homology and concerted evolution at the $\alpha 1$ and $\alpha 2$ loci of human α-globin, *Nature* **290:**26–29.

Lomedico, P., Rosenthal, N., Efstratiadis, A., Gilbert, W., Kolodner, R., and Tizard, R., 1979, The structure and evolution of the two nonallelic rat preproinsulin genes, *Cell* **18:**545–558.

Long, E. O., and Dawid, I. B., 1980, Repeated genes in eukaryotes, *Annu. Rev. Biochem.* **49:**727–764.

MacIntyre, R. J., 1976, Evolution and ecological value of duplicate genes, *Annu. Rev. Ecol. Syst.* **7:**721–748.

Margoliash, E., 1963, Primary structure and evolution of cytochrome *c*, *Proc. Natl. Acad. Sci. USA* **50:**672–679.

Min Jou, W., and Fiers, W., 1976, Studies on the bacteriophage MS2, XXXIII. Comparison of the nucleotide sequences in related bacteriophage RNAs, *J. Mol. Biol.* **106**:1047–1060.

Min Jou, W., Haegeman, G., Ysebaert, M., and Fiers, W., 1972, Nucleotide sequence of the gene coding for the bacteriophage MS2 coat protein, *Nature* **237**:82–88.

Minty, A., and Newmark, P., 1980, Gene regulation: New, old, and remote controls, *Nature* **288**:210–211.

Miyata, T., and Yasunaga, T., 1980, Molecular evolution of mRNA: A method for estimating evolutionary rates of synonymous and amino acid substitution from homologous sequences and its application, *J. Mol. Evol.* **16**:23–26.

Miyata, T., and Yasunaga, T., 1981, Rapidly evolving mouse α-globin-related pseudogene and its evolutionary history, *Proc. Natl. Acad. Sci. USA* **78**:450–453.

Miyata, T., Yasunaga, T., and Nishida, T., 1980, Nucleotide sequence divergence and functional constraint in mRNA evolution, *Proc. Natl. Acad. Sci. USA* **77**:7328–7332.

Miyata, T., Hayashida, H., Kikuno, R., Hasegawa, M., Kobayashi, M., and Koike, K., 1982, Molecular clock of silent substitution: At least six-fold preponderance of silent changes in mitochondrial genes over those in nuclear genes, *J. Mol. Evol.* **19**:28–35.

Modiano, G., Battistuzzi, G., and Motulsky, A. G., 1981, Nonrandom pattern of codon usage and of nucleotide substitution in human α- and β-globin genes: An evolutionary strategy reducing the rate of mutations with drastic effects? *Proc. Natl. Acad. Sci. USA* **78**:1110–1114.

Moore, G. W., Goodman, M., and Barnabas, J., 1973, An iterative approach from the standpoint of the adaptive hypothesis to the dendrogram problem posed by molecular data sets, *J. Theor. Biol.* **38**:423–457.

Muller, U. R., and Fitch, W. M., 1982, Evolutionary selection for perfect hairpin structures in viral DNAs, *Nature* **298**:582–585.

Murgola, E. J., Pagel, F. T., and Hijazi, K. A., 1984, Codon context effects in missense suppression, *J. Mol. Biol.* **174**:1–9.

Nagylaki, T., 1984, The evolution of multigene families under intrachromosomal gene conversion, *Genetics* **106**:529–548.

Nagylaki, T., and Petes, T. D., 1982, Intrachromosomal gene conversion and the maintenance of sequence homogeneity among repeated genes, *Genetics* **100**:315–337.

Nambiar, K. P., Stackhouse, J., Stauffer, D. M., Kennedy, W. P., Eldredge, J. K., and Banner, S. A., 1984, Total synthesis and cloning of a gene coding for the ribonuclease S protein, *Science* **223**:1299–1301.

Nei, M., and Tateno, Y., 1978, Nonrandom amino acid substitution and estimation of the number of nucleotide substitutions in evolution, *J. Mol. Evol.* **11**:333–347.

Nishimura, S., 1978, Modified nucleotide and isoaccepting tRNA, in: *Transfer RNA* (S. Altman, ed.), MIT Press, Cambridge, Massachusetts, pp. 168–195.

Nomura, M., Post, L. E., and Jinks, C. S., 1980, New aspects of regulation of ribosomal protein genes in *Escherichia coli*, in: *RNA-Polymerase, tRNA and Ribosomes* (S. Osawa, ed.), University of Tokyo Press, Tokyo, Japan, pp. 315–328.

Nussinov, R., 1981, The universal dinucleotide asymmetry rules in DNA and the amino acid codon choice, *J. Mol. Evol.* **17**:237–244.

O'Farrell, P. H., 1978, The suppression of defective translation by ppGpp and its role in the stringent response, *Cell* **15**:545–547.

Ohno, S., 1970, *Evolution by Gene Duplication*, Springer, Berlin.

Ohta, T., 1976, A simple model for treating the evolution of multigene families, *Nature* **263**:74–76.

Ohta, T., 1980, *Evolution of Multigene Families* (Lecture Notes in Biomathematics 37), Springer, Berlin.

Ohta, T., 1982, Further study on the genetic correlation between gene numbers of a multigene family, *Genetics* **99:**555–571.

Ohta, T., 1983, On the evolution of multigene families, *Theor. Popul. Biol.* **23:**216–240.

Ohta, T., 1984, Some models of gene conversion for treating the evolution of multigene families, *Genetics* **106:**517–528.

Ohta, T., and Dover, G. A., 1983, Population genetics of multigene families that are dispersed into two or more chromosomes, *Proc. Natl. Acad. Sci. USA* **80:**4079–4083.

Ohta, T., and Kimura, M., 1971, On the constancy of the evolutionary rate of cistrons, *J. Mol. Evol.* **1:**18–25.

Ohta, T., and Kimura, M., 1981, Some calculations on the amount of selfish DNA, *Proc. Natl. Acad. Sci. USA* **78:**1129–1132.

Orgel, L. E., and Crick, F. H. C., 1980, Selfish, DNA: The ultimate parasite, *Nature* **284:**604–607.

Orgel, L. E., Crick, F. H. C., and Sapienza, C., 1980, Selfish DNA, *Nature* **228:**645–646.

Parker, J., Polland, J., Friesen, J. D., and Stanners, C. P., 1978, Stuttering: High level mistranslation in animal and bacteria cells, *Proc. Natl. Acad. Sci. USA* **75:**1091–1095.

Parker, J., Johnston, T. C., Borgia, P. T., Holtz, G., Remaut, E., and Fiers, W., 1983, Codon usage and mistranslation: *In vivo* basal level misreading of the MS2 coat protein message, *J. Biol. Chem.* **258:**10007–10012.

Perler, R., Efstratiadis, A., Lomedico, P., Gilbert, W., Klodner, R., and Dodgson, J., 1980, The evolution of genes: The chicken preproinsulin gene, *Cell* **20:**555–566.

Pieczenik, G., 1980, Predicting coding function from nucleotide sequence or survival of "fitness" of tRNA, *Proc. Natl. Acad. Sci. USA* **77:**3539–3543.

Post, L. E., Strycharz, G. D., Nomura, M., Lewis, H., and Dennis, P. P., 1979, Nucleotide sequence of the ribosomal protein gene cluster adjacent to the gene for RNA polymerase subunit β in *Escherichia coli, Proc. Natl. Acad. Sci. USA* **76:**1697–1701.

Proudfoot, N., 1984, The end of the message and beyond, *Nature* **307:**412–413.

Radding, C. M., 1978, Genetic recombination: Strand transfer and mismatch repair, *Annu. Rev. Biochem.* **47:**847–880.

Razin, A., and Riggs, A. D., 1980, DNA methylation and gene function, *Science* **210:**604–610.

Rice, N. R., 1972, Change in repeated DNA in evolution, in: *Evolution of Genetic Systems* (H. H. Smith, ed.), Gordon and Breach, New York, pp. 44–79.

Ritossa, F. M., Atwood, K. C., and Speigelman, S., 1966, On the redundancy of DNA complementary to amino acid transfer RNA and its absence from the nucleolus organizer region of *Drosophila melanogaster, Genetics* **54:**663–676.

Rogers, J., and Wall, R., 1980, A mechanism for RNA splicing, *Proc. Natl. Acad. Sci. USA* **77:**1877–1879.

Romer, A. S., 1966, *Vertebrate Paleontology*, University of Chicago Press, Chicago.

Rosenberg, A. H., Simon, M. N., and Studier, F. W., 1979, Survey and mapping of restriction cleavage sites in bacteriophage T7 DNA, *Proc. Natl. Acad. Sci. USA* **135:**907–915.

Salser, W., and Issacson, J. S., 1976, Mutation rates in globin genes: The genetic load in Haldane's dilemma, *Progr. Nucleic Acids Res. Mol. Biol.* **19:**205–220.

Sanger, F., Air, G. M., Barrell, B. G., Brown, N. L., Coulson, A. R., Hutchinson III, C. A., Slocombe, P. M., and Smith, M., 1977, Nucleotide sequence of bacteriophage ϕX174, *Nature* **265:**687–695.

Sarich, V. M., 1972, Generation time and albumin evolution, *Biochem. Genet.* **7:**205–212.

Sarich, V. M., and Wilson, A. C., 1967, Immunological time scale for hominid evolution, *Science* **158:**1200–1203.

Sarich, V. M., and Wilson, A. C., 1973, Generation time and genomic evolution in primates, *Science* **179:**1144–1147.

Scherer, S., and Davis, R. W., 1980, Recombination of dispersed repeated DNA sequences in yeast, *Science* **209:**1380–1384.

Schon, E. A., Cleary, M. L., Haynes, J. R., and Lingrel, J. B., 1981, Structure and evolution of goat γ-, β^{C}- and β^{A}-globin genes: Three developmentally regulated genes contain inserted elements, *Cell* **27:**359–369.

Scott, A. F., Heath, P., Trusko, S., Boyer, S. H., Prass, W., Goodman, M., Czelusniak, J., Chang, L.-Y. E., and Slightom, J. L., 1984, The sequence of the gorilla fetal gamma globin genes: Evidence for multiple gene conversions in human evolution, *Mol. Biol. Evol.* **1:**371–389.

Shapiro, S. G., Schon, E. A., Townes, T. M., and Lingrel, J. B., 1983, Sequence and linkage of the goat ge^{I} and ϵ^{II} β-globin genes, *J. Mol. Biol.* **169:**31–52.

Sharp, P. M., Rogers, M. S., and McConnel, D. J., 1985, Selection pressures on codon usage in the complete genome of bacteriophage T7, *J. Mol. Evol.*, **21:**150–160.

Shepherd, J. C. W., 1981, Method to determine the reading frame of protein from the purine/pyrimidine genome sequence and its possible evolutionary justification, *Proc. Natl. Acad. Sci. USA* **78:**1596–1600.

Sinha, N. K., and Haimes, M. D., 1980, Probing the mechanism of transition and transversion mutagenesis using the phage T4 DNA replication apparatus *in vitro*, in: *Mechanistic Studies of DNA Replication and Genetic Recombination* (B. Alberts, ed.), Academic Press, New York, pp. 707–723.

Slightom, J., Blechl, A. E., and Smithies, O., 1980, Human fetal $^{G}\gamma$- and $^{A}\gamma$-globin genes: Complete nucleotide sequences suggest that DNA can be exchanged between these duplicated genes, *Cell* **21:**627–638.

Smith, G. P., 1974, Unequal crossover and the evolution of multigene families, *Cold Spring Harbor Symp. Quant. Biol.* **38:**507–513.

Smith, G. P., 1976, Evolution of repeated DNA sequences by unequal crossover, *Science* **191:**528–535.

Sneath, P. H. A., and Sokal, R. R., 1973, *Numerical Taxonomy*, Freeman, San Francisco.

Steitz, J. A., 1979, Genetic signals and nucleotide sequences in mRNA, in: *Biological Regulation and Development*, Vol. 1 (R. F. Goldberger, ed.), Plenum Press, New York, pp. 349–399.

Szostak, J. W., and Wu, R., 1980, Unequal crossing over in the ribosomal DNA of *Saccharomyces cerevisiae*, *Nature* **284:**426–430.

Szostak, J. W., Orr-Weaver, T. L., Rothstein, R. J., and Stahl, F. W., 1983, The double-stranded-break repair model for recombination, *Cell* **33:**25–35.

Tajima, F., and Nei, M., 1982, Biases of the estimates of DNA divergence obtained by the restriction enzyme technique, *J. Mol. Evol.* **18:**115–120.

Tajima, F., and Nei, M., 1984, Estimation of evolutionary distance between nucleotide sequences. *Mol. Biol. Evol.* **1:**269–285.

Takahata, N., and Kimura, M., 1981, A model of evolutionary base substitutions and its application with special reference to rapid change of pseudogenes, *Genetics* **98:**641–657.

Tateno, Y., and Nei, M., 1978, Goodman *et al*'s method for augmenting the number of nucleotide augmenting substitutions, *J. Mol. Evol.* **11:**67–73.

Tateno, Y., Nei, M., and Tajima, F., 1982, Accuracy of estimated phylogenetic trees from molecular data: I. Distantly related species, *J. Mol. Evol.* **18:**387–404.

Topal, M. D., and Fresco, J. R., 1976, Complementary base pairing and the origin of substitution mutations, *Nature* **263:**285–289.

Van Ooyen, A., Van den Berg, J., Manter, N., and Weissmann, C., 1979, Comparison of total sequence of a cloned rabbit β-globin gene and its flanking regions with a homologous mouse sequence, *Science* **206:**337–344.

Villafranca, J. E., Howell, E. E., Voet, D. H., Strobel, M. S., Ogden, R. C., Abelson, J. N., and Krant, J., 1983, Directed mutagenesis of dihydrofolate reductase, *Science* **222**:782–788.

Vogel, F., and Kopun, M., 1977, Higher frequencies of transitions among point mutations, *J. Mol. Evol.* **9**:159–180.

Wallace, R. B., Schold, M., Johnson, M. J., Dembek, P., and Itakura, K., 1981, Oligonucleotide directed mutagenesis of the human β-globin gene: A general method for producing specific point mutation in cloned DNA, *Nucleic Acids Res.* **9**:3647–3657.

Weiss, E. H., Mellor, A., Golden, L., Fahrner, K., Simpson, E., Hurst, J., and Flavell, R. A., 1983, The structure of a mutant H-2 gene suggests that the generation of polymorphism in H-2 genes may occur by gene conversion-like events, *Nature* **301**:671–674.

Weissenbach, J., and Dirheimer, G., 1978, Pairing properties of the methylester of 5-carboxymethyl uridine in the wobble position of yeast tRNA, *Biochim. Biophys. Acta* **518**:530–534.

Wellauer, P. K., Reader, R. H., Dawid, I. B., and Brown, D. D., 1976, The arrangement of length heterogeneity in repeating units of amplified and chromosomal DNA from *Xenopus laevis*, *J. Mol. Biol.* **105**:447–505.

Wilde, C. D., Growther, E., Cripe, T. P., Lee, M. G.-S., and Cowan, N. J., 1982, Evidence that a human β-tubulin pseudogene is derived from its corresponding mRNA, *Nature* **297**:83–84.

Wilson, A. C., Carlson, S. S., and White, T. J., 1977, Biochemical evolution, *Annu. Rev. Biochem.* **46**:573–639.

Wright, S., 1969, *Evolution and the Genetics of Populations*, Vol. II, University of Chicago Press, Chicago.

Wu, C.-I., and Li, W.-H., and 1985, Evidence for higher rates of nucleotide substitution in rodents than in man. *Proc. Natl. Acad. Sci. USA* **82**:1741–1745.

Yarus, M., 1979, The accuracy of translation, *Progr. Nucleic Acids Res. Mol. Biol.* **23**:195–225.

Zimmer, E. A., Artin, S. L., Beverley, S. M., Kan, Y. W., and Wilson, A. C., 1980, Rapid duplication and loss of genes coding for the α chains of hemoglobin, *Proc. Natl. Acad. Sci. USA* **77**:2158–2162.

Zuckerkandl, E., and Pauling, L., 1962, Molecular disease, evolution and genic heterogeneity, in: *Horizons in Biochemistry* (M. Kasha and B. Pullman, eds.), Academic Press, New York, pp. 189–225.

Zuckerkandl, E., and Pauling, L., 1965, Evolutionary divergence and convergence in proteins, in: *Evolving Genes and Proteins* (V. Bryson and H. J. Vogel, eds.), Academic Press, New York, pp.97–116.

Zuckerkandl, E., Derancourt, J., and Vogel, H., 1971, Mutational trends and random process in the evolution of informational macromolecules, *J. Mol. Biol.* **59**:473–490.

CHAPTER 2

The Mitochondrial Genome of Animals

WESLEY M. BROWN

1. Introduction

A major goal of molecular genetic research is to provide an accurate description of how DNA sequences and sequence organization change with time. This requires an enumeration of the kinds of changes that can occur and an accurate estimate of how frequently each occurs. In order to assess the effects of and constraints on the changes that are observed, knowledge of the structural and functional significance of the DNA sequences must be acquired. The qualitative and quantitative aspects of the enzymatic mechanisms that are responsible for changing and for preventing changes in the DNA (e.g., those for replication, recombination, and repair) must also be described. Finally, the way that all of these parameters vary among organisms, both within and between taxonomic groups, and how (or whether) the parameters can be influenced by environmental conditions must be known. Because of the large and complex nature of organismal genomes, even such limited tasks as describing and measuring the relative frequencies of changes are difficult to perform and interpret. Many of these tasks can be simplified by using less complex genomes, such as those found in mitochondria and chloroplasts, as model systems.

The mitochondrial system of animals has many characteristics that make it a useful one in which to study genome evolution. It appears to

WESLEY M. BROWN • Division of Biological Sciences and Museum of Zoology, University of Michigan, Ann Arbor, Michigan 48109.

be much simpler than the mitochondrial system of single-celled eukaryotes and the eukaryotic nuclear or prokaryotic systems with regard to the number and variety of enzymes present that carry out nucleic acid metabolism. For example, there are at least three distinct eukaryotic nuclear DNA polymerases, whereas there appears to be only one in the mitochondrion; animal mitochondrial DNA (mtDNA) replication is continuous, asymmetric, and unidirectional, and thus requires many fewer enzymes than the discontinuous, symmetric, and bidirectional DNA replication of nuclear and prokaryotic systems; both eukaryotic nuclear and prokaryotic systems have several complex enzymatic systems for carrying out DNA repair, whereas none have been documented for animal mitochondria; the animal mitochondrial system, with 22 tRNAs, requires only half as many aminoacyl synthetases as the eukaryotic cytoplasmic system, with over 40 tRNAs. Many other examples of the relative simplicity of the mitochondrial system could be listed.

With regard to the genome itself, the mtDNA of animals is both smaller and simpler than either eukaryotic nuclear genomes or the mtDNAs of plants and single-celled eukaryotes. Animal mtDNA is 25,000 times smaller than the smallest animal nuclear genome. It varies in size among different taxa by a factor of 1.3, whereas the nuclear genome varies by a factor of 1000. It consists of a single, duplex, closed circular DNA molecule, whereas the haploid nuclear genome consists of from 2 to >128 DNA molecules (assuming one molecule per chromosome). The sequences contained in animal mtDNA occur, with few exceptions, only once. In nuclear DNA, by contrast, many sequences (perhaps most) occur in multicopy form and range in copy number from two to several hundred thousand per haploid genome. In nuclear DNA, both intergenic sequences and intervening sequences within transcribed genes (introns) are usually present and often occupy a larger portion of the DNA than the actual coding sequences for the structural genes. In animal mtDNA, however, intergenic sequences are either very small or absent and introns do not occur. Sequence rearrangements (e.g., inversions and transpositions) appear to be extremely rare in animal mtDNA, probably due to the absence of repetitive, intergenic, and intron sequences. This is in marked contrast to the mtDNA of single-celled eukaryotes and to nuclear DNA, in which rearrangements occur at a high frequency. Finally, animal mtDNA appears to be uniparentally (maternally) transmitted, and is thus free from the genetic complexities that accompany a biparental mode of transmission.

The evolutionary scenario presented in this chapter is based almost exclusively on molecular considerations. This reflects the present state of development in this area of evolutionary biology rather than a disregard for the potential importance of transmission and population genetic considerations, both of which have been the subject of recent summaries and reviews (Birky, 1983a,b; Birky *et al.*, 1982, 1983; Avise and Lansman, 1983). For a truly comprehensive picture of animal mtDNA evolution, many more studies at all levels of biological organization must be performed and, ultimately, integrated.

2. Structure, Content, and Function in Animal Mitochondrial DNA

This section is a brief summary of the molecular biology of animal mtDNA and is intended only to provide a basis for understanding the material presented in subsequent sections of this chapter. For more detailed information about the molecular biology of animal mtDNA, the paper by Anderson *et al.* (1982) and several recent reviews (Attardi, 1985; Clayton, 1982, 1984; Wolstenholme and Clary, 1985) may be consulted.

Multicellular animal mtDNA possesses a unique structure. Unlike the mtDNAs of protists and plants, which exhibit a high degree of variation in size, structure, and (perhaps) gene content, the mtDNA of multicellular animals is relatively uniform. It is a duplex, covalently closed circular DNA molecule that exhibits little variation in size and none in gene content. This shared uniqueness is consistent with the hypothesis of a monophyletic origin for the metazoa. However, because there are few mtDNA data for most metazoan groups, present generalizations are based almost entirely on studies of mtDNAs from vertebrate and *Drosophila* species, and are therefore preliminary.

2.1. Gene Content

Although few mtDNAs have been thoroughly analyzed, the complete sequences of those from three mammals [human and cow (Anderson *et al.*, 1981, 1982); house mouse (Bibb *et al.*, 1981)], one amphibian [*Xenopus laevis* (B. Rowe, personal communication)], and one insect [*Drosophila yakuba* (Clary and Wolstenholme, 1983a,b, 1984; Clary *et al.*, 1982, 1983, 1984)] have been determined. Because vertebrates and insects represent the two major divisions of the metazoa (deuterostomes and protostomes), it is likely that they also represent the extremes in mtDNA diversity among multicellular animals. Features held in common between them are thus likely to be general for metazoan mtDNAs. One such feature is gene content, which is summarized in Table I. Genes for 13 proteins, two ribosomal RNAs (rRNAs), and 22 transfer RNAs (tRNAs) are present, as well as a region (hereafter called the control region) that although

Table I
Human (H), Mouse (M), Cow (C), and *Drosophila* (D) Mitochondrial Gene Comparisons[a]

	Gene size (base pairs)				Protein size (amino acids)				Percent conservation							
									Nucleotides				Amino acids			
Gene	H	M	C	D	H	M	C	D	H/M	H/C	M/C	D/M	H/M	H/C	M/C	D/M
srRNA	954	956	955	789	—	—	—	—	75	77	79	—	—	—	—	—
lrRNA	1559	1582	1571	1326	—	—	—	—	74	76	77	—	—	—	—	—
tRNAs	59–75	59–75	60–75	63–72	—	—	—	—	68–90	69–96	—	32–70	—	—	—	—
COI	1542	1545	1545	1540	513	514	514	512	77	79		72	90	91	93	75
COIII	784	784	784	789	261	261	261	262	76	79		67	86	87	86	64
Cyt *b*	1141	1144	1140	1137	380	381	379	378	74	73		65	78	79	84	67
ND1	956	946	956	975	318	315	318	324	71	73		52	78	78	78	46
ATPase 6	680	680	680	675	226	226	226	224	72	73		49	75	78	78	36
COII	684	684	682	685	227	227	227	228	71	70		64	71	73	90	56
ND4L	297	294	297	291	98	97	98	96	69	75		34	67	74	68	40
ND4	1378	1378	1378	1340	459	459	459	446	68	72		51	67	74	72	42
ND3	346	345	347	354	115	114	115	117	66	70		52	65	74	68	42
ND5	1812	1824	1821	1721	603	607	606	573	65	71		42	63	70	68	32
ND2	1042	1036	1042	860	347	345	347	286	61	66		47	57	63	56	35
URF 6	525	519	528	525	174	172	175	174	63	65		38	52	63	59	17
A6L	207	204	201	162	68	67	66	53	65	63		—	46	52	58	30

[a]Gene abbreviations: srRNA and lrRNA, small (12S) and large (16S) ribosomal RNAs; tRNA, transfer RNA; COI, COII, COIII, subunits I–III of the cytochrome oxidase complex; Cyt *b*, apoenzyme of cytochrome *b* complex; ATPase 6, subunit 6 of the mitochondrial ATP synthetase complex; A6L, (probable) subunit 8 of the mitochondrial ATP synthetase complex (Devlin, 1982; Macreadie *et al.*, 1983; Mariottini *et al.*, 1983); ND1–5, subunits 1–5 of the NADH dehydrogenase complex (Chomyn *et al.*, 1985); URF6, reading frame for a functionally unidentified protein. Mammalian mtDNA sequence sources: human, Anderson *et al.* (1981); mouse, Bibb *et al.* (1981); cow, Anderson *et al.* (1982). *Drosophila* (= *D. yakuba*) sequences are from Clary and Wolstenholme (1983a,b, 1984), Clary *et al.* (1982, 1983, 1984), Wolstenholme and Clary (1985), and Clary and Wolstenholme (1985).

generally devoid of structural genes, contains sites for the initiation of both mtDNA replication and RNA transcription. In addition to these sites, the control region probably contains other sequences that are important in the regulation of mtDNA replication and gene expression.

Of the 13 mitochondrially encoded proteins (Table I), 12 are subunits of enzymatic complexes that carry out electron transport and ATP synthesis; the function of the remaining protein is unknown [Macreadie *et al.* (1983); Michael *et al.* (1984); Chomyn *et al.*, 1985]. The minimum number of enzymes required for mitochondrial DNA replication, transcription, and transcript processing is in excess of 100. Moreover, at least several hundred enzymes are required to carry out the numerous metabolic functions that take place in the mitochondrion. With so few proteins encoded by mtDNA itself, most of the enzymes found in mitochondria are encoded by nuclear genes and must be imported from the cytoplasm.

2.2. Genome Structure

Four structural characteristics are hallmarks of metazoan mtDNAs and, when taken together, separate them as a group from all other mtDNAs. These are: (1) a highly compact gene arrangement, in which intergenic sequences are reduced or absent and adjacent genes occasionally overlap; (2) the absence of intervening sequences (introns) within structural genes; (3) the regular interspersion of tRNA genes between most of the remaining genes, resulting in a dispersed distribution for the tRNA genes; and (4) a concentration of most of the essential regulatory sequences in a single region of the genome (the control region). It should be noted that the control region, unlike the remainder of the mtDNA, also contains sequences that appear to be functionally nonessential. This will be discussed in detail in Section 3.4.2.

2.3. Gene Order

The sequential order of the mitochondrial genes is identical between frogs and mammals. Because their respective lineages diverged over 350 million years ago (mya), mitochondrial gene order appears to be a highly conserved feature. However, this conservation is not absolute. The gene order differs between insect and vertebrate mtDNAs (Fig. 1). A third gene order, based on data from transfer-hybridization experiments, has been reported for a sea urchin [Echinodermata: *Strongylocentrotus* (Roberts *et al.*, 1983)], and a fourth, based on sequence data, has been observed in a roundworm [Nematoda: *Ascaris* (D. R. Wolstenholme, personal communication)]. Because only one genus from each of the invertebrate groups

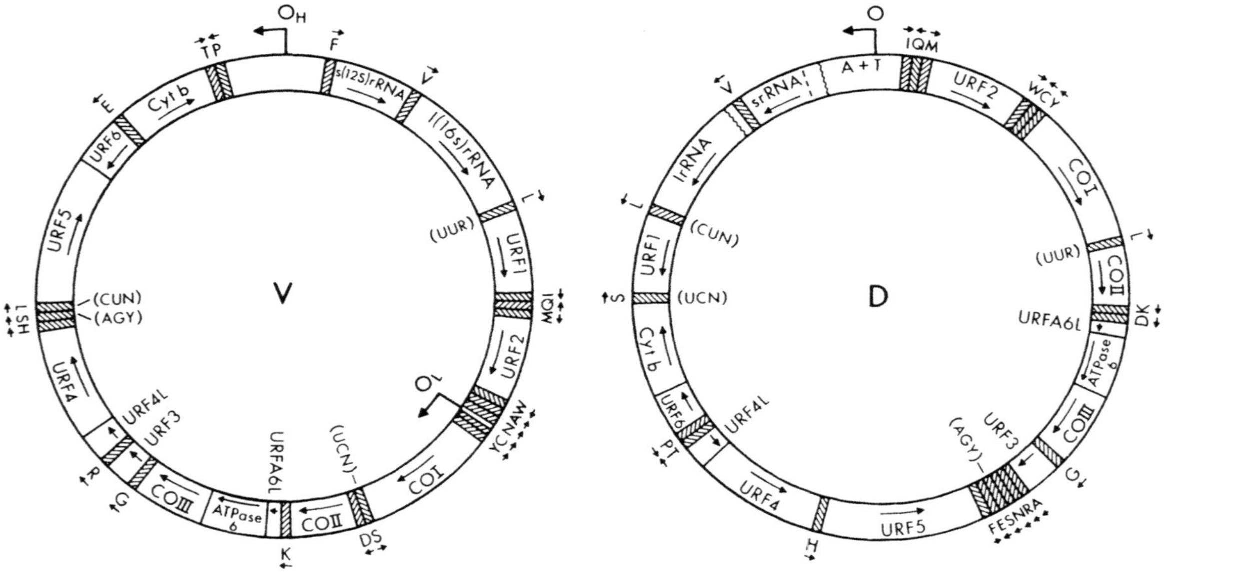

Figure 1. Gene order in (V) vertebrate and (D) *Drosophila* mitochondrial DNAs. The genes are abbreviated as explained in footnote *a* of Table I. The tRNA genes (cross-hatched) are designated by the single letter code for amino acids (see Table II); where the same amino acid is carried by two different tRNAs, the anticodons are given in parentheses. O_H and O_L are, respectively, the origins of heavy- and light-strand replication; the directions are indicated by the arrows. A small arrow associated with each gene indicates the direction in which it is transcribed. The vertebrate mtDNA map is modified from Fig. 1 of W. M. Brown (1983) and the *Drosophila* map from Fig. 1 of Wolstenholme and Clary (1985) URF1–5 correspond to ND1–5 of Table I.

has been investigated, it is not possible at present to distinguish between a hypothesis of gene order conservation within major groups with diversity between them, and one of gene order conservation within the vertebrates with nonconservation in other groups. Resolution of this will come when the mtDNA diversity within other groups has been assessed.

2.4. Genome Size

Metazoan mtDNAs range in size from 15,700 to 19,500 base pairs (bp) (Fauron and Wolstenholme, 1976; W. M. Brown, 1976, 1981, 1983). No correlation between genome size and taxonomic group appears to exist: vertebrate mtDNAs range from 16,300 to 19,200 bp (W. M. Brown, 1981, 1983) and, among invertebrate groups, mtDNAs from *Drosophila* species alone range from 15,700 to 19,500 bp (Fauron and Wolstenholme, 1976). In all cases examined, major size variation occurs exclusively in the control region, presumably reflecting changes in the functionally nonessential sequences. Size variation also occurs in other regions of the genome, but is minor in extent (see Sections 3 and 4.2).

2.5. Control Region

Although present evidence is meager, the data that exist suggest that the control region of animal mtDNA is structurally different but functionally similar between major groups. Structurally, the control region is characterized by the presence of a displacement loop (D-loop) structure in vertebrates and echinoderms (Fig. 2) and by the presence of an A + T-rich region in insects (at least in *Drosophila*). Functionally, the control region contains the points at which both replication and transcription of mtDNA are initiated.

The vertebrate D loop (Fig. 2) is formed by the synthesis of a short piece of DNA (the D-loop DNA) that is complementary to the mitochondrial light strand (L strand). This displaces the parental heavy strand (H strand) in the region of D-loop DNA synthesis. The points of initiation of D-loop DNA synthesis and the origin of H-strand replication (see below) are identical. This identity has given rise to the long-standing (and long-unsubstantiated) hypothesis that the D-loop DNA primes daughter H-strand synthesis. Both initiation and termination of D-loop DNA synthesis occur at discrete points in the control region, and there can be more than one point for either of these (W. M. Brown *et al.*, 1978; Gillum and Clayton, 1978; Clayton, 1982). The sequences and secondary structures at these points vary greatly—not only between species, but even within the same molecule (Walberg and Clayton, 1981). Chang and Clayton (1985)

Figure 2. Electron micrograph of an animal mitochondrial DNA molecule. The circular DNA duplex crosses itself at six points. The D loop appears at the extreme right extention of the molecule; its single-stranded branch (which appears both shorter and thinner than the duplex branch) lies inside a larger loop that is formed by two consecutive crossovers. The molecule shown is from a lizard (genus *Cnemidophorus*).

have demonstrated the presence of two discrete RNA species whose 3′ ends terminate adjacent to the initiation points of two minor D-loop DNA species in human mtDNA, and suggest that these RNAs may serve as primers for the initiation of D-loop DNA synthesis. No covalent linkage between the RNA and DNA moieties was detected, however, so their relationship remains inferential. The basis for D-loop DNA initiation and termination is not understood, and is likely to remain unknown until the enzymatic components that are responsible for these processes have been identified and characterized.

Sea urchin mtDNA exhibits D loops that resemble those of vertebrates in both their structure and their association with the origin of replication (Matsumoto *et al.*, 1974). Structures resembling D loops have also been reported in *Drosophila* mtDNA (Zakian, 1976). However, extensive data from three other studies has led to the tentative conclusion that *Drosophila* mtDNA probably lacks D loops (Klukas and Dawid, 1976; Rubenstein *et al.*, 1977; Wolstenholme *et al.*, 1979) and D-loop DNA (Rubenstein, *et al.*, 1977). This issue deserves to be clarified, since it involves questions of fundamental interest to both molecular biology and molecular evolution (see Section 5).

2.6. Replication

Replication of animal mtDNA has been characterized for several vertebrates [mostly mammals; see Clayton (1982) for a review and references], for *Drosophila* species [see Wolstenholme *et al.*, (1979) for a review and references], and for a sea urchin (Matsumoto *et al.*, 1974). In all of these cases replication initiates independently on each of the two complementary strands, proceeds by a continuous mode of DNA synthesis, and is highly asymmetric with respect to the time of initiation for each of the strands. These features give rise to the observation that animal mtDNA replicates unidirectionally from a fixed origin.

Among tetrapod vertebrates (fishes have not been studied), there is one fixed origin of replication for each mtDNA strand; these are separated by several genes and approximately 5000 bp (see Fig. 1). MtDNA replication begins with daughter H-strand synthesis, which is initiated at the origin located in the control region (O_H of Fig. 1), and extends in the direction away from the adjacent rRNA genes (counterclockwise in Fig. 1). When H-strand synthesis has reached the position of the origin of L-strand synthesis(O_L of Fig. 1), daughter L-strand synthesis is initiated, using the displaced parental H strand as a template, and proceeds in the opposite direction. Daughter molecules are separated (presumably by an enzyme with a topoisomerase-like activity) when daughter H-strand synthesis is complete. This occurs prior to the completion of daughter L-strand synthesis, which takes place after the daughter molecules have separated.

The patterns of mtDNA replication seen in *Drosophila* and sea urchin are notably different than the vertebrate pattern. The asymmetric replication of the two strands is even more pronounced than among vertebrates, and synthesis of the initiating daughter strand is frequently >90% complete before synthesis of the second daughter strand has begun. There appear to be multiple sites at which the replication of the second strand can be initiated, rather than the single site found in vertebrate mtDNA. Finally, although mtDNA replication is initiated in the control region, extension in *Drosophila* is toward the adjacent rRNA genes, the opposite of the vertebrate case; the direction of extension relative to the rRNA genes in sea urchin mtDNA is unknown.

2.7. Transcription

The genomic locations and some properties of mitochondrial gene transcripts are well characterized for several vertebrate and invertebrate systems (e.g., Rastl and Dawid, 1979a,b; Battey and Clayton, 1980; Ojala *et al.*, 1980; Merten and Pardue, 1981). However, details of the transcriptional process are known only for mammalian systems and have come primarily from studies conducted in the laboratories of G. Attardi and D. Clayton, both of whom have reviewed this topic in detail (Attardi, 1985;

Clayton, 1984). In brief, transcription of the H strand appears to be initiated at a promoter site in the control region and, possibly, at a nearby second site within the directly-adjacent $tRNA^{Phe}$ gene; initiation of L-strand transcription appears to take place at a single site located in the control region (see Fig. 1). Transcript elongation results in the production of polycistronic RNAs, which must be processed by a hypothetical enzyme (or enzymes) into mRNAs, tRNAs, rRNAs, and nonstructural RNAs. The latter appear to be rapidly degraded. The interspersed distribution of tRNA genes in vertebrate mtDNA led to the very reasonable hypothesis that all or part of their structure serves as a signal for transcript processing (Ojala *et al.*, 1980; Battey and Clayton, 1980). The similar tRNA gene distribution in *Drosophila* mtDNA supports an extension of this hypothesis to all multicellular animals. Given the compact nature of animal mtDNA, it would not be surprising if the vertebrate pattern of transcription were general. However, it should be noted that a yeast species with an unusually small mtDNA [18.9 kb (Sriprakash and Clark-Walker (1983)] nevertheless contains three distinctly different and well-separated transcriptional promotors, and that Wolstenholme and co-workers (Clary *et al.*, 1983) have speculated that a sequence present in the larger intergenic sequences of *Drosophila* mtDNA (see Section 3.4.1) may function as a promoter.

2.8. Strand Distribution of Genes and Nucleotides

The distribution of both genes and nucleotides between the respective DNA strands is markedly different when vertebrate and *Drosophila* mtDNAs are compared (Table II). There is a distinct asymmetry in the strand distribution of the 37 structural genes in the mtDNA of vertebrates. Only one protein gene (URF6) and eight tRNA genes (see Fig. 1 and Table II) are encoded by (i.e., transcribed from) the L strand. The sequences of these nine genes comprise only 7% of the total sequence devoted to structural genes in animal mtDNA. This informational asymmetry is not seen in *Drosophila* mtDNA, where 23 genes, comprising 52% of the portion of the genome devoted to structural genes, are transcribed from the "minus" strand (see Fig. 1 and Tables I and II).

Regarding the strand distribution of nucleotides, a difference analogous to (and possibly correlated with) the distribution of genes also exists between vertebrate and invertebrate mtDNAs. Among vertebrates there is a pronounced bias in the G + T composition of the H and L strands, which gives rise to a pronounced difference in the buoyant densities of the strands in CsCl gradients (thus their designation as heavy and light strands). For example, the H:L ratios of G and T, respectively, are 2.27 and 1.25 in human mtDNA and 1.92 and 1.20 in mouse mtDNA

Table II
Strand Distribution of Genes in Animal Mitochondrial DNA[a]

tRNA gene	V	D	tRNA gene	V	D	Protein gene	V	D	rRNA gene	V	D
F Phe	+	+	C Cys	−	+	COI	+	−	Small (12S)	+	+
H His	+	+	P Pro	−	+	COII	+	−	Large (16S)	+	+
L Leu [CUN]	+	+	Q Gln	−	+	COIII	+	−			
V Val	+	+	Y Try	−	+	Cyt b	+	−			
D Asp	+	−	A Ala	−	−	ATPase 6	+	−			
G Gly	+	−	E Glu	−	−	A6L	+	−			
I Ile	+	−	N Asn	−	−	ND2	+	−			
K Lys	+	−	S Ser [UCN]	−	−	ND3	+	−			
L Leu [UUR]	+	−				ND1	+	+			
M Met	+	−				ND4L	+	+			
R Arg	+	−				ND4	+	+			
S Ser [AGY]	+	−				ND5	+	+			
T Thr	+	−				URF 6	−	−			
W Trp	+	−									

[a]The strands of vertebrate (V) and *Drosophila* (D) mtDNAs from which the rRNA genes are transcribed have been designated as the "plus" (+) strands. Both single- and three-letter abbreviat ons for amino acids are given. The protein genes are abbreviated as explained in the footnote to Table I.

(W. M. Brown, 1976, 1981; Anderson *et al.*, 1981; Bibb *et al.*, 1981). No strong interstrand bias in base composition has been reported among invertebrates, but the data base supporting this generality is very limited [see Borst and Flavell (1976) and W. M. Brown (1981) for tabulations and references]. In general, the base compositions of animal mtDNAs appear to be lower in G + C content than those of their nuclear DNAs, and range from a low, among *Drosophila* species, of 20–25%, to a high, among avian species, of 45–50% [see W. M. Brown (1981, 1983) for references]. The reasons for the maintenance of a G + T strand bias among vertebrate mtDNAs or for the relatively low G + C content among animal mtDNAs are unknown.

2.9. Genetic Code and Codon Usage

The animal and fungal mitochondrial genetic codes differ slightly from the "universal" (i.e., eukaryotic nuclear and prokaryotic) code, and also from one another (Barrell *et al.*, 1979, 1980; Bonitz *et al.*, 1980; Heckman *et al.*, 1980; Bibb *et al.*, 1981; Kochel *et al.*, 1981; Clary and Wolstenholme, 1983a,b; de Bruijn, 1983) [see Fig. 1 in Attardi (1985) for a compilation and comparison]. Translation of the mitochondrial codes is carried out in a way that requires fewer tRNA species than are required by eukaryotic nuclear and prokaryotic systems. None of the genetic codes differ radically from one another, however, and this close resemblance suggests that the mitochondrial codes are probably altered descendants of the "universal" code, rather than the vestiges of an ancient, more primitive code.

Codon usage in the mitochondrial protein genes of six animal species is summarized in Table III. There is a bias favoring the use of codons ending in A and T (= U) and against those ending in G that goes beyond the bias in the base compositions of the coding strands. Among vertebrates, this biased usage is not extreme, and all codons are employed. However, in *Drosophila* the bias toward A or T in the third position is extreme, and there is an accompanying dramatic shift in the relative usage frequencies for the two leucine codons between *Drosophila* and the vertebrates, which also reflects the shift in mtDNA base composition between these taxa.

3. Sequence Variation in Animal Mitochondrial DNA

Both inter- and intraspecific mtDNA sequence variation have been assessed by a variety of methods. Early studies, employing buoyant den-

Table III

Codon Usage in Mitochondrial Protein Genes of Five Animal Species[a]

		HU	CO	MO	DY	DM			HU	CO	MO	DY	DM			HU	CO	MO	DY	DM			HU	CO	MO	DY	DM
TTT	Phe	77	107	109	120	128	TCT	Ser	32	51	43	49	42	TAT	Tyr	46	64	65	45	50	TGT	Cys	5	6	11	7	8
TTC		141	133	132	13	8	TCC		99	66	47	1	3	TAC		89	74	59	14	10	TGC		17	16	20	1	0
TTA	Leu	73	110	131	198	193	TCA		83	99	148	49	53	TAA	Ter	2	6	7	3	2	TGA	Trp	93	92	97	48	50
TTG		16	16	16	6	3	TCG		7	5	4	2	2	TAG		2	0	0	0	1	TGG		11	12	7	2	0
CTT	Leu	65	62	87	19	16	CCT	Pro	41	42	30	37	35	CAT	His	18	30	34	43	42	CGT	Arg	7	7	8	3	4
CTC		167	95	65	2	1	CCC		119	63	34	2	4	CAC		79	63	63	6	8	CGC		25	11	18	0	0
CTA		276	285	266	6	11	CCA		52	135	89	25	22	CAA	Gln	81	79	79	37	36	CGA		29	42	36	23	22
CTG		45	29	26	0	0	CCG		7	3	2	2	4	CAG		9	8	3	0	0	CGG		2	3	3	0	0
ATT	Ile	125	165	234	140	137	ACT	Thr	51	44	58	49	46	AAT	Asn	33	60	60	81	81	AGT	Ser	14	12	14	10	10
ATC		196	160	140	4	10	ACC		155	96	80	2	3	AAC		131	102	108	6	5	AGC		39	43	35	1	0
ATA	Met	167	218	218	68	84	ACA		133	153	157	47	47	AAA	Lys	85	90	100	23	25	AGA	Ter/Ser*	1	1	0	23*	23*
ATG		40	44	31	11	4	ACG		10	15	6	1	1	AAG		10	11	2	2	2	AGG		1	0	0	0	0
GTT	Val	30	40	53	36	30	GCT	Ala	43	52	47	57	64	GAT	Asp	15	22	31	24	27	GGT	Gly	24	29	36	23	16
GTC		49	48	34	1	1	GCC		124	91	82	7	2	GAC		51	46	43	7	4	GGC		88	62	39	0	1
GTA		70	87	76	45	46	GCA		80	103	97	21	18	GAA	Glu	64	78	83	36	35	GGA		67	97	109	69	80
GTC		18	15	10	3	5	GCG		8	2	7	1	1	GAG		24	17	11	0	0	GGG		34	31	29	7	2

[a]Compiled from Table 1 of Bibb *et al.* (1981), Table 3 of Anderson *et al.* (1982), and Table 5 of Wolstenholme and Clary (1985).

[b]Species abbreviations: HU, human; CO, cow; MO, mouse; DY; *Drosophila yakuba*; DM, *D. melanogaster*.

sity shift and thermal stability analyses of homo- and heteroduplexes, provided evidence that mtDNA sequences were not highly conserved and that variation was accumulating throughout the genome at a rate at least comparable to that in nuclear DNA (Dawid and Wolstenholme, 1968; Dawid, 1972) [see Borst (1972) for a comprehensive review]. Electron microscopy and restriction endonuclease cleavage site mapping, especially in combination, provided a large gain in resolution and a means of aligning animal mtDNAs for comparison (W. M. Brown and Vinograd, 1974; Robberson *et al.*, 1974). These techniques were used to confirm Dawid's (1972) report that nucleotide substitutions occurred in all portions of mammalian mtDNA, including the rRNA genes, to discover that the absolute size of mtDNA varied (W. M. Brown, 1976, 1981; Fauron and Wolstenholme, 1976), to demonstrate that this size variation was associated with the control region (Fauron and Wolstenholme, 1976; Upholt and Dawid, 1977; W. M. Brown, 1981), and to reveal that, among tetrapod vertebrates, substitutions occurred at a much higher rate in mtDNA than in single-copy nuclear DNA (W. M. Brown *et al.*, 1979).

The development of techniques for DNA sequencing (Maxam and Gilbert, 1977; Sanger *et al.*, 1977, 1978) provided the ultimate in resolution. With sequence data it was possible to test directly the inferences drawn in earlier studies, and to begin to investigate and understand the mechanistic bases that underlie animal mtDNA variation and evolution. One unexpected, possibly general, and still unexplained result is that differences due to base substitutions are predominantly transitions when closely related taxa are compared (W. M. Brown *et al.*, 1982; G. G. Brown and Simpson, 1982; Greenberg *et al.*, 1983; Aquadro and Greenberg, 1983; Wolstenholme and Clary, 1984). The observed ratio of transitions to transversions decreases with increasing divergence time. This time-dependent decrease probably reflects a gradual accumulation of transversions, which occur at the same nucleotide positions as transitions, but at a much lower frequency (W. M. Brown *et al.*, 1982; W. M. Brown, 1983).

3.1. Protein Genes

The general features of mitochondrial protein gene evolution have been reviewed recently (W. M. Brown, 1983; Attardi, 1985; Wolstenholme and Clary, 1984). Except for the rates at which they occur in vertebrates, most of these features appear to be similar in mitochondrial and nuclear systems. Silent substitutions (which do not result in amino acid replacements) occur several times more frequently than replacement substitutions. The substitution frequency at codon positions is in the order third >> first > second, in accordance with the facts that (1) most third-position substitutions are silent (the structure of the codes makes the proportion of silent third positions even greater in the mitochondrial than in the nuclear system); and (2) first-position substitutions result in less radical amino acid replacements (and thus a lower probability of functional impairment of the protein) than do second-position substitutions. Interspecific comparisons of the relative extents of divergence among mitochondrial protein genes indicate that those with COI-III functions are generally more conserved than with ND and ATPase functions (Table I). With one notable exception (discussed in Section 4.1), the relative order of sequence conservation among the 13 genes is the same in species representing three mammalian orders; however, an altogether different order of conservation appears to occur in *Drosophila* (Table I).

Codon-sized additions and deletions also occur in the mitochondrial protein genes, but at an extremely low frequency. These events appear to be most common at the 3′ and 5′ termini, but internal codons are occasionally involved, especially when the comparisons are between the sequences of protein genes from distantly related taxa, such as from mammals of different orders or *Drosophila* and vertebrates (see Table I).

3.2. Transfer RNA Genes

The sequences of animal mitochondrial tRNA genes differ radically from those of their nuclear counterparts in a number of respects (de Bruijn *et al.*, 1980; Anderson *et al.*, 1981, 1982; Bibb *et al.*, 1981; W. M. Brown *et al.*, 1982; Cantatore *et al.*, 1982; Clary *et al.*, 1982, 1983, de Bruijn, 1983; Attardi, 1985). They are smaller, their G + C contents are much lower, and many more nonstandard base pairs (A:C and G:U) occur in their stem regions; all of these factors lower the stability of their secondary structures. Perhaps the most striking difference, however, is in the degree of tRNA sequence (and resulting structural) variation that is tolerated in the mitochondrial system. Nuclear (and prokaryotic) tRNA genes exhibit many invariant features, such as lengths of stems and loops and the presence of specific nucleotides or oligonucleotide sequences at specific tRNA positions (Gauss and Sprinzl, 1984ab). Most of these features are variable in animal mitochondrial tRNAs [see Attardi (1985) for a concise summary].

Both substitutions and small (1–2 bp) deletions/additions occur in mitochondrial tRNA genes. In stem regions, substitutions on one strand are highly correlated with compensating substitutions on the complementary strand so that base pairing (and therefore tRNA structure) is maintained. Transitions are the predominant form of substitution observed, even in comparisons between distantly related species. Since tRNA stems with mismatched bases must be intermediates in tRNA gene evolution, the destabilizing effect of a mismatch should strongly affect the probability of fixation of the substitution causing it. Uncompensated transitions result in A:C and G:U pairs, both of which decrease the stability of tRNA structure less than the A:A, G:G, U:U, and C:C pairs that result from uncompensated transversions. The persistence of a bias favoring transitions is probably a reflection of the greater stability of their uncompensated form. Deletions/additions of 1–2 bp occur, but are much less frequent than substitutions. They occur more frequently in loops than in stems, and most often in runs of identical bases.

3.3. Ribosomal RNA Genes

At present the only animal mitochondrial rRNA sequences published are for vertebrate species (Eperon *et al.*, 1980; Van Etten *et al.*, 1980; Kobayashi *et al.*, 1981; Saccone *et al.*, 1981). The small (12S) and large (16S) rRNAs of vertebrate mitochondria consist of approximately 950 and 1550 nucleotides, respectively. These are significantly smaller than the corresponding 16S and 23S rRNAs of prokaryotes, at 1550 and 2900 nucleotides, or the 18S and 28S eukaryotic nuclear rRNAs, at 1900 and 4100 nucleotides. No mitochondrial rRNA gene corresponding to the prokaryotic 5S or eukaryotic nuclear 5.8S rRNA is present, although Nierlich (1982) has proposed that a sequence corresponding to the 5S rRNA gene is present at the 3′ end of the large mitochondrial rRNA gene. The general features of sequence variation and evolution in vertebrate mitochondrial rRNA genes are very similar to those in the tRNA genes. Base substitutions and small (1–5 bp) deletions/additions occur. Transitions predominate in all comparisons, probably because of the same kinds of structural constraints on uncompensated substitutions that were just discussed for the tRNA genes. Deletions/additions generally occur in loop structures, usually amid runs of identical bases. Runs of C appear to be especially susceptible among mammalian rRNAs (Hixson and Brown, 1985). There is much more tolerance of sequence variation in mitochondrial than in prokaryotic or eukaryotic nuclear rRNA genes, but the basic secondary structural features of the mitochondrial rRNAs are conserved [see Maly and Brimacombe (1983) for a comprehensive review of rRNA structure].

3.4. Other Regions

In addition to sequences that encode tRNAs, rRNAs, and proteins, animal mtDNA contains noncoding sequences. Most are quite small (<10 bp) and are found between immediately adjacent structural genes. These are often called "spacers." However, no function (including "spacing") is known for them, and they are hereafter referred to simply as intergenic sequences. In the mtDNA of vertebrates (but not in *Drosophila* mtDNA) a 31- to 34-bp sequence occurs between the genes for $tRNA^{Asp}$ and $tRNA^{Cys}$ (Fig. 1) within which daughter L-strand replication originates. This particular sequence is referred to as O_L. Finally, in the mtDNA of all animals examined there is an extensive region that contains sequences that have been associated with a variety of functions, particularly with the initiation of mtDNA replication and (among vertebrates) with the initiation of transcription. This region has been called the D-loop region, the noncoding region, and (in *Drosophila*) the A+T-rich region. Because of the nature of the functions that are localized in this region and the strong possibility that any mtDNA sequence-specific components of their regulation also reside here, this region will be referred to as the control region.

3.4.1. Intergenic Sequences

Complete or extensive (i.e., >5000 contiguous bp) sequences have been published for mtDNAs of human (Anderson *et al.*, 1981), house mouse (Bibb *et al.*, 1981), cow (Anderson *et al.*, 1982), norway rat (Saccone *et al.*, 1981; Pepe *et al.*, 1983), *Drosophila yakuba* (Clary *et al.*, 1982, 1983, 1984; Clary and Wolstenholme, 1983a,b, 1984; Wolstenholme and Clary, 1984), and *D. melanogaster* (de Bruijn, 1983). Collectively, these sequences contain 188 intergenic junctions. Of these, 89 (47%) have intergenic sequences. In the remaining 53%, the adjacent genes either abut directly or overlap with each other. Of 140 vertebrate junctions examined, only three contain intergenic sequences of more than 5 bp. Intergenic sequences average somewhat larger in *Drosophila*, where 11 of 47 junctions include intergenic sequences of more than 6 bp. Substitutions occur with approximately the same frequency in the intergenic sequences as at the third positions of codons in protein genes, and additions/deletions occur at a higher frequency than in any other region of the mtDNA [except, possibly, parts of the control region (Cann and Wilson, 1983; Cann *et al.*, 1984)]. These data suggest strongly that all intergenic sequences in vertebrate mtDNAs and most in *Drosophila* mtDNA are without function. However, several of the largest intergenic sequences in *Drosophila* mtDNA contain the sequence 5'TTTATTAT or 5'TATATTAT upstream from tRNA genes. This correlation has led to the suggestion that this intergenic sequence may be of functional significance (Clary *et al.*, 1983; Wolstenholme and Clary, 1984).

3.4.2. The Origin of L-Strand Replication in Vertebrates

The O_L sequence is located on the H strand between the asparagine and cysteine tRNA genes (Fig. 1). This sequence differs from other intergenic sequences by its greater size and by the conservation of several of its features. O_L sequences are available for four mammals, representing three different orders, and for an amphibian (Table IV). Among the mammalian species, the two stem sequences, each of 11 nucleotides, are perfectly complementary, and are identical in three. The two substitutions observed (in mouse) are compensating transversions. This level of primary sequence conservation is at least as high as that observed in the most conserved stem sequences of mammalian mitochondrial tRNAs. The loop sequences do not exhibit this conservation. A minimum of two single-base deletion/addition events and 11 substitutions (five transitions and six transversions) have occurred in this 12- to 13-nucleotide sequence among the four mammals. It is noteworthy that the greatest pairwise difference observed is between the two most closely related species, mouse and rat (one deletion/addition, two transitions, and three transversions; Table IV).

Addition of the amphibian (*Xenopus laevis*) sequence to these comparisons requires only minor changes in the above parameters: The stem regions in *Xenopus* are each two nucleotides shorter than in the mammals, but are identical in sequence to the basal nine nucleotides of the mammalian stems; the loop sequence in *Xenopus* is longer, and consists of 19 nucleotides. One loop feature has been conserved among the five species and may prove to be of functional significance. This is the high proportion of T's (50–85%), most of which occur as a tandem repeat in the portion of the loop sequence proximal to the $tRNA^{Cys}$ gene (Table IV).

3.4.3. The Control Region

From both an evolutionary and a molecular biological perspective this region is the most complex, most interesting, and least understood in the mitochondrial genome. At present there are no published sequence data for the control regions of invertebrate mtDNAs. The following discussion of sequence features is therefore based entirely on comparisons among eight vertebrate species: rat (Sekiya *et al.*, 1980), human and cow (Anderson *et al.*, 1981, 1982), mouse (Bibb *et al.*, 1981), frog (*Xenopus* (Wong *et al.*, 1983), and three ape species (pygmy and common chimpanzees, and gorilla (J. Hixson and W. Brown, unpublished data).

At the top of Fig. 3 is a schematic depiction of the human mtDNA control region, and below it are depictions of its sequence similarity with each of seven other vertebrate species. The most obvious difference is in size, which varies from about 900 bp (in mouse, rat, and cow) to over 2100 bp (in frog). From the difference between total mitochondrial genome size (among animals, 15.7–19.5 kb) and the combined sizes of the protein and RNA genes (approximately 15.5 kb, including intergenic sequences)

Table IV
Sequences of the L-Strand Origin of Replication for Five mtDNAs[a]

	tRNA^{Cys} Gene	Proposed L-strand origin structure			tRNA^{Asn} Gene
Species		Stem	Loop	Stem	
HU	CCTGCCGGGG	CTTCTCCCGCC	TTTTTT-CCCGGC	GGCGGGAGAAG	TAGATTGAAGCCA
CO	TCTGCCGGGG	CTTCTCCCGCC	TTTTTTTTCTTGC	GGCGGGAGAAG	TAGATTGAAGCCA
MO	TCTACTAAG	ACTTCTACCGCC	ATTTTTTTTTTTC	GGCGGTAGAAGT	AGATTGAAGCCA
RA	TCTACTAAGG	CTTCTCCCGCCT	CTTTT-CTGAT	AGGCGGGAGAAG	TAGATTGAAGCCA
XL	TTTGCCGGGG	CTTCTCCCG	GTTTTTTTGGCTTAATAAA	CGGGAGAAG	TAGAATGAAGCTC

[a] The 3′ end of the tRNA^{Cys} and the 5′ end of the tRNA^{Asn} genes, which overlap or abut with the O_L sequence, are also shown. The template (i.e., H-strand) sequences are shown, in 5′ to 3′ orientation. Species abbreviations and published sequence sources are: human (HU), Anderson *et al.* (1981); cow (CO), Anderson *et al.* (1982); mouse (MO), Bibb *et al.* (1981); rat (RA), Saccone *et al.* (1981); *Xenopus laevis* (XL), Wong *et al.* (1983).

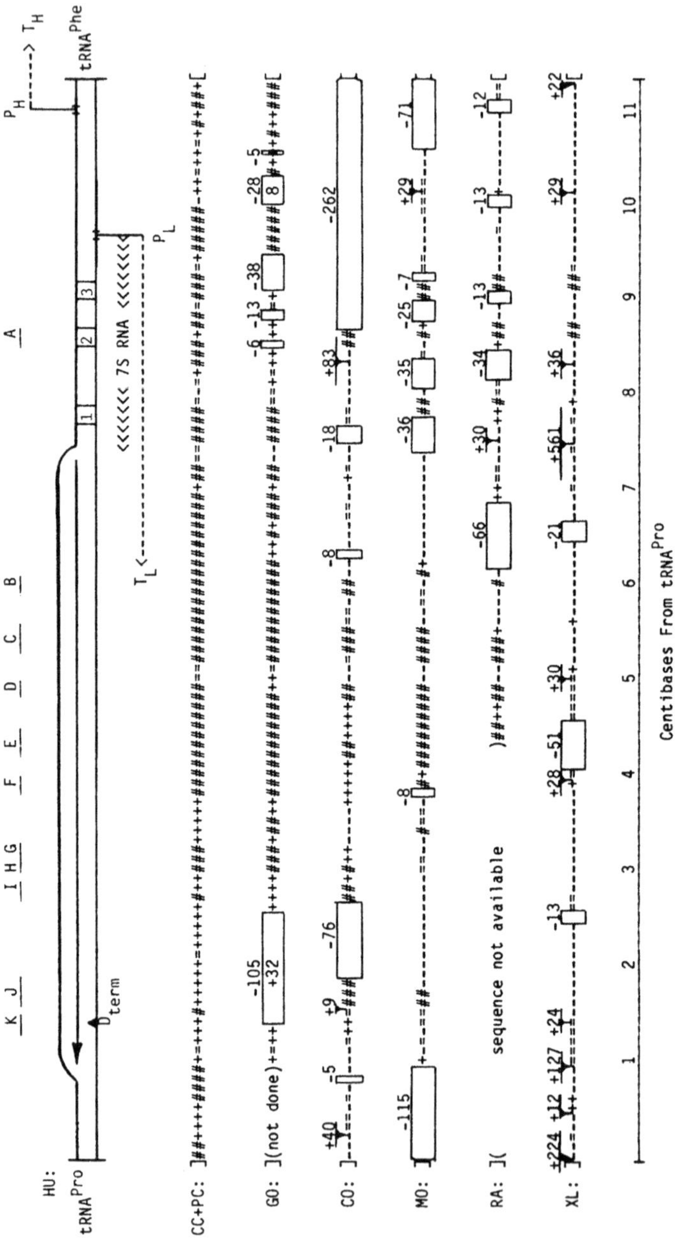
HU:
tRNAPro
tRNAPhe
CC+PC:
GO:
](not done)
CO:
MO:
RA:
sequence not available
XL:
7S RNA
Centibases From tRNAPro

Figure 3. Comparison of the control region sequence of human mitochondrial DNA with those of seven other vertebrates. The diagram of the human control region is shown at the top, with the D loop and the sequences of known functional importance indicated. Lines A–K above the diagram indicate sequences that are highly conserved between human and cow (Anderson *et al.*, 1982). P_H and P_L are, respectively, the heavy- and light-strand promoters in human mitochondrial DNA (Chang and Clayton, 1984; Bogenhagen *et al.*, 1984); the arrows T_H and T_L show their directions of transcription. D_{term} indicates the position of the putative D-loop DNA termination sequence, and boxes 1–3 the postions of conserved sequence blocks 1–3 (Walberg and Clayton, 1981; Clayton, 1982). The extent of the mitochondrial 7S RNA transcript [see Attardi (1985) and Clayton (1984) for references] is shown above T_L. The lines below the control region diagram represent the control region sequences of common and pygmy chimpanzees (CC, PC), gorilla (GO), cow (C), mouse (MO), rat (RA), and the frog *Xenopus laevis* (XL). The approximate sequence similarity is: #, >90%; +, >70%, =, >50%; −, <50%. Open boxes in these lines represent regions that are deleted relative to the human sequence; additions are shown as pedestals. Relative extents are indicated by the negative and positive numbers, respectively. In gorilla, an unalignable sequence is contained in each of two deleted regions. The extent of the deletion end points is given as a negative number above the box; that of the contained unalignable sequence is given within the box. Sequence sources: HU, Anderson *et al.* (1981); CC, PC, GO, J. Hixson and W. Brown (unpublished data); CO, Anderson *et al.* (1982); MO, Bibb *et al.* (1981); RA, Sekiya *et al.* (1980); XL, Wong *et al.* (1983).

it is estimated that the size range of the control region among animals is 200–4100 bp. It could be argued that this size difference is due to a difference in gene content. However, even in the greatly expanded control region of *Xenopus* (Fig. 3), no tRNA-like and/or protein coding sequences have been found (Wong *et al.*, 1983). When compared to the control region of human, that of gorilla (a close relative) exhibits four deletions, totalling 170 bp (J. J. Hixson and W. M. Brown, unpublished data). Rat, mouse, and cow sequences also show several deletions and size reductions of about 300 bp relative to the human sequence (Fig. 3). These direct sequence comparisons confirm and extend the results of many other studies. Data from quantitative electron microscopy has shown that the mtDNA size range for most *Drosophila* species is 16.2–16.7 kb; however, among the closely related species in the *melanogaster* subgroup the range is 15.7–19.5 kb, with the control region contributing (within experimental error) all of the variation (Fauron and Wolstenholme, 1976). In sequence comparisons of *Drosophila* mtDNAs by heteroduplex analysis the control regions remained single-stranded, even though the remainder of the genome formed a stable duplex structure (Wolstenholme *et al.*, 1979; Fauron and Wolstenholme, 1980a,b). High-resolution electron microscopy combined with restriction endonuclease analysis indicated that size differences in the control regions of sheep and goats, a closely related species pair, occurred primarily at or near the D-loop forks (Upholt and Dawid, 1977). Restriction endonuclease analysis of mtDNA sampled from individual lizards in the same local population demonstrated frequent and extensive (up to 400 bp) size variation in the control region (Densmore *et al.*, 1985).

An examination of the pairwise sequence comparisons in Fig. 3 indicates that the degree of sequence similarity to the human control region decreases with increasing evolutionary distance. The most closely related species pair, pygmy and common chimpanzee, differ from each other by approximately 8%, and each differs from human by about 12% (J. E. Hixson and W. M. Brown, unpublished data). A similar amount of divergence, 13%, is seen between gorilla and human and, in addition to this divergence by substitution, more than 15% of the gorilla sequence has been deleted. The remaining mammalian sequences show patterns in which small portions of the control region exhibit strong similarity while the remainder shows little or none. This "patchy" pattern of sequence similarity was noted in the initial reports of these sequences (Bibb *et al.*, 1981; Walberg and Clayton, 1981; Anderson *et al.*, 1982). Finally, the comparison of the *Xenopus* and human sequences indicates virtual dissimilarity over all but a tiny portion of the region.

The extensive divergence and size variation exhibited by most of the sequences of the control region make it difficult to entertain a hypothesis that they are of functional importance. Attempts to find conservation of secondary structure have been without success. Although several conserved sequence blocks have been identified (see Fig. 3), only one has been inferentially associated with a function (Chang and Clayton, 1985), and no other functionally identified position has been associated with a strongly conserved sequence or secondary structure. Based on the data, the most conserved blocks are those numbered 1, 2, and 3 in Fig. 3. Number 1 is present in all, but has been reduced to a 6-bp sequence in *Xenopus*. Numbers 2 and 3 are partially deleted from gorilla mtDNA, and number 3 is completely deleted from cow mtDNA. Only in the context of the even more labile and expendable sequences found elsewhere in this region could such sequences be considered conserved.

4. Rates of Change in Animal Mitochondrial DNA

From the preceding discussions it is evident that three kinds of changes have been well documented in animal mtDNA: (1) nucleotide substitutions; (2) additions and deletions; and (3) sequence rearrangements (inversions and transpositions). The manner in which mtDNA changes with time is determined by both the absolute and relative rates of occurrence of these different kinds of changes. For several years after the discovery of animal mtDNA it was assumed that it would prove to be a highly conserved molecule. In addition to being uniform in size and (probably) gene content (Borst, 1972), it was the genome of an organelle that was central to cellular metabolism. It was assumed that the functional constraints on mtDNA were so stringent that most changes would be lethal. The first indication that this very reasonable assumption was incorrect came from a comparison between two species of *Xenopus* (Dawid, 1972). The *Xenopus* mtDNAs exhibited as much, and possibly more, change than their nuclear DNAs, and the entire mtDNA (including the tRNA and rRNA genes) was affected. Studies by Fauron and Wolstenholme (1976) and Upholt and Dawid (1977) demonstrated that the control region of mtDNA was changing at an extremely high rate, and that additions and deletions as well as substitutions were involved. Finally, definitive evidence for major genomic rearrangements was first provided by comparisons between *Drosophila* and vertebrate mtDNA sequences (Clary *et al.*, 1982).

4.1. Rates of Nucleotide Substitution

Based on restriction endonuclease cleavage map comparisons and thermal stability analysis, the substitution rate in mammalian mtDNA was estimated to be 5–10 times greater than in single-copy nuclear DNA (W. M. Brown *et al.*, 1979). This estimate is now supported by direct sequence comparisons as well as by additional cleavage map comparisons [reviewed in W. M. Brown (1981, 1983)]. Although data for other tetrapod groups are less extensive than those for mammals, there is support for a rapid substitution rate in these groups as well. Based on the current minimum and maximum estimates for the times of divergence of hominoids (human and apes) provided, respectively, by Wilson *et al.* (1977) and by Pilbeam (1984), the average rate of mtDNA divergence in these primates is approximately 1% or 0.5% per million years per lineage (W. M. Brown *et al.*, 1979, 1982). The validity of this rate estimate for other vertebrate groups is unknown, and additional comparisons of both mtDNA and single-copy nuclear DNA among sets of taxa with well-characterized divergence times are needed.

The rate of substitution is greatest in the control region. Because of the high probability of multiple events at the same site and the high frequency of deletions and additions in this region, it is impossible to extract a reliable rate estimate from the sequence comparisons depicted in Fig. 3. A better estimate could be obtained by high-resolution cleavage map comparisons among conspecific individuals. In such a study, the observed rate in the control region was 1.4 times greater than the rate in the (combined) protein genes, and the estimated rate calculated on the basis of a parsimony analysis of the data was 2.8 times greater (Cann *et al.*, 1984). However, the best estimate available, from a phylogenetic and statistical analysis of seven human control region sequences (Aquadro and Greenberg, 1983), indicates that the rate is at least five times greater than that in the remainder of the genome.

The rate of substitution is lowest in the tRNA and rRNA genes, as originally inferred from comparative thermal stability analysis of mtDNA:rRNA hybrids of *Xenopus* species (Dawid, 1972) and recently confirmed, using primate mtDNAs, by high-resolution cleavage map comparisons (Cann *et al.*, 1984) and sequencing (W. M. Brown *et al.*, 1982; Hixson and Brown, 1985). The observed rates for primate tRNA and rRNA genes were approximately one-half the rate for protein genes. However, the rates in these genes are on the order of 100 times higher than the rates in their nuclear counterparts (W. M. Brown *et al.*, 1982; Cann *et al.*, 1984; Hixson and Brown, 1985).

Each of the 13 mitochondrial protein genes exhibits its own characteristic rate of change; the relative rates among these genes appear, with one exception, to be the same among vertebrates. The exception is the COII gene, in which the rates of both nucleotide substitution and amino acid replacement differ significantly among mammalian orders.

Pairwise comparisons of COII gene sequences from human, rat, mouse, and cow (G. G. Brown and Simpson, 1982; Cann *et al.*, 1984) and human and chimpanzee (R. Honeycutt, J. Levy, and W. Brown, unpublished data) indicate a significantly higher rate of change in the primate than in the rodent or artiodactyl lineages. A higher rate of change in the primate lineage has also been noted for a nuclear-encoded protein, cytochrome *c* (Carlson *et al.*, 1977), which is known to interact directly with COII in the mitochondrion (Tzagoloff, 1982). In a comparative analysis of the reactions between homologous and heterologous cytochromes *c* and cytochrome oxidases, Osheroff *et al.* (1983) observed an increase in the compatability of heterologous components with increased evolutionary distance from human, again suggesting a rate discontinuity specific to the primate lineage. Coevolution between COII and cytochrome *c* has been suggested as the basis for this correlated rate acceleration (Cann *et al.*, 1984), but the basis for the acceleration itself and the extent to which it is exhibited by primate groups other than hominoids (apes and human) is unknown. A similar coevolutionary rate correlation appears to exist for the mitochondrial rRNA genes and the 85 mitochondrial ribosomal proteins, most (all?) of which must be encoded by nuclear genes (Schieber and O'Brien, 1982; Matthews *et al.*, 1982). Unlike the case for the COII gene, however, the question of between-lineage rate variation in mitochondrial rRNA gene evolution has not been explored.

The rate of mtDNA substitution has not been investigated among invertebrates, although preliminary data exist for *Drosophila*. Shah and Langley (1979) proposed that the rate of *Drosophila* mtDNA sequence divergence, exclusive of the control region, was no greater, and possibly less, than the rate in its nuclear protein genes. Although the data supporting this were meager, the proposal was plausible and could be supported inferentially by data from studies of the DNA polymerases in *Drosophila* [see W. M. Brown (1981) for discussion and references]. Direct sequence comparisons are now available for a contiguous 5-kb segment of mtDNA between *Drosophila yakuba* and *D. melanogaster* (Clary *et al.*, 1982, 1983, 1984; Clary and Wolstenholme, 1983a,b; Wolstenholme and Clary, 1985; de Bruijn, 1983), and for the nuclear alcohol dehydrogenase (ADH) gene between several *Drosophila* species, including *D. melanogaster* and *D. orena* (Bodmer and Ashburner, 1984). Of the species whose ADH genes were compared, *orena* is least related to *melanogaster*. However, *melanogaster* is more closely related to *orena* than to *yakuba* [see Fig. 1 in Bodmer and Ashburner (1984)]. Comparisons of the mitochondrial and nuclear protein genes (Table V) show very similar amounts of nucleotide substitution, and the amount of amino acid replacement in the ADH protein is greater than that of all but one of the mitochondrial

Table V
Drosophila Mitochondrial and Nuclear Gene Sequence Comparisons[a]

Protein	Size[b] (bp)	Sequence divergence (%)		Substitutions (%)	
		Amino acid	Nucleotide	Silent	Transition
URF 2	860	5.6	7.0	70	42
COI	1540	1.0	6.9	95	44
COII	685	1.8	6.4	87	61
A6L	162	3.8	6.2	80	60
ATPase 6	675	2.2	7.0	87	47
COIII	789	1.9	7.2	88	51
ADH	768	3.9	4.2	76	63
tRNA[c]	**Size (bp)**	**Substitutions**		**Deletions/additions**	
		Total	Transition	Number	Size (bp)
Trp	66	8	7	1	+1
Cys	63	2	0	1	+1
Tyr	66	2	0	0	—
Leu	66	2	2	0	—
Lys	69	1	0	0	—
Asp	67	2	1	1	+1
Gly	66	1	0	0	—
Totals[d]	455	17 (3.9%)	9 (53%)	3	+3
IGS[e]	44	4	1 (25%)	3	7+3+12
NCR[f]	741	62 (8.4%)	30 (48%)	>9	1 to >12

[a] Published sequence sources: *D. melanogaster* mtDNA, Clary *et al.* (1983), de Bruijn (1983); *D. yakuba* mtDNA, Clary *et al.* (1982, 1983, 1984), Clary and Wolstenholme (1983a,b, 1984), Wolstenholme and Clary (1984); *D. melanogaster* and *D. orena* ADH genes, Bodmer and Ashburner (1984).
[b] Size includes coding region and encoded portion of termination codon. Only 860 bp from the URF2 gene of *D. melanogaster* mtDNA have been reported.
[c] The tRNA sequences are referenced to those of *D. melanogaster* mtDNA.
[d] Totals reflect the 8-bp overlap between the $tRNA^{Trp}$ and $tRNA^{Cys}$ genes.
[e] IGS, Intergenic sequences of mtDNA, referenced to *D. melanogaster*.
[f] NCR, Noncoding regions of the ADH gene (3′ and 5′ flanking sequences plus introns 1–3).

proteins. Assuming that the above evolutionary relationships are correct and that the degree of sequence dissimilarity is roughly correlated with increasing evolutionary distance, it is reasonable to expect that an ADH gene comparison between *melanogaster* and *yakuba* would show an even greater amount of divergence than the comparison between *melanogaster* and *orena*. These data provide support for Shah and Langley's (1979) hypothesis that the rates of nucleotide substitution and amino acid replacement in mitochondrial genes and proteins are no faster than those in nuclear genes. However, until the kinetics of substitution has been determined by sequence comparisons within and among several, more closely related species of *Drosophila*, no firm conclusions regarding the dynamics of *Drosophila* mtDNA evolution can be drawn.

4.2. Additions and Deletions

Early studies indicated that addition/deletion events were frequent, although less so than substitutions. In these early studies the methods employed (usually electron microscopy) were insufficient for the detection of small events, and therefore only those taking place in the control region were of sufficient size for detection. When the sequences of mouse, human, and cow were available for comparison, it became clear that deletion/addition events also occurred frequently in the intergenic sequences and, although less frequently, in the tRNA and rRNA genes. Two quantitative estimates for the frequency of occurrence of addition and deletion events are presently available. From sequence comparisons of a 900-bp portion of the control region among seven individual human mtDNAs, Aquadro and Greenberg (1983) estimated that addition/deletion events were about one-fifth as frequent in occurrence as substitutions. From high-resolution restriction endonuclease cleavage map comparisons among 112 mtDNA samples from individual humans, Cann and Wilson (1983) estimated that addition/deletion events were approximately one-half as frequent in occurrence as substitutions in the combined intergenic and control regions of human mtDNA. Deletions and additions appear to be generated at an even higher frequency in a homopolymer (poly C) tract in the control region of cattle (Hauswirth *et al.*, 1984) and in a tandemly repeated sequence in the control region of the mtDNA from *Cnemidophorus* lizards (Densmore *et al.*, 1985). Deletions and additions were also most frequent within poly C tracts in sequence comparisons among 12S rRNA genes of great apes and human (Hixson and Brown, 1985).

4.3. Rates of Genome Rearrangements

At present, the most quantitative statement that can be made about the frequency of occurrence of genome rearrangements is that they are rare. At least three rearrangement events separate *Drosophila* and vertebrate mtDNAs, and at least one appears to separate the mtDNAs of sea urchin and vertebrates. This provides a tentative rate estimate between 100 and 1000 times lower than those for substitutions or additions/deletions. If this low rate estimate holds, then mitochondrial gene order may be a useful character in inferring evolutionary relationships between very distantly related taxa, such as many classes and most phyla.

5. Speculations, Problems, and Prospects

The mitochondrial genome of animals evolves by the classic modes of nucleotide substitution, sequence additions/deletions, and sequence rearrangements. While these general modes of change were predictable, the uniquely mitochondrial aspects of them were not. The most striking of these aspects are the high rate of substitutions, the extreme kinetic bias toward transitions, and the low rate of genome rearrangements. The latter aspect is the easiest to understand. Given the closely packed gene organization in animal mtDNA, any rearrangement would result in the creation of nonsense sequences unless its end points coincided with the boundaries of genes. In addition, inversions would require transcription and processing to take place from the opposite mtDNA strand, which might impose a barrier to proper mitochondrial gene expression. At present it is impossible to do more than speculate about the mechanism(s) by which rearrangements arise. Given the paucity, small size, and sequence dissimilarity of intergenic regions, it is unlikely that they could serve on any but a random basis as end points for rearrangement; however, the sequence similarity found in some intergenic regions of *Drosophila* mtDNA (Clary *et al.*, 1983) should be noted in this regard. If rearrangements proceed by nonhomologous pairing between similar oligonucleotide sequences, one set of candidate sequences that might serve as successful end points are the tRNA genes, since these occur between most of the other mitochondrial genes (see Fig. 1). Mispairing and recombination between tRNA genes might occasionally result in the production of impaired but functional tRNAs, given the apparently high tolerance to variation in tRNA structure possessed by the animal mitochondrial system. A much larger sample of total animal mtDNA diversity is needed before reasonable hypotheses about the manner in which rearrangements arise and before their rate of occurrence may be accurately estimated. Data from invertebrate groups will be especially important in this regard.

Speculations about the mechanistic basis of the high rate of substitution in vertebrate mtDNA have been discussesd at length in recent reviews and papers (W. M. Brown, 1981, 1983; W. M. Brown *et al.*, 1982; Cann and Wilson, 1983; Cann *et al.*, 1984). These reports summarize the circumstantial evidence in support of the hypothesis of an elevated mutation rate in vertebrate mitochondria and the theoretical reasons why a higher probability of fixation of mutations should occur in very small genomes, such as mtDNA. However, the arguments presented are purely speculative, and no data exist that provide a direct correlation between the high rate of substitution and a biochemical mechanism or mechanisms. The chief reason for this is that almost nothing is known about the enzymes that carry out nucleic acid metabolism in animal mitochondria. For almost a decade it has been assumed that γ DNA polymerase is the enzyme responsible for vertebrate mtDNA replication [see Weissback (1977), Kornberg (1980), and Clayton (1982) for reviews and references], yet there is no *direct* evidence that this enzyme is mitochondrial or that mtDNA is its natural substrate. At present, the only nucleic acid metabolic activity that has been isolated and is demonstrably mitochondrial is one for transcription of mitochondrial DNA (Walberg and Clayton, 1983; Chang and Clayton, 1984). Lack of information about the enzymology of mtDNA replication and gene expression is presently the single greatest barrier to a better understanding of the evolution and molecular biology of mtDNA.

More information about the occurrence of additions and deletions in animal mtDNA is needed. Even though only limited data exist, there is already a clear indication from sequence comparisons among conspecific individuals (Greenberg *et al.*, 1983; Aquadro and Greenberg, 1983; G. G. Brown and DesRosiers, 1983; Hauswirth *et al.*, 1984) and closely related species (W. M. Brown *et al.*, 1982; Hixson and Brown, 1985) that slippage and mispairing in tandemly repeated runs of the same nucleotide during DNA replication is one mechanism by which additions and deletions frequently arise (Streisinger *et al.*, 1966). It is unlikely that this mechanism also accounts for the more extensive addition and deletion events that take place in the control region, however, and other mechanisms must be sought. The interspecific sequence comparisons that have been made have not provided data that are useful in formulating additional mechanistic hypotheses, probably because the rate of sequence evolution in the control region is so high that changes have obscured any sequence-related clues that may have existed when additions and deletions were formed. However, the existence of extensive mtDNA size variation between conspecific individuals has been documented for several species (Fauron and Wolstenholme, 1976; W. M. Brown, 1981; Solignac *et al.*, 1983; Densmore *et al.*, 1985; Harrison *et al.*, 1985). MtDNA sequence determinations from such individuals provide the means for making comparisons between sequences that differ by large additions or deletions but are otherwise identical. These comparisons are needed not only for the information that they may provide about potential mechanisms of sequence addition and deletion, but also because they will result in the precise identification of the sequences that are involved in addition/deletion events, thereby providing substrates that can be used in *in vitro* assays for the proteins that catalyze these events.

Another major question that must be addressed and that can be addressed with present techniques is how the mitochondrial and nuclear genomes coevolve in a functionally coordinated way. The mitochondrial gene products, whether protein or RNA, must each interact with many proteins whose genes are encoded in nuclear DNA. Given the rapid evolution of vertebrate mitochondrial genes and gene products, the associated nuclear genes might also be expected to show an elevated rate of change. The data from comparative studies of mitochondrial ribosomal proteins and cytochrome *c* (see Section 4.1) provide evidence that this expectation is met. However, it is far from clear how this coevolutionary process is orchestrated. Are the maximum potential rates of change for nuclear genes high enough to keep pace with the changes in the mitochondrial genes, or must mechanisms for hypervariability of the nuclear genes be invoked? How is it possible to coordinate and select for the *appropriate* changes in the multitude of nuclear genes whose products must interact in a functionally efficient way with mitochondrial gene products? If there is a strong component in the rapid evolution of vertebrate mtDNA that is due primarily to mutation pressure, questions such as these will have to be addressed.

Alternatively, the high rate of change exhibited by vertebrate mtDNA may be an evolved response to changes in the associated nuclear genes. Phrased in another way, changes in nuclear genes may be the driving force for the rapid evolution of mtDNA. Many recent studies have demonstrated that the mtDNA population within individuals is polymorphic (Hauswirth and Laipis, 1982; Olivo *et al.,* 1983; Greenberg *et al.,* 1983; G. G. Brown and DesRosiers, 1983; Solignac *et al.,* 1983; Hauswirth *et al.,* 1984; Densmore *et al.,* 1985). This situation, coupled with the presence of several hundred mitochondria per cell, would provide a mutant enzyme such as an RNA polymerase with an opportunity to select a substrate (i.e., a promoter sequence) with which it could best interact. Since RNA priming has been strongly implicated in the initiation of mtDNA replication, a rapid restructuring of the mitochondrial population would probably result. The characterization and comparison of nuclear genes whose products interact with mtDNA or its gene products is a necessary first step in addressing these issues.

References

Anderson, S., Bankier, A. T., Barrell, B. G., de Bruijn, M. H. L., Coulson, A. R., Drouin, J., Eperon, I. C., Nierlich, D. P., Roe, B. A., Sanger, F., Schreier, P. H., Smith, A. J. H., Staden, R., and Young, I. G., 1981, Sequence and organization of the human mitochondrial genome, *Nature* **290**:457–465.

Anderson, S., de Bruijn, M. H. L., Coulson, A. R., Eperon, I. C., Sanger, F., and Young, I. G., 1982, Complete sequence of bovine mitochondrial DNA, *J. Mol. Biol.* **156**:683–717.

Aquadro, C. F., and Greenberg, B. D., 1983, Human mitochondrial DNA variation and evolution: Analysis of nucleotide sequences from seven individuals, *Genetics* **103**:287–312.

Attardi, G., 1985, Animal mitochondrial DNA: An extreme example of genetic economy in: Genome evolution in procaryotes and eukaryotes (D. C. Reamey ed.), *Int. Rev. Cytol.* **93.**

Avise, J. C., and Lansman, R. A., 1983, Genetic polymorphism and the role of mutation in evolution, in: *Evolution of Genes and Proteins* (M. Nei and R. K. Koehn, eds.), Sinauer, Sunderland, Massachusetts, pp. 165–190.

Barrell, B. G., Bankier, A. T., and Drouin, J., 1979, A different genetic code in human mitochondria, *Nature* **282:**189–194.

Barrell, B. G., Anderson, S., Bankier, A. T., de Bruijn, M. H. L., Chen, E., Coulson, A. R., Drouin, J., Eperon, I. C., Nierlich, D. P., Roe, B. A., Sanger, F., Schreier, P. H., Smith, A. J. H., Staden, R., and Young, I. G., 1980, Different patterns of codon recognition by mammalian mitochondrial tRNAs, *Proc. Natl. Acad. Sci. USA* **77:**3164–3166.

Battey, J., and Clayton, D. A., 1980, The transcription map of human mitochondrial DNA implicates transfer RNA excision as a major processing event, *J. Biol. Chem.* **255:**11599–11606.

Bibb, M. J., Van Etten, R. A., Wright, C. T., Walberg, M. W., and Clayton, D. A., 1981, Sequence and gene organization of mouse mitochondrial DNA, *Cell* **26:**167–180.

Birky, C. W., Jr., 1983a, Relaxed cellular and organelle heredity, *Science* **222:**468–475.

Birky, C. W., Jr., 1983b, The partitioning of cytoplasmic organelles at cell division, *Int. Rev. Cytol.* **1983**(Suppl. 15):49–89.

Birky, C. W., Jr., Acton, A. R., Dietrich, R., and Carver, M., 1982, Mitochondrial transmission genetics: Replication, recombination, and segregation of mitochondrial DNA and its inheritance in crosses, in: *Mitochondrial Genes* (P. Slonimski, P. Borst, and G. Attardi, eds.), Cold Spring Harbor Laboratory, Cold Spring Harbor, New York, pp. 333–348.

Birky, C. W., Jr., Maruyama, T., and Fuerst, P., 1983, An approach to population and evolutionary genetic theory for genes in mitochondria and chloroplasts, and some results, *Genetics* **103:**513–527.

Bodmer, M., and Ashburner, M., 1984, Conservation and change in the DNA sequences coding for alcohol dehydrogenase in sibling species of *Drosophila*, *Nature* **309:**425–430.

Bogenhagen, D. F., Applegate, E F., and Yoza, B. K., 1984, Identification of a promoter for transcription of the heavy strand of human mtDNA: *In vitro* transcription and deletion mutagenesis, *Cell* **36:**1105–1113.

Bonitz, S. G., Bertani, R., Coruzzi, G., Li, M., Macino, G., Nobrega, F. G., Nobrega, M. P., Thalenfeld, B. E., and Tzagoloff, A., 1980, Codon recognition rules in yeast mitochondria, *Proc. Natl. Acad. Sci. USA* **77:**3167–3170.

Borst, P., 1972, Mitochondrial nucleic acids, *Annu. Rev. Biochem.* **41:**333–376.

Borst, P., and Flavell, R. A., 1976, Properties of mitochondrial DNAs, in: *Handbook of Biochemistry and Molecular Biology,* 3rd ed., *Nucleic Acids,* Vol. II (G. D. Fasman, ed.), CRC Press, Cleveland, Ohio, pp. 363–374.

Brown, G. G., and DesRosiers, L. J., 1983, Rat mitochondrial DNA polymorphism: Sequence analysis of a hypervariable site for insertions/deletions, *Nucleic Acids Res.* **11**(19):6699–6708.

Brown, G. G., and Simpson, M. V., 1982, Novel features of animal mtDNA evolution as shown by sequences of two rat cytochrome oxidase subunit II genes, *Proc. Natl. Acad. Sci. USA* **79:**3246–3250.

Brown, W. M., 1976, Interspecific Comparisons of Animal Mitochondrial DNAs, Ph.D. thesis, California Institute of Technology, Pasadena, California.

Brown, W. M., 1980, Polymorphism in mitochondrial DNA of humans as revealed by restriction endonuclease analysis, *Proc. Natl. Acad. Sci. USA* **77**:3605–3609.

Brown, W. M., 1981, Mechanisms of evolution in animal mitochondrial DNA, *Ann. N.Y. Acad. Sci.* **361**:119–134.

Brown, W. M., 1983, Evolution of mitochondrial DNA, in: *Evolution of Genes and Proteins* (M. Nei and R. K. Koehn, eds.), Sinauer, Sunderland, Massachusetts, pp. 62–88.

Brown, W. M., and Vinograd, J., 1974, Restriction endonuclease cleavage maps of animal mitochondrial DNAs, *Proc. Natl. Acad. Sci. USA* **71**:4617–4621.

Brown, W. M., Shine, J., and Goodman, H. M., 1978, Human mitochondrial DNA: Analysis of 7S DNA from the origin of replication, *Proc. Natl. Acad. Sci. USA* **75**:735–739.

Brown, W. M., George, M., Jr., and Wilson, A. C., 1979, Rapid evolution of mitochondrial DNA, *Proc. Natl. Acad. Sci. USA* **76**:1967–1971.

Brown, W. M., Prager, E. M., Wang, A., and Wilson, A. C., 1982, Mitochondrial DNA sequences of primates: Tempo and mode of evolution, *J. Mol. Evol.* **18**:225–239.

Cann, R. L., and Wilson, A. C., 1983, Length mutations in mitochondrial DNA, *Genetics* **104**:699–711.

Cann, R. L., Brown, W. M., and Wilson, A. C., 1984, Polymorphic sites and the mechanism of evolution in human mitochondrial DNA, *Genetics* **106**:479–499.

Cantatore, P., De Benedetto, C., Gadaleta, G., Gallerani, R., Kroon, A. M., Holtrop, M., Lanave, C., Pepe, G., Quagliariello, C., Saccone, C., and Sbisa, E., 1982, The nucleotide sequences of several tRNA genes from rat mitochondria: Common features and relatedness to homologous species, *Nucleic Acids Res.* **10**(10):3279–3289.

Carlson, S. S., Mross, G. A., Wilson, A. C., Mead, R. T., Wolin, L. D., Bowers, S. F., Foley, N. T., Muijsers, A. O., and Margoliash, E., 1977, Primary structure of mouse, rat, and guinea pig cytochrome *c*, *Biochemistry* **16**:1437–1442.

Chang, D., and Clayton, D. A., 1984, Precise identification of individual promoters for transcription of each strand of human mitochondrial DNA, *Cell* **36**:635–643.

Chang, D., and Clayton, D. A., 1985, Priming of human mitochondrial DNA replication occurs at the light strand promoter, *Proc. Natl. Acad. Sci. USA* **82**:351–355.

Chomyn, A., Mariottini, P., Cleeter, M., Ragan, F., Matsuno-Yagi, A., Hatefi, Y., Doolittle, R., and Attardi, G., 1985, Six unidentified reading frames of human mitochondrial DNA encode components of the respiratory-chain NADH dehydrogenase, *Nature* **314**:592–597.

Clary, D. O., and Wolstenholme, D. R., 1983a, Nucleotide sequence of a segment of *Drosophila* mitochondrial DNA that contains the genes for cytochrome *c* oxidase subunits II and III and ATPase subunit 6, *Nucleic Acids Res.* **11**(12):4211–4227.

Clary, D. O., and Wolstenholme, D. R., 1983b, Genes for cytochrome *c* oxidase subunit I, URF2, and three tRNAs in *Drosophila* mitochondrial DNA, *Nucleic Acids Res.* **11**(19):6859–6872.

Clary, D. O., and Wolstenholme, D. R., 1984, A cluster of six tRNA genes in *Drosophila* mitochondrial DNA that includes a gene for an unusual $tRNA^{ser}$[AGY], *Nucleic Acids Res.* **12**(5):2367–2379.

Clary, D. O., and Wolstenholme, D. R., 1985, The ribosomal RNA genes of *Drosophila* mitochondrial DNA, *Nucleic Acids Res.* **13**(11):4029–4045.

Clary, D. O., Goddard, J. M., Martin, S. C., Fauron, C. M.-R., and Wolstenholme, D. R., 1982, *Drosophila* mitochondrial DNA: A novel gene order, *Nucleic Acids Res.* **10**(21):6619–6637.

Clary, D. O., Wahleithner, J. A., and Wolstenholme, D. R., 1983, Transfer RNA genes in *Drosophila* mitochondrial DNA: Related 5′ flanking sequences and comparisons to mammalian mitochondrial tRNA genes, *Nucleic Acids Res.* **11**(8):2411–2425.

Clary, D. O., Wahleithner, J. A., and Wolstenholme, D. R., 1984, Sequence and arrangement of the genes for cytochrome b, URF1, URF4L, URF4, URF5, URF6 and five tRNAs in *Drosophila* mitochondrial DNA, *Nucleic Acids Res.* **12**(9):3747–3762.

Clayton, D. A., 1982, Replication of animal mitochondrial DNA, *Cell* **28**:693–705.

Clayton, D. A., 1984, Transcription of the mammalian mitochondrial genome, *Annu. Rev. Biochem.* **53**:573–594.

Dawid, I. B., 1972, Evolution of mitochondrial DNA sequences in *Xenopus*, *Dev. Biol.* **29**:139–151.

Dawid, I. B., and Wolstenholme, D. R., 1968, Renaturation and hybridization studies of mitochondrial DNA, *Biophys. J.* **1968**:65–81.

De Bruijn, M. H. L., 1983, *Drosophila melanogaster* mitochondrial DNA, A novel organization and genetic code, *Nature* **304**:234–241.

De Bruijn, M. H. L., Schreier, P. H., Eperon, I. C., Barrell, B. G., Chen, E. Y., Armstrong, P. W., Wong, J. F. H., and Roe, B. A., 1980, A mammalian mitochondrial serine transfer RNA lacking the 'dihydrouridine' loop and stem, *Nucleic Acids Res.* **8**:5213–5222.

Devlin, R. B., 1982, Biogenesis of the mitochondrial ATPase from sea urchin embryos, *J. Biol. Chem.* **257**:9711–9716.

Densmore, L. D., Wright, J. W., and Brown, W. M., 1985, Length variation and heteroplasmy are frequent in mitochondrial DNA from parthenogenetic and bisexual lizards (genus *Cnemidophorus*), *Genetics* (in press).

Eperon, I. C., Anderson, S., and Nierlich, D. P., 1980, Distinctive sequence of human mitochondrial ribosomal RNA genes, *Nature* **286**:460–467.

Fauron, C. M.-R., and Wolstenholme, D. R., 1976, Structural heterogeneity of mitochondrial DNA molecules within the genus *Drosophila*, *Proc. Natl. Acad. Sci. USA* **73**:3623–3627.

Fauron, C. M.-R., and Wolstenholme, D. R., 1980a, Extensive diversity among *Drosophila* species with respect to nucleotide sequences within the adenine + thymine-rich region of mitochondrial DNA molecules, *Nucleic Acids Res.* **8**(11):2439–2452.

Fauron, C. M.-R., and Wolstenholme, D. R., 1980b, Intraspecific diversity of nucleotide sequences within the adenine + thymine-rich region of mitochondrial DNA molecules of *Drosophila mauritiana*, *Drosophila melanogaster*, and *Drosophila simulans*, *Nucleic Acids Res.* **8**(17):5391–5410.

Gauss, D. H., and Sprinzl, M., 1984a, Compilation of tRNA sequences, *Nucleic Acids Res.* **12**:1r–r58.

Gauss, D. H., and Sprinzl, M., 1984b, Compilation of sequences of tRNA genes, *Nucleic Acids Res.* **12**:r59–r132.

Gillum, A. M., and Clayton, D. A., 1978, Displacement loop replication initiation sequence in animal mitochondrial DNA exists as a family of discrete lengths, *Proc. Natl. Acad. Sci. USA* **75**:677–681.

Greenberg, B. D., Newbold, J. E., and Sugino, A., 1983, Intraspecific nucleotide sequence variability surrounding the origin of replication in human mitochondrial DNA, *Gene* **21**:33–49.

Harrison, R. G., Rand, D. M., and Wheeler, W. C., 1985, Mitochondrial DNA size variation within individual crickets, *Science* **228**:1446–1448.

Hauswirth, W. W., and Laipis, P. J., 1982, Mitochondrial DNA polymorphism in a maternal lineage of Holstein cows, *Proc. Natl. Acad. Sci. USA* **79**:4686–4690.

Hauswirth, W. W., and Laipis, P. J., 1982, Mitochondrial DNA polymorphism in a maternal lineage of Holstein cows, *Proc. Natl. Acad. Sci. USA* **79**:4686–4690.

Hauswirth, W. W., Van de Walle, M. J., Laipis, P. H., and Olivo, P. D., 1984, Heterogeneous mitochondrial DNA D-loop sequences in bovine tissue, *Cell* **37**:1001–1007.

Heckman, J. E., Sarnoff, J., Alzner-DeWeerd, B., Tin, S., and RajBhandary, U. L., 1980, Novel features in the genetic code and codon reading patterns in *Neurospora crassa* mitochondria based on sequences of six mitochondrial tRNAs, *Proc. Natl. Acad. Sci. USA* **77**:3159–3163.

Hixson, J. E., and Brown, W. M., 1985, A comparison of the small ribosomal RNA genes from the mitochondrial DNA of the great apes and human: Sequence, structure, evolution and phylogenetic implications (submitted).

Klukas, C. K., and Dawid, I. B., 1976, Characterization and mapping of mitochondrial ribosomal RNA and mitochondrial DNA in *Drosophila melanogaster, Cell* **9**:615–625.

Kobayashi, M., Seki, T., Yaginuma, K., and Koike, K., 1981, Nucleotide sequences of small ribosomal RNA and adjacent transfer RNA genes in rat mitochondrial DNA, *Gene* **16**:297–307.

Kochel, H. G., Lazarus, C. M., Basak, N., and Kuntzel, H., 1981, Mitochondrial tRNA gene clusters in *Aspergillus nidulans:* Organization and nucleotide sequence, *Cell* **23**:625–633.

Kornberg, A., 1980, *DNA Replication,* Freeman, San Francisco.

Macreadie, I. G., Novitski, C. E., Maxwell, R. J., John, U., Ooi, B.-G., McMullen, G. L., Lukins, H. B., Linnane, A. W., and Nagley, P., 1983, Biogenesis of mitochondria: The mitochondrial gene (*aapl*) coding for mitochondrial ATPase subunit 8 in *Saccharomyces cerevisiae, Nucleic Acids Res.* **11**(13):4435–4451.

Maly, P., and Brimacombe, R., 1983, Refined secondary structure models for the 16S and 23S ribosomal RNA of *Escherichia coli, Nucleic Acids Res.* **11**(21):7263–7286.

Mariottini, P., Chomyn, A., and Attardi, G., 1983, Antibodies against synthetic peptides reveal that the unidentified reading frame A6L, overlapping the ATPase 6 gene, is expressed in human mitochondria, *Cell* **32**:1269–1277.

Matsumoto, L., Kasamatsu, H., Piko, L., and Vinograd, J., 1974, Mitochondrial DNA replication in sea urchin oocytes, *J. Cell Biol.* **63**:146–159.

Matthews, D. E., Hessler, R. A., Denslow, N. D., Edwards, H. S., and O'Brien, T. W., 1982, Protein composition of the bovine mitochondrial ribosome, *J. Biol. Chem.* **257**(15):8788–8794.

Maxam, A. M., and Gilbert, W., 1977, A new method for sequencing DNA, *Proc. Natl. Acad. Sci. USA* **74**:560–564.

Merten, S. H., and Pardue, M. L., 1981, Mitochondrial DNA in *Drosophila:* An analysis of genome organization and transcription in *Drosophila melanogaster* and *Drosophila virilis, J. Mol. Biol.* **153**:1–21.

Michael, N. L., Rothbard, J. B., Shiurba, R.A., Linke, H. K., Schoolnik, G. K., and Clayton, D. A., 1984, All unassigned reading frames of mouse mitochondrial DNA are expressed, *EMBO J.* **3**:3165–3175.

Nierlich, D. P., 1982, Fragmentary 5S rRNA gene in the human mitochondrial genome, *Mol. Cell. Biol.* **2**(2):207–209.

Ojala, D., Merkel, C., Gelfand, R., and Attardi, G., 1980, The tRNA genes punctuate the reading of genetic information in human mitochondrial DNA, *Cell* **22**:393–403.

Olivo, P. D., Van de Walle, M. J., Laipis, P. J., and Hauswirth, W. W., 1983, Nucleotide sequence evidence for rapid genotypic shifts in the bovine mitochondrial DNA D-loop, *Nature* **306**:400–402.

Osheroff, N., Speck, S. H., Margoliash, E., Veerman, E. C. I., Wilms, J., Konig, B. W., and Muijsers, A. O., 1983, The reaction of primate cytochromes *c* with cytochrome oxidase: Analysis of the polarographic assay, *J. Biol. Chem.* **258**:5731–5738.

Pepe, G., Holtrop, M., Gadaleta, G., Kroon, A. M., Cantatore, P., Gallerani, R., De

Benedetto, C., Quagliariello, C., Sbisa, E., and Saccone, C., 1983, Non-random patterns of nucleotide substitutions and codon strategy in the mammalian mitochondrial genes coding for identified and unidentified reading frames, *Biochem. Int.* **6**:553–563.

Pilbeam, D. R., 1984, The descent of hominoids and hominids, *Sci. Am.* **250**:84–96.

Rastl, E., and Dawid, I. B., 1979a, Structure and evolution of animal mitochondrial DNA, in: *Extrachromosomal DNA* (D. J. Cummings, P. Borst, I. B. Dawid, S. M. Weissman, and C. F. Fox, eds.), Academic Press, New York, pp. 395–407.

Rastl, E., and Dawid, I. B., 1979b, Expression of the mitochondrial genome in *Xenopus laevis:* A map of transcripts, *Cell* **18**:501–510.

Robberson, D. L., Clayton, D. A., and Morrow, J. F., 1974, Cleavage of replicating forms of mitochondrial DNA by *Eco*RI endonuclease, *Proc. Natl. Acad. Sci. USA* **71**:4447–4451.

Roberts, J. W., Grula, J. W., Posakony, J. W., Hudspeth, R., Davidson, E. H., and Britten, R. J., 1983, Comparison of sea urchin and human mtDNA: Evolutionary rearrangement, *Proc. Natl. Acad. Sci. USA* **80**:4614–4618.

Rubenstein, J. L. R., Brutlag, D., and Clayton, D. A., 1977, The mitochondrial DNA of *Drosophila melanogaster* exists in two distinct and stable superhelical forms, *Cell* **12**:471–482.

Saccone, C., Cantatore, P., Gadaleta, G., Gallerani, R., Lanave, C., Pepe, G., and Kroon, A. M., 1981, The nucleotide sequence of the large ribosomal RNA gene and the adjacent tRNA genes from rat mitochondria, *Nucleic Acids Res.* **9**(16):4139–4148.

Sanger, F., and Coulson, A. R., 1978, The use of thin acrylamide gels for DNA sequencing, *FEBS Lett.* **87**:107–110.

Sanger, F., Nicklen, S., and Coulson, A. R., 1977, DNA sequencing with chain-terminating inhibitors, *Proc. Natl. Acad. Sci. USA* **74**:5463–5467.

Schieber, G. L., and O'Brien, T. W., 1982, Extraction of proteins from the large subunit of bovine mitochondrial ribosomes under nondenaturing conditions, *J. Biol. Chem.* **257**(15):8781–8787.

Sekiya, T., Kobayashi, M., Seki, T., and Koike, K., 1980, Nucleotide sequence of a cloned fragment of rat mitochondrial DNA containing the replication origin, *Gene* **11**:53–62.

Shah, D. M., and Langley, C. H., 1979, Inter- and intraspecific variation in restriction maps of *Drosophila* mitochondrial DNAs, *Nature* **281**:696–699.

Solignac, M., Monnerot, M., and Mounolou, J.-C., 1983, Mitochondrial DNA heteroplasmy in *Drosophila mauritiana*, *Proc. Natl. Acad. Sci. USA* **80**:6942–6946.

Sriprakash, K. S., and Clark-Walker, G. D., 1983, Multiple "start-transcription" sites in *Torulopsis glabrata* mitochondrial DNA, in: *Manipulation and Expression of Genes in Eukaryotes* (P. Nagley, A. Linnane, W. Peacock, and J. Pateman, eds.), Academic Press, Sydney, Australia, pp. 303–306.

Streisinger, G., Okada, Y., Emrich, J., Newton, J., Tsugita, A., Terzaghi, E., and Inouye M., 1966, Frameshift mutations and the genetic code, *Cold Spring Harbor Symp. Quant. Biol.* **31**:77–84.

Tzagoloff, A., 1982, *Mitochondria,* Plenum Press, New York.

Upholt, W. B., and Dawid, I. B., 1977, Mapping of mitochondrial DNA of individual sheep and goats: Rapid evolution in the D-loop region, *Cell* **11**:571–583.

Van Etten, R. A., Walberg, M. W., and Clayton, D. A., 1980, Precise localization and nucleotide sequence of the two mouse mitochondrial rRNA genes and three immediately adjacent novel tRNA genes, *Cell* **22**:157–170.

Walberg, M. W., and Clayton, D. A., 1981, Sequence and properties of the human KB cell and mouse L cell D-loop regions of mitochondrial DNA, *Nucleic Acids Res.* **9**(20):5411–5421.

Walberg, M. W., and Clayton, D. A., 1983, *In vitro* transcription of human mitochondrial DNA: Identification of specific light strand transcripts from the displacement loop region, *J. Biol. Chem.* **258**:1268–1275.

Weissback, A., 1977, Eukaryotic DNA polymerases, *Annu. Rev. Biochem.* **46**:25–47.

Wilson, A. C., Carlson, S. S., and White, T. J., 1977, Biochemical evolution, *Annu. Rev. Biochem.* **46**:573–639.

Wolstenholme, D. R., and Clary, D. O., 1985, Sequence evolution of *Drosophila* mitochondrial DNA, *Genetics* **109**:725–744.

Wolstenholme, D. R., Goddard, J. M., and Fauron, C. M.-R., 1979, Structure and replication of mitochondrial DNA from the genus *Drosophila*, in: *Extrachromosomal DNA* (D. J. Cummings, P. Borst, I. B. Dawid, S. M. Weissman, and C. F. Fox, eds.), Academic Press, New York, pp.409–425.

Wong, J. F. H., Ma, E. P., Wilson, R. K., and Roe, B. A., 1983, DNA sequence of the *Xenopus laevis* mitochondrial heavy and light strand replication origins and flanking tRNA genes, *Nucleic Acids Res.* **11**(14):4977–4995.

Zakian, V. A., 1976, Electron microscopic analysis of DNA replication in main band and satellite DNAs of *Drosophila virilis*, *J. Mol. Biol.* **108**:305–331.

CHAPTER 3

Evolution of Chloroplast and Mitochondrial DNA in Plants and Algae

JEFFREY D. PALMER

1. Introduction

Plants and algae are unique among eukaryotes in possessing three DNA-containing organelles—nucleus, mitochondrion, and plastid. The amount and size of the DNA in the cytoplasmic organelles are small relative to the nucleus, and thus, in spite of early genetic evidence that genes reside in the cytoplasm (Correns, 1909; Bauer, 1909), it was only 20 years ago that convincing physical evidence was obtained demonstrating that chloroplasts (Sager and Ishida, 1963; Chun *et al.*, 1963) and mitochondria (Luck and Reich, 1964) contain their own DNA.

The powerful molecular techniques available today have recently allowed unparalleled insight into the organization, gene content, and gene structure of organelle genomes, culminating in the complete sequence description of three mammalian mitochondrial genomes (Bibb *et al.*, 1981; Anderson *et al.*, 1981, 1982). Since these molecular investigations have often been conducted in parallel on a variety of phylogenetically diverse organisms, it has been possible to deduce many features of the evolution of organelle DNAs, and, conversely, to exploit organelle DNA variation in order to assess species relationships and population structures. The depth and sophistication of these evolutionary studies of organelle DNA has been greatest with the smallest and simplest genomes, the mitochon-

JEFFREY D. PALMER • Division of Biological Sciences, University of Michigan, Ann Arbor, Michigan 48109.

drial genomes of animals [reviewed in Avise and Lansman (1983) and W. M. Brown (1983, and this volume, Chapter 2)]. Much is also known about mitochondrial DNA (mtDNA) diversity and evolution in fungi (M. W. Gray, 1982; Wallace, 1982; Sederoff, 1984), particularly with regard to variation in the structure of mitochondrial introns (Mahler, 1983).

The purpose of this chapter is to review the available information on the comparative structure and sequence of chloroplast DNA (cpDNA) and mitochondrial DNA (mtDNA) in plants and eukaryotic algae. As will become apparent, the extent of our present understanding of both organelle DNAs is decidedly lopsided from a phylogenetic standpoint. Certain cpDNAs, from a number of angiosperms and two algae (*Chlamydomonas reinhardtii and Euglena gracilis*), have been studied in great molecular detail and it is possible to describe many features of their evolution. Unfortunately, cpDNA has been studied in only a limited detail in a few nonchlorophytic* algae, so that we are almost completely ignorant of how their evolution compares with that of the green plant–green algal lineage. Our understanding of plant mtDNAs, which are the largest and most complex organelle DNAs known, lags far behind that of plant cpDNAs and also animal and fungal mtDNAs. The properties of algal mtDNAs are even more poorly understood. Nonetheless, one of the major aims of this review is to compare and contrast the evolution of cpDNA and mtDNA in plants, with the ultimate goal being to deduce any commonality in their evolution that might result from their coexistence in the same cell.

Several recent review articles are particularly relevant to the topics considered here. These articles are, for chloroplasts, Bohnert *et al.* (1982a), Whitfeld and Bottomley (1983), Sears (1983), Crouse *et al.* (1985), and Gillham *et al.* (1985), and for mitochondria, Leaver and Gray (1982), M. W. Gray (1982), Wallace (1982), and Sederoff (1984).

2. Chloroplast Genome Evolution

This section will discuss the large-scale evolutionary changes that have resulted in variation in the inheritance, cytological structure, and

*Throughout this chapter the term "chlorophyte" will be used for green algae and land plants (bryophytes, ferns, gymnosperms, angiosperms); "euglenophyte" for members of the genus *Euglena;* "rhodophyte" for red algae; and "chromophyte" for dinoflagellates, diatoms, cryptomonads, brown algae, golden algae, and yellow-green algae [cf. algal classification schemes based on plastid endosymbioses as discussed in Whatley and Whatley (1981), Whatley (1983), and Section 5.1]. The only other chloroplast-containing organism discussed here is *Cyanophora paradoxa*, which is of an uncertain classification relative to these other groups (Jaynes and Vernon, 1982).

molecular structure of cpDNA among different plants and algae. Particular attention will be given to a description of the kinds and frequencies of mutational events that lead to rearrangement of cpDNA sequences. Specific features of chloroplast gene evolution will be presented in the next section; individual chloroplast genes will be discussed here only in the context of their presence, *vis-à-vis* gene transfer between cellular compartments, and arrangement on the genome.

2.1. Chloroplast DNA Inheritance

The inheritance of the plastid and its genome has been considered from an evolutionary perspective in three recent and comprehensive review articles (Sears, 1980, 1983; Whatley, 1982) and also in a monograph (Kirk and Tilney-Bassett, 1978). It should be kept in mind that most studies of plastid inheritance have relied primarily if not entirely on ultrastructural observations, although similar conclusions have usually been reached in those instances where parallel genetic or molecular studies have been carried out on the same organism. The primary conclusion reached in the above review papers is that there is an extreme diversity of modes and mechanisms of plastid inheritance in most phyla of plants and algae. The pattern of plastid inheritance covers almost the complete spectrum from strictly maternal to predominantly paternal. Nonetheless, maternal plastid transmission is the dominant mode, strong paternal inheritance being known only in a few species of gymnosperms. Even where inheritance is biparental, the maternal contribution is usually much greater than the paternal. Elimination of plastids from one parent, usually the paternal one, can occur in several ways: exclusion from the male gamete during spermatogenesis, loss from the motile sperm, exclusion during fertilization, or degradation within the zygote (Sears, 1980). Furthermore, fluorescent microscopic observations indicate that maternal inheritance of cpDNA in *Chlamydomonas* results from the specific degradation of cpDNA of the paternal parent occurring after fusion of the cell walls of the gametes (zygote formation) but before fusion of the gametic plastids (Kuroiwa *et al.*, 1982; Coleman and Maguire, 1983; Coleman, 1984a).

The great variation in patterns and mechanisms of plastid inheritance found in green algae and in vascular plants, and in certain cases even among closely related angiosperm species, suggests that any given mode of inheritance may have evolved independently on many occasions in different lineages (Kirk and Tilney-Bassett, 1978; Sears, 1980, 1983; Whatley, 1982). It is thus difficult to discern any clear-cut trends in the evolution of plastid inheritance, and also to understand what the selective pressures might be that could cause such a reticulate and convergent pattern of evolution for this trait. In particular, it is difficult to imagine a selection that acts directly on the plastid genetic component itself. For example, uniparental plastid inheritance could be viewed as a mechanism for preventing recombination between parental genomes. However, it seems unnecessary to postulate such selection, since, with the notable exception of the algal genus *Chlamydomonas* (Gillham, 1978; Lemieux *et al.*, 1981, 1984), there is little evidence for plastid fusion (Sears, 1980, 1983) and none at all for plastid DNA recombination (e.g., Kutzelnigg and Stubbe, 1974) even in those species that exhibit biparental plastid inheritance. On the other hand, it is possible to envision selection operating without regard to the plastid as a genetic unit on the basis of physiological factors, as in the development of a streamlined, highly motile sperm in many species with maternal inheritance, or, alternatively, the development of a large, metabolically long-lived spermatozoid in gymnosperms that feature paternal inheritance. In addition, it is entirely possible that in certain cases changes in the mode of plastid inheritance may operate outside of any direct selection, as "the more or less accidental result of other evolutionary events" (Whatley, 1982).

2.2. Cytological Organization of Chloroplast DNA

Algae and higher plants almost always contain multiple copies of the plastid genome in each plastid (Possingham and Lawrence, 1983). The only known exception to this rule is that in several species of the alga *Acetabularia* as many as 75% of the chloroplasts do not contain any detectable DNA (Woodcock and Bogorad, 1970; Coleman, 1979; Luttke and Bonotto, 1982). Although there are no obvious evolutionary trends in the amount of cpDNA per plastid, per cell, or relative to nuclear DNA content and ploidy level (Possingham and Lawrence, 1983), there do appear to be some very broad relationships between the cytological organization of DNA within plastids and the phylogenetic position of many plants and algae. In general, the multiple copies of the plastid genome are organized into discrete structures within the plastid called nucleoids, each of which may contain between one and 25 copies of the genome (Possingham and Lawrence, 1983). Electron and fluorescence microscope studies have revealed three general patterns of nucleoid organization [reviewed in Coleman (1985b) and Gillham *et al.* (1985); also see Possingham and Lawrence (1983)]: (1) In the most distinctive pattern, characteristic of five classes of chromophytic algae, cpDNA appears as a continuous peripheral ring of variable thickness. (2) In certain small algae and in proplastids and chromoplasts of higher plants, cpDNA is present in a single, more or less centrally located nucleoid. (3) Variable numbers of discrete cpDNA nucleoids are found in chloroplasts of flowering plants and chlorophytic algae, rhodophytes, and certain chromophytes. These nucleoids are often scattered in the stroma in such a way as to appear associated with the thylakoids. A somewhat different pattern is found in certain grasses, where the nucleoids are relegated to peripheral stromal regions (Possingham and Lawrence, 1983). In spite of our ability to recognize and classify distinct plastid DNA configurations, the ultimate physiological and phylogenetic significance of this variation is unclear (Coleman, 1985b).

2.3. Physicochemical Properties of Chloroplast DNA

An impressive uniformity exists in the size and conformation of land plant chloroplast DNAs. Analysis by electron microscopy and restriction endonuclease digestion has shown that in six nonangiospermous land plants and in every one of over 200 species of angiosperms examined cpDNA is organized as a single size class of circular molecules (Table I). The multiple circles of cpDNA found in each plant cell are homogeneous and indistinguishable from one another in all respects, except for the case of those genomes that exist as two genetically identical, but physically distinct inversion isomers (see Section 2.4.1a).

The total extent of size variation (120–217 kb) observed for angiosperm cpDNAs is in a sense misleading. The lower extreme of this variation occurs in a single group of legumes as the result of the deletion of one copy of a large two-copy inverted repeat sequence present in all other angiosperms (see Section 2.4.1a), while the upper 55 kb of the size spectrum is represented by only two species, *Spirodela oligorrhiza* [genome size 180 kb (Van Ee *et al.*, 1980)] and *Pelargonium hortorum* [genome size 217 kb (J. Palmer and J. Nugent, unpublished data)]. Furthermore, most of the increase in *Pelargonium* results from an enlarged inverted repeat of over 75 kb, as compared to the 20- to 30-kb inverted repeat found in most angiosperms. Thus, among the great majority of angiosperms examined cpDNA varies in size over the relatively narrow range of 135–160 kb. Moreover, the total range of sequence complexities encountered among *all* angiosperm cpDNAs is only 115–150 kb.

Considerably greater variation in size (85–195 kb) and sequence complexity (from at most 85 kb in *Codium* to 170 kb in *Chlamydomonas*) is found among the unicircular cpDNAs of three green algae (Table I). A seemingly unique case is presented by the green algal genus *Acetabularia*, whose chloroplast genome is estimated from reassociation kinetics to be on the order of 2000 kb in size (Green *et al.*, 1977; Padmanabhan and Green, 1978) and whose cpDNA is isolated as large, linear molecules of up to 600 kb in size (Green, 1976; Luttke and Bonotto, 1982). Since this

Table I
Physicochemical Properties of Chloroplast DNAs

Taxon	GC Content (%)	Size (kb)	Conformation	Reference
Angiosperms (210 species, 78 genera, 33 families)	37–39	120–217	Circular	—[a]
Gymnosperms				
Ginkgo biloba	—	158	Circular	J. Palmer and W. Thompson (unpublished data)
Ferns				
Asplenium nidus	38	150	Circular	Herrmann *et al.* (1980)
Pteris vittata	38	150	Circular	Hermann *et al.* (1980)
Osmunda cinnamomea	—	144	Circular	Palmer and Stein (1982)
Bryophytes				
Sphaerocarpus donnellii	31	125	Circular	Hermann *et al.* (1980)
Marchantia polymorpha	—	121	Circular	Ohyama *et al.* (1983)
Green algae				
Chlamydomonas reinhardtii	36	169–235	Circular	—[b]
Chlorella ellipsoidea	36	174	Circular	Yamada (1983)
Codium fragile	37	85	Circular	Hedberg *et al.* (1981)
Acetabularia (two species)	47	~2000	Unknown	Padmanabhan and Green (1978)
Polytoma obtusum[c]	17	~200	Unknown	Siu *et al.* (1975)

Euglenophyta				
Euglena gracilis	24–28	130–152	Circular	See Section 2.4.1b
Classification uncertain				
Cyanophora paradoxa	33–36	127–138	Circular	Bohnert and Loffelhardt (1982), Loffelhardt *et al.* (1983)
Chromophyta				
Olisthodiscus luteus	31	150	Circular	Aldrich and Cattolico (1981)
Vaucheria sessilis	—	125	Circular	Hennig and Kowallik[d]
Plylciella littorali	34	140	Multicircular?	Dalmon *et al.* (1983)
Sphacelaria sp.	33	156	Multicircular?	Dalmon *et al.* (1983)

[a] Data summarized from tables presented in Bohnert *et al.* (1982a) and Whitfeld and Bottomley (1983), plus the following recent studies: Chu and Tewari (1982), Coates and Cullis (1982), Poulsen (1983), Spielmann *et al.* (1983), Mubumbila *et al.* (1983), Palmer *et al.* (1983a,b), and J. Palmer and W. Thompson (unpublished data).

[b] Wild-type *Chlamydomonas* cpDNA is 195 kb in size (Rochaix, 1978), while a series of mutant cpDNAs range in size from 169 to 235 kb (Myers *et al.*, 1982).

[c] A colorless alga probably descended from green algae (see Section 2.4.4).

[d] Cited in Herrmann *et al.* (1980).

result is at complete variance with those for all other chloroplast genomes, including that of another siphonaceous greeen alga (Hedberg *et al.*, 1981), the identity of this *Acetabularia* DNA as *bona fide* cpDNA should be confirmed by challenging it with hydridization probes for several well-characterized chloroplast genes.

Euglena gracilis resembles green algae and land plants in terms of its chlorophyll composition and photosynthetic characteristics, but is likely to have acquired its chloroplast by a separate endosymbiotic event from the green algal–land plant lineage (see Section 5.1). Nonetheless, its chloroplast genome also consists of a single circular species, which shows interstrain size variation of between 130 and 152 kb due to a variable number of ribosomal RNA operons (see Section 2.4.1b).

The cpDNAs of three other nonchlorophytic algae also exist as a single circular species of the usual size (Table I). Interestingly, one of these organisms, *Cyanophora paradoxa*, is thought to represent an intermediate step in the endosymbiotic invasion of a unicellular protist by a cyanobacterium in that its chloroplast (often termed "cyanelle") is surrounded by a thin peptidylglycan cell wall and is located within the cell's vacuole (Jaynes and Vernon, 1981). Nonetheless, on the basis of size (Table I), its circular chloroplast genome appears to contain no more genetic information than the normal chloroplast genome. Furthermore, *Cyanophora* is so far indistinguishable from red algae in particular in terms of the distribution of genes between chloroplast and nuclear genomes (Grossman *et al.*, 1983), (Section 2.4.4).

In contrast to the structural uniformity of those plant and algal cpDNAs discussed thus far, a recent report (Dalmon *et al.*, 1983) indicates that two brown algal cpDNAs may be organized in a heterogeneous pattern found heretofore only among angiosperm mtDNAs (Section 4.2). Electron microscopic analysis suggests that *Pylaiella littorali* cpDNA may exist as two discrete circular classes of molecules, 54 and 114 kb in size, and possibly a further size class represented by large linear molecules over 200 kb in size. Summation of restriction fragments (without taking into account extensive stoichiometry differences among fragments) indicates a minimal genome size of 140 kb, greater than can be accommodated by either circular size class alone (Dalmon *et al.*, 1983). *Sphacelaria* sp. cpDNA exhibits in the electron microscope a fairly continuous range of circular molecules of from 96 to 160 kb in size, while a minimal genome size estimate of 156 kb is obtained by restriction analysis (Dalmon *et al.*, 1983). Although these findings are preliminary and should be confirmed by restriction mapping, they clearly emphasize the need to examine many more nonchlorophytic algae in order to appreciate the full extent of chloroplast genome differentiation.

Consistent with many other aspects of their conservative evolution, angiosperm cpDNAs span a relatively narrow range (37–39% GC) in average base composition [reviewed in Bohnert *et al.* (1982a)], whereas a much greater range (25–47% GC) of base compositions is found among algal cpDNAs (Table I). Intramolecular heterogeneity in base composition is greatest among the most AT-rich genomes. For example, *Euglena gracilis* cpDNA, which at 24–28% GC is the most AT-rich photosynthetic plastid genome known, consists of stretches comprising 30% of the genome that are only 12% GC, whereas 10% of the genome is greater than 60% GC (Schmitt *et al.*, 1981). The plastid DNA of *Polytoma obtusum*, a unicellular alga that appears to have sustained the permanent loss of photosynthetic function, is even more AT-rich (17% GC) than *Euglena* cpDNA, and is also extremely heterogeneous in base composition (Siu *et al.*, 1975). The extremely AT-rich sequences in *Polytoma* and *Euglena* plastid DNA are unlikely to have coding function and probably represent spacer and intron sequences (Koller and Delius, 1984). The high proportion of such sequences in *Polytoma* may be related to the fact that it is no longer photosynthetically competent and thus is likely to have lost function in a significant number of its plastid genes, which now may be evolving freely as pseudogenes (see Section 2.4.4).

Chloroplast DNA of all plants and algae examined lacks completely any modified bases, such as 5-methylcytosine [reviewed in Bohnert *et al.* (1982a) and Loiseau and Dalmon (1983)]. The interesting exception to this rule is that cpDNA from *Chlamydomonas reinhardtii* shows a low level of cytosine methylation in vegetative cells, much higher levels in gametic cells, and still higher levels in zygotes (Burton *et al.*, 1979; Bolen *et al.*, 1982; Feng and Chiang, 1984). Since at all stages, cpDNA of maternal (mating type$^+$) strains of *C. reinhardtii* is much more extensively methylated than that of paternal (mating type$^-$) strains, it has been hypothesized that the maternal inheritance of cpDNA in *C. reinhardtii* is governed by a methylation-restriction system analogous to that in bacteria (Burton *et al.*, 1979; Sager and Grabowy, 1983). However, this hypothesis has been challenged based on recent findings that the pattern of cpDNA inheritance in *C. reinhardtii* is essentially unchanged despite extensive cpDNA hypermethylation (Bolen *et al.*, 1982) or hypomethylation (Feng and Chiang, 1984). It shall be interesting to see whether similar forms of base modification occur during gametogenesis in other organisms that feature uniparental plastid DNA inheritance.

2.4. Molecular Organization and Rearrangement of Chloroplast DNA

It should now be clear that cpDNAs of land plants and green algae, with the possible exception of *Acetabularia*, are highly conserved in basic structure (unicircular) and size (85–210 kb). In contrast, initial studies suggest that the cpDNA of certain brown algae may be distributed among several molecules of different size and sequence composition. In this section, we explore how genes and other sequences are arranged in different cpDNAs and seek to characterize those events that lead to evolutionary change in the linear order of chloroplast sequences. Data that bear on these questions are quite limited from a phylogenetic standpoint: details of cpDNA sequence organization are known for a large number of angiosperms, three nonangiospermous land plants, and five algae.

2.4.1. Repeated Sequences in Chloroplast DNA

Organelle DNAs characteristically have very few repeated sequences compared to most eukaryotic nuclear genomes. Although this generalization also holds true for cpDNA, a large inverted repeat sequence that is present in most chloroplast genomes serves, by virtue of its size, gene content, phylogenetic conservation, and associated recombination properties, as a landmark feature of these genomes. Aside from this inverted repeat, there are only three other known cpDNA repeat families whose individual members are larger than 100 bp in size. Each of these repeat families appears to be confined to a single genus of plants or algae.

2.4.1a. The Inverted Repeat. The only angiosperms lacking a large rRNA-encoding inverted repeat (e.g., Fig. 1) are confined to one section of the subfamily Papilionoideae of the legume family Fabaceae (Table II). Species in the tribes Phaseoleae (*Vigna, Phaseolus, Glycine*) and Genisteae (*Lupinus*) have retained the repeat, while all those examined from four other tribes have lost the repeat (Table II). Given that all species examined from 32 other angiosperm families contain the inverted repeat, the phylogenetic direction of the mutation responsible for the difference among legumes is clear; i.e., the inverted repeat is ancestral to the legume family as a whole and the repeat structure has been lost in one section of the family. The loss of the inverted repeat appears to have occurred without the loss of any sequence information; i.e, although one entire segment of the repeat was deleted, sequences representing the entirety of the other segment are still present in these legume chloroplast genomes (Palmer and Thompson, 1981a; 1982).

A similar inverted repeat has been found in a single gymnosperm, a fern, and a liverwort (Table II). The three features that are common to this large inverted repeat among all land plants are its location in the genome (i.e, the two repeat segments are always positioned asymmetrically, dividing the genome into small and large single-copy regions of 12–28 kb and 80–100 kb, respectively) [Fig. 1; see Table 2 in Bohnert *et al.* (1982a)]; the presence of a complete set of ribosomal RNA genes within

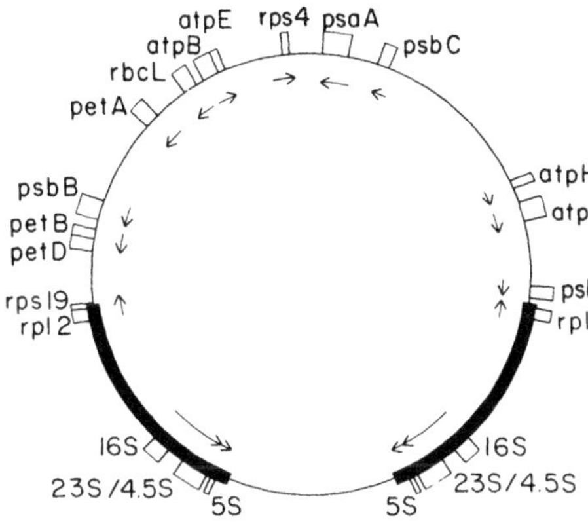

Figure 1. Gene map of spinach chloroplast DNA. Arrows indicate direction of transcription of genes and operons coding for ribosomal RNA and 15 proteins on the circular map of the 150-kb spinach chloroplast genome. Transfer RNA gene positions (Driesel *et al.*, 1979) are not shown. Heavy lines indicate the extent of the inverted repeat (Kolodner and Tewari, 1979). Note that only one of the two possible orientations of the genome are shown (cf. Fig. 3 and Section 2.4.1a). All genes shown have been mapped on the spinach genome except for *rps4*, which has been mapped only in maize (Subramanian *et al.*, 1983). The orientation and approximate position of this gene on the spinach genome have been inferred based on the colinearity of the spinach and maize genomes in this region (Palmer and Thompson, 1982). Gene nomenclature is that used by Whitfeld and Bottomley (1983), which was recently codified by workers in the field (Hallick and Bottomley, 1983). Gene designations and sources of mapping data are as follows: in counterclockwise order starting at the bottom of the map 5S, 23S/4.5S, 16S, ribosomal RNA genes (Whitfeld *et a.*, 1978); *rpl2*, gene for a chloroplast ribosomal protein homologous to *E. coli* L2 (Zurawski *et al.*, 1984b); *psbA*, gene for the "32-kD" herbicide-binding, photosystem II polypeptide (Driesel *et al.*, 1980); *atpA*, gene for the α subunit of chloroplast ATPase (Westhoff *et al.*, 1981); *atpH*, gene for the DCCD-binding proteolipid of ATPase (Atl *et al.*, 1983b); *psbC*, gene for a 44-kD chlorophyll a-binding photosystem II polypeptide (Westhoff *et al.*, 1983a); *psaA*, gene for the 67-kD P_{700} chlorophyll a apoprotein of photosystem I (Westhoff *et al.*, 1983b); *rps4*, maize gene for a chloroplast ribosomal protein homologous to *E. coli* S4 (Subramanian *et al.*, 1983); *atpE*, gene for the ϵ subunit of chloroplast ATPase (Westhoff *et al.*, 1981); *atpB*, gene for the β subunit of chloroplast ATPase (Westhoff *et al.*, 1981); *rbcL*, gene for the large subunit of ribulose-1,5-bisphosphate carboxylase (Whitfeld and Bottomley, 1980); *petA*, gene for cytochrome *f* (Alt *et al.*, 1983a); *psbB*, gene for a 51-kD chlorophyll a-binding photosystem II polypeptide (Westhoff *et al.*, 1983a); *petB*, gene for cytochrome *b*6 (Alt *et al.*, 1983a); *petD*, gene for subunit 4 of the cytochrome b_6-f complex (Alt *et al.*, 1983a); *rps19*, gene for a chloroplast ribosomal protein homologous to *E. coli* S19 (Zurawski *et al.*, 1984b).

each repeat unit; and the orientation of the rRNA genes (i.e., the 23S gene is always found closer to the small single-copy region and the 16S gene closer to the large single-copy region) (Fig. 1). Based on these common features it is likely that the inverted repeat was present in the chloroplast genome of the common ancestor of all land plants.

Table II
Distribution of a Large, Ribosomal DNA-Containing Inverted Repeat in cpDNAs of Plants and Algae[a]

Taxon	Repeat size[b] (kb)
Species containing the inverted repeat	
Angiosperms (33 families, 70 genera, 200 species)	10–76
Gymnosperms	
Ginkgo biloba	17 (11.9–21.7)
Ferns	
Osmunda cinnamomea	11 (7.7–13.4)
Bryophytes	
Marchantia polymorpha	11 (9.8–11.7)
Green algae	
Chlamydomonas reinhardtii	22 (4–65)
Chlorella ellipsoidea	23
Classification uncertain	
Cyanophora paradoxa	10
Species lacking the inverted repeat	
Angiosperms	
One family (Fabaceae)	
Four tribes (Vicieae, Cicereae, Trifolieae, Tephrosieae)	
Eight genera (*Pisum, Vicia, Lathyrus, Cicer, Medicago, Trifolium, Melilotus, Wisteria*)	
Euglenophyta	
Euglena gracilis	
Chromophyta	
Olisthodiscus luteus[c]	

[a] Data taken from references given in Table I; *Olisthodiscus* information is from the unpublished work of M. Reith and R. Cattolico.
[b] The range of inverted repeat sizes encountered in different species is given for angiosperms. Gymnosperm, fern, and bryophyte repeat sizes are approximations based on restriction mapping; numbers in parentheses give the minimum and maximum possible sizes of the repeat in these species. The algal repeat sizes are exact determinations made by electron microscopy. Numbers in parentheses for the *Chlamydomonas* repeat indicate the range of repeat sizes encountered in different laboratory mutants (Myers *et al.*, 1982; Palmer, *et al.*, 1985b).
[c] CpDNA from *Olisthodiscus luteus* possesses an unrelated inverted repeat, which encodes a different set of genes (see Section 2.4.1a).

Four of the five mapped algal chloroplast genomes contain a large inverted repeat with varying degrees of similarity to that of land plants (Table II and below). CpDNA from *Euglena gracilis* lacks any large inverted repeat and instead features between one and five complete rRNA operons repeated in tandem (see Section 2.4.1b). The uncertain, probably very distant, phylogenetic relationship of *Euglena* to land plants and green algae (see Section 5.1) precludes any conclusions as to the ancestral state of its chloroplast genome. Although cpDNA from the chromophyte *Olisthodiscus luteus* does have a large inverted repeat, the gene content of this repeat is distinctly different from that of the usual chlorophytic inverted repeat. The *Olisthodiscus* inverted repeat contains cotranscribed genes for the large and small subunits of ribulose-1,5-bisphosphate carboxylase, while only a single set of rRNA genes is present on the *Olisthodiscus* chloroplast chromosome (M. Reith and R. Cattolico, unpublished data). Given these differences, and also the probably separate endosymbiotic origin of the plastid in *Olisthodiscus* compared to chlorophytes (Section 5.1), it is highly likely that their inverted repeats are of independent origin.

CpDNAs from *Cyanophora paradoxa* (Bohnert and Loffelhardt, 1982) and two green algae, *Chlamydomonas reinhardtii* (Rochaix, 1978) and *Chlorella ellipsoidea* (Yamada, 1983), contain a large inverted repeat encoding a complete set of rRNA genes. It is intriguing that such a repeat exists in *Cyanophora,* since it is almost certain that this alga gained its plastid via a separate endosymbiotic event from the land plant–green algal lineage (Jaynes and Vernon, 1982). If this is the case, then either the inverted repeat structure originated independently in the two plastid lineages, or else it was present in the genome of the common ancestor of the photosynthetic prokaryotes that established these separate endosymbioses. In this regard, it is noteworthy that the cyanobacterium *Anacystis nidulans,* whose rRNA genes have extensive sequence homology with chloroplast rRNA genes (Section 5.1), also contains two sets of rRNA genes (Tomioka *et al.*, 1981), although their relative orientation in the genome is not known. [This compares with nonphotosynthetic eubacteria, which generally have between five and ten sets of rRNA genes (Henner and Hock, 1980; Bachmann, 1983; Sanderson and Roth, 1983).] On the other hand, the fact that inverted arrangements of duplicated rRNA genes appear to have evolved independently on several occasions in mitochondria (Table III; Section 4.2) makes the former possibility seem plausible.

The inverted repeat segments are asymmetrically located in *Chlorella* and *Cyanophora,* as in land plants, while the *Chlamydomonas reinhardtii* repeat units are almost 180° apart, producing single-copy regions nearly equal in size (74 and 80 kb). The enlargement of the small single-copy region in *C. reinhardtii* relative to all other plants and algae may be the result of two processes: (1) It is likely that extra sequences have been added to the *C. reinhardtii* genome, given that it has the largest sequence complexity of any mapped chloroplast genome (Table I) (Rochaix, 1978). (2) Preexisting cpDNA sequences may have been redistributed between the unique regions in *C. reinhardtii.* For example, the *atpB* and *rbcL* genes are located adjacent to one another within the large single-copy

Table III
Repeats in Organelle DNAs[a]

Taxon	Repeat size (kb)	Repeat orientation	Repeat active in recombination	Genes on repeat	Reference
Chloroplast DNA					
Chlorophytes and *Cyanophora*	10–76	Inverted	Yes	rRNA operon[b]	See Tables II and IV
Euglena gracilis	6	Direct	No	rRNA operon[c]	See Section 2.4.1b
Olisthodiscus luteus	22	Inverted	Unknown	*rbcL*, *rbcS*	M. Reith and R. Cattolico (unpublished data)
Mitochondrial DNA					
Angiosperms					
Turnip	2	Direct	Yes	Cytochrome oxidase subunit II	Palmer and Shields (1984), Stern and Palmer (1984b)
Maize	10	Direct	No		Lonsdale *et al.* (1984)
Maize	14	Indirect	Yes		Lonsdale *et al.* (1984)
Maize	3	Direct	Yes		Lonsdale *et al.* (1984)
Wheat	4	Unknown	Yes	18S, 5S rRNA	Falconet *et al.* (1984)
Wheat	6	Unknown	Yes	26S rRNA	Stern and Palmer (1984b)
Spinach	5	Unknown	Yes	26S rRNA	Stern and Palmer (1984b)
Pokeweed	>1	Unknown	Yes	26S rRNA	Stern and Palmer (1984b)
Fungi					
Achyla ambisexualis	11	Inverted	Yes	rRNA operon	Hudspeth *et al.* (1983)
Kloeckera africana	4	Inverted	Unknown	Large rRNA (part)	Clark-Walker *et al.* (1981)
Protist					
Tetrahymena pyriformis	2	Inverted	Unknown	Large rRNA (part?)	Goldbach *et al.* (1978)

[a] Only those repeats larger than 1 kb are included. See Section 2.4.1b for smaller cpDNA repeats.
[b] Includes the entire rRNA operon, including spacer tRNA genes, and also various other tRNA and protein genes, depending on the repeat size (Section 2.4.1a).
[c] Includes the entire rRNA operon, including spacer tRNA genes, and a pseudogene cluster (Section 2.4.1b).

regions in land plants (e.g., Fig. 1), but in *C. reinhardtii* the two genes are present on opposite sides of the inverted repeat (Woessner *et al.*, 1984).

The orientation of the ribosomal RNA genes has not been established in *Chlorella*, but in *C. reinhardtii* and *Cyanophora* these genes appear to be inverted relative to land plants. *In Cyanophora* this conclusion is based on the fact that the 16S and 23S rRNA genes are inverted relative to the small and large unique regions compared to land plants (Bohnert and Loffelhardt, 1982). In *C. reinhardtii* the 16S gene is also closer to the smaller of the two unique regions than is the 23S gene (Rochaix, 1978), although the significance of this observation is questionable, given that the unique regions are almost equal in size. More consistent with this inversion hypothesis is the fact that of the four single-copy genes mapped in angiosperms and in *C. reinhardtii*, all of which are located in the large single-copy region in angiosperms, three are also present in the larger unique region in *C. reinhardtii* (*Woessner et al.*, 1984), the exception being the *atpB* gene, as discussed above. It is tempting to speculate that the putative inversion of the rRNA genes (possibly of the entire inverted repeat) and the transfer of the *atpB* gene between unique regions may have occurred as part of the same large inversion. In this regard, it is worth noting that large symmetric inversions encompassing the rDNA genes present within the inverted repeat have been found in laboratory mutants of *Chlamydomonas reinhardtii* (Myers *et al.*, 1982).

Although early work indicated that the inverted repeat is highly conserved in size and information content (Kolodner and Tewari, 1979; Bohnert *et al.*, 1982a), more extensive investigation has revealed wide variation in both these parameters, particularly among angiosperms (Table II). Evolutionary variation in size of the inverted repeat is in a sense linked with the convergence or copy-correction mechanism that is postulated to maintain sequence identity between the two repeat units in a given genome. Evidence for this correction is of two types: (1) The two repeats are always found to be identical within an individual plant and molecule as judged by homoduplex analysis of the entire repeat (Kolodner and Tewari, 1979) and sequence analysis of portions of the repeat (Kato *et al.*, 1981; Spielmann and Stutz, 1983; Sugita *et al.*, 1984; Zurawski *et al.* 1984b). (2) All naturally occurring (Fluhr and Edelman, 1981; Gordon *et al.*, 1982) as well as induced (Myers *et al.*, 1982) mutations completely within the inverted sequence occur symmetrically in both segments. In other words, deletions or insertions that have both boundaries within the repeat are found in both repeat units. Conversely, it has been found that deletions that overlap one of the four repeat–single-copy junctions can occur asymmetrically (Palmer, 1981; Myers *et al.*, 1982).

Electron microscopic studies have shown that the 19.3-kb inverted repeat of *Nicotiana tabacum* cpDNA is bounded at its large single-copy end by a 3.5-kb region containing two short inverted repeats interspersed with unique sequences (Fig. 2). Approximately 30 other species of *Nicotiana* have an inverted repeat of this same approximate size (Shen *et al.*, 1982), while in *N. acuminata* the inverted repeat is enlarged by precisely the same amount and in the same position as the interspersion region in *N. tabacum* (Fig. 2). Thus, it appears that in *N. acuminata* the inverted repeat has spread through this region via copy correction. It should be pointed out that successive rounds of copy-corrected shrinkage and engulfment of the inverted repeat have the potential to not only manipulate the copy number of sequences in or near the ends of the repeat, but also to effect the "transposition" of these sequences from one side of the chloroplast chromosome to the other.

That changes in the size of the inverted repeat do actually alter the copy number of chloroplast genes can be indirectly inferred from the fact that the chloroplast genome as a whole and the inverted repeat specifically are tightly packed with coding sequences (Poulsen, 1983). In addition, certain chloroplast genes have been mapped in or near the ends of the inverted repeat (Fig. 1). One of these genes, *psbA*, is found in the large single-copy region immediately adjacent to the inverted repeat in all land plants examined (Whitfeld and Bottomley, 1983). In *Chlamydomonas*, however, *psbA* is located within the inverted repeat (J. Erickson and

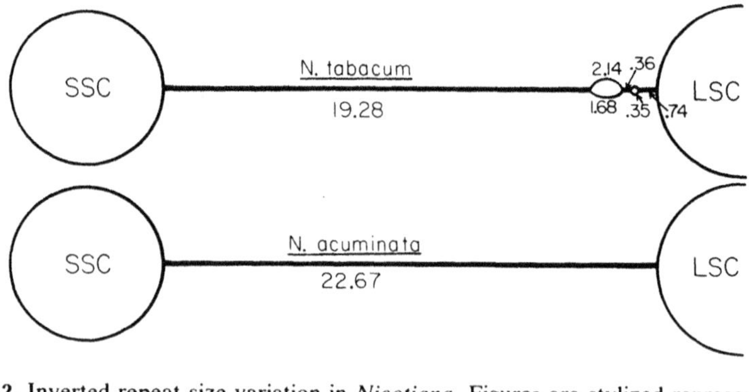

Figure 2. Inverted repeat size variation in *Nicotiana*. Figures are stylized representations of self-renatured cpDNA molecules from *N. tabacum* and *N. acuminata*. Thick lines represent double-stranded DNA (i.e., inverted repeats) and thin lines single-stranded DNA. The small single-copy region is indicated as SSC and the large single-copy region as LSC. Sizes of inverted repeats and interspersed single-copy regions are given in kb. [Redrawn and modified from Shen *et al.* (1982).]

J. Rochaix, unpublished data). In maize, the two genes for $tRNA^{His}$ are located within the large single-copy end of the inverted repeat (Schwarz *et al.*, 1981), while in spinach (Driesel *et al.*, 1979; Zurawski *et al.*, 1984b), soybean (Spielmann and Stutz, 1983), common bean (Mubumbila *et al.*, 1983), and tobacco (Sugita *et al.*, 1984) the single $tRNA^{His}$ gene is in the large single-copy region just between the end of the inverted repeat and *psbA*. Selden *et al.* (1983) have hypothesized that this difference may reflect movement of the ends of the inverted repeat, whereas Zurawski *et al.* (1984b) have presented sequence comparisons suggesting that this tRNA gene may have been moved by a small inversion near the inverted-repeat–single-copy junction. In any case, given that the ends of the inverted repeat can "move" through regions of the genome, it is perhaps not surprising that sequence analysis has not revealed any significant sequence elements or motifs at the four ends of the inverted repeat (Spielmann *et al.*, 1983; Sugita *et al.*, 1984; Zurawski *et al.*, 1984b).

Although the inverted repeat can vary by at least a factor of six in size, with concomitant changes in gene content, it is striking that it always contains a complete set of ribosomal RNA genes. This is particularly remarkable, given that across a wide phylogenetic spectrum (Table II) the inverted repeat is just large enough (8–13 kb) to encode a complete set of chloroplast rRNA genes (7 kb) and not much else. Although rRNA genes are present in multiple copies in nuclear (Long and Dawid, 1980) and most eubacterial (Bachmann, 1983; Sanderson and Roth, 1983; Amikan *et al.*, 1984) genomes, they usually are present only once in animal and fungal mtDNAs (M. W. Gray, 1982; Wallace, 1982). Nonetheless, almost all the major repeated sequences that have been found in organelle genomes contain all or part of a rRNA gene (Table III). Thus, as noted by Hudspeth *et al.* (1983), there may be a selective advantage to having multiple, or at least duplicate, rRNA genes in certain organelle genomes. Evidence against this selection hypothesis in the case of the chloroplast inverted repeat rRNA genes comes from the work of Myers *et al.* (1982), who found normal levels (measured relative to the level of cytoplasmic ribosomes) of chloroplast ribosomes in two *Chlamydomonas* mutants that had deleted one entire set of rRNA genes. [Although both of these mutants are nonphotosynthetic and grow slowly relative to wild-type *Chlamydomonas*, these properties are almost certainly unrelated to the absence of one set of rRNA genes (Myers *et al.*, 1982; Woessner *et al.*, 1984)].

It may be significant that almost all duplicated organelle rRNA genes whose relative orientation has been determined are present as inverted repeats, which are not susceptible to deletion as in the case of direct repeats. Indeed, the one organelle genome that possesses tandemly repeated rRNA genes, the cpDNA of *Euglena gracilis*, also shows interstrain variation of between one and five copies of the repeat (see next section).

In addition to the conversion activity that is postulated to maintain homogeneity between segments of the inverted repeat, two other recombination activities are associated with its presence in the chloroplast genome. Head-to-head circular dimers are found in spinach and lettuce cpDNAs, both of which contain the inverted repeat, but not in pea cpDNA, which lacks the repeat (Kolodner and Tewari, 1979). It is hypothesized (Kolodner and Tewari, 1979) that these dimers arise via intermolecular recombination within the inverted repeat, although the possibility that they form as replicative intermediates has not been ruled out. Since these dimers are found in two distantly related dicots, this may be a general recombination property common to all inverted repeat-containing cpDNAs.

Recombination also appears to occur intramolecularly within the inverted repeat. This conclusion is based on the recent findings that cpDNAs from a number of diverse plants and algae (Table IV) all exist as an equimolar population of two molecules differing only in the relative orientation of their single-copy sequences (Fig. 3). Although alternative explanations are possible (Palmer, 1983), it is highly likely that these inversion isomers arise via reciprocal recombination between intramolecularly paired inverted repeat segments (Fig. 3). The data suggest that this high-frequency inversion reaction is universal among inverted repeat-containing cpDNAs (Table IV). That it has not yet been found in a greater proportion of those cpDNAs that have been restriction mapped is merely a reflection of the technical difficulty of finding restriction enzymes that do not cleave within the large inverted repeat (Palmer, 1983; Palmer *et al.*, 1984). No known functions can be ascribed to this high-frequency

Table IV
Inverted Repeat-Containing CpDNAs in Which Inversion Heterogeneity Has Been Demonstrated

Taxon	Reference
Angiosperms	
Phaseolus vulgaris	Palmer (1983), Mubumbila *et al.* (1983)
Glycine max	Palmer *et al.* (1984)
Coriandrum sativum	J. Palmer (unpublished data)
Ferns	
Osmunda (three species)	J. Palmer, D. Stein, and W. Thompson (unpublished data)
Green algae	
Chlamydomonas reinhardtii	Aldrich *et al.* (1985), Palmer *et al.* (1985b)
Classification uncertain	
Cyanophora paradoxa	Bohnert and Loffelhardt (1982)

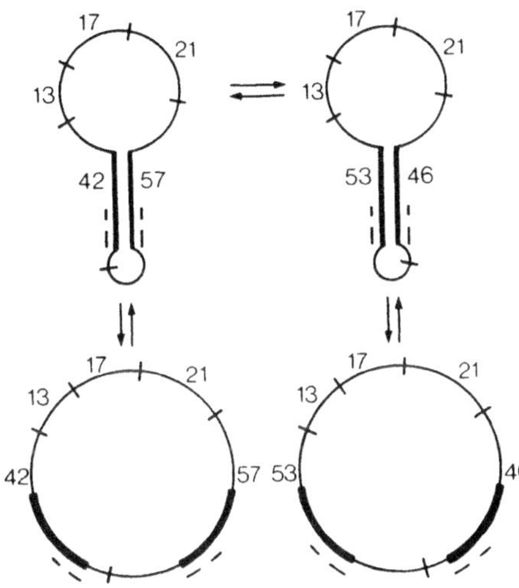

Figure 3. Model for cpDNA isomerization via intramolecular recombination between segments of the inverted repeat. Bottom: *Sal*I restriction maps of the two inversion isomers of *Phaseolus vulgaris* cpDNA. These molecules differ only with respect to the relative orientation of their single-copy DNA regions. Top: Dumbbell-shaped molecules postulated to occur as intermediates in the interconversion, via reciprocal recombination within intramolecularly paired inverted repeat segments, of the two inversion isomers shown at bottom. Long, heavy lines indicate the position of the inverted repeat. Short lines at the small single-copy end of the inverted repeat indicate positions of the 16S and 23S rRNA genes (cf. Fig. 1). Short radial lines on the circles represent *Sal*I sites and numbers give the size of the *Sal*I fragments in kb. [Reprinted with permission from Palmer (1983).]

recombination system and thus it is difficult to imagine any selective pressures that might be responsible for its retention over such a long period of evolutionary time. Perhaps this type of recombination is an unavoidable physical consequence of having such a large region of intramolecular homology. Alternatively, both the inverted repeat and its associated inversion activity may have originated and also be maintained by the recruitment and modified operation of a transposon encoding a site-specific recombination activity [see discussions in Szekeley and Simon (1983) and Plasterk *et al.*, (1983)].

2.4.1b. Other Repeats. In addition to the near-universal large inverted repeat, one tandem repeat and two dispersed repeat families have been found in certain cpDNAs. The only one of these repeats for which evolutionary information is available is a tandemly repeated ribosomal RNA gene cluster in *Euglena gracilis* cpDNA. Initial studies revealed the presence of three tandemly repeated complete sets of rRNA genes in cpDNA from both the Z strain (P. W. Gray and Hallick, 1978; Jenni and Stutz, 1978; Rawson *et al.*, 1978) and *bacillaris* strain (Helling *et al.*, 1979) of *E. gracilis*. The repeats are very similar, but not identical in each strain. Restriction analysis has revealed the existence of length polymorphisms in the spacer between rDNA operons that distinguish the three *bacillaris* repeats from one another and from the Z-strain repeats (Helling *et al.*,

1979). Extensive sequence analysis of the three, approximately 1-kb-long rDNA spacers of the *bacillaris* strain has revealed a 292-bp deletion in the middle spacer relative to the two outer ones and a minimum of 17 single base pair substitutions among the three spacers (El-Gewely *et al.*, 1984). Sequence analysis has also revealed a number of base pair substitutions that distinguish between at least two of the three 5S rRNA genes in the Z-strain repeats (Karabin *et al.*, 1983). Repeated sequences, especially those organized in tandem arrays, often evolve in concert as the result of such homogenization processes as gene conversion and unequal crossover [reviewed in Arnheim (1983)]. The degree of sequence divergence among the *Euglena* rDNA repeats suggests that these correction processes must occur fairly infrequently (if at all) at the *Euglena* rDNA locus, particularly when compared to the rapid homogenization of the two segments of the rDNA-containing inverted repeat present in chlorophytic cpDNAs (see previous section).

The leader region of the *Euglena* chloroplast rDNA operon contains a partial duplication of a portion of the transcribed rDNA, including the 3′ end of the 16S gene, the adjacent spacer, and a $tRNA^{Ile}$ gene (Orozco *et al.*, 1980; El-Gewely *et al.*, 1984). This partial duplication region extends, beginning at a segment that lies within the duplicated $tRNA^{Ile}$ gene, into a region of homology to a portion of a $tRNA^{Met}$–$tRNA^{Trp}$–$tRNA^{Glu}$ cluster located elsewhere in the *Euglena* chloroplast genome (Hollingsworth and Hallick, 1982; El-Gewely *et al.*, 1984). El-Gewely *et al.* (1984) have therefore proposed that the duplication region within the rDNA leader is the result of two insertion events, one involving a segment from the middle of the rRNA operon and the second involving the tRNA cluster, with the second occurring by recombination in a region of $tRNA^{Ile}$–RNA^{Met} homology. On the basis of sequence analysis (Orozco *et al.*, 1980) of the Z-strain rDNA leader, Miyata *et al.*, (1982) have proposed that this partial duplication region is evolving at an accelerated rate as a pseudogene cluster free from selective constraints. In contrast, based on their sequence analysis of the *bacillaris*-strain rDNA leader, El-Gewely *et al.* (1984) have hypothesized that the structural gene segments that were inserted in the rDNA leader have assumed a regulatory role in rDNA transcription, and that, although they may initially have gone through a period of very rapid evolution, these sequences are now evolving relatively slowly and under strong functional constraints. This latter argument is predicated on the assumption that the Z and *bacillaris* strains have diverged for more than 150 million years; i.e. since the two continents from which they were isolated last came in contact (El-Gewely *et al.*, 1984). However, this assumption seems somewhat questionable given the complete identity (aside from single *Eco*RI and *Xho*I site difference in the rDNA locus) of the Z and *bacillaris* cpDNAs at 65 restriction sites (Hallick, 1985).

The Z and *bacillaris* strains contain a partial rRNA gene set, including an entire fourth 16S rRNA gene, which precedes the three complete and apparently functional rRNA operons (Jenni and Stutz, 1979; Koller and Delius, 1982a–c; El-Gewely *et al.*, 1984; Koller *et al.*, 1984b). Although the extra 16S gene shows 98% sequence homology with the functional gene, it does not appear to be transcribed into stable rRNA and may therefore represent an incipient or newly formed pseudogene (Roux *et al.*, 1983).

Investigation of additional *E. gracilis* laboratory strains has revealed variation in the number of complete rDNA operons: one operon is found in the Z-S strain (Wurtz and Buetow, 1981; Koller *et al.*, 1984a); two in the permanently bleached mutant Y_3BUD (Ravel-Chapuis *et al.*, 1984); and five in the strain *bacillaris*-ATCC (Koller and Delius, 1982a). The *bacillaris*-ATCC strain also contains a second extra 16S gene. Based on their restriction mapping studies, Flamant *et al.*, (1984) have postulated that the *bacillaris*-ATCC cpDNA is derived by unequal crossover between two molecules of the Z/*bacillaris* type, each possessing three intact operons and one supplementary 16S gene (Fig. 4). Similarly, the two-operon genome found in the Y_3BUD mutant is postulated to result from unequal crossover occurring at a different position between two molecules of the three-operon genome (Ravel-Chapuis *et al.*, 1984).

Dispersed repeated sequences larger than 100 bp in size have been found in the chloroplast genomes of *Chlamydomonas reinhardtii* and *Trifolium subterraneum* (subclover). In *C. reinhardtii* these sequences occur as short (0.1–0.3 kb) inverted repeats present in 25–40 copies and scattered throughout the chloroplast genome (Rochaix, 1972, 1978; Rochaix and Malnoe, 1978; Gelvin and Howell, 1979). Clover cpDNA contains at least five copies of a 0.2- to 0.8-kb repeat element dispersed throughout at least

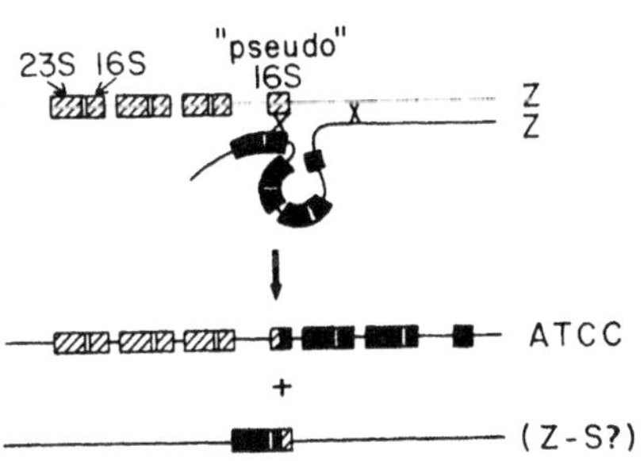

Figure 4. Unequal crossover in the chloroplast rDNA of *Euglena gracilis*. Top: Double reciprocal cross over between two identical cpDNA molecules of the Z strain of *Euglena gracilis*. Each molecule contains three tandemly repeated complete rDNA operons and an additional "pseudo" 16S rRNA gene. Bottom: Outcome of the Z-strain crossover shown at top, leading to one molecule (*bacillaris*-ATCC strain) with five complete operons and two extra 16S genes and one molecule with one complete operon. [Redrawn and modified from Flamant *et al.* (1984).]

one-fourth of the genome (J. Palmer and W. Thompson, unpublished data). Three aspects of the evolution of these repeats, and the genome as a whole, should be addressed by comparing related species and strains of *C. reinhardtii* and *T. subterraneum*: (1) Do these dispersed repeats "move" about the chloroplast genome in a fashion similar to transposable elements? (2) Does recombination between these repeated elements lead to cpDNA rearrangement? (3) Do the repeats evolve in concert; i.e., is there a significantly greater degree of sequence homogeneity between different repeat elements of the same genome or species than between the same repeat element present in two different species?

2.4.2. Deletions and Additions

Processes of deletion and addition must play an important role in chloroplast genome evolution, given the size variation found among angiosperms and green algae thus far (even discounting the effects of variation in the size of the large inverted repeat; see Section 2.3). In fact, however, the great majority of cpDNA length mutations appear to be very short and unlikely to contribute significantly to the observed variation in genome size. For example, a minimum of 40 deletions/additions were found in a sequence comparison of the 2294-bp transcribed rDNA spacer of tobacco and maize cpDNAs (Takaiwa and Sugiura, 1982a). One mutation is 229 bp in length, a second is 103 bp in length, while the remaining 38 are all between 1 and 10 bp in size. Similar results have been obtained by comparative sequence analysis of various other regions of the chloroplast genome (e.g., Zurawski *et al.*, 1982a,b, 1984a; Takaiwa and Sugiura, 1982b; Tohdoh and Sugiura, 1982; Shinozaki and Sugiura, 1982), leading to the general conclusion that most cpDNA length mutations, in both coding and noncoding sequences, are small deletions and insertions 1–10 bp in size. Many of these small length mutations are flanked by or close to small (at least an order of magnitude smaller than the repeats discussed in Section 2.4.1b) direct repeats (e.g., Takaiwa and Sugiura, 1982a; Zurawski *et al.*, 1984a). Experimental results and models from other systems suggest that these may occur primarily as deletions arising through slippage and mispairing of DNA strands during replication or repair (Streisinger *et al.*, 1966; Efstratiadis *et al.*, 1980; de Jong and Ryden, 1981; Albertini *et al.*, 1982; Jones and Kafatos, 1982; Jones *et al.*, 1982). Zurawski *et al.* (1984a) have also noted a tendency for small length mutations to occur preferentially in AT-rich spacer regions of the chloroplast genome.

Moderate-sized length mutations, on the order of 50–1200 bp in size, although infrequent compared to 1- to 10-bp size changes, appear to provide an important component of the total size variation observed among cpDNAs. For example, depending on the exact experimental approach used, these events can produce a significant proportion, sometimes even all, of the variation detected in studies on cpDNA restriction site variation [see Palmer *et al.* (1985a) for discussion; also see Fluhr and Edelman (1981), Gordon *et al.* (1982), Kung *et al.* (1982), Bowman *et al.* (1983), and Palmer *et al.* (1983a)]. Comparison with other genomes and systems suggests that these larger changes might occur primarily as deletions between short direct repeats (Albertini *et al*, 1982; Jones *et al.*, 1982), or, alternatively, as both deletions and insertions through unequal crossing over within tandem repeats (G. P. Smith, 1976; Long and Dawid, 1980; Szostak and Wu, 1980).

Restriction analysis has revealed two cases of length polymorphisms that probably occur by unequal crossover. CpDNA from the alga *Euglena gracilis* contains approximately 15 size forms of a single region of the genome (Jenni *et al.*, 1981; Schlunegger *et al.*, 1983). The regular spacing of these size forms on agarose gels suggests that the size heterogeneity is due to a variable number of copies of a 55-bp repeat element (Schlunegger *et al.*, 1983). It may be significant that this polymorphic region is quite AT-rich (Schlunegger *et al.*, 1983) and maps very close to the origin of replication of *Euglena* cpDNA (Koller and Delius, 1982d; Ravel-Chapuis *et al.*, 1982). A variably sized, AT-rich region is also close to the origin of replication of *Drosophila* mtDNA (Fauron and Wolstenholme, 1980), and most of the size variation found in the structurally constrained mtDNA of mammals is also located near the origin of replication (W. M. Brown, 1983). Size variation in a highly polymorphic region of the pea chloroplast genome may also reflect the presence of a tandemly repeated, approximately 50-bp sequence element (Palmer *et al.*, 1985a). The spatial relationship of this element to the origin of replication of pea cpDNA (Kolodner and Tewari, 1975a,b) is not known.

Length mutations larger than 1 or 2 kb in size are extremely rare in plant and green algal cpDNAs. Data were reviewed in Section 2.4.1a indicating that the distinctive, large inverted repeat has been lost, through the deletion of one entire segment that is duplicated, only once in several hundred million years of chlorophytic chloroplast evolution. Furthermore, only a single instance is known of a significantly enlarged inverted repeat [76 kb in *Pelargonium hortorum* (J. Palmer and J. Nugent, unpublished data) as compared to 20–30 kb in the great majority of angiosperms]. The only known major length mutation occurring outside the inverted repeat is found in *Nicotiana acuminata,* whose chloroplast genome is 11–13 kb larger than that of 30 other species in *Nicotiana* (Shen *et al.*, 1982) and also larger by the same amount than the genomes of numerous species in other genera in the Solanaceae (Bovenberg *et al.*, 1981; Fluhr and Edelman, 1981; Palmer and Zamir, 1982). The weight of this phylogenetic evidence strongly argues that this increase in genome size in *N. acuminata* represents a fairly recent insertion in this species [rather than a deletion in all other *Nicotiana* species, a view originally stated, erroneously, in my opinion, by Shen *et al.* (1982) and more recently expounded by Tassopulu and Kung (1984)].

Although large length mutations occur very rarely in nature, they can be readily produced in the green alga *Chlamydomonas reinhardtii* by growing cells in the presence of 5-fluorodeoxyuridine (Myers *et al.*, 1982). These mutant cpDNAs contain deletions and duplications of as much as 40 kb of the genome, some of which are accompanied by inversions. All of these physical alterations extend at least partially into the inverted repeat; indeed, the range of inverted repeat sizes found in these single-step *C. reinhardtii* mutants encompasses nearly the entire range of naturally occurring inverted repeat sizes found among all plants and algae examined (Table II). To explain the frequent involvement of the inverted repeat in large structural mutations, Myers *et al.* (1982) have hypothesized that intramolecular pairing between repeat segments allows double-stranded free ends resulting from damage to be joined and repaired, whereas broken ends in single-copy regions will often not be religated, yielding linear molecules unable to survive. All of these mutant cpDNAs were defective in function; indeed, they were selected on the basis of a nonphotosynthetic phenotype (Myers *et al.*, 1982).

Since it is clear that the capability for producing large length mutations in cpDNAs must exist, one wonders why these mutations do not occur more often in nature. That large deletions are rare may be a simple consequence of the relatively compact organization of the chloroplast genome into transcribed and otherwise functional sequences (Oishi *et al.*, 1981; Poulsen, 1983). That large insertions appear to be equally rare may be a consequence of several forces: (1) These events may simply not occur very often, either endogenously, by the duplication of preexisting chloroplast sequences, or exogenously, by the import of sequences from other genomes or DNAs. However, since DNA sequences appear to leave the chloroplast and enter other genomes fairly regularly (Table V; Sections 2.4.4 and 4.6), it is hard to believe that the reverse does not also occur. (2) There may be strong selection to maintain a fairly uniform and compact genome, so that once they occur, large insertions are soon pruned down in size by deletion events. What the basis for such a selection might be is totally unclear. (3) In the absence of any strong selection against increases in genome size, deletion forces may yet be so overwhelming that large insertions generally will survive as such for only a short time.

The ratio of length mutations to base substitutions in cpDNA approximates that in other genomes (Cann and Wilson, 1983; Zurawski *et al.*, 1984a; Palmer *et al.*, 1985a). Since it is now clear that the rate of cpDNA base substitutions is extremely low compared to other genomes (Section 3.2.4), one can indirectly infer that length mutations also occur at a relatively low rate in cpDNA.

As is the case with many aspects of its organization, expression, and evolution, *Euglena gracilis* cpDNA differs significantly from land plant and green algal cpDNAs in its ability to accommodate large size mutations, although at least in this instance the reasons for these differences are apparent. As discussed in the preceding section, the tandem nature of repeated rDNA operons in *Euglena* cpDNA is likely to be responsible for the observed variation in their copy number by virtue of the opportunities for unequal crossover this arrangement provides. In addition, Heizmann and coworkers (Heizmann *et al.*, 1981, 1982a,b; Hussein *et al.*, 1982) have shown that permanently bleached mutants of *Euglena* contain nonstoichiometric proportions of different regions of the chloroplast genome. In particular, the ribosomal RNA genes are generally found amplified relative to other sequences, and in at least one mutant, rDNA sequences are found on small extrachromosomal molecules (Heizmann *et al.*, 1982b). The origin of replication of *Euglena* cpDNA maps very close to the rDNA locus in wild-type cells (Koller and Delius, 1982d; Ravel-Chapuis *et al.*, 1982). Thus these defective *Euglena* genomes resemble mitochondrial genomes from most petite mutants of yeast, which also contain replication origins (Bernardi, 1982). As discussed in Section 2.4.4, *Euglena*'s unique ability to tolerate such defective chloroplast genomes appears to be a consequence of a differential cellular partitioning of various nonphotosynthetic functions relative to chlorophytes.

2.4.3. Inversions

Inversions appear to occur very rarely during chloroplast genome evolution. The great majority of the 200 angiosperm species that have been examined have the same linear order of chloroplast sequences, except of course for small deletions and additions, as discussed in the preceding section. This conclusion comes from four types of studies. The first involves comparison of restriction fragment patterns and maps among closely related species (usually within a genus). Approximately 160 species distributed among ten genera of flowering plants have been examined in this way (Gordon *et al.*, 1982; Kung *et al.*, 1982; Kumar *et al.*, 1982; Palmer, 1982; Palmer and Zamir, 1982; Palmer *et al.*, 1982b, 1983a, 1985a; Bowman *et al.*, 1983; Tsunewaki and Ogihara, 1983; Clegg *et al.*, 1984)

Table V
Interorganellar DNA Sequence Homologies

Taxon	Size of homology	Genes present on homology region	Reference
CpDNA–mtDNA sequence homologies			
Maize	12 kb	16S rRNA, tRNAs	Stern and Lonsdale (1982)
Maize	2 kb	*rbcL*	Lonsdale *et al.* (1983)
Eight species[a]	Many homologies of undetermined size	*rbcL*	Stern *et al.* (1983)
Four species[b]	Many homologies of undetermined size	*rbcL*, *atpB*	Stern and Palmer (1984a)
CpDNA–nuclear DNA sequence homologies			
Spinach	Many homologies of undetermined size	*rbcL*, *atpB*, 23S rRNA	Timmis and Scott (1983), Scott and Timmis (1984)
Chlamydomonas reinhardtii[c]	One homology of undetermined size	Beta subunit of RNA polymerase	Watson and Surzycki (1983)
MtDNA–nuclear DNA sequence homologies			
Maize	Many homologies of undetermined size	S_1 plasmid	Kemble *et al.* (1983a)

Neurospora crassa	Several hundred bp	ATPase subunit 9	Van den Boogaart *et al.* (1982)
Yeast	Several hundred bp	*var 1*, cytochrome *b*	Farrelly and Butow (1983)
Locusta migratoria	Several kb	Large and small rRNAs	Gellissen *et al.* (1983)
Podospora anserina	12.4 kb	Cytochrome oxidase subunits I and II	Wright and Cummings (1983)
Sea urchin	Several kb	16S rRNA, cytochrome oxidase subunit I	Jacobs *et al.* (1983)
Human	Several homologies of undetermined size	16S rRNA	Tsuzuki *et al.* (1983)
Rat	Less than 3 kb	rRNA	Hadler *et al.* (1983)

[a] Spinach cpDNA sequences were used to identify homologous sequences in the mitochondrial genomes of spinach, mung bean, pea, maize, watermelon, zucchini, cucumber, and muskmelon.

[b] Spinach and mung bean cpDNA sequences were used to identify homologous sequences in the mitochondrial genomes of spinach, mung bean, pea, and maize.

[c] A probe for the β subunit gene of *E. coli* RNA polymerase hybridized to both cpDNA and nuclear DNA of *Chlamydomonas*, although it was not determined whether the same or different portions of the probe were responsible for the observed hybridizations.

and the only inversion found is one of 2.2–5.2 kb in the wild ancestor (*Pisum humile*) of the garden pea, *P. sativum* (Palmer *et al.*, 1985a).

Three experimental approaches—comparative denaturation mapping (Kolodner and Tewari, 1979), heteroduplex analysis (Lamppa and Bendrich, 1981), and heterologous filter hybridization—have been used to compare the organization of distantly related chloroplast genomes. Comparative filter hybridization studies fall into two classes according to the kind of hybridization probe used. One set of studies has used as probes various large (5–20 kb), but poorly defined (in terms of gene content) restriction fragments, which together often encompass an entire chloroplast genome, while the other set has used various small (0.1–2 kb), but well-defined gene-specific molecules (either restriction fragments or individual tRNAs). The great majority of the approximately 30 family and 60 genus representatives examined by filter hybridization studies of the first type do not contain any detectable inversions or any other major rearrangements (Fluhr and Edelman, 1981; Palmer and Thompson, 1981a, 1982, and unpublished data; Palmer *et al.*, 1983a,b; de Heij *et al.*, 1983). The best documented inversion among these species is a 50-kb inversion, located within the large single-copy DNA region, that was postulated to explain the observed hybridization pattern between mung bean and spinach cpDNA restriction fragments (Fig. 5). Additional support for the inversion hypothesis in this particular case came from the demonstration that the *rbcL* and *psbA* genes, which in spinach are 50 kb apart and are

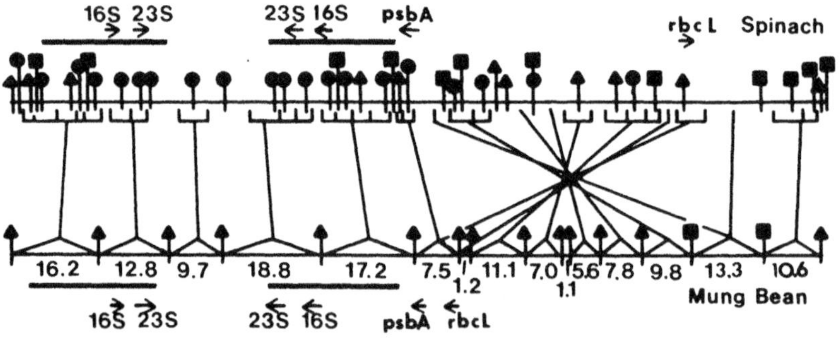

Figure 5. Mapping of a 50-kb inversion between mung bean and spinach cpDNAs. Each of 14 cloned (Palmer and Thompson, 1981b) mung bean fragments (sizes given in kb) was hybridized to a nitrocellulose filter containing spinach cpDNA restriction fragments. The spinach fragments to which the mung bean probes hybridize are indicated by the lines leading from the mung bean fragments to the spinach fragments. See Fig. 1 for sources of spinach mapping data. Mung bean mapping data are from Palmer and Thompson (1981a) and Palmer *et al.* (1982a). Restriction sites shown: (↑) *Pst*I; (↑) *Sal*I; (↑) *Sma*I. [Reprinted with permission from Palmer and Thompson (1982).]

transcribed in opposite directions (Fig. 1), are only 5 kb apart and are transcribed in the same direction in mung bean (Fig. 5) (Palmer *et al.*, 1982a). More extensive examination has shown that the spinach genome represents the ancestral angiosperm chloroplast genome, as found in the majority of monocots and dicots, and that the 50-kb inversion is shared (and derived) by all species examined from 12 genera and six tribes of the subfamily Papilionoideae of the legume family Fabaceae (Palmer and Thompson, 1982, and unpublished data; Palmer *et al.*, 1983b; Spielmann *et al.*, 1983).

Gene mapping studies indicate the presence of a large inversion of similar size and location to the legume inversion in the genus *Oenothera*, also when compared to spinach (Herrmann *et al.*, 1983). However, since cpDNA from *Fuchsia*, a member of the same family (Onagraceae) as *Oenothera*, is colinear with spinach cpDNA (J. Palmer and W. Thompson, unpublished data), is is likely that this represents a second, independent occurrence of an inversion in the same region of the genome. Other documented inversions in chloroplast genomes that feature the inverted repeat include the following: a 20-kb inversion located within the large single-copy region and characteristic of three genera of the Poaceae (Palmer and Thompson, 1982, and unpublished data; Howe *et al.*, 1983; Poulsen, 1983; de Heij *et al*, 1983); a 30-kb inversion also located within the large single-copy region and characteristic of six tribes of the Asteraceae (J. Palmer and W. Thompson, unpublished data); and between one and three rearrangements (probably inversions, although transpositions and duplications have not been ruled out) of undetermined size and location in each of the families Ranunculaceae, Geraniaceae, and Campanulaceae (J. Palmer and W. Thompson, unpublished data).

Transfer RNA mapping studies (Selden *et al.*, 1983) indicate that a small rearrangement, perhaps an inversion [see discussion in Zurawski *et al.* (1984b)], involving the gene for $tRNA^{His}$ may have occurred at the junction of the inverted-repeat–large-single-copy region in maize relative to spinach and other dicots. Aside from this one difference, the inverted repeats of maize, spinach, and lettuce appear highly similar in size and linear organization as judged by comparative denaturation mapping (Kolodner and Tewari, 1979). Transfer RNA mapping studies of inverted repeat-containing cpDNAs from spinach (Driesel *et al.*, 1979), maize (Selden *et al.*, 1983), common bean (Mubumbila *et al.*, 1983), and tobacco (Bergmann *et al.*, 1984) suggest that a few other small sequence rearrangements (transpositions?) may have occurred during the evolution of these species. The specific nature of these rearrangements needs further investigation by sequencing and by more extensive filter hybridizations using sequences adjacent to the tRNAs genes as probes.

The rarity of major sequence inversions in angiosperms that contain the large inverted repeat contrasts with a greater frequency of rearrangement in several of those legume genomes that have lost the repeat structure (Palmer and Thompson, 1981a, 1982, and unpublished data). As an example, Fig. 6 shows that sequences common to two species (pea and broad bean) that lack the inverted repeat have become extensively scrambled since their divergence from one another. This comparison is particularly striking because pea and broad bean are classified in the same taxonomic tribe and because their cpDNAs are quite highly related at the nucleotide sequence level (Palmer and Thompson, 1982). Additional evidence that extensive rearrangements have occurred in pea and broad bean cpDNAs comes from mapping studies of protein genes in pea (Huttly and Gray, 1984; Phillips and Gray; 1984a; Willey *et al.*, 1984a) and tRNA genes in broad bean (Mubumbila *et al.*, 1984), and from the analysis of heteroduplexes formed between pea and maize cpDNAs (Lamppa and Bendich, 1981).

Further studies (Palmer *et al.*, 1983b; J. Palmer and W. Thompson, unpublished data) on these and other legume cpDNAs have led to the following conclusions regarding the structural evolution of legume cpDNAs: (1) Instability is not a general feature of all legume cpDNAs, but is confined to only certain of those genomes that have lost the inverted

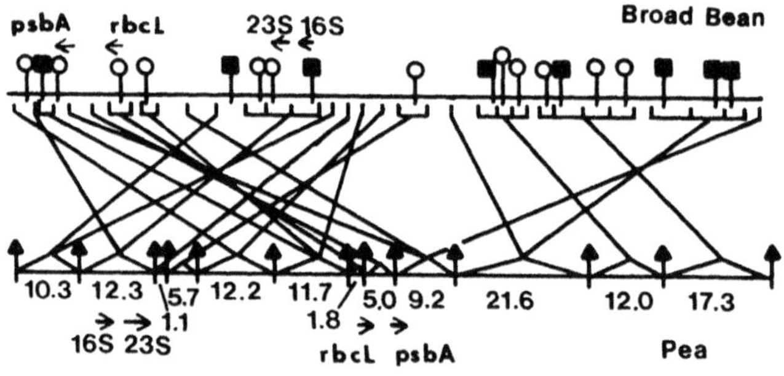

Figure 6. Scrambled arrangement of sequences in pea and broad bean cpDNAs. Each of 12 cloned (Palmer and Thompson, 1981b) pea fragments (sizes given in kb) was hybridized to a nitrocellulose filter containing broad bean cpDNA restriction fragments. The broad bean fragments to which the pea probes hybridize are indicated by the lines leading from the pea fragments to the broad bean fragments. Broad bean mapping data are from Koller and Delius (1980) and J. Palmer and W. Thompson (unpublished data). Pea mapping data are from Palmer and Thompson (1981a) and Palmer *et al.* (1982a). Restriction sites shown: (⚲) *Kpn*I; (↑) *Pst*I; (⬆) *Sal*I. [Reprinted with permission from Palmer and Thompson, (1982).]

repeat. (2) The loss of the inverted repeat probably occurred only once in the legume family, *prior* to all other rearrangements observed in species such as pea and broad bean. (3) No further rearrangements have occurred in two separate legume lineages, represented by alfalfa and wisteria, following the loss of the inverted repeat. (4) Two large inversions occurred in the broad bean genome subsequent to the loss of the repeat. These appear to be independent of any of the pea rearrangements. (5) Many (~12) rearrangements have occurred in the pea genome. The only one whose origin and nature is clear is the small inversion in *P. humile* described earlier. Although the nature of the remaining rearrangements is unclear, the hybridization data are not inconsistent with the majority of the changes being inversions. (6) A third, independent lineage of rearranged legume genomes is represented by clover. The specific nature of the rearrangements in clover is unclear, although they include small sequence duplications (see Section 2.4.1b), and thus possibly transpositions, and probably inversions as well.

Very little is known about sequence rearrangements in cpDNAs of nonangiosperms. The inverted repeat and four protein genes located in the large single-copy region (*rbcL, psbA, atpA, atpB*) are in similar positions in the fern *Osmunda cinnamomea* (Palmer and Stein, 1982), the gymnosperm *Ginkgo biloba* (J. Palmer and W. Thompson, unpublished data), and the angiosperm spinach (Fig. 1). The inverted repeat and *rbcL* genes also have similar locations in the liverwort, *Marchantia polymorpha* (Ohyama *et al.*, 1983). Thus, the basic uniformity of chloroplast genome arrangement found among angiosperms may extend throughout land plants, although more complete comparisons are clearly needed.

As discussed in Section 2.4.1a, the rRNA genes appear to be inverted and sequences, among them *atpB*, redistributed between the two single-copy regions in *Chlamydomonas reinhardtii* relative to land plants (Woessner *et al*, 1984). In addition, there are positional differences with respect to the *psbA, atpA,* and *atpH* genes (Woessner *et al.*, 1984; J. Erickson and J. Rochaix, unpublished data). These data suggest at least a moderate amount of cpDNA rearrangement in *Chlamydomonas* relative to land plants, although the precise nature of the changes, and also when and in what particular lineages they occurred, are completely open issues.

Why are major sequence rearrangements so rare among most land plant cpDNAs, particularly those with the large inverted repeat? Several theories have been advanced by Roth and Schmid (1981; Schmid and Roth, 1983a,b) to explain a similar evolutionary conservation in the linear arrangement of bacterial chromosomes and to explain why bacterial inversions are so rare even when there is genetic selection for them. The following discussion will include several of these theories as well as others more particular to the chloroplast genome (Palmer and Thompson, 1982; Palmer *et al.*, 1984).

One set of theories invokes the disruptive effects of rearrangements on gene function. For example, genes whose products interact closely may become coadapted in a way that selects for their close linkage, as protection against the breakup of these advantageous combinations by outcrossing. This obviously makes no sense for land plant chloroplast genes, where sexual recombination is unknown (see Section 2.1). In addition, relatively few chloroplast genes are grouped according to function (Fig. 1; see Section 3.2.1).

Gene dosage effects might also constrain genomic rearrangement. CpDNAs from angiosperms (Kolodner and Tewari, 1975a, b) and *Chlamydomonas* (Waddell *et al.*, 1984) replicate bidirectionally from two closely spaced and oppositely oriented fixed origins. Therefore, a gradient of gene dosages must exist according to chromosomal position relative to the replication origins. A fine tuning of expression levels of chloroplast genes may exist that takes into account these dosage effects and would be disrupted by chromosomal rearrangements. While this argument makes a certain amount of sense for bacterial cells, with their predictable growth cycles and regular rounds of DNA replication, it is harder to apply to chloroplasts, which are biosynthetically quite active for a prolonged period of leaf and cellular expansion during which cell division and cpDNA replication have ceased (Possingham and Lawrence, 1983). On the other hand, gene replication/dosage effects may be important during leaf cell division and at the onset of the cell expansion phase, when a threefold overreplication of cpDNA occurs relative to nuclear DNA (Possingham and Lawrence, 1983). If gene dosage does constrain genomic rearrangement, then "movement" of the ends of the inverted repeat through genes should be even more constrained, since this produces complete gene duplication, or the reverse, and we would not expect to find the sevenfold variation in inverted repeat size observed among angiosperms (Section 2.4.1a). However, it should be stressed that in most angiosperms the inverted repeat falls into a relatively narrow size range of between 20 and 30 kb and that this level of size variation could be accomplished by internal deletions/additions, without "moving" the repeat ends relative to fixed genes. For example, sequence studies (Schwarz *et al.*, 1981; Spielmann and Stutz, 1983; Sugita *et al.*, 1984; Zurawski *et al.*, 1984b) reveal that the large single-copy end of the inverted repeat is situated within the same 2- to 3-kb region of the genome in three dicots and one monocot that diverged on the order of 50–150 mya (Doyle, 1978; Muller, 1981).

A second set of theories involves the disruptive effects of genome rearrangement on chromosome structure and function, such as its ability to fold properly within the cell. There is evidence that the *E. coli* chromosome is folded into about 40 independently supercoiled domains (Sinden and Pettijohn, 1981) and it is thought that major rearrangements that disturb the spacing of sites that determine this folding might be deleterious (Roth and Schmid, 1981). Ultrastructural studies suggest that the chloroplast chromosome has a folded, knotted structure reminiscent of that of the bacterial chromosome (Yosida *et al.*, 1978; Yoshida, 1981; Briat *et al.*, 1982b). Also, a protein component of this folded chloroplast chromosome shows strong size and immunological similarities with the histone-like protein HU of bacterial chromosomes (Briat *et al.*, 1984). So far, however, there is no evidence as to whether folding of the chloroplast chromosome is in any way sequence-specific.

Schmid and Roth (1983b) found that most large bacterial inversions include the origin (or sometimes terminus) of DNA replication. They speculate that this may indicate a need to maintain a particular orientation of certain sequences with respect to the replication fork. In this regard, it is perhaps significant that all four well-mapped inversions in inverted repeat-containing cpDNAs are located within the same one-third of the genome; i.e., the 50-kb domain encompassed by the mung bean inversion shown in Fig. 5. On the other hand, rearrangements occur throughout the chromosome in genomes that lack the inverted repeat (Fig. 6); this may indicate a fundamental difference in rearrangement properties of these two genome types (see below).

Mechanistic arguments can also be advanced to explain the relative infrequency of chloroplast genome rearrangement. One could argue that the recombination systems that promote these changes are poorly developed in chloroplasts. However, it is clear that processes of intra- and intermolecular recombination involving the inverted repeat are extremely well developed in angiosperm chloroplasts (Fig. 3; Section 2.4.1a). Furthermore, recombination appears to occur throughout the chloroplast genome in species of *Chlamydomonas* (Lemieux *et al.*, 1981, 1984), and given appropriate mutagenic and selective conditions, it is possible to induce large inversions, deletions, and duplications in *Chlamydomonas* cpDNA in a single-step fashion (Myers et al., 1982).

Bacterial rearrangements, in particular inversions, appear to occur by homologous recombination between short (~100 bp) sequences repeated in several locations around the chromosome (Schmid and Roth, 1983b). Perhaps chloroplast rearrangements are rare because such sequences are rare. In fact, self-hybridization of cloned restriction fragments has failed to reveal any repeated sequences (aside from the large inverted repeat) in cpDNAs of nine divergent angiosperms (Palmer and Thompson,

1981a, 1982, and unpublished data; Coates and Cullis, 1982; de Heij *et al.*, 1983; Spielmann *et al.*, 1983; Ko *et al.* 1983; Fluhr *et al.*, 1983a). The only exceptional angiosperm is clover, where the presence of several dispersed repetitive sequences (see Section 2.4.1b) may be related to the extensive rearrangements seen in its cpDNA (J. Palmer and W. Thompson, unpublished data). It should be pointed out, however, that the filter hybridization approach may fail to detect very short dispersed repeats that could still be important as sites for recombination.

The relatively compact organization of plant chloroplast genomes with respect to coding sequences (Oishi *et al.*, 1981; Poulsen, 1983; Dix and Rawson, 1983; Palmer *et al.*, 1984) should place severe constraints on the allowable recombination and rearrangement events. While this factor must certainly contribute to a general lowering of rearrangement frequencies for the chloroplast genome, it is also true that sequencing studies (see Section 3 for specific references) have revealed that reasonably sized spacer sequences (of a few hundred base pairs) are present between many chloroplast genes. By way of comparison, major sequence rearrangements are essentially unknown in mammalian mtDNA, where noncoding sequences are almost totally absent (Bibb *et al.*, 1981, Anderson *et al.*, 1981, 1982; for a review see W. M. Brown, 1983).

The relative lack of rearrangements in inverted repeat-containing cpDNAs, coupled with an apparent increase in their frequency in genomes lacking the repeat (Fig. 6), led to a hypothesis that the inverted repeat might, in some unknown fashion, stabilize the chloroplast genome against rearrangements (Palmer and Thompson, 1982). Given the recent observation that intramolecular recombination occurs at a high frequency between inverted repeat segments (Fig. 3; Section 2.4.1a), one can now imagine in more specific terms how such a stabilization might result. The large size of the inverted repeat and the extended opportunities for intramolecular base pairing it provides, may, whenever any significant denaturation of the duplex occurs, constrain the chromosome into the dumbbell-shaped conformation shown in Fig. 3. In other words, conformational opportunities for intramolecular recombination and rearrangement within the entire genome may be effectively suppressed by the dominance of pairing and recombination within the repeat. According to this idea, loss of the inverted repeat should relax the genome and encourage rearrangements throughout the molecule. While this prediction appears to be borne out in certain genomes (from pea, broad bean, and clover) that lack the repeat and are considerably rearranged, it is at odds with the finding that alfalfa and wisteria also lack the repeat but are otherwise unrearranged (see above). For a further elaboration of how the inverted repeat might act as a stabilizing force in chloroplast genome evolution see Palmer *et al.* (1984).

In conclusion, there are no clear answers as to why inversions and other rearrangements are so rare in most chloroplast genomes. However, we can at least point to certain factors—most prominently the compact gene organization of cpDNA and a lack of repeated sequences to serve as substrates for recombination, but also gene dosage effects, constraints on chromosome folding and replication, and structural and recombinational constraints imposed by the large inverted repeat—which in combination may be responsible for the structural stability of chloroplast chromosomes.

2.4.4. Gene Content and Gene Transfer

The chloroplast genome, at an average size of 150 kb, has the potential to encode more than 100 different proteins and stable RNAs. Measurements of the total proportion of the chloroplast genome transcribed into RNA indicate that a large portion of this coding potential is realized in such diverse organisms as *Euglena gracilis* (Chelm and Hallick, 1976; Rawson and Boerma, 1976; Dix and Rawson, 1983; Koller and Delius, 1984), *Chlamydomonas reinhardtii* (Howell and Walker, 1977), and angiosperms (Oishi *et al.*, 1981; Poulsen, 1983). Furthermore, isolated chloroplasts from pea may synthesize as many as 80 different soluble proteins (Ellis *et al.*, 1980), although the identity of most of these apparent translation products with authentic chloroplast polypeptides has not yet been established and it is possible that a large number of these might be artifactual products resulting from premature termination of translation. Chloroplast gene products identified thus far include all the chloroplast rRNAs, probably all the chloroplast tRNAs, and 15 polypeptides, with another 20 or so polypeptides known to be synthesized within the chloroplast (Fig. 1) (Bohnert *et al.*, 1982a; Bottomley and Bohnert, 1982; Whitfeld and Bottomley, 1983; Crouse *et al.*, 1985). With one exception—chloroplast elongation factor G, which appears to be encoded by cpDNA in spinach (Ciferri *et al.*, 1979) and *Chlorella vulgaris* (Ciferri and Tiboni, 1976), but by nuclear DNA in *Euglena gracilis* (Breitenberger *et al.*, 1979) and perhaps *Chlamydomonas reinhardtii* (Carbonera *et al.*, 1981)—the same complement of these genes has been found in all chlorophytic and euglenophytic chloroplast genomes (see above references). The converse to this conclusion is that there exists a similar phylogenetic uniformity with regard to those chloroplast proteins that have been identified as encoded by the nuclear genome (Kirk and Tilney-Bassett, 1978; Gillham, 1978; Bottomley and Bohnert, 1982).

The apparent constancy of the division of genetic labor between chloroplast and nucleus suggests that the original pace of transfer of genes between cellular compartments (regardless of whether chloroplasts originated endosymbiotically or autogenously) has slackened almost to zero in these organisms. It is difficult to imagine how or why chloroplast genomes have always retained the same set of genes. The stable RNAs (ribosomal and transfer) are known to be encoded within all organelle genomes, and since there is no precedent for the import of any RNA molecules into organelles (Bottomley and Bohnert, 1982), it can be argued that the lack of effective mechanisms for their intracellular transport has prevented the transfer of these genes. One hypothesis to explain the dispersal of polypeptide genes between genomes is that developed by Bogorad (1975). His argument contains two major points: (1) Certain polypeptides may remain produced by the chloroplast because they are untransportable, that is, mechanisms for their transport from the cytoplasm into the chloroplast could not evolve fast enough to accommodate the transfer of their genes into the nucleus. (2) Certain genes might also be retained in the chloroplast because of some sort of fundamental requirement that their polypeptide products be initially present in order to "lock-in" nuclear-encoded, cytoplasmically synthesized polypeptides by any of a variety of mechanisms (Bogorad, 1975). This latter argument is consistent with the observation that one or more polypeptide components of most of the major protein complexes involved in chloroplast gene expression and photosynthesis—the chloroplast ribosome, ribulose-1,5-bisphosphate carboxylase, the chloroplast ATPase, the cytochrome b_6-f complex, and the photosystem I and II complexes—are known to be encoded in the chloroplast genome, while at least one other component is encoded in the nuclear genome.

An alternative hypothesis is that at some point during chloroplast evolution the entire complement of chloroplast gene products became stabilized in the chloroplast genome as a result of some profound divergence in the machineries for chloroplast and nuclear gene expression that rendered the expression of a gene from one compartment impossible in the other. There is, however, little evidence in support of this idea. Chloroplast mRNAs can be expressed faithfully in eukaryotic translation systems, both chloroplast and nuclear genes can be transcribed and translated in *E. coli* systems, and both use the same genetic code [reviewed in Petes (1980), Whitfeld and Bottomley (1983), and Crouse *et al.* (1985)].

A selectionist hypothesis to explain patterns of intergenomic gene dispersal could be advanced on the basis of known differences in the way the two genomes evolve, i.e., chloroplast genes normally evolve in an asexual environment without any recombination between genes of different lineages (Gillham, 1978; Birky, 1983) and at a rate that appears to be significantly lower than for most nuclear genes (see Section 3.2.4). Thus, one could imagine that the conservative evolution of the chloroplast genome selects for the retention of genes whose products are highly constrained in evolution. Evidence for this idea comes from the observation that the major catalytic and structural polypeptides of several chloroplast complexes—the catalytic subunit (large subunit) of ribulose-1,5-bisphosphate carboxylase, the catalytic (β subunit) and proton-translocating subunits of the ATPase, the major chlorophyll-binding polypeptides of the two reaction centers, and several of the rRNA-binding ribosomal proteins—are encoded by the chloroplast genome (see Section 3.1). On the other hand, certain polypeptides, such as the ϵ subunit of the ATPase, of less obvious critical function and whose evolution is generally more rapid (see Section 3.1.2b), are also encoded by all known cpDNAs.

Although there is no evidence for any recent transfer of gene function from the chloroplast into the nucleus among chlorophytes and euglenophytes, it should be emphasized that only a small sampling of these organisms, as well as of their chloroplast genes, have been examined. Indeed, there is now ample physical evidence that DNA sequences can and do become transferred between cellular compartments and their genomes. Cross-hybridizations between cloned chloroplast sequences and purified nuclear and mtDNAs indicates the existence of extensive interorganellar DNA sequence homologies in a variety of angiosperms (Table V). There is, however, no evidence that any of these chloroplast-homologous sequences, several of which appear to represent recently transferred chloroplast genes (see Section 4.6), are actually expressed where present in nuclear or mtDNA.

Relatively little is known about the chloroplast gene content of organisms other than chlorophytes and *Euglena gracilis*. CpDNA (cyanelle DNA) of *Cyanophora paradoxa* appears to encode a complete set of tRNAs, many of which have strong sequence homologies with angiosperm chloroplast tRNAs (Kuntz *et al.*, 1984). More limited similarities have been noted in the positional arrangement of tRNA genes in angiosperm and cyanelle cpDNAs. Most prominently, the rDNA spacer of *C. paradoxa* cpDNA (Kuntz *et al.*, 1984) contains the same two tRNA genes as do all other chloroplast rDNA spacers (Fig. 7). *Cyanophora paradoxa* cpDNA also hybridizes strongly with probes for spinach chloroplast genes *psbA*, *atpA*, *atpB*, and *atpE* and those for ribosomal RNA (unpublished results of H. Bohnert and W. Loffelhardt, cited in Bohnert *et al.*, 1982a).

Heinhorst and Shively (1983) have found that the cyanelle genome of *C. paradoxa* contains tightly linked genes for both the small and large subunits of ribulose-1,5-bisphosphate carboxylase, whereas in all chlorophytes and euglenophytes the small subunit gene(s) is located in the nucleus [reviewed in Ellis (1981) and Bottomley and Bohnert (1982)]. Furthermore, the small and large subunit genes are located adjacent to

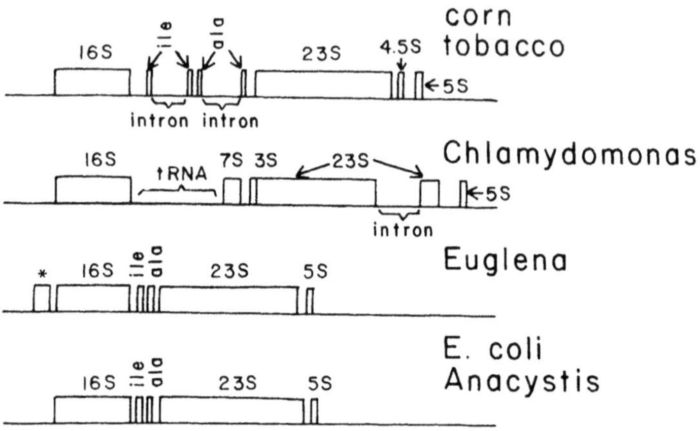

Figure 7. Ribosomal RNA transcription units of chloroplasts and eubacteria. Corn (maize) data are from Kossel *et al.* (1982), tobacco data are from Takaiwa and Sugiura (1982a), *Chlamydomonas* data are from Rochaix and Malnoe (1978) and Rochaix and Darlix (1982), *Euglena* data are from Karabin *et al.* (1983), *E. coli* data are for the *rrnA*, *rrnD*, and *rrnH* operons and are from Morgan *et al.* (1977), and *Anacystis* data are from Williamson and Doolittle (1983) and Tomioka and Sugiura (1984). The asterisk denotes a pseudogene cluster in the *Euglena* rDNA leader (see Section 2.4.1b). [Redrawn and modified from Bohnert *et al.* (1982a).]

one another and are apparently cotranscribed in cpDNA of the chromophyte *Olisthodiscus luteus* (M. Reith and R. Cattolico, unpublished data). These findings are phylogenetically consistent with two other recent observations. First, Steinmuller *et al.* (1983) have shown that in two red algae, *Porphyridium aerugineum* and *Cyanidium caldarium*, the small subunit is translated *in vitro* from mRNA that lacks poly A tails, a condition generally characteristic of chloroplast and not nuclear mRNAs (Ellis, 1981; Grossman *et al.*, 1982). More cogently, the small-subunit and large-subunit genes appear to be cotranscribed as a single operon in the cyanobacteria *Anacystis nidulans* (Shinozaki and Sugiura, 1983) and *Anabaena* (Nierzwicki-Bauer *et al.*, 1984). At this early juncture, these results establish one significant link among the genomes of *Cyanophora*, chromophytes, rhodophytes, and cyanobacteria relative to those of chlorophytes and euglenophytes.

The presence of the complex light-harvesting structures known as phycobilisomes is another feature that distinguishes *Cyanophora*, cyanobacteria, rhodophytes, and certain chromophytes (the cryptomonads) from chlorophytes and euglenophytes. Grossman and collaborators (Egelhoff and Grossman, 1983; Grossman, *et al.*, 1983) have shown that many of the polypeptide components, including the major pigmented polypeptides and at least one structural polypeptide (the anchor protein), of phycobilisomes from *Cyanophora paradoxa* and three rhodophytes are synthesized within the chloroplast. Since all chloroplast-synthesized polypeptides have so far been found to be encoded by cpDNA, it is suggested that the genes for these phycobilisome polypeptides are also encoded in these algal cpDNAs. Since chlorophytes and *Euglena* do not possess phycobilisomes, it can be safely assumed that this represents another fundamental difference in the coding content of chloroplast genomes. Furthermore, because the major pigmented polypeptides in phycobilisomes comprise a group of closely related polypeptides, the corresponding plastid genes may constitute a multigene family (Egelhoff and Grossman, 1983), a feature hitherto unknown to organelle genomes.

Two cases of altered cpDNA gene content can be predicted based on a known reduction in physiological function of the plastid. Numerous algae and plants are permanently nonphotosynthetic, but have retained plastids (Kirk and Tilney-Bassett, 1978; Droop, 1974; Menke and Wolfersdorf, 1968). In at least two of these algae, *Polytoma obtusum* and *Chilomonas paramecium*, the starch-containing plastids contain DNA and ribosomes (Sepsenwol, 1973; Siu *et al.*, 1976b). Furthermore, *Polytoma* plastid DNA is the same size as *Chlamydomonas* cpDNA and has significant sequence homologies with it (Siu *et al.*, 1975). Since a number of chloroplast gene products function in photosynthesis (Fig. 1; Section 3.1.2), one should expect to find a considerable reduction in the functional gene content of *Polytoma* plastid DNA; i.e., in the absence of continued selection for their function, photosynthetic genes may either be entirely deleted or be evolving as dead pseudogenes (Li, 1983). As noted in Section 2.3, the extreme AT-richness of *Polytoma* plastid DNA (Siu *et al.*, 1975) is consistent with a large portion of the genome having no function.

That plastid genomes are retained and expressed (plastid ribosomes are present) in nonphotosynthetic organisms strongly suggests that the genome must provide other indispensable functions to the plastid and the cell. It is well known, for example, that plastids carry out many other metabolic functions in addition to photosynthesis. These include steps in the biosynthesis of fatty acids and essential amino acids, nitrate reduction, sulfate reduction, and starch metabolism and storage (Kirk and Tilney-Bassett, 1978; Jensen, 1980; Weeden, 1981; Sears, 1983). So far, however, there is no direct evidence that the plastid genome encodes components controlling any of these processes (Kirk and Tilney-Bassett, 1978; Bottomley and Bohnert, 1982).

Euglena gracilis may represent the other side of the coin from organisms such as *Polytoma*. *Euglena* is unique among photosynthetic organisms in being able to survive, even thrive, after "bleaching" with various agents that cause the loss of most of the plastid structure [reviewed in Gillham (1978) and Schiff (1980)] and also large parts of the plastid genome (see Section 2.4.2). This suggests that *Euglena* cpDNA probably does not encode those nonphotosynthetic functions postulated in the previous paragraph to be under plastid control in *Polytoma* and perhaps all other plastid-containing organisms. Consistent with this speculation, various nonphotosynthetic functions (e.g., assimilatory sulfate reduction; Brunold and Schiff, 1976) that occur within the plastid in plants and green algae are known to occur external to the plastid in *Euglena*. Thus, we can recognize three basic kinds of plastid genomes in terms of their gene content and function. In most organisms, the plastid genome provides both photosynthetic and putatively nonphotosynthetic functions, in colorless plants and algae such as *Polytoma* it may provide only nonphotosynthetic functions, and in *Euglena* it appears to be responsible only for photosynthesis.

3. Chloroplast Gene Evolution

The study of the structure, sequence, and function of chloroplast genes is a rapidly growing field. A significant number of chloroplast genes have now been sequenced in several different plants and algae. This sequence information is revealing not only how these genes are evolving in different chloroplast lineages, but also how they have changed relative to their eubacterial ancestors, or at least the extant descendants of their ancestors. Each chloroplast gene or set of related genes for which DNA sequence information is available will first be discussed individually, in order to provide a clear picture of how that specific gene is evolving. Then, our composite knowledge of how chloroplast genes are evolving will be summarized with reference to general features of their organization, expression, and rate of sequence change.

3.1. Evolution of Individual Chloroplast Genes

As already discussed, chloroplast gene products thus far identified fall into two classes—those that function in chloroplast protein synthesis and those that function in photosynthesis. The plastid performs a great variety of other metabolic functions, and, as discussed in Section 2.4.4, there is some reason to believe that the plastid genome might encode some of the components necessary for these other processes. However, direct evidence that it does, much less any evidence as to the identity of these putative gene products, is completely lacking.

3.1.1. Genes for Chloroplast Protein Synthesis

Chloroplast genomes encode all of the ribosomal and transfer RNAs, about one-third of the ribosomal proteins, and certain of the elongation factors found in the chloroplast. In contrast, all of the chloroplast amino acyl-tRNA synthetases appear to be encoded by the nucleus (Weil and Parthier, 1982). Much is known about the evolution of the stable RNAs, particularly relative to those of prokaryotes, but comparatively little about the evolution of chloroplast ribosomal proteins and elongation factors.

3.1.1a. Ribosomal RNA Genes. Much of the large body of comparative information on chloroplast rRNA genes has recently been reviewed by Whitfeld and Bottomley (1983) and Gillham *et al.* (1985), and therefore I will summarize here only the most cogent evolutionary details. All chloroplast rRNA operons are very similar in organization to both rRNA operons of *Anacystis nidulans* and to three operons (*rrnA*, *rrnD*, and *rrnH*) of the seven found in *E. coli* (Fig. 7). In eubacteria and chloroplasts the DNA region encompassing the 16S gene, the 23S gene, and the intergenic spacer is cotranscribed into a single large precursor RNA. In *E. coli* and *Anacystis* (Douglas and Doolittle, 1984a) this precursor includes the 5S rRNA, while in angiosperms the chloroplast 5S gene is probably transcribed independently from a promoter located in the 4.5S–5S rRNA spacer [reviewed in Whitfeld and Bottomley (1983) and Crouse *et al.* (1985)]. It is not known how the 5S gene is transcribed in *Chlamydomonas*, while there is only indirect evidence that it might be cotranscribed with the 16S and 23S rRNAs in *Euglena* (Karabin *et al.*, 1983).

Several sequence motifs similar to the "-35" and "-10" (Pribnow box) sequences of *E. coli* promoters are found 5′ to the 16S rRNA gene of angiosperm chloroplasts. A $tRNA^{Val}$ gene is located in the midst of these "promoters." *In vitro* transcription experiments suggest that a promoter 5′ to this tRNA gene may be the functional one for ribosomal RNA transcription in tobacco; i.e., the $tRNA^{Val}$ might be cotranscribed with the rRNA operon (Tohdoh *et al.*, 1981). On the other hand, analysis of rRNAs isolated from chloroplasts of maize (Kossel *et al.*, 1982), spinach (Briat *et al.*, 1982a), and duckweed (Keus *et al.*, 1983a) suggests that rRNA transcription initiates 3′ to the $tRNA^{Val}$ gene; i.e., the $tRNA^{Val}$ and rRNA operon are transcribed independently. Briat *et al.* (1983) recently postulated that transcription of angiosperm chloroplast rDNA may be regulated according to bacterial models, with premature termination (attenuation) occurring at stem-loop structures that can form within the rDNA leader transcript.

Three types of 23S rRNA genes are found in chloroplasts. Those in *Euglena* are coextensive with eubacterial ones, in *Chlamydomonas* there are two small rRNAs (7S and 3S RNA), which appear to have split off from the 5′ end of the 23S gene (Rochaix and Darlix, 1982), and in angiosperms and the one fern examined a small RNA (4.5S RNA) has split off from the 3′ end of the 23S gene (Fig. 7) [reviewed in Whitfeld and Bottomley (1983) and Crouse *et al.* (1985)].

A large intron of 870 bp is unique to the 23S gene of *Chlamydomonas* (Rochaix and Malnoe, 1978; Allet and Rochaix, 1979). The 5′ ends of the 7S and 3S rRNAs that are also unique to *Chlamydomonas* have partial homology with the 23S rRNA sequences that flank the intron. It has therefore been suggested that these small rRNAs might act as guide RNAs in the excision of the intron from the 23S precursor (Rochaix and Darlix, 1982). Waring *et al.* (1983) have noted the existence of sequence and secondary structure similarities between the *Chlamydomonas* 23S intron (Allet and Rochaix, 1979), a *Tetrahymena* nuclear rDNA intron, and "group I" introns of fungal mitochondrial genes, and have proposed that similar splicing mechanisms may operate in these systems.

Parts or all of the chloroplast rDNA have been sequenced from *Chlamydomonas, Euglena,* and four angiosperms—maize, tobacco, spinach, and duckweed. Detailed discussions of the comparative features of these sequences and their relationships to eubacterial sequences are found in Whitfeld and Bottomley (1983) and Gillham, *et al.* (1985). The most salient evolutionary conclusions from these sequence studies are: (1) Chloroplast and eubacterial rRNAs are highly conserved at the primary sequence level. 16S gene homologies range from 96% between the dicot tobacco and the monocot maize to 83% between tobacco and the cyanobacterium *Anacystis nidulans,* while the 23S gene is slightly less conserved at 92% tobacco–maize and 78% tobacco–*Anacystis* (Tomioka and Sugiura, 1983; Kumano *et al.*, 1983; Douglas and Doolittle, 1984b). (2) Chloroplast and eubacterial rRNA genes are also highly conserved at the secondary structure level. For example, analysis of the secondary structure of 23S rRNA reveals over 450 compensating base changes between maize and *E. coli* (Glotz *et al.*, 1981). (3) Many of the conserved regions correlate with known functional regions of the *E. coli* rRNAs. These include sites of base modification, ribosomal protein-binding sites, regions of interaction between ribosomal subunits, a sequence at the 3′ end of 16S rRNA that is complementary to the Shine–Delgarno sequence on mRNAs, and a site in the 23S rRNA involved in peptidyl transferase activity. (4) Deletion and insertion differences of up to 78 bp in size distinguish chloroplast and *E. coli* rRNAs. Some of these are similar in structure to bacterial insertion elements (Edwards and Kossel, 1981). The possibility that some of these sequences (those that are insertions in chloroplasts relative to *E. coli*) may be introns and therefore not present in the mature chloroplast rRNAs has not been investigated. (5) Sequences flanking the ends of both the 16S and 23S rRNAs of chloroplasts can form large hairpin structures similar to those that in *E. coli* are substrates for RNAse III in the processing of precursor rRNAs into their mature forms. In summary, chloroplast and eubacterial rRNAs are overwhelmingly similar; given these similarities, it is hardly surprising that the endosymbiotic theory for the origins of chloroplasts is now almost universally accepted (Section 5.1).

3.1.1b. Ribosomal Protein Genes. Approximately 20 of the 60–65 chloroplast ribosomal proteins are synthesized, and presumably encoded, within the chloroplast of chlorophytes and euglenophytes, while the remainder are synthesized in the cytoplasm (Freyssinet, 1978; Eneas-Filho *et al.*, 1981; Schmidt *et al.*, 1983; Posno *et al.*, 1984; Dorne *et al.*, 1984a). Given the pronounced sequence and structural similarities of chloroplast and eubacterial rRNAs, it is not surprising that a number of chloroplast ribosomal proteins show immunological similarities with *E. coli* ribosomal proteins (Bartsch *et al.*, 1982; Dorne *et al.*, 1984b; Schmidt *et al.*, 1984). Interestingly, chloroplast-synthesized ribosomal proteins from *Chlamydomonas reinhardtii* have substantially more antigenic, size, and charge similarities with *E. coli* ribosomal proteins than do cytoplasmically synthesized chloroplast ribosomal proteins (Schmidt *et al.*, 1984). The *E. coli* homologues of these conserved chloroplast-synthesized ribosomal proteins and also those of two chloroplast ribosomal proteins thought to be chloroplast gene products in angiosperms (Sugita and Sugiura, 1983; Subramanian *et al.*, 1983) all interact strongly with rRNA in the *E. coli* ribosome [see discussion in Schmidt *et al.* (1984)]. This raises the possibility that the genes for these ribosomal proteins might have been retained within the chloroplast on the basis of this functional interaction and perhaps also as key components in the process of ribosomal subunit assembly (Schmidt *et al.*, 1984; Dorne *et al.*, 1984a).

Five putative genes for ribosomal proteins have recently been identified on the chloroplast chromosomes of angiosperms (Fig. 1) and *Euglena gracilis* (Montandon and Stutz, 1984) based on deduced amino acid homologies with *E. coli* ribosomal proteins. These homologies are 39% for *rps4* (Subramanian *et al.*, 1983), 43% for *rpl2* and 55–58% for *rps19* (Sugita and Sugiura, 1983; Zurawski *et al.*, 1984b), and 38% for *rps7* and 68% for *rps12* (Montandon and Stutz, 1984). In addition, there is 50% amino acid homology between *E. coli* L12 and an antigenically related ribosomal protein from spinach chloroplasts (Bartsch *et al.*, 1982).

In *E. coli* the *rps4* gene is cotranscribed with genes for three other ribosomal proteins and an RNA polymerase subunit (Lindahl and Zengel, 1982). However, *rps4* does not appear to be part of any large operon in maize cpDNA, and instead is closely flanked by separately transcribed tRNA genes (Subramanian *et al.*, 1983). The *rps19* and *rpl2* genes of *E. coli* are among 11 ribosomal protein genes contained within the S10 operon (Lindahl and Zengel, 1982). It is likely that their homologues in spinach and tobacco cpDNA are also cotranscribed, but as part of a smaller, dicistronic operon (Zurawski *et al.*, 1984b). The *rpl2* from spinach lacks a 666-bp intron found in *rpl2* from tobacco (Zurawski *et al.*, 1984b). As pointed out by the authors, given the great homology (97%) found in the coding regions of these two genes, it should be relatively easy to trace the origin of this intron deletion/insertion through the dicots. The *rps7, rps12,* and *tufA* genes, all of which are part of the *str* operon in *E. coli* (Lindahl and Zengel, 1982), are adjacent in *Euglena gracilis,* but *rps7* and *rps12* appear to be cotranscribed as a dicistronic cluster (Montandon and Stutz, 1984) that is separate from the *tufA* transcription unit (Montandon and Stutz, 1983).

3.1.1c. Transfer RNA Genes. Studies on a number of different angiosperms and algae indicate that all chloroplast genomes probably encode a complete set of tRNA genes. Over 20 different chloroplast tRNA genes have already been sequenced in a variety of organisms and several excellent reviews discussing the evolution of these genes have recently appeared (Weil and Parthier, 1982; Bohnert *et al.*, 1982a; Whitfeld and Bottomley, 1983; Hallick *et al.*, 1983; Crouse *et al.*, 1985; Gillham *et al.*, 1985). The most important conclusions from these comparisons are summarized in the following paragraphs.

Transfer RNA genes have two distinct patterns of organization in the chloroplast genome. Those in *Euglena* cpDNA are clustered in groups of two to six genes. Short AT-rich spacers separate genes within a cluster, and most clusters, or subclusters, appear to be transcribed into polycistronic precursor RNAs, which are then processed into component tRNAs (also see Gruissem *et al.*, 1982, 1983a,b). In contrast, except for the two tRNAs present in the transcribed rDNA spacer (Fig. 7), angiosperm chloroplast tRNA genes are nonclustered and transcribed into monocistronic RNAs. In maize, the gene for $tRNA^{His}$ overlaps for a few nucleotides at its 5′ end the divergently transcribed *rps19* gene (Schwarz *et al.*, 1981) [see discussion in Zurawski *et al.* (1984b)]. Whether the expression of these overlapping genes is in any way linked is not known; however, the same genes do not overlap in dicots (Spielmann and Stutz, 1983; Sugita *et al.*, 1984; Zurawski *et al.*, 1984b).

Chloroplast tRNA genes are highly conserved in primary sequence among plants and algae [see Table 1s of Karabin and Hallick (1983) and Table 2 of Gillham *et al.* (1985); also see references in Table 3 of Crouse *et al.* (1985)]. Furthermore, chloroplast tRNAs always show much more sequence homology with their cognate prokaryotic tRNAs than they do with cognate cytoplasmic tRNAs from the same plant. Most aspects of secondary structure are also highly conserved between chloroplast and eubacterial tRNAs.

The genetic code in chloroplasts, unlike in mitochondria, appears to be read according to the rules of the wobble hypothesis. The most exceptional chloroplast tRNA gene in this regard is that for $tRNA^{Ile}$ from spinach (Francis and Dudock, 1982; Kashdan and Dudock, 1982). This gene has the Met anticodon CAT, but reads isoleucine, probably as a result of modification of the C residue. Furthermore, this tRNA shows little homology to other $tRNAs^{Ile}$, but is approximately equally related to $tRNA^{Met}$ and $tRNA^{Ile}$ from *E. coli*.

The most striking difference between chloroplast tRNA genes and those from *E. coli* is that the former lack the 3′-terminal CCA found in the mature tRNAs. Thus, this sequence must be added posttranscriptionally in the chloroplast. It has recently been shown that the gene for $tRNA^{Ile}$ from the cyanobacterium *Anacystis nidulans*, which is highly homologous with the corresponding chloroplast gene, also does not encode the 3′-terminal CCA, although this sequence is encoded in the *Anacystis* $tRNA^{Ala}$ gene (Williamson and Doolittle, 1983; Tomioka and Sugiura, 1984).

All 17 sequenced *Euglena* chloroplast tRNA genes lack introns (Hallick *et al.*, 1983), whereas five of 20 sequenced angiosperm genes have introns (Koch *et al.*, 1981; Takaiwa and Sugiura, 1982a; Steinmetz *et al.*, 1982; Deno *et al.*, 1982; Deno and Sugiura, 1984; Zurawski and Clegg, 1984; Krebbers *et al.*, 1984; Bonnard *et al.*, 1984). Included among the *Euglena* genes are cognates for four of the five intron-containing angiosperm genes. The chloroplast tRNA gene introns range from 451 to 949 bp in length, far larger than the 15- to 60-bp introns found in yeast and *Drosophila* tRNA genes. Moreover, the chloroplast introns are found at varying positions within the anticodon loop, whereas nuclear and archaebacterial introns are invariably located one nucleotide away from the anticodon triplet toward the 3′ end (Steinmetz *et al.*, 1982; Bonnard *et al.*, 1984).

Michel and Dujon (1983) and Bonnard *et al.* (1984) have observed that three of these chloroplast tRNA introns [from tRNA genes for Ile and Ala from maize and tobacco (Koch *et al.*, 1981; Takaiwa and Sugiura, 1982a) and Val from tobacco (Deno *et al.*, 1982)] have distinctive sequence and potential RNA secondary structure homologies with so-called "group II" introns of yeast mitochondrial genes. Their secondary structure models would bring these conserved sequence elements and also intron–exon junctions into close proximity. Furthermore, Bonnard *et al.* (1984) find that a fourth chloroplast tRNA intron [(from the tRNA gene for Leu from maize and broad bean (Steinmetz *et al.*, 1982; Bonnard *et al.*, 1984)] has strong sequence and secondary structure homologies with "group I" introns present in fungal mitochondrial genes and in nuclear rDNA genes from *Tetrahymena* and *Physarum*. At present, there is no direct evidence as to whether excision of this chloroplast type I intron requires the help of other proteins, as in the case of mitochondrial type I introns (Michel and Dujon, 1983), or whether it is self-directed by the precursor tRNA, as in the case of the self-splicing rRNA precursor from *Tetrahymena* (Cech, 1983). However, Bonnard, *et al.* (1984) note that the chloroplast $tRNA^{Leu}$ intron resembles the *Tetrahymena* rDNA intron in size (451–458 bp vs. 407–413 bp, respectively) and in lacking any significant open reading frame, whereas many fungal mitochondrial introns contain long open reading frames encoding splicing enzymes ("maturases"). The fifth chloroplast tRNA intron, from the gene for Gly (Deno and Sugiura, 1984), has not yet been compared to group I and group II introns.

Chloroplast tRNA introns differ markedly in size among plant species, although most of the size differences in the rDNA spacer $tRNA^{Ile}$ and $tRNA^{Ala}$ introns of maize and tobacco are due to single large deletions of 229 and 103 bp, respectively, in each intron (Koch *et al.*, 1981; Takaiwa and Sugiura, 1982a). These two introns have considerable sequence homology with one another, suggestive of a possible common origin or of more recent sequence "cross-talk" between the two. Although each of the introns from both plant species contains open reading frames (36–123 amino acids long), those from cognate tRNAs are of different size and location, casting doubt on their functional significance. Similarly, short open reading frames found in the intron of the tobacco gene for $tRNA^{Val}$ (Deno *et al.*, 1982) are not conserved in the homologous genes from barley (Zurawski and Clegg, 1984) and maize (Krebbers *et al.*, 1984). However, an extensive analysis of the pattern of nucleotide substitution between the barley and tobacco $tRNA^{Val}$ introns suggests that selection may be constraining the evolution of certain intron sequences, although the significance of these conserved sequences is unclear (Zurawski and Clegg, 1984).

3.1.1d. Elongation Factor Genes. Of the two chloroplast elongation factors, EF-Tu and EF-G, known to be chloroplast gene products in most plants and algae (see Section 2.4.4) (Bottomley and Bohnert, 1982), only the gene for EF-Tu (*tufA*) from *Euglena* (Montandon and Stutz, 1983) has been isolated and sequenced. EF-Tu is highly conserved among bacteria (Bosch *et al.*, 1983), and *tufA* sequences from *E. coli* hybridize strongly with cpDNA from *Chlamydomonas* (Watson and Surzycki, 1982) and *Euglena* (Passavant *et al.*, 1983). It is therefore not too surprising that relatively high homology (70%) was found between the *E. coli* and *Euglena* chloroplast EF-Tu proteins compared, for example, to bacterial–chloroplast homologies of several ATPase polypeptides (cf. Table VI and section 3.1.1b). *Euglena* EF-Tu is translated from a mRNA that is spliced together from four exons (Montandon and Stutz, 1983). Surprisingly, EF-Tu is completely encoded by the first three exons and exons 3 and 4 appear to contain an additional, nonoverlapping open reading frame of 250 amino acids.

3.1.2. Genes for Photosynthesis

Chloroplasts encode many of the critical polypeptides that make up four supramolecular complexes—the ATPase complex, the cytochrome b_6-*f* complex, and the photosystem I and II complexes—of the thylakoid membranes. Evolutionary comparisons of these genes from three of the complexes will be presented below. [Insufficient information is available for chloroplast genes of the photosystem I complex (Westhoff *et al.*, 1983b; A. G. Smith and Gray, 1984) to warrant further discussion here.] Among Calvin cycle enzymes, only the large subunit of ribulose-1,5-bisphosphate carboxylase is encoded by cpDNA in chlorophytes and euglenophytes, while in certain rhodophytes and chromophytes both the large and small carboxylase subunits are chloroplast gene products (see Section 2.4.4).

3.1.2a. Gene for the Large Subunit of Ribulose-1,5-bisphosphate. The chloroplast gene (*rbcL*) for the large subunit of ribulose-1,5-bisphosphate carboxylase (RuBPC) has now been sequenced from five angiosperms [maize (McIntosh *et al.*, 1980), spinach (Zurawski *et al.*, 1981), tobacco (Shinozaki and Sugiura, 1982), barley (Zurawski *et al.*, 1984a), and pea (Zurawski et al., 1985)], the green alga *Chlamydomonas reinhardtii* (Dron *et al.*, 1982), and in part from *Euglena gracilis* (Koller *et al.*, 1984). In addition, prokaryotic *rbcL* sequences are available from the cyanobacteria *Anabaena* (Curtis and Haselkorn, 1983) and *Anacystis nidulans* (Shinozaki *et al.*, 1983a; Reichelt and Delaney, 1983), and the nonsulfur purple bacterium *Rhodospirillum rubrum* (Nargang *et al.*, 1984). The *rbcL* gene is clearly highly conserved among chlorophytes and cyanobacteria. Deduced sequence homologies for the 472- to 477-amino acid protein are

86–95% among chlorophytes (spinach, tobacco, maize, and *Chlamydomonas*), 79–85% between chlorophytes and cyanobacteria, and 84% between two cyanobacteria. In contrast, the *Rhodospirillum* large-subunit protein shows only about 25% homology with the large subunit polypeptide of these other enzymes (Nargang *et al.*, 1984), reinforcing the concept of a close evolutionary relationship of chloroplasts and cyanobacteria.

Not surprisingly, amino acid conservation is greatest in regions of the large-subunit protein that make up the active site and that bind the activating CO_2 molecule (Miziorko and Lorimer, 1983). The most notable difference in these regions is that a COOH-terminal Cys residue that is photoaffinity-labeled in the spinach enzyme and is conserved in *rbcL* of all chloroplasts and *Anabaena* is replaced by Leu in *Anacystis*. Whether this difference is in any way responsible for the lowered carboxylation/oxygenation ratio found in certain cyanobacterial enzymes [unfortunately, neither the *Anacystis* nor *Anabaena rbcL* enzymes have been examined in this regard (Jordan and Ogren, 1981, 1983)], thereby nominating this site as an attractive candidate for genetic modification (Shinozaki *et al.*, 1983a), or whether this means simply that the site is not an essential component of the active site (Curtis and Haselkorn, 1983) remains to be seen.

Many hopes for improving plant productivity rest on the idea of improving the efficiency of RuBPC, primarily by lowering its oxygenase activity (Somerville and Ogren, 1982). Thus, we should expect to see continued emphasis on nucleotide sequence comparisons of *rbcL* as a means of assessing conserved regions of functional importance, followed by attempts to modify these regions by *in vitro* mutagenesis. An important first step in this latter process is provided by the ability to express properly the chloroplast *rbcL* gene from maize and wheat in *E. coli* (Gatenby *et al.*, 1981; Gatenby and Castleton, 1982). In the case of the *Rhodospirillum* large-subunit gene, both expression and a fully functional enzyme were obtained (Somerville and Somerville, 1984). (In most photosynthetic organisms RuBPC is a heteromultimeric enzyme composed of two nonidentical subunits, and thus expression of the large subunit alone is unlikely to yield a functional enzyme, whereas the *Rhodospirillum* enzyme is a homodimer of two large-subunit polypeptides.)

In those angiosperms where *rbcL* has been sequenced and also in *Chlamydomonas* the gene encodes a single, unspliced transcript approximately 1.6 kb in length. This situation appears to prevail in most angiosperms, with the notable exception of one subtribe of the legume family, where three *rbcL* transcripts are found (Palmer *et al.*, 1982a). In these legumes the two larger transcripts (2.6 and 2.4 kb in size) are the major RNA species and the normal transcript of 1.6 kb is reduced to only a small fraction of total *rbcL* RNA. Recent studies (Palmer *et al.*, 1984, and unpublished data) indicate that the larger transcripts in these legumes contain 3′ extensions of the 1.6-kb *rbcL* mRNA, and preliminary sequence analysis suggests that this transcriptional readthrough may result from mutational alterations within the hairpin loop that normally (Section 3.2.2) serves to terminate *rbcL* transcription.

The *Euglena rbcL* gene differs fundamentally from all other characterized *rbcL* genes in containing no less than nine introns, each approximately 0.5 kb in length (Stiegler *et al.*, 1982; Koller *et al.*, 1984b). In spite of these structural differences, *Euglena* exon sequences are highly conserved (85–90% deduced amino acid homology over 40% of the coding region) compared to other *rbcL* genes. The DNA sequences of five 3′-intron/5′-exon and four 3′-exon/5′-intron boundaries are highly conserved relative to one another and show partial homology with nuclear mRNA splice junctions (Koller *et al.*, 1984b). The homology is much greater at the 5′-intron consensus sequence. Since U1 small nuclear ribonucleoproteins bind selectively to the 5′-splice site of nuclear pre-mRNA *in vitro* (Mount *et al.*, 1983; also see Cech, 1983), Koller *et al.* (1984b) speculate that an analogous class of ribonucleoproteins might be found in *Euglena* chloroplasts. They also speculate that the U1-like RNA would probably be a chloroplast gene product, since all known chloroplast RNAs are encoded by the chloroplast genome.

General features of *rbcL* genes, including promoter and terminator sequences, codon usage, and patterns of nucleotide substitution, that appear common to most chloroplast protein genes will be discussed in Section 3.2.

3.1.2b. ATPase Genes. Five of the eight subunits of the chloroplast ATPase are synthesized in the chloroplast (Bottomley and Bohnert, 1982; Whitfeld and Bottomley, 1983). All but one (the gene for subunit I of the membrane component CF_0) of the five chloroplast genes has been sequenced from one or more angiosperms. Different ATPase genes are evolving at different rates, not only within the chloroplast lineage (70–100% deduced amino acid homologies between dicots and monocots, depending on the gene), but also between chloroplasts and prokaryotes (23–67% amino acid homologies; Table VI). Surprisingly, however, relative rates of evolution of the various ATPase genes do not appear to be constant over long periods of time. The *atpB* gene is highly conserved both among chloroplasts (91% amino acid homology between monocots and dicots) and also between chloroplasts and *E. coli* (67%), while *atpE* is quite divergent in both comparisons (70 and 25%, respectively; Table VI). In striking contrast, *atpH* is only poorly related between chloroplasts and

E. coli (25–35% amino acid homology), but is completely conserved (Table VI) in amino acid sequence between two angiosperms (wheat and spinach) that diverged on the order of 100–150 mya (Doyle, 1978; Muller, 1981). Thus, relative to *atpB* and *atpE*, *atpH* has experienced a marked slowdown in sequence evolution among plant chloroplasts.

The *atpE* gene is evolving much more rapidly than *atpB* and *rbcL* based on amino acid homologies calculated from two pairwise comparisons among dicots, one comparison between dicots and monocots, and three comparisons between angiosperms and *E. coli* (Table VI). In contrast, *atpE* is slightly more conserved between two monocots, maize and barley (both members of the family Poaceae), than *atpB* or *rbcL* (Table VI). Thus *atpE* sequence evolution has drastically slowed among monocots, or at least the cereals.

Much more comparative sequence data and information on ATPase structure–function relationships will be needed before we understand the functional significance of these alterations in evolutionary rates of the *atpH* and *atpE* genes. In this regard, it should be noted that the ϵ subunit of the chloroplast ATPase (the *atpE* gene product) may serve somewhat different regulatory and structural roles in the different kinds of ATPases (mitochondrial, chloroplast, bacterial) (Nelson, 1980; Harris, 1981). Moreover, the ϵ subunits of spinach and *Euglena* chloroplasts display different inhibitory activities in *in vitro* studies (Kahn, 1982).

Table VI
Deduced Amino Acid Homologies of *rbcL* and ATPase Genes in Chloroplasts and *E. coli* (%)[a]

	rbcL	*atpB*	*atpE*	*atpH*	*atpA*
Dicot–dicot					
Pea–spinach	94	92	82	—	—
Tobacco–spinach	93	96	87	—	—
Monocot–dicot					
Maize–tobacco	90	91	70	—	—
Wheat–spinach	—	—	—	100	—
Monocot–monocot					
Maize–barley	95	97	98	—	—
Angiosperm–*E. coli*					
Tobacco–*E. coli*	—	62	27	—	54
Spinach–*E. coli*	—	67	26	30	—
Maize–*E. coli*	—	58	23	—	—

[a] Data compiled from Howe *et al.* (1982), Zurawski *et al.* (1982, 1984a, 1985c), Krebbers *et al.* (1982), Deno *et al.* (1983), Shinozaki *et al.* (1983b), Alt *et al.* (1983b), and Zurawski and Clegg (1984).

Of some relevance to these chloroplast gene comparisons is the observation that subunit II of cytochrome oxidase (a mtDNA gene product) is evolving about five times more rapidly in primates than in other mammals (G. G. Brown and Simpson, 1982; Cann *et al.*, 1984). Cytochrome *c*, a nuclear-encoded protein that interacts directly with the oxidase II subunit, also is evolving much more rapidly in primates (Carlson *et al.*, 1977), suggesting a specific coevolution of the two genes. By analogy, it will be interesting to see whether the ATPase γ and δ subunits, which are nuclear-encoded and interact structurally with the ϵ subunit, also have altered evolutionary rates among the cereals.

In *E. coli* all eight ATPase genes are organized into a single operon (Futai and Kanazawa, 1983). In contrast, in *Rhodopseudomonas blastica* the F_0 component genes are separate from the F_1 component genes, although the latter form an operon with the same gene order as in *E. coli* (Walker *et al.*, 1984). Preliminary studies of the cyanobacterium *Anabaena* (S. Curtis, unpublished data) suggest even further dispersion of ATPase genes: The *atpB* and *atpE* genes appear to be separate from the remaining F_1 component genes. Remnants of these eubacterial ATPase operons can be discerned in the chloroplast genome (Fig. 1). The *atpB* and *atpE* genes, which are adjacent to one another in the above eubacteria, are also adjacent and cotranscribed in five angiosperm chloroplast genomes [from maize, barley, spinach, tobacco, and pea (Zurawski *et al.*, 1982a, 1985; Krebbers *et al.*, 1982; Shinozaki *et al.*, 1983b; Zurawski and Clegg, 1984)]. Moreover, in four of these five angiosperms the coding regions of *atpB* and *atpE* overlap by 4 bp. In these species the TGA stop codon of *atpB* is preceded by an A, which forms the ATG initiator methionine of *atpE*. In pea, the *atpB* and *atpE* coding regions no longer overlap, but instead are separated by a 22-bp untranslated region similar in size to the 20-bp *atpBE* intercistronic sequence in *E. coli* (Zurawski *et al.*, 1985). However, the pea *atpB* and *atpE* genes are still transcribed into a dicistronic mRNA. In contrast to the cotranscription of *atpB* and *atpE*, *atpA* and *atpH*, although only 2 kb apart from one another, appear to be transcribed independently in cpDNA of spinach, wheat, petunia, and pea (Fig. 1) (Alt *et al.*, 1983b; Howe *et al.*, 1983; Bovenberg *et al.*, 1984; Huttly and Gray, 1984). The two genes, atpD and *atpG*, that separate *atpA* and *atpH* in the *E. coli* ATPase operon (Futai and Kanazawa, 1983) are present in the nuclear genome in plants (Westhoff *et al.*, 1981).

3.1.2c. Photosystem II Genes. The chloroplast *psbA* gene encodes the so-called "32-kD" protein, a rapidly metabolized polypeptide of apparent molecular weight 32,000, which is associated with photosystem II and is responsible for binding of herbicides such as atrazine and diuron

(Steinback *et al.*, 1981; Mattoo *et al.*, 1981; Pfister *et al.*, 1981). This gene is currently one of the prime targets of plant genetic engineers, since a single amino acid change abolishes atrazine binding and renders plants resistant to the herbicide (Hirschberg and McIntosh, 1983). The *psbA* gene is extraordinarily highly conserved among eukaryotes. The *psbA* gene product differs at between zero and three residues among five distantly related dicots [spinach, tobacco, soybean, mustard, and amaranth (Zurawski *et al.*, 1982b; Spielmann and Stutz, 1983; Hirschberg and McIntosh, 1983; Link and Langridge, 1984)] that diverged on the order of 50–100 mya (Muller, 1981). Furthermore, 87% homology exists between the 32-kD polypeptides of *Euglena* and angiosperms (Hollingsworth *et al.*, 1984) and also between those of the cyanobacterium *Anabaena* and angiosperms (Curtis and Haselkorn, 1984). In contrast to the single *psbA* gene located in angiosperm cpDNAs, *Anabaena* contains multiple (2–4) related copies of the *psbA* gene, only one of which appears to be transcribed (Curtis and Haselkorn, 1984).

The *psbA* gene encodes a single transcript of 1.2 kb, colinear with the gene, in the five angiosperms where it has been sequenced. A single *psbA* transcript of the same size was found in a survey of a large number of additional angiosperms, with the exception of sweetpea, where a second *psbA* transcript of 1.4 kb was also observed (Palmer *et al.*, 1982a). The molecular basis for this transcript variation is not known. Unlike angiosperm *psbA* genes, those of algae are interrupted by introns. The *psbA* gene of *Euglena gracilis* contains four introns (Hollingsworth *et al.*, 1984) with conserved boundary sequences (G. Karabin, M. Farley, J. Narita, and R. Hallick, unpublished data) similar to those found in the *Euglena rbcL* gene (Koller *et al.*, 1984). However, whereas only a single, fully spliced *rbcL* transcript is detected in *Euglena* (Stiegler *et al.*, 1982; Koller *et al.*, 1984), five, presumably unspliced or only partially spliced precursor RNAs are found for the 1.2-kb *psbA* mRNA (Hollingsworth *et al.*, 1984). The *psbA* gene of *Chlamydomonas reinhardtii*, which is located within the inverted repeat (see Section 2.4.3), also contains four introns (J. Erickson and J. Rochaix, unpublished data). It will be interesting to compare the positions and sequences of the *psbA* introns in *Chlamydomonas* and *Euglena;* a primary prediction of separate endosymbiotic origins for their chloroplasts (Section 5.1) is that they should not contain the same introns.

The two chlorophyll a-conjugated polypeptides of photosystem II, of apparent molecular weight 51,000 and 44,000, are encoded by cpDNA in spinach (Westhoff *et al.*, 1983a). The two genes, *psbB* and *psbC*, respectively, for these polypeptides are quite far apart from each other and from the *psbA* gene on the spinach chloroplast genome (Fig. 1). Interestingly, however, the *psbB* gene is adjacent to and cotranscribed with the *petB* and *petD* genes of the cytochrome b_6-*f* complex (Fig. 1) (Westhoff *et al.*, 1983a; Morris and Herrmann, 1984). At least 11 discrete RNA species, ranging in size from 0.9 to 6 kb, encode between one and all three of these proteins (Morris and Herrmann, 1984). Sequence studies show that the largest transcripts, encoding all three proteins, probably arise by read-through at two stem-loop, putative transcription termination sequences found between the genes (Morris and Herrmann, 1984).

3.1.2d. Cytochrome b_6-*f* Genes. Three of the four polypeptide components of the cytochrome b_6-*f* complex are encoded by cpDNA in flowering plants (Alt *et al.*, 1983a; Phillips and Gray, 1984a, b; Willey *et al.*, 1983, 1984a, b). In spinach, the chloroplast gene for cytochrome *f* (*petA*) maps 15 kb away from the genes for cytochrome b_6 (*petB*) and subunit 4 of the complex (*petD*) (Fig. 1) (Alt *et al.*, 1983a). Moreover, as described in the preceding section, these latter two genes are cotranscribed together with the *psbB* gene. The *petA* genes of pea and wheat have 85% nucleic acid homology and 88% amino acid homology (Willey *et al.*, 1984a,b). Especially conserved regions include the heme-binding site and the hydrophobic COOH-terminal sequence. An intriguing hypothesis, stemming from recent sequence comparisons, is that the functions of the single mitochondrial polypeptide cytochrome *b* are distributed between two polypeptides (the *petB* and *petD* gene products) of the chloroplast cytochrome complex (Widger *et al.*, 1984; Phillips and Gray, 1984a).

3.2. General Features of Chloroplast Gene Evolution

3.2.1. Transcriptional Organization and Introns

Two major conclusions can be drawn concerning the origin and evolution of chloroplast transcription units. First, starting from a typical prokaryotic operon built around the concept of common, cooperative function of products from linked genes, chloroplasts have evolved a variety of organizational units of gene function. At one extreme, both ribosomal RNA operons of *Anacystis nidulans* and three of seven operons of *E. coli* are found almost perfectly conserved in the chloroplast genome (Fig. 7). All identified chloroplast rRNAs are transcribed from cpDNA as a single large precursor, which is processed into mature rRNAs in a typically prokaryotic fashion (Crouse *et al.*, 1985). The one major difference, compared to both *Anacystis* and *E. coli* (Douglas and Doolittle, 1984a), is that in vascular plants at least, chloroplast 5S rRNA, although tightly linked physically to the rest of the rRNA operon, probably has its own promoter and is transcribed independently. At an intermediate level, other prokaryotic operons have been fragmented into smaller pieces, several of which are still recognizably prokaryotic in organization (Table VII). For the most part, however, functionally related chloroplast proteins are encoded by genes whose ancient linkages have been obliterated, foremost by their transfer from the plastid to the nucleus (Section 2.4.4) and secondarily by the dispersion of genes throughout the chloroplast chromosome in a fairly random fashion. So far, genes for proteins that make up chloroplast ribosomes, the ATPase, photosystem II, and cytochrome b_6-*f* have been found to be scattered widely in two or more regions of the chloroplast genome (Fig. 1). A second conclusion is that the chloroplast genome now appears to be locked into a set pattern of gene arrangements and transcription units, particularly among angiosperms. This has already been discussed in the context of the relatively fixed dispersal of genes between chloroplast and nucleus and the rarity of major structural rearrangements within the chloroplast genome. In the context of individual gene position and transcription, this constrained organization extends to the level of conserved chloroplast operons encoding rRNA genes (all land plants and algae), *atpB* and *atpE* [(angiosperms and probably *Chlamydomonas* (Woessner *et al.*, 1984)], and *rps19* and *rpl2* (tobacco and spinach).

Euglena gracilis departs significantly from plants and other algae in certain aspects of its gene organization and transcription. This is not surprising given that it probably derived its plastid by a separate endosymbiotic event (Section 5.1). *Euglena* tRNA genes differ in two important ways: They lack the large introns present in one-fourth of the sequenced angiosperm tRNAs and they are clustered tightly into polycistronic arrays, as compared to the monocistronic organization of almost all angiosperm tRNA genes. In contrast to these prokaryotic features of *Euglena* tRNA genes, certain *Euglena* protein genes contain multiple introns (nine, four, and three introns in *Euglena rbcL, psbA,* and *tufA,* respectively). Moreover, comprehensive examination of the entire *Euglena* genome by electron microscopy reveals that almost all transcribed regions contain introns (at least 50 introns, accounting for over 20% of the genome, were detected), whereas parallel studies reveal only four spliced transcripts and a minimum total of only six introns in cpDNA from broad bean, an angiosperm (Koller and Delius, 1984). In addition, only one gene (*rpl2*) among 15 characterized chloroplast protein genes from angiosperms (Fig. 1) has even a single intron, and even this intron is absent from certain angiosperm species. Although relatively few genes have been characterized in the *Chlamydomonas reinhardtii* chloroplast genome, it appears to be somewhat intermediate in intron frequency. The *Chlamydomonas rbcL* gene, like those of angiosperms, lacks introns, while the *psbA* gene contains multiple introns and the 23S rRNA gene has a single intron.

Table VII
Polycistronic Operons in Chloroplast DNA

Taxon	Cotranscribed genes	Prokaryotic homologues	Reference
All chloroplast DNAs	rRNA operon[a]	Yes	Section 3.1.1a
Five angiosperms	*atpB*, *atpE*[a]	Yes	Section 3.1.2b
Spinach	*petB*, *petD*, *psbB*[a]	?	Alt *et al.* (1983a), Morris and Herrmann (1984)
Spinach, tobacco	*rps19*, *rpl2*[b]	Yes	Zurawski *et al.* (1984b)
Euglena gracilis	tRNA operons[a]	No[c]	Gruissem *et al.* (1982, 1983a)
Euglena gracilis	*tufA*, 250-amino acid ORF[d]	?	Montandon and Stutz (1983)
Euglena gracilis	*rps7*, *rps12*[a]	Yes	Montandon and Stutz (1984)
Cyanophora paradoxa and *Olisthodiscus luteus*	*rbcL*, *rbcS*[e]	Yes	Heinhorst and Shively (1983), M. Reith and R. Cattolico (unpublished data)

[a] Polycistronic RNAs demonstrated *in vivo*.

[b] Cotranscription is inferred from their close proximity, the lack of transcriptional regulatory sequences in the short spacer between the genes, and because the homologous genes are cotranscribed in *E. coli*.

[c] Except for the $tRNA^{Ala}$ and $tRNA^{Ile}$ genes, which are part of all rRNA transcription units in chloroplast genomes and some units in eubacteria, none of the polycistronic tRNA operons in *Euglena* cpDNA (Hallick, 1985) correspond to those identified in *E. coli* (Bachmann, 1983).

[d] It has not yet been shown that the 250-amino acid open reading frame (ORF) present on *tufA* mRNA is translated *in vivo*.

[e] In the case of *Cyanophora*, cotranscription is inferred by analogy to the cyanobacteria *Anacystis* and *Anabaena*, where good evidence exists that the similarly closely located *rbcL* and *rbcS* are cotranscribed (Shinozaki and Sugiura, 1983; Nierzwicki-Bauer *et al.*, 1984).

Chloroplasts appear to utilize at least two of the three general classes of RNA splicing mechanisms recognized by Cech (1983). Nuclear mRNA splice sites and apparently those of *Euglena* chloroplast mRNAs (Koller *et al.*, 1984b) seem to be determined largely by short, highly conserved sequences at the intron–exon junctions. In this respect, *Euglena* split genes are unique among all characterized organelle genes. In contrast, in the chloroplast genomes of chlorophytes, the secondary structure of the intron is postulated to direct splicing of several split tRNAs (Michel and Dujon, 1983; Bonnard *et al.*, 1984) and, quite tentatively, the split 23S rRNA of *Chlamydomonas reinhardtii* (Waring *et al.*, 1983). Furthermore, these structured chloroplast introns fall into two distinct subgroups based on homologies with structured group I and group II mitochondrial and nuclear introns. So far, chloroplast genes lack the introns characteristic of certain nuclear tRNA genes, whose splicing is determined primarily by exon structure (Cech, 1983).

3.2.2. Regulatory Sequences

Highly conserved sequences closely resembling the "-10" and "-35" promoter regions of prokaryotic genes are found in a region 5′ to essentially all chloroplast genes (Table VIII) (Whitfeld and Bottomley, 1983; Crouse *et al.*, 1985). As in the case of *E. coli*, the 5′ end of chloroplast transcripts is usually found only a few base pairs downstream from these putative promoters (Table VIII). Direct evidence for or against a promoter function for these chloroplast "-10" and "-35" sequences has not yet been obtained. Such evidence will require the continued application of homologous systems developed for the *in vitro* transcription of chloroplast genes (Bogorad *et al.*, 1983; Gruissem *et al.*, 1983a,b) and ultimately will

Table VIII
Consensus Sequences for Putative Chloroplast Promoters and for Bacterial Promoters[a]

	"-35 region"		"-10 region"	
Chloroplast consensus sequence	TTGaNa	9–23 bp	TAtaaT	2–34 bp
Bacterial consensus sequence	TTGaca	16–19 bp	TAtaaT	7–10 bp

[a] Chloroplast data are from Table VI of Crouse *et al.* (1984). [Also see Fig. 3 of Whitfeld and Bottomley (1983).] Bacterial data are from Hawley and McClure (1983). An upper case letter in a consensus sequence indicates that the given nucleotide is found in that position in the vast majority of the sequences examined, while a lower case letter indicates a somewhat less, although still predominant, occurrence of the given nucleotide. Numbers between the "-35 region" and the "-10 region" indicate the number of nucleotides separating these regions. Numbers following the "-10 region" give the number of nucleotides between this region and the assumed or demonstrated start site of transcription.

require the development of chloroplast transformation systems. Termination of chloroplast transcription also appears distinctly prokaryotic, since complementary sequences capable of forming short hairpin loops are found at the end of both chloroplast and eubacterial genes [reviewed in Whitfeld and Bottomley (1983) and Crouse *et al.* (1985)]. In at least one case, the presence of multiple, unexpectedly large transcripts from a single gene appears to be the consequence of a mutation that destroys such a hairpin structure (see Section 3.1.2a). Furthermore, it is hypothesized that chloroplasts may prematurally terminate rDNA transcription by a process analogous to bacterial attenuation (Briat *et al.*, 1983) (Section 3.1.1a). Many, but not all chloroplast mRNAs contain identifiable "Shine–Delgarno" sequences at their 5′ end that are complementary to sequences at the 3′ end of chloroplast 16S rRNA (Kozak, 1983; Whitfeld and Bottomley, 1983). In summary, although dispersive forces have greatly disrupted the prokaryotic operon structure in chloroplast genomes, the basic regulatory sequences found in prokaryotic genes are highly conserved in individual chloroplast genes. It thus follows that a necessary part of the operon disruption process has been the recruitment of regulatory sequences by newly "orphaned" genes.

3.2.3. Genetic Code and Codon Use

In contrast to mitochondria, chloroplasts utilize the universal genetic code (McIntosh *et al.*, 1980; for review, see Whitfeld and Bottomley, 1983). All possible codons are used, including the three stop codons, UAA, UAG, and UGA. The high AT content of cpDNA (Section 2.3.2) is reflected by a strong AT bias (69% AT) in third-position codon usage (Whitfeld and Bottomley, 1983). Angiosperm chloroplast genes are slightly (4–5%) more GC-rich than the genome as a whole, while spacer sequences tend to be AT-rich (Whitfeld and Bottomley, 1983). The greater AT-richness of the *Chlamydomonas* and *Euglena* genomes is reflected by an even greater bias in AT usage in coding sequences and by a more restricted codon usage (Dron *et al.*, 1982; Montandon and Stutz, 1983, 1984). For example, only 40 different codons are used in *rbcL* of *Chlamydomonas,* compared to 55 and 58 in spinach and maize, respectively (Dron *et al.*, 1982). Although codon usage in *rbcL* of the cyanobacterium *Anacystis* differs from that of chloroplast *rbcL,* most of these differences reflect a tendency toward synonymous third-position GC usage in *Anacystis* (Shinozaki *et al.*, 1983a). Moreover, *rbcL* from *Anacystis* and chlorophytic chloroplasts has normal frequencies (relative to random expectations) of certain codons that are rarely used in eukaryotic nuclear genes (Shinozaki *et al.*, 1983a).

3.2.4. Rates of Sequence Evolution

Global studies of two types—solution hybridizations among distantly related angiosperms (Lamppa and Bendich, 1979, 1981; Bisaro and Siegel, 1980; Palmer and Thompson, 1982) and restriction site comparisons among closely related species (Gordon *et al.*, 1982; Kung *et al.*, 1982; Palmer and Zamir, 1982; Palmer *et al.*, 1983a, 1985a; Bowman *et al.*, 1983)—have indicated that nucleotide substitution rates are relatively low over the entire chloroplast genome. In addition, DNA sequence analysis has provided direct evidence that many individual chloroplast genes are also changing very slowly (Section 3.1; Tables VI and IX).

The lack of a well-documented angiosperm fossil record (particularly for those plants whose chloroplast genomes have been well studied) makes it difficult to calculate an absolute base substitution rate for cpDNA and also to compare directly evolutionary rates among the various cellular genomes. Zurawski *et al.*, (1984a) have estimated that the synonymous rate of sequence change of the *rbcL* gene among grasses is 1×10^{-9} nucleotide substitutions per year, compared to estimates of $(2\text{–}3) \times 10^{-9}$ for animal nuclear genes (Kimura, 1983), 5×10^{-9} for pseudogenes (Li, 1983), and 100×10^{-9} for animal mtDNA (W. M. Brown *et al.*, 1982;

Table IX
Sequence Comparisons of Chloroplast Genes of Tobacco (T), Spinach (S), and Maize (M)[a]

	Protein size (amino acids)[b]			Percentage amino acid conservation		Percentage nucleotide conservation	
Gene	T	S	M	T vs. S	T or S vs. M	T vs. S	T or S vs. M
16S rRNA	—	—	—	—	—	—	96
23S rRNA	—	—	—	—	—	—	92
5S rRNA	—	—	—	—	—	100	99
psbA	353	353	—	100	—	96	—
rbcL	477	475	475	93	90	91	85
atpB	498	498	498	96	91	—	—
atpE	133	134	137	87	70	—	—
atpH	—	81	81[c]	—	100[c]	—	93[c]
rpl2	266	286	—	97	—	95	—
rps19	92	92	—	92	—	89	—

[a] Data compiled from Tohdoh and Sugiura (1982), Takaiwa and Sugiura (1982b), Keus *et al.* (1983b), Zurawski *et al.* (1982b, 1984b), Shinozaki and Sugiura (1982), Shinozaki *et al.* (1983b), and Alt *et al.* (1983b).
[b] Amino acid sequences are inferred from nucleotide sequences.
[c] The *AtpH* sequence is from wheat (Howe *et al.*, 1982) rather than maize.

Aquadro and Greenberg, 1983). While emphasizing the tentative nature of these calculations, Zurawski *et al.*, 1984a) nonetheless concluded that cpDNA is a rather conservative molecule in its sequence evolution, especially compared to animal mtDNA. Similar conclusions have been reached from restriction site studies (Palmer and Zamir, 1982; Palmer *et al.*, 1983a, 1985a), where interspecific cpDNA divergence values are calculated that are well below those from interspecific comparisons of animal mtDNA, and even at the low end of intraspecific population comparisons for these DNAs (Avise *et al.*, 1979a,b; G. G. Brown and Simpson, 1981; Avise and Lansman, 1983). Comparisons of evolutionary rates among the three plant cell genomes, while preliminary, suggest that cpDNA is changing slowly relative to single-copy nuclear genes in *Atriplex* (Belford and Thompson, 1981; Palmer *et al.*, 1982b) but at about the same rate as mtDNA in *Brassica* (Section 4.7).

Relatively little information is available about the distribution of nucleotide sequence change in the chloroplast genome. The relatively large size of the chloroplast genome and low density of restriction sites examined in comparative studies preclude analysis of the entire genome on a gene-by-gene basis, as best exemplified by the study of Cann *et al.* (1984) on human mtDNA. At best, restriction site studies have indicated a somewhat lowered rate of sequence evolution for the inverted repeat relative to the single-copy regions (Palmer and Zamir, 1982; Kung *et al.*, 1982; Clegg *et al.*, 1984). Much of this differential conservation can be accounted for by the slow rate of evolution of the inverted repeat rRNA genes (Palmer *et al.*, 1983a,b).

Much more precise and complete information is available on relative rates of sequence change for those few chloroplast genes that have been sequenced in two or more species (Tables VI and IX). A range of derived amino acid conservation values is evident in dicot–dicot comparisons of between 100 and 87%, while monocot–dicot conservations range between 100 and 70% (Table IX). Given the limited data pool, one can tentatively conclude that relative rates of evolution of different genes are fairly constant in different chloroplast lineages (Table IX). The major exception in this regard is *atpE*, which appears to be evolving relatively slowly among cereals (see Section 3.1.2b and Table VI). The genes listed in Table IX comprise no more than perhaps 10 to 15% of those present in the chloroplast genome, and thus we have much to learn about evolutionary rates of individual chloroplast genes, as well as the functional constraints controlling their evolution.

Zurawski and co-workers (Zurawski *et al*, 1984a; Zurawski and Clegg, 1984) have analyzed in detail the pattern of nucleotide substitution within *rbcL*, *atpB*, *atpE*, and flanking noncoding and nontranscribed regions in barley and maize cpDNA. They observed a twofold excess of transitions over transversions relative to random expectations for all three genes. As they point out, this bias is an order of magnitude smaller than the transition bias in animal mtDNA (W. M. Brown *et al.*, 1982; G. G. Brown and Simpson, 1982; Aquadro and Greenberg, 1983) [reviewed in W. M. Brown (1983)]. Not surprisingly, most nucleotide substitutions (71 and 75% for *rbcL* and *atpB*, respectively) occurred in the third position of codons as silent changes; i.e., without leading to amino acid substitutions. Silent substitutions were randomly distributed over *atpB* and *atpE* coding regions, while missense substitutions were clustered at the ends of the molecules (Zurawski and Clegg, 1984). In contrast, both types of changes appear to have occurred nonrandomly in *rbcL* (Zurawski *et al.*, 1984a). Even including insertions and deletions, which are confined to noncoding regions in these genes, the frequency of mutations in the 5′ noncoding leader sequence of *rbcL* is significantly below the third-position substitution rate (Zurawski *et al.*, 1984a). The authors therefore suggest that functional constraints on promoter and ribosome binding sequences may be responsible for this reduced rate of leader sequence evolution.

4. Mitochondrial DNA Evolution

Research on plant and algal mtDNA has lagged far behind that on other organelle DNAs. Several reasons for this lag are apparent. Plants and algae generally contain less mtDNA than other organisms and its rigorous purification free of nuclear DNA and cpDNA is often difficult. Early work indicated that plant mitochondrial genomes are quite large and heterogeneous, making than less easy to characterize than animal and fungal mtDNAs. Unlike the chloroplast genome of *Chlamydomonas* and the mitochondrial genome of *Saccharomyces*, no model genetic systems are available for plant and algal mitochondria. Finally, the presence of the chloroplast and the fundamental importance of photosynthesis have tended to steer plant molecular biologists toward this organelle rather than the mitochondrion, whereas the latter organelle has been intensively studied in many fungi and animals.

A catalytic event that greatly stimulated an interest in the molecular biology of flowering plant mitochondria was the discovery that the genetic determinants for cytoplasmic male sterility (CMS) in maize are carried by the mitochondrial genome [for review, see Leaver and Gray (1982), Levings (1983), Laughnan and Gabay-Laughnan (1983), and Sederoff (1984)]. One consequence of this discovery, as will become apparent in the following, is that we are at last beginning to understand the structure of angiosperms mtDNAs, their coding properties, and their evolution. At the same time, however, we are still almost totally ignorant of the characteristics of mtDNA from algae and plants other than angiosperms.

4.1. Genome Size

The early literature indicates a general confusion about the size of angiosperm mtDNAs. This confusion stems from the complexity and apparent heterogeneity of mitochondrial DNA restriction profiles and the broad size range of mtDNA molecules (primarily linear with a low percentage of circles) visualized in the electron microscope [reviewed in Leaver and Gray (1982) and Lonsdale (1984)]. The landmark study of Ward *et al.* (1981) was the first step in resolving this confusion and in showing definitively that angiosperm mtDNAs differ radically in size from all other mtDNAs. Ward *et al.* (1981) estimated mitochondrial genome sizes of six species of angiosperms by optical reassociation kinetics and by restriction analysis. MtDNA kinetic complexities were found to range from 340 to 2500 kb for pea, maize, and four species in the Cucurbitaceae (Table X). Similar genome size estimates were obtained by restriction analysis for four of the species (the extremely complex mtDNA restriction patterns of muskmelon and cucumber precluded effective analysis). This was the first and is still the only determination of angiosperm mitochondrial genome size by two independent measurements. Furthermore, the accuracy of the estimate of 500 kb for the size of the maize mitochondrial genome has since been confirmed by complete physical mapping of the genome (see next section and Table X).

The large and highly variable size of angiosperm mtDNAs is without parallel. The smallest well-characterized angiosperm mtDNA is 218 kb in size (Palmer and Shields, 1984). Thus, there is over a sevenfold range in genome size within a single family (Cucurbitaceae) of flowering plants, at least an 11-fold range in genome size among angiosperms, and, including the 16-kb mitochondrial genome of *Chlamydomonas reinhardtii* (see next paragraph), over a 150-fold range in mitochondrial genome size among chlorophytes. In contrast, there is only a sevenfold range in mitochondrial genome size among the rest of the eukaryotes, and in these the largest genome is only half the size of the smallest angiosperm mitochondrial genome (Mahler, 1983; Sederoff, 1984).

The evidence suggests that algal mitochondrial genomes, although still quite variable in size, are much smaller than those of angiosperms, and that at least one of these, from *Chlamydomonas reinhardtii*, may be predominantly linear in form (Table X). However, it should also be recognized that only two of the characterized algal "mitochondrial" DNAs

Table X
Physicochemical Properties of Plant and Algal Mitochondrial DNAs[a]

Taxon	GC Content (%)	Size (kb)	Conformation	Reference
Angiosperms				
Muskmelon	41	2500	—	Ward *et al.* (1981)
Cucumber	43	1500	—	Ward *et al.* (1981)
Zucchini	42	870	—	Ward *et al.* (1981)
Watermelon	44	340	—	Ward *et al.* (1981)
Pea	44	370	—	Ward *et al.* (1981)
Maize (normal, fertile)	41	500	—	Ward *et al.* (1981)
Maize (normal, fertile)	—	570	Multicircular	Lonsdale *et al.* (1984)
Maize (CMS-S cytoplasm)	—	570	Multilinear	Schardl *et al.* (1984)
Turnip	—	218	Three circles	Palmer and Shields (1984)
Green algae				
Chlamydomonas reinhardtii	47	16	Linear	Ryan *et al.* (1978), Grant and Chiang (1980)
Chlorella pyrenoidosa	40	80	—	Bayen and Rode (1973)
Euglenophyta				
Euglena gracilis	25	60	—	Fonty *et al.* (1975)
Classification uncertain				
Cyanophora paradoxa	55	200	—	Bohnert *et al.* (1982b)
Chromophyta				
Olisthodiscus luteus	35	40	Circular	Aldrich and Cattolico (1981), Aldrich *et al.* (1982)

[a] Angiosperm data are included only for those genomes whose sizes have been estimated rigorously, either by reassociation kinetics, usually in combination with restriction analysis, or by complete restriction mapping. Not included are a large number of earlier studies that produced conflicting and often unsubstantiated results [see discussions in Ward *et al.* (1981) and Leaver and Gray (1982)]. All published data on algal mtDNAs are cited, including several studies in which the mitochondrial origin of the DNA studied was not rigorously confirmed.

listed in Table X, those of *C. reinhardtii* and *Euglena gracilis*, have been shown conclusively to be of mitochondrial origin. Furthermore, there is no published evidence that even those molecules of true mitochondrial origin actually constitute "the mitochondrial genome" (i.e., code for genes normally found in mtDNA), although recent studies indicate that the 16-kb linear molecule found in mitochondria of *C. reinhardtii* (Ryan *et al.*, 1978; Grant and Chiang, 1980) does indeed contain several *bona fide* mitochondrial genes (P. Boer, L. Bonen, R. Lee, and M. Gray, unpublished data). Caution is also warranted, given that mitochondria from numerous angiosperms and fungi contain extrachromosomal plasmids of various sizes and conformations, which in some cases are of much higher copy number than the main mitochondrial genome (Sederoff, 1984) (see Section 4.5).

4.2. Genome Organization and Intragenomic Recombination

In the last year it has become clear that much of the previously described "heterogeneity" of plant mtDNAs, especially those DNAs isolated from native plant tissues, can be understood in terms of specific recombination events that resolve a single circular "master" chromosome into various smaller circles and sometimes even linear molecules. For example, the simplest model for the mitochondrial genome of turnip (*Brassica campestris*) is that it is organized into three discrete circular chromosomes (Fig. 8). A master chromosome contains the entire 218-kb se-

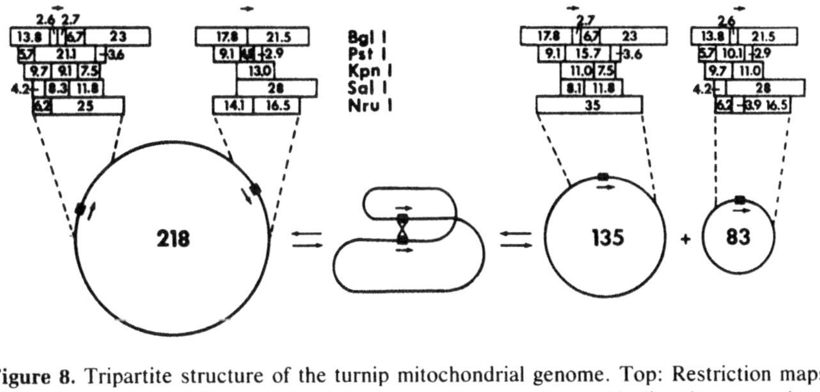

Figure 8. Tripartite structure of the turnip mitochondrial genome. Top: Restriction maps of the four regions of the mitochondrial genome that contain a 2-kb directly repeated sequence. Bottom: Predicted organization of the turnip mitochondrial genome into three circular molecules and their interconversion via recombination within the 2-kb repeat. Solid boxes indicate the position of the repeat and arrows its relative orientation. [Reprinted with permission from Palmer and Shields (1984).]

quence complexity of the mitochondrial genome, including two copies of a 2-kb sequence element present as direct repeats separated by 135 and 83 kb (Palmer and Shields, 1984). The two smaller circles are also 135 and 83 kb in size, contain one copy each of the 2-kb repeat element, and are postulated to interconvert with the master chromosome via a cointegration-resolution pathway mediated by reciprocal recombination within the 2-kb repeat (Fig. 8). Approximately two molecules of each of the two smaller circles are present for each molecule of the master chromosome (Palmer and Shields, 1984).

The basic recombinational elements of the turnip genome are also found in the mitochondrial genome of normal, male-fertile maize (Lonsdale *et al.*, 1983b, 1984; Lonsdale, 1984). However, the maize genome is larger (570 kb, compared to 218 kb in turnip) and features six pairs of large repeated sequences (compared to one pair in turnip), at least five pairs of which appear to be recombinationally active. Lonsdale and coworkers (1984) suggest that two of the pairs of repeats are preferred sites of recombination, and that the bulk of the mtDNA exists as four circles, of 67, 250, 253, and 503 kb. The circular organization of the mitochondrial genome in male-fertile maize contrasts with the situation in mitochondria from the CMS-S cytoplasm. Recombination between the linear S_1 and S_2 plasmids unique to the S cytoplasm and homologous sequences in chromosomal mtDNA appears to convert a high proportion of the S genome into linear molecules with S_1 and S_2 at their termini (Schardl *et al.*, 1984).

Repeated sequences that mediate intragenomic recombination have also been found in several other plant mtDNAs (Table III). In most of these less extensively mapped genomes the orientation of these recombination repeats (and therefore the outcome of the recombination event, inversion or deletion) has not been determined. As judged by filter hybridization homologies (Stern and Palmer, 1984b), it appears that different sequences are repeated and serve as sites for recombination in various angiosperm mitochondrial genomes. For example, strong sequence homologies were found to the recombination repeats from turnip and spinach in each of 13 angiosperms mtDNAs tested. Yet those sequences homologous to the turnip repeat appear to function as a recombination repeat only in turnip, while sequences homologous to the spinach repeat appear to be recombinationally active only in spinach, two species of pokeweed, and wheat (Table III) (Stern and Palmer, 1984b). In contrast, a major recombination repeat from maize [the "3-kb" repeat (Lonsdale et al., 1983b)] did not even hybridize to any other mitochondrial genome (Stern and Palmer, 1984b). It should be stressed that these filter hybridization studies do not rule out the possibility that a common, but relatively short recombination site is embedded within these nominally different recombination repeats.

Five of the six well-characterized recombination repeats encode part or all of a plant mitochondrial gene of known function (Table III). The frequent association of genes with repeated sequences seems especially peculiar for plant mitochondrial genomes in view of their large size, limited repeat content, and limited gene content (only 2–20% of their sequences are likely to be genic; see Section 4.6). Four of the five gene-containing recombination repeats in angiosperm mtDNAs encode one or more ribosomal RNAs (Table III). This seems to be part of a larger pattern noted in Section 2.4.1a, in which almost all known major repeated sequences in organelle genomes contain rRNA genes (Table III). The presence of a 26S rRNA-encoding recombination repeat in three diverse angiosperms (wheat, spinach, and pokeweed), together with the absence of such a repeat in nine other diverse species that occupy intermediate taxonomic positions relative to these three (Stern and Palmer, 1984b), strongly suggests that the 26S gene and associated sequences have either gained or lost the organization and function of a recombination repeat on multiple independent occasions during the evolution of angiosperm mtDNAs.

4.3. Rearrangement in Evolution

The 11-fold size range of angiosperm mtDNAs is *prima facie* evidence that extensive sequence rearrangements, at the very least large insertions, deletions, and duplications, must have occurred during their evolution. Interestingly, the sevenfold range in genome size encountered within the Cucurbitaceae is unaccompanied by gross changes in average base composition of mtDNA or by significant changes in the proportion of repeated sequences—5 to 10% in the smallest genome and 5% in the largest (Ward *et al.*, 1981). This suggests that large increases in mitochondrial genome size occur either by the uptake of single-copy sequences from foreign sources or by the amplification of preexisting mitochondrial sequences, followed by very rapid sequence drift and/or extensive rearrangement on a very fine scale [see Ward *et al.* (1981) for further discussion]. There is no correlation between mitochondrial genome size in these cucurbits and the amount of cpDNA-homologous sequences they contain (Stern *et al.*, 1983, and unpublished data). Therefore, one should look to the nucleus as a possible source of the extra sequences found in the very large mitochondrial genomes.

Sederoff *et al.* (1981) and Spruill *et al.* (1981) used filter hybridization techniques to study the evolution of DNA sequence organization in mitochondrial genomes of various species, races, and cultivars of *Zea* (maize and teosinte). Since only very limited restriction mapping data were then available for maize mtDNA, their conclusions are necessarily indirect and imprecise. Nonetheless, their data suggest that rearrangement events, of an undetermined nature, are frequent relative to nucleotide substitutions in *Zea*.

A similar study (J. Palmer and L. Herbon, unpublished data), in which cloned restriction fragments encompassing the mapped mitochondrial genome of turnip (*Brassica campestris*) (Palmer and Shields, 1984) were hybridized to filters containing mtDNAs from five other species in the genus *Brassica*, has allowed a direct comparison of the evolution of mtDNA and cpDNA (Palmer *et al.*, 1983a) within the same cytoplasm. Several large sequence rearrangements (probably inversions) were found in the six *Brassica* mtDNAs, whereas no rearrangements were found in the cpDNAs (cf. Section 2.4.3). Although major rearrangements are important in the evolution of *Brassica* mtDNAs, small deletions/additions of the type described in Section 2.4.2 occur no more often in these mtDNAs than in cpDNAs of the same species (J. Palmer and L. Herbon, unpublished data).

A special class of mtDNA rearrangements is associated with two linear, transposon-like plasmid DNAs (S_1 and S_2) that are found in the CMS-S cytoplasm of maize (see Section 4.5). It was originally shown by Levings *et al.* (1980) that in spontaneous cytoplasmic revertants to fertility from the CMS-S cytoplasm Vg, S_1 and S_2 disappear as autonomous replicons and that at least one of these, S_2, may become integrated into the main mitochondrial genome. More recently, Kemble and Mans (1983) have provided evidence that upon reversion most or all of S_2 and parts of S_1 transpose into variable target sites within the main mitochondrial genome, or, alternatively, that rearrangement occurs in the organization of preexisting S-homologous sequences in the main genome. Although sequences homologous to S_1 and S_2 are present in the high-molecular weight mtDNA of fertile (N) cytoplasms (Thompson *et al.*, 1980; Lonsdale *et al.*, 1981; McNay *et al.*, 1983; Kemble and Mans, 1983), reversion of the CMS-S plants to fertility does not restore the organization of the genome to that of N plants (Kemble and Mans, 1983).

4.4. Rearrangement in Cultured Cells and Somatic Crosses

The plasticity of the plant mitochondrial genome has also become evident from studies on plant cells cultured *in vitro* and plants derived by somatic fusion of protoplasts. Plants regenerated from fused protoplasts of species in either *Nicotiana* or *Petunia* have consistently exhibited mtDNA restriction patterns that differ from those of either fusion parent (Belliard *et al.*, 1979; Nagy *et al.*, 1981, 1983; Galun *et al.*, 1982; Boeshore *et al.*, 1983). Moreover, individual plants regenerated from the same fusion usually have mtDNAs that differ from one another. These alterations are not due to culture or selection conditions, or alterations in nuclear background, and are not detected when protoplasts with identical mtDNA restriction patterns are fused (Nagy *et al.*, 1983; Boeshore *et al.*, 1983). The complexity of the restriction patterns and a lack of information on the organization of the mitochondrial genomes in these species has hindered definitive molecular explanations for these alterations. The most detailed published analysis of altered mtDNAs is that of Boeshore *et al.* (1983), who used cloned probes and filter hybridizations to analyze mtDNAs of interspecific somatic hybrids in *Petunia*. They suggest two nonexclusive explanations for their results: (1) Intermolecular recombination of parental mtDNA molecules may occur following fusion. (2) Separate mtDNA molecules from the two parents may assort independently following fusion. Recent analysis of cosmid clones containing large segments of the mtDNA from these *Petunia* somatic hybrid regenerants has provided direct evidence in favor of the recombination hypothesis (M. Rothenberg, C. Snyder, B. Moeshore, M. Hanson, and S. Izhar, unpublished data).

The evolutionary ramifications of the apparent recombination or reassortment of mitochondrial genomes in somatic cell crosses is unclear. Since cpDNA and mtDNA are both maternally inherited in *Nicotiana* and *Petunia*, somatic fusion alterations in mtDNA are probably of no consequence to the evolution of these species. However, an important, but so far unexplored area, is the state of the mitochondrial genome in plants, such as *Oenothera* or *Pelargonium* in which plastids are known to be biparentally inherited. Thus far, there is little evidence that plastids actually fuse in these species and none at all that their cpDNAs recombine (Kutzelnigg and Stubbe, 1974; Sears, 1983). However, the fact that mitochondrial and chloroplast behavior is so different in somatic cell crosses—cpDNAs do not recombine in the *Nicotiana* and *Petunia* protoplast fusions in which mtDNAs apparently do recombine, and chloroplasts and mitochondria also assort independently of one another (Belliard *et al.*, 1978; Aviv *et al.*, 1980, 1984; Scowcroft and Larkin, 1981; Schiller *et al.*, 1982; Galun *et al.*, 1982; Kumar *et al.*, 1982; Fluhr *et al.*, 1980, 1984)—raises the possibility that mitochondria might fuse and their genomes recombine in species with biparental inheritance. This of course assumes that mitochondrial inheritance is biparental in those species where plastid inheritance is known to be biparental. Unfortunately, very few data are available on this question, and in fact limited ultrastructural studies (Vaughn, 1981; Vaughn *et al.*, 1981) suggest that mitochondrial inheritance may be uniparental in species with biparental plastid inheritance.

As reviewed in Section 4.2, there is growing evidence that the bulk of the mtDNA prepared from intact plants is organized into fairly large circular molecules. This conclusion is consistent with electron microscopic studies in which this DNA has been visualized as consisting of a broad size range of primarily linear molecules, which presumably result from degradation of the large circles during isolation [reviewed in Leaver and Gray (1982)]. In contrast, as much as 45% of total mtDNA can be isolated as supercoiled, circular molecules from a variety of plant cell suspension culture lines (Sparks and Dale, 1980; Dale, 1981; Dale *et al.*, 1981, 1983) and tissue cultures (Brennicke and Blanz, 1982). The circles are generally quite small, ranging in size from 1.5 to 60 kb, and the smallest circles are often present in extremely high copy number. Different size classes of small circles often show sequence homology with one another (Dale, 1981), in certain cases resulting from simple dimer and trimer relationships (Dale *et al.*, 1983). Different cultures established from the same variety sometimes have radically different populations, both qualitatively and quantitatively, of these small circles (Dale *et al.*, 1981).

These observations, together with the failure to find circles of these sizes in anywhere near these levels in native plant tissues, suggest that processes of intra- and intermolecular recombination and deletion of mtDNA may be considerably intensified when plant cells are cultured *in vitro*. Whether this occurs by an intensification of processes that normally occur at a low frequency in native tissues or by entirely new modes of recombination and rearrangement remains to be seen. In contrast to the instability of mtDNA in culture, cpDNA prepared from long-term (1000 generations) cell suspension cultures of tobacco is indistinguishable in circle size and restriction patterns from tobacco leaf cpDNA (Seyer *et al*, 1982).

A different form of tissue culture variation in mitochondrial genome composition has been reported in maize. Certain maize plants regenerated from tissue cultures containing the T cytoplasm, which confers male sterility and sensitivity to *Helminthosporium* toxin, phenotypically have the normal (N) cytoplasm; i.e., are male-fertile and toxin-resistant (Gengenbach *et al.*, 1977; Brettell *et al.*, 1980). The mtDNA restriction patterns of individual regenerants of this type differ from one another and from mtDNA of both the T and N cytoplasms (Gengenbach *et al.*, 1981; Kemble *et al.*, 1982). No differences were observed in the regenerants in terms of their composition of the small plasmid DNAs (Table XI) found in many maize cytoplasms (Kemble *et al.*, 1982). Tissue culture variation in maize mtDNA composition has also been reported in the N cytoplasm (McNay *et al.*, 1984) and S cytoplasm (Chourey and Kemble, 1982). In none of these cases has the molecular basis of these mtDNA alterations been determined. As pointed out by Kemble *et al.* (1982), the fact that mtDNA changes occur readily during culture and/or regeneration contrasts with the constancy of the genome during conventional sexual propagation.

4.5. Extrachromosomal Elements in Plant Mitochondria

Mitochondria isolated from native tissues of a number of plant species contain small (1–11 kb) circular and linear plasmid DNAs that replicate autonomously from the large and complex chromosomal genome (Table XI). In many of these species, most prominently maize, the presence of or alterations in the number and/or size of these small plasmid DNAs show various correlations with cytoplasmic male sterility [see references in Table XI and Laughnan and Gabay-Laughnan (1983)]. However, aside from their possible involvement in CMS, there are no other phenotypic effects associated with these molecules.

Table XI
Plasmids in Plant Mitochondria[a]

Taxon	Plasmid size and conformation[b]	Reference
Zea	7.5 kb, linear, ITR	Weissinger *et al.* (1982)
	6.4 kb, linear, ITR	Pring *et al.* (1977)
	5.4 kb, linear, ITR	Pring *et al.* (1977)
	2.3 kb, linear; 1.9, 1.6, and 1.4 kb, circular	Kemble and Bedbrook (1980), Kemble *et al.* (1980, 1983b)
Sorghum bicolor	5.7 and 5.3 kb, linear ITR	Dixon and Leaver (1982), Pring *et al.* (1982)
Triticum aestivum	0.3–6.0 kb, circular	Handa *et al.* (1984)
Beta vulgaris	Four circles, 1.3–1.5 kb, and linear molecules, 2–10 kb	Powling (1981)
Vicia faba	Three circles, 1.4–1.7 kb	Negruk *et al.* (1982), Boutry and Briquet (1982), Goblet *et al.* (1983)
Vicia villosa	Multiple plasmids	Nikiforova and Negruk (1983)
Medicago sativa	Multiple plasmids	Nikiforova and Negruk (1983)
Brassica	11.3 kb, linear	Palmer *et al.* (1983c)

[a]This listing only includes low-molecular weight, extrachromosomal DNAs isolated from mitochondrial of native plant tissues; mitochondria from cultured plant cells contain additional small circular molecules (Section 4.4).
[b]ITR, Inverted terminal repeats.

In the present context, it is of interest to examine the origins of these plasmids and to see how they vary among related species. Much of this information concerns the S_1 and S_2 plasmids introduced in Section 4.3. The S_1 and S_2 plasmids are linear molecules, 6.4 and 5.4 kb in length (Pring *et al.*, 1977). They share 1.5 kb in sequence homology at one end and have terminal inverted repeats 0.2 kb in length (Kim *et al.*, 1982). The S_2 plasmid has been sequenced and contains two unidentified open reading frames of 339 and 1098 amino acids (Levings and Sederoff, 1983). Weissinger, *et al.* (1982) found that 12 out of 81 Latin American maize races surveyed contain two mtDNA plasmids, termed R_1 and R_2. The R_2 is apparently identical to S_2 and R_1 is largely homologous with, but 1 kb larger than, S_1. Further analysis has led to the hypothesis that S_1 may have arisen by recombination between R_1 and R_2 (S_2) (Levings *et al.*, 1983). In contrast to the correlation between the presence of free S_1 and S_2 molecules and sterility in the S cytoplasm, all 12 races that contain R_1 and R_2 show normal male fertility (Weissinger *et al.*, 1982). In surveys of 31 inbred maize lines and 25 Mexican maize races Kemble *et al.* (1980, 1983b) found tremendous variation in the content of several mitochondrial plasmid molecules that are smaller and less abundant than S_1 and S_2 and their relatives.

Two CMS lines of sorghum (*Sorghum bicolor*) contain large amounts of two linear plasmid DNAs of similar sizes, 5.7 and 5.3 kb, to S_1 and S_2 (Dixon and Leaver, 1982; Pring *et al.*, 1982). These two molecules have extensive cross-homology with one another and possess terminal inverted repeats, both features of S_1 and S_2. In addition, the *Sorghum* plasmids show weak sequence homology with S_1 and S_2 in filter hybridizations. The diverse array of S_1- and S_2-related plasmid DNAs found in only a limited cross section of the large number of wild and cultivated lines from *Zea* and *Sorghum* examined thus far, together with the demonstration that sequences largely homologous to these molecules are present in the main mitochondrial genome of fertile maize lines (Section 4.3), further emphasizes the dynamic nature of plant mtDNAs in terms of their propensity to transpose and rearrange, and to lose and gain sequences.

Although S_1 and S_2 and their relatives may have originated from the main mitochondrial genome, such is apparently not the case for other small mtDNA plasmids from *Zea* (Kemble and Bedbrook, 1980), and also for mtDNA plasmids from *Beta* (Powling and Ellis, 1983) and *Brassica* (Palmer *et al.*, 1983c), all of which fail to hybridize to the main genome of their respective species. The absence of main genome homology for the linear *Brassica* plasmid is consistent with indirect genetic evidence that this molecule, although mitochondrially located, is not transmitted through the cytoplasm and may even originate from the nucleus (Palmer *et al.*, 1983c). In this context, it is noteworthy that multiple sequence homologies have been found between the maize S_1 plasmid and the maize nuclear genome (Kemble *et al.*, 1983a) (Sections 2.4.4 and 4.6 and Table V).

The diversity and abundance of plasmid DNAs in plant mitochondria contrasts with their absence from chloroplasts in over 200 species of angiosperms examined. Among algae, minicircular cpDNA molecules have been observed in at least one bleached mutant of *Euglena*, although these appear to be remnants of the main chloroplast genome (Heizmann *et al.* 1982b and references therein) (see Section 2.4.2). The only evidence for DNA plasmids in chloroplasts of normal function comes from two independent studies on the alga *Acetabularia* (Green, 1976; Mazza *et al.*, 1980). The nature and origin of these molecules, and also of the spectacularly large putative chloroplast genome of these species (Section 2.3), need further investigation.

4.6. Gene Content and Gene Transfer

There is no published information about mitochondrial genes of algae. The information content of angiosperm mtDNAs has recently been reviewed by Leaver and Gray (1982) and Leaver *et al.* (1983). So far, angiosperm mtDNAs have been shown to encode 26S, 18S, and 5S rRNAs, an undetermined number of tRNAs, and three identified polypeptides—subunits I and II of cytochrome *c* oxidase and apocytochrome *b*. Isolated plant mitochondria synthesize 20–30 polypeptides [also reviewed in Leaver and Gray (1982) and Leaver *et al.* (1983)]. Six of these have been identified and include the three known to be encoded by mtDNA, plus the α subunit and the DCCD-binding proteolipid of the mitochondrial ATPase, and a putative ribosomal protein (probably homologous with the *var 1* protein of yeast mitochondrial ribosomes). Although some of these translation products may be artifactual, it nonetheless appears that angiosperm mitochondria may synthesize, and presumably encode, a few additional polypeptides beyond the 13–20 synthesized and encoded in animal and fungal mtDNAs. Even so, at most only 20% of the smallest angiosperm mitochondrial genome, from turnip, is likely to consist of coding sequences, and less than 2% of the largest, from muskmelon.

There is no indication of any variation in the coding content of the mitochondrial genome among different plants, although only a limited number of genes—for rRNAs (Bonen and Gray, 1980; Stern *et al.*, 1982; Iam and Sinclair, 1982; Huh and Gray; 1982) and subunit II of cytochrome *c* oxidase (T. D. Fox and Leaver, 1981; Hiesel and Brennicke, 1983)—or mitochondrially synthesized polypeptides—the α subunit of ATPase

(Boutry *et al.*, 1983; Hack and Leaver, 1983)—have been examined. In addition, the same spectrum of polypeptides is synthesized in isolated mitochondria from fertile cytoplasms of a range of angiosperms (Leaver and Gray, 1982). Specific variations do occur in mitochondrial translation products in association with CMS in maize (Forde *et al.*, 1978; Forde and Leaver, 1980) and sorghum (Dixon and Leaver, 1982). However, these probably result from alterations in the structure or expression of certain mitochondrial genes, and are unlikely to represent differences in gene content (Leaver and Gray, 1982; Leaver *et al.*, 1983).

Angiosperm mtDNAs contain at least two genes not found in animal and fungal mtDNAs [reviewed in Leaver *et al.* (1983)]. Only angiosperm mitochondria are known to synthesize the α subunit of ATPase and to encode 5S rRNA (in fact, only angiosperms are known to even contain a 5S rRNA in their mitochondrial ribosomes). Angiosperm and yeast mtDNAs encode the DCCD-binding proteolipid (ATPase subunit 9), while in animals and *Neurospora* the active gene is encoded by nuclear DNA.

Angiosperm mtDNAs are quite adept at incorporating sequences from other genomes, in particular the chloroplast genome (Table V). A number of plant mtDNAs have now been examined and each contains a species-specific pattern of chloroplast DNA-homologous sequences. Because many of these sequence homologies are to *bona fide* chloroplast genes (Table V), some of which are present with essentially identical sequence in the mtDNA and cpDNA of a given species (Stern and Lonsdale, 1982; Lonsdale *et al.*, 1983), it is inferred that the direction of sequence transfer is primarily from chloroplast to mitochondrion, and that sequence transfer is a continuing and ongoing evolutionary process [see discussion in Stern and Palmer (1984a)]. Although there is only a single reported instance of apparent traffic between plant mitochondrial and nuclear genomes (Kemble *et al.* 1983b), it would be very surprising if further investigation did not show more extensive homology between the two genomes. So far, nothing is known about the mechanisms whereby sequences are exchanged among the three cellular genomes. Among the possibilities that have been discussed are DNA transformation, both with or without organelle contact and fusion, organelle fusion and direct recombination between different genomes, and transfer mediated by cytoplasmic vectors, such as plasmids, transposons, or viruses (Stern and Lonsdale, 1982; Stern and Palmer, 1984a; Scott and Timmis, 1984).

Unfortunately, the sheer profligacy of genomic intercourse between mitochondria and other organelles is likely to obscure those relatively rare events in which actual gene function is transferred. In none of the reports of sequence homology involving plant mtDNA and cpDNA is there any evidence that the foreign sequences are actually functional within the mitochondrion. Furthermore, of the large number of recent studies that report sequence homologies between animal or fungal mtDNAs and nuclear DNA (Table V), only one (Van den Boogaart *et al.*, 1982) provides clear evidence for the actual transfer of gene function from the mitochondrion to the nucleus.

4.7. Gene Evolution and Genetic Code

Mitochondrial rRNA genes have a uniform and fairly unusual arrangement in a variety of angiosperms (Bonen and Gray, 1980; Huh and Gray, 1982; Stern *et al.*, 1982; Iams and Sinclair, 1982). The 18S and 26S rRNA genes are always found unlinked, at least 15 kb apart, while the 5S gene is always found closely linked to the 18S gene. Aside from angiosperms, only the yeast *Saccharomyces cerevisiae* has its small- and large-subunit mitochondrial rRNA genes located so far apart (M. W. Gray, 1982; Wallace, 1982). Moreover, the 5S gene arrangement is unique compared to chloroplasts and prokaryotes, whose 5S genes are always found closely linked to large-subunit rRNA genes (Fig. 7), and also compared to eukaryotic nuclei (except in the case of certain fungi), whose 5S genes are unlinked to either the 18S or 28S genes (Long and Dawid, 1980).

Complete sequence analysis of the 18S rRNA gene from wheat mitochondria (M. W. Gray, 1983; Spencer *et al.*, 1984) and the 26S rRNA gene from maize mitochondria (Dale *et al.*, 1984) reveals extensive structural and sequence homologies with the corresponding rRNA genes of *E. coli* and chloroplasts. The most notable homologies are for the small-subunit rRNA genes. For example, within six conserved regions that are strict secondary structure counterparts of the same regions in eubacterial 16S rRNA there is almost as much homology (76%) between wheat mitochondrial 18S rRNA and *E. coli* 16S rRNA as between maize chloroplast and *E. coli* 16S rRNAs (78%) (Spencer *et al.*, 1984). These homologies are much greater than those between small-subunit rRNAs of eubacteria and those of any fungal or animal mitochondrial genome, and provide the strongest evidence yet in favor of an endosymbiotic, specifically eubacterial, origin for any mitochondrial genome (M. W. Gray, 1983) (see Section 5.1). The strongly eubacterial nature of wheat 18S rRNA indicates that, similar to cpDNA (Section 3.2.4), plant mtDNA is evolving very slowly at the sequence level, especially compared to fungal and animal mtDNA (W. M. Brown, 1983; M. W. Gray, 1983). Consistent with this notion is the fact that only one sequence difference exists between the 3′-terminal 295 nucleotides of wheat and maize 18S mitochondrial rRNAs (Chao *et al.*, 1983; Spencer *et al.*, 1984).

Wheat mitochondrial 5S rRNA has an ambiguous sequence and structural relationship to both prokaryotic and eukaryotic 5S rRNAs (Spencer *et al.*, 1981). It shares only marginally more sequence homology with *E. coli* 5S rRNA than with wheat cytoplasmic 5S rRNA, and displays secondary structure features characteristic of both prokaryotic and eukaryotic rRNAs. In addition, it possesses unique features found only in mitochondrial 5S rRNAs of angiosperms, and hence has been considered a distinct class of 5S rRNA (Spencer *et al.*, 1981; M. W. Gray and Spencer, 1981). Wheat mitochondrial 5S rRNA is, however, closely related (97% nucleotide sequence homology) to maize mitochondrial 5S rRNA (Chao *et al.*, 1983). One possible explanation for these observations is that mitochondrial 5S rRNA may have gone through a period of rapid evolution, perhaps as a consequence of changes in its function [see Spencer *et al.*, (1981) for discussion], during an early period of plant evolution. Now, with its function stabilized, it is evolving more slowly, in tune with the rest of the highly conserved (at the sequence level) angiosperm mitochondrial genome. Indeed, the fact that among mitochondrial ribosomes only those of angiosperms even contain 5S rRNA may be taken as a general indication of alterations in its function and/or selective constraints on its evolution within mitochondria.

The only sequenced plant mitochondrial tRNA and tRNA gene is a $tRNA^{Met}$ gene from wheat (M. W. Gray and Spencer, 1983). Three observations stand out from the sequence data. First, this tRNA gene is separated by only 1 bp from the 5′ end of the 18S rRNA and appears to be cotranscribed with it. If this is the case, then the $tRNA^{Met}$ may serve as a signal for processing of the primary transcript, as is postulated for mammalian mitochondria (Ojala *et al.*, 1981). Second, wheat mitochondrial $tRNA^{Met}$ shows considerable structural and sequence homologies with eubacterial and chloroplast $tRNAs^{Met}$ (M. W. Gray and Spencer, 1983). These homologies are dramatically stronger than those between any animal or fungal mitochondrial tRNAs and eubacterial ones. Interestingly, the wheat mitochondrial tRNA does share a few structural deviations found thus far only in three fungal mitochondrial tRNAs and is also closer in sequence to these mitochondrial tRNAs than they are to eubacterial tRNAs. Thus, plant mitochondrial tRNAs may provide a crucial evolutionary link between eubacterial and fungal mitochondrial tRNAs (M. W. Gray and Spencer, 1983). A third observation is that the wheat mitochondrial $tRNA^{Met}$ gene does not encode the CCA terminus found in the mature tRNA (M. W. Gray and Spencer, 1983), as is also the case for tRNA genes of all other mitochondria (Wallace, 1982) and chloroplasts (Section 2.4.3), and for certain tRNA genes of cyanobacteria (Williamson and Doolittle, 1983; Tomioka and Sugiura, 1984).

The only sequenced plant mtDNA protein gene is that for cytochrome oxidase subunit II from maize (T. D. Fox and Leaver, 1981) and *Oenothera berteriana* (Hiesel and Brennicke, 1983). The two genes have 89% nucleotide sequence homology and 88% deduced amino acid homology, which is on a par with monocot–dicot homologies for chloroplast genes (Tables VI and IX). Only 47% of the nucleotide substitutions in the coding regions of the two oxidase II genes are transitions; thus, like cpDNA, plant mtDNA does not appear to suffer from the extreme transition bias found in animal mtDNA evolution (see Section 3.2.4). Unlike the *Oenothera* gene and all homologous yeast and animal mitochondrial genes (Mahler, 1983), the maize gene contains a single, centrally located 794-bp intron. The intron contains two open reading frames, of 118 and 111 codons. However, neither open reading frame contains an AUG initiation codon or is in phase with the exons, and thus their significance is unclear. Michel and Dujon (1983) have recently shown that the oxidase II intron in maize has significant sequence and potential secondary structure homologies, suggestive of a common splicing mechanism, with "group II" introns of yeast mitochondrial genes and also with certain introns of chloroplast tRNA genes (see Section 3.1.1c).

Only a single, 1.2-kb transcript was found from the *Oenothera* gene (Hiesel and Brennicke, 1983), while multiple transcripts were found in maize (T. D. Fox and Leaver, 1981). The smaller, abundant maize transcripts lack intron sequences found in the larger transcripts and thus intron splicing is clearly implicated. However, there are surprisingly many transcripts from the maize gene, leading T. D. Fox and Leaver (1981) to postulate that some combination of multiple transcription start or stop signals and multistep processing may be involved.

The maize and *Oenothera* cytochrome oxidase II gene sequences (T. D. Fox and Leaver, 1981; Hiesel and Brennicke, 1983) provide the only available evidence concerning the genetic code used by plant mitochondria. In animal and fungal mitochondria, TGA, a stop codon in the standard genetic code, is used as a Trp codon, and in fact is favored overwhelmingly over the standard TGG in coding for Trp. However, no TGA codons are found within either plant gene and thus TGA may well be used as a nonsense codon in plant mitochondria. Both genes contain CGG (normally an Arg codon) in three highly conserved positions in which homologous mitochondrial genes from other organisms encode Trp. Thus, the genetic code in plant mitochondria may differ from the universal code and all other mitochondrial codes in using both CGG and TGG as Trp codons. In support of this inference, a fourth CGG codon is found in *Oenothera* in the same position as a TGG in maize.

Mitochondrial rRNA genes, a tRNA gene, and a polypeptide gene are highly conserved in sequence among angiosperms. However, since genes comprise only a small portion of plant mitochondrial genomes, one must ask whether the bulk of the genome is evolving at an equally slow rate as these select genes. Several groups have studied variation in restriction patterns of total mtDNA from normal and male-sterile maize cytoplasms (Pring and Levings, 1978; Borck and Walbot, 1982) and from wild races and species of *Zea* (Levings and Pring, 1977; Timothy *et al.*, 1979; Kemble *et al.*, 1983b; Weissinger *et. al.*, 1983). However, the complexity of these restriction patterns, together with the prevalence of rearrangements in these DNAs (Sederoff *et al.*, 1981; Spruill *et al.*, 1981), has precluded the precise quantitation of point mutation rates on the basis of restriction fragment differences alone. More recently, restriction mapping and filter hybridization comparisons have revealed a striking degree of interspecific restriction site conservation over the entirety of the mapped, ~220-kb mitochondrial genomes of six species in the genus *Brassica* (J. Palmer and L. Herbon, unpublished data). In fact, the *Brassica* mitochondrial genome appears to be evolving with a slow rate of sequence change similar to that of the highly conserved chloroplast genome (Palmer *et al.*, 1983a) of the same species.

5. Phylogenetic Implications of Organelle DNA Variation

The record of change contained within organelle genomes has been used chiefly to study two quite different kinds of evolutionary histories—the first relating to the primary lines of origin and descent of the organelles themselves, and the second concerning phylogenetic relationships among recently diverged species of flowering plants.

5.1. Endosymbiotic Origins of Chloroplasts and Mitochondria

The question of how chloroplasts originated—either autogenously, via compartmentalization of a single ancestral protoeukaryotic cell and its genome to form both chloroplast and nucleus, or xenogenously (endosymbiotically), through the symbiotic association of a photosynthetic prokaryote and a primitive protoeukaryote—now seems largely settled in favor of the latter theory. The logical basis of the relevant arguments and the nature of the molecular evidence have been reviewed in detail by M. W. Gray and Doolittle (1982) and Wallace (1982), and more recently summarized by M. W. Gray (1983). Suffice it to say that in essentially every aspect of genome organization, gene arrangement (see Fig. 7 and Table VII), and gene structure and sequence (Section 3), plastid and eubacterial genomes are much more closely related to each other than either is to the nuclear genome. Hence, the argument goes, plastid and eubacterial genomes almost certainly shared a common ancestor more recently than did plastid and nuclear genomes, and thus the present-day occurrence of plastid and nuclear genomes in the same cell must reflect horizontal evolution; i.e., endosymbiosis.

Some of the strongest evidence in favor of the endosymbiotic theory is provided by the striking similarities in sequence and structure of the rRNA genes of the eubacterium *E. coli* and those of chloroplasts of various plants and algae (Section 3.1.1a). More recently, sequence analysis of rRNA genes from the cyanobacterium *Anacystis nidulans* (Tomioka and Sugiura, 1983; Kumano *et al.*, 1983; Douglas and Doolittle, 1984a, b) has provided even more compelling support for a eubacterial origin of plastids in general, and a cyanobacterial origin in particular: Each of the 16S, 23S, and 5S rRNA genes is significantly more closely related between angiosperm chloroplasts and *Anacystis* (83, 78, and 66% homology, respectively) than between chloroplasts and *E. coli* (74, 68, and 54% homology, respectively). Similar sequence homology relationships exist for the $tRNA^{Ile}$ and $tRNA^{Ala}$ genes located in the transcribed rDNA spacer. Furthermore, as noted in Section 3.1.1c, the *Anacystis* $tRNA^{Ile}$ gene, like all characterized chloroplast tRNA genes, but unlike the *Anacystis* $tRNA^{Ala}$ gene and all tRNA genes from *E. coli* and *B. subtilis*, does not encode the 3′-terminal CCA sequence found in the mature tRNA (Williamson and Doolittle, 1983; Tomioka and Sugiura, 1984). Extremely strong sequence homologies have also been found between the *rbcL* genes of various cyanobacteria and chloroplasts (Section 3.1.2a). Thus, as more sequence information becomes available, the evolutionary ties between chloroplasts and eubacteria, specifically cyanobacteria, appear stronger and stronger.

Accepting an endosymbiotic origin for chloroplasts, we can now turn our attention to several much more complex and interesting problems relating to the origin and establishment of plastids: i.e., how many separate endosymbioses occurred in the lines of descent leading to extant plants and algae; which extant photosynthetic prokaryotes are likely to be the most direct descendants of the organisms that established these endosymbioses; and how do the evolutionary fates of the resulting plastid genomes compare in terms of gene sequence, gene arrangement, and gene transfer to the nucleus? Consideration of the great variety of photosynthetic (pigment composition and thylakoid arrangement) and ultrastructural (number and type of envelope membranes) features found in extant algae has led most current authors to postulate at least two (Cavalier-Smith, 1982) and as many as six or seven (Gibbs, 1981a,b; Whatley and Whatley, 1981; Whatley, 1983) separate plastid endosymbioses. The more complicated schemes postulate single-step endosymbioses leading to chloroplasts of chlorophytes (from a chlorophyll a- and b-containing prokaryote such as *Prochloron*) and those of *Cyanophora* and rhodophytes (from chloropyll a- and phycobilin-containing prokaryotes such as cyanobacteria), two successive endosymbioses leading to the euglenophyte chloroplast (the second being the uptake of a green algal chloroplast or cell), and several independent two-step endosymbioses, all starting with a cyanobacterium, leading to the various chloroplast types found among chromophytes.

Molecular studies have so far not supported the idea that chlorophytic plastids originated from a *Prochloron*-type prokaryote rather than a cyanobacterium; i.e., plastid genes of chlorophytes appear equally similar to those of *Prochloron* and cyanobacteria (Seewaldt and Stackebrandt, 1982; MacKay *et al.*, 1982). However, the comparative data base is limited to a single, very short (~120 bp) sequenced gene [for 5S rRNA (MacKay *et al.*, 1982)] and a partial sequence (an oligonucleotide catalogue) of a larger gene [16S rRNA (Seewaldt and Stackebrandt, 1982)]. In addition, the phylogenetic interpretation of even these limited data is controversial (Val Valen, 1982). Much more sequence information (e.g., of entire rRNA and ATPase genes) is therefore needed before *Prochloron* can be dismissed (Cavalier-Smith, 1982) as just a derivative cyanobacterium.

Although molecular studies do not thus far support a *Prochloron*-type origin for chlorophytic plastids, they are consistent with a specifically cyanobacterial origin for plastids of red algae (rhodophytes). That is, oligonucleotide catalogues reveal somewhat greater sequence identity between 16S rRNAs of a red alga (*Porphyridium cruentum*) and cyanobacteria than between either type and those of *Euglena* and the angiosperm *Lemmna minor* (G. E. Fox *et al.*, 1980; Doolittle and Bonen, 1981). Again, however, complete sequence analysis, preferably of several different genes, is needed before these results can be accepted at full face value.

Perhaps the strongest molecular evidence in favor of multiple plastid endosymbioses concerns the relationship between plastids of *Euglena* and chlorophytes. As summarized in Section 6, *Euglena* cpDNA differs significantly from chlorophytic cpDNA in a number of important parameters. For example, *Euglena* chloroplast 16S rRNA is somewhat *less* homologous in sequence to angiosperm 16S rRNA [80% homology (Graf *et al.*, 1982)] than are 16S rRNAs of angiosperms and cyanobacteria to each other [83% homology (Tomioka and Sugiura, 1983)]. Although unequal rates of 16S evolution could certainly affect this comparison, it nonetheless suggests a very deep branching for chloroplasts of chlorophytes and euglenophytes relative to cyanobacteria. Much more convincing molecular evidence that the *Euglena* chloroplast is of an independent endosymbiotic origin actually comes from sequence analysis of two cytoplasmic RNA species (5S rRNA and $tRNA^{Phe}$). These RNAs from *Euglena* are much more similar to those of animals and protists (particularly trypanosomes) than to those of chlorophytes (Chang *et al.*, 1981; Delihas *et al.*, 1981; Kumazaki *et al.*, 1982). In other words, the *Euglena* plastid is present in a nonchlorophytic nuclear background and must have arisen by a separate endosymbiosis. The *Euglena* plastid resembles those of chlorophytes ultrastructurally and photosynthetically, but is surrounded by an extra outer membrane. It has therefore been suggested that it arose by the uptake (in some protozoan line) of a green algal cell (Gibbs, 1981a, b) or green algal chloroplast (Whatley and Whatley, 1981; Whatley, 1983).

No sequence data are available that bear on the origin of the diverse plastid types found among chromophytic algae. As discussed in Section 2.4.4, chromophytic algae (at least one representative, *Olisthodiscus luteus*) appear to resemble *Cyanophora paradoxa* and rhodophytic algae in encoding the gene for the small subunit of ribulose-1,5-bisphosphate carboxylase in their plastid genomes, apparently in a single operon together with the large-subunit gene, whereas in *Euglena gracilis* and all chlorophytes the small-subunit gene(s) is nuclear-encoded. Although this result does not relate directly to the question of multiple plastid origins, it clearly points the way to future sequence and structural studies on such key plastid gene complexes as those of ribosomal RNAs, ribosomal proteins, ATPase subunits, and carboxylase subunits. Over the next few years, these studies should greatly increase our understanding of the origins and relationships of algal plastids.

The eubacterial origin of plant mitochondria now seems largely settled by the recent sequencing of the 18S rRNA gene of wheat mitochondria (Spencer *et al.*, 1984) (also see Section 4.7 and discussion in M. W. Gray (1983)]. Biochemical and structural considerations suggest an organism resembling present-day nonsulfur purple bacteria as the likely endosymbiotic precursor to mitochondria, but rRNA oligonucleotide cataloging studies do not support this inference [reviewed in M. W. Gray and Doolittle (1982)]. Although plant mitochondrial rRNA is very clearly eubacterial in sequence (as is at least one plant mitochondrial tRNA; see Section 4.7), fungal and animal mitochondrial rRNAs are much less so, raising the possibility that these mitochondria may have independent origins or else vastly different evolutionary rates [see discussions in M. W. Gray and Doolittle (1982), M. W. Gray (1983), and Spencer *et al.* (1984)]. Unfortunately, no sequence data are available for any plant or algal mitochondrial genes other than those of angiosperms.

5.2. Phylogenetic Relationships among Plants and Algae

The phylogenetic uses and merits of studying cpDNA variation have recently been reviewed (Palmer, 1985a,b; Giannasi and Crawford, 1985) and only a brief summary of this rapidly growing field will be presented here. The analysis of cpDNA restriction fragment patterns is a very rapid means of indirectly sampling a large sequence pool (up to 3000 bp) from a number of closely related species. Restriction site mutations and length mutations can usually be inferred by direct inspection of restriction patterns, although some sort of restriction mapping analysis is often needed to confirm these inferences. As a consequence of the very slow rate of sequence change in cpDNA, and also its relative freedom from an extreme transition/transversion bias, the incidence of parallelism and convergence in cpDNA restriction site mutations is very low at the interspecific level (Palmer and Zamir, 1982; Palmer *et al.*, 1983a, 1985a). This results in molecular phylogenies that are essentially free of ambiguity. So far, cpDNA restriction fragment studies have provided a wealth of phylogenetic information at the interspecific level, particularly in such agriculturally important genera as *Lycopersicon*, *Nicotiana*, *Brassica*, *Triticum*, and *Pisum* (Palmer 1985a,b; Giannasi and Crawford, 1985). The fact that cpDNA is maternally inherited in many angiosperms (Sears, 1980, 1983) has proved to be critical in allowing important qualitative and quantitative insights into the origin and ancestry of the interspecific hybrids and polyploid species that are extremely common among plants (e.g. Kung *et al.*, 1982; Palmer *et al.*, 1983a; Erickson *et al.*, 1983; Bowman *et al.*, 1983; Tsunewaki and Ogihara, 1983).

The direct analysis of complete cpDNA sequences has so far not been attempted with the explicit purpose of exploring phylogeny. Nonetheless, certain phylogenetic conclusions, e.g., concerning the endosymbiotic origins of chloroplasts (see preceding section), can already be extracted from the large number of accumulated chloroplast gene sequences. A second conclusion supported by these sequence data is that the monocots appear to be a very early, probably monophyletic branch of evolution among the angiosperms (cf. Tables VI and IX) (Palmer, 1985a,b). With ever-continuing increases in the ease and efficiency of DNA sequencing, it is expected that cpDNA sequence analysis will become widely used to investigate major questions of plant phylogeny and evolution, particularly above the genus and family levels, where restriction studies are difficult and often inappropriate. Sequence analysis may ultimately be of the greatest value in establishing a rational phylogenetic outline for eukaryotic algae, whose simple morphology often impedes conventional taxonomic analysis.

The analysis of major structural rearrangements in chloroplast genome architecture constitutes a third, decidedly more limited approach for assessing plant relationships. As individual mutations, the large inversions described in Section 2.4.3 and the inverted repeat deletion described in Section 2.4.1.1 are extremely dramatic and phylogenetically powerful. For example, three of the described cpDNA inversions each circumscribe one of the so-called "great" plant families (Fabaceae, Poaceae, Asteraceae). If more extensive phylogenetic examination were to reveal that certain other families or orders also possessed one of these inversions, this would constitute strong evidence for common descent of these taxa and the corresponding "great" family. Unfortunately, the very rarity of these major rearrangements places severe limitations on their general usefulness for most phylogenetic questions.

Since plant mtDNA has the same pattern of inheritance and a similarly slow rate of sequence evolution as cpDNA, one might expect it to be equally useful for phylogenetic studies. Technical and methodological considerations, however, suggest otherwise. Plant tissues, particularly leaves, contain much less mtDNA than cpDNA, and the purification of mtDNA in sufficient quantity for comparative analysis is a major problem for many groups of plants. The extremely large size of most plant mitochondrial genomes effectively precludes the critical analysis of whole genome restriction patterns, as is quite feasible with cpDNA. In addition, the increased prevalence of rearrangements, which require very extensive molecular analysis for careful and complete diagnosis, makes it much harder to analyze satisfactorily mtDNA restriction pattern variation. Technically, mtDNA and cpDNA are on a more equal footing in terms of gene sequencing for systematic purposes. However, given that cpDNA encodes more genes, thus providing a greater range of sequences to choose from; that chloroplast gene sequencing has a large headstart on plant mitochondrial gene sequencing; and that certain chloroplast genes (e.g., *rbcL* and *psbA*) will probably continue to be sequenced widely for practical reasons relating to genetic engineering; it seems likely that chloroplast genes will be a more natural substrate for phylogenetic studies. In summary, cpDNA possesses a number of advantages relative to plant mtDNA in terms of phylogenetic analysis. Although there will undoubtedly be certain specific cases (as in the evolution of maize; see references in Section 4.7) where mtDNA will be of great phylogenetic value, most future systematic studies using plant organelle DNA will probably focus on the chloroplast genome.

6. Summary Comparison of Chloroplast and Mitochondrial DNA Evolution

The data reviewed in this chapter reveal a striking dichotomy in the evolution of the two cytoplasmic genomes of angiosperms (Table XII).

Table XII
Comparison of cpDNA and mtDNA Evolution in Angiosperms

Similarities
Slow rate of nucleotide sequence evolution
Constancy of base composition
Constancy of gene content and operon structure
Variable (optional) introns
Slow rate of small deletions/additions
Differences
Genome size: relatively constant in chloroplasts; highly variable in mitochondria
Uptake of foreign sequences: rare in chloroplasts; frequent in mitochondria
Genome conformation: single circular genetic unit in chloroplasts, no plasmids; variable number of circular (rarely linear) genetic units in mitochondria, plasmids common
Genome arrangement: relatively constant in chloroplasts; highly variable in mitochondria

Both genomes are quite conservative at the gene level and nucleotide sequence level, whereas plant mtDNA is changing much more rapidly in terms of genome structure and arrangement than is cpDNA. Solution hybridizations, restriction site comparisons, and gene sequencing all document a very low rate of nucleotide sequence evolution for the chloroplast genome as a whole and for many individual chloroplast genes. This sequence conservation is reflected at a very gross level by the conservation of average base composition of angiosperm cpDNA. Although relatively few plant mitochondrial gene sequences have been determined and although whole genome restriction site comparisons are still at an early stage, it nonetheless appears that plant mtDNA sequences are also changing very slowly, perhaps equally as slowly as chloroplast sequences. Different species of angiosperms all contain the same spectrum of organelle genes, although again the data base is much smaller for mitochondria than for chloroplasts. In addition, wherever any degree of operon structure (polycistronic mRNAs) is found in one angiosperm, it is found in all. For example, all chloroplast genes for 16S, 23S, and 4.5S rRNA are transcribed coordinately, but separately from that for 5S rRNA, while all plant mitochondrial genes for 18S and 5S rRNA are tightly linked physically (and probably cotranscribed), and are relatively far apart from that for 26S rRNA. A limited number of introns are found in both organelle genomes (in mRNA and tRNA genes in angiosperm chloroplasts; so far only in mRNA genes in angiosperm mitochondria). Furthermore, these introns can be optional in the sense that each of the mRNA introns found in the two genomes is not universally present in all angiosperms. The large number of sequenced angiosperm chloroplast genes all appear to utilize the same (universal) genetic code, while the one mitochondrial protein gene sequenced in two different plants also uses the same (albeit nonuniversal) code. Restriction site comparisons and sequencing studies reveal that small deletions/additions (of several hundred base pairs or less) are relatively infrequent in chloroplast genomes compared to other genomes. Initial studies suggest a similar trend in plant mtDNA.

Data on mitochondrial genome structure and arrangement are much more extensive than for mitochondrial sequence evolution, and here it is apparent that mitochondria and chloroplasts differ dramatically in their evolution. Among angiosperms, cpDNA ranges in size from 120 to 210 kb. Factoring out the sequence repetition represented by the large inverted repeat gives a total range of sequence complexities of only 115 to 150 kb. In contrast, the sequence complexity of angiosperm mtDNA ranges from approximately 200 to over 2000 kb. One reflection, although apparently not directly related in any proportional sense, of the size variation of plant mtDNAs is their ability to readily incorporate cpDNA sequences, while no evidence has yet been obtained for the reverse phenomenon. Angiosperm cpDNAs exist as a single circular genetic unit, and intragenomic recombination, where present, occurs within the same repeat sequence in all genomes. Angiosperm mtDNAs, on the other hand, exist as a collection of different sized circular molecules (sometimes even linear ones), resulting from intragenomic recombination within a variety of repeat sequences. In addition, plant mitochondria contain many diverse circular and linear extrachromosomal plasmids, both with and without homology to the main genome, whereas plasmids have not been found in angiosperm chloroplasts. A large body of evidence suggests that major sequence rearrangements (inversions, transpositions, and deletions/additions larger than 1 kb) are extremely rare in most angiosperm cpDNAs, although the frequency of some of these events, particularly inversions, seems to be somewhat higher in certain legumes that have lost the inverted repeat structure that is the hallmark of most cpDNAs. Plant mitochondrial genomes appear to sustain internal rearrangements, often independent of any large size changes, quite frequently. These alterations are apparent not only in evolutionary comparisons between species, but also in studies on mtDNA rearrangement and recombination in cultured cells and somatic cell fusions.

In a very general sense, we can recognize a different mode of genome evolution for each of four well-characterized organelle genomes (Table XIII). These comparisons clearly show a complete uncoupling between rates of sequence evolution and structural evolution in organelle DNA. The one intriguing exception to this generalization is that the rate of small

Table XIII
Classification Scheme of Organelle Genome Evolution[a]

Genome	Sequence evolution	Structural evolution
Angiosperm cpDNA	Slow	Slow
Angiosperm mtDNA	Slow	Rapid
Mammalian mtDNA	Rapid	Slow
Fungal mtDNA	Rapid	Rapid

[a]See W. M. Brown (1983) for data on animal mtDNA evolution, and M. W. Gray (1982), Wallace (1982), Mahler (1983), and Sederoff (1984) for data on fungal mtDNA evolution.

length mutations seems to be more or less coupled to the rate of point mutations (i.e. both rates are low in cpDNA and mtDNA of angiosperms) but are high in animal mtDNA (Cann and Wilson, 1983). The high rate of sequence evolution of animal mtDNA has been explained in terms of general factors affecting the entire genome—such as an enhanced mutation pressure resulting from unfaithful DNA replication, faulty or deficient repair, rapid DNA turnover, and high exposure to mutagens, and also a relative lack of a recombinational mechanism that could eliminate mildly deleterious mutations linked to advantageous ones—and also a specific relaxation of constraints on genes encoding the translational apparatus as a result of the reduced translational capacity (~13 polypeptides) of animal mitochondria (W. M. Brown, 1983; Cann and Wilson, 1983; Cann *et al.*, 1984). There are essentially no data that bear on the possible contribution of these factors to the slow rate of sequence change in plant cytoplasmic genomes. Perhaps the easiest to test experimentally are the accuracy of DNA replication and repair processes. These processes may also be particularly relevant, given that animal mitochondria are well known to be seriously deficient in repair and faulty in replication (W. M. Brown, 1983).

The hypothesis that relaxed translational constraints allow more rapid evolution of genes encoding translational components (Cann *et al.*, 1984), while perhaps true for animal and even fungal mitochondrial genomes, clearly cannot hold for plant mtDNA. This is in spite of the fact that plant mitochondrial genomes do not appear to encode substantially more polypeptides than other mitochondrial genomes. The low rate of sequence evolution of the *entire* plant mitochondrial genome is even more perplexing, given that only a small portion of this DNA is likely to have coding functions. The fact that both genes and nongenic sequences of plant mitochondria are highly conserved, more or less equally with chloroplast sequences, raises the possibility that the same set of factors, presumably all under the common control of the nucleus, regulates rates of sequence evolution in the two plant organelles.

Factors responsible for the conserved sequence arrangement of cpDNA were discussed at length in an earlier section. In the present context, it is relevant that certain of these factors—a reduced mutation pressure in terms of the absence of dispersed repeats that could serve as substrates for recombination and rearrangement, and a reduced fixation pressure in terms of the close packing of organelle genes—appear common both to the chloroplast genome and to the highly structurally constrained genome of animal mitochondria (W. M. Brown, 1983). A facile explanation for the much more dynamic structural arrangement of plant mitochondrial genomes is that the high proportion of noncoding sequences allows fixation of most rearrangements that occur. This simply raises the fundamental question of angiosperm mtDNA evolution: Why are these genomes so large and so variable in size, especially relative to the fairly invariant size of chloroplast genomes? This question becomes even more vexing if we extend our consideration to green algae, where the *Chlamydomonas reinhardtii* mitochondrial genome is over 100 times smaller than the largest angiosperm mitochondrial genome, even though the *C. reinhardtii* chloroplast genome is essentially the same size (actually slightly larger) as angiosperm cpDNAs. Unfortunately, no ready answers are forthcoming to explain this mitochondrial *C*-value paradox.

We have no clues yet about how chloroplast and mitochondrial genomes are evolving within any plant or algal lineages other than angiosperms. However, we can at least begin to compare the end products of chloroplast genome evolution in two algal lineages (as represented by the chlorophyte *Chlamydomonas reinhardtii* and the euglenophyte *Euglena gracilis*) and in angiosperms. The high levels of sequence homology between algal and angiosperm chloroplast rRNA genes and *rbcL* suggest that rates of sequence evolution are comparably slow in all three lineages. Whether the same holds true for rhodophytic and chromophytic cpDNAs remains to be seen. On the other hand, the *Euglena* genome is fundamentally different in terms of its dispensability, as witnessed by the ready formation of permanently bleached mutants that lack large portions of the plastid genome. This dispensability appears to be functionally related to the fact that many metabolic functions that normally occur inside the plastid in chlorophytes occur elsewhere in the cell in *Euglena*. It is likely that both of these features reflect in a fundamental sense the different endosymbiotic origins of these two plastid types. At least one difference in chloroplast gene content (for elongation factor G) is apparent between *Euglena* and chlorophytes for those functions that are present in both plastids. Furthermore, *Euglena* chloroplasts are distinguished by the multitude and unusual nature (for organelles) of the introns found in those genes encoding proteins, and by the polycistronic organization and lack of introns in those genes encoding tRNA. The organization of rRNA operons into a tandem repeat structure is a distinctive feature of the *Euglena* chloroplast genome that is clearly of evolutionary significance among laboratory strains, and presumably among natural species as well. In spite of the above differences, it should be emphasized that *Euglena* and chlorophytic chloroplasts are overwhelmingly similar in such functions as photosynthesis and protein synthesis.

Although cpDNAs of *Chlamydomonas reinhardtii* and angiosperms clearly have the same endosymbiotic origin and are overall very similar in sequence and function, they nonetheless differ in several important ways. *Chlamydomonas* cpDNA is 25 kb larger in sequence complexity than the largest known angiosperm cpDNA, is considerably different in sequence organization, features a distinctive family of dispersed repetitive sequence elements, has introns (in *psbA* and 23S rRNA genes) not found in angiosperms, and has a distinctive mosaic 23S rRNA gene structure (Fig. 7). How these differences relate to present-day evolutionary processes in green algae is not known.

We are clearly still at a primitive state in understanding the diversity of organelle genomes present in plants and algae, as well as the evolutionary forces responsible for this diversity. A tremendous amount of basic descriptive molecular biology is needed in order to determine the fundamental outlines of the multiple origins and lines of descent of plastids (and perhaps mitochondria, too), as well as the present course of evolution of these organelle genomes. At the same time, we are now in a position to begin to investigate some of the basic evolutionary mechanisms and pressures underlying the variation documented thus far in the cytoplasmic genomes of angiosperms.

ACKNOWLEDGMENTS. I thank J. E. Boynton and N. W. Gillham for their generous support and encouragement throughout the writing of this manuscript, A. Rosemond and J. E. Boynton for producing the figures, J. E. Boynton, N. W. Gillham, R. B. Helling, and M. E. Zolan for critical reading of the manuscript, and numerous colleagues for preprints and communication of unpublished studies.

References

Albertini, A. M., Hofer, M., Calos, M. P., and Miller, J. H., 1982, On the formation of spontaneous deletions: The importance of short sequence homologies in the generation of large deletions, *Cell* **29:**319–328.

Aldrich, J., and Cattolico, R. A., 1981, Isolation and characterization of chloroplast DNA from the marine chromophyte, *Olisthodiscus luteus:* Electron microscopic visualization of isomeric molecular forms, *Plant Physiol.* **68:**641–647.

Aldrich, J., Gelvin, S., and Cattolico, R. A., 1982, Extranuclear DNA of a marine chromophytic alga. Restriction endonuclease analysis, *Plant Physiol.* **69:**1189–1195.

Aldrich, J., Cherney, B., Merlin, E., Williams, C., and Mets, L., 1985, Recombination within the inverted repeat sequences of the *Chlamydomonas reinhardii* chloroplast genome produces two orientation isomers, *Curr. Genet.* **9:**233–238.

Allet, B., and Rochaix, J. D., 1979, Structure analysis at the ends of the intervening DNA sequences in the chloroplast 23S ribosomal genes of *C. reinhardtii*, *Cell* **18:**55–60.

Alt, J., Westhoff, P., Sears, B. B., Nelson, N., Hurt, E., Hauska, G., and Herrmann, R. G., 1983a, Genes and transcripts for the polypeptides of the cytochrome *b6/f* complex from spinach thylakoid membranes, *EMBO J.* **2:**979–986.

Alt, J., Winter, P., Sebald, W., Moser, J. G., Schedel, R., Westhoff, P., and Herrmann, R. G., 1983b, Localization and nucleotide sequence of the gene for the ATP synthase proteolipid subunit of the spinach plastid chromosome, *Curr. Genet.* **7:**129–138.

Amikan, D., Glaser, G., and Razin, S., 1984, Mycoplasmas (Mollicutes) have a low number of rRNA genes, *J. Bacteriol.* **158:**376–378.

Anderson, S., Bankier, A. T., Barrell, B. G., de Bruijn, M. H. L., Coulson, A. R., Drouin, J., Eperon, I. C., Nierlich, D. P., Roe, B. A., Sanger, F., Schreier, P. H., Smith, A. J. H., Staden, R., and Young, I. G., 1981, Sequence and organization of the human mitochondrial genome, *Nature* **290:**457–464.

Anderson, S., de Bruijn, M. H. L., Coulson, A. R., Eperon, I. C., Sanger, F., and Young, I. G., 1982, Complete sequence of bovine mitochondrial DNA. Conserved features of the mammalian mitochondrial genome, *J. Mol. Biol.* **156:**683–717.

Aquadro, J. C., and Greenberg, P. B., 1983, Human mitochondrial DNA variation and evolution: Analysis of nucleotide sequences from seven individuals, *Genetics* **103:**287–312.

Arnheim, N., 1983, Concerted evolution of multigene families, in: *Evolution of Genes and Proteins* (M. Nei and R. K. Koehn, eds.), Sinauer, Sunderland, Massachusetts, pp. 38–61.

Avise, J. C., and Lansman, R. A., 1983, Polymorphism of mitochondrial DNA in populations of higher animals, in: *Evolution of Genes and Proteins* (M. Nei and R. K. Koehn, eds.), Sinauer, Sunderland, Massachusetts, pp. 147–164.

Avise, J. C., Giblin-Davidson, C., Laerm, J., Patton, J. C., and Lansman, R. A., 1979a, Mitochondrial DNA clones and matriarchal phylogeny within and among geographic populations of the pocket gopher *Geomys pinetis*, *Proc. Natl. Acad. Sci. USA* **76:**6694–6698.

Avise, J. C., Lansman, R. A., and Shade, R. O., 1979b, The use of restriction endonucleases to measure mitochondrial DNA sequence relatedness in natural populations. I. Population structure and evolution in the genus *Peromyscus*, *Genetics* **92:**279–295.

Aviv, D., Fluhr, R., Edelman, M., and Galun, E., 1980, Progeny analysis of the interspecific somatic hybrids: *Nicotiana tabacum* ((CMS) and *Nicotiana sylvestris* with respect to nuclear and chloroplast markers, *Theor. Appl. Genet.* **56:**145–150.

Aviv, D., Bleichman, S., Arzee-Gonen, P., and Galun, E., 1984, Intersectional cytoplasmic hybrids in *Nicotiana*. Identification of plastomes and chondriomes in *N. sylvestris* + *N. rustica* cybrids having *N. sylvestris* nuclear genomes, *Theor. Appl. Genet.* **67:**499–504.

Bachmann, B. J., 1983, Linkage map of *Escherichia coli* K-12, Edition 7, *Microbiol. Rev.* **47:**180–230.

Bartsch, M., Kimura, M., and Subramanian, A. R., 1982, Purification, primary structure, and homology relationships of a chloroplast ribosomal protein, *Proc. Natl. Acad. Sci. USA* **79:**6871–6875.

Baur, E., 1909, Das wesen und die Erblichkeitsverhaltnisse der "Varietates albomarginatae hort" von *Perlargonium zonale*, *Z. Vererbungsl.* **1:**330–351.

Bayen, M., and Rode, A., 1973, The 1.700 DNA of *Chlorella pyrenoidosa:* Heterogeneity and complexity, *Plant Sci. Lett.* **1**:385–389.

Belford, H. S., and Thompson, W. F., 1981, Single copy DNA homologies in *Atriplex*. II. Hybrid thermal stabilities and molecular phylogeny, *Heredity* **46**:102–122.

Belliard, G., Pelletier, G., Vedel, F., and Quetier, F., 1978, Morphological characteristics and chloroplast DNA distribution in different cytoplasmic parasexual hybrids of *Nicotiana tabacum*, *Mol. Gen. Genet.* **165**:231–237.

Belliard, G., Vedel, F., and Pelletier, G., 1979, Mitochondrial recombination in cytoplasmic hybrids of *Nicotiana tabacum* by protoplast fusion, *Nature* **281**:401–403.

Bergmann, P., Seyer, P., Burkard, G., and Weil, J. H., 1984, Mapping of transfer RNA genes on tobacco chloroplast DNA, *Plant Mol. Biol.* **3**:29–36.

Bernardi, G., 1982, The origins of replication of the mitochondrial genome of yeast, *Trends Biochem. Sci.* **7**:404–408.

Bibb, M. J., Van Etten, R. A., Wright, C. T., Walberg, M. W., and Clayton, D. A., 1981, Sequence and gene organization of mouse mitochondrial DNA, *Cell* **26**:167–180.

Birky, C. W., Jr., 1983, Relaxed cellular controls and organelle heredity, *Science* **222**:468–475.

Bisaro, D., and Siegel, A., 1980, Sequence homology between chloroplast DNAs from several higher plants, *Plant Physiol.* **65**:234–237.

Boeshore, M. L., Lifshitz, I., Hanson, M. R., and Izhar, S., 1983, Novel composition of mitochondrial genomes in *Petunia* somatic hybrids derived from cytoplasmic male sterile and fertile plants, *Mol. Gen. Genet.* **190**:459–467.

Bogorad, L., 1975, Evolution of organelles and eukarotic genomes, *Science* **188**:891–898.

Bogorad, L., Gubbins, E. J., Jolly, S. O., Krebbers, E. T., Larrinua, I. M., Muskavitch, K. M. T., Rodermel, S. R., Subramanian, A., and Steinmetz, A., 1983, Maize plastid genes: Structure and expression, in: *Gene Structure and Regulation in Development*, Alan R. Liss, New York, pp. 13–32.

Bohnert, H. J., and Loffelhardt, W., 1982, Cyanelle DNA from *Cyanophora paradoxa* exists in two forms due to intramolecular recombination, *FEBS Lett.* **150**:403–406.

Bohnert, H. J., Crouse, E. J., and Schmitt, J. M., 1982a, Chloroplast genome organization and RNA synthesis, *Encyclopedia Plant Physiol.* **14B**:475–530.

Bohnert, H. J., Crouse, E. J., Pouyet, J., Mucke, H., and Loffelhardt, W., 1982b, The subcellular localization of DNA components from *Cyanophora paradoxa*, a flagellate containing endosymbiotic cyanelles, *Eur. J. Biochem.* **126**:381–388.

Bolen, P. L., Grant, D. M., Swinton, D., Boynton, J. E., and Gillham, N. W., 1982, Extensive methylation of chloroplast DNA by a nuclear gene mutation does not affect chloroplast gene transmission in *Chlamydomonas*, *Cell* 28:335–343.

Bonen, L., and Gray, M. W., 1980, Organization and expression of the mitochondrial genome of plants I. The genes for wheat mitochondrial ribosomal and transfer RNA: Evidence for an unusual arrangement, *Nucleic Acids Res.* **8**:319–335.

Bonnard, G., Michel, F., Weil, J. H., and Steinmetz, A., 1984, Nucleotide sequence of the split $tRNA^{Leu}$-UAA gene from *Vicia faba* chloroplasts: Evidence for structural homologies of the chloroplast $tRNA^{Leu}$ intron with the intron from the autosplicable *Tetrahymena* ribosomal RNA precursor, *Mol. Gen. Genet.* **194**:330–336.

Borck, K. S., and Walbot, V., 1982, Comparison of the restriction endonuclease digestion patterns of mitochondrial DNA from normal and male sterile cytoplasms of *Zea mays* L., *Genetics* **102**:109–128.

Bosch, L., Kraal, B., Van der Meide, P. H., Duisterwinkel, F. J., and Van Noort, J. M., 1983, The elongation factor EF-Tu and its two encoding genes, *Prog. Nucleic Acid Res. Mol. Biol.* **30**:91–124.

Bottomley, W., and Bohnert, H. J., 1982, The biosynthesis of chloroplast proteins, *Encyclopedia Plant Physiol.* **14B**:531–596.

Boutry, M., and Briquet, M., 1982, Mitochondrial modifications associated with the cytoplasmic male sterility in faba beans, *Eur. J. Biochem.* **127**:129–135.

Boutry, M., Briquet, M., and Goffeau, A., 1983, The alpha subunit of a plant mitochondrial Fl-ATPase is translated in mitochondria, *J. Biol. Chem.* **258**:8524–8526.

Bovenberg, W. A., Kool, A. J., and Nijkamp, H. J. J., 1981, Isolation, characterization and restriction endonuclease mapping of the *Petunia hybrida* chloroplast DNA, *Nucleic Acids Res.* **9**:503–517.

Bovenberg, W. A., Howe, C. J., Kool, A. J., and Nijkamp, H. J. J., 1984, Physical mapping of genes for chloroplast DNA encoded subunit polypeptides of the ATP synthase complex from *Petunia hybrida*, *Curr. Genet.* **8**:283–290.

Bowman, C. M., Bonnard, G., and Dyer, T. A., 1983, Chloroplast DNA variation between species of *Triticum* and *Aegilops*. Location of the variation on the chloroplast genome and its relevance to the inheritance and classification of the cytoplasm, *Theor. Appl. Genet.* **65**:247–262.

Breitenberger, C. A., Graves, M. C., and Spremulli, L. I., 1979, Evidence for the nuclear location of the gene for chloroplast elongation factor G, *Arch. Biochem. Biophys.* **194**:265–270.

Brennicke, A., and Blanz, P., 1982, Circular mitochondrial DNA species from *Oenothera* with unique sequences, *Mol. Gen. Genet.* **187**:461–466.

Brettell, R. I. S., Thomas, E., and Ingram, D. S., 1980, reversion of Texas male-sterile cytoplasm maize in culture to give fertile, T-toxin resistant plants, *Theor. Appl. Genet.* **58**:55–58.

Briat, J. F., Dron, M., Loiseaux, S., and Mache, R., 1982a, Structure and transcription of the spinach chloroplast rDNA leader region, *Nucleic Acids Res.* **10**:6865–6878.

Briat, J. F., Gigot, C., Laulhere, J. P., and Mache, R., 1982b, Visualization of a spinach plastid transcriptionally active DNA–protein complex in a highly condensed structure, Plant Physiol. **69**:1205–1211.

Briat, J. F., Dron, M., and Mache, R., 1983, Is transcription of higher plant chloroplast ribosomal operons regulated by premature termination?, *FEBS Lett.* **163**:1–5.

Briat, J. F., Letoffe, S., Mache, R., and Rouviere-Yaniv, J., 1984, Similarity between the bacterial histone-like protein HU and a protein from spinach chloroplasts, *FEBS Lett.* **172**:75–79.

Brown, G. G., and Simpson, M. V., 1981, Intra- and interspecific variation of the mitochondrial genome in *Rattus norvegicus* and *Rattus rattus:* Restriction enzyme analysis of variant mitochondrial DNA molecules and their evolutionary relationships, *Genetics* **97**:125–143.

Brown, G. G., and Simpson, M. V., 1982, Novel features of animal mtDNA evolution as shown by sequences of two rat cytochrome oxidase subunit II genes, *Proc. Natl. Acad. Sci. USA* **79**:3246–3250.

Brown, W. M., 1983, Evolution of animal mitochondrial DNA, in: *Evolution of Genes and Proteins* (M. Nei, and R. K. Koehn, eds.), Sinauer, Sunderland, Massachusetts, pp. 62–88.

Brown, W. M., Prager, E. M., Wang, A., and Wilson, A. C., 1982, Mitochondrial DNA sequences of primates: Tempo and mode of evolution, *J. Mol. Evol.* **18**:225–239.

Brunold, C., and Schiff, J. A., 1976, Studies of sulfate utilization by algae. 15. Enzymes of assimilatory sulfate reduction in *Euglena* and their cellular localization, *Plant. Physiol.* **57**:430–436.

Burton, W. G., Grabowy, C. T., and Sager, R., 1979, Role of methylation in the modification and restriction of chloroplast DNA in *Chlamydomonas, Proc. Natl. Acad. Sci. USA* **76:**1390–1394.

Cann, R. L., and Wilson, A. C., 1983, Length mutations in human mitochondrial DNA, *Genetics* **104:**699–711.

Cann, R. L., Brown, W. M., and Wilson, A. C., 1984, Polymorphic sites and the mechanism of evolution in human mitochondrial DNA, *Genetics* **106:**479–499.

Carbonera, D., Sora, S., Riccardi, G., Camerino, G., and Ciferri, O., 1981, Characterization of a mutant of *Chlamydomonas reinhardtii* resistant to fusidic acid, *FEBS Lett.* **132:**227–230.

Carlson, S. S., Mross, G. A., Wilson, A. C., Mead, R. T., Wolin, L. D., Bowers, S. F., Foley, N. T., Muijsers, A. O., and Margoliash, E., 1977, Primary structure of mouse, rat and guinea pig cytochrome c, *Biochemistry* **16:**1437–1442.

Cavalier-Smith, T., 1982, The origins of plastids, *Biol. J. Linn. Soc.* **17:**289–306.

Cech, T. R., 1983, RNA splicing: Three themes with variations, *Cell* **34:**713–716.

Chang, S. H., Hecker, L. I., Brum, C. K., Schnabel, J. J., Heckman, J. E., Silberkland, M., Rajbhandary, U. L., and Barnett, W. E., 1981, The nucleotide sequence of *Euglena* cytoplasmic phenylalanine transfer RNA. Evidence for possible classification of *Euglena* among the animal rather than the plant kingdom, *Nucleic Acids Res.* **9:**3199–3204.

Chao, S., Sederoff, R. R., and Levings, C. S., III, 1983, Partial sequence analysis of the 5S to 18S rRNA gene region of the maize mitochondrial genome, *Plant Physiol.* **71:**190–193.

Chelm, B. K., and Hallick, R. B., 1976, Changes in the expression of the chloroplast genome of *Euglena gracilis* during chloroplast development, *Biochemistry* **15:**593–599.

Chourey, P. S., and Kemble, R. J., 1982, Transposition event in tissue cultured cells of maize, in: *5th International Congress Plant Tissue Cell Culture,* Tokyo, Japan, pp. 425–426.

Chu, N. M., and Tewari, K. K., 1982, Arrangement of the ribosomal RNA genes in chloroplast DNA of Leguminosae, *Mol. Gen. Genet.* **186:**23–32.

Chun, E. H. L., Vaughan, M. H., and Rich, A., 1963, The isolation and characterization of DNA associated with chloroplast preparations, *J. Mol. Biol.* **7:**130–141.

Ciferri, O., and Tiboni, O., 1976, Evidence for the synthesis in the chloroplast of elongation factor G, *Plant Sci. Lett.* **7:**455–466.

Ciferri, O., Di Pasquale, G., and Tiboni, O., 1979, Chloroplast elongation factors are synthesized in the chloroplast, *Eur. J. Biochem.* **102:**331–335.

Clark-Walker, G. D., McArthur, C. R., and Sriprakash, K. S., 1981, Partial duplication of the large ribosomal RNA sequence in an inverted repeat in circular mitochondrial DNA from *Kloeckera africana.* Implications for mechanisms of the petite mutation, *J. Mol. Biol.* **147:**399–415.

Clegg, M. T., Rawson, J. R. Y., and Thomas, K., 1984, Chloroplast DNA variation in pearl millet and related species, *Genetics* **106:**449–461.

Coates, D., and Cullis, C. A., 1982, The chloroplast DNAs of flax genotrophs, *Plant Mol. Biol.* **1:**183–189.

Coleman, A. W., 1979, Use of the fluorochrome 4′6-diamidino-2-phenylindole in genetic and developmental studies of chloroplast DNA, *J. Cell Biol.* **82:**299–305.

Coleman, A. W., 1984a, The fate of chloroplast DNA during cell fusion, zygote maturation and zygote germination in *Chlamydomonas reinhardi* as revealed by DAPI staining, *Exp. Cell Res.* **152:**528–540.

Coleman, A. W., 1985, Diversity of plastid DNA configuration among classes of eukaryotic algae, *J. Phycol.,* **21:**1–16.

Coleman, A. W., and Maguire, M. J., 1983, Cytological detection of the basis of uniparental inheritance of plastid DNA in *Chlamydomonas moewusii, Curr. Genet.* **7:**211–218.

Correns, C., 1909, Zur kenntnis der rolle von kern and plasma bei der verebung, *Z. Vererbungsl.* **2**:331–340.

Crouse, E. J., Bohnert, H. J., and Schmitt, J. M., 1985, Chloroplast RNA synthesis, in: *Chloroplast Biogenesis* (R. J. Ellis, ed.), Cambridge University Press, pp. 83–136.

Curtis, S. E., and Haselkorn, R., 1983, Isolation and sequence of the gene for the large subunit of ribulose-1,5-bisphosphate carboxylase from the cyanobacterium *Anabaena* 7120, *Proc. Natl. Acad. Sci. USA* **80**:1835–1839.

Curtis, S. E., and Haselkorn, R., 1984, Isolation, sequence and expression of two members of the 32 kd thylakoid membrane protein gene family from the cyanobacterium *Anabaena* 7120, *Plant Mol. Biol.*, **3**:249–258.

Dale, R. M. K., 1981, Sequence homology among different size classes of plant mtDNAs, *Proc. Natl. Acad. Sci. USA* **78**:4453–4457.

Dale, R. M. K., Duesing, J. H., and Keene, D., 1981, Supercoiled mitochondrial DNAs from plant tissue culture cells, *Nucleic Acids Res.* **9**:4583–4593.

Dale, R. M. K., Wu, M., and Kiernan, M. C. C., 1983, Analysis of four tobacco mitochondrial DNA size classes, *Nucleic Acids Res.* **11**:1673–1685.

Dale, R. M. K., Mendu, N., Ginsburg, H., and Kridl, J. C., 1984, Sequence analysis of the maize mitochondrial 26S rRNA gene and flanking regions, *Plasmid* **11**:141–150.

Dalmon, J., Loiseaux, S., and Bazetoux, S., 1983, Heterogeneity of plastid DNA of two species of brown algae *Plant Sci. Lett.* **29**:243–253.

De Heij, H. T., Lustig, H., Moeskops, D. J. M., Bovenberg, W. A., Bisanz, C., and Groot, G. S. P., 1983, Chloroplast DNAs of *Spinacia, Petunia* and *Spirodela* have a similar gene organization, *Curr. Genet.* **7**:1–6.

De Jong, W. W., and Ryden, L., 1981, Causes of more frequent deletions than insertions in mutations and protein evolution, *Nature* **290**:157–159.

Delihas, N., Andersen, J., Andresini, W., Kaufman, L., and Lyman, H., 1981, The 5S ribosomal RNA of *Euglena gracilis* cytoplasmic ribosomes is closely homologous to the 5S RNA of the trypanosomatid protozoa, *Nucleic Acids Res.* **9**:6627–6633.

Deno, H., and Sugiura, M., 1984, Chloroplast $tRNA^{Gly}$ gene contains a long intron in the D stem: Nucleotide sequences of tobacco chloroplast genes for $tRNA^{Gly}$(UCC) and $tRNA^{Arg}$(UCU), *Proc. Natl. Acad. Sci. USA* **81**:405–408.

Deno, H., Kato, A., Shinozaki, K., and Sugiura, M., 1982, Nucleotide sequences of tobacco chloroplast genes for elongator $tRNA^{Met}$ and $tRNA^{Val}$ (UAC): The $tRNA^{Val}$ (UAC) gene contains a long intron, *Nucleic Acids Res.* **10**:7511–7520.

Deno, H., Shinozaki, K., and Sugiura, M., 1983, Nucleotide sequence of tobacco chloroplast gene for the alpha subunit of proton-translocating ATPase, *Nucleic Acids Res.* **11**:2185–2191.

Dix, K. P., and Rawson, J. R. Y., 1983, *In vivo* transcriptional products of the chloroplast DNA of *Euglena gracilis, Curr. Genet.* **7**:265–272.

Dixon, L. K., and Leaver, C. J., 1982, Mitochondrial gene expression and cytoplasmic male sterility in sorghum, *Plant Mol. Biol.* **1**:89–102.

Doolittle, W. F., and Bonen, L., 1981, Molecular sequence data indicating an endosymbiotic origin for plastids, *Ann. N.Y. Acad. Sci.* **361**:248–256.

Dorne, A. M., Lescure, A. M., and Mache, R., 1984a, Site of synthesis of spinach chloroplast ribosomal proteins and formation of incomplete ribosomal particles in isolated chloroplasts, *Plant Mol. Biol.* **3**:83–90.

Dorne, A. M., Eneas-Filho, J., Heizmann, P., and Mache, R., 1984b, Comparisons of ribosomal proteins of chloroplast from spinach and of *E. coli, Mol. Gen. Genet.* **193**:129–134.

Douglas, S. E., and Doolittle, W. F., 1984a, Nucleotide sequence of the 5S rRNA gene and flanking regions in the cyanobacterium *Anacystis nidulans*, *FEBS Lett.* **166**:307–310.

Douglas, S. E., and Doolittle, W. F., 1984b, Complete nucleotide sequence of the 23S rRNA gene of the cyanobacterium *Anacystis nidulans*, *Nucleic Acids Res.* **12**:3373–3386.

Doyle, J. A. 1978, Origin of angiosperms, *Annu. Rev. Ecol. Syst.* **9**:365–392.

Driesel, A. J., Crouse, E. J., Gordon, K., Bohnert, H. J., Herrmann, R. G., Steinmetz, A., Mubumbila, M., Keller, M., Burkard, G., and Weil, J. H., 1979, Fractionation and identification of spinach chloroplast transfer RNAs and mapping of their genes on the restriction map of chloroplast DNA, *Gene* **6**:285–306.

Driesel, A. J., Speirs, J., and Bohnert, N. J., 1980, Spinach chloroplast mRNA for a 32,000 dalton polypeptide: Size and localization on the physical map of the chloroplast DNA, *Biochim. Biophys. Acta* **610**:297–310.

Dron, M., Rahire, M., and Rochaix, J. D., 1982, Sequence of the chloroplast DNA region of *Chlamydomonas reinhardtii* containing the gene of the large subunit of ribulose bisphosphate carboxylase and parts of its flanking genes, *J. Mol. Biol.* **162**:775–793.

Droop, M. R., 1974, Heterotrophy of algae, in: *Algal Physiology and Biochemistry* (W. A. P. Stewart, ed.), University of California Press, Berkeley, California, pp. 530–559.

Edwards, K., and Kossel, H., 1981, The rRNA operon from *Zea mays* chloroplasts: Nucleotide sequence of 23S rDNA and its homology with *E. coli* rDNA, *Nucleic Acids Res.* **9**:2853–2869.

Efstratiadis, A., Posakony, J. W., Maniatis, T., Lawn, R. M., O'Connell, C., Spritz, R. A., DeRiel, J. K., Forget, B. G., Weissman, M., Slightom, J. L., Blechl, A. E., Smithies, O., Baralle, F. E., Shoulders, C. C., and Proudfoot, N. J., 1980, The structure and evolution of the human beta-globin gene famiy, *Cell* **21**:653–668.

Egelhoff, T., and Grossman, A., 1983, Cytoplasmic and chloroplast synthesis of phycobilisome polypeptides, *Proc. Natl. Acad. Sci. USA* **80**:3339–3343.

El-Gewely, M. R., Helling, R. B., and Dibbits, J. G. T., 1984, Sequence and evolution of the regions between the *rrn* operons in the chloroplast genome of *Euglena gracilis bacillaris*, *Mol. Gen. Genet.* **194**:432–443.

Ellis, R. J., 1981, Chloroplast proteins: Synthesis, transport and assembly, *Annu. Rev. Plant Physiol.* **32**:111–137.

Ellis, R. J., Smith, S. M., and Barraclough, R., 1980, Synthesis, transport and assembly of chloroplast proteins, in: *Genome Organization and Expression in Plants* (C. J. Leaver, ed.), Plenum Press, New York, pp. 321–335.

Eneas-Filho, J. Hartley, M. R., and Mache, R., 1981, Pea chloroplast ribosomal proteins: Characterization and site of synthesis, *Mol. Gen. Genet.* **184**:484–488.

Erickson, L. R., Straus, N. A., and Beversdorf, W. D., 1983, Restriction patterns reveal origins of chloroplast genomes in *Brassica* amphidiploids, *Theor. Appl. Genet.* **65**:201–206.

Falconet, D., Lejeune, B., Quetier, F., and Gray, M. W., 1984, Evidence for homologous recombination between repeated sequences containing 18S and 5S ribosomal RNA genes in wheat mitochondrial DNA, *EMBO J.* **3**:297–302.

Farrelley, F., and Butow, R. A., 1983, Rearranged mitochondrial genes in the yeast nuclear genome, *Nature* **301**:296–301.

Fauron, C. M. R., and Wolstenholme, D. R., 1980, Intraspecific diversity of nucleotide sequences within the adenine + thymine-rich region of mitochondrial DNA molecules of *Drosophila mauritania*, *Drosophila melanogaster* and *Drosophila simulans*, *Nucleic Acids Res.* **8**:5391–5404.

Feng, T. Y., and Chiang, K. S., 1984, The persistence of maternal inheritance in *Chlamydomonas* despite hypomethylation of chloroplast DNA induced by inhibitors, *Proc. Natl. Acad. Sci. USA* **81**:3438–3442.

Flamant, F., Heizmann, P., and Nigon, V., 1984, Rearrangement of chloroplast ribosomal cistrons by unequal crossing-over in *Euglena gracilis, Curr. Genet.* **8:**9–13.

Fluhr, R., and Edelman, M., 1981, Conservation of sequence arrangement among high plant chloroplast DNAs: Molecular cross hybridization among the Solanaceae and between *Nicotiana* and *Spinacia, Nucleic Acids Res.* **9:**6841–6853.

Fluhr, R., Fromm, H., and Edelman, M., 1983a, Clone bank of *Nicotiana tabacum* chloroplast DNA: Mapping of the alpha, beta and epsilon subunits of the ATPase coupling factor, the large subunit of ribulosebisphosphate carboxylase, and the 32-kDal membrane protein, *Gene* **25:**271–280.

Fluhr, R., Aviv, D., Edelman, M., and Galun, E., 1983b, Cybrids containing mixed and sorted-out chloroplasts following interspecific somatic fusions in *Nicotiana, Theor. Appl. Genet.* **65:**289–294.

Fluhr, R., Aviv, D., Galun, E., and Edelman, M., 1984, Generation of heteroplastidic *Nicotiana* cybrids by protoplast fusion: Analysis for plastid recombinant types, *Theor. Appl. Genet.* **67:**491–497.

Fonty, G., Crouse, E. J. Stutz, E., and Bernardi, G., 1975, The mitochondrial genome of *Euglena gracilis, Eur. J. Biochem.* **54:**367–372.

Forde, B. G., and Leaver, C. J., 1980, Nuclear and cytoplasmic genes controlling synthesis of variant mitochondrial polypeptides in male sterile maize, *Proc. Natl. Acad. Sci. USA* **77:**418–422.

Forde, B. G., Oliver, R. J. C., and Leaver, C. J., 1978, Variation in mitochondrial translation products associated with male-sterile cytoplasms in maize, *Proc. Natl. Acad. Sci. USA* **75:**3841–3845.

Fox, G. E., Stackebrandt, E., Hespell, R. B., Gibson, J., Maniloff, J., Dyer, T. A., Wolfe, R. S., Balch, W. E., Tanner, R. S., Magrum, L. J., Zablen, L. B., Blakemore, R., Gupta, R., Bonen, L., Lewis, B. J., Stahl, D. A., Luehrsen, K. R., Chen, K. N., and Woese, C. R., 1980, The phylogeny of prokaryotes, *Science* **209:**457–463.

Fox, T. D., and Leaver, C. J., 1981, The *Zea mays* mitochondrial gene coding cytochrome oxidase subunit II has an intervening sequence and does not contain TGA codons, *Cell* **26:**315–323.

Francis, M. A., and Dudock, B. S., 1982, Nucleotide sequence of a spinach chloroplast isoleucine tRNA, *J. Biol. Chem.* **257:**11195–11198.

Freyssinet, G., 1978, Determination of the site of synthesis of some *Euglena* cytoplasmic and chloroplast ribosomal proteins, *Exp. Cell Res.* **115:**207–219.

Futai, M., and Kanazawa, H., 1983, Structure and function of proton-translocating adenosine triphosphatase (F_0F_1): Biochemical and molecular biological approaches, *Microbiol. Rev.* **47:**285–312.

Galun, E., Arzee-Gonen, P., Fluhr, R., Edelman, M., and Aviv, D., 1982, Cytoplasmic hybridization in *Nicotiana:* Mitochondrial DNA analysis in progenies resulting from fusion between protoplasts having different organelle constitutions, *Mol. Gen. Genet.* **186:**50–56.

Gatenby, A. A., and Castleton, J. A., 1982, Amplification of maize ribulose bisphosphate carboxylase large subunit synthesis in *E. coli* by transcriptional fusion with the lambda N operon, *Mol. Gen. Genet.* **185:**424–429.

Gatenby, A. A., Castleton, J. A., and Saul, M. W., 1981, Expression in *E. coli* of maize and wheat chloroplast genes for large subunit of ribulose bisphosphate carboxylase, *Nature* **291:**117–121.

Gellissen, G., Bradfield, J. Y., White, B. N., and Wyatt, G. R., 1983, Mitochondrial DNA sequences in the nuclear genome of a locust, *Nature* **301:**631–634.

Gelvin, S. B., and Howell, S. H., 1979, Small repeated sequences in the chloroplast genome of *Chlamydomonas reinhardtii*, *Mol. Gen. Genet.* **173**:315–322.

Gengenbach, B. G., Green, C. E., and Donovan, C. D., 1977, Inheritance of selected pathotoxin resistance in maize plants regenerated from cell cultures, *Proc. Natl. Acad. Sci. USA* **74**:5113–5117.

Gengenbach, B. G., Connelly, J. A., Pring, D. R., and Conde, M. F., 1981, Mitochondrial DNA variation in maize plants regenerated during tissue culture selection, *Theor. Appl. Genet.* **59**:161–167.

Giannasi, D. E., and Crawford, D. J., 1985, Plant chemosystematics, *Evol. Biol.* (in press).

Gibbs, S. P., 1981a, The chloroplast endoplasmic reticulum: Structure, function, evolutionary significance, *Int. Rev. Cytol.* **72**:49–99.

Gibbs, S. P., 1981b, Chloroplasts of some groups may have evolved from endosymbiotic eukaryotic algae, *Ann. N.Y. Acad. Sci.* **361**:193–207.

Gillham, N. W., 1978, *Organelle Heredity*, Raven Press, New York.

Gillham, N. W., Boynton, J. E., and Harris, E. H., 1985, Evolution of plastid DNA, in: *DNA and Evolution: Natural Selection and Genome Size* (T. Cavalier-Smith ed.), Wiley, New York, pp. 299–351.

Glotz, C., Zweib, C., Brimacombe, R., Edwards, K., and Kossel, H., 1981, Secondary structure of the large subunit ribosomal RNA from *Escherichia coli*, *Zea mays* chloroplast, and human and mouse mitochondrial ribosomes, *Nucleic Acids Res.* **9**:3287–3306.

Goblet, J. P., Boutry, M., Duc, G., and Briquet, M., 1983, Mitochondrial plasmid-like molecules in fertile and male sterile *Vicia faba* L., *Plant Mol. Biol.* **2**:305–309.

Goldbach, R. W., Borst, P., Bollen-de Boer, J. E., and van Bruggen, E. F. J., 1978, The organization of ribosomal RNA genes in the mitochondrial DNA of *Tetrahymena pyriformis* strain ST, *Biochim. Biophys. Acta* **521**:169–186.

Gordon, K. H. J., Crouse, E. J., Bohnert, H. J., and Herrmann, R. G., 1982, Physical mapping of differences in chloroplast DNA of the five wild-type plastomes in *Oenothera* subsection *Euoenothera*, *Theor. Appl. Genet.* **61**:373–384.

Graf, L., Roux, E., Stutz, E., and Kossel, H., 1982, Nucleotide sequence of a *Euglena gracilis* chloroplast gene coding for the 16S rRNA: Homologies to *E. coli* and *Zea mays* chloroplast 16S rRNA, *Nucleic Acids Res.* **10**:6369–6381.

Grant, D., and Chiang, D. S., 1980, Physical mapping and characterization of *Chlamydomonas* mitochondrial DNA molecules: Their unique ends, sequence homogeneity, and conservation, *Plasmid* **4**:82–96.

Gray, M. W., 1982, Mitochondrial genome diversity and the evolution of mitochondrial DNA, *Can. J. Biochem.* **60**:157–171.

Gray, M. W., 1983, The bacterial ancestry of plastids and mitochondria, *Bioscience*, **33**:693–699.

Gray, M. W., and Doolittle, W. F., 1982, Has the endosymbiont hypothesis been proven?, *Microbiol. Rev.* **46**:1–42.

Gray, M. W., and Spencer, D. F., 1981, Is wheat mitochondrial 5S ribosomal RNA prokaryotic in nature?, *Nucleic Acids Res.* **9**:3523–3529.

Gray, M. W., and Spencer, D. F., 1983, Wheat mitochondrial DNA encodes a eubacteria-like initiator methionine transfer RNA, *FEBS Lett.* **161**:323–327.

Gray, P. W., and Hallick, R. B., 1978, Physical mapping of the *Euglena gracilis* chloroplast DNA and ribosomal RNA gene region, *Biochemistry* **17**:284–289.

Green, B. R., 1976, Covalently closed minicircular DNA associated with *Acetabularia* chloroplasts, *Biochim. Biophys. Acta* **447**:156–166.

Green, B. R., Muir, B. L., and Padmanabhan, U., 1977, The *Acetabularia* chloroplast genome: Small circles and large kinetic complexity, in: *Progress in Acetabularia Research* (C. F. L. Woodcock, ed.), Academic Press, New York, pp. 107–122.

Grossman, A. G., Bartlett, S. G., Schmidt, G. W., Mullet, J. E., and Chua, N. H., 1982, Optimal conditions for post-translational uptake of proteins by isolated chloroplasts. *In vitro* synthesis and transport of plastocyanin, ferredoxin-NADP$^+$ oxidoreductase, and fructose-1,6-bisphosphatase, *J. Biol. Chem.* **257**:1558–1563.

Grossman, A., Talbot, L., and Egelhoff, T., 1983, Biosynthesis of phycobilisome polypeptides of *Porphyridium aerugineum* and *Cyanophora paradoxa*, *Carnegie Inst. Washington Yearb.* **82**:112–116.

Gruissem, W., Prescott, D. M., Greenberg, B. M., and Hallick, R. B., 1982, Transcription of *E. coli* and *Euglena* chloroplast tRNA gene clusters and processing of polycistronic transcripts in a HeLa cell-free system, *Cell* **30**:81–92.

Gruissem, W., Greenberg, B. M., Zurawski, G., Prescott, D. M., and Hallick, R. B., 1983a, Biosynthesis of chloroplast transfer RNA in a spinach chloroplast transcription system, *Cell* **35**:815–828.

Gruissem, W., Narita, J. O., Greenberg, B. M., Prescott, D. M., and Hallick, R. B., 1983, Selective *in vitro* transcription of chloroplast genes, *J. Cell. Biochem.* **22**:31–46.

Hack, E., and Leaver, C. J., 1983, The alpha subunit of the maize F_1-ATPase is synthesized in the mitochondrion, *EMBO J.* **2**:1783–1789.

Hadler, H. I., Dimitrijevic, B., and Mahalingam, R., 1983, Mitochondrial and nuclear DNA from normal rat liver have a common sequence, *Proc. Natl. Acad. Sci. USA* **80**:6495–6499.

Hallick, R. B., 1985, Chloroplast DNA, in: *The Biology of Euglena*, Vol. IV (D. E. Buetow, ed.), Academic Press, New York (in press).

Hallick, R. B., and W. Bottomley, 1983, Proposal for naming of chloroplast genes, *Plant Mol. Rep.* **1**:38–43.

Hallick, R. B., Greenberg, B. M., Gruissem, W., Hollingsworth, M. J., Karabin, G. D., Narita, J. O., Nickoloff, J. A., Passavant, C. W., and Stiegler, G. L., 1983, Organization and expression of the chloroplast genome of *Euglena gracilis*, in: *Structure and Function of Plant Genomes* (O. Ciferri and L. Dure III, eds.), Plenum Press, New York, pp. 155–166.

Handa, H., Tsunewaki, K., Kunisada, T., and Yamagishi, H., 1984, Small circular DNA molecules in wheat mitochondria, *Mol. Gen. Genet.* **194**:368–372.

Harris, D. A., 1981, The coupling ATPase complex: An evolutionary view, *Biosystems* **14**:113–121.

Hawley, D. K., and McClure, W. R., 1983, Compilation and analysis of *Escherichia coli* promoter DNA sequences, *Nucleic Acid Res.* **11**:2237–2255.

Hedberg, M. F., Huang, Y. S., and Hommersand, M. H., 1981, Size of the chloroplast genome in *Codium fragile*, *Science* **213**:445–447.

Heinhorst, S., and Shively, J. M., 1983, Encoding of both subunits of ribulose-1,5-bisphosphate carboxylase by organelle genome of *Cyanophora paradoxa*, *Nature* **304**:373–374.

Heizmann, P., Doly, J., Hussein, Y., Nicolas, P., Nigon, V., and Bernardi, G., 1981, The chloroplast genome of bleached mutants of *Euglena gracilis*, *Biochim. Biophys. Acta* **653**:412–415.

Heizmann, P., Hussein, Y., Nicolas, P., and Nigon, V., 1982a, Modifications of chloroplast DNA during streptomycin induced mutagenesis in *Euglena gracilis*, *Curr. Genet.* **5**:9–15.

Heizmann, P., Ravel-Chapuis, P., and Nigon, V., 1982b, Minicircular DNA having sequence homologies with chloroplast DNA in a bleached mutant of *Euglena gracilis*, *Curr. Genet.* **6**:119–122.

Helling, R. B., El-Gewely, M. R., Lomax, M. I., Baumgartner, J. E., Schwartzbach, S. D., and Barnett, W. E., 1979, Organization of the chloroplast ribosomal RNA genes of *Euglena gracilis bacillaris*, *Mol. Gen. Genet.* **174**:1–10.

Henner, D. J., and Hock, J. A., 1980, The *Bacillus subtilis* chromosome, *Microbiol. Rev.* **44**:57–82.

Herrmann, R. G., Palta, H. K., and Kowallik, K. V., 1980, Chloroplast DNA from three archegoniates, *Planta* **148**:319–327.

Herrmann, R. G., Westhoff, P., Alt, J., Winter, P., Tittgen, J., Bisanz, C., Sears, B. B., Nelson, N., Hurt, E., Hauska, G., Viebrock, A., and Sebald, W., 1983, Identification and characterization of genes for polypeptides of the thylakoid membrane, in: *Structure and Function of Plant Genomes* (O. Ciferri and L. Dure, III, eds.), Plenum Press, New York, pp. 143–153.

Hiesel, R., and Brennicke, A., 1983, Cytochrome oxidase subunit II gene in mitochondria of *Oenothera* has no intron, *EMBO J.* **2**:2173–2178.

Hirschberg, J., and McIntosh, L., 1983, Molecular basis of herbicide resistance in *Amaranthus hybridus*, *Science* **222**:1346–1349.

Hollingsworth, M. J., and Hallick, R. B., 1982, *Euglena gracilis* chloroplast transfer RNA transcription units. Nucleotide sequence analysis of a $tRNA^{Tyr}$-$tRNA^{His}$-$tRNA^{Met}$-$tRNA^{Trp}$-$tRNA^{Glu}$-$tRNA^{Gly}$ gene cluster, *J. Biol. Chem.* **257**:12795–12799.

Hollingsworth, M. J., Johanningmeier, U., Karabin, G. D., Stiegler, G. L., and Hallick, R. B., 1984, Detection of multiple, unspliced precursor mRNA transcripts for the M_r 32,000 thylakoid membrane protein from *Euglena gracilis* chloroplasts, *Nucleic Acids Res.* **12**:2001–2017.

Howe, C. J., Auffret, A. D., Doherty, A., Bowman, C. M., Dyer, T. A., and Gray, J. C., 1982, Location and nucleotide sequence of the gene for the proton-translocating subunit of wheat chloroplast ATPase synthase, *Proc. Natl. Acad. Sci. USA* **79**:6903–6907.

Howe, C. J., Bowman, C. J., Dyer, T. A., and Gray, J. C., 1983, The genes for the alpha and proton-translocation subunits of wheat chloroplast ATP synthase are close together on the same strand of chloroplast DNA, *Mol. Gen. Genet.* **190**:51–55.

Howell, S. H., and Walker, L. L., 1977, Transcription of the nuclear and chloroplast genomes during the vegetative cell cycle in *Chlamydomonas reinhardtii, Dev. Biol.* **56**:11–23.

Hudspeth, M. E. S., Shumard, D. S., Bradford, C. J. R., and Grossman, L. I., 1983, Organization of *Achlya* mtDNA: A population with two orientations and a large inverted repeat containing the rRNA genes, *Proc. Natl. Acad. Sci. USA* **80**:142–146.

Huh, T. Y., and Gray, M. W., 1982, Conservation of ribosomal RNA gene arrangement in the mitochondrial DNA of angiosperms, *Plant Mol. Biol.* **1**:245–249.

Hussein, Y., Heizmann, P., Nicolas, P., and Nigon, V., 1982, Quantitative estimates of chloroplast DNA in bleached mutants of *Euglena gracilis, Curr. Genet.* **6**:111–117.

Huttly, A. K., and Gray, J. C. 1984, Localisation of genes for four ATP synthase subunits in pea chloroplast DNA, *Mol. Gen. Genet.* **194**:402–409.

Iams, K. P., and Sinclair, J. H., 1982, Mapping the mitochondrial DNA of *Zea mays:* Ribosomal gene localization, *Proc. Natl. Acad. Sci. USA* **79**:5926–5929.

Jacobs, H. T., Posakony, J. W., Grula, J. W., Roberts, J. W., Xin, J. H., Britten, R. J., and Davidson, E. H., 1983, Mitochondrial DNA sequences in the nuclear genome of *Strongylocentrotus purpuratus, J. Mol. Biol.* **165**:609–632.

Jaynes, J. M., and Vernon, L. P., 1982, The cyanelle of *Cyanophora paradoxa:* Almost a cyanobacterial chloroplast, *Trends Biochem. Sci.* **7**:22–24.

Jenni, B., and Stutz, E., 1978, Physical mapping of the ribosomal DNA region of *Euglena gracilis* chloroplast DNA, *Eur. J. Biochem.* **88**:127–134.

Jenni, B., and Stutz, E., 1979, Analysis of *Euglena gracilis* chloroplast DNA. Mapping of a DNA sequence complementary to 16S rRNA outside of the three rRNA gene sets, *FEBS Lett.* **102**:95–99.

Jenni, B., Fasnacht, M., and Stutz, E., 1981, The multiple copies of the *Euglena gracilis* chloroplast genome are not uniform in size, *FEBS Lett.* **125:**175–179.

Jensen, R. G., 1980, Biochemistry of the chloroplast, in: *The Biochemistry of Plants*, Vol. 1, Academic Press, New York, pp. 273–313.

Jones, C. E., and Kafatos, F. C., 1982, Accepted mutations in a gene family: Evolutionary diversification of duplicated DNA, *J. Mol. Evol.* **19:**87–103.

Jones, I. A., Primrose, S. B., and Ehrlich, S. D., 1982, Recombination between short direct repeats and a *RecA* host, *Mol. Gen. Genet.* **188:**486–489.

Jordan, D. B., and Ogren, W. L., 1981, Species variation in the specificity of ribulose bisphosphate carboxylase/oxygenase, *Nature* **291:**513–515.

Jordan, D. B., and Ogren, W. L., 1983, Species variation in kinetic properties of ribulose 1,5-bisphosphate carboxylase/oxygenase, *Arch. Biochem. Biophys.* **227:**425–433.

Kahn, J. S., 1982, The epsilon subunit of the chloroplast coupling factor-1 from *Euglena gracilis*—A possible role in controlling ATPase activity, *Plant Physiol.* **70:**451–455.

Karabin, G. D., and Hallick, R. B., 1983, *Euglena gracilis* chloroplast transfer RNA transcription units. Nucleotide sequence analysis of a $tRNA^{Thr}$-$tRNA^{Gly}$-$tRNA^{Met}$-$tRNA^{Ser}$-$tRNA^{Gln}$ gene cluster, *J. Biol. Chem.* **258:**5512–5518.

Karabin, G. D., Narita, J. O., Dodd, J. R., and Hallick, R. B., 1983, *Euglena gracilis* chloroplast ribosomal RNA transcription units. Nucleotide sequence polymorphism in 5S rRNA genes and 5S rRNAs, *J. Biol. Chem.* **258:**14790–14796.

Kashdan, M. A., and Dudock, B. S., 1982, The gene for a spinach chloroplast isoleucine tRNA has a methionine anticodon, *J. Biol. Chem.* **257:**11191–11194.

Kato, A., Shimada, H., Kusuda, M., and Sugiura, M., 1981, The nucleotide sequences of two $tRNA^{Asn}$ genes from tobacco chloroplasts, *Nucleic Acids Res.* **9:**5601–5607.

Kemble, R. J., and Bedbrook, J. R., 1980, Low molecular weight circular and linear DNA in mitochondria from normal and male-sterile *Zea mays* cytoplasm, *Nature* **284:**565–566.

Kemble, R. J., and Mans, R. J., 1983, Examination of the mitochondrial genome of revertant progeny from S cms maize with cloned S-1 and S-2 hybridization probes, *J. Mol. Appl. Genet.* **2:**161–171.

Kemble, R. J., Gunn, R. E., and Flavell, R. B., 1980, Classification of normal and male-sterile cytoplasms in maize. II. Electrophoretic analysis of DNA species in mitochondria, *Genetics* **95:**451–458.

Kemble, R. J., Flavell, R. B., and Brettell, R. I. S., 1982, Mitochondrial DNA analyses of fertile and sterile maize plants derived from tissue culture with the Texas male sterile cytoplasm, *Theor. Appl. Genet.* **62:**213–217.

Kemble, R. J., Mans, R. J., Gabay-Laughnan, S., and Laughnan, J. R., 1983a, Sequences homologous to episomal mitochondrial DNAs in the maize nuclear genome, *Nature* **304:**744–747.

Kemble, R. J., Gunn, R. E., and Flavell, R. B., 1983b, Mitochondrial DNA variation in races of maize indigenous to Mexico, *Theor. Appl. Genet.* **65:**129–144.

Keus, R. J. A., Dekker, A. F., van Roon, M. A., and Groot, G. S. P., 1983a, The nucleotide sequences of the regions flanking the genes coding for 23S, 16S and 4.5S ribosomal RNA on chloroplast DNA from *Spirodela oligorhiza*, *Nucleic Acids Res.* **11:**6465–6474.

Keus, R. J. A., Roobers, D. J., Dekker, A. F., and Groot, G. S. P., 1983b, The nucleotide sequence of the 4.5S and 5S rRNA genes and flanking regions from *Spirodela oligorhiza* chloroplasts, *Nucleic Acids Res.* **11:**3501–3510.

Kim, B. D., Mans, R. J., Conde, M. F., Pring, D. R., and Levings, C. S., III, 1982, Physical mapping of homologous segments of mitochondrial episomes from S male-sterile maize, *Plasmid*, **7:**1–14.

Kimura, M., 1983, The neutral theory of molecular evolution, in: *Evolution of Genes and*

Proteins (M. Nei and R. K. Koehn, eds.), Sinauer, Sunderland, Massachusetts, pp. 208–233.

Kirk, J. T. O., and Tilney-Bassett, R. A. E., 1978, *The Plastids. Their Chemistry, Structure, Growth and Inheritance,* 2nd ed., Elsevier, Amsterdam.

Ko, K., Straus, N. A., and Williams, J. P., 1983, Mapping the chloroplast DNA of *Vicia faba, Curr. Genet.* **7:**255–263.

Koch, W., Edwards, K., and Kossel, H., 1981, Sequencing of the 16S–23S spacer in a ribosomal RNA operon of *Zea mays* chloroplast DNA reveals two split tRNA genes, *Cell* **25:**203–213.

Koller, B., and Delius, H., 1980, *Vicia faba* chloroplast DNA has only one set of ribosomal RNA genes as shown by partial denaturation mapping and R-loop analysis, *Mol. Gen. Genet.* **178:**261–269.

Koller, B., and Delius, H., 1982a, A chloroplast DNA of *Euglena gracilis* with five complete rRNA operons and two extra 16S rRNA genes, *Mol. Gen. Genet.* **188:**305–308.

Koller, B., and Delius, H., 1982b, Electron microscopic analysis of the extra 16S rRNA gene and its neighbourhood in chloroplast DNA from *Euglena gracilis* strain Z, *FEBS Lett.* **139:**86–92.

Koller, B., and Delius, H., 1982c, Parts of the sequence between the complete rRNA operons are repeated on either side of the extra 16S rRNA gene in chloroplast DNA of *Euglena gracilis* strain Z, *FEBS Lett.* **140:**198–202.

Koller, B., and Delius, H., 1982d, Origin of replication in chloroplast DNA of *Euglena gracilis* located close to the region of variable size, *EMBO J.* **1:**995–998.

Koller, B., and Delius, H., 1984, Intervening sequences in chloroplast genomes, *Cell* **36:**613–622.

Koller, B., Delius, H., and Helling, R. B., 1984a, Structure and rearrangement of rRNA genes in chloroplast DNA in two strains of *Euglena gracilis, Plant Mol. Biol.* **3:**127–136.

Koller, B., Gingrich, J. C., Stiegler, G. L., Farley, M. A., Delius, H., and Hallick, R. B., 1984b, Nine introns with conserved boundary sequences in the *Euglena gracilis* chloroplast ribulose-1,5-bisphosphate carboxylase gene, *Cell,* 36:545–553.

Kolodner, R., and Tewari, K. K., 1975a, Presence of displacement loops in the covalently closed circular chloroplast deoxyribonucleic acid from higher plants, *J. Biol. Chem* **250:**8840–8847.

Kolodner, R. D., and Tewari, K. K., 1975b, Chloroplast DNA from higher plants replicates by both the Cairns and the rolling circle mechanism, *Nature* **256:**708–711.

Kolodner, R., and Tewari, K. K., 1979, Inverted repeats in chloroplast DNA from higher plants, *Proc. Natl. Acad. Sci. USA* **76:**41–45.

Kossel, H., Edwards, K., Koch, W., Langridge, P., Sciefermayr, E., Schwarz, Z., Strittmatter, G., and Zenke, G., 1982, Structural and functional analysis of an rRNA operon and its flanking tRNA genes from *Zea mays* chloroplast, *Nucleic Acids Res. Symp. Ser.* **11:**117–120.

Kozak, M., 1983, Comparison of initiation of protein synthesis in procaryotes, eucaryotes, and organelles, *Microbiol. Rev.* **47:**1–45.

Krebbers, E. T., Larrinua, I. M., McIntosh, L., and Bogorad, L., 1982, The maize chloroplast genes for the beta and epsilon subunits of the photosynthetic coupling factor CF_1 are fused, *Nucleic Acids Res.* **10:**4985–5002.

Krebbers, E. T., Steinmetz, A., and Bogorad, L., 1984, DNA sequences for the *Zea mays* tRNA genes tV-UAC and tS-UGA: tV-UAC contains a large intron, *Plant Mol. Biol.* **3:**13–20.

Kumano, M., Tomoika, N., and Sugiura, M., 1983, The complete nucleotide sequence of a 23S rRNA gene from a blue-green alga, *Anacystis nidulans, Gene* **24:**219–225.

Kumar, A., Cocking, E. C., Bovenberg, W. A., and Kool, A. J., 1982, Restriction endonuclease analysis of chloroplast DNA in interspecies somatic hybrids of *Petunia*, *Theor. Appl. Genet.* **62**:377–383.

Kumazaki, T., Hori, H., and Osawa, S., 1982, Nucleotide sequence of cytoplasmic 5S ribosomal RNA from *Euglena gracilis*, *J. Mol. Evol.* **18**:293–296.

Kung, S. D., Zhu, Y. S., and Shen, G. F., 1982, *Nicotiana* chloroplast genome. III. Chloroplast DNA evolution, *Theor. Appl. Genet.* **61**:73–79.

Kuntz, M., Crouse, E. J., Mubumbila, M., Burkard, G., Weil, J. H., Bohnert, H. J., Mucke, H., and Loffelhardt, W., 1984, Transfer RNA gene mapping studies on cyanelle DNA from *Cyanophora paradoxa*, *Mol. Gen. Genet.* **194**:508–512.

Kuroiwa, T., Kawano, S., and Nishibayashi, S., 1982, Epifluorescent microscopic evidence for maternal inheritance of chloroplast DNA, *Nature* **298**:481–483.

Kutzelnigg, H., and Stubbe, W., 1974, Investigations on plastome mutants in *Oenothera*. I. General considerations, *Sub-Cell. Biochem.* **3**:73–89.

Lamppa, G. K., and Bendich, A. J., 1979, Chloroplast DNA sequence homologies among vascular plants, *Plant Physiol.* **63**:660–668.

Lamppa, G. K., and Bendich, A. J., 1981, Fine scale interspersion of conserved sequences in the pea and corn chloroplast genomes, *Mol. Gen. Genet.* **182**:310–320.

Laughnan, J. R., and Gabay-Laughnan, S., 1983, Cytoplasmic male sterility in maize, *Annu. Rev. Genet.* **17**:27–48.

Leaver, C. J., and Gray, M. W., 1982, Mitochondrial genome organization and expression in higher plants, *Annu. Rev. Plant Physiol.* **33**:373–402.

Leaver, C. J., Dixon, L. K., Hack, E., Fox, T. D., and Dawson, A. J., 1983, Mitochondrial genes and their expression in higher plants, in: *Structure and Function of Plant Genomes* (O. Ciferri and L. Dure, eds.), Plenum Press, New York, pp. 347–361.

Lemieux, C., Turmel, M., and Lee, R. W., 1981, Physical evidence for recombination of chloroplast DNA in hybrid progeny of *Chlamydomonas eugametos* and *C. moewusii*, *Curr. Genet.* **3**:97–103.

Lemieux, C., Turmel, M., Seligy, V. L., and Lee, R. W., 1984, Chloroplast DNA recombination in interspecific hybrids of *Chlamydomonas:* Linkage between a nonmendelian locus for streptomycin resistance and restriction fragments coding for 16S rRNA, *Proc. Natl. Acad. Sci. USA* **81**:1164–1168.

Levings, C. S., III, 1983, The plant mitochondrial genome and its mutants, *Cell* **32**:659–661.

Levings, C. S., III, and Pring, D. R., 1977, Diversity of mitochondrial genomes among normal cytoplasms of maize, *J. Hered.* **68**:350–354.

Levings, C. S., III, and Sederoff, R. R., 1983, Nucleotide sequence of the S-2 mitochondrial DNA from the S cytoplasm of maize, *Proc. Natl. Acad. Sci. USA* **80**:4055–4059.

Levings, C. S., III, Kim, B. D., Pring, D. R., Conde, M. F., Mans, R. J., Laughnan, J. R., and Gabay-Laughnan, S. J., 1980, Cytoplasmic reversion of cms-S in maize: Association with a transpositional event, *Science* **209**:1021–1023.

Levings, C. S., III, Sederoff, R. R., Hu, W. W. L., and Timothy, D. H., 1983, Relationships among plasmid-like DNAs of the maize mitochondria, in: *Structure and Function of Plant Genomes* (O. Ciferri and L. Dure III, eds.), Plenum Press, New York, pp. 363–371.

Li, W. H., 1983, Evolution of duplicate genes and pseudogenes, in: *Evolution of Genes and Proteins*, (M. Nei and R. K. Koehn, eds.), Sinauer, Sunderland, Massachusetts, pp. 14–37.

Lindahl, L., and Zengel, J. M., 1982, Expression of ribosomal genes in bacteria, *Adv. Genet.* **21**:53–121.

Link, G., and Langridge, U., 1984, Structure of the chloroplast gene for the precursor of the M_r 32,000 photosystem II protein from mustard (*Sinapis alba* L.), *Nucleic Acids Res.* **12**:945–958.

Loffelhardt, W., Mucke, H., Crouse, E. J., and Bohnert, H. J., 1983, Comparison of the cyanelle DNA from two different strains of *Cyanophora paradoxa*, *Curr. Genet.* **7**:139–144.

Loiseaux, S., and Dalmon, J., 1983, Some characteristics of plastid DNA from different algae, *Physiol. Veg.* **21**:775–784.

Long, E. O., and Dawid, I. B., 1980, Repeated genes in eukaryotes, *Annu. Rev. Biochem.* **49**:727–764.

Lonsdale, D. M., 1984, A review of the structure and organization of the mitochondrial genome of higher plants, *Plant Mol. Biol.* **3**:201–206.

Lonsdale, D. M., Thompson, R. D., and Hodge, T. P., 1981, The integrated forms of the S1 and S2 DNA elements of maize male sterile mitochondrial DNA are flanked by a large repeated sequence, *Nucleic Acids Res.* **9**:3657–3669.

Lonsdale, D. M., Hodge, T. P., Howe, C. J., and Stern, D. B., 1983, Maize mitochondrial DNA contains a sequence homologous to the ribulose-1,5-bisphosphate carboxylase large subunit gene of chloroplast DNA, *Cell* **34**:1007–1014.

Lonsdale, D. M., Hodge, T. P., and Fauron, C. M. R., 1984, The physical map and organization of the mitochondrial genome from the fertile cytoplasm of maize, *Nucleic Acids Res.* **12**:9251–9261.

Luck, D. J. L., and Reich, E., 1964, DNA in mitochondria of *Neurospora crassa*, *Proc. Natl. Acad. Sci. USA* **52**:931–937.

Luttke, A., and Bonotto, S., 1982, Chloroplasts and chloroplast DNA of *Acetabularia mediterranea:* Facts and hypotheses, *Int. Rev. Cytol.* **77**:205–242.

MacKay, R. M., Salgado, D., Bonen, L., Stackebrandt, E., and Doolittle, W. F., 1982, The 5S ribosomal RNAs of *Paracoccus denitrificans* and *Prochloron*, *Nucleic Acids Res.* **10**:2963–2970.

Mahler, H. R., 1983, The exon:intron structure of some mitochondrial genes and its relation to mitochondrial evolution, *Int. Rev. Cytol.* **82**:1–98.

Mattoo, A. K., Pick, U., Hoffman-Falk, H., and Edelman, M., 1981, The rapidly metabolized 32,000 dalton polypeptide of the chloroplast is the "proteinaceous shield" regulating photosystem II electron transport and mediating diuron herbicide sensitivity, *Proc. Natl. Acad. Sci. USA* **78**:1572–1576.

Mazza, A., Casale, A., Sassone-Corsi, P., and Bonotto, S., 1980, A minicircular component of *Acetabularia acetabulum* chloroplast DNA replicating by the rolling circle, *Biochem. Biophys. Res. Commun.* **93**:668–674.

McIntosh, L., Poulsen, C., and Bogorad, L., 1980, Chloroplast gene sequence for the large subunit of ribulose bisphosphate carboxylase of maize, *Nature 288*:556–560.

McNay, J. W., Pring, D. R., and Lonsdale, D. M., 1983, Polymorphism of mitochondrial DNA 'S' regions among normal cytoplasms of maize, *Plant Mol. Biol.* **2**:177–187.

McNay, J. W., Chourey, P. S., and Pring, D. R., 1984, Molecular analysis of genomic stability of mitochondrial DNA in tissue cultured cells of maize, *Theor. Appl. Genet.* **67**:433–437.

Menke, W., and Wolfersdorff, B., 1968, Uber die plastiden von *Neottia nidus-avis*, *Planta* **78**:134–143.

Michel, F., and Dujon, B., 1983, Conservation of RNA secondary structures in two intron families including mitochondrial-, chloroplast- and nuclear-encoded members, *EMBO J.* **2**:33–38.

Miyata, T., Kikuno, R., and Ohshima, Y., 1982, A pseudogene cluster in the leader region of the *Euglena* chloroplast 16S–23S rRNA genes, *Nucleic Acids Res.* **10**:1771–1780.

Miziorko, H. M., and Lorimer, G. H., 1983, Ribulose-1,5-bisphosphate carboxylase oxygenase, *Annu. Rev. Biochem.* **52**:507–535.

Montandon, P. E., and Stutz, E., 1983, Nucleotide sequence of a *Euglena gracilis* chloroplast genome region coding for the elongation factor Tu; Evidence for a spliced mRNA, *Nucleic Acids Res.* **11**:5877–5892.

Montandon, P. E., and Stutz, E., 1984, The genes for the ribosomal proteins S12 and S7 are clustered with the gene for the EF-Tu protein on the chloroplast genome of *Euglena gracilis*, *Nucleic Acids Res.* **12**:2851–2859.

Morgan, E. A., Ikemura, T., and Nomura, M., 1977, Identification of spacer tRNA genes in individual ribosomal RNA transcription units of *Eschericia coli*, *Proc. Natl. Acad. Sci. USA* **74**:2710–2714.

Morris, J., and Herrmann, R. G., 1984, Nucleotide sequence of the gene for the P_{680} chlorophyll a apoprotein of the photosystem II reaction center from spinach, *Nucleic Acids Res.* **12**:2837–2850.

Mount, S. M., Pettersson, I., Hinterberger, M., Karmas, A., and Steitz, J. A., 1983, The U1 small nuclear RNA–protein complex selectively binds a 5′-splice site *in vitro*, *Cell* **33**:509–518.

Mubumbila, M., Gordon, K. H. J., Crouse, E. J., Burkard, G., and Weil, J. H., 1983, Construction of the physical map of the chloroplast DNA of *Phaseolus vulgaris* and localization of ribosomal and transfer RNA genes, *Gene* **21**:257–266.

Mubumbila, M., Crouse, E. J., and Weil, J. H., 1984, Transfer RNAs and tRNA genes of *Vicia faba* chloroplasts, *Curr. Genet.* **8**:379–385.

Muller, J., 1981, Fossil pollen records of extant angiosperms, *Bot. Rev.* **47**:1–142.

Myers, A. M., Grant, D. M., Rabert, D. K., Harris, E. H., Boynton, J. E., and Gillham, N. W., 1982, Mutants of *Chlamydomonas reinhardtii* with physical alterations in their chloroplast DNA, *Plasmid* **7**:133–151.

Nagy, F., Torok, I., and Maliga, P., 1981, Extensive rearrangements in the mitochondrial DNA in somatic hybrids of *Nicotiana tabacum* and *Nicotiana knightiana*, *Mol. Gen. Genet.* **183**:437–439.

Nagy, F., Lazar, G., Menczel, L., and Maliga, P., 1983, A heteroplasmic state induced by protoplast fusion is a necessary condition for detecting rearrangements in *Nicotiana* mitochondrial DNA, *Theor. Appl. Genet.* **66**:203–207.

Nargang, F., McIntosh, L., and Somerville, C., 1984, Nucleotide sequence of the ribulose-bisphosphate carboxylase gene from *Rhodospirillum rubrum*, *Mol. Gen. Genet.* **193**:220–224.

Negruk, V. I., Cherny, D. I., Nikiforova, I. D., Aleksandrov, A. A., and Butenko, R. G., 1982, Isolation and characterization of minicircular DNAs found in mitochondrial fraction of *Vicia faba*, *FEBS Lett.* **142**:115–117.

Nelson, N., 1980, Proton ATPase of chloroplasts, *Curr. Top. Bioenergetics* **11**:1–33.

Nierzwicki-Bauer, S. A., Curtis, S. E., and Haselkorn, R., 1984, Cotranscription of genes encoding the small and large subunits of ribulose-1,5-bisphosphate carboxylase in the cyanobacterium *Anabaena* 7120, *Proc. Natl. Acad. Sci. USA* **81**:5961–5965.

Nikiforova, I. D., and Negruk, V. I., 1983, Comparative electrophoretical analysis of plasmid-like mitochondrial DNAs in *Vicia faba* and in some other legumes, *Planta* **157**:81–84.

Ohyama, K., Yamano, Y., Fukuzawa, H., Komano, T., Yamagishi, H., Fujimoto, S., and Sugiura, M., 1983, Physical mappings of chloroplast DNA from liverwort *Marchantia polymorpha* L. cell suspension cultures, *Mol. Gen. Genet.* **189**:1–9.

Oishi, K., Sumnicht, T., and Tewari, K. K., 1981, Messenger ribonucleic acid transcripts of pea chloroplast deoxyribonucleic acid, *Biochemistry* **20**:5710–5717.

Ojala, D., Montoya, J., and Attardi, G., 1981, tRNA punctuation model of RNA processing in human mitochondria, *Nature* **290**:470–474.

Orozco, E. M., Jr., Rushlow, K. E., Dodd, J. R., and Hallick, R. B., 1980, *Euglena gracilis* chloroplast ribosomal RNA transcription units. II. Nucleotide sequence homology between the 16S–23S ribosomal RNA spacer and the 16S ribosomal RNA leader regions, *J. Biol. Chem.* **255**:10997–11003.

Padmanabhan, U., and Green, B. V., 1978, The kinetic complexity of *Acetabularia* chloroplast DNA, *Biochim. Biophys. Acta.* **521**:67–73.

Palmer, J. D., 1981, Chloroplast DNA Evolution: Molecular and Phylogenetic Studies, Ph.D. Thesis, Stanford University, Stanford, California.

Palmer, J. D., 1982, Physical and gene mapping of chloroplast DNA from *Atriplex triangularis* and *Cucumis sativa*, *Nucleic Acids Res.* **10**:1593–1605.

Palmer, J. D., 1983, Chloroplast DNA exists in two orientations, *Nature* **301**:92–93.

Palmer, J. D., 1985a, Phylogenetic analysis of chloroplast DNA variation, *Ann. Missouri Bot. Garden* (in press).

Palmer, J. D. 1985b, Chloroplast DNA and phylogenetic relationships, in: *DNA Systematics: Evolution* (S. K. Dutta, ed.), CRC Press, Boca Raton, Florida (in press).

Palmer, J. D., and Shields, C. R., 1984, Tripartite structure of the *Brassica campestris* mitochondrial genome, *Nature*, **307**:437–440.

Palmer, J. D., and Stein, D. B., 1982, Chloroplast DNA from the fern *Osmunda cinnamomea: Physical organization, gene localization and comparison to angiosperm chloroplast DNA*, *Curr. Genet.* **5**:165–170.

Palmer, J. D., and Thompson, W. F., 1981a, Rearrangements in the chloroplast genomes of mung bean and pea, *Proc. Natl. Acad. Sci. USA* **78**:5533–5537.

Palmer, J. D., and Thompson, W. F., 1981b, Clone banks of mung bean, pea and spinach chloroplast DNA, *Gene* **15**:21–26.

Palmer, J. D., and Thompson, W. F., 1982, Chloroplast DNA rearrangements are more frequent when a large inverted repeat sequence is lost, *Cell* **29**:537–550.

Palmer, J. D., and Zamir, D., 1982, Chloroplast DNA evolution and phylogenetic relationships in *Lycopersicon*, *Proc. Natl. Acad. Sci. USA* **79**:5006–5010.

Palmer, J. D., Edwards, H., Jorgensen, R. A., and Thompson, W. F., 1982a, Novel evolutionary variation in transcription and location of two chloroplast genes, *Nucleic Acids Res.* **10**:6819–6832.

Palmer, J. D., Osorio, B., Thompson, W. F., and Nobs, M. A., 1982b, Chloroplast DNA evolution in *Atriplex*, *Carnegie Inst. Washington Yearb.* **81**:96–97.

Palmer, J. D., Shields, C. R., Cohen, D. B., and Orton, T. J., 1983a, Chloroplast DNA evolution and the origin of amphidiploid *Brassica* species, *Theor. Appl. Genet.* **65**:181–189.

Palmer, J. D. Singh, G. P., and Pillay, D. T. N., 1983b, Structure and sequence evolution of three legume chloroplast DNAs, *Mol. Gen. Genet.* **190**:13–19.

Palmer, J. D., Shields, C. R., Cohen, D. B., and Orton, T. J., 1983c, An unusual mitochondrial DNA plasmid in the genus *Brassica*, *Nature* **301**:725–728.

Palmer, J. D., Osorio, B., Watson, J. C., Edwards, H., Dodd, J., and Thompson, W. F., 1984, Evolutionary aspects of chloroplast genome expression and organization, in: *Biosynthesis of the Photosynthetic Apparatus: Molecular Biology, Development and Regulation*, UCLA Symposia on Molecular and Cellular Biology, New Series, Vol. 14, (J. P. Thornber, L. A. Staehelin, and R. Hallick, eds.), Alan R. Liss, New York, pp. 273–283.

Palmer, J. D., Jorgensen, R. A., and Thompson, W. F., 1985a, Chloroplast DNA variation and evolution in *Pisum:* Patterns of change and phylogenetic analysis, *Genetics*, **109**:195–213.

Palmer, J. D., Boynton, J. E., Gillham, N. W., and Harris, E. H., 1985b, Evolution and recombination of the large inverted repeat in *Chlamydomonas* chloroplast DNA, in:

Molecular Biology of the Photosynthetic Apparatus, (C. Arntzen, L. Bogorad, S. Bonitz, and K. Steinback, eds.), Cold Spring Harbor Press, New York (in press).

Passavant, C. W., Stiegler, G. L., and Hallick, R. B., 1983, Location of the single gene for elongation factor Tu on the *Euglena gracilis* chloroplast chromosome, *J. Biol. Chem.* **258**:693–695.

Petes, T. D., 1980, Molecular genetics of yeast, *Annu. Rev. Biochem.* **49**:845–876.

Pfister, K., Steinback, K. E., Gardner, G., and Arntzen, C. J., 1981, Photoaffinity labeling of an herbicide receptor in chloroplast membranes, *Proc. Natl. Acad. Sci. USA* **78**:981–985.

Phillips, A. L., and Gray, J. C., 1984a, Location and nucleotide sequence of the gene for the 15.2 kDa polypeptide of the cytochrome *b-f* complex from pea chloroplasts, *Mol. Gen. Genet.* **194**:477–484.

Phillips, A. L., and Gray, J. C., 1984b, Synthesis of components of the cytochrome *b-f* complex by isolated pea chloroplasts, *Eur. J. Biochem.* **138**:591–595.

Plasterk, R. H. A., Brinkman A., and Van de Putte, P., 1983, DNA inversions in the chromosome of *Escherichia coli* and in bacteriophage Mu: Relationship to other site-specific recombination systems, *Proc. Natl. Acad. Sci. USA* **80**:5355–5358.

Posno, M., van Noort, M., Debise, R., and Groot, G. S. P., 1984, Isolation, characterization, phosphorylation and site of synthesis of *Spinacia* chloroplast ribosomal proteins, *Curr. Genet.* **8**:147–154.

Possingham, J. V., and Lawrence, M. E., 1983, Controls to plastid division, *Int. Rev. Cytol.* **84**:1–56.

Poulsen, C. R., 1983, The barley chloroplast genome: Physical structure and transcriptional activity *in vivo*, *Carlsberg Res. Commun.* **48**:57–80.

Powling, A., 1981, Species of small DNA molecules found in mitochondria from sugarbeet with normal and male sterile cytoplasms, *Mol. Gen. Genet.* **183**:82–84.

Powling, A., and Ellis, T. H. N., 1983, Studies on the organelle genomes of sugarbeet with male-fertile and male-sterile cytoplasms, *Theor. Appl. Genet.* **65**:323–328.

Pring, D. R., and Levings, C. S., III, 1978, Heterogeneity of maize cytoplasmic genomes among male-sterile cytoplasms, *Genetics* **89**:121–136.

Pring, D. R., Levings, C. S., III., Hu, W. W. L., and Timothy, D. H., 1977, Unique DNA associated with mitochondria in the "S"-type cytoplasm of male-sterile maize, *Proc. Natl. Acad. Sci. USA* **74**:2904–2908.

Pring, D. R., Conde, M. F., Schertz, K. F., and Levings, C. S., III, 1982, Plasmid-like DNAs associated with mitochondria of cytoplasmic male-sterile *Sorghum*, *Mol. Gen. Genet.* **186**:180–184.

Ravel-Chapuis, P., Heizmann, P., and Nigon, V., 1982, Electron microscopic localization of the replication origin of *Euglena gracilis* chloroplast DNA, *Nature* **300**:78–81.

Ravel-Chapuis, P., Flamant, F., Nicolas, P., Heizmann, P., and Nigon, V., 1984, Diversity of the ribosomal structures in the *Euglena gracilis* chloroplast genome: Description of a mutant with two ribosomal operons and possible mechanism for its production, *Nucleic Acids Res.* **12**:1039–1048.

Rawson, J. R. Y., and Boerma, C. L., 1976, A measurement of the fraction of chloroplast DNA transcribed during chloroplast development in *Euglena gracilis*, *Biochemistry* **15**:588–592.

Rawson, J. R. Y., Kushner, S. R., Vapnek, D., Kirby, N., and Boerma, C. L., 1978, Chloroplast ribosomal RNA genes in *Euglena gracilis* exist as three clustered tandem repeats, *Gene* **3**:191–209.

Reichelt, B., and Delaney, S., 1983, The nucleotide sequence for the large subunit of ribulose-1,5-bisphosphate carboxylase from a unicellular cyanobacterium, *Synechococcus* PCC6301, *DNA* **2**:121–130.

Rochaix, J. D., 1972, Cyclization of chloroplast DNA fragments of *Chlamydomonas reinhardtii*, *Nature* **238**:76–78.

Rochaix, J. D., 1978, Restriction endonuclease map of the chloroplast DNA of *Chlamydomonas reinhardtii*, *J. Mol. Biol.* **126**:597–617.

Rochaix, J. D., and Darlix, J. L., 1982, Composite structure of the chloroplast 23S ribosomal RNA genes in *Chlamydomonas reinhardtii:* Evolutionary and functional implications, *J. Mol. Biol.* **159**:383–395.

Rochaix, J. D., and Malnoe, P., 1978, Anatomy of the chloroplast ribosomal DNA of *Chlamydomonas reinhardtii*, *Cell* **15**:661–670.

Roth, J. R., and Schmid, M. B., 1981, Arrangement and rearrangement of the bacterial chromosome, *Stadler Symp.* **13**:53–70.

Roux, E., Graf, L., and Stutz, E., 1983, Nucleotide sequence of a 'truncated rRNA operon' of the *Euglena gracilis* chloroplast genome, *Nucleic Acids Res.* **11**:1957–1968.

Ryan, R., Grant, D., Chiang, K. S., and Swift, H., 1978, Isolation and characterization of mitochondrial DNA from *Chlamydomonas reinhardtii*, *Proc. Natl. Acad. Sci. USA* **75**:3268–3272.

Sager, R., and Grabowy, C., 1983, Differential methylation of chloroplast DNA regulates maternal inheritance in a methylated mutant of *Chlamydomonas*, *Proc. Natl. Acad. Sci. USA* **80**:3025–3029.

Sager, R., and Ishida, M. R., 1963, Chloroplast DNA in *Chlamydomonas*, *Proc. Natl. Acad. Sci. USA* **50**:2725–2730.

Sanderson, K. E., and Roth, J. R., 1983, Linkage map of *Salmonella typhimurium*, Edition VI, *Microbiol. Rev.* **47**:410–453.

Schardl, C. L., Lonsdale, D. M., Pring, D. R., and Rose, K. R., 1984, Linearisation of maize mitochondrial chromosomes by recombination with linear episomes, *Nature*, **310**:292–296.

Schiff, J. A., 1980, Development, inheritance and evolution of plastids and mitochondria, *Biochem. Plants* **1**:209–272.

Schiller, B., Herrmann, R. G., and Melchers, G., 1982, Restriction endonuclease analysis of plastid DNA from tomato, potato and some of their somatic hybrids, *Mol. Gen. Genet.* **186**:453–459.

Schlunegger, B., Fasnacht, M., Stutz, E., Koller, B., and Delius, H., 1983, Analysis of a polymorphic region of the *Euglena gracilis* chloroplast genome, *Biochim. Biophys. Acta* **739**:114–121.

Schmid, M. B., and Roth, J. R., 1983a, Genetic methods for analysis and manipulation of inversion mutations in bacteria, *Genetics* **105**:517–537.

Schmid, M. B., and Roth, J. R., 1983b, Selection and endpoint distribution of bacterial inversion mutations, *Genetics* **105**:539–557.

Schmidt, R. J., Richardson, C. B., Gillham, N. W., and Boynton, J. E., 1983, Sites of synthesis of chloroplast ribosomal proteins in *Chlamydomonas*, *J. Cell Biol.* **96**:1451–1463.

Schmidt, R. J., Myers, A. M., Gillham, N. W., and Boynton, J. E., 1984, Immunological similarities between specific chloroplast ribosomal proteins from *Chlamydomonas reinhardtii* and ribosomal proteins from *Escherichia coli*, *Mol. Biol. Evol.* **1**:317–334.

Schmitt, J. M., Bohnert, H. J., Gordon, K. H. J., Herrmann, R., Bernardi, G., and Crouse, E. J., 1981, Compositional heterogeneity of the chloroplast DNAs from *Euglena gracilis* and *Spinacia oleracea*, *Eur. J. Biochem.* **117**:375–382.

Schwarz, Z., Jolly, S. O., Steinmetz, A. A., and Bogorad, L., 1981, Overlapping divergent genes in the maize chloroplast chromosome and *in vitro* transcription of the gene for $tRNA^{His}$, *Proc. Natl. Acad. Sci. USA* **78**:3423–3427.

Scott, N. S., and Timmis, J. N., 1984, Homologies between nuclear and plastid DNA in spinach, *Theor. Appl. Genet.* **67**:279–288.

Scowcroft, W. R., and Larkin, P. J., 1981, Chloroplast DNA assorts randomly in intraspecific somatic hybrids of *Nicotiana debneyi*, *Theor. Appl. Genet.* **60**:179–184.

Sears, B. B., 1980, Elimination of plastids during spermatogenesis and fertilization in the plant kingdom, *Plasmid* **4**:233–255.

Sears, B. B., 1983, Genetics and evolution of the chloroplast, *Stadler Symp.* **15**:119–139.

Sederoff, R. R., 1984, Structural variation in mitochondrial DNA, *Adv. Genet.* **22**:1–108.

Sederoff, R. R., Levings, C. S., III, Timothy, D. H., and Hu, W. W. L., 1981, Evolution of DNA sequence organization in mitochondrial genomes of *Zea*, *Proc. Natl. Acad. Sci. USA* **78**:5953–5957.

Seewaldt, E., and Stackebrandt, E., 1982, Partial sequence of 16S ribosomal RNA and the phylogeny of *Prochloron*, *Nature* **295**:618–620.

Selden, R. F., Steinmetz, A., McIntosh, L., Bogorad, L., Burkard, G., Mubumbila, M., Kuntz, M., Crouse, E. J., and Weil, J. H., 1983, Transfer RNA genes of *Zea mays* chloroplast DNA, *Plant Mol. Biol.* **2**:141–153.

Sepsenwol, S., 1973, Leucoplast of the cryptomonad, *Chilomonas paramecium*, *Exp. Cell. Res.* **76**:395–409.

Seyer, P., Herrmann, R. G., and Lescure, A. M., 1982, Purification of plastid DNA from tobacco cell suspensions, *Plant Sci. Lett.* **25**:345–352.

Shen, G. F., Chen, K., Wu, M., and Kung, S. D., 1982, *Nicotiana* chloroplast genome IV. *N. accuminata* has larger inverted repeats and genome size, *Mol. Gen. Genet.* **187**:12–18.

Shinozaki, K., and Sugiura, M., 1982, The nucleotide sequence of the tobacco chloroplast gene for the large subunit of ribulose-1,5-bisphosphate carboxylase/oxygenase, *Gene* **20**:91–102.

Shinozaki, K., and Sugiura, M., 1983, The gene for the small subunit of ribulose-1,5-bisphosphate carboxylase/oxygenase is located close to the gene for the large subunit in the cyanobacterium *Anacystis nidulans* 6301, *Nucleic Acids Res.* **11**:6957–6964.

Shinozaki, K., Yamada, C., Takahata, N., and Sugiura, M., 1983a, Molecular cloning and sequence analysis of the cyanobacterial gene for the large subunit of ribulose-1,5-bisphosphate carboxylase/oxygenase, *Proc. Natl. Acad. Sci. USA* **80**:4050–4054.

Shinozaki, K., Deno, H., Kato, A., and Sugiura, M., 1983b, Overlap and cotranscription of the genes for the beta and epsilon subunits of tobacco chloroplast ATPase, *Gene* **24**:147–155.

Sinden, R. R., and Pettijohn, D. E., 1981, Chromosomes in living *Escherichia coli* cells are segregated into domains of supercoiling, *Proc. Natl. Acad. Sci. USA* **78**:224–228.

Siu, C. H., Chiang, K. S., and Swift, H., 1975, Characterization of cytoplasmic and nuclear genomes in the colorless alga *Polytoma*. V. Molecular structure and heterogeneity of leucoplast DNA, *J. Mol. Biol.* **98**:369–391.

Siu, C. H., Swift, H., and Chiang, K. S., 1976a, Characterization of cytoplasmic and nuclear genomes in the colorless alga *Polytoma*. I. Ultrastructural analysis of organelles, *J. Cell Biol.* **69**:352–370.

Siu, C. H., Swift, H., and Chiang, K. S., 1976b, Characterization of cytoplasmic and nuclear genomes in the colorless alga *Polytoma*. II. General characterization of organelle nucleic acids, *J. Cell Biol.* **69**:371–382.

Smith, A. G., and Gray, J. C., 1984, Localization of the gene for P700 chlorophyll a protein in pea chloroplast DNA, *Mol. Gen. Genet.* **194**:471–476.

Smith, G. P., 1976, Evolution of repeated DNA sequences by unequal crossover, *Science* **191**:528–535.

Somerville, C. R., and Ogren, W. L., 1982, Genetic modification of photorespiration, *Trends Biochem. Sci.* **7:**171–174.

Somerville, C. R., and Somerville, S. C., 1984, Cloning and expression of the *Rhodospirillum rubrum* ribulosebisphosphate carboxylase gene in *E. coli, Mol. Gen. Genet.* **193:**214–219.

Sparks, R. B., Jr, and Dale, R. M. K., 1980, Characterization of ^{3}H-labeled supercoiled mitochondrial DNA from tobacco suspension culture cells, *Mol. Gen. Genet.* **180:**351–355.

Spencer, D. F., Bonen, L., and Gray, M. W., 1981, Primary sequence of wheat mitochondrial 5S ribosomal ribonucleic acid: Functional and evolutionary implications, *Biochemistry* **20:**4022–4029.

Spencer, D. F., Schnare, M. N., and Gray, M. W., 1984, Pronounced structural similarities between the small subunit ribosomal RNA genes of wheat mitochondria and *Escherichia coli, Proc. Natl. Acad. Sci. USA* **81:**493–497.

Spielmann, A., and Stutz, E., 1983, Nucleotide sequence of soybean chloroplast DNA regions which contain the *psbA* and *trnH* genes and cover the ends of the large single copy region and one end of the inverted repeats, *Nucleic Acids Res.* **11:**7157–7167.

Spielmann, A., Ortiz, W., and Stutz, E., 1983, The soybean chloroplast genome: Construction of a circular restriction site map and location of DNA regions encoding the genes for rRNAs, the large subunit of the ribulose-1,5-bisphosphate carboxylase and the 32 kd protein of the photosystem II reaction center, *Mol. Gen. Genet.* **190:**5–12.

Spruill, W. M., Jr., Levings, C. S., III, and Sederoff, R. R., 1981, Organization of mitochondrial DNA in normal and Texas male sterile cytoplasms of maize, *Dev. Genet.* **2:**319–336.

Steinback, K. E., McIntosh, L., Bogorad, L., and Arntzen, C. J., 1981, Identification of the triazine receptor protein as a chloroplast gene product, *Proc. Natl. Acad. Sci. USA* **78:**7463–7467.

Steinmetz, A., Gubbins, E. J., and Bogorad, L., 1982, The anticodon of the maize chloroplast gene for $tRNA^{Leu}$ (UAA) is split by a large intron, *Nucleic Acids Res.* **10:**3027–3037.

Steinmuller, K., Kaling, M., and Zetsche, K., 1983, *In vitro* synthesis of phycobiliproteins and ribulose-1,5-bisphosphate carboxylase by non-polyadenylated-RNA of *Cyanidium caldarium* and *Porphyridium aerugineum, Planta* **159:**308–313.

Stern, D. B., and Lonsdale, D. M., 1982, Mitochondrial and chloroplast genomes of maize have a 12-kilobase DNA sequence in common, *Nature* **299:**698–702.

Stern, D. B., and Palmer, J. D., 1984a, Extensive and widespread homologies between mitochondrial DNA and chloroplast DNA in plants, *Proc. Natl. Acad. Sci. USA* **81:**1946–1950.

Stern, D. B., and Palmer, J. D., 1984b, Recombination sequences in plant mitochondrial genomes: Diversity and homologies to known mitochondrial genes, *Nucleic Acids Res.*, **12:**6141–6157.

Stern, D. B., Dyer, T. A., and Lonsdale, D. M., 1982, Organization of the mitochondrial ribosomal RNA genes of maize, *Nucleic Acids Res.* **10:**3333–3340.

Stern, D. B., Palmer, J. D., Thompson, W. F., and Lonsdale, D. M., 1983, Mitochondrial DNA sequence evolution and homology to chloroplast DNA in angiosperms, in:*Plant Molecular Biology* (R. B. Goldberg, ed.), Alan R. Liss, New York, pp. 467–477.

Stiegler, G. L., Matthews, H. M., Bingham, S. E., and Hallick, R. B., 1982, The gene for the large subunit of ribulose-1,5-bisphosphate carboxylase in *Euglena gracilis* chloroplast DNA: Location, polarity, cloning and evidence for an intervening sequence, *Nucleic Acids Res.* **10:**3427–3444.

Streisinger, G., Okada, Y., Emrich, J., Newton, J., Tsugita, A., Terzhagi, E., and Inouye, M., 1966, Frameshift mutations and the genetic code, *Cold Spring Harbor Symp. Quant. Biol.* **31:**77–84.

Subramanian, A. R., Steinmetz, A., and Bogorad, L., 1983, Maize chloroplast DNA encodes a protein sequence homologous to the bacterial ribosome assembly protein S4, *Nucleic Acids Res.* **11**:5277–5286.

Sugita, M., and Sugiura, M., 1983, A putative gene of tobacco chloroplast coding for ribosomal protein similar to *E. coli* ribosomal protein S19, *Nucleic Acids Res.* **11**:1913–1918.

Sugita, M., Kato, A., Shimada, H., and Sugiura, M., 1984, Sequence analysis of the junctions between a large inverted repeat and single-copy regions in tobacco chloroplast DNA, *Mol. Gen. Genet.* **194**:200–205.

Szekely, E., and Simon, M., 1983, DNA sequence adjacent to flagellar genes and evolution of flagellar-phage variation, *J. Bacteriol.* **155**:74–81.

Szostak, J. W., and Wu, R., 1980, Unequal crossing over in the ribosomal DNA of *Saccharomyces cerevisiae*, *Nature* **284**:426–430.

Takaiwa, F., and Sugiura, M., 1982a, Nucleotide sequence of the 16S–23S spacer region in an rRNA gene cluster from tobacco chloroplast DNA, *Nucleic Acids Res.* **10**:2665–2676.

Takaiwa, F., and Sugiura, M., 1982b, The complete nucleotide sequence of a 23S rRNA gene from tobacco chloroplasts, *Eur. J. Biochem.* **124**:13–19.

Tassopulu, D., and Kung, S. D., 1984, *Nicotiana* chloroplast genome. 6. Deletion and hot spot—A proposed origin of the inverted repeats, *Theor. Appl. Genet.* **67**:185–193.

Thompson, R. D., Kemble, R. J., and Flavell, R. B., 1980, Variations in mitochondrial DNA organization between normal and male-sterile cytoplasms in maize, *Nucleic Acids Res.* **8**:1999–2008.

Timmis, J. N., and Scott, N. S., 1983, Sequence homology between spinach nuclear and chloroplast genomes, *Nature* **305**:65–67.

Timothy, D. H., Levings, C. S., III, Pring, D. R., Conde, M. F., and Kermicle, J. L., 1979, Organelle DNA variation and systematic relationships in the genus *Zea:* Teosinte, *Proc. Natl. Acad. Sci. USA* **76**:4220–4224.

Tohdoh, N., and Sugiura, M., 1982, The complete nucleotide sequence of a 16S ribosomal RNA gene from tobacco chloroplasts, *Gene* **17**:213–218.

Tohdoh, N., Shinozaki, K., and Sugiura, M., 1981, Sequence of a putative promoter region for the rRNA genes of tobacco chloroplast DNA, *Nucleic Acids Res.* **9**:5399–5406.

Tomioka, N., and Sugiura, M., 1983, The complete nucleotide sequence of a 16S ribosomal RNA gene from a blue-green alga, *Anacystis nidulans*, *Mol. Gen. Genet.* **191**:46–50.

Tomioka, N., and Sugiura, M., 1984, Nucleotide sequence of the 16S–23S spacer region in the *rrnA* operon from a blue-green alga, *Anacystis nidulans*, *Mol. Gen. Genet.* **193**:427–430.

Tomioka, N., Shinozaki, K., and Sugiura, M., 1981, Molecular cloning and characterization of ribosomal RNA genes from a blue-green alga, *Anacystis nidulans*, *Mol. Gen. Genet.* **184**:359–363.

Tsunewaki, K., and Ogihara, Y., 1983, The molecular basis of genetic diversity among cytoplasms of *Triticum* and *Aegilops* species. II. On the origin of polyploid wheat cytoplasms as suggested by chloroplast DNA restriction fragment patterns, *Genetics* **104**:155–171.

Tsuzuki, T., Nomiyama, H., Setoyama, C., Maeda, S., and Shimada, K., 1983, Presence of mitochondrial-DNA-like sequences in the human nuclear DNA, *Gene* **25**:223–229.

Van den Boogaart, P., Samallo, J., and Agsteribbe, E., 1982, Similar genes for a mitochondrial ATPase subunit in the nuclear and mitochondrial genomes of *Neurospora crassa*, *Nature* **298**:187–189.

Van Ee, J. H., Vos, Y. J., and Planta, R. J., 1980, Physical map of chloroplast DNA of *Spirodela oligorrhiza;* analysis by the restriction endonucleases *Pst*I, *Xho*I and *Sac*I, *Gene* **12**:191–200.

Van Valen, L. M., 1982, Phylogenies in molecular evolution: *Prochloron*, *Nature* **298**:493–494.

Vaughn, K. C., 1981, Organelle transmission in higher plants: Organelle alteration vs. physical exclusion, *J. Hered.* **72**:335–337.

Vaughn, K. C., Kimpel, D. L., and Wilson, K. G., 1981, Control of organelle transmission in *Chlorophytum*, *Curr. Genet.* **3**:105–108.

Waddell, J., Wang, X. M., and Wu, M., 1984, Electron microscopic localization of the chloroplast DNA replicative origins in *Chlamydomonas reinhardtii*, *Nucleic Acids Res.* **12**:3843–3856.

Walker, J. E., Falk, G., Gay, N. J., and Tybulewicz, V. L. J., 1984, Genes for bacterial and mitochondrial ATP synthase, *Biochem. Soc. Trans.* **12**:234–235.

Wallace, D. C., 1982, Structure and evolution of organelle genomes, *Microbiol. Rev.* **46**:208–240.

Ward, B. L., Anderson, R. S., and Bendich, A. J., 1981, The mitochondrial genome is large and variable in a family of plants (Cucurbitaceae), *Cell* **25**:793–803.

Waring, R. B., Scazzocchio, C., Brown, T. A., and Davies, R. W., 1983, Close relationship between certain nuclear and mitochondrial introns. Implications for the mechanism of RNA splicing, *J. Mol. Biol.* **167**:595–605.

Watson, J. C., and Surzycki, S. J., 1982, Extensive sequence homology in the DNA coding for elongation factor Tu from *Escherichia coli* and the *Chlamydomonas reinhardtii* chloroplast, *Proc. Natl. Acad. Sci. USA* **79**:2264–2267.

Watson, J. C., and Surzycki, S. J., 1983, Both the chloroplast and nuclear genomes of *Chlamydomonas reinhardtii* share homology with *Escherichia coli* genes for transcriptional and translational components, *Curr. Genet.* **7**:201–210.

Weeden, N. F., 1981, Genetic and biochemical implications of the endosymbiotic origin of the chloroplast, *J. Mol. Evol.* **17**:133–139.

Weil, J. H., and Parthier, B., 1982, Transfer RNA and aminoacyl-tRNA synthetases in plants, *Encyclopedia Plant Physiol.* **14A**:65–112.

Weissinger, A. K., Timothy, D. H., Levings, C. S., III, Hu, W. W. L., and Goodman, M. M., 1982, Unique plasmid-like mitochondrial DNAs from indigenous maize races of Latin America, *Proc. Natl. Acad. Sci. USA* **79**:1–5.

Weissinger, A. K., Timothy, D. H., Levings, C. S., III, and Goodman, M. M.,, 1983, Patterns of mitochondrial DNA variation in indigenous maize races of Latin America, *Genetics* **104**:365–379.

Westhoff, P., Nelson, N., Bunemann, H., and Herrmann, R. G., 1981, Localization of genes for coupling factor subunits on the spinach plastid chromosome, *Curr. Genet.* **4**:109–120.

Westhoff, P., Alt, J., and Herrmann, R. G., 1983a, Localization of the genes for the two chlorophyll a-conjugated polypeptides (mol. wt. 51 and 44 kd) of the photosystem II reaction center on the spinach plastid chromosome, *EMBO J.* **2**:2229–2237.

Westhoff, P., Alt, J., Nelson, N., Bottomley, W., Bunemann, H., and Herrmann, R. G., 1983b, Genes and transcripts for the P_{700} chlorophyll a apoprotein and subunit 2 of the photosystem I reaction center complex from spinach thylakoid membranes, *Plant Mol. Biol.* **2**:95–107.

Whatley, J. M., 1982, Ultrastructure of plastid inheritance: Green algae to angiosperms, *Biol. Rev.* **57**:527–569.

Whatley, J. M., 1983, Plastids—Past, present, and future, *Int. Rev. Cytol. Suppl.*, **14**:329–374.

Whatley, J. M., and Whatley, F. R., 1981, Chloroplast evolution, *New Phytol.* **87**:233–247.

Whitfeld, P. R., and Bottomley, W., 1980, Mapping of the gene for the large subunit of ribulose bisphosphate carboxylase on spinach chloroplast DNA, *Biochem. Int.* **1**:172–178.

Whitfeld, P. R., and Bottomley, W., 1983, Organization and structure of chloroplast genes, *Annu. Rev. Plant Physiol.* **34**:279–310.

Whitfeld, P. R., Herrmann, R. G., and Bottomley, W., 1978, Mapping of the ribosomal RNA genes of spinach chloroplast DNA, *Nucleic Acids Res.* **5**:1741–1751.

Widger, W. R., Cramer, W. A., Herrmann, R. G., and Trebst, A., 1984, Sequence homology and structural similarity between cytochrome *b* of mitochondrial complex III and the chloroplast b_6-*f* complex: Position of the cytochrome *b* hemes in the membrane, *Proc. Natl. Acad. Sci. USA* **81**:674–678.

Willey, D. L., Huttly, A. K., Phillips, A. L., and Gray, J. C., 1983, Localization of the gene for cytochrome *f* in pea chloroplast DNA, *Mol. Gen. Genet.* **189**:85–89.

Willey, D. L., Howe, C. J., Auffret, A. D., Bowman, C. M., Dyer, T. A., and Gray, J. C., 1984a, Location and nucleotide sequence of the gene for cytochrome *f* in wheat chloroplast DNA, *Mol. Gen. Genet.* **194**:416–422.

Willey, D. L., Auffret, A. D., and Gray, J. C., 1984b, Structure and topology of cytochrome *f* in pea chloroplast membranes, *Cell* **36**:555–562.

Williamson, S. E., and Doolittle, W. F., 1983, Genes for $tRNA^{Ile}$ and $tRNA^{Ala}$ in the spacer between the 16S and 23S rRNA genes of a blue-green alga: Strong homology to chloroplast tRNA genes and tRNA of the *E. coli rrnD* gene cluster, *Nucleic Acids Res.* **11**:225–235.

Woessner, J. P., Masson, A., Harris, E. H., Bennoun, P., Gillham, N. W., and Boynton, J. E., 1984, Molecular and genetic analysis of the chloroplast ATPase of *Chlamydomonas*, *Plant Mol. Biol.* **3**:177–190.

Woodcock, C. L. F., and Bogorad, L., 1970, Evidence for variation in the quantity of DNA among plastids of *Acetabularia*, *J. Cell Biol.* **44**:361–375.

Wright, R. M., and Cummings, D. J., 1983, Integration of mitochondrial gene sequences within the nuclear genome during senescence in a fungus, *Nature* **302**:86–88.

Wurtz, E. A., and Buetow, D. E., 1981, Intraspecific variation in structural organization and redundancy of chloroplast ribosomal DNA cistrons in *Euglena gracilis*, *Curr. Genet.* **3**:181–187.

Yamada, T., 1983, Characterization of inverted repeat sequences and ribosomal RNA genes of chloroplast DNA from *Chlorella ellipsoidea*, *Curr. Genet.* **7**:481–487.

Yoshida, Y., 1981, Configurations indicative of functional sites in spinach chloroplast DNA, *Cell. Biol. Int. Rep.* **5**:115–123.

Yoshida, Y., Laulhere, J. P., Rozier, C., and Mache, R., 1978, Visualization of folded chloroplast DNA from spinach, *Biol. Cell.* **32**:187–190.

Zurawski, G., and Clegg, M. T., 1984, The barley chloroplast DNA *atpBE*, *trnM2*, and *trnV1* loci, *Nucleic Acids Res.* **12**:2549–2559.

Zurawski, G., Perrot, B., Bottomley, W., and Whitfeld, P. R., 1981, The structure of the gene for the large subunit of ribulose 1,5-bisphosphate carboxylase from spinach chloroplast DNA, *Nucleic Acids Res.* **9**:3251–3270.

Zurawski, G., Bottomley, W., and Whitfeld, P. R., 1982a, Structures of the genes for the beta and epsilon subunits of spinach chloroplast ATPase indicate a dicistronic mRNA and an overlapping translation stop/start signal, *Proc. Natl. Acad. Sci. USA* **79**:6260–6264.

Zurawski, G., Bohnert, H. J., Whitfeld, P. R., and Bottomley, W., 1982b, Nucleotide sequence of the gene for the M_r 32,000 thylakoid membrane protein from *Spinacia oleracea* and *Nicotiana debneyi* predicts a totally conserved primary translation product of M_r 38,950, *Proc. Natl. Acad. Sci. USA* **79**:7699–7703.

Zurawski, G., Clegg, M. T., and Brown, A. H. D., 1984a, The nature of nucleotide sequence divergence between barley and maize chloroplast DNA, *Genetics* **106**:735–749.

Zurawski, G., Bottomley, W., and Whitfeld, P. R., 1984b. Junction of the large single copy region and the inverted repeats in *Spinacia oleracea* and *Nicotiana debneyi* chloroplast DNA: Sequence of the genes for $tRNA^{His}$ and the ribosomal proteins S19 and L2. *Nucleic Acids Res.* **12**:6547–6558.

Zurawski, G., Bottomley, W., and Whitfeld, P. R., 1985, Molecular analysis of adjacent transcriptional units from pea chloroplast DNA: *rbcL*, the gene for the large subunit of ribulose bisphosphate carboxylase, and *atpBE*, the genes for the beta and epsilon subunits of ATPase (in preparation).

CHAPTER 4

Localized Highly Repetitive DNA Sequences in Vertebrate and Invertebrate Genomes

GEORGE L. GABOR MIKLOS

1. Introduction

Most eukaryotes carry repetitive DNA sequences, which occur in tandem arrays in the vicinity of centromeres, telomeres, or, less frequently, in interstitial regions of chromosomes. These localized arrays generally constitute the heterochromatic regions of chromosomes. They are easily visualized as prominent DNA fragments following restriction endonuclease digestions of the genomic DNA of a species, or as the classical satellite peaks in buoyant density gradients.

Some species of American kangaroo rats can have more than half their genomic DNA invested in these localized tandem arrays. For example, the MS, α, and β components of the genome of *Dipodomys ordii* consist of variants of the sequences AAG, TTAGGG, and ACACAGCGGG (Hatch *et al.*, 1976; Salser *et al.*, 1976; Fry and Salser, 1977). These constitute approximately 2400, 2200, and 1200 million base pairs (bp) of DNA, respectively, so that in a diploid nucleus there is nearly 6000 million bp of nontranscribed simple sequence DNA. It is this blatant profligacy of repeated sequences within some eukaryotic genomes that fascinates many investigators, and had led to two general and mutually exclusive conclusions concerning their status in the genome. One is that

GEORGE L. GABOR MIKLOS • Department of Population Biology, and Center For Recombinant DNA Research, Research School of Biological Sciences, Australian National University, Canberra, A. C. T. 2600, Australia.

these sequences have functions; the other is that they represent the natural outcomes of genomic turnover processes that involve selfish, ignorant, or junk DNA. The functionalists rationalize the existence of these sequences have been held to function, for example, in determining centromere strength, chromosome pairing, recombination, the three-dimensional architecture of the nucleus, genomic reorganization, and speciation.

On the other hand the proponents of the second class of hypotheses propose that these sequences are inevitable byproducts of molecular mechanisms involved in DNA replication and recombination, and that such genomic turnover is not supervised by natural selection acting through the phenotype.

These viewpoints accurately reflect other deep divisions in contemporary biology, such as those concerning selection versus neutrality and function versus junk. An overt unease is now apparent in studies on highly repetitive DNAs, since, despite the focusing of the awesome power of recombinant DNA technology onto the structure of these sequences, the expected biological roles have still proved elusive.

The properties and evolutionary aspects of highly repetitive DNA sequences have been reviewed in quite diverse contexts (John and Milkos, 1979; Kurnit, 1979; Bostock, 1980; Brutlag, 1980; Doolittle and Sapienza, 1980; Hilliker *et al.*, 1980; Orgel and Crick, 1980; Schopf, 1981; Dover, 1982; Flavell, 1982; Miklos, 1982; Singer, 1982; Rose and Doolittle, 1983). At the conclusion of many of these reviews, however, the major competing hypotheses concerning the potential functional attributes of these DNA sequences have, in general, emerged unscathed. Why is it that the structural data have not impinged on what localized tandem arrays of highly repetitive sequences actually do, or do not do, at the cellular, organismal, populational, or evolutionary levels? In an attempt to answer this question, I first show that a great deal can be learned from the historical perspectives in this field. Second, I describe the DNA sequence data base in the context of biological phenomena. Finally, I point out the directions in which relevant data might now be sought.

1.1. Historical Perspectives: *Audi alteram partem*

Cytogenetic analyses in the 1930s revealed that the X chromosome of *Drosophila melanogaster* had a sizeable heterochromatic region that was genetically inert. Since that time the heterochromatic regions of chromosomes have been studied intensely and many hypotheses have been erected for their functional attributes. In essence most of these are still with us, unconnected in any real sense with the DNA sequence data. An important technical advance was made in the late 1960s when analytical ultracentrifugation techniques revealed DNA sequences of such unusual base composition that they appeared as distinct satellite peaks in cesium chloride gradients. These satellite sequences were found to be highly repeated and to hybridize to the heterochromatic regions of chromosomes.

Two other events occurred at about this time that might have been expected to cause significant soul searching, but went largely unnoticed. First, as early as 1970, Ohno published an essay entitled, "So much 'junk' DNA in our genome" (Ohno, 1970). Second, Kimura's ideas on neutrality were providing an alternative viewpoint to that of the selectionists [summarized in Kimura (1979)]. It also became clear that satellite DNA was not transcribed, which implied that if it was to face meaningful biological constraints, these were unlikely to be of the type acting on genic sequences or multigene families.

When the first nucleotide sequences became available, the concept of conservation swept onto the stage. DNA sequences in one species were found to be present in others and this was interpreted to mean that they *must* have been functionally constrained. Little serious consideration was given to other obvious alternatives. Were there, for example, mechanisms that periodically amplified DNA in the absence of selective forces? If selection did occur, how did it act? It was postulated that specific proteins bound to particular repetitive sequences and these were important in chromosome pairing. If this process was defective, it created infertility barriers between organisms, which led to speciation. Thus the library hypothesis was born, in which particular DNA sequences were envisaged as books in a library from whence they were periodically withdrawn and amplified. This hypothesis was readily accepted by many because it reflected a number of popular beliefs: (1) that satellite sequences were the basis of chromosome recognition phenomena; (2) that tandem DNA arrays were highly homogeneous within and between species and that they were retained over long evolutionary time periods for functional reasons; (3) that natural selection eliminated some DNA variants while promoting others; and (4) that significant evolutionary changes, such as speciation processes, could be rationalized in a molecular milieu through the study of tandemly repeated DNA sequences and their associated protein complexes.

DNA sequence conservation between species was seen as a natural outcome of selection acting on tandem arrays; how else could an array remain so pure? However, the extrapolation from selection acting at a single protein coding locus to selection acting on a tandem array of sequences is a brave one indeed. Just consider for a moment how selection acting through the phenotype could detect a single variant copy in an array of 100,000 copies if these sequences are used in chromosome pairing. Selection could only survey an entire family, not its individual members. It is unrealistic to imagine that two homologous chromosomes each with 100,000 perfect tandem copies of a sequence would fare any better in the rigors of meiosis or in the metabolic arena than two homologous chromosomes one of which has only 99,999 good copies and a lone variant.

Until the advent of cloning and sequencing, good estimates of within-species variability based on cloned sequences were unavailable. How much sequence change exists within a single individual for a particular array, and how much variation is there between individuals? Without such estimates of variability, hypotheses relating to conservation are severely constrained.

The advent of recombinant DNA technology placed the structural data for the first time on the only acceptable footing for the evaluation of hypotheses concerning the significance of DNA sequence structure. This data base for nucleotide sequence variation within and between individuals is, however, still inchoate, since the bulk of it deals with consensus sequences for a given species.

The whole problem of whether there was a function at all for highly repetitive sequences exploded into the harsh glare of criticism when the concept of selfish DNA was propounded. Doolittle and Sapienza (1980), presenting arguments based on properties of mobile elements, and Orgel and Crick (1980), using similar arguments pertinent to highly repetitive DNAs, argued that sequences having little effect on the phenotype could be spread throughout a genome simply on the basis of their preferential replicative properties.

A further novel development came from the concept of molecular drive (Dover, 1982), an attempt to examine genome flux and its evolutionary consequences in terms of DNA turnover processes of multicopy families within the genome. This concept is based on processes that, it is argued, could have important phenotypic effects without necessarily invoking natural selection. Indeed, molecular drive utilizes known molecular mechanisms and so has forced the adoption of different attitudes to the introduction of genetic variation into a genome and a reappraisal of the types and frequencies of DNA sequence variants in natural populations.

It has also been argued that one major evolutionary function for highly repetitive DNA sequences has been in providing an inert substrate for chromosomal rearrangements, which in turn led to infertility problems and hence to speciation. There has been insufficient cognizance, however, of how many of the known rearrangements are due to the presence of nomadic elements rather than to highly repetitive sequences. It is important to keep in mind that mobile elements are recombinogenic, and hence rearrangements occurring within heterochromatic regions may be due to mobile element excision, and not to the occurrence of highly repeated DNA sequences *per se*.

There have also been attempts to utilize these repetitive DNA sequences for the construction of phylogenetic trees and to determine rates of evolution. The construction of phylogenies can certainly be satisfactorily attempted using either morphological characteristics or the nucleotide sequences from *bona fide* single-copy genes, since these can be used in the conventional sense to determine divergence parameters. However, since a tandem pack of DNA sequences can be subject to unequal sister chromatid exchange, conversion phenomena, and amplification and deletion events, as well as insertion of nonhomologous sequences, it becomes hazardous to determine accurate *rates* of change or meaningful phylogenetic relationships.

It is important to realize the extent to which sequence data will or will not impinge upon what this class of DNA molecule can possibly do. The broad classes into which repetitive DNA functions have historically been classed are: chromosome organization, cell metabolism, chromosome pairing, speciation, evolution, and the newer one of nonfunctional DNA. Except for the last, these have been discussed extensively in terms of heterochromatin for many years, and biological data already exist that can provide meaningful directions in which to gather sequence data. Nearly 50 years ago it was shown that the addition of extra inert regions had no measurable effect on the phenotype, the viability, or the fertility of *D. melanogaster* (Gershenson, 1933, 1940). Gershenson specifically and perceptively addressed himself to what roles this inert region might have in chromosome pairing, meiotic recombination, cell size, and somatic phenotypic effects. While he recognized that there was an effect on pairing in the male, there were no other identifiable effects on other processes. We now know that the inert region in question, the heterochromatin of the X, is a pastiche of different repetitive DNA sequences, a variety of different nomadic element families, 200 or so ribosomal RNA genes, and a number of pairing sites utilized in male meiosis. All of these together, when duplicated in a genome, have negligible effects on somatic development and metabolism.

Since Gershenson's work the properties of "heterochromatin" have been reviewed almost every decade, but it was Cooper (1959) who drew attention to its inconsistent galaxy of attributes. He summarized many of the suggestions that had been made for this genetically inert material and the data upon which they were based. As he pointed out, it has been assumed to modify gene expression, to bring about or prevent chromosome pairing, to control meiotic recombination, to influence cell size, to regulate growth and differentiation, and to play a role in speciation via chromosomal rearrangements, and these assumptions have been re-echoed in the years that followed.

I have spelled out these classical hypotheses because they can be compared with the themes of the 1970s and 1980s, which are expressed in terms of DNA sequences. Thus satellite or highly repetitive DNAs have been invoked in chromosome pairing (Brutlag *et al.*, 1977), chromosomal rearrangements (Hatch *et al.*, 1976), speciation (Fry and Salser, 1977; Arnason *et al.*, 1978), germ line processes (John and Miklos, 1979; Bostock, 1980), alteration of regulatory pathways via genomic reorganizations (Donehower and Gillespie, 1979), and the three-dimensional structure of the interphase nucleus (Hilliker *et al.*, 1980; Manuelidis, 1982a). Highly repetitive DNA sequences may, alternatively, be thought of as examples of selfish DNA (Doolittle and Sapeinza, 1980; Orgel and Crick, 1980) or ignorant DNA (Dover and Doolittle, 1980; Maresca and Singer, 1983). Thus highly repetitive DNAs are considered to be subject to natural selection (Fry and Salser, 1977; Brutlag, 1980), to be neutral (Miklos and Gill, 1981, 1982), or to be subject to molecular drive (Dover, 1982).

This litany of functional explanations led Doolittle and Sapienza (1980) and Orgel and Crick (1980) to formulate their concept of selfishness, in which no phenotypic or evolutionary function was called for. These authors summarized what they considered to be unproductive avenues in the searches for functionality in the following terms (Doolittle and Sapienza, 1980):

> The phenotype paradigm underlies attempts to explain genome structure. There is a hierarchy of types of explanations we use in efforts to rationalize, in neo-Darwinian terms, DNA sequences which do not code for proteins . . . Non transcribed DNA, and in particular repetitive sequences, are thought of as regulatory or somehow essential to chromosome structure or pairing. When all attempts to assign to a given sequence class of DNA functions of immediate phenotypic benefit to the organism fail, we resort to evolutionary explanations. The DNA is there because it facilitates genetic rearrangements which increase evolutionary versatility (and hence long term phenotypic benefit), or because it is a repository from which new functional sequences can be recruited or, at worst, because it is the yet-to-be eliminated by-product of past chromosomal rearrangements of evolutionary significance.

In summary, the main discussions on the significance of repetitive sequences have centered on natural selection, the library hypothesis, selfish, ignorant, or neutral DNA, germ line phenomena, genomic rearrangements, three-dimensional nuclear architecture, and molecular drive. The field keeps on returning to the same hypotheses or to the same major pairs of alternatives. Few of these have been falsified using DNA sequence data and that is the fountainhead of our present dilemma. A dispassionate comparison of the *theories* indicates little significant progress. The situation in terms of hypotheses is reminiscent of the "Glass Bead Game" in Herman Hesse's (1943) novel of that name, where scholars play a purely internal and interminable game with the pieces of theories. The field has essentially oscillated from hypothesis to hypothesis over half a century and it badly needs *data* that impinge *directly* on such hypotheses.

Recombinant DNA technology gave biologists the wherewithal to participate in a veritable feast of genomic DNA dissection, and current fashion dictates that the avenues of research be strictly confined to DNA sequence data. However, such data in isolation from relevant biological contexts are as impotent as biological data without DNA sequences. In some instances DNA sequence data are now available to render certain cherished hypotheses inviable, and to indicate that further data accumulation is largely unnecessary. Similarly, biological data exist that render some current avenues of recombinant DNA research already obsolete.

What I have attempted to do in this chapter is to choose organisms that illustrate the general principles involved in relevant phenomena at the cellular, organismal, and populational levels, and to meld the biological with the nucleotide sequence data.

2. The Molecular and Biological Data Base: Vertebrates

2.1. Kangaroo Rats and the Concept of a Library

Rodents of the genus *Dipodomys* can have over half their genomes in highly repetitive DNAs (Hatch *et al.*, 1976). The genomic components of one species, *D. ordii,* have three major components termed MS, HS-α, and HS-β, which make up 22, 19, and 11%, respectively of this genome. Although these components have been neither cloned nor sequenced by conventional methods, their average DNA sequences have been deduced using ribosubstitution methods and have been found to be variants of AAG, TTAGGG, and ACACAGCGGG (Salser, *et al.*, 1976; Fry and Salser, 1977). Closely related species that show few morphological differences can have such reduced quantities of these same sequences that *Dipodomys* species differ in genome size by almost a factor of two. Thus *D. ordii monoensis* has nearly 4000 million bp of nuclear DNA in excess of *D. heermani tularensis*. Clearly neither the developmental, metabolic, nor morphological spheres are much perturbed by such a huge difference in repetitive DNA content [reviewed in John and Miklos (1979); and Miklos and Gill (1982)].

The significance of the molecular data from the *Dipodomys* species is manifold. First, at least one major variant sequence, TTAGGG, is the same as the major variant of a guinea pig repetitive DNA family (Southern, 1970). Since kangaroo rats were thought to share a common ancestor with guinea pigs 50 million years ago, it was proposed that these sequences were conserved over long evolutionary time periods by protein–DNA interactions that were important in chromosome pairing and hence to speciation events (Salser *et al.*, 1976; Fry and Salser, 1977). A second proposal was that this particular DNA sequence was conserved between species by natural selection. Yet the heterogeneity at the nucleotide sequence level, even from these uncloned DNAs, is substantial. The variant sequences of just the HS-α satellite are shown in Table I, and the range is indicative of extensive variation within the sequence pool.

More recently, whenever a phylogeny has been examined and *similar* sequences have been found by DNA hybridization, restriction endonuclease digest patterns, or cloned DNA sequences, conservation has almost invariably been invoked. The use of this word nearly always implies that there has been conservation for functional reasons by natural selection. If indeed conservation has occurred, then we need to examine the biological context in which it has been interpreted.

The library hypothesis was erected to account for the apparent conservation of DNA sequences between species. Satellite sequences were envisaged as being like books in a library, which were periodically withdrawn and amplified, resulting in the satellite peaks seen in various lineages. The DNA sequence data were tied to three biological tenets:

Table I
Uncloned DNA Sequences and Their Relative Yields from the HS-α Satellite of *Dipodomys ordii*[a]

Sequence	Relative yield
T T A G G G	1.0
T T A G A G	0.5
T T A G G T	0.5
T G A G G G	0.5
T T A G G A	0.3
T T A G T G	0.3
T A A G G G	0.2
T T A G T T	0.2
G A A G G G	0.2
T T A G G C	0.2
T G A G G A	0.1
T T A A G G	0.1

[a] Summarized from Fry and Salser (1977).

(1) The maintenance of high fidelity at the DNA sequence level due to DNA-(1)binding proteins; (2) The induction of meiotic pairing problems due to differences in the amounts of localized repeated sequences on homologous chromosomes; and (3) Changes in the amounts of repetitive sequences due to speciation "genes."

This complex hypothesis has a number of predictions. In its common usage, however, many authors have laid aside the biological properties with which it was endowed, and consider it merely as involving a library of sequences that are amplified from time to time. The original hypothesis, however, was much more than this, even at the cellular level, since not only were "candidate" satellite sequences created *de novo,* but having acquired a function via a particular DNA binding protein, each sequence became a *bona fide* library member. In fact there were now *two* libraries, one for DNA sequences and another for their individual binding proteins.

There are a number of data bases upon which the library hypothesis can be tested. One is to examine if there has been conservation using nucleotide sequence data. Another is to ask if satellite DNA differences between homologues give rise to any meiotic pairing problems, and yet another is to determine if there are indeed speciation "genes" of the kind postulated by Fry and Salser (1977).

In order to appreciate the complexities of interpreting the subsequent cloned data, consider the following example. Consider three individuals, each with two chromosomes that have identical repeating units in their repeated arrays (Fig. 1A). If DNA is extracted from these individuals and pooled, and random clones are sequenced, there will be no base changes or other rearrangements between clones, and the sequence would be *homogeneous*. If, on the other hand, the individual repeat units differ by base changes, for example, then pooling, cloning, and sequencing individual clones will reveal a certain proportion of variants. Thus the pool of clones is heterogeneous, but it is not possible to partition the source of the variation, since it is not known if the clones came from chromosome 1 or 2 (Fig. 1B), or whether all the variation was from one chromosome in one individual (Fig. 1C).

When clones are isolated from two different species, the degree of heterogeneity *within* each species needs to be known, since a similar array of variants in both species does not necessarily allow us to distinguish among selection, neutrality, or molecular drive as a source of variation.

2.2. Amphibians and the Transcriptional Capacities of Tandem Arrays

Amphibians are a providential group from the viewpoint of studying highly repetitive DNA sequences because they reveal unambiguous evi-

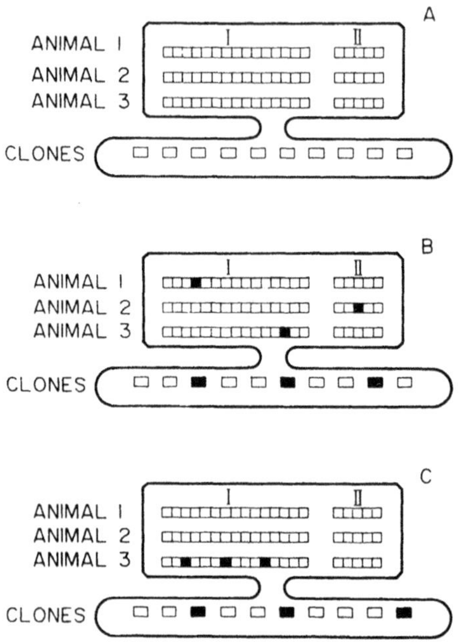

Figure 1. Three animals depicted carrying tandem arrays of highly repetitive sequences on chromosomes I and II. (A) Individual repeating units, denoted □, are identical within and between individuals; (B) some repeating units carry mutational perturbations, denoted ■, in one or more chromosomes in all individuals; (C) mutant repeating units occur on only one chromosome in one individual.

dence of the type and significance of RNA transcripts emanating from tandem arrays. Varley *et al.* (1980) originally demonstrated that satellite DNA fractions from the newt *Triturus cristatus carnifex* bound *in situ* to some actively transcribing loops in lampbrush chromosomes. These authors suggested that satellite DNA transcription was the consequence of random "overrun" transcription. Recombinant DNA analyses in another amphibian placed these findings in a more exact context. A DNA fraction termed satellite I has been isolated from the newt *Notophthalamus viridescens* (Diaz *et al.*, 1981; Stephenson *et al.*, 1981). The basic repeating unit is 222 bp in length and hybridizes *in situ* predominantly to the centromeric heterochromatin of all chromosomes and to two minor regions on chromosomes 2 and 6. It is these minor arrays of 222-bp units that are in fact transcribed, since the cytological localization of transcription is *not* from the centromeric heterochromatin, but from these two regions at the noncentromeric sites. These "orphon" satellite sequences, which are no longer sequestered within the bosom of their centromeric families, occur as 50- to 100-kb spacers between histone gene clusters. Following transcriptional initiation of these histone genes, termination sometimes fails to occur, leading to readthrough into the satellite DNA. Since the genes in the histone cluster are of opposite polarity, both strands of the satellite sequence are transcribed. This is transcription by default.

These results highlight the problems of other RNA analyses that are carried out on total RNA from different tissues and where there is no information of its chromosomal origin. Because of the favorable cytology of this newt, it is clear that transcription emanates from noncentromeric regions. Thus, whenever RNA transcripts of satellite DNAs are found, it will be necessary to show that they are not the adventitious products of a small cluster of repetitive DNA sequences that have been transposed to near coding regions of the genome, or due to initiation sequences that have been transposed into a tandem family.

There have been reports of satellite DNA transcripts in mouse and rat tissue culture cells (A. K. Cohen *et al.*, 1973; Sealy *et al.*, 1981), but it is well documented that cultured cells undergo chromosomal reorganization, resulting in the appearance of satellite DNA sequences in novel interstitial regions (Holmquist and Comings, 1975; White *et al.*, 1975). If any of these newly located sequences fall next to a region that has adventitious readthrough, then RNA transcripts will again be formed.

The reverse process may also occur, since nomadic elements are nearly always interdigitated into centromeric heterochromatin. The *copia* mobile elements, for example, can be found juxtaposed to the 359-bp satellite sequences of *D. melanogaster* (Carlson and Brutlag, 1978a,b). Furthermore, most nomadic elements of *D. melanogaster* hybridize *in situ* to the chromocenter (Dowsett and Young, 1982) and thus probably occur adjacent to satellite DNA sequences of different kinds. In the African green monkey, members of the *Kpn*I family of dispersed repeated sequences can be found flanked by α-satellite DNA and have probably immigrated to such sites by transposition (Grimaldi and Singer, 1983).

It is thus likely that any transcriptional activity of repeated sequences is due to readthrough from external promoters. Caution therefore needs to be exercised before functional interpretations of satellite DNA transcripts are invoked.

In *Xenopus laevis*, a 741-bp monomer sequence has been cloned and is the first satellite to be analyzed at this level in this organism (Lam and Carroll, 1983). It has a copy number of about 30,000 and hybridizes *in situ* to the centromeric heterochromatin of about half of the chromosomes. Computer analysis reveals that it has no internal subrepeats. It does, however, have a number of open reading frames, which potentially allow it to code for a protein, and it has sequence motifs (TATAA) that are normally found in the landscapes surrounding common gene sequences. The biological significance of these motifs in tandem arrays is not known, but they do occur in other complex satellite DNAs of both vertebrates and invertebrates. In the Hawaiian *Drosophila* where individual clones have been analyzed for such motifs, these sequences show variability among different clones from the one laboratory stock (Miklos and Gill, 1981).

Some idea of the variation present in satellite I of *X. laevis* is available from four clones that have been examined at the level of restriction endonuclease sites (Fig. 2). Clone 7, for example, harbors the basic repeating unit of this satellite, a 741-bp sequence that differs from the 741-bp monomers of clone 8 by the internal spacing and type of restriction endonuclease sites. I refer to these sequences as 741a and 741b types. Clone 6 differs yet again in that it has a 741c-type sequence into which a 310-bp foreign DNA has been inserted. The composite 1050-bp unit has then been amplified, a situation that occurs in the satellite DNAs of the calf. Finally, clone 12 harbors a mixture of 741e-type sequences, interdigitated with 1000-bp-type sequences. The 1000-type consists of a 260-bp insertion of nonhomologous DNA into a 741d-type sequence. These sequences are then present in the order . . .$(741)_4$ 1000 $(741)_2$ 1000 $(741)_4$ (Fig. 2). The cloned data attest to the large amount of variation that is apparent within a satellite, variation that is cryptic when 741-bp fragments are visualized on gels. What is clear is that amplification of sequences can occur following an insertion event. Thus, although there are as yet no sequence estimates of variability between adjacent repeating units in any one clone, there is nevertheless extensive organizational variation among clones.

Two further important points relate to the library hypothesis and to transcription. First, satellite I of *X. laevis* is not found in the genomes of the related species *X. mulleri* and *X. borealis* under standard stringency conditions. Second, no transcripts were detected from this satellite DNA

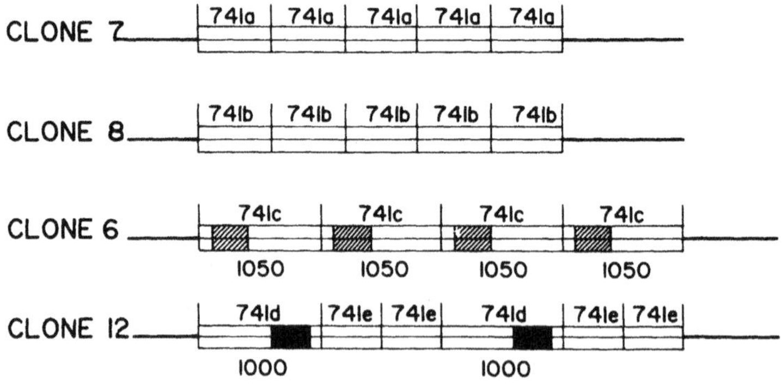

Figure 2. Diagrammatic illustration of four clones containing inserts of *X. laevis* satellite I DNA. The different 741-bp monomer sequences are as shown. Inserts of foreign DNA into the monomer sequences of clones 6 and 12 are hatched. [Redrawn from Lam and Carroll (1983).]

in Northern blot analyses using RNA from adult liver, embryos, and oocytes.

Lam and Carroll (1983) have pointed out that the nucleotide sequence of the 741-bp unit *per se* provides scant insight into its possible functions, and that the prevalence of certain variants may be due either to random processes or to natural selection. This candidly highlights the limits to which sequence data can be pushed in terms of evolutionary interpretations. While the structural data are obligatory for estimates of homogeneity within a species and for evaluating possible conservation of sequences between species, other types of information are necessary to determine if sequence structure *per se* does or does not have selective value. Certainly the *Xenopus* data indicate that some sequences are heterogeneous within, and not conserved between, species.

2.3. The Calf and the Concept of DNA Reamplification

The calf (*Bos taurus*) has at least eight satellites, four of which have been subjected to cloning and sequencing analyses (Pech *et al.*, 1979a; Poschl and Streeck, 1980; Gaillard *et al.*, 1981; Streeck, 1981; Pzucienniczak *et al.*, 1982; Taparowksy and Gerbi, 1982). Summaries of the properties and evolutionary characteristics of the calf highly repetitive DNA families have recently been compiled (Miklos and Gill, 1982; Roizes and Pages, 1982). All four satellites hybridize to the centromeric regions of most autosomes, but none give hydridization to the sex chromosomes (Kurnit *et al.*, 1973). Thus, unlike the autosomes, the X and Y chromosomes are both likely to contain only very small amounts of repetitive sequences in their centromeric regions.

The consensus sequence of the 1.720 satellite has been determined from uncloned DNA and found to be 46 bp in length (Poschl and Streeck, 1980). The 46-bp repeat is in fact a 23-bp dimer, so that its amplification can be formally considered as a 46-bp "bite" from a DNA landscape originally consisting of 23-bp repeating units, which themselves were at various levels of sequence divergence. Since there are as yet no individual clones of this satellite, estimates of the heterogeneity among 46-bp units are unavailable.

The 1.711a satellite has also been sequenced (Streeck, 1981). It is 1413 bp in length and is a mosaic at the sequence level, consisting of 23-bp units into which a 611-bp foreign piece of DNA has been inserted. The entire new unit was then amplified, this time as a 1413-bp "bite" containing the 611-bp insertion flanked by its 23-bp primordial units (Fig. 3). The basic structure of this satellite is particularly noteworthy, because it illustrates the principle of alien DNA insertion into a repeated array and

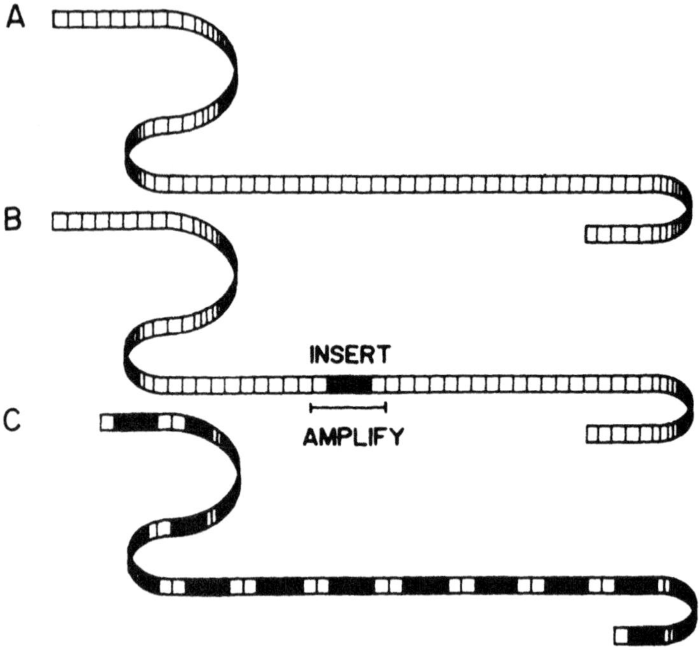

Figure 3. Three tandem arrays of highly repetitive sequences. (A) Primordial monomer unit; (B) insertion of foreign DNA into the array; (C) amplification of the inserted repeating unit plus some flanking monomers.

the subsequent amplification of the new unit. It will be informative to determine the *in situ* localization of this amplified satellite using only its 611-bp insert as a probe. It should thus be possible to gauge the extent to which the new sequence has spread throughout the genome following amplification.

A number of characteristics of this insertion draw attention to properties that can be bestowed upon a repeated sequence once an insertion has occurred and amplification has taken place. This particular 611-bp insertion has a short open reading frame and a number of motifs that potentially would allow transcription. It is obvious that any sequence with initiation properties that becomes inserted into a tandem array may cause transcriptional activity that proceeds into the adjacent sequences. Thus these sequences can be transcribed by default, and there is no need in the first instance to ascribe functional properties to them. Any mobile element with long terminal repeats could induce such transcription.

The 1.715 satellite of 1402 bp has also been cloned and sequenced in its entirety and it reveals evidence of a 31-bp repeating unit. This unit has homology to the previous two satellites because it has arisen as a 31-bp "bite" from an array of 23-bp sequences. A subsequent 1402-bp "bite" has then been taken from a tandem array of 31-bp units and amplified to yield the present mature satellite. The satellite contains a low level of 5-methylcytosine (Gautier *et al.*, 1977; Pages and Roizes, 1982), the patterns of which are tissue-specific, but the biological significance of which is unknown.

The 1.706 satellite has a basic repeating unit 2350 bp in length, which is one of the longest basic repeats known. It is constructed of four unequal segments terms A, B, C, and D. Each of these segments has a substructure based on either a 23-bp or a 46-bp unit, which itself may even have been derived from shorter sequences (Pech *et al.*, 1979a). Segment B has one region that could code for a protein of 42 amino acids. The evolution of the 2350-bp unit thus involved recruitment of four segments, each of which had undergone significant base changes. Further amplification involved a 2350-bp "bite" and its subsequent amplification (Pech *et al.*, 1979b; Roizes and Pages, 1982).

Some measure of the variation at the nucleotide sequence level in the 1.706 satellite can be estimated from eight clones derived from the 2350-bp repeats (Fig. 4). The base changes are prominent and affect nearly all of the possible 23 positions. The clones, however, are from an unknown number of individuals, and so the overall variation cannot be subdivided in the manner one would have wished in the theoretical example given in Fig. 1. In order to obtain an estimate of the variation among 2350-bp units, it will be necessary to clone these long units individually and subject each one to sequence analysis. In the absence of these estimates, the only available cloned measure of variation concerns the one that exists among 23-bp repeats. This is 12% (total number of base changes and deletions divided by the total number of nucleotides).

The positions of the perturbations in the 23-bp unit are nonrandom, there being a higher probability of a deletion or a base change in the middle of the unit. Over all eight clones, only three positions escape some form of change. If there are DNA-binding proteins that utilize the 23-bp monomer as a recognition platform, then either there are a very large number of such proteins, or the binding must tolerate a lot of leeway. Alternatively, the sequences have diverged nonrandomly without functional constraints.

In summary, the data from the calf beautifully illustrate a number of points:

1. Tandem arrays with long basic repeating units have arisen from smaller tandem arrays by processes of amplification involving pro-

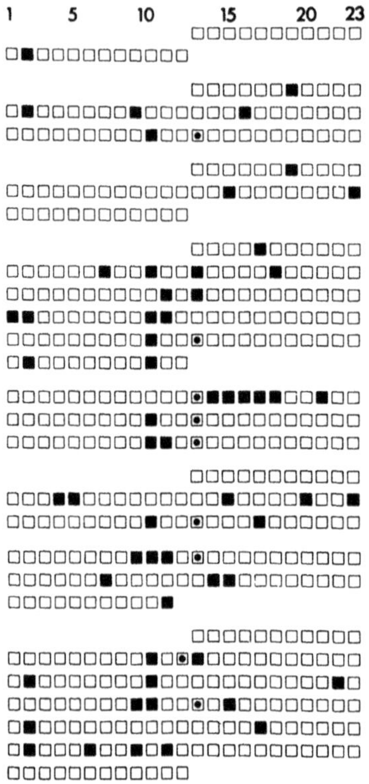

Figure 4. Diagrammatic illustration of the within-species variation in eight clones of the calf 1.706 satellite. Each clone consists of one or more contiguous 23-bp repeating units. The deviations from a concensus sequence are as follows: ■, base changes: ⊡, deletions. [Data from Pech *et al.* (1979a).]

gressively larger repeating units. In fact, all mature satellites of the calf can be considered as arising from a primordial pool based on a 23-bp sequence, which itself was probably a composite of smaller units.

2. Units of amplification can harbor sequences that have been inserted into them from elsewhere in the genome.
3. There is extensive heterogeneity among clones carrying multiple 23-bp units, as well as variation among different 23-bp units within a clone. The overall heterogeneity, however, is not easily partitioned, since different individuals were pooled to provide the initial source of DNA. The variation within 23-bp units, irrespective of their origin, is nonrandom, with most perturbations being near the middle of the repeat unit.
4. If there are biologically significant satellite DNA-binding proteins that recognize the 23-bp units, then either their binding constants are very variable, or else there are an inordinately large number of such proteins to match the large number of DNA variants.

5. The CpG doublets in the satellites can be modified as 5-methylcytosine, and this modification exhibits tissue specificity in its patterns.
6. If cross-hybridization is found to occur between the calf satellites and the genomic DNA of related species, then in order for the library hypothesis to be of biological relevance, each variant (if it is indeed a book in a library) will have to be shown to have been maintained by selective forces due to a specific DNA-binding protein. This does challenge one's credulity.

It should be noted that none of the six points allow the DNA sequence data to be easily put into a biological context, nor is it possible to use the available structural information as such in testing (falsifying) any of the six major themes raised in the introduction.

2.4. Primate Repeats and DNA Sequence Variation

2.4.1. Humans

The highly repetitive DNAs of primates have recently been reviewed (Singer, 1982). The four classical human satellites I–IV (Mitchell *et al.*, 1979) were originally isolated from buoyant density gradients and mapped *in situ* necessarily employing uncloned probes (Gosden *et al.*, 1975). They hybridize predominantly to the centromeric heterochromatin of different chromosomes. An additional uncloned AT-rich sequence as well as a more complex repeat, also have heterochromatic localities (Manuelidis, 1978; summarized in Miklos and John, 1979). When a cloned DNA representative of satellite III was used for hybridization, however, it effectively singled out only that centromeric heterochromatin of chromosome I (Gosden *et al.*, 1980), reinforcing the notion that only cloned probes are now acceptable for *in situ* analyses, a point stressed by Singer (1982). A quite different clone derived from satellite III, however, had a major location on the Y chromosome (Bostock *et al.*, 1978; Cooke and McKay, 1978). Thus satellite III is a mixture of sequences, deriving from different chromosomal locations within the genome, whose common buoyant density results in their copurification.

The Y-chromosome clone belongs to a family of sequences with a 3.4-kb repeating unit and it is most useful because it allows examination of these sequences in a biological context. It is well known that human chromosomes exhibit considerable inherited polymorphisms, which have little effect on phenotype. Males, for example, can have Y chromosomes of different lengths in which the polymorphic region generally involves the heterochromatic long arm. This variation also has no known clinical manifestations.

McKay *et al.* (1978a,b) examined a number of males with different sized Y chromosomes by restriction endonuclease digestions of their genomic DNAs, and found that the 3.4-kb *Hae*III band showed an almost direct proportionality to the length of the distal heterochromatin (Fig. 5; Table II). This does not mean that the differences in length between the Y chromosomes are due solely to their 3.4-kb fragments, only that its spacing among other sequences reflects the average composition of the distal heterochromatin.

Another male-specific repetitive DNA family of 2.1 kb has also been examined in two further males. One individual possessed hundreds of copies of the 2.1-kb fragment, whereas in the other they were below the level of detection (McKay *et al.*, 1978a,b).

The conclusions from this work are that:

1. Radical differences in DNA amounts among human Y chromosomes are mirrored by differences attributable to particular repetitive DNA sequence families.

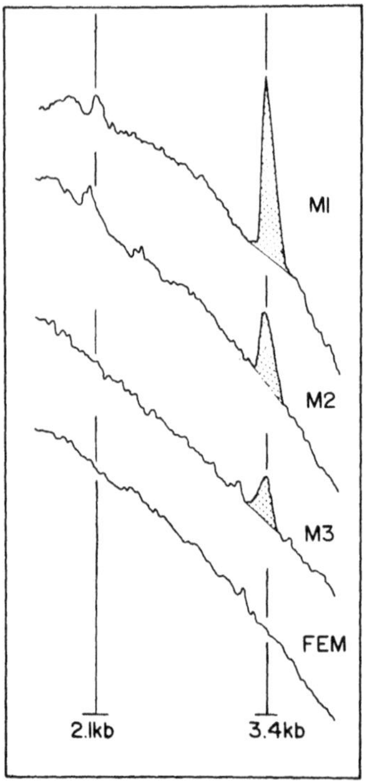

Figure 5. Microdensitometer tracings of restriction endonuclease digests of human genomic DNA from three males, M1, M2, and M3, and a control female, FEM, showing variation in the content of the 3.4- and 2.1-kb Y-chromosome-specific repetitive sequences. [Redrawn from McKay *et al.* (1978a).]

Table II
Relative Amounts of Heterochromatin and 3.4-kb-type Sequences in Different Y Chromosomes of Humans[a]

	Ratio of heterochromatin to nonheterochromatin	Relative amounts of 3.4-kb sequences
Male 1	5.6	5.5
Male 2	1.7	2.0
Male 3	1.0	1.0

[a] Data from McKay *et al.* (1978b).

2. Some sequences are below the level of resolution in one individual but in hundreds of copies in another.
3. There are no known clinical manifestations of this variation. Just as in *Drosophila*, where the heterochromatin of the X chromosome is devoid of conventional genes and graciously endowed with simple sequence DNA as well as nomadic elements, the addition or deletion of predominantly repetitive DNA sequences has negligible effects on phenotype.

In order to pursue the analysis at a finer level of resolution, it is necessary to examine the DNA sequences of individual chromosomes. Prior to recombinant DNA technology, single (or a few) chromosomes were examined by first constructing human–rodent hybrid cell lines from which most human chromosomes were absent (Beauchamp *et al.*, 1979). Restriction endonuclease cleavage of DNA from these cell lines revealed that human chromosomes possessed characteristic patterns of repetitive sequence arrays of satellite III. Thus, chromosome 7 carries a predominant 2.6-kb *Hae*III band, whereas chromosome 22 has a 1.9-kb *Hae*III fragment that is not found in other tested chromosomes. Uncloned material also revealed that the sequence 5′-TTCCA-3′ was a major component of satellite III (Frommer *et al.*, 1982).

This chromosomal signature becomes even more apparent when nucleotide sequences are examined (Cooke and Hindley, 1979; Deininger *et al.*, 1981). Three DNA sequences are shown in Fig. 6A–6C, which originate from a clone derived from the heterochromatin of chromosome I. While there is an internal substructure based on the sequence TTCCA, it is also apparent that there may be a higher order periodicity of 26 bp, and probably periodicities even higher than this. The clearest point is that within each sequence run there is considerable heterogeneity of at least 20% (base changes and deletions divided by the total number of nucleotide positions). Thus for one part of one human chromosome there is enormous inhomogeneity at the DNA sequence level.

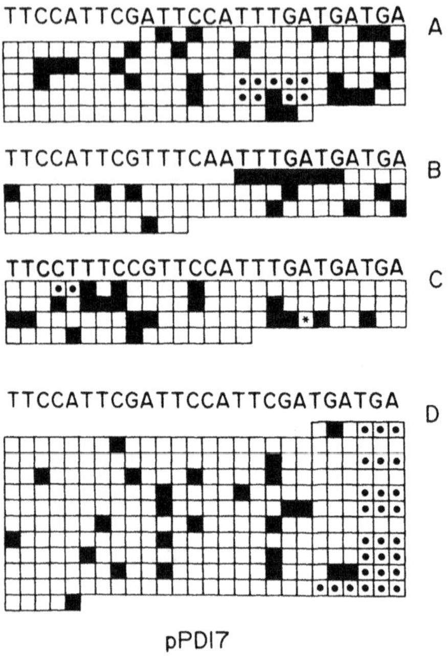

Figure 6. Variation at the nucleotide sequence level in four clones related to human satellite III DNA. ■, Base changes, [•], deletions, and [*] additions relative to a consensus sequence. [Based on Cooke and Hindley (1979) and Deininger *et al.* (1981).]

Deininger *et al.*, (1981) also provided sequence data for individual clones, one of which is inferred to occur on the Y chromosome, and two others whose location is less securely known. The two putatively Y-specific clones, PD9 and PD18, are based on a clean TTCCA repeating unit (Fig. 7), whereas clone PD17 (Fig. 6D) shows more affinities to the sequences on chromosome I isolated by Cooke and Hindley (1979). All three clones are inhomogeneous to the extent of nearly 20%. All these sequences are related to satellite III and so the extent of the heterogeneity within a single satellite is clear, as is the heterogeneity in a single chromosome. Both are extensive.

In order to ask meaningful questions about the conservation of these inhomogeneous sequences between primate species, it will be necessary to compare the corresponding cloned areas of a number of individuals of different species. An optimal and objective approach would be to compare a repetitive DNA landscape within the heterochromatin that is flanked by two unique sequences. In this way one would be certain that the same array was being cloned each time.

In summary, the data from humans reveal that:

1. Repetitive families with long periodicities can be constructed from simpler primordial repeats.

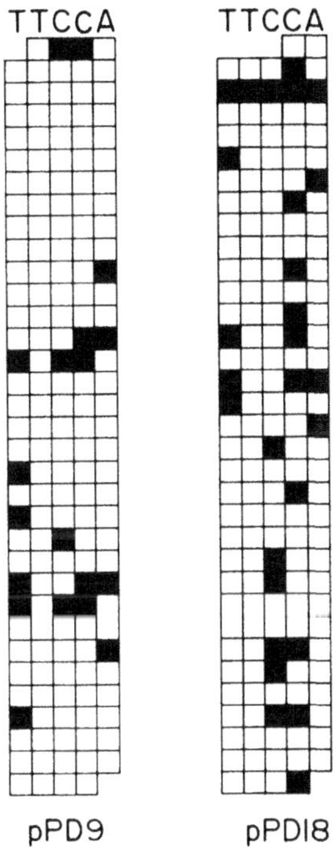

Figure 7. Variation at the nucleotide sequence level in two clones related to human satellite III DNA. ■ Base changes relative to the canonical sequence TTCCA are denoted. [Based on Deininger *et al.* (1981).]

2. There is variation among individuals in the copy number of family members.
3. There is extensive variation at the nucleotide sequence level on single chromosomes.
4. Chromosomes are individualized by their patterns of sequence variants.
5. There are no phenotypic effects associated with Y-chromosome polymorphisms, which mainly involve repetitive DNA sequence families.

2.4.2. The African Green Monkey: α-Satellite Sequences

This organism has been extensively characterized in terms of its repetitive DNA content (Maio, 1971; Kurnit and Maio, 1974; Rosenberg *et al.*, 1978; F. L. Brown *et al.*, 1979: Donehower and Gillespie, 1979; Graf *et al.*, 1979; Horz *et al.*, 1983; Zhang *et al.*, 1983; reviewed by Singer, 1982). It contains several million copies of a 172-bp sequence, which constitutes a significant proportion of the α satellite that can make up nearly 25% of the genome, and which is located in the centromeric heterochromatin of most chromosomes. In a cell line it reduces to about 7% of the genome, indicating that much of it can be banished from the metabolic arena (Rosenberg *et al.*, 1978; Thayer *et al.*, 1981). Furthermore, individual clones have been isolated and sequenced, and differences have been described among them. While most of the changes are single-base perturbations, one clone of 150 bp is particularly aberrant (Thayer *et al.*, 1981). Length variation of this type is also found within the 359-bp sequence family of *D. melanogaster*, where one variant is only 254 bp in length.

The best insights into the organization and variation of the α component derive, as in the human case, from analyses of a single chromosome present in a rodent–African green monkey hybrid cell line (Lee and Singer, 1982). The structural characteristics of clones derived from this single chromosome are shown in Fig. 8, which illustrates restriction endonuclease cleavage site summaries from λ clones 1, 2, 5, 6, 8, and 20. Each clone is distinctly different from all others as judged by the distribution of *Hin*dIII and *Eco*R1 sites. Phage 1, for example, has *Hin*dIII monomers and dimers, and most monomers carry *Eco*R1 sites. Phage 2, on the other hand, consists virtually of *Eco*R1 trimers and *Hin*dIII monomers. Phage

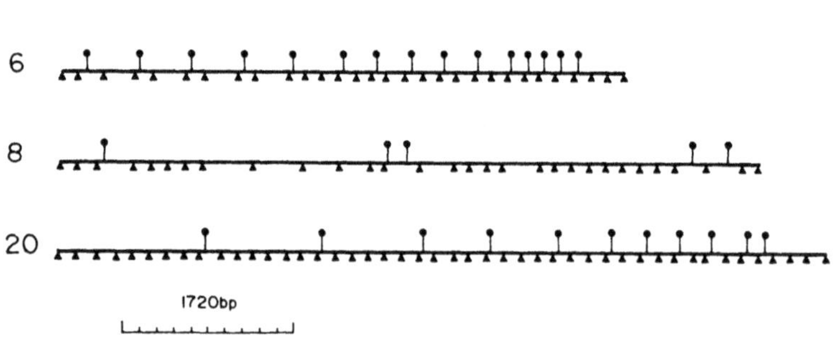

Figure 8. Distribution of restriction endonuclease cleavage sites (not restriction maps) in six clones derived from a single chromosome of the African green monkey and containing α-satellite sequences. ↑, *Eco*R1 sites; ▲, *Hind*III sites. [Redrawn from Lee and Singer (1982).]

5 has a paucity of *Eco*R1 sites, and also contains material nonhomologous to α satellite. Phage 20 consists almost exclusively of *Hin*dIII monomers and *Eco*R1 higher order oligomers. Since these data all derive from a single chromosome, they indicate a considerable heterogeneity in sequence organization.

A similar but less sensitive data set has been obtained by S. D. M. Brown and Dover (1980a) in the mouse *Mus musculus*. Here a single X chromosome was examined for its mouse satellite DNA organization in a mouse–Chinese hamster hybrid cell line, and found to differ from bulk satellite DNA patterns. In these examples of the African green monkey and the mouse, as well as the human data discussed previously, the distribution of sequence variants on individual chromosomes provides a marker of chromosome individuality. There is little doubt that each chromosome will have a tandemly repeated sequence signature. The critical question, however, is whether the signatures have any functional significance or are the byproducts of genomic turnover processes of no functional significance whatsoever.

Clones containing African green monkey α-satellite DNA can also harbor other sequences, such as *Kpn*I and *Alu* nomadic elements (McCutchan *et al.*, 1982; Grimaldi and Singer, 1983; Maresca and Singer, 1983). Both these intruders are members of families that are dispersed throughout primate genomes. In one such clone the *Kpn*I insertion is flanked by α-satellite sequences and thus has been inserted into a tandem array. The African green monkey *Kpn*I sequences are probably mobile elements, and they are related to the human *Kpn*I sequences described previously (Manuelidis, 1982b; Shafit-Zagardo *et al.*, 1983). Thus in the three clones analyzed by Grimaldi and Singer (1983), *Kpn*I and *Alu* insertions (Schmid and Jelinek, 1982) have been found in close proximity to α-satellite arrays. Similar nestlings of foreign DNA sequences in proximity to satellite components have been described for human repeated sequences (Potter and Jones, 1983) and *Drosophila* satellites (Carlson and Brutlag, 1978a,b; Dowsett and Young 1982).

2.4.3. The African Green Monkey: *deca* Satellite Sequences

Putatively mobile or mobilizable sequences are not the only ones found adjacent to satellite DNA. Clones isolated from the African green monkey genome contain sequences termed *deca* satellite, which consist of variations on a 10-bp sequence, 5′-AAACCGGNTC-3′ (Maresca and Singer, 1983). This sequence hybridizes *in situ* to the centromeric heterochromatin of about half the chromosomes of the genome. Furthermore, unlike the α sequences, which have homologies to other primates, the *deca* satellite is not found in the human genome at standard stringencies.

The sequences of two separate clones (denoted p2 and p4) containing *deca* sequences are shown in Fig. 9. There is again enormous variation between adjacent 10-bp units, of the order of 25% in clone p2 and 32% in clone p4. While these changes are nonrandom, they again illustrate the extensive sequence alterations that occur in the genome and must be taken into consideration when functional interpretations are sought.

There is also enormous variation among individuals and among different cell lines in the genomic hybridization patterns of *deca* satellite. When liver DNAs from four African green monkeys and kidney DNAs from four others were compared to two cell lines, *all gave different hybridization patterns*. Sometimes the intensity of hybridization was different for a particular genomic fragment; at other times a particular fragment was either absent or below the resolution of the technique. Within the tissues of any one individual, however, the patterns were quite constant. Since no two genomic DNAs yielded the same pattern of hybridization in Southern blots, extensive rearrangements of *deca* satellite must exist among different individuals.

The African green monkey data are particularly instructive in showing that:

1. There is extensive inhomogeneity within α satellite at the sequence level, and this variation is visible on a single chromosome.
2. α-Satellite sequences can be interrupted or adjacent to other DNA sequences.
3. The amount of α satellite can be reduced from 25% of the genome to 7% without any impairment of viability.

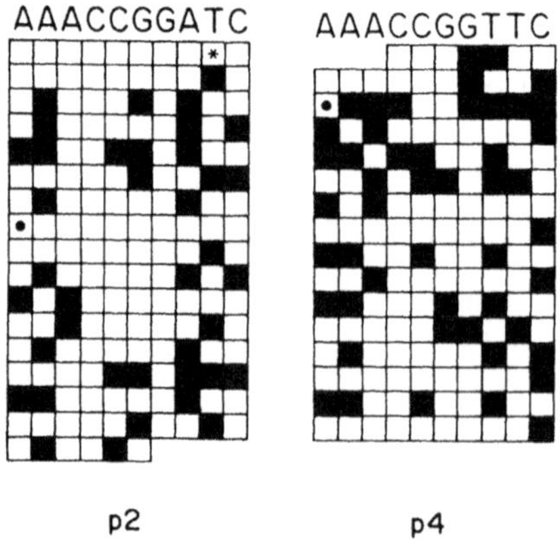

Figure 9. Nucleotide sequence variation within two clones containing *deca* satellite from the African green monkey. Perturbations from the 10-bp consensus sequence are denoted as follows: ■, base changes; ⊡, deletions, ⊠, additions. [Based on Maresca and Singer (1983).]

4. The amount of variation within the *deca* satellite and the amount among *different individual* African green monkeys is staggering.
5. *Deca*-satellite sequences are not found in the human genome at standard stringencies of hybridization.
6. *Deca* and α satellites can be found adjacent to each other in centromeric heterochromatin, as is the case for different *D. melanogaster* sequences.

2.4.4. Primate Alphoid Sequences

Most primates contain families of sequences (termed alphoid DNAs) related to the 172-bp α repeat of the African green monkey (Donehower and Gillespie, 1979; Musich *et al.*, 1980; Maio *et al.*, 1981). These generally occur as variations on a length of $170n$, where $n = 1, 2$, or 4. Thus the west African baboon (*Papio papio*) has a repeat length of 343 bp made up of two nonidentical sections of 172 and 171 bp, respectively (Donehower *et al.*, 1980). Human beings (*Homo sapiens*) possess a 340-bp sequence, which consists of unequal sections 171 and 169 bp in length (Manuelidis and Wu, 1978; Wu and Manuelidis, 1980). The bonnet monkey (*Macaca mulatta*) has a 343-bp sequence with unequal sections of 172 and 171 bp (Rubin *et al.*, 1980). The Colobus monkey (*Colobus badius*) has a repeat length based on 680 bp (Donehower and Gillespie, 1979). Positive blot hybridizations of alphoid sequences to a black lemur (*Lemus macaca*) indicate that these sequences are of reasonable antiquity (Musich *et al.*, 1980). The molecular taxonomy also reveals that the three major groupings that include baboons, guenons, and colobus are readily distinguished by their predominant repeating units in restriction endonuclease digestions (Donehower and Gillespie, 1979).

Most of the following sequence data for these primate repeats stem from uncloned DNAs. Thus restriction fragments containing a population of DNA fragments of 170 and 340 bp were eluted from gels and the sequences determined. This provides a consensus sequence where the predominant base is given at each position. Since the 340-bp human, baboon, and bonnet monkey sequences are constructed of unequal sections of approximately 170 bp, each line in Figure 10 denotes these sections as human 1, human 2, African green monkey, baboon 1, baboon 2, bonnet 1, and bonnet 2, respectively. It should be realized that this comparison is between *means* of populations of sequences drawn from many chromosomes of the genome.

This comparison (Fig. 10) reveals that a large number of base changes distinguishes not only the respective sections of each 170-bp sequence within a species, but that there is significant variation among species.

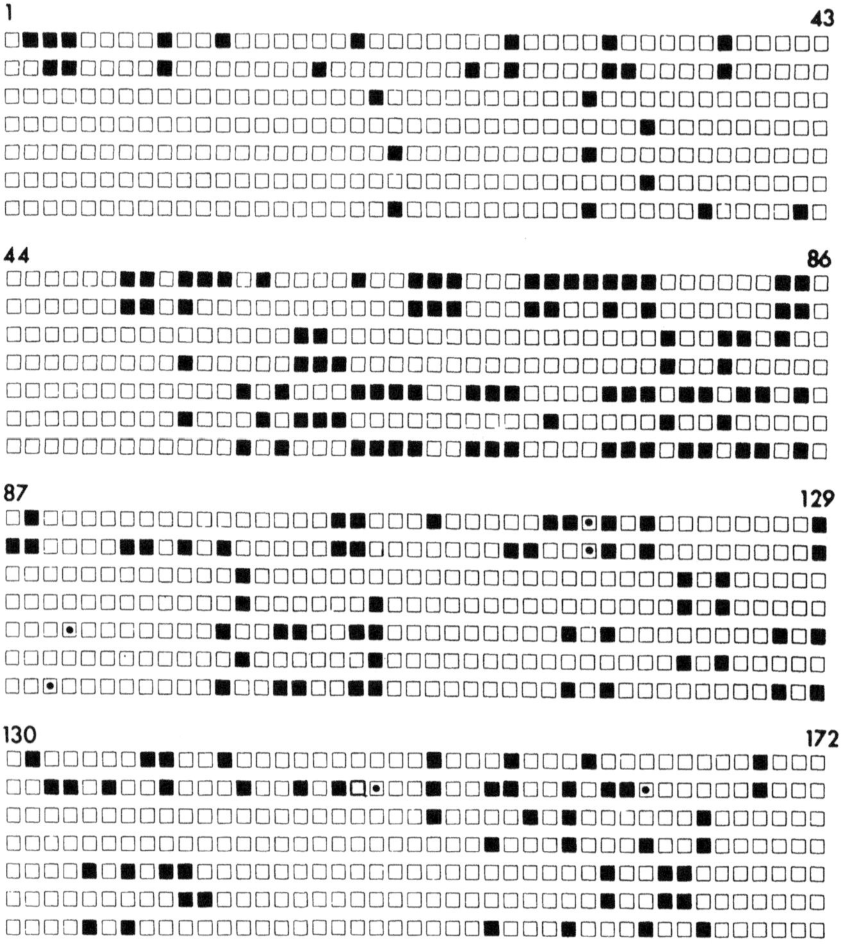

Figure 10. Nucleotide sequence variation in the basic 172-bp units making up the human, African green monkey, African baboon, and bonnet monkey repetitive sequences. The seven lines in each group represent, respectively, human subunit 1, human subunit 2, African green monkey, baboon subunit 1, baboon subunit 2, bonnet monkey subunit 1, and bonnet monkey subunit 2. ■, Base changes; ⊡, deletions. [Redrawn from Miklos and Gill (1982); data compiled from Wu and Manuelidis (1980), Donehower *et al.* (1980),and Rubin *et al.* (1980).]

Thus more than 60% of the average sequence has been perturbed. The sequence changes appear to be nonrandom, but the biological significance of this nonrandomness remains to be determined.

Donehower *et al.* (1980) have pointed out that the core oligonucleotides, 5′-CTTT-3′ and 5′-CCTT-3′, are overrepresented in the baboon sequence and may have provided exchange points for recombination in the evolution of this repeated sequence from a smaller unit. At the very least, the 340-bp unit has evolved from a 170-bp one by dimerization, followed by subsequent amplification. While the original 170-bp sequence may have been constructed from simpler repeats, the evidence for this is not convincing (Wu and Manuelidis, 1980).

I have summarized a particular dyadic symmetry among the various primate repeats (Fig. 11) to see how well structural features might be *conserved* among species. Dyadic symmetries are of value in protein DNA interactions, and this is the reason for examining them in satellite sequences. Once again there is extensive variation among species. In the bonnet monkey, for example (Fig. 11f), there are 15 out of 18 paired bases, whereas the corresponding run in the human sequence has only nine out of 18 matches and the hairpin is not particularly stable. In the Hawaiian *Drosophila* data, where individual clones from within the same species are available for comparison, the dyadic symmetries are inhomogeneous even within a laboratory strain (Miklos and Gill, 1981).

Having seen the variation in consensus sequences among different primate species, we can now return to the manner in which such alphoid sequences has been treated in evolutionary contexts. The reader will recall that in the kangaroo rat–guinea pig comparison, the existence of the shared sequence TTAGGG was used as an argument both for sequence conservation by selection and for the role of such DNAs in speciation phenomena. Donehower and Gillespie (1979) have argued that the primate data provide support for certain aspects of the library hypothesis, since the alphoid sequences arose from a pool of ancestral sequences. However, they point out that since speciation in these primates clearly took place subsequent to the latest major amplification, the newly amplified sequences probably facilitated diversification, rather than providing the means for species barriers. It should be stressed that in this account, the original library hypothesis has been biologically emasculated, since it neither deals with the postulated specific protein–DNA interactions nor with a role in speciation. It is now reduced to a pool of sequences that can be added

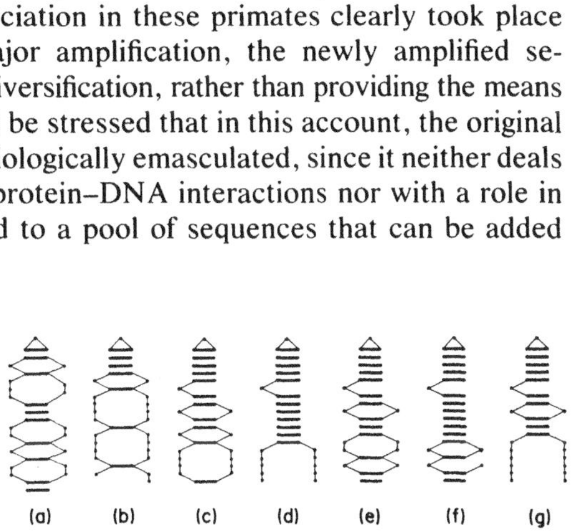

Figure 11. Potential long dyadic symmetries in the basic 172-bp primate units: (a) human subunit 1; (b) human subunit 2; (c) African green monkey; (d) baboon subunit 1; (e) baboon subunit 2; (f) bonnet monkey subunit 1; and (g) bonnet monkey subunit 2. [Based on Miklos and Gill (1982); data compiled from the same sources as in Fig. 10.]

to or drawn from. In this form the sequence pool has no biological context and is merely shackled to the whims of DNA turnover mechanisms in the genome.

Once again an important point needs to be stressed. Sequence similarities among species may be due to a number of forces, of which selection is only one. The finding of common sequences among species does not automatically imply that selection has been operative. Finally, since the *deca* satellite sequences are absent from the human genome, hypotheses on their conservation would seem ill conceived. Similarly, mouse satellite sequences are absent from the Chinese hamster (S. D. M. Brown and Dover, 1980a), and a large repeated sequence family of *D. orena* is absent from other sibling species of the *Drosophila* group (Strachan *et al.*, 1985). Detailed evaluation of the minimal requirements for investigating libraries of conserved sequences that are framed in testable biological frameworks are given in John and Miklos (1979) and Miklos and Gill (1981).

In summary, the primate data are much like data from any other source, be it crabs, fruitflies, or rats. There are repetitive sequences of short or long repeat length, which may have homologies between some species but not others. All sequences display heterogeneity at the DNA sequence level and the sequence data provide no obvious functional or evolutionary contexts into which they can be placed. Despite attempts to force them into intellectually satisfying biological categories, these sequences are excellent candidates for genomic flotsam and jetsam.

2.5. The Mouse, DNA Polymorphisms, and Chromosome Pairing

One of the the very first mammalian highly repeated DNAs to be characterized was that of the mouse *Mus musculus*. The mouse satellite, which can make up 10% of the genome, is a 234-bp unit with an internal substructure based on 28- and 30-bp units (Horz and Altenburger, 1981). As with other satellites, it apparently evolved from smaller repeating units (Southern, 1975). *In situ* hybridization of uncloned satellite DNAs indicates that it hybridizes to the centromeric heterochromatin of all chromosomes except the Y (Pardue and Gall, 1970). It is apparent that the X and Y are unlikely to utilize most of these sequences as pairing entities, since the X has substantial amounts of this repetitive DNA, whereas its homologue has them below the level of detection. A similar situation also applies in the autosomes. For example, *Mus spretus* is a relative of *M. musculus*, but it has only 1% of the amount of mouse satellite DNA (S. D. M. Brown and Dover, 1980b). The female *M. spretus*–*M. musculus* hybrids are fertile, indicating that enormous satellite DNA discrepancies between homologues are no hindrance to normal mitotic or meiotic events.

This example is instructive in another sense. When these two species are compared by restriction endonuclease and blotting techniques, similar DNA variants are shared between them, even though *M. spretus* has such a reduced amount of these sequences. Either there has been wholesale reduction of mouse satellite DNA sequences near every *M. spretus* centromere, or a concomitant amplification has occurred in *M. musculus*. Whatever the true situation, the amplification and deletion mechanisms in this case are addressing themselves to existing sequence sets and are not randomly generating sequences *de novo*. Similar conclusions can be drawn from data on species of *Apodemus* (S. D. M. Brown and Dover, 1979).

Other results from mouse hybrids also support the conclusion that differences in satellite DNA between homologous chromosomes are not directly implicated in meiotic chromosome pairing. *Mus musculus musculus* and *M. m. molossinus* also differ in their satellite DNA content, *M. m. molossinus* having approximately 40% less than the other subspecies (Rice and Straus, 1973). This discrepancy is not evenly spread throughout the karyotype, so that some chromosomes have large blocks of centromeric heterochromatin whereas others have very small ones. Hybrids between *M. m. musculus* and *M. m. molossinus* are phenotypically normal and fertile, and at meiosis, chromosome pairing is perfectly normal even though some bivalents are highly asymmetric in terms of their satellite DNA content (Dev *et al.*, 1975). A similar situation obtains in populations of wild Bohemian mice, where natural polymorphisms of the centromeric heterochromatin are found (Forejt, 1973). Thus in rodents, meiotic pairing regimes are not sensibly attributable to satellite DNA, a conclusion also borne out by direct experimental manipulations in *Drosophila*. These findings in rodents invalidate the first and most important biological tenet of the library hypothesis.

S. D. M. Brown and Dover (1980a) have also examined the satellite sequences on a single mouse X chromosome by utilizing a mouse–Chinese hamster hybrid cell line into which the *Mus musculus* X chromosome has been sequestered. The control Chinese hamster cell line (without the X) was totally devoid of mouse satellite sequences at normal conditions of stringency, indicating that a library of conserved sequences is unlikely to exist between these two rodents. The experimental cell line, however, revealed that while the overall satellite patterns on the X were similar to those of the bulk satellite patterns, the relative band intensities were different, showing that there are organizational differences between the X and other chromosomes.

Finally, the question of what causes mouse satellite DNA-containing regions to be heterochromatic has begun to be addressed (Zhang and Horz, 1982). Blumenfeld *et al.* (1978), Gurley *et al.* (1978), and Billings *et al.*

(1979) suggested that phosphorylation of histones may be an important factor in satellite DNA compaction in *Drosophila virilis* and in the deer mouse, *Peromyscus*. The isolation of purified satellite DNA chromatin from the mouse, however, has shown that there is no H1 histone phosphorylation of note. Furthermore, bulk chromatin and mouse satellite chromatin also differ insignificantly in terms of their histone composition. One difference between these two broad chromatin classes is that the HMG proteins are effectively absent from satellite chromatin (Zhang and Horz, 1982).

In summary, the mouse DNA data place severe constraints on two of the backbone hypotheses of satellite DNA function. The involvement of this satellite DNA in chromosome recognition phenomena is most unlikely, and the sequences themselves are not substantially held in common between the mouse and the Chinese hamster.

2.6. Rats and Rearrangements

2.6.1. Introduction

The genome of *Rattus norvegicus* contains a number of repetitious sequences, both dispersed and localized, which together constitute about 10% of the genome (Lapeyre and Becker, 1980; Miklos *et al.*, 1980). Predominant 92- and 93-bp *Eco*R1 fragments are repeated about one million times in the haploid genome of *R. norvegicus*, but are undectable in other species, such as *R. villosissimus*.

Satellite I of *R. norvegicus* consists of a 370-bp *Hin*dIII monomer unit constructed from four alternating 93-, 92-, 93-, and 92-bp *Eco*R1 sequences. The sequence of the 370-bp repeat has been determined from uncloned DNA (Pech *et al.*, 1979b) and when the subrepeats are compared against a consensus sequence, the overall variation is 19%. A better estimate of the inhomogeneity is available from individual clones that have been sequenced (Sealy *et al.*, 1981). When these 93- and 92-bp inserts are compared *inter se*, extensive sequence heterogeneity is obvious (Table III). Sealy *et al.* (1981) draw attention to the extensive perturbations in this satellite, with only 47 of the 93 positions being common to all seven clones. Those regions that are homologous are short and dispersed throughout the repeating unit.

In terms of their structural properties, clones 2A and 15A have termination codons in all possible reading frames of both strands. The remaining five clones have stop codons in five out of six possible reading frames, and these sequences thus do not engender translational excitement. The number of inverted repeat sequences is also low and none

Table III
Percent Homology between 93-bp Cloned Repetitive Sequences from *Rattus norvegicus*[a]

	2A	2B,15B	10A	15A	15C	10B
2A	—	90	86	69	65	63
2B,15B	—	—	89	67	63	62
10A	—	—	—	63	60	65
15A	—	—	—	—	83	72
15C	—	—	—	—	—	55
10B	—	—	—	—	—	—

[a]Summarized from Sealy *et al.* (1981).

exceed 4 bp in length, indicating that the secondary structure potential of the 93-bp monomer is not high. In this sense it is analogous to the human complex repeat sequenced by Wu and Manuelidis (1980).

When *in situ* hybridizations are carried out on tissue culture cells with cloned probes, hybridization occurs to telomeric, centromeric, and some interstitial locations. Nuclear RNA sequences homologous to such probes can also be detected, indicating that these repetitive sequences are transcribed, but their significance has yet to be evaluated. It may be similar to the overrun transcription seen in amphibia, or it may be important. The fact that it is from tissue culture cells suggests caution, since it is known that extensive chromosomal rearrangements occur in such cells. If RNA transcripts are found in tissues of whole organisms by Northern blot or cDNA analyses, then the search for their transcriptional identity can be widened.

The cloned data focus attention on the significance of consensus sequence comparisons between species. Sequence homologies of varying length are found between the 93-bp rat units and repetitive DNA sequences of the mouse, the 1.706 calf satellite, the 359-bp *D. melanogaster* family, and the 170-bp units of humans and African green monkeys (Pech *et al.*, 1979b). It has been argued that these homologies could be due either to evolutionary relatedness or perhaps to protein-binding sites of functional significance. The meaning of these homologies should become clear when DNA-binding proteins are tested against these satellites, and when the variation among individual clones from different species are known. Until these tests are finalized, statistical homologies are probably inadequate indicators of biological significance.

Highly repetitive DNA has also been isolated from the roof rat, *Rattus rattus,* and here the predominant repeating unit is 185 bp in length and has been denoted satellite 1′. It is related to the 370-bp unit of *R. norvegicus* satellite 1, and consists of an asymmetric 185-bp "bite" from a diverged 370-bp monomer unit (Witney and Furano, 1983). *Rattus rattus* also harbors within its genome 370-bp units that are homologous to the 370-bp units of *R. norvegicus,* but these are much reduced in frequency relative to the 185-bp units. Thus these two rat species share variant sequence pools, which are amplification derivatives of 92/93-bp building blocks. Because of the commonality of these sequence pools, the authors have pointed out their consistency with the library hypothesis. It should also be kept in mind that since these two species supposedly diverged from a common ancestor only about 2 million years ago, it is to be expected that some sequence variants should still be shared in the absence of any functional constraints.

2.6.2. *Rattus sordidus, R. villosissimus,* and Centric Rearrangements

Comparisons of different rat species from a biological viewpoint allow us to ask questions about one of the recurring hypotheses of satellite DNA function, namely that satellite DNA in large amounts facilitates chromosome rearrangements. This hypotheses was originally erected for heterochromatin (H. J. Muller, 1940), and later applied to highly repeated DNA. As with many of the hypotheses on satellite DNA, it seeks to place this DNA in an evolutionary context, usually with an emphasis on its contribution to reproductive isolation and hence speciation. The details of each variant of this class of hypotheses have been evaluated and summarized by Miklos *et al.* (1980).

The hypothesis itself has been experimentally examined in a group of closely related rats that are phenotypically almost indistinguishable, yet have undergone massive genome reorganizations involving centric fusions (or fissions) (Miklos *et al.*, 1980). Eleven major centric rearrangements have been fixed, so that the genomes of *R. sordidus* and *R. villosissinus* now consist of 32 and 50 chromosomes. *Rattus sordidus* has almost all its chromosomes in a biarmed configuration (Fig. 12).

Analyses of these present-day genomes by restriction endonuclease cleavages, thermal denaturation studies, and cytogenetic techniques reveal that there are negligible quantities of centromerically located repetitive DNA sequences relative to *R. norvegicus,* which itself is not exactly well endowed with large amounts of localized tandem arrays. Thus the formation and fixation of centric rearrangements can go on in the virtual absence of highly repetitive DNAs. This conclusion clearly has obvious restrictions. Large amounts of repetitive DNA may well have been present when these rearrangements took place, and the amounts may have been reduced during evolution to their present levels. This is an inescapable

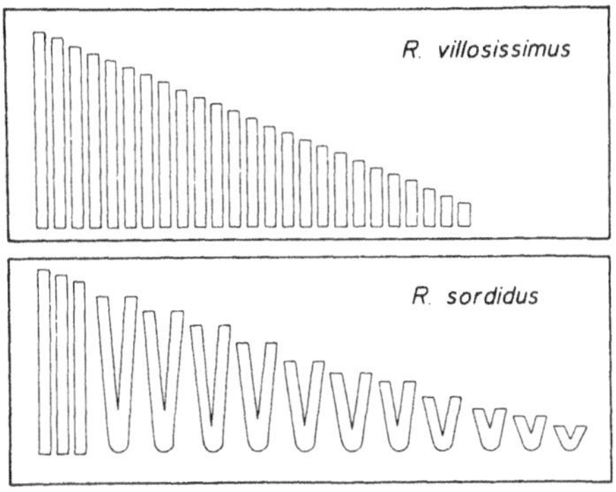

Figure 12. Diagrammatic representation of the major karyotypic differences between two species of *Rattus*.

difficulty and is implicit in every argument based on comparisons of extant species.

Some of the attempts to place satellite DNA in an evolutionary framework via genomic reorganizations stem in part from mammalian groups such as the Carnivora, which have "stable" karyotypes and very little C-banding (Pathak and Wurster-Hill, 1977). However, within the Carnivora, this argument would only have validity for groups such as the Felidae (Matthews *et al.*, 1980), since other carnivores can show quite remarkable departures from this generality. Foxes, for example, can have chromosome numbers of $2n = 38, 48, 50, 64$, and 78.

The other side of the coin is seen in groups such as *Peromyscus* (Hazen *et al.*, 1977), where significant amounts of satellite DNA occur, but where hardly any centric rearrangements have been either formed or fixed. This situation is not restricted to mammals. In the Hawaiian *Drosophila*, where satellite sequences can in some instances account for 70% of the genome, centric rearrangements are rare, despite the fact that in extant species the centromere regions are loaded with repetitive DNA (Carson *et al.*, 1967; Clayton, 1971; Miklos and Gill, 1981).

Thus the seductive appeal of a causal relationship between chromosome rearrangements and centromerically located, highly repetitive DNA sequences is not borne out by the data from natural populations. It should also be noted that the next step in the hypothesis hierarchy is to argue that centric rearrangements were causal factors in speciation processes. In order to demonstrate this, one needs to separate the effects of the rearrangements on fertility from any effects of the background genotype. This requires disproving the null hypothesis that *any* of the changes found at the molecular level anywhere in the genomes of the species being compared were instrumental in the speciation process itself. This is a daunting and so far unresolved issue. It requires a decision on whether macrokaryotypic changes are incidental or causal components of speciation processes.

In summary, the molecular data from different rat species conform very much to the rules and regulations obvious in other mammalian groups. In a biological context the molecular data severly restrict the generality of the satellite DNA–chromosome rearrangement hypothesis.

2.7. Sheep, Goats, Satellite Homologies, and Rearrangements

The only DNA sequence data in this group stemming from cloned sources concern the 816-bp monomer sequence of the 1.714 satellite of the sheep (Reisner and Bucholtz, 1983). This sequence has indications of being internally repetitious, and two sections of it have statistically significant homologies to the 1.715 sequence of the calf. In these two portions of 67 and 59 bp, greater than 70% matching is achieved. This homology has led these authors to conclude that the sequence is important and that its function is sequence-related. Whether such homologies are functional or are the nonfunctional remnants of past ancestral sequences is difficult to ascertain. One avenue is to determine the level of variation between individual monomers and to examine if the heterogeneity is at acceptable levels. Another is to perform experiments to examine their protein-binding properties. If insertions of alien DNA sequences are a common occurrence in tandem arrays, homologies should occasionally be found to protein-coding regions, to nomadic elements, and to noncoding sequences. The sheep 816-bp monomer in fact exhibits homology to a human class 1 transplantation antigen. The statistical homologies are guideposts, but whether they are likely to lead toward or away from significant research avenues is not known.

The remainder of the molecular data in this group concern restriction endonuclease digestions and Southern blotting experiments on sequence homologies between different species (Maio *et al.*, 1977; Forstova *et al.*, 1979; Buckland, 1983). These data are no different to those I have previously discussed. Satellite I of the domestic sheep, for example, shares homology with an 820-bp *Eco*R1 fragment of Saanan goats, Scottish blackface sheep, Barbary sheep, European mouflon sheep, Afghan urial sheep, and the Siberian ibex. Other fragment sizes, however, can vary among these organisms. Thus, sequence variants are held in common among these groups, but it is well to recall the amphibian data, where monomeric units consisted of many different families while still retaining their 741-bp unit length. The evolutionary history of the Caprini is difficult to assess, and the taxonomic relationships between sheep and goats is very much in dispute. Hence this group is not a particularly useful one for evaluating repetitive DNA sequence relationships in terms of phylogenies.

Since the sheep has six large metacentric chromosomes and the goat has one, it has been proposed that sheep satellite II may have been responsible for the production of matacentrics by the fusion of other chromosomes (Forstova *et al.*, 1979). Once again the evolutionarily popular satellite DNA–chromosomal rearrangement hypothesis has been invoked in the usual context of reduced fertility. However, the biological data already pose severe difficulties for such a proposition. Valdez *et al.* (1978), in analyzing the evolution of wild sheep in Iran, found divergent chromosome numbers of from $2n = 54$ to 58. In northern Iran, where sheep with chromosome numbers of $2n = 55$, 56, 57, and 58 all coexist, the fecundity of a limited number of hybrid forms is as good as the nonhybrids, and the fetuses are morphologically normal. A similar situation has been described in domestic sheep (Bruere and Chapman, 1974). In these cases translocation heterozygotes do not have reduced fertility even when heterozygous for two different translocations. Similarly, natural populations of the vole mouse, *Akodon molinae* (Bianchi *et al.*, 1973) are polymorphic for *metacentric/two acrocentric* situations, and this seems to be no impediment to normalcy. Clearly, the widespread occurrence of this heterozygosity is not causing sterility barriers. This cautions against immediately embracing centric translocations as the primary events in reproductive isolation and hence speciation. Bruere and Chapman (1974) long ago pointed out that the correlation between centric fusions and reduced fertility was inconclusive. The actual causes of any infertility first need to be ascertained, rather than assuming that it is the translocation *per se* that is the causal factor.

These biological data, as well as the rat data described previously, all render the satellite DNA–centric rearrangement–speciation theories quite untenable. In this regard, see also O'Brien *et al.* (this volume, Chapter 8, Section 4.3.1), who also find no evidence that satellite DNA is involved in chromosomal rearrangements.

3. The Molecular and Biological Data Base: Invertebrates

3.1. *Drosophila* Species

Drosophila melanogaster and its dipteran relatives are the most useful organisms for evaluating nearly all hypotheses on highly repetitive DNAs. They have simple sequence as well as complex highly repetitive DNAs, and a satellite DNA binding protein has been found in *D. melanogaster*. The evolutionary history of the *Drosophila* group is adequately known and a plethora of chromosomal rearrangements are available for shuffling genomic components. Except for the lack of 5-methylcytosine in their tandem arrays, *Drosophila* species deviate in no major ways from the properties of vertebrate highly repetitive sequences. *Drosophila* is thus a paradigm for other species.

The haploid genome is approximately 165 million bp and is organized chromosomally as shown in Fig. 13. Blocks of heterochromatin engulf the centromeres and the Y chromosome is almost totally heterochromatic. The heterochromatic regions account for about 50 million bp of DNA, at least 35 million bp of which is made up of satellite sequences (Brutlag *et al.*, 1977; Barnes *et al.*, 1978; Peacock *et al.*, 1978; Brutlag, 1980). Some of these heterochromatic blocks, such as those of chromosome II, are also inhabited by a low density of genetic complementation groups. Others, such as the blocks in chromosome III, apparently lack such genic residents (Hilliker *et al.*, 1980). There are also sundry nomadic elements (Carlson and Brutlag, 1978a,b; Dowsett and Young, 1982) and at least one major multigene family, the ribosomal RNA genes, in the heterochromatin. The ribosomal RNA genes are embedded in the heterochromatin of the X and Y chromosomes, but there is nothing sacrosanct about these sex-chromosomal addresses. In the closely related species *D. simulans*, the ribosomal genes are absent from the Y, and in some Hawaiian *Drosophila*, the ribosomal genes can be autosomally located. Finally, there are also repetitive DNA sequences within the centromeric hetero-

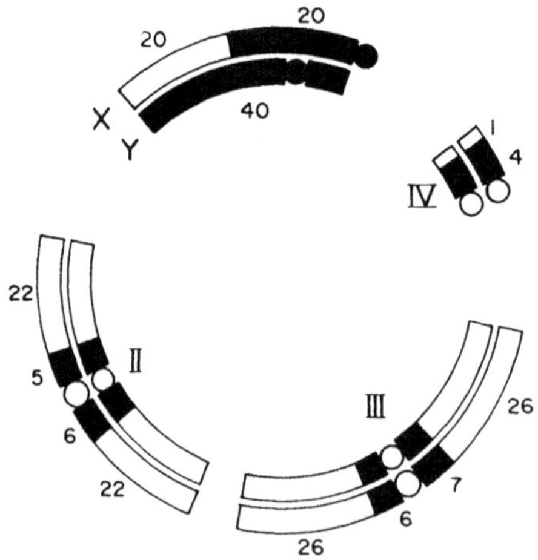

Figure 13. Chromatic subdivisions of the male *D. melanogaster* genome in million bp. Heterochromatic regions are solid, euchromatic areas are open.

chromatin that share homology with telomeric sequences (Young *et al.*, 1983).

The simple sequence satellite data for *D. melanogaster* all derive from uncloned DNAs. The consensus sequences of the different satellites are 5′-AATAT-3′, 5′-AATATAT-3′, 5′-AAGAG-3′, 5′-AAGAGAG-3′, and 5′-AATAACATAG-3′. One particular cloned satellite is 359 bp in length but has variants that are only 254 bp in length (Brutlag, 1980). These various uncloned DNAs have been used as probes for *in situ* hybridization reactions and all hybridize to the heterochromatic regions of the chromosomes (Peacock *et al.*, 1978; Steffensen *et al.*, 1981; Hilliker and Appels, 1982). Since each chromosome had its own particular arrangement of satellites, Brutlag *et al.* (1977) and Peacock *et al.* (1978) argued that these patterns were critical to recognition processes between homologous as well as between nonhomologous chromosomes. This notion was reinforced when Hsieh and Brutlag (1979) isolated an embryonic DNA-binding protein that preferentially bound to a part of the 359-bp satellite. Since this satellite was predominantly located on the X and Y chromosomes, these authors proposed that it had a function in either sex-chromosome meiotic pairing or postmeiotic processes.

The genetic data, however, indicate that such pairing functions are unlikely to be a property of such repetitive DNA sequences. First, the pairing properties differ between the sexes. In females the pairing sites are predominantly euchromatic and enormous deficiencies and rearrangements of the heterochromatic regions have little effect on chromosome pairing (Yamamoto and Miklos, 1978; Hawley, 1980). In the male, pairing sites are highly localized in the heterochromatin of the X and Y chromosomes (Cooper, 1964; Yamamoto and Miklos, 1977, 1978), whereas in the autosomes, they occupy euchromatic locations (Yamamoto, 1979a). Thus in all situations where homologues share euchromatic regions, the pairing sites are euchromatic. In the one case where they do not, as in the totally heterochromatic Y and its homologue the X, the pairing sites are necessarily located within these heterochromatic regions but are not a general property of them.

When deletions of the heterochromatic regions of the X and Y are constructed and tested in males, chromosome pairing can be severely upset, but this is due to the deletion of pairing sites in specific regions of the heterochromatin. When the molecular data are superimposed on the genetic data, this lack of correspondence between particular satellites and pairing sites becomes obvious. For example, it is known that the uncloned AATAT and AAGAG sequences are distributed in at least eight regions of the Y chromosome, whereas they are cramped very close to the centromere of the X (Peacock *et al.*, 1978; Steffensen *et al.*, 1981). The region of the X in which these particular sequences are located is totally devoid of male pairing sites, indicating also that the centromere itself is without pairing attributes. Furthermore, the two distal regions of the Y chromosome also lack these sites even though these satellites occur there. Hence there is no reason to conclude that the interstitial satellite sites on the Y are in any way involved in the interstitial male pairing sites.

The *in situ* results from the 359-bp complex satellite are equally informative. This satellite was originally mapped to the heterochromatic regions of *all* chromosomes of the genome. However, the use of cloned probes has now shown that it is localized exclusively in the X-chromosome heterochromatin (Hilliker and Appels, 1982). Since this sequence is restricted to the X, its involvement in any pairing regime with the Y on the basis of primary sequence is precluded. A similar situation will later be seen to hold for *D. hydei,* where one particular satellite occupies almost all the heterochromatin of the X and is not found elsewhere in the genome (Renkawitz, 1978).

The description of the heterochromatic landscapes of the various *D. melanogaster* chromosomes will not only have to be reassessed with cloned probes, but these will need to be sequenced first, to ensure that no other sequences are contained within the insert. The problems inherent in the use of uncloned probes have been clearly illustrated by the differences in hybridizations using uncloned versus cloned DNAs of human satellite III (Gosden *et al.*, 1980).

This brings us to a critical point. How relevant can a detailed molecular description of heterochromatic landscapes be in terms of the putative functions of these DNAs? We have already seen in the case of vertebrates that certain sequences oscillate wildly in amount and arrangement between different individuals—for example, the *deca* satellite of African green monkey. In the case of the *Drosophila* AAGAG sequences, stocks can vary in the amounts of this sequence on both chromosomes III and the X (Steffensen *et al.*, 1981). Heterochromatic regions are also interdigitated with nomadic elements, so much so that probably every nomadic family has members in the heterochromatin (Dowsett and Young, 1982). It has been argued by Brutlag (1980) that significant progress about functions of satellite sequences will only be forthcoming when the detailed molecular architecture and the protein components of these regions are known. This, however, is the structure–function approach that has led to the present imbroglio. Since most of the heterochromatin can be deleted in both the mouse and *Drosophila* and the organism is still unimpaired in terms of its viability and fertility, is a description of the underlying molecular landscape a sensible route of enquiry? It may well be that the products of genomic change are such that a chromosome cannot avoid having some kind of repetitious DNA sequences near the centromeres. These sequences are continually changing in amount and composition and descriptions of each and every variant landscape would seem to be an imprudent line of enquiry.

Satellite DNA-binding proteins are essential to revealing important properties of DNA–protein interactions, as well as to contributing to knowledge of chromatin structure and to the understanding of how these complexes interact with nuclear scaffolding, for example. However, this structural line of enquiry is very distinct from, and not necessarily germane to, enquiries about functional relevance at the phenotypic level. The hope that function will become apparent when a constantly changing molecular landscape is described in its entirety is a misguided one for localized highly repetitive sequences. I shall illustrate the point with a number of examples.

Drosophila nasutoides is a Samoan fly, 60% of whose genome exists as a number of satellite sequences (Cordeiro-Stone and Lee, 1976; Wheeler and Altenburg, 1977; Lee, 1981). These sequences occur on one giant pair of chromosomes (Fig. 14) and constitute the bulk of this enormous entity. There is no detectable *in situ* hybridization of the satellites to the ordinary members of the complement, which show very little conventional heterochromatin. Thus the normal chromosomes perform mitotic and meiotic recognition programs with an insignificantly small amount of localized highly repetitive DNA. Is it prudent to argue in the face of such data that one chromosome sequesters 60% of the genomic sequences to perform such a task? Surely it is foolish to think so. The important message from this example is that if highly repetitive DNA sequences do provide a basis for chromosome recognition phenomena, then the minimum amount to do so is but a minute fraction of the amount on any chromosome. This was also the conclusion from the comparison of *M. musculus* and *M. spretus* (S. D. M. Brown and Dover, 1980b).

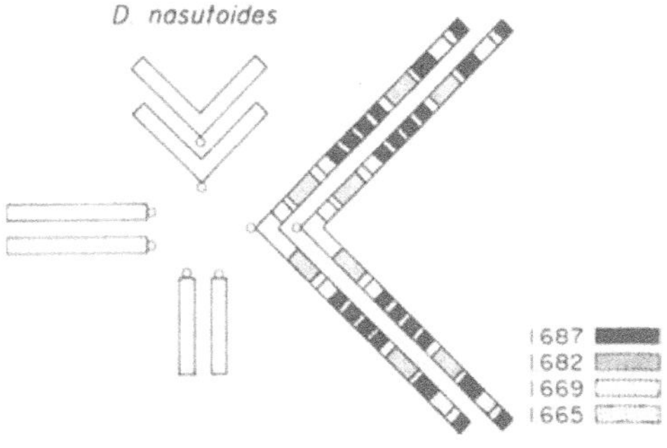

Figure 14. Approximate *in situ* hybridization sites of the four major satellite DNAs of *D. nasutoides*. [Data from Wheeler and Altenburg (1977) and Wheeler *et al.* (1978).]

It would seem inappropriate to characterize the underlying molecular architecture and protein-binding capacities of this giant chromosome as a means of asking about its functional capabilities. A more direct approach would seem to involve analyses of whether this chromosome has any effects on phenotype, or whether related species have homologous chromosomes of this size. If the *D. nasutoides* chromosome is an amplificatory accident of nature, as it may well be, molecular descriptions of it will be oblique in a functional sense. A much more harmonious interpretation of the *D. melanogaster* and *D. nasutoides* results is that tandem arrays of particular sequences have accumulated on various chromosomes due to mechanisms of genomic flux. To determine whether their nonrandom distribution in the genome is due to selective or nonselective regimes is a challenge for the future.

3.2. Manipulation of Heterochromatic Regions

In *D. melanogaster* both induced and spontaneous rearrangements involving the heterochromatic regions of most chromosomes, but especially those of the X and the Y, have been used to evaluate hypotheses on function. If a homozygous deficiency for most of the satellite sequences is viable and fertile, then the underlying organization is sufficiently irrelevant not to impinge upon cellular metabolism and development. When a deletion of 80% of the X heterochromatin is analyzed in females, chromosome behavior as measured genetically is quite normal (Yamamoto and Miklos, 1978). Not only is an enormous deletion tolerated at the level of chromosome mechanics, but viability, fertility, and developmental characteristics are unperturbed.

There are two further points from this example. First, these deleted chromosomes also form a normal chromocenter (Fig. 15), so whatever somatic molecular interactions are involved in bringing nonhomologues together, their efficacy is not impaired by massive deletions and rearrangement of tandem packs of sequences. Furthermore, centromere activity is not inhibited when nearly 90% of the adjacent repetitive sequences are removed from its proximity.

The major repetitive DNA of *D. hydei* is also germane to recognition phenomena involving the sex chromosomes. Here a 1.696, highly repetitive DNA comprises 13% of the diploid genome (Renkawitz, 1978); its distribution on the sex chromosomes of *D. hydei* is depicted in Fig. 16. Using the uncloned 1.696 DNA as a probe, Renkawitz (1978) demonstrated that it localized exclusively to the X-chromosome heterochromatin. Fortunately, here it does not matter that the probes were uncloned DNAs, since there was no cross-hybridization to other chromosomes.

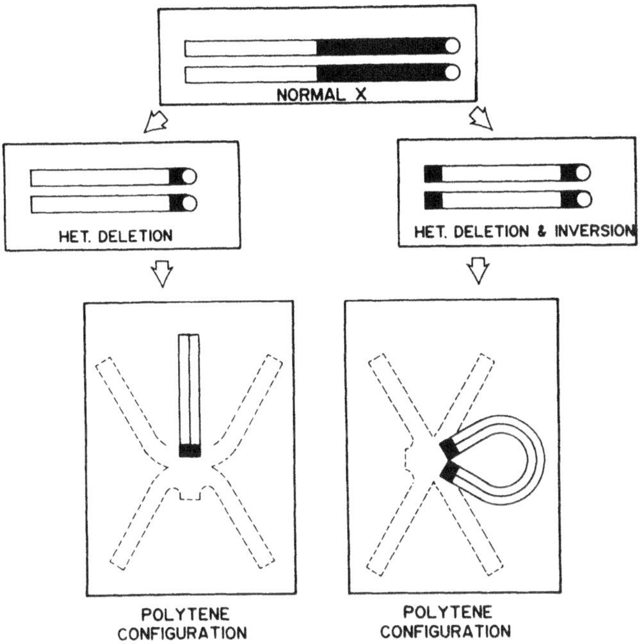

Figure 15. Experimentally constructed X chromosomes and their meiotic and somatic association relationships in *D. melanogaster*.

The amount of X heterochromatin in *D. hydei* is about 13% of the haploid genome, which is the same as the amount of the 1.696 satellite. If we allow for the DNA content of the ribosomal genes that are in the X heterochromatin and also for the usual bevy of nomadic elements that should be found there, the heterochromatin of the X is effectively full. Thus if satellite DNA sequence homology *per se* were important for sex-chromosome recognition in the male of *D. hydei,* then it cannot be utilizing the 1.696 satellite, which is effectively absent from the Y. A similar argument holds for the 1.688 satellite of *D. melanogaster* (Fig. 16).

The second point from the *D. melanogaster* manipulations is that these massive deficiencies, or even larger ones, such as the removal of the Y chromosome to yield XO males, have no known phenotypic effects directly attributable to highly repetitive sequences *per se*. Similarly, while duplications that yield XYY and XYYY males contribute to sterility, these males are nevertheless somatically unperturbed. The upsets that do occur are more likely to be attributable to genic or ribosomal DNA dosages in these genotypes. Thus the *D. melanogaster* genome can tolerate removal

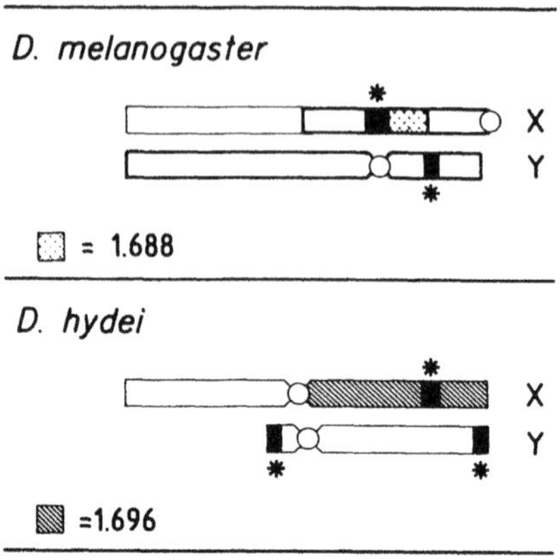

Figure 16. Localization of specific satellite DNAs in *D. melanogaster* and *D. hydei*. Asterisk denotes the ribosomal gene cluster. [Data of Hilliker and Appels (1982) and Renkawitz (1978).]

of at least 40 million bp of its repetitive DNA content or have 90 million bp added, and its somatic developmental capabilities are unhindered.

This is very important in a metabolic context because it sets the boundary conditions within which repetitive DNA sequences can act as functional sinks for nonspecific protein–DNA binding. In *E. coli,* for example, many *bona fide* regulatory proteins exhibit very tight binding to their specific targets but bind nonspecifically around the remainder of the genome. This latter binding influences the concentration of unbound repressors and may have an important role in regulatory activities (Lin and Riggs, 1975; Kao-Huang *et al.*, 1977; Berg *et al.*, 1982; Takeda *et al.*, 1983). Lin and Riggs (1975) found that the specific binding of the *lac* repressor was quite sensitive to total DNA content, and they postulated that junk DNA in eukaryotes had a role in the maintenance of total genomic DNA concentration for all regulatory proteins. If this concentration were to be significantly altered, suboptimal function of regulatory circuits would ensue. The *Drosophila* data set the limits within which development can proceed normally. So also do the kangaroo rat data, where related species can differ by 4000 million bp per diploid nucleus and still allow the functioning of a similar developmental network.

We have already seen that huge deficiencies of the repetitive DNA sequences of the X chromosome of *D. melanogaster* females are irrelevant to chromosome mechanics in a homozygous state. *Drosophila* is also similar to the mouse hybrids in that heterozygous differences are tolerated with ease (Fig. 17A and 17B).

We can continue in this vein and determine whether the autosomal repetitive sequences of *D. melanogaster* resemble the sex-chromosomal ones in their irrelevance to chromosome mechanics. When a chromosome

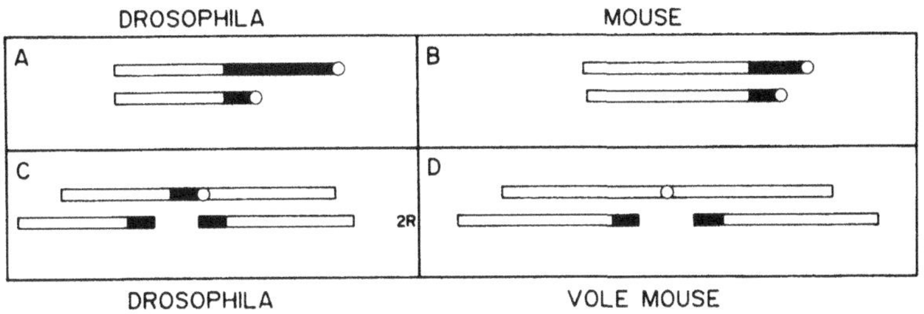

Figure 17. Variant homologous chromosome combinations, which nevertheless lead to normal homologous pairing in *D. melanogaster, Mus musculus,* and *Akodon molinae.*

II with an almost total deficiency of heterochromatin on its right arm is challenged with two acrocentric derivatives, chromosome pairing as assayed cytologically in males is again normal (Fig. 17C). Thus a chromosome arm that has very little heterochromatin pairs normally with one that does (Yamamoto, 1979a). Exactly the same result is obtained in natural populations of the rodent *Akodon molinae* (Bianchi *et al.*, 1973), where the population is polymorphic for a particular metacentric/acrocentric chromosomal setup (Fig. 17D).

These experiments are open to a number of criticisms, the chief one being that the heterochromatin of each and every manipulated chromosome has not been characterized molecularly, and hence it is not assuredly known just what sequences do occur in these heterochromatic regions (Brutlag, 1980; Steffensen *et al.*, 1981). These tentative fears can be soothed and allayed by utilizing chromosomes that do not contain heterochromatic rearrangements. For example, consider a euchromatic translocation between chromosomes IV and I (the X) in which the heterochromatin of both chromosomes has been left unruffled, but parts of their euchromatin have been interchanged. There are now two chromosome IV's, which share total heterochromatic commonalities but lack euchromatic homologies (Fig. 18). There are two possible avenues of action; the IV's pair either on the basis of euchromatic or on heterochromatic homologies. In fact the euchromatic option is exercised (Fig. 18A). Thus, whatever undisturbed tandem arrays are in the heterochromatin of the IV's, they are insufficient for chromosome recognition phenomena (Fig. 18B). Clearly, euchromatic homology is obligatory in autosomal recognition processes (Yamamoto, 1979a). Results as clear and pristine as these should dispel any ungrounded fears.

If doubts still linger about *D. melanogaster* as a paradigm for these processes, then a similar experiment in the mouse is instructive in this regard. Here the pairing properties of a particular translocation T(1;13)70H

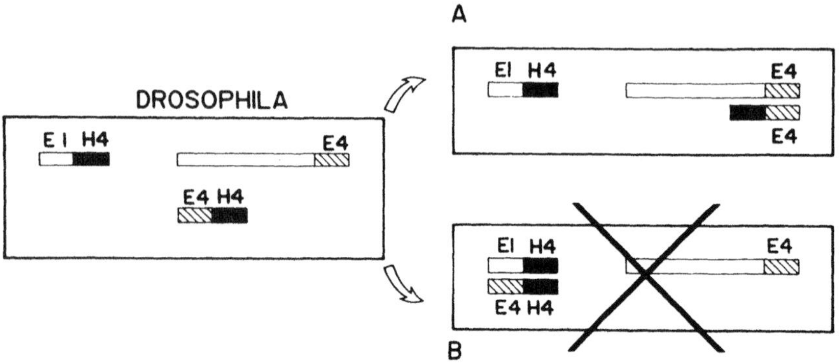

Figure 18. Experimental protocol and resultant pairing configuration in a *D. melanogaster* male containing a translocation between chromosomes 1 and 4. H4, Heterochromatin of chromosome 4; E4, euchromatin of chromosome 4; E1, euchromatin of chromosome 1.

have been analyzed in which the tester chromosome is given a choice; does it pair on the basis of its heterochromatic homology with chromosome I or its euchromatic homology with chromosome 13 (De Boer and Groen, 1974)? Just as in the *Drosophila* example, it turns out to do so on its euchromatic homologies (Fig. 19).

If overkill is considered desirable in this area, then further experiments with autosomes II and III of *D. melanogaster* lead to exactly the same conclusions. In the case of chromosome II, deletion of most of the euchromatic sequences leaves a mini II that is almost totally heterochromatic (Fig. 20). When a pair of these mini II's are made homozygous,

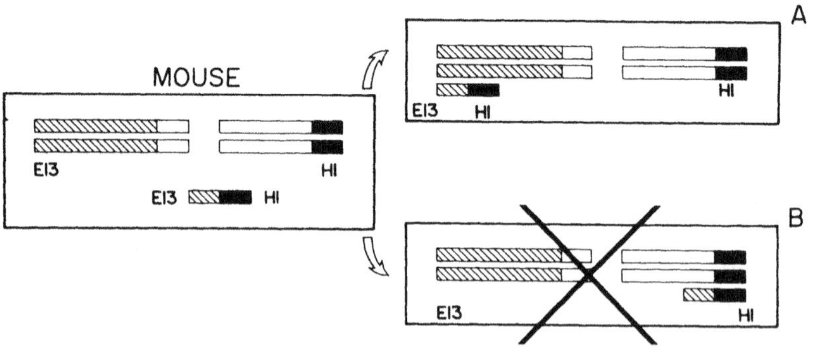

Figure 19. Experimental protocol and resultant pairing configuration in a *Mus musculus* partial trisomic. H1, Heterochromatin of chromosome one; E13, euchromatin of chromosome 13.

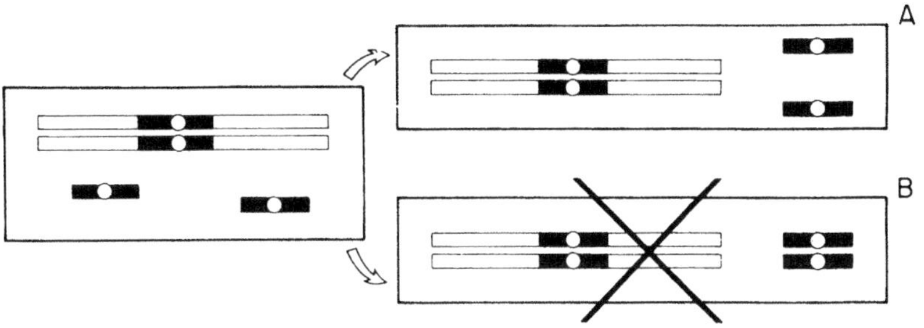

Figure 20. Experimental protocol and resultant pairing configuration in a *D. melanogaster* male homozygous for an almost totally heterochromatic mini chromosome II and the two normal chromosome II's. Heterochromatin shown solid, euchromatin shown open.

they do not recognize each other and stay as univalents (Yamamoto, 1979a).

The conclusions from these manipulative systems in *Drosophila* and the corresponding examples from rodents are quite unambiguous. In the autosomes euchromatic homology is paramount and heterochromatic homology is irrelevant. Thus the genetic data show that heterochromatic regions are *not* a sensible place in which to look for recognition phenomena in either flies, mice, or men (Gosden *et al.*, 1975; Yamamoto and Miklos, 1977, 1978; John and Miklos, 1979; Yamamoto, 1979a; S. D. M. Brown and Dover 1980b; Miklos, 1982). The molecular data have in many instances been unfortunately tied to this incorrect satellite DNA–chromosome pairing dogma. It should now be clear that both the vertebrate and invertebrate examples lead to the same conclusions.

3.3. Satellite DNA and Developmental Timing

One of the long-standing notions about excessive genomic DNA is that is can have important effects on cell size and cell cycle time and hence on developmental timing (Bennett, 1971, 1982; Cavalier-Smith, 1978, 1980). These are all variants of the concept of nucleotype (Bennett, 1971), in which DNA mass, rather than its sequence content, can have important phenotypic effects, which are assumed to be of evolutionary significance.

If such hypotheses had validity and generality, then the perturbation of a genome using genetically depauperate DNAs should show significant effects. The X heterochromatin of *Drosophila* is almost totally lacking in conventional loci, and is an ideal vehicle for such purposes. The bulk of the heterochromatin consists of simple sequence DNAs, the 359-bp family, about 250 ribosomal RNA genes, representatives of most of the nomadic element families, and the compensatory responder (cr^+) locus (Procunier and Tartof, 1978). What happens when all or most of these sequences are subtracted from or added to a genome? Is early embryogenesis upset? Figure 21 illustrates the *Drosophila* genome and the mini X chromosomes used in a study of such a situation. The bulk of the euchromatic genic sequences of the X have been deleted, resulting in these small and almost totally heterochromatic chromosomes. The genetic crossing program is critically important, since by mating X.Y males to X.X.mini X females, four different progeny genotypes are available from the *same* cross; X.Y males, X.Y.mini X males, X.X females and X.X.mini X females (Fig. 22).

The length of time taken to complete embryogenesis was measured from egg laying to larval hatching, with the initial expectation that development may have been longer in genotypes with additional repetitive sequences. It should also be noted that each experimental genotype has its own internal control, thus X.Y males and X.Y.mini X males are derived from the same mothers, as are X.X and X.X.mini X females. The results of such experiments are depicted in Fig. 23; additional data are given in Miklos (1982).

The average time to complete embryogenesis and hatch from the egg was 1250 min (21 hr) and the time differences between control and experimental genotypes did not exceed 2% of this figure. Interestingly, these small effects are *not* due to chromosomes carrying the most heterochromatin, but to the few euchromatic sequences that serve as phenotypic markers for these various chromosomes. Thus, mini 1337 is relatively more savage in its effects than the Y chromosome, and the genic content

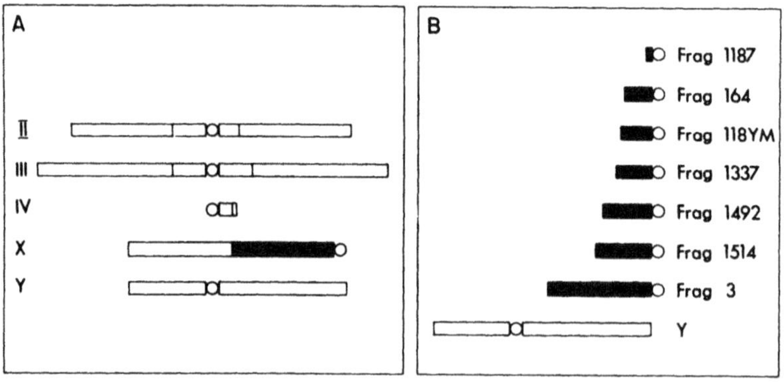

Figure 21. Diagrammatic illustration of the *D. melanogaster* genome and a series of experimental mini X chromosomes. Only heterochromatic regions of the X chromosome are shown solid.

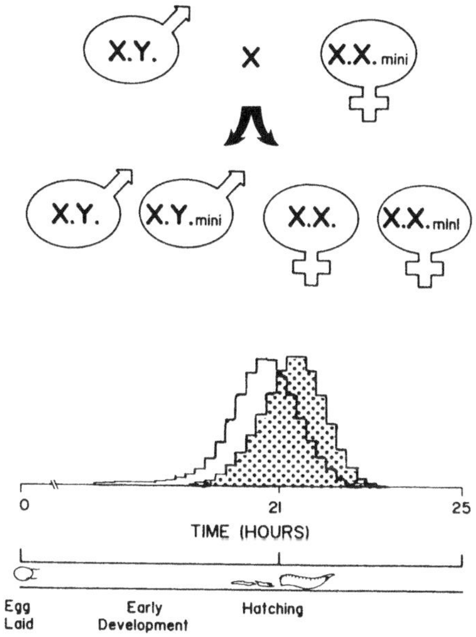

Figure 22. Genotypes, genetic protocol, and emergence times expected from total DNA content of eggs in control (unshaded) and experimental (shaded) progeny, from a crossing program involving the series of mini X chromosomes in Fig. 21.

of a chromosome can have effects out of all proportion to its size. When DNA is removed from a genome to produce XO males, these, too, develop no differently from their XY controls (timing difference of 1%).

Thus, the overall conclusions from these experiments are that when heterochromatic sequences of known average content, but almost totally lacking conventional genic sequences, are added to or subtracted from a genome, there are negligible somatic effects on embryonic processes, which one might expect to be particularly sensitive to such changes. *All* of the small timing differences could be due to the marker genes carried on these chromosomes and *none* to the motley assemblage of repetitive sequences carried within the heterochromatin.

It is noteworthy that in a careful examination of plant hybrids in which the parental genomes differ by 40% there were no detectable effects of this supplementary DNA on viability, fertility, or somatic phenotypic characteristics (Hutchinson *et al.*, 1979; Rees *et al.*, 1982).

In order to keep these *Drosophila* results in perspective, it should be remembered that there is no doubt whatsoever that in many systems there are excellent correlations between cell cycle time and the nuclear DNA content of different genomes. However, this is only the beginning of the issue. The crux of the matter is whether the correlations are due to the quality or the quantity of the DNA. It still needs to be demonstrated that

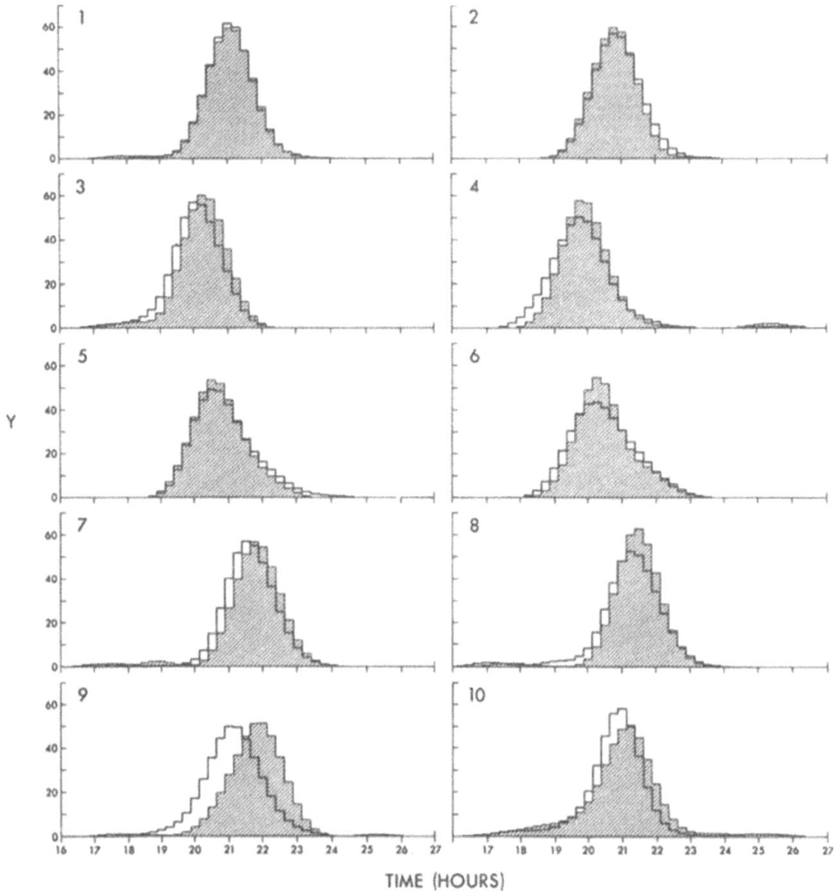

Figure 23. Duration of embryogenesis in various genotypes of *D. melanogaster*. Individuals carrying a mini X chromosome are shaded, normal XY or XX controls are unshaded. (1, 2) Mini 1187 males and females; (3, 4) mini 1492 males and females; (5, 6) mini 1514 males and females; (7, 8) mini 3 males and females; (9, 10) Y-chromosome-carrying males and females. [Reprinted from Miklos (1982) by permission.]

such correlations have coding or noncoding DNA as their basis. While these tests are difficult to perform, the DNA used for such purposes must be characterized to some extent. Thus, if the addition of a chromosome or a heterochromatic block causes a particular effect, it needs to be determined how much of this entity is devoid of conventional coding sequences; otherwise, any phenotypic effects may be due to the coding sequences sequestered within tandem arrays of noncoding sequences.

This is precisely why the heterochromatin of the *Drosophila* X is such a useful landscape. Except for the ribosomal genes and the compensatory responder locus, it is genetically empty.

3.4. Evolutionary Considerations

3.4.1. Hawaiian *Drosophila*

Three Hawaiian species have been examined for variation at the DNA sequence level with respect to their highly repetitive DNAs. These species, *D. grimshawi, D. gymnobasis,* and *D. silvarentis,* share a similar satellite DNA, which constitutes at least 40% of their genomes. This family of DNAs has been characterized by sequencing individual clones and the sequences have been found to consist of variations on a 189-bp repeat, which can have relatives of 185, 179, and 177 bp (Miklos and Gill, 1981, 1982). The sequences of individual clones and the variation among them is shown in Fig. 24. The clones fall into at least four different classes, on the basis of deficiencies as well as base changes, but it is not known whether these classes distinguish individual chromosomes or whether all can occur on a single chromosome. Within a class the variation is low, being less than 2% in all cases. Among classes, however, the amount of variation is approximately 12%.

The 189-bp sequence has a substructure in which the sequence TTAA occurs at intervals of 35, 37, 35, and 71 bp. Thus the present sequence, like many mammalian ones, may have been amplified from a shorter repeating unit. There are also a number of sequence motifs present in the basic repeat, including two potential RNA polymerase recognition sites, TATAAG and TATAAA. The first lies within the deletion zone and the second is not maintained between *D. gymnobasis* and *D. grimshawi.* There are five stop codons in each reading frame and at least four substantial inverted repeats spread throughout the sequence. The sequence can also fold up into a variety of distorted cloverleaf shapes, but none of these is differentially conserved between species. Thus, there is no reason to attach anything but low probabilities of functional significance to the sequence structure itself. This conclusion is similar to that of Wu and Manuelidis (1980), who argued from their human sequencing data that individual unit repeat lengths and secondary structure potential were not compelling features.

When the three species are compared in all possible 183 positions, only about half of the positions are free of base changes or deletion/addition events and this is not a very impressive figure for the concept of conservation. If, on the other hand, the variation among individual

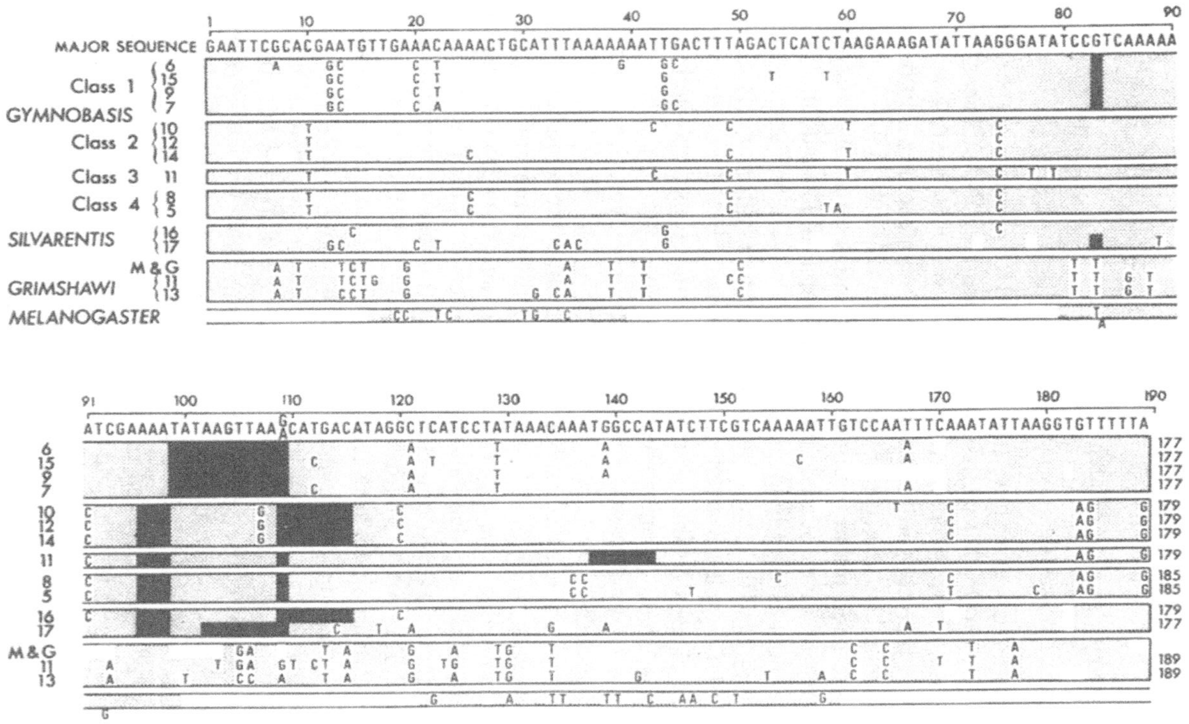

Figure 24. Nucleotide sequence variations expressed as deviations from a consensus sequence for ten Hawaiian *Drosophila* satellite clones from *D. gymnobasis* (6, 15, 9, 7, 10, 12, 14, 11, 8, and 5), two clones from *D. silvarentis* (16 and 17), and two clones from *D. grimshawi* (11 and 13). Base sequence changes are as shown and deletions are shown solid. Open areas refer to undetermined bases. Regions showing homology to the *D. melanogaster* 359-bp sequence are as indicated. [Reprinted from Miklos and Gill (1981) by permission.]

clones is ignored and consensus sequences are compared for all the species, *D. gymnobasis* and *D. silvarentis* are almost indistinguishable.

Furthermore, *D. gymnobasis* and *D. silvarentis* are *homosequential* in their polytene banding patterns. So close are they in external phenotype that the females of the species are not easily classified as being different, and examination of males is necessary in order to make this decision. These two species retain 40% of their genomes as a single satellite, which in terms of consensus sequences is virtually identical. Thus, speciation between *D. gymnobasis* and *D. silvarentis* has occurred in the absence of significant changes in either the amount or sequence to this class of highly repetitive sequences (Miklos and Gill, 1981). For the Hawaiian *Drosophila* this is not surprising, since much of the isolation between species is behavioral. In spite of newer methods for assaying differences between species, the old problem remains; how are the changes that gave rise to the species to be sorted from those that have accrued since that time and now distinguish the present-day taxa? The information on satellite DNA sequences in terms of their transcriptive reticence, and particularly their closeness in sequence and amount in *D. gymnobasis* and *D. silvarentis,* seems to provide no evidence in favor of their pivotal role in speciation processes within this group.

3.4.2. Other *Drosophila* Species

Drosophila virilis has at least four satellites, whose consensus sequences based on uncloned DNAs are 5′-ACAAACT-3′, 5′-ATAAACT-3′, 5′-ACAAATT-3′, and 5′-AATATAG-3′ (Gall and Atherton, 1974; Mullins and Blumenfeld, 1979). These are again located in the centromeric heterochromatin and together constitute over 40% of the genome in diploid cells. However, in cells that are polytene or polyploid the satellite sequences can be under- or overeplicated, indicating that somatic functions can be totally normal in certain tissues when nearly half the genome has been removed by underreplication (Endow and Gall, 1975).

In an evolutionary perspective, some relatives of *D. virilis,* such as *D. littoralis* and *D. ezoana,* are almost totally devoid of the particular repeated sequences found in *D. virilis,* as well as having very small amounts of highly repetitive DNA in general (E. H. Cohen and Bowman, 1979). The *D. virilis* phylogeny and its peculiarities have been previously summarized and evaluated (John and Miklos, 1979) and the salient conclusions from these studies are that:

1. Some species, such as *D. littoralis* and *D. ezoana,* can indeed perform all the basic cellular functions involved in chromosome

structure, three-dimensional nuclear architecture, and meiotic and mitotic recognition programs in the virtual absence of highly repeated DNA.
2. Neither viability nor fertility is correlated with satellite DNA amounts in interspecific hybrids.
3. There can be gain and loss of satellite sequences without major karyotypic changes.
4. When major chromosomal changes do occur, they can take place or be fixed in the absence of significant quantities of satellite DNA.

3.4.3. Molecular Evolution in the *melanogaster* Species Subgroup

The most extensive array of clones and sequences for one particular highly repetitive family, as well as extensive comparisons of other genomic components, is available from the seven species of the *melanogaster* subgroup (Barnes *et al.*, 1978; Coen *et al.*, 1982; Dover *et al.*, 1982; Strachan *et al.*, 1982, 1985). The phylogeny of this group has been documented using polytene comparisons (Lemeunier and Ashburner, 1976), and the number and abundance of different repetitive sequences have been determined. Two major DNA sequence families are of immediate interest, termed, respectively, 180 and 360 (Strachan *et al.*, 1982, 1985). When restriction endonuclease and blotting experiments using uncloned DNAs are carried out within this subgroup, it is found that the 180 family appears only in *D. orena,* where it constitutes nearly 15% of the genome. This DNA sequence family is distributed in the centromeric heterochromatin of all chromosomes, which suggests not only that it is of recent evolutionary origin in *D. orena,* but that following amplification it was a successful invader of all chromosomes of the genome. There is yet another sequence family that might be exclusive to *D. orena,* and this localizes to the X and Y chromosomes, and is apparently excluded from autosomal excursions.

The X-chromosome-specific 360 family, on the other hand, has representatives in all seven groups, except that in *D. teissieri* it has suffered a large deletion, which reduces its size to 196 bp. It does not cross-hybridize under relaxed conditions to a distantly related species, *D. busckii.*

The existence of the 360 family in all seven species indicates that it was probably amplified and present in progenitors prior to the separation of these seven species. It has been inherited and undergone certain sequence alterations, which are examinable using clones and sequencing techniques.

The within- and between-species variation, for the seven sibling species in the 360 family are presented in Table IV. The data are quite

Table IV
Intra- and Interspecific Variation in the *360* Family of Repetitive Sequences, Expressed to the Nearest Percent, for Seven Species of the *melanogaster* Subgroup[a]

Intraspecific variation			
melanogaster	4	*erecta*	ND
mauritiana	3	*orena*	3
simulans	3	*teissieri*	5
		yakuba	1

Interspecific variation

	mel	*mau*	*sim*	*ere*	*ore*	*tei*	*yak*
melanogaster	—	33	35	ND	21	31	35
mauritiana		—	3	ND	32	33	35
simulans			—	ND	32	33	35
erecta				—	ND	ND	ND
orena					—	29	35
teissieri						—	35
yakuba							—

[a]Summarized from Strachan *et al.* (1985). ND, Not determined.

straightforward, in that the between-species comparisons based on consensus sequences are all tenfold higher than the within-species estimates (Strachan *et al.*, 1985). The one exception is the *D. simulans–D. mauritiana* comparison, where the two species are known to be very closely related on other criteria.

What does this large differential mean? Dover (1982) has drawn attention to the different expectations involved when amplified sequence arrays are inherited and undergo changes imposed by two different regimes. In one, mutations are free to accumulate over time, whereas in another, homogenization of sequences occurs due to a variety of inbuilt genomic mechanisms that impinge on these arrays. Random sampling of sequences in the first case should provide equivalent measures of variation within and between species. In the second, variation within a species should be lower than that between species due to homogenization of sequences. The finding that the data conform to the second alternative provides evidence in favor of homogenizing mechanisms. Molecular drive challenges natural selection as a prime mover in evolution, and it frees us from the constraints of selection, which as far as multigene or multisequence families are concerned, is very restricted in both a pragmatic and a predictive molecular sense.

The results then could mean that homogenizing mechanisms operate within a species, and it is the fixation of sequence variants that leads to

DNA sequences diagnostic for a species (Dover *et al.*, 1982; Dover, 1982; Strachan *et al.*, 1985). Mutation and turnover may well be largely unconstrained and be time-dependent phenomena.

3.4.4. Structural Characteristics

As with other highly repeated DNAs, the differences between species in the 360 family are due mainly to base substitution and deletion/addition events. However, *D. teissieri* has a large deletion as well, reducing its repeating unit to 196 bp. Since it is known that the 360 family in *D. melanogaster* is bound to by a specific protein in a region of dyadic symmetry (Brutlag, 1980), it is now possible to compare homologous consensus sequences in the different species groups. These comparisons reveal that there is substantial leeway in the secondary structure of the stem and loop homologous to *D. melanogaster*. For example, *D. mauritiana* has a particularly perturbed stem and in *D. teissieri* one-half of the potential stem falls into the large deletion (Fig. 25). A stem can still be formed, however, in *D. teissieri* by recruiting a sequence from elsewhere in the repeating unit. Whatever the final outcome of stability, the pro-

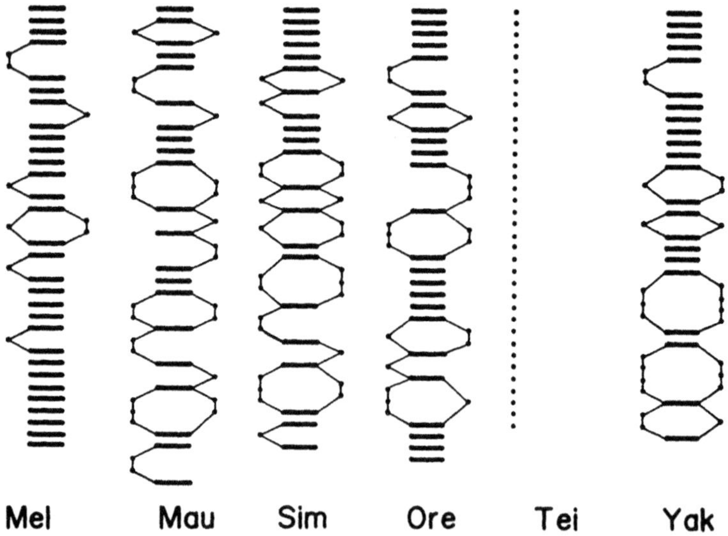

Figure 25. Potential dyadic symmetries in the 360 DNA sequence family of six species of *Drosophila*. The *D. teissieri* sequence for the complementary region is contained within a large deficiency and so the formal dyadic symmetry is aborted. [Data from Strachan *et al.* (1985).]

tein–DNA binding assays are now open to experimental tests. It may well be that the same protein binds to all sequences, or that different proteins bind to the different sequences. This second alternative, if it occurred, would be particularly noteworthy, since it could mean that molecular drive within a sequence family has forced a protein-coding gene elsewhere in the genome to change in concert with the changing repeats.

The protein-binding capacities of different repeated sequences have begun to be analyzed in mammals and invertebrates (Wiche *et al.*, 1978; Will and Bautz, 1980; Schmidt and Keyl, 1981), but the *Drosophila* species subgroup, with its localization of the 360 family on a single chromosome, and known sequences in related species, as well as the P-factor transformation capability, offers the most viable present alternative for investigations.

3.4.5. Centromere Sequence Homologies

The second structural characteristic of the 360 family is that there are regions of homology between it and the centromere sequences of yeast (Fitzgerald-Hayes *et al.*, 1982). This has again raised the issue of functionality, namely that the 360 family may somehow help in centromere activity. The only direct observation impinging on this point biologically is that huge deletions of the heterochromatin of the X chromsome, which totally delete the 360 family, do not cause meiotic impairment of chromosome segregation in females. The *Mus musculus–M. spretus* data also indicate that centromeres can function with less than 1% of the mouse repetitive DNA landscape in their neighborhood.

Finally, it is possible to compare the consensus sequences of the seven species of the *melanogaster* subgroup with element 1 of the centromere of yeast chromosome 3. This comparison reveals that extensive base changes have occurred in this sequence (Fig. 26). In the case of *D. teissieri*, the sequence in question again falls into the deletion, but sequences elsewhere in the unit can be found to compensate for it.

This absence of strict conservation unfortunately does not negate the hypothesis that this sequence is important for centromere activity. Strachan *et al.* (1985) have pointed out that function and conservation are not mutually required. Thus it may be more important to have high levels of family homogeneity within a species than conservation of a given sequence between species. The distinction between these two possibilities has all the hallmarks of being experimentally quite troublesome. It may be more pragmatic in the first instance to isolate the 360 binding proteins from the seven species and then readdress any questions that relate to putative functionality.

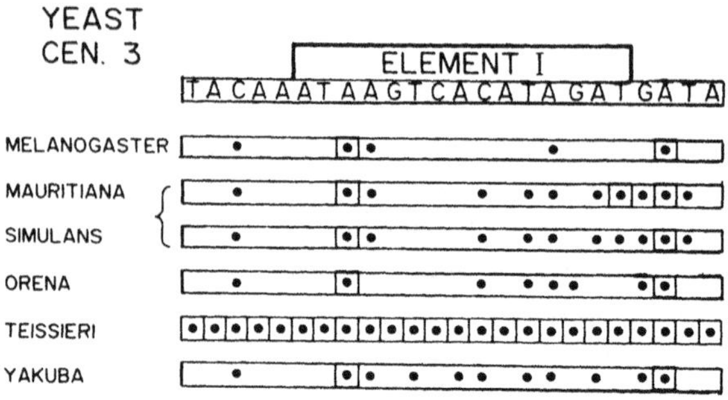

Figure 26. Nucleotide sequence comparisons between element 1 of yeast centromere 3 and homologous regions in six species of the *melanogaster* species subgroup. ●, Base changes; ⊡, deletions. [Data from Fitzgerald-Hayes *et al.* (1982) and Strachan *et al.* (1985).]

3.5. Germ Line Effects

It has been pointed out that one of the highest probabilities of repetitive sequence effects are in the germ line (John and Miklos, 1979; Bostock, 1980; F. Muller *et al.*, 1982). Study of grasshoppers of the genus *Atractomorpha*, for example, provides a connection between heterochromatin consisting of satellite DNA and some of its biological effects. In *Atractomorpha similis*, *A. species 1*, and *A. australis*, the genome sizes are in the ratio of 1.4:1.1:1.0 and these differences are largely due to the presence of centromeric and telomeric blocks of heterochromatin (Miklos and Nankivell, 1976). The presence of these large blocks of highly repetitive DNA has characteristic meiotic effects. Recombination events in bivalents that contain these telomeric sequences occur with a lower probability and with altered distributions. Thus, one effect of a telomeric block is to render regions near to it *rec*$^{-}$ (Fig. 27).

When populations of *A. similis* are surveyed for their C-band variation (John and King, 1983), a stunning variety of polymorphisms is found (Fig. 28). Thus, within one species, the array of chromosomal differences is truly enormous, with every chromosome exhibiting variation in length. In this case, it is likely that the cytological variation will reflect the variation in the repeated DNA content, and this is certainly a group in which this formal possibility can be rigorously tested between different populations.

The other major biological conclusion that emerges from these data is that because individuals are polymorphic for the presence or absence of telomeric blocks, these blocks are unlikely to have any significant somatic effects on metabolism or development. The situation resembles a scaled-up version of the polymorphisms on the human Y chromosome.

These germ line effects on recombinational probabilities and spatial

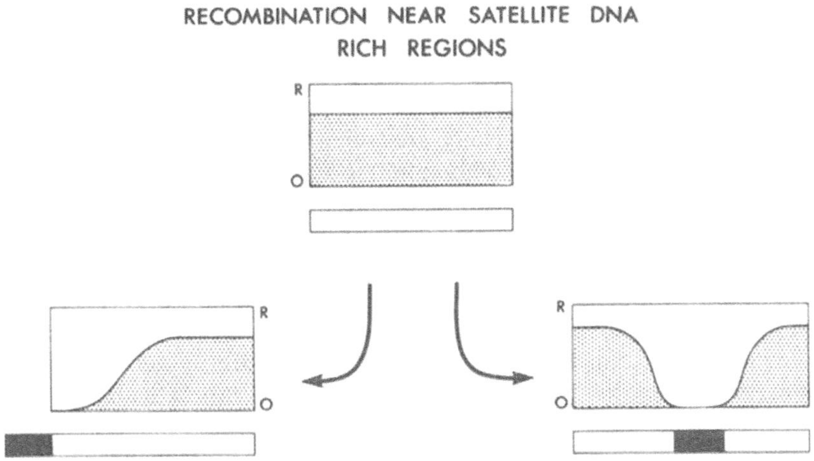

Figure 27. Diagrammatic illustration of recombination probabilities (from zero to *R*) along chromosomes containing either telomeric or interstitial blocks of heterochromatin. For illustrative purposes only the control chromosome is given a uniform exchange probability along its length.

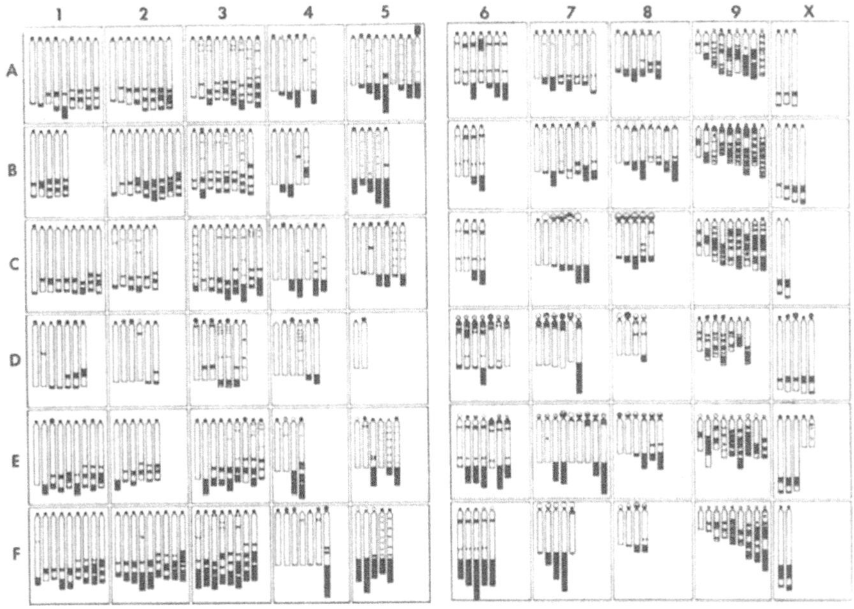

Figure 28. Diagrammatic representation of the C-band variation within and among the individual chromosomes (1–9 and the X) of six populational groups (A–F) of the grasshopper *A. similis*. [Reprinted from John and King (1983) by permission.]

distributions do not occur only in grasshoppers, but are paralleled by the effects of naturally occurring telomeric blocks of heterochromatin in hedgehogs, where the DNA is of unknown molecular composition. An experimentally produced situation in *Drosophila,* where repetitive sequences from the Y chromosome are added to the telomere of the X, also results in a reduction in recombination in this region. The results of such experiments are illustrated in Fig. 27 and are dealt with in more detail in John and Miklos (1979) and Miklos and John (1979).

A similar effect of these heterochromatic regions also obtains when they are interstitially located. Only in *Drosophila,* however, are the average genetic and molecular composition of such blocks even roughly known. In antelope squirrels of the genus *Ammospermophilus* (Mascarello, 1980) interstitial blocks of heterochromatin can be asynapsed, and while a detailed analysis needs to be done, at least in some individuals regions next to such blocks participate in recombination events at a lower probability (Fig. 27).

What is now clearly required is to investigate the underlying molecular architecture giving rise to such effects. In the *Drosophila* examples the DNA is predominantly simple sequence in content together with nomadic elements and their immovable relics. In the hedgehog and squirrel cases the contents of the block remain to be determined. A further critical point is to find out if heterochromatinization *per se,* irrespective of the underlying architecture, can also indulge in such perturbations. Thus, if normally euchromatic regions were able to be reversibly heterochromatinized and still have these same effects, then the problem would move into the domain of chromatin condensation phenomena.

Finally, these germ line effects are not restricted to a single chromosome. If the repetitious sequences on the *Drosophila* X chromosome are perturbed by deletions, so that recombination near the X heterochromatin is reduced, then recombination probabilities on other chromosomes are *increased,* generally in regions near heterochromatic blocks (Fig. 29) (Yamamoto, 1979b). These are variations on the classic interchromosomal effects in *Drosophila.* Thus the genome is not static in terms of its recombinational characteristics; if one part is affected by heterochromatic polymorphisms, another part may compensate.

These germ line effects are useful in thinking about particular sequence arrays at different levels, since the usual preoccupation is only at the *cellular* level. However, the effects of meiotic processes do not intrude on the phenotype of the organism in question. It is irrelevant to an animal in a somatic sense whether it is *rec*$^{-}$ or *rec*$^{+}$ in its germ cells. The significance of such events only becomes apparent in the next generation, where the genetic variability is open to scrutiny via the phenotype.

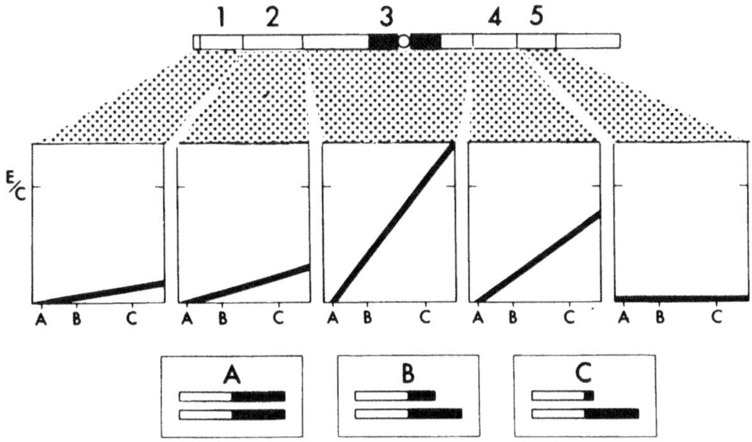

Figure 29. The ratio of recombination levels (experimental/control) measured in five regions of chromosome III of *D. melanogaster* in the presence of three different X-chromosome genotypes. (A) Control X chromosome; (B) heterozygosity for a medium-sized deletion of the X heterochromatin; (C) heterozygosity for a very large deletion of the X heterochromatin. [Reprinted from Miklos (1982) by permission; data from Yamamoto (1979b).]

The major problem with this body of data, as with the molecular data, is to *prove* that the gross influences on the meiotic system have any adaptive or evolutionary significance. The challenge is no different for the proponents of the more topical ways of introducing variation into a genome via nomadic element excursions. One has to demonstrate that the genetic variability has a phenotypic component that is in some sense adaptive or has contributed to evolutionary change. This is still the crux of the problem, and there are no hard data impinging upon it in an experimental context.

3.6. Crabs: From the Simple to the Complex

3.6.1. Introduction

The remaining molecular data source stems from the Crustacea, where the complexities of particular highly repeated DNAs vary from 2 to at least 2000 bp (Skinner, 1977; Skinner *et al.*, 1982). This enormous range in complexity is reminiscent of the kangaroo rat and calf data, indicating that repeat unit complexities are not the exclusive domain of vertebrates.

In *Cancer borealis* the major repeating unit is AT, and when 500 million bp of the genome is almost unadulterated ATATATATAT. . . with only the lone G or C occasionally blighting the landscape, does this leave much scope for its functional utilization at the cellular level? Surely such a purity of sequences would hinder, rather than help, homologous chromosome association, for example. Is this the stuff of important chromosome architecture?

In the hermit crab, *Pagurus pollicaris,* the canonical repeats are TAGG and variants of $(CAG)_n$CTGCACT, so that the wearisome monotony of ATs is somewhat relieved. The most frequent length is 31 bp, so that the repeat is of the form $(CAG)_8$CTGCACT (Chambers *et al.*, 1978). In the red crab, *Geryon quinquedens,* the basic uncloned unit has been sequenced and found to be 81 bp (Christie and Skinner, 1980a,b) and constitutes about 2% of the genome. In the Bermuda land crab, *Gecarcinus lateralis,* the average unit is 2070 bp in length (Skinner *et al.*, 1982) and makes up 3% of the genome.

Analyses of uncloned satellite DNAs using restriction endonucleases have, as expected, yielded levels of heterogeneity on a basic sequence. In the 81-bp repeat of the red crab, the dimers exhibit heterogeneity and are divisible into at least three subclasses. The conclusion from analyses of oligomeric forms is that mutation alone is insufficient to generate the diversity; what is also required is selective amplification of particular variants (Christie and Skinner, 1980a,b).

3.6.2. Cloned Data

Sixteen clones have been examined for their variation from the Bermuda land crab and the variation is extensive (Skinner *et al.*, 1982). Not only do most of the clones range in size from 2060 to 2090 bp, indicating heterogeneity in length, but restriction endonuclease analyses show that almost every clone is different from every other, which is similar to the situation for the α-satellite of the African green monkey. If this were to be the case, then the Bermuda land crab would have an astonishing assemblage approaching 16,000 different variant sequences.

The sequence analyses of two further clones of heretical length have also been completed. One clone is short (1690 bp), whereas the other is long (2700 bp), and both have distinctive characteristics that can be compared to the more normal 2070-bp units. The gross sequence architecture of the normal clone is depicted in Fig. 30, with its substructure laid out in 100-bp divisions (Skinner *et al.*, 1982). The differences among the short,normal, and long variants are manifold and are scrupulously listed below in order to allow the reader to appreciate the extensive type and form of the sequence variations:

In section 0.4, there is a G tract of 20 bases. In the short variant

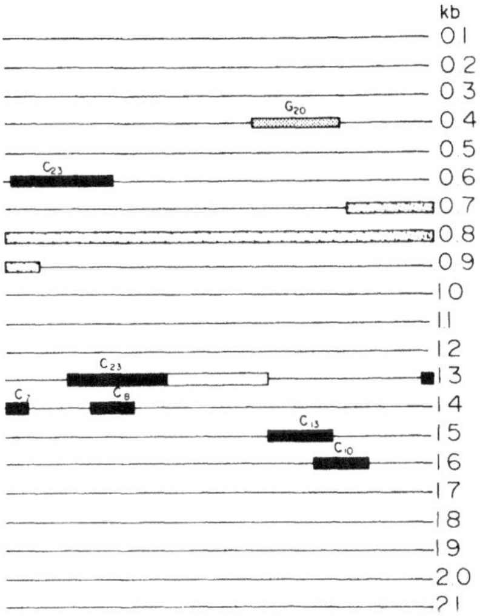

Figure 30. Diagrammatic illustration of the variation present in the DNA landscape of an average clone of the Bermuda land crab satellite. Stretches of G or C residues are as indicated; the large hypervariable region in line 0.8 is hatched, and a short region of internal repetition in line 1.3 is unshaded. [Redrawn from Skinner *et al.* (1982).]

there are only 17 Gs, whereas in the long one there are two G tracts of 23 and 18 bases.

In section 0.6, there is a C tract of 23 bases. Both the short and the long variants have 16- and 29-bp deletions in this general area.

In sections 0.7, 0.8, and 0.9 the long variant has six tandem repeats of a 142-bp sequence and this accounts almost entirely for its heretical length. Skinner *et al.* (1982) argue that unlike the insertion seen in the calf 1.711 satellite DNA, these five extra copies are due to internal replicative events.

In sections 1.2 and 1.3 the long variant has a 110-bp deletion.

In section 1.3, there are five tandem repeats of the sequence CGCAC following the C23 tract.

In section 1.5 downstream from the C13 tract, there are 5- and 10-bp inserts in the long and short variants.

The sequence in general also has an abundance of alternating purines and pyrimidines, GpC and CpG, which, it has been argued, could adopt Z-DNA configurations. As with most invertebrates, there is no methylation of cytosines or at best only minute amounts. This is not a peculiarity of satellite sequences, since the remainder of the genome is also unmethylated, but it does appear to be a property of many invertebrates. It may well be that satellite sequences in vertebrates are methylated by default, simply because the methylation pathways cannot avoid carrying out the modifications.

The satellite has no obvious internal substructure to indicate whetherit might have arisen from a primordial simple sequence variant. It does have potential open reading frames, which could give rise to proteins of 163, 97, and 49 amino acids. No analyses by Northern blot hybridizations are available, however, and hence whether it is transcribed or not has yet to be determined. Earlier indications yielded suggestive evidence that the sequence was homologous to RNA at a certain time in development, and this can now be exained using sensitive modern blotting techniques.

I have described this G- and C-rich crab sequence in what must seem to be excruciating detail, but it is an unrivalled example of sequence inhomogeneity *par excellence*. It would require some fortitude to propose that these sequences are conserved by natural selection between species, although the hybridization data do indicate that homologies exist between the Bermuda land crab satellites and the DNA of other species. This could mean no more than that it was inherited from a common ancestor and has undergone the vicissitudes attributable to DNA amplification, deletion, base changes, unequal exchange, and conversion phenomena. If indeed there has been conservation between species, it would require that variants be held in common, and this is certainly testable using cloned data from related species.

4. Data Summary

The molecular and biological data on highly repetitive sequences have been especially chosen from ten groups of organisms, each of which interlocks to produce an information network. The salient features of each group are briefly summarized below, so that the underlying themes will be at hand when their relevance is evaluated in the next section.

4.1. Kangaroo Rats

(a) The repeat units consist predominantly of simple sequence DNAs; (b) there are enormous differences in satellite DNA amounts among species in the face of developmental and morphological similarities; (c) common sequence variants are shared between kangaroo rats and guinea pigs; (d) extensive heterogeneity exists at the DNA sequence level in uncloned DNAs.

4.2. Amphibia

(a) Transcription of highly repetitive DNA sequences is due to readthrough from foreign promoters; (b) alien sequences insert into tandem arrays and are reamplified in newer repeating lengths; (c) there is significant inhomogeneity among clones.

4.3. The Calf

(a) Repeating units of short and very long lengths are based on a 23-bp (or lower) primordial unit; (b) insertion of alien sequences into tandem arrays can be followed by reamplification; (c) heterogeneity is found among cloned sequences and there is nonrandom positioning of perturbations within a repeat; (d) potential protein-coding sequences and transcriptional motifs are found in repeating units; (e) 5-methylcytosine occurs in repeating units.

4.4. Primates

(a) Repeating units are of both short and long length; (b) polymorphisms of sequence families are found among individuals without measurable phenotypic effects; (c) specific patterns of satellite DNA arrangements are found on individual chromosomes; (d) there is extensive heterogeneity at the DNA sequence level on individual chromosomes; (e) there is enormous variation among individuals for *deca*-satellite sequences of the African green monkey; (f) *deca*-satellite sequences are not detectable in the human genome and hence are not conserved among species; (g) short and long nomadic elements interrupt tandem arrays; (h) variants of α-satellite sequence types are found among most primate groups; (i) dyadic symmetries within α-satellite repeating units are not well conserved among species.

4.5. The Mouse

(a) The repeating unit is long, with a shorter internal substructure; (b) there is a huge reduction of satellite DNA amounts in *Mus spretus*, but similar sequence patterns to *M. musculus;* (c) chromosome pairing and meiosis are normal in fertile hybrids in which there are large differentials in the content of centromeric satellite DNA amounts; (d) *Mus* satellite sequences are absent from the Chinese hamster and hence are not conserved in certain rodent genomes; (e) some high mobility group proteins are absent from satellite DNA chromatin.

4.6. Rats

(a) There can be extensive genomic reorganizations in centromeric regions in the presence of minute quantities of centromerically located repeated DNA sequences; (b) there is heterogeneity among cloned repeats; (c) the amplification of larger units is from smaller ones; (d) similar sequences but of differing repeat length are found in a related species, *R. rattus;* (e) short consensus sequences are found between parts of *R. norvegicus* repeats and parts of vertebrate and invertebrate repeating units.

4.7. Sheep and Goats

(a) Translocation heterozygotes carrying centric rearrangements can be normal and fertile; (b) consensus sequence homologies are found between sheep and calf satellites.

4.8. *Drosophila*

(a) There are short and long repeating units within as well as between species; (b) chromosome-specific arrangements of different satellites are found in *D. melanogaster,* whereas all satellites are on only one giant chromosome in *D. nasutoides;* (c) the chromosome pairing–satellite DNA dogma is totally refuted by extensive experimental data; (d) a DNA-binding protein occurs for the 360 family of repeats; (e) deletion and addition of large amounts of repetitive DNA sequences leads to no impairment of early developmental events nor to somatic perturbations; (f) in Hawaiian *Drosophila* the consensus sequences are almost indistinguishable between very closely related species and there is no evidence for satellite DNA involvement in speciation events; (g) the internal transcriptional motifs in repeating units are inhomogeneous in Hawaiian *Drosophila;* (h) some species (*D. littoralis* and *D. ezoana*) have very small amounts of highly repetitive DNA, while others in the same group have investments totalling nearly half the genome (*D. virilis*); (i) the within-species variation for the 360 family in the *melanogaster* group is tenfold less than the between-species variation and provides evidence in favor of molecular drive; (j) the 180 family is restrited to *D. orena* and has already spread over all chromosomes; it is not conserved between species.

4.9. Grasshoppers

(a) The polymorphic variation within and between natural populations is almost unimaginable and is without overt phenotypic effects; (b) there are germ line effects due to satellite DNA-rich heterochromatin, which cause adjacent euchromatic regions to become *rec*$^-$; (c) equivalent meiotic recombination reductions are found in *Drosophila,* hedgehogs, and squirrels.

4.10. Crabs

(a) The repeat units vary form 2 to 2000 bp; (b) there is enormous variation among cloned repeats; (c) no 5-methycytosine residues occur in the large repeating units.

5. Epilogue

The general thrust of the research avenues concerned with highly repetitive sequences relate to two basic courts of appeal, the cellular and the evolutionary.

5.1. Cellular, Structural, and Mechanistic Approaches

This avenue has been the most fruitful because almost every hypothesis pertaining to cellular phenomena has been falsifiable in both vertebrate and invertebrate systems. The central emergent realization from the data on highly repetitive sequences is one of overwhelming variation at all levels. In general the power of the DNA sequence data has been in the absolute precision that it alone has brought to estimates of variation. These data make it difficult to assign cellular functions to classes of DNA sequence that tolerate almost every conceivable mutational insult. Furthermore, the polymorphisms of DNA amounts between *individual* humans, and the extensive sequence heterogeneity between *individual* African green monkeys, are extraordinarily crucial data sets. The enormous differences in satellite DNA amounts among species of kangaroo rats that are morphologically and developmentally very similar indicate that these sequences are oblique to developmental phenomena. The large experimental perturbations in *Drosophila melanogaster* also do not effect hard-wired developmental processes. Since the constraints of developmental biology hold many of the keys to macroevolutionary change, then if these DNAs do not impinge on development, they are unlikely to be pertinent to, let alone causal factors in, macroevolution.

Similarly, other variable aspects of these sequences lead to identical conclusions. Differences in repeating unit length, such as the basic 7-bp repeat in *D. virilis* and the 189-bp one in *D. grimshawi*, do not impinge on phenotype even though both constitute nearly half the genome. These two organisms are effectively equivalent in a developmental sense. Furthermore, it makes little difference to a final developmental package how these sequences are parceled out in a genome: *D. nasutoides* has them all on one giant chromosome; whereas *D. melanogaster* is catholic in its genomic distributions.

There is also metabolic normalcy when the amounts of these sequences vary in the genome. Thus African green monkey cell lines are viable when their α-satellite content is reduced from nearly 25% to 7% of the genome. *Mus spretus* is unperturbed with only 1% of the *M. musculus* satellite sequences. *Drosophila* males are somatically normal when over one-third of their highly repetitive sequences are experimentally excised from the genome. The polymorphisms in natural populations of grasshoppers also drive home these facts.

Most structural characteristics of these sequences, however, cannot be generalized. Thus, methylation of cytosines in CpG doublets occurs extensively in vertebrates in both coding and noncoding DNA, whereas in invertebrates it is almost nonexistent. Thus, methylation of satellite sequences in vertebrates, no matter how tissue-specific, is probably an epiphenomenon that highly repetitive DNA sequences cannot avoid. Clearly if there are common attributes of highly repetitive sequences between these two major divisions of organisms, they are unlikely to operate via methylation regimes.

Similarly, the invocation of higher order chromatin foldings, as in potential Z-DNA configurations and their associated effects, is seriously compromised in chromosome polymorphisms. If some individuals are totally missing tandem arrays on a chromosome, whereas others have millions of base pairs of a DNA family, then their importance to developmental, metabolic, or other phenotypic endpoints must be minimal.

We are effectively left with utilizing tandem arrays of sequences for purposes other than direct functional attacks. These sequences provide barometric indicators of genomic processes such as nucleosome phasing, which may be difficult to examine in other genomic compartments. It is noteworthy that it has been the molecular biology of internal genomic processes, together with the manipulative ability and genetic paradise provided by *D. melanogaster*, that has allowed the most progress at this level. These data have mainly eliminated general hypotheses and have left only the germ line as a potential sphere of influence (John and Miklos, 1979; Bostock, 1980; F. Muller *et al.*, 1982).

5.2. Evolutionary Concepts and Tandem Arrays: *Anguillam cauda tenes*

In contrast to these previous hypotheses, the evolutionary ones remain impervious to direct experimental test and there are a number of reasons for this. Evolution is conventionally thought to manifest itself in two major ways, by leading to adaptations and to new species. While adaptation is self-evident from even a cursory view of nature, it is very difficult to measure and evaluate (Lewontin, 1978; Gould and Lewontin, 1979). In speciation, on the other hand, it has not been possible to disentangle the events that gave rise to species during their formation from those that were irrelevant to these processes but have accumulated since species formation. This truism has been discussed for over 100 years without meaningful resolution. With these constraints in the foregound, let us reexamine the concepts of selection, neutrality, selfishness, and molecular drive and determine if there has been a significant interplay between the expansive DNA data base and these concepts.

5.2.1. Conventional Viewpoints: 1. Selection

In the neo-Darwinian view of evolution, genomes within a species produce variation that is acted upon by selection. Thus those individuals whose genomes produce phenotypes better adapted to the environment will be more successful than their less well adapted compatriots. The evolution of a species is seen as the accumulation of adaptive mutant genes and the characters that spring from them. When viewed in this light, a new alternative at a locus is better or less well adapted than the form form which it arises. Whether the new form makes it into the population depends on natural selection.

It has been enormously difficult to demonstrate convincingly for any protein-coding locus that positive selection has in fact occurred (MacIntyre, 1982). The problem is an even more demanding one for a tandem array of nontranscribed sequences. How is a variant among tens of thousands of other copies to be viewed by selection acting through the phenotype, especially when the data base shows that these arrays have negligible phenotypic effects and are readily tolerated as polymorphisms in natural situations? It is glib to discuss protein–DNA interactions in tandem arrays and to argue that the fixation of this or that variant has been achieved by selection. The *existence* of a homogeneous tandem array does not tell us that selection has been active. When a sequence arises and spreads through a population, one way in which it can do so is if it has a selective advantage due to its expression in phenotype. However, what happens if a newly arisen sequence is neither more or less advantageous than the old one?

5.2.2. Conventional Viewpoints: 2. Neutrality

The concept of neutrality was developed for genic sequences, but the principles are applicable to tandem arrays (Kimura, 1979). The first point is that neutral genes need *not* be functionless, only that different alleles are equally effective in promoting the survival and reproduction of the individual. One essential difference between selectionist and neutralist viewpoints is mirrored by the processes by which variants are substituted in a species. If a new sequence is selectively equivalent to ones that already exist, then the frequency of the new variant will fluctuate over time; some variants will be lost, while others will be fixed by drift. In terms of highly repetitive sequences it is necessary to determine if a new variant within an array modifies it in such a way as to promote the survival and reproduction of the individual. This is an awesome task even before we consider the spread of this variant through a population. Furthermore, the existence of functional differences between two forms is not evidence for the operation of natural selection, something that has to be rightly determined by survival rates and fecundity. More important, there is a real distinction between positive Darwinian selection and negative selection. Negative selection simply removes deleterious mutants and by and large does not impinge upon gene substitutions of evolution. Negative selection does not contradict the concepts of neutralism.

The neutral theory proposes that as functional constraint diminishes, the rate of evolution converges to the maximum set by the total mutation rate. In terms of the data on tandem arrays, can we negate neutrality? This is a critical question, because if we cannot distinguish among neutral, selective, and other modes of evolution for nontranscribed tandemly arranged sequences, then we are left with nucleotide sequence data that cannot be easily placed into an evolutionary framework.

5.2.3. Unconventional Viewpoints: 1. Selfish DNA

Doolittle and Sapienza (1980), Doolittle (1982), and Orgel and Crick (1980) have drawn attention to what has bogged down our evolutionary concepts of the genome. We have been wedded to thinking of it in functional terms and have considered natural selection as an omnipotent reaper that would rapidly slash any nonproductive DNA from the genome. The existence of highly repetitive sequences has inevitably led to a search for their function, and this in turn has revolved around a role in chromosome mechanics or chromosome architecture. The data, however, do not support either of these theories. Consequently, evolutionary avenues have been explored. Thus, highly repetitive sequences have been invoked in facilitating chromosome rearrangements, but once again the data base does not favor such excursions. Similarly, a role has been sought for them in speciation processes, but again no unbroken chain of data leads to this as a likely possibility. Doolittle and Sapienza (1980), Doolittle (1982), stressed that most "explanations" are unrealistic. If there are DNA sequences without effects on phenotype, then they can arise, spread, and be maintained in a functional vacuum. The data indicate that satellite sequences *per se* have no known phenotypic effects.

Every theory must have predictions and those put forward by Orgel and Crick (1980) for selfish DNA concern determinations of (1) the genomic positions of DNAs with no obvious functions; (2) the molecular mechanisms involved in DNA amplification and transposition; and (3) an evaluation of the nonspecific effects of such DNAs on cellular behavior. The first two of these "predictions" are known for tandemly repeated arrays, but they are insufficient for deciding among the actions of forces such as selection, neutrality, and selfishness for any given sequence array. The third prediction stresses *cellular* phenomena and since highly repetitive sequences have mainly germ line effects, their *consequences* are in the next generation, *not* on the cell or the organism itself.

How then does one decide among the various alternatives? Orgel and Crick (1980) suggested that DNA sequence information should help in this regard. As we have witnessed in the data base, there is now an excellent assemblage of cloned DNA sequence data from both vertebrate and invertebrate groups and it is insufficient for deciding among selection, neutrality, and selfishness. The selfish DNA hypothesis as applied to highly repetitive sequences (in contrast to nomadic elements) requires proving that different sequences within the same genome *compete on the basis of replicative efficiencies*. Nothing less will do, since there is yet another possibility, namely that turnover events simply address themselves to existing parts of tandem arrays. In this case the DNA is being manipulated by genomic mechanisms that it cannot avoid, and it is effectively "ignorant" (Dover, 1980; Dover and Doolittle, 1980).

We do not yet know in any eukaryotic genome how many of the repetitive DNA sequences arose by sequence-dependent or sequence-independent phenomena and the data base does not help in distinguishing these alternatives. The beneficial effect of this theory has been that certain cherished concepts on the superefficiency of genomes have been called into question and the mandatory search for functions for every single piece of genomic DNA has been courageously challenged. What is still required is to design experiments to differentiate among hypotheses, not to produce another consensus sequence or similar biological data from another part of a phylogeny. Whether the hypothesis of selfish DNA will ultimately prove to be a falsifiable hypothesis is still too early to tell. Its major stimulus has been in providing a novel and unconventional viewpoint about genome evolution.

5.2.4. Unconventional Viewpoints: 2. Molecular Drive

Molecular drive is an entirely different concept to selection, neutrality, and selfish DNA since it concerns cohesive *patterns* of change and their potential evolutionary consequences. Molecular drive is *not* differential survival in the conventional sense, since the population *as a whole* goes either toward survival or extinction. In molecular drive, the internal genomic turnover processes contrast with the external forces of selection and neutrality. The fixation of mutations in a population may occur as a result of a number of molecular processes *within the genome*. In general, we already know a fair amount at the moelcular level about three processes, transposition, unequal exchange, and conversion, which can all lead to nonreciprocal transfer of sequences between and within chromosomes. This transfer destroys the nexus between the behavior of chromosomes and the behavior of genes. Molecular drive is *not* just biased gene conversion; any processes that give rise to basic gain and loss events in a family of sequences can have consequences for variant fixation in a population. Molecular drive is even more than this, because turnover processes affecting multigene families can increase variants more or less simultaneously in each individual of a population. This is the concerted part of the process. What effects, however, can these internal processes acting on multigene or multisequence families have on phenotype? In the case of the ribosomal RNA genes, the genes themselves remain conserved between species but the spacers are changed. Since the spacers contain the promoter batteries, it is easy to see how concerted evolution of spacers could lead to differential expression of these genes during development. Significant phenotypic effects could be brought about by internal molecular genomic meddlings. Thus Dover (1984) has evaluated the example of the bat-eared foxes, in order to highlight the differences between the acquisition of characters under molecular drive of multigene families and the adaptionist viewpoint under natural selection.

The major problem with nontranscribed highly repetitive arrays, however, is quite different to that of multigene families. There is no quibble with the notion that turnover processes in *genic* families can have important phenotypic consequences. In our case we have to restrict ourselves to tandemly repeated arrays and ask questions about their homogeneity and their effects on phenotype. Highly repeated arrays are not in general transcribed, and in contrast to the *multigene* families, it is difficult to conceive of them in a phenotypic context.

Thus, while the predictions of molecular drive are concordant with the extensive sequence data from tandem arrays as well as from that of many multigene families, the role that turnover processes in satellite arrays have played in adaptation and speciation is speculative.

5.2.5. Speciation: *Ignotum per ignotius*

I have left the most intractable and emotive problem until the end, because it, too, has had highly repetitive sequences thrust upon it. To date, however, very few components of the genome, coding or noncoding, have been unequivocally demonstrated to have been the causative agent in either speciation processes or in adaptations (Doolittle, 1982; Dover, 1982; Rose and Doolittle, 1983; Dover, 1984).

In the Hawaiian *Drosophila* it was seen that two very closely related species, *D. gymnobasis* and *D. silvarentis*, shared the same amount and effectively the same sequence of their satellite DNAs. In this case at least, speciation had occurred in the absence of any major perturbations to the satellite DNAs at the nucleotide sequence level. Thus the assertion that repetitive DNAs, which have no known phenotypic effects, are prime movers in speciation processes is an ill-founded concept. It has already come to grief on the satellite DNA–chromosome pairing, as well as the satellite DNA–phenotypic effects, data bases. Unless another compelling data base provides a chain of testable events connecting satellite DNA to speciation events, this class of hypotheses remains to me an intellectually unacceptable refuge. It is not only impossible to falsify, but it has little predictive value. Most accept that speciation processes are the result of a plurality of causes, but it is difficult to see how we are going to progress past the generalities.

6. Conclusions

The cloned sequence data together with the biological data are impressive in their implications; they have negated *every* functional test to which satellite DNA has been put in a mechanistic cellular domain and they have seriously compromised many of the evolutionary concepts pertaining to this class of DNA. In this field there has been an almost overpowering reluctance to lay aside hypotheses that were erected in the absence of relevant data. There has been a desire to avoid the unpalatable, or the unthinkable—that tandem arrays of highly repetitive sequences have no functional characteristics whatsoever but are the ignorant playthings of genomic turnover processes. There is a fear that if not one hypothesis is left standing, then little of substance will remain in this field. However, highly repetitive DNAs have provided a vanguard for the problems that must be encountered in interfacing molecular biology with developmental and evolutionary biology at any level. These sequences have revealed many of the hurdles likely to be encountered in both coding and noncoding parts of the genome. These problems will not go away, nor will our understanding of genomic relevance be helped if the hypotheses that are erected are of little direct predictive value.

I believe that an acceptable state of affairs has been reached for highly repetitive localized arrays at the cellular level. Most hypotheses have been dispensed with, and we are confrontd with the stark realism that the obsessive hunt for function has been tangential and illusory.

The evolutionary hypotheses that have been left unscathed have rarely been tainted by direct data sources. It was not until the formulation of the concepts of selfish DNA and molecular drive that attention was focused onto genomic turnover processes and their evolutionary implications. Now fresh viewpoints are available on this rather unsalubrious state of affairs and it is the multigene families that have risen to prominence.

Nontranscribed highly repetitive DNA (satellite DNA) in a functional sense is dead. It has been replaced by an awareness that genomic flux is an everyday event and by an adage of the physicists, which is directly applicable to DNA and its retinue of enzymes: *if it is not biochemically forbidden, it will occur.* The important point for biologists is that given the internal mechanisms of genomic turnover, we need to determine their consequences on phenotype, or in the case of germ line phenomena, the consequences of their actions on genetic variability in succeeding generations. Our evolutionary impasse stems from being unable to assign the basis of macroevolutionary changes to specific genomic compartments, such as changes in single-copy or multigene families, or to known developmental circuits. At least the localized tandem arrays can be banished from this arena.

ACKNOWLEDGMENTS. I would like to thank those scientists whose ideas have led me to question the entrenched dogmas of eukaryotic genome organization and their biological relevance: S. Ohno, M. Kimura, G. Dover, R. Flavell, F. Doolittle, A. Bendich, and J. H. Campbell. I also wish to thank those colleagues who provided preprints, especially Maxine Singer and the members of her laboratory whose African green monkey data bore so critically on many hypotheses.

I thank all the members of the illustration and photographic sections of the Research School of Biological Sciences for their efficient and cheerful help, as well as Marilyn Miklos for her preparation of the manuscript.

Finally, I express deep gratitude to my colleague Bernard John, whose forbearance provided an atmosphere conducive to unshackled thinking and whose knowledge of biological phenomena greatly enriched my thoughts.

References

Arnason, U., Purdom, I. F., and Jones, K. W., 1978, Conservation and chromosomal localization of DNA satellites in Balenopterid whales, *Chromosoma (Berl.)* **66:**141–159.

Barnes, S. R., Webb, D. A., and Dover, G., 1978, The distribution of satellite and main-band DNA components in the *melanogaster* species subgroup of *Drosophila:* I. Fractionation of DNA in actinomycin D and distamycin A density gradients, *Chromosoma (Berl.)* **67:**341–363.

Beauchamp, R. S., Mitchell, A. R., Buckland, R. A., and Bostock, C. J., 1979, Specific arrangements of human satellite III DNA sequences in human chromosomes, *Chromosoma (Berl.)* **71:**153–166.

Bennett, M. D., 1971, The duration of meiosis, *Proc. R. Soc. Lond. B* **178:**277–299.

Bennett, M. D., 1982, Nucleotypic basis of the spatial ordering of chromosomes in eukaryotes and the implications of the order for genome evolution and phenotypic variation, in: *Genome Evolution* (G. A. Dover, and R. B. Flavell, eds.), Academic Press, London, pp. 239–261.

Berg, O. G., Winter, R. B., and Von Hippel, P. H., 1982, How do genome-regulatory proteins locate their DNA target sites?, *Trends Biochem. Sci.* **7:**52–55.

Bianchi, N. O., Vidal-Rioja, L., and Bianchi, M. S., 1973, Constitutive heterochromatin G-bands and Robertsonian rearrangements in the chromosomes of *Akodon molinae* (Rodentia–Cricetidae), *Can. J. Genet. Cytol.* **15:**855–861.

Billings, P. C., Orf, J. W., Palmer, D. K., Talmage, D. A., Pan, C. G., and Blumenfeld, M., 1979, Anomalous electrophoretic mobility of *Drosophila* phosphorylated H1 histone: Is it related to the compaction of satellite DNA into heterochromatin?, *Nucleic Acids Res.* **6:**2151–2164.

Blumenfeld, M., Orf, J. W., Sina, B. J., Kreber, R. A., Callahan, M. A., Mullins, J. I., and Snyder, L. A., 1978, Correlation between phosphorylated H1 histones and satellite DNAs in *Drosophila virilis, Proc. Natl. Acad. Sci. USA* **75:**866–870.

Bostock, C., 1980, A function for satellite DNA?, *Trends Biochem. Sci.* **5:**117–119.

Bostock, C. J., Gosden, J. R., and Mitchell, A. R., 1978, Localization of a male-specific DNA fragment to a sub-region of the human Y chromosome, *Nature* **272:**324–328.

Brown, S. D. M., and Dover, G. A., 1979, Conservation of sequences in related genomes of *Apodemus:* Constraints on the maintenance of satellite DNA sequences, *Nucleic Acids Res.* **6:**2423–2434.

Brown, S. D. M., and Dover, G. A., 1980a, The specific organization of satellite DNA sequences on the X-chromosome of *Mus musculus:* Partial independence of chromosome evolution, *Nucleic Acids Res.* **8:**781–792.

Brown, S. D. M., and Dover, G. A., 1980b, Conservation of segmental variants of satellite DNA of *Mus musculus* in a related species: *Mus spretus, Nature* **285:**47–49.

Brown, F. L., Musich, P. R. and Maio, J. J., 1979, The repetitive sequence structure of component α DNA and its relationship to the nucleosomes of the African green monkey, *J. Mol. Biol.* **131:**777–799.

Bruere, A. N., and Chapman, H. M., 1974, Double translocation heterozygosity and normal fertility in domestic sheep, *Cytogenet. Cell. Genet.* **13**:342–351.

Brutlag, D. L., 1980, Molecular arrangement and evolution of heterochromatic DNA, *Annu. Rev. Genet.* **14**:121–144.

Brutlag, D., Appels, R., Dennis, E. S., and Peacock, W. J., 1977, Highly repeated DNA in *Drosophila melanogaster, J. Mol. Biol.* **112**:31–47.

Buckland, R. A., 1983, Comparative structure and evolution of goat and sheep satellite I DNAs, *Nucleic Acids Res.* **11**:1349–1360.

Carlson, M., and Brutlag, D., 1978a, A gene adjacent to satellite DNA in *Drosophila melanogaster, Proc. Natl. Acad. Sci. USA* **75**:5898–5902.

Carlson, M., and Brutlag, D., 1978b, One of the *copia* genes is adjacent to satellite DNA in *Drosophila melanogaster, Cell* **15**:733–742.

Carson, H. L., Clayton, F. E., and Stalker, H. D., 1967, Karyotypic stability and speciation in Hawaiian *Drosophila, Proc. Natl. Acad. Sci. USA* **57**:1280–1285.

Cavalier-Smith, T., 1978, Nuclear volume control by nucleo-skeletal DNA, selection for cell volume and cell growth rate, and the solution of the DNA *C*-value paradox, *J. Cell. Sci.* **34**:247–278.

Cavalier-Smith, T., 1980, How selfish is DNA?, *Nature* **285**:617–618.

Chambers, C. A., Schell, M. P., and Skinner, D. M., 1978, The primary sequence of a crustacean satellite DNA containing a family of repeats, *Cell* **13**:97–110.

Christie, N. T., and Skinner, D. M., 1980a, Evidence for nonrandom alteration in a fraction of the highly repetitive DNA of a eukaryote, *Nucleic Acids Res.* **8**:279–298.

Christie, N. T., and Skinner, D. M., 1980b, Selective amplification of variants of a complex repeating unit in DNA of a crustacean, *Proc. Natl. Acad. Sci. USA* **77**:2786–2790.

Clayton, F. E., 1971, VIII Additional karyotypes of Hawaiian Drosophilidae, *Univ. Texas Publ.* **7103**:171–181.

Coen, E., Strachan, T., and Dover, G., 1982, Dynamics of concerted evolution of ribosomal DNA and histone gene families in the *melanogaster* species subgroup of *Drosophila, J. Mol. Biol.* **158**:17–35.

Cohen, E. H., and Bowman, S. C., 1979, Detection and location of three simple sequence DNAs in polytene chromosomes from *virilis* group species of *Drosophila, Chromosoma (Berl.)* **73**:327–355.

Cohen, A. K., Huh, T. Y., and Helleiner, C. W., 1973, Transcription of satellite DNA in mouse L-cells, *Can. J. Biochem.* **51**:529–532.

Cooke, H. J., and Hindley, J., 1979, Cloning of human satellite III DNA: Differnt components are on different chromosomes, *Nucleic Acids Res.* **6**:3177–3197.

Cooke, H. J., and McKay, R. D. G., 1978, Evolution of a human Y chromosome-specific repeated sequence, *Cell* **13**:453–460.

Cooper, K. W., 1959, Cytogenetic analysis of major heterochromatic elements (especially Xh and Y) in *Drosophila melanogaster* and the theory of "heterochromatin," *Chromosoma (Berl.)* **10**:535–588.

Cooper, K. W., 1964, Meiotic conjunctive elements not involving chiasmata, *Proc. Natl. Acad. Sci. USA* **52**:1248–1255.

Cordeiro-Stone, M., and Lee, C. S., 1976, Studies on the satellite DNAs of *Drosophila nasutoides:* Their buoyant densities, melting temperatures, reassociation rates and localizations in polytene chromosomes, *J. Mol. Biol.* **104**:1–24.

De Boer, P., and Groen, A., 1974, Fertility and meiotic behavior of male *T70H* teriary trisomics of the mouse (*Mus musculus*), *Cytogenet. Cell Genet.* **13**:489–510.

Deininger, P. L., Jolly, D. J., Rubin, C. M., Friedmann, T., and Schmid, C. W., 1981, Base sequence studies of 300 nucleotide renatured repeated human DNA clones, *J. Mol. Biol.* **151**:17–33.

Dev. V. G., Miller, D. A., Tantravahi, R., Schreck, R. R., Roderick, T. H., Erlanger, B. F., and Miller, O. J., 1975, Chromosome markers in *Mus musculus*: Differences in C-banding between the subspecies *M. m. musculus* and *M. m. molossinus, Chromosoma (Berl.)* **53**:335–344.

Diaz, M. O., Barsacchi-Pilone, G., Mahon, K. A., and Gall, J. G., 1981, Transcripts from both strands of a satellite DNA occur on lampbrush chromosome loops of the newt *Notophthalmus, Cell* **24**:649–659.

Donehower, L., and Gillespie, D., 1979, Restriction site periodicities in highly repetitive DNA of primates, *J. Mol. Biol.* **134**:805–834.

Donehower, L., Furlong, C., Gillespie, D., and Kurnit, D., 1980, DNA sequence of baboon highly repeated DNA: Evidence for evolution by nonrandom unequal crossovers, *Proc. Natl. Acad. Sci. USA* **77**:2129–2133.

Doolittle, W. F., 1982, Selfish DNA after fourteen months, in: *Genome Evolution* (G. A. Dover, and R. B. Flavell, eds.), Academic Press, London, pp. 3–28.

Doolittle, W. F., and Sapienza, C., 1980, Selfish genes, the phenotype paradigm and genome evolution, *Nature* **284**:601–603.

Dover, G., 1980, Ignorant DNA?, *Nature* **285**:618–619.

Dover, G., 1982, Molecular drive: A cohesive mode of species evolution, *Nature* **299**:111–117.

Dover, G. A., 1984, Forces of evolution. An origin of bat-eared foxes by means of molecular drive, *Kos* **5**:129–136.

Dover, G., and Doolittle, W. F., 1980, Modes of genome evolution, *Nature* **288**:646–647.

Dover, G., Brown, S., Coen, E., Dallas, J., Strachan, T., and Trick, M., 1982, The dynamics of genome evolution and species differentiation, in: *Genome Evolution* (G. A. Dover and R. B. Flavell, eds.), Academic Press, London, pp. 343–372.

Dowsett, A. P., and Young, M. W., 1982, Differing levels of dispersed repetitive DNA among closely related species of *Drosophila, Proc. Natl. Acad. Sci. USA* **79**:4570–4574.

Endow, S. A., and Gall, J. G., 1975, Differential replication of satellite DNA in polyploid tissues of *Drosophila virilis, Chromosoma (Berl.)* **50**:175–192.

Fitzgerald-Hayes, M., Clarke, L., and Carbon,J., 1982, Nucleotide sequence comparisons and functional analysis of yeast centromere DNAs, *Cell* **29**:235–244.

Flavell, R., 1982, Sequence amplification, deletion and rearrangement: Major sources of variation during species divergence, in: *Genome Evolution* (G. A. Dover, and R. B. Flavell, eds.), Academic Press, London, pp. 301–323.

Forejt, J., 1973, Centromeric heterochromatin polymorphism in the house mouse. Evidence from inbred strains and natural populations. *Chromosomes (Berl.)* **43**:187–201.

Forstova, J., Votavova, H., Guttmann, T., Pivec, L., and Doskocil, J., 1979, A comparative study of G+C rich satellite components of sheep and goat DNA, *Nucleic Acids. Res.* **6**:57–70.

Frommer, M., Prosser, J., Tkachuk, D., Reisner, A. R., and Vincent, P. C., 1982, Simple repeated sequences in human satellite DNA, *Nucleic Acids Res.* **10**:547–563.

Fry, K., and Salser, W., 1977, Nucleotide sequences of HS-α satellite DNA from kangaroo rat *Dipodomys ordii* and characterization of similar sequences in other rodents, *Cell* **12**:1069–1084.

Gaillard, C., Doly, J., Cortadas, J., and Bernardi, G., 1981, The primary structure of bovine satellite 1.715, *Nucleic Acids. Res.* **9**:6069–6082.

Gall, J. G., and Atherton, D. D., 1974, Satellite DNA sequences in *Drosophila virilis, J. Mol. Biol.* **85**:633–664.

Gautier, F., Bunemann, H., and Grotjahn, L., 1977, Analysis of calf-thymus satellite DNA: Evidence for specific methylation of cytosine in C–G sequences, *Eur. J. Biochem.* **80**:175–183.

Gershenson, S., 1933, Studies on the genetically inert region of the X-chromosome of

Drosophila. 1. Behavior of an X-chromosome deficient for a part of its insert region, *J. Genet.* **28**:297–313.

Gershenson, S., 1940, The nature of the so-called genetically inert parts of chromosomes, *Vid. Akad. Nauk USSR* **1940**:3–116 (in Ukrainian, with English summary).

Gosden, J. R., Mitchell, A. R., Buckland, R. A., Clayton, R. P., and Evans, H. J., 1975, The location of four human satellite DNAs on human chromosomes, *Exp. Cell Res.* **92**:148–158.

Gosden, J. R., Lawrie, S. S., and Cooke, H. J., 1980, A cloned repeated DNA sequence in human chromosome heteromorphisms, *Cytogenet. Cell Genet.* **29**:32–39.

Gould, S. J., and Lewontin, R. C., 1979, The spandrels of San Marco and the Panglossian paradigm: A critique of the adaptionist programme, *Proc. R. Soc. Lond. B* **205**:581–598.

Graf, H., Fittler, F., and Zachau, H. G., 1979, Studies on the organization of the α-satellite DNA from African green monkey cells using restriction nucleases and molecular cloning, *Gene* **5**:93–110.

Grimaldi, G., and Singer, M. F., 1983, Members of the *Kpn*I family of long interspersed repeated sequences join and interrupt α-satellite in the monkey genome, *Nucleic Acids Res.* **11**:321–338.

Gurley, C. R., Walters, R. A., Barham, S. S., and Deaven, L. L., 1978, Heterochromatin and histone phosphorylation, *Exp. Cell Res.* **111**:373–383.

Hatch, F. T., Bodner, A. J., Mazrimas, J. A., and Moore, D. H., 1976, Satellite DNA and cytogenetic evolution. DNA quantity, satellite DNA and karyotypic variations in kangaroo rats (*Genus Dipodomys*), *Chromosoma (Berl.)* **58**:155–168.

Hawley, R. S., 1980, Chromosomal sites necessary for normal levels of meiotic recombination in *Drosophila melanogaster*. 1. Evidence for and mapping of the sites, *Genetics* **94**:625–646.

Hazen, M. W., Kuo, M. T., and Arrighi, F. E., 1977, Genome analysis of *Peromyscus* (Rodentia–Cricetidae). VII Localization of satellite DNA sequences of *P. eremicus* on metaphase chromosomes, *Chromosoma (Berl.)* **64**:133–142.

Hesse, H., 1943, *The Glass Bead Game*, Penguin Books.

Hilliker, A. J., and Appels, R., 1982, Pleiotropic effects associated with the deletion of heterochromatin surrounding rDNA on the X chromosome of *Drosophila*, *Chromosoma (Berl.)* **86**:469–490.

Hilliker, A. J., Appels, R., and Schalet, A., 1980, The genetic analysis of *D. melanogaster* heterochromatin, *Cell* **21**:607–619.

Holmquist, G. P., and Comings, D. E., 1975, Sister chromatid exchange and chromosome organization based on a bromodeoxyuridine Giemsa-C-banding technique (TC-banding), *Chromosoma (Berl.)* **52**:245–259.

Horz, W., and Altenburger, W., 1981, Nucleotide sequence of mouse satellite DNA, *Nucleic Acids Res.* **9**:683–696.

Horz, W., Fittler, F., and Zachau, H. G., 1983, Sequence specific cleavage of African green monkey α-satellite DNA by micrococcal nuclease, *Nucleic Acids Res.* **11**:4275–4285.

Hsieh, T., and Brutlag, D. L., 1979, A protein that preferentially binds *Drosophila* satellite DNA, *Proc. Natl. Acad. Sci. USA* **76**:726–730.

Hutchinson, J., Rees, H., and Seal, A. G., 1979, An assay of the activity of supplementary DNA in *Lolium*, *Heredity* **43**(3):411–421.

John B., and King, M., 1983, Population cytogenetics of *Atractomorpha similis*. 1. C-Band variation, *Chromosoma (Berl.)* **88**:57–68.

John, B., and Miklos, G. L. G., 1979, Fucntional aspects of satellite DNA and heterochromatin, *Int. Rev. Cytol.* **58**:1–114.

Kao-Huang, Y., Revzin, A., Butler, A. P., O'Conner, P., Noble, D. W., and Von Hippel,

P., 1977, Non specific DNA binding of genome-regulating proteins as a biological control mechanism: Measurement of DNA-bound *Escherichia coli lac* repressor *in vivo, Proc. Natl. Acad. Sci. USA* **74**:4228–4232.

Kimura, M., 1979, The neutral theory of molecular evolution, *Sci. Am.* **241**:94–104.

Kurnit, D. M., 1979, Satellite DNA and heterochromatin variants: The case for unequal mitotic crossing over, *Hum. Genet.* **47**:169–186.

Kurnit, D. M., and Maio, J. J., 1974, Variable satellite DNA's in the African green monkey *Cercopithecus aethiops, Chromosoma (Berl.)* **45**:387–400.

Kurnit, D. M., Shafit, B. R., and Maio, J. J., 1973, Multiple satellite deoxyribonucleic acids in the calf and their relation to the sex chromosomes, *J. Mol. Biol.* **81**:273–284.

Lam, B. S., and Carroll, D., 1983, Tandemly repeated DNA sequences from *Xenopus laevis*. 1. Studies on sequence organization and variation in satellite I DNA (741 base-pair repeat), *J. Mol. Biol.* **165**:567–585.

Lapeyre, J., and Becker, F. F., 1980, Analysis of highly repeated DNA sequences of rat with *Eco*R1 endonuclease, *Biochim. Biophys. Acta* **607**:23–35.

Lee, C. S., 1981, Restriction enzyme analysis of a highly diverged satellite DNA from *Drosophila nasutoides, Chromosoma (Berl.)*. **83**:367–379.

Lee, T. N. H., and Singer, M. F., 1982, Structural organization of α-satellite DNA in a single monkey chromosome, *J. Mol. Biol.* **161**:323–342.

Lemeunier, F., and Ashburner, M., 1976, Relationships within the *melanogaster* species subgroup of the genus *Drosophila (Sophophora), Proc. Roy. Soc. Lond. B* **193**:275–294.

Lewontin, R. C., 1978, Adaptation, *Sci. Am.* **239**(3):213–230.

Lin, S., and Riggs, A. D., 1975, The general affinity of *lac* repressor for *E. coli* DNA: Implications for gene regulation in procaryotes and eucaryotes, *Cell* **4**:107–111.

MacIntyre, R. J., 1982, Regulatory genes and adaptation: Past, present and future, in: *Evolutionary Biology*, Vol. 15 (M. K. Hecht, B. Wallace, and G. T. Prance, eds.), Plenum Press, New York, pp. 247–285.

Maio, J. J., 1971, DNA strand reassociation and polyribonucleotide binding in the African green monkey, *Cercopithecus aethiops, J. Mol. Biol.* **56**:579–595.

Maio, J. J. Brown, F. L., and Musich, P. R., 1977, Subunit structure of chromatin and the organization of eukaryotic highly repetitive DNA: Recurrent periodicities and models for the evolutionary origins of repetitive DNA, *J. Mol. Biol.* **117**:637–655.

Maio, J. J., Brown, F. L., and Musich, F. R., 1981, Toward a molecular paleontology of primate genomes. I The *Hin*dIII and *Eco*R1 dimer families of alphoid DNAs, *Chromosoma (Berl.)* **83**:103–125.

Manuelidis, L., 1978, Chromosomal localization of complex and simple repeated human DNAs, *Chromosoma (Berl.)* **66**:23–32.

Manuelidis, L., 1982a, Repeated DNA sequences and nuclear structure, in: *Genome Evolution* (G. A. Dover, and R. B. Flavell, eds.), Academic Press, London, pp. 263–285.

Manuelidis, L., 1982b, Nucleotide sequence definition of a major human repeated DNA, the *Hin*dIII 1.9 kb family, *Nucleic Acids Res.* **10**:3211–3219.

Manuelidis, L., and Wu, J. C., 1978, Homology between human and simian repeated DNA, *Nature* **276**:92–94.

Maresca, A., and Singer, M. F., 1983, *Deca*-satellite: A highly polymorphic satellite that joins α-satellite in the African green monkey genome, *J. Mol. Biol.* **164**:493–511.
and molecular analyses of three rat genomes with special reference to chromosome rearrangement and speciation problems, *Chromosoma (Berl.)* **76**:339–363.

Matthews, H. R., Pearson, M. D., and MacLean, N., 1980, Cat satellite DNA. Isolation using netropsin with CsCl gradients, *Biochim. Biophys. Acta* **606**:228–235.

McCutchan, T., Hsu, H., Thayer, R. E., and Singer, M. F., 1982, Organization of African green monkey DNA at junctions between α-satellite and other DNA sequences, *J. Mol. Biol.* **157:**195–211.

McKay, R. D. G,, Bobrow, M., and Cooke, H. J., 1978a. The identification of a repeated DNA sequence involved in the karyotype polymorphism of the human Y chromosome, *Cytogenet. Cell Genet.* **21:**19–32.

McKay, R. D. G., Heritage, J., Bobrow, M., and Cooke, H. J., 1978b, Endonuclease analysis of Y chromosome DNA, *Cytogenet. Cell Genet.* **22:**357–358.

Miklos, G. L. G., 1982, Sequencing and manipulating highly repeated DNA, in: *Genome Evolution* (G. A. Dover and R. B. Flavell, eds.), Academic Press, London, pp. 41–68.

Miklos, G. L. G., and Gill, A. C., 1981, The DNA sequences of cloned complex satellite DNAs from Hawaiian *Drosophila* and their bearing on satellite DNA sequence conservation, *Chromosoma (Berl.)* **82:**409–427.

Miklos, G. L. G., and Gill, A. C., 1982, Nucleotide sequences of highly repeated DNAs: Compilation and comments, *Genet. Res. Camb.* **39:**1–30.

Miklos, G. L. G., and John B., 1979, Heterochromatin and satellite DNA in man: Properties and prospects, *Am. J. Hum. Genet.* **31:**264–280.

Miklos, G. L. G., and Nankivell, R. N., 1976, Telomeric satellite DNA functions in regulating recombination, *Chromosoma (Berl.)* **56:**143–167.

Miklos, G. L. G., Willcocks, D. A., and Baverstock, P. R., 1980, Restriction endonuclease and molecular analyses of three rat genomes with special reference to chromosome rearrangement and speciation problems, *Chromosoma (Berl.)* **76:**339–363.

Mitchell, A. R., Beauchamp, R. S., and Bostock, C. J., 1979, A study of sequence homologies in four satellite DNAs of man, *J. Mol. Biol.* **135:**127–149.

Muller, H. J., 1940, Bearings of the '*Drosophila*' work on systematics, in: *The New Systematics* (J. Huxley, ed.), Oxford University Press, pp. 185–268.

Muller, F., Walker, P., Aeby, P., Neuhaus, H., Felder, H., Back, E., and Tobler, H., 1982, Nucleotide sequence of satellite DNA contained in the eliminated genome of *Ascaris lumbricoides, Nucleic Acids Res.* **10:**7493–7510.

Mullins, J. I., and Blumenfeld, M., 1979, Satellite 1c: A possible link between the satellite DNAs of *D. virilis* and *D. melanogaster, Cell* **17:**615–621.

Musich, P. R., Brown, F. L., and Maio, J. J., 1980, Highly repetitive component α and related alphoid DNAs in man and monkeys, *Chromosoma (Berl.)* **80:**331–348.

Ohno, S., 1970, So much "junk" DNA in our genome, in: *Evolution of Genetic Systems,* Vol. 23 (H. H. Smith, ed.), Gordon and Breach, New York, pp. 366–370.

Orgel, L. E., and Crick, F. H. C., 1980, Selfish DNA: The ultimate parasite, *Nature* **284:**604–607.

Pages, M., and Roizes, G., 1982, Tissue specificity and organization of CpG methylation in calf satellite DNA I, *Nucleic Acids Res.* **10:**565–576.

Pardue, M. L., and Gall, J. G., 1970, Chromosomal localization of mouse satellite DNA, *Science* **168:**1356–1358.

Pathak, S., and Wurster-Hill, D. H., 1977, Distribution of constitutive heterochromatin in carnivores, *Cytogenet. Cell. Genet.* **18:**245–254.

Peacock, W. J., Lohe, A. R., Gerlach, W. L., Dunsmuir, P., Dennis, E. S., and Appels, R., 1978, Fine structure and evolution of DNA in heterochromatin, *Cold Spring Harbor Symp. Quant. Biol.* **42:**1121–1135.

Pech, M., Streeck, R. E., and Zachau, H. G., 1979a, Patchwork structure of a bovine satellite DNA, *Cell* **18:**883–893.

Pech, M., Igo-Kemenes, T., and Zachau, H. G., 1979b, Nucleotide sequence of a highly repetitive component of rat DNA, *Nucleic Acids Res.* **7:**417–432.

Poschl, E., and Streeck, R. E., 1980, Prototype sequence of bovine 1.720 satellite DNA, *J. Mol. Biol.* **143:**147–153.
Potter, S. S., and Jones, R. S., 1983, Unusual domains of human alphoid satellite DNA with contiguous non-satellite sequences: Sequence analysis of a junction region, *Nucleic Acids Res.* **11:**3137–3153.
Procunier, J. D., and Tartof, K. D., 1978, A genetic locus having *trans* and *contiguous cis* functions that control the disproportionate replication of ribosomal RNA genes in *Drosophila melanogaster, Genetics* **88:**67–79.
Pzucienniczak, A., Skowronski, J., and Jaworski, J., 1982, Nucleotide sequence of bovine 1.715 satellite DNA and its relation to other bovine satellite sequences, *J. Mol. Biol.* **158:**293–304.
Rees, H., Jenkins, G., Seal, A. G., and Hutchinson, J., 1982, Assays of the phenotypic effects of changes in DNA amounts, in: *Genome Evolution* (G. A. Dover and R. B. Flavell, eds.), Academic Press, London, pp. 287–297.
Reisner, A. H., and Bucholtz, C. A., 1983, Apparent relatedness of the main component of ovine 1.714 satellite DNA to bovine 1.715 satellite DNA, *EMBO J.* **2**(7)**:**1145–1149.
Renkawitz, R., 1978, Two highly repetitive DNA satellites of *Drosophila hydei* localized within the α-heterochromatin of specific chromosomes, *Chromosoma (Berl.)* **66:**237–248.
Rice, N. R., and Straus, N. A., 1973, Relatedness of mouse satellite deoxyribonucleic acid to deoxyribonucleic acid of various *Mus* species, *Proc. Natl. Acad. Sci. USA* **70:**3546–3550.
Roizes, G. P., and Pages, M., 1982, Conserved and divergent sequences of bovine satellite DNAs, in: *Genome Evolution* (G. A. Dover and R. B. Flavell, eds.), Academic Press, London, pp. 93–111.
Rose, M. R., and Doolittle, W. F., 1983, Molecular biological mechanisms of speciation, *Science* **220:**157–162.
Rosenberg, H., Singer, M. F., and Rosenberg, M., 1978, Highly reiterated sequences of SIMIANSIMIANSIMIANSIMIANSIMIAN, *Science* **200:**394–402.
Rubin, C. M., Deininger, P. L., Houck, C. M., and Schmid, C. W., 1980, A dimer satellite sequence in bonnet monkey DNA consists of distinct monomer subunits, *J. Mol. Biol.* **136:**151–167.
Salser, W., Bowen, S., Browne, D., El Adli, F., Fedoroff, N., Fry, K., Heindell, H., Paddock, G., Poon, R., Wallace, B., and Whitcome, P., 1976, Investigation of the organization of mammalian chromosomes at the DNA sequence level, *Fed. Proc.* **35**(1)**:**23–35.
Schmid, C. W., and Jelinek, W. R., 1982, The Alu family of dispersed repetitive sequences, *Science* **216:**1065–1070.
Schmidt, E. R., and Keyl, H. G., 1981, *In situ* binding of AT-rich repetitive DNA to the centromeric heterochromatin in polytene chromosomes of Chironomids, *Chromosoma (Berl.)* **82:**197–204.
Schopf, T. J. M., 1981, Evidence from findings of molecular biology with regard to the rapidity of genomic change: Implications for species durations, in: *Paleobotany, Paleoecology and Evolution,* Vol. 1, (K. J. Niklas, ed.), Praeger, New York, pp. 135–192.
Scaly, L., Hartley, J., Donelson, J., Chalkley, R., Hutchison, N., and Hamkalo, B., 1981, Characterization of a highly repetitive sequence DNA family in rat, *J. Mol. Biol.* **145:**291–318.
Shafit-Zagardo, B., Brown, F. L., Zavodny, P. J., and Maio, J. J., 1983, Transcription of the *Kpn*I families of long interspersed DNAs in human cells, *Nature* **304:**277–280.
Singer, M. F., 1982, Highly repeated sequences in mammalian genomes, *Int. Rev. Cytol.* **76:**67–112.

Skinner, D. M., 1977, Satellite DNAs, *BioScience* **27**:790–796.

Skinner, D. M., Bonnewell, V., and Fowler, R. F., 1982, Sites of divergence in the sequence of a complex satellite DNA and several cloned variants, *Cold Spring Harbor Symp. Quant. Biol.* **47**:1151–1157.

Southern, E. M., 1970, Base sequence and evolution of guinea-pig α-satellite DNA, *Nature* **227**:794–798.

Southern, E. M., 1975, Long range periodicities in mouse satellite DNA, *J. Mol. Biol.* **94**:51–69.

Steffensen, D. M., Appels, R., and Peacock, W. J., 1981, The distribution of two highly repeated DNA sequences within *Drosophila melanogaster* chromosomes, *Chromosoma (Berl.)* **82**:525–541.

Stephenson, E. C., Erba, H. P., and Gall, J. C., 1981, Histone gene clusters of the newt *Notophthalmus* are separated by long tracts of satellite DNA, *Cell* **24**:639–647.

Strachan, T., Coen, E., Webb, D., and Dover, G., 1982, Modes and rates of change of complex DNA families of *Drosophila, J. Mol. Biol.* **158**:37–54.

Strachan, T., Webb, D., and Dover, G., 1985, Transition stages during sequence turnover in multiple-copy DNA families in *Drosophila, EMBO J.* (submitted).

Streeck, R. E., 1981, Inserted sequences in bovine satellite DNA's *Science* **213**:443–445.

Takeda, Y., Ohlendorf, D. H., Anderson, W. F., and Matthews, B. W., 1983, DNA-binding proteins, *Science* **221**:1020–1026.

Taparowsky, E. J., and Gerbi, S. A., 1982, Sequence analysis of bovine satellite I DNA (1.715 gm/cm^3), *Nucleic Acids Res.* **10**:1271–1281.

Thayer, R. E., Singer, M. F., and McCutchan, T. F., 1981, Sequence relationships between single repeat units of highly reiterated African green monkey DNA, *Nucleic Acids Res.* **9**:169–181.

Valdez, R., Nadler, C. F., and Bunch, T. D., 1978, Evolution of wild sheep in Iran, *Evolution* **32**:56–72.

Varley, J. M., Macgregor, H. C., Nardi, I., Andrews, C., and Erba, H. P., 1980, Cytological evidence of transcription of highly repeated DNA sequences during the lampbrush stage in *Triturus cristatus carnifex, Chromosoma (Berl.)* **80**:289–307.

Wheeler, L. L., and Altenburg, L. C., 1977, Hoechst 33258 banding of *Drosophila nasutoides* metaphase chromosomes, *Chromosoma (Berl.)* **62**:351–360.

Wheeler, L. L., Arrighi, F., Cordeiro-Stone, M., and Lee, C. S., 1978, Localization of *Drosophila nasutoides* satellite DNAs in metaphase chromosomes, *Chromosoma (Berl.)* **70**:41–50.

White, R., Pasztor, L. M., and Hu, F., 1975, Mouse satellite DNA in noncentromeric heterochromatin of cultured cells, *Chromosoma (Berl.)* **50**:275–283.

Wiche, G., Corces, V. G., and Avila, J., 1978, Preferential binding of hog brain microtubule-associated proteins to mouse satellite versus bulk DNA preparations, *Nature* **273**:403–405.

Will, H., and Bautz, E. K. F., 1980, Immunological identification of a chromocenter-associated protein in polytene chromosomes of *Drosophila, Exp. Cell Res.* **125**:401–410.

Witney, F. R., and Furano, A. V., 1983, The independent evolution of two closely related satellite DNA elements in rats (*Rattus*), *Nucleic Acids Res.* **11**:291–304.

Wu, J. C., and Manuelidis, L., 1980, Sequence definition and organisation of a human repeated DNA, *J. Mol. Biol.* **142**:363–386.

Yamamoto, M., 1979a, Cytological studies of heterochromatin function in the *Drosophila melanogaster* male: Autosomal meiotic pairing, *Chromosoma (Berl.)* **72**:293–328.

Yamamoto, M., 1979b, Interchromosomal effects of heterochromatic deletions on recombination in *Drosophila melanogaster, Genetics* **93**:437–448.

Yamamoto, M., and Miklos, G. L. G., 1977, Genetic dissection of heterochromatin in

Drosophila: The role of basal X heterochromatin in meiotic sex chromosome behavior, *Chromosoma (Berl.)* **60:**283–296.
Yamamoto, M., and Miklos, G. L. G., 1978, Genetic studies on heterochromatin in *Drosophila melanogaster* and their implications for the functions of satellite DNA, *Chromosoma (Berl.)* **66:**71–98.
Young, B. S., Pession, A., Traverse, K. L., French, C., and Pardue, M. L., 1983, Telomere regions in *Drosophila* share complex DNA sequences with pericentric heterochromatin, *Cell* **34:**85–94.
Zhang, X. Y., and Horz, W., 1982, Analysis of highly purified satellite DNA containing chromatin from the mouse, *Nucleic Acids Res.* **10:**1481–1494.
Zhang, X. Y., Fittler, F., and Horz, W., 1983, Eight different highly specific nucleosome phases on α-satellite DNA in the African green monkey, *Nucleic Acids Res.* **11:**4287–4306.

CHAPTER 5

The Evolution of Interspersed Repetitive DNA Sequences in Mammals and Other Vertebrates

CARL W. SCHMID and CHE-KUN JAMES SHEN

1. Introduction

The major class of interspersed repeats in the DNA of higher organisms closely resembles the structure of processed RNA pseudogenes. These repetitive sequences result from a reinsertion of transcribed sequences into the genome and are shown to be mobile at least on an evolutionary time scale. We term such sequences as inserted repeats.

This review primarily describes the inserted repeats in rodent (mostly mouse) and primate (mostly human), which are the most extensively studied vertebrates. As judged by either mass or number, there are two major families of inserted repeats, which dominate the rodent and primate genomes. The rodent and primate versions of each of these two families are ancestrally related. These two families differ in a variety of properties, such as repetition frequency, length, transcriptional activity and most importantly primary structure. However, they share a number of features that indirectly demonstrate that they are dispersed throughout the genome by very similar mechanisms.

Most of these features as well as other properties of repetitive DNAs are exemplified by the human *Alu* family. We summarize available human *Alu* sequences to derive an accurate genomic consensus. *Alu* family mem-

CARL W. SCHMID • Department of Chemistry, University of California, Davis, California 95616. *CHE-KUN JAMES SHEN* • Department of Genetics, University of California, Davis, California 95616.

bers are found to insert preferentially at genomic sites that are A-rich and homologous to the extreme 5′ end of the *Alu* sequence.

A second, less abundant class of interspersed repeats consists of tandemly repeated oligonucleotides. This second class of repeats probably results from a different mechanism than the one that generates the inserted repeats. Possible biological functions for both classes of interspersed repeats are briefly considered. Although the biological functions of these sequences are still unknown, we now recognize a number of significant genetic effects attributable to interspersed repeats. In particular cases, the presence or insertion of either class of interspersed repeats results in genetic instability in flanking regions.

2. Background

2.1. DNA Renaturation Studies

Repetitive sequences were first identified in the seminal DNA renaturation rate study of Britten and Kohne (1968). Until the advent of molecular cloning techniques, subsequent studies on repetitive sequences continued to rely on renaturation techniques. This older literature, which has been well reviewed (Davidson *et al.*, 1975; Davidson and Britten, 1973; Britten and Davidson, 1971), is now largely superseded by modern cloning methods and will not be considered in detail here. However, a few generalizations from this literature provide the background for our present considerations.

The fraction of the vertebrate genome that is composed of repetitive DNA is quite varied, 40–80% for amphibians and 20–40% for primates (Britten and Davidson, 1971). There are two broad organizational classes of repetitive DNA, namely tandem repeats, which are commonly called satellite DNAs, and repeats that are interspersed with single-copy sequences. Satellite DNAs undergo rapid amplifications in evolution, so that they account for much of the variation in the repetitive sequence content of closely related species. For example, 25% of the African green monkey genome is composed of α-satellite DNA (Kurnit and Maio, 1973), whereas only 1–2% of the human genome is composed of this sequence (Manuelidis, 1976).

Because they are present in multiple copies, repetitive sequences are excellent candidates to perform any of a number of possible biological functions. Since single-copy DNA includes many structural genes, this suggests an association between interspersed repeats and the expression of any flanking single-copy genes. In many cases (e.g., human, *Xenopus*, and sea urchin,) this interspersed DNA sequence arrangement consists of short repetitive sequences (typically 300 bp in length), which are interspersed with roughly "gene-size" single-copy sequences (1000–2000 bp in length) (Schmid and Deininger, 1975 Deininger and Schmid, 1976; Davidson *et al.*, 1975). In typical examples over half of an organism genome is occupied by this close interspersion of repetitive and single-copy sequences. The close interspersion pattern does not preclude other types of interspersed sequence arrangements, including, as we shall see below, the presence of extremely long repeated sequences. Nor is this close interspersion pattern found in all eukaryotic genomes. Exceptions are *Drosophila*, in which long repeats are interspersed with long, single-copy sequences (Manning *et al.*, 1975), and *Achlya*, which is essentially devoid of interspersed repeats (Hudspeth *et al.*, 1977). Surprisingly, the short interspersion pattern is not even shared by all vertebrate. In chicken DNA short repeated sequences are interspersed with rather long single-copy DNA sequences (Stumph *et al.*, 1981). We are permitted only the weak generalization that all vertebrate genomes that have been studied contain short interspersed repeats.

The repetition frequencies of interspersed repeats within a genome span a tremendous range. DNA renaturation rate studies suggest that some families of repeats contain as few as 30 members (Davidson *et al.*, 1975), whereas the human *Alu* family is present in excess of 500,000 copies (Rinehart *et al.*, 1981). The lower limit approximates the size of many multigene families and it is now recognized that many interspersed repeats are in fact members of multigene families (see below). The higher value of 500,000 copies is also noteworthy. Such a prolific family of sequences is arithmetically sufficient to account for much of the interspersion of repetitive DNA within the human genome (Schmid and Jelinek, 1982; Jelinek and Schmid, 1982). Hyperabundant repeat families have not been detected in such lower eukaryotes as sea urchin (Davidson *et al.*, 1975) and, as discussed below, hyperabundant repeats may be restricted to mammals.

Repetitive sequences that cross-hybridize to form stable duplex structures are not necessarily identical sequences. The melting temperature of renatured repetitive DNA is lower than the melting temperature of perfectly paired duplex molecules (Britten and Dvaidson, 1971). This difference in melting temperature has been correctly attributed to the base pair mismatching between inexact sequence complements. It is therefore useful to regard a group of repeats that cross-hybridize as a family of related but not identical indivduals. In an example presented here, members of one particular family of repeats differ by about 20% in base sequence.

The diversity of members of a family of repeats might suggest that these sequences are poorly conserved, the implication being that they are not subject to selective pressure and are not functional. In this case we would expect families of sequences to be poorly conserved between divergent species. Alternatively such sequence differences could provide functional diversity within a family of otherwise similar sequence. In this case the differences in sequence among members of a family might be ancient and predate the divergence of related species.

These ideas have been tested by cross-renaturing the repetitive sequences of related species to form interspecies DNA heteroduplexes. One finding of such studies is that relative to single-copy sequences repetitive sequence families are conserved between such related species as mouse and rat (Rice, 1971) and human and monkey (Deininger and Schmid, 1979). A second finding is that these renatured DNA heteroduplexes exhibit interspecies base pair mismatching in addition to the intraspecies mismatching (Deininger and Schmid, 1979). The members of a family of repeats must therefore be homogenized for species-specific sequence features.

2.2. Summary of Renaturation Studies

Based on this older literature it is interesting to ask how members of a repeat family differ in sequence. Are mutations among the member sequences random or are some regions relatively well conserved? Do different members of a family form closely related subfamilies, suggesting a specialized function? In what ways do new members of a family or even new families appear in the genome? To what extent are similar families shared by divergent species and how do these families differ?

Modern recombinant DNA techniques have permitted substantial progress toward answering these as well as other questions that we did not know enough to raise. The remainder of this review will concentrate on these newer findings.

3. RNA Pseudogenes

3.1. Pseudogenes Resulting from Processed mRNA

Many interspersed repeats are now recognized as either being RNA pseudogenes or resembling RNA pseudogenes in various details. For this reason we will briefly review those features that identify processed pseudogenes resulting from structural genes prior to considering interspersed repeats.

Processed pseudogenes resemble the mature mRNA more closely than the parent gene in that their intervening sequences have been removed by processing, their 3′ end is polyadenylated, and their 5′ end is defined by the transcriptional start site. A hallmark of processed genes is that they are flanked by short direct repeats, which result from a duplication of existing chromosomal sequences upon insertion of the pseudogene. The flanking short direct repeats are typically 10–20 nucleotides in length. In examples discussed below (Section 4.2.1) there is direct evidence that the direct repeats result from a duplication of the insertion site. The structure of processed pseudogenes can thus be schematically depicted as

5′. . . direct repeat, transcriptional start, coding sequence, poly A, direct repeat. . . 3′

Processed pseudogenes that exhibit some or all of the features outlined above include human genes for λ immunoglobulin, β tubulin, and metallothionein, rat genes for cytochrome *c* and α tubulin, and a mouse α-globin gene (Hollis *et al.*, 1982; Nishioka *et al.*, 1980; Wilde *et al.*, 1982; Karin and Richards, 1982; Lemishka and Sharp, 1982; Scarpulla and Wu, 1983; Vanin *et al.*, 1980). The polyadenylation, processing of intervening sequences, and, most importantly, the correspondence of the 5′ end of the pseudogene to the transcriptional start site are compelling evidence that these pseudogenes are formed through an mRNA intermediate. Studies on small RNA pseudogenes suggest that the RNA intermediate is converted to a complementary DNA (cDNA) prior to its insertion into the genome.

3.2. Small RNA Pseudogenes Are Recognizably Interspersed Repeats

The genes and pseudogenes for a number of small RNAs (e.g., U1, U2, U3, U4, and U6) fill all of the required properties of interspersed repeats (Van Ardsell *et al.*, 1981; Hammarström *et al.*, 1982, Hayashi, 1981). They are sequences present in high copy number (~1000–10,000) broadly distributed throughout the genome and correspond in size to typical short interspersed repeats [reviewed in Van Ardsell *et al.* (1981)]. There are a number of distinct types of small RNA pseudogenes (Denison and Weiner, 1982). Here we are only concerned with those pseudogenes that result from RNA intermediates. These may be recognized by combinations of some or all of the following features: 5′ ends that correspond to the transcriptional start site; a polyadenylated 3′ end; and flanking short direct repeats (Van Ardsell *et al.*, 1981; Jagadeeswaran *et al.*, 1981).

Usually these pseudogenes are full-length sequences, whereas one class of U3 RNA pseudogenes has severely truncated 3′ ends (Bernstein *et al.*, 1983). This 3′ truncation has been attributed to the folded secondary structure that the U3 RNA adopts upon self-priming for reverse transcription (Bernstein *et al.*, 1983). Full-length molecules generated by reverse transcription could initiate at the 3′ end of the RNA intermediate. For polyadenylated pseudogenes, the A-rich 3′ ends that ultimately result from this process are variable in both length and sequence; in many cases they resemble a long tandem repetition of a short, A-rich oligonucleotide. A suggestion is that reverse transcriptase stutters when it initiates on the relatively unstable priming interaction.

Given that some small RNA pseudogenes are in effect a class of interspersed repeats, to what extent do they typify other families of interspersed repeats? To distinguish between other possible classes of interspersed repeats and those that are inserted into the genome by an RNA-dependent mechanism, we shall refer to the latter as "inserted repeats." Examples of these repeats that are inserted by way of an RNA intermediate include the RNA pseudogenes described above and the *Alu* family and other known groups of interspersed repeats described below.

4. The *Alu* Family and Its Allies

4.1. Structure of the Human *Alu* Family

Renaturation rate kinetics predicted the existence of an extremely abundant family of 300-bp interspersed repeats in human DNA. This prediction has been confirmed by the identification of the *Alu* family of repeats (Houck *et al.*, 1979), which consists of at least 500,000 member sequences (Rinehart *et al.*, 1981). As anticipated from the results of renaturation studies, individual members of this family are related but not identical sequences. A consensus sequence for the *Alu* family has been derived from clones constructed from renatured members of the family that had been isolated by an S1 digestion procedure (Fig. 1) (Deininger *et al.*, 1981). The use of S1 nuclease in this experiment nibbled the 5′ ends of the sequences and could conceivably lead to a bias against exceptionally divergent members of the family. Further, any one clone resulting from this procedure is an unknown average of the two heteroduplexed strands and might not be typical of any one *Alu* family member. For these reasons we have constructed a genomic consensus sequence based on the many *Alu* family sequences in the literature. This is shown in Fig. 1.

The 25 *Alu*'s reported in Fig. 1 are full-length sequences. We are also aware of three partial *Alu*'s that have suffered a 5′ truncation (Houck,

```
Position--------  1  2                                                            51  52                                                                         101
                     **    *          *                                               *                                         *
BLUR consensus         GGTGGGCGTG  GTGGCTCACA  CCTGTAATCC  CAGCACTTTG  GGAGGCCGAG  GTGGGTG  GAT  CACCTGAGGT  CAGGAGTTCA  AGACCAGCCT  GGCCAACATG
GENOMIC consensus G    GCcGGGCGcG  GTGGCTCACg  CCTGTAATCC  CAGCACTTTG  GGAGGCCGAG  GCGGGcg  GAT  CACCTGAGGT  CAGGAGTTCG  AGACCAGCCT  GGCCAACATG
LAMBDA 4               ..T.T...T.. ...T.....A  ...A......  .....T....  ....A.T...  C....T.  ...  ...T....CC  ..........  ..........  .-........
GROWTH HORMONE         .AA...A..A  ....T.....  ..........  .....A....  .......A..  .T...TA  ...  .........A.  T.......G.  ..........  ......T...
HGH V2                 .AA....C.A  ........TA  ..........  .....A....  ..........  .......  ...  T.........  ......A.G.  ........A..  ......T...
ACTH a             taG ..G....A.A  ..........  ..........  .......C..  ......T...  T.....A  ...  ...--.....  ..A...A.G.  ....ATT...  .......G...
ACTH b             tgG ..T...T.T.  ......A..A  ..........  ........CA  ...TA.T..T  .T.----  ...  T..T....CC  .......C.T.  ..........  A.C......A
ACTH c             ctG ......G...  .......G..  .........G.  ...A......  ..G....A..  .......  ...  .......TT..  .G......T.  ..........  .A........
ACTH d             gaG A..A....A.  A.......T.  G.........  .........T  A..T..TA..  .T...A.  ...  TG.T....CC  ....GA...A  ....A.....  ..G......A
ACTH f             agG .......AT.  .........A  ...A......  ..........  .......A..  ..A...A  ...  ...A......  .........A  ..........  .......TATA
INSULIN            gaG ..T...T...  ........T.  ...A......  .........A  ......T...  ......A  ...  ..........  .G.......A  ..........  .A......G.
G gamma            gaG ..T...A.T.  ..........  ..........  ...A.T....  .......A..  ..A...A  ...  ..........  ..A......A  .......A...  ..........
delta globin       ctG ..T..AT...  ........G.  .T.....A..  ..........  .......A..  ..A...A  ...  ...T......  .........A  ..........  .A........
delta A gamma      ttG ..T...T.T.  .........CG .....G....  .........C  A.........  .A...T.  ...  ........T..  T.CC......  ...T......  .A........
alpha 1            tcC .A.A...A..  ..........  ..........  ..........  ..........  .T...A.  ...  ..........  .G......T.  .......C...  .AT.......
alpha 2            tcC .A.A...A..  .........A  ..........  ...T......  ......T...  ...A.A.  ...  ..........  CG......T.  ..........  .A....T...
alpha 2 dimer 1    aaG ..........  T........-  G.........  ......AT..  .A....T...  ..T...A  ...  GC---.....  .......A...  .......T...  C..T..TGCA
alpha 2 dimer 2    ttG ...A...---  TC...G....  ....G.....  ..........  ..........  .T...T.G ...  ..G...T...  .G........  ........G..  .A........
pseudo alpha       ggG ..T...T.T.  ..........  .....C....  ..........  ....A..-..  .A....A  ...  ...T......  ........T.  ..........  A........A
3 prime alpha      ttG ..........  .........A  ........TG  ..........  CC....TT..  ..A..T.  ...  CA......A..  ....G.....  ..........  A........A
RANDOM             tgG ...A..T.T.  .........A  ..........  ..........  ..........  .T...T.  ...  ..........  ...A.....A  ...-T...A.  ..........
PROTHROMBIN 1      caG .T.A..T..A  .........C  ..........  ..........  .......A..  A.A..AA  ...  ...T.....G  ..........  ..G.....T.  ..GA..ATG.
PROTHROMBIN 2      tgG ......T.T.  ........T.  ..........  .G........  ......T...  ..A..T.  ...  ...T......  .........A  ..........  ..........
PROTHROMBIN 3      ggG ...A.....A  ......T..A  ..........  ...A......  ..........  CT....A  ...  ....G..ACA  -........A  ..T.AG..AA  ..........
PROTHROMBIN 4      atG ..T......A  ...C.....C  .........T  .....G....  ..........  .....A.  ...  ...--A....  ........T.  ...T......  ..........
pJP53              agG ..TA......  --..T.....  ..........  .....T....  ......T...  A....T.  ...  ..--......  .......A...  ......T...  ...T......
Globin Region      ttG A---..T...  .........T  .........A  .....G....  ...C...A..  .T....A  ...  ...T......  ........G.  .C.......C  A.........
```

(continued)

Figure 1. Genomic consensus derived from 0.005% of the human *Alu's*. The Blur consensus, which results from S1-treated sequences (Deininger *et al.*, 1981), is constructed from nine cloned sequences, not all of which are full length. Consequently, this consensus is somewhat less certain over the first 20 nt and does not define the 5′ end. The genomic consensus represents the most frequent base of the 25 clones listed below. Positions of consensus uncertainty are designated by a lowercase letter, undefined positions by the letter n, deletions by a hyphen, and insertions by either a number of word. The names of individual *Alu's* usually refer to neighboring genes found on the same clone as the repeat. These are as follows: λ4 (C. Houck, unpublished); growth hormone (Seeburg, 1982); ACTH corticotropin β-lipotropin (Tsukada *et al.*, 1982); insulin (Bell *et al.*, 1980); γ^G globin (Duncan, 1981); δγ globin (Jagadeeswaran *et al.*, 1982); α1, α2, and a dimeric pair 5′ to the α2 globin (Hess *et al.*, 1983), pseudo α globin (Sawada *et al.*, 1983), 3′ α globin (Shen and Maniatis, 1982), random (E. Paulson, unpublished results), pJP53 (Elder *et al.*, 1981), globin region (Shen and Smithies, 1982), prothrombin (Friezner Degen *et al.*, 1983).

```
Position--------     102                                                     150      151                                                          200
                                                                *                               **
BLUR consensus       GTGAAACCCC CTCTCTACTA AAAATACAAA AA TTAGCCG GGCTTGGTGG      CG---CGCGCCTG TAATCCCAGC TACTCGGGAG GCTGAGGCAG GAGAATCGCT
GENOMIC consensus    GTGAAACCCC GTCTCTACTA AAAATACAAA AA TTAGCTG GGcGTGGTGG      cg---CATGCCTG TAATCCCAGC TACTCGGGAG GCTGAGGCAG GAGAATCGCT
LAMBDA 4             .CA......T .......... .......... A.. ....... AA.A......     .A---........ .GG.T..... ....A.A... .......... A.....T...
GROWTH HORMONE       .......... ........C. ....A..... .. ....... A..C....CA      T.---........ G.......A. A......... .......... ..........
HGH V2               .......... ........C. .......... .. ....... A..C....CT      T.---........ G.......A. ..T....... A......... ..........
ACTH a               .C........ .......... .......... .. ....... ..........      ..---..CA.... ..G....... .........T. .......... .C........
ACTH b               .........T ........C. ....C(INSERT). ....... ...A......     .A---TGCAT... .GG.T....G .....A.... .....A...T ...G...A..
ACTH c               .A.....T.. ...A...... .......... .. .....C. ..........      ..---........ .......... .......... .....A.... ..........
ACTH d               ....G...T. CACT.....A. ......A... .G .....CA .AT.......     .A---TG...... ..G...T... ....T..... .TT------- ------.A..
ACTH f               ......TA.. A......G.. T.....A.T. .. .A...T. ..T.......      ..---GGC..... .......... .....T.... C......... .....C....
INSULIN              .A........ A...T..... ...C...... -. ....... ..T.......      .A---........ .........A ..T....... .......... ..........
G gamma              ......T... A.......A. -......... .. ....A.A ...A..A...      .A---AG...... .......... ....T..... ........A.. ......T...
delta globin         .......... A......... .......... -. .C...C. ......-...      T.---........ C.G....... ..T..A..T. .......... ....C.T...
dalta A gamma        .......... C.A....... ......G... -. .....CA ...A......      T.GCA......-. ..G.G..... ....T..... .......... ..........
alpha 1              TA........ A...A..... .......... -. .C..... .AT.GT.GCC      -----........ ...A...... .....C.... .......... .......AT.
alpha 2              .A........ AGT.A..... .......... -. ....... ..T....GC.      -----........ ......T... ....A..A.. .......... ..........
alpha 2 dimer 1      .......T.T ........T.. ....A.G... A.. ......AC .A.TA.TG..GGGC GT.---T.GC..A. CT.-.T.... G......... ........G.. ......G..G
alpha 2 dimer 2      .A........ .......... .......... .. ....... -.T.......      T.---..C..... .......... .......... .TGG....T. .......CT.
pseudo alpha         .......TGT ........T.. .........T. .. .....CA ..AA...C..     .C---........ .......... ...A...... A.A....... ..........
3 prime alpha        .........T .......... .......... .. ..GT.CA ..T...A..A      .T---........ ...A..TG.. -...TT..GAG ..G...---- ----------
RANDOM               .....G...T ........-. .......... .. ....... ...A..A..C      G.---GA-..... ....T..... .......... .......... ..........
PROTHROMBIN 1        ..A....... .......... .....G.... .. .....CA ..T..A.CA.      .A---TG.C.... ..G....... ....AA.... .........G. ...G.....C
PROTHROMBIN 2        .........T .........T .......... .. .....C.. ..T.......     ..---....T... ...ATT.... .....A.... ..........A ........T.
PROTHROMBIN 3        AA........ .......... .......... -. .....AA ..........      ..---TG...... .......... ..TAT.A... .........- .........C
PROTHROBMIN 4        .......... .......... .C.TA.A... A.. .....CA ..A.......     .A---..C..... ......T... .....A.... .....A.... .G........
pPJ53                .......... .......... .......... .. CC...C. A.T.......      T.---GGCA.... ..G....... .....A.... .......... .......GA..
Globin Region        ........TG ......G..TA ......T... .. ....... .A.......A      ..---GGA..... .......-.A. ....T..... .......... ..........
```

```
Position--------     201                                                  246  247                                                     296
                          *          *                           * *                              *                        *
BLUR consensus       TGAACCCAGG  AGGTGGAGGT  TGCAGTGAGC  CGA-GATCGC-  ---GCCACTG  CACTC---CAECC  TGGGCAACA-  GAGCGAGACT  CCATCTCAAA  AAAAAAAAAAA
GENOMIC consensus    TGAACCcgGG  AGGCgGAGGT  TGCAGTGAGC  CgA-GATTGn-  ---GCCACTG  CACTC---CACCC  TGGGtaACA-  GAGcgAGACT  CCGTCTCAAA  AAAAAAAAAAA
LAMBDA 4             ......TT...  ...T......  ..........  TA.-.....AT  GGA.....A.  ...C.---.....  .A...G...-  ..ATA.....  .T.C......  ..........G
GROWTH HORMONE       .......A..  .........A.  ..........  .A.-....GT-  ---.......  .....---....T  ...T.CC.-G  A.T--....C  ...C
HGH V2               ..GG...A..  .........A.  .....C....  .A.-..C.---  ---.......  .....---.....  ...T.CC.-G  AAT--....C  .T.
ACTH a               ......T...  ..........  ..........  .A.-...C.C-  ---.......  .....---.....  ...-CG...-  C...A.....  ..A...A...  ........
ACTH b               ...G...A..  ...T.....C  ..........  A.T-GT.CAC-  ---A.TG...  ..T..---.....  ....C....-  ...T.....C  .TAAT.....  ..
ACTH c               ..........  .....A....  ...G......  A..-...CAC-  ---C..TT..  .T...---....A  ....C....A  ......A...  ..........  ........
ACTH d               ...GG..A..  ..T.T..C.A  CA....A...  TAT-...CA.-  ---A...T..  .....---...T.  .........-  ..AT.....C  TT........  .C....C....T
ACTH f               ......T...  ...TA.....  ..........  .A.-...CAC-  ---A......  .....---..Ɛ..  A...C....-  ....A.....  ..........  ....T..
INSULIN              ......T...  .A..A.....  ...GC.....  ...-...G.C-  ---A...T..  .....---.....  ....C...GA  ......A...  ..........  ....C.....
G gamma              .A....T..A  ....A.GA..  ..........  ...-...CAT-  ---A......  .....---.....  .....G...-  ..A.A.....  .T........  .........G
delta globin         .......A..  ....A.....  ...G......  .T.-.....C-  ---A...T..  .....---T...T  ....C..T.G  .GAT..A...  ..A.....G.  .G.G....G.
delta A gamma        .C....T...  ....A..T..  ..........  .TG-...CAT-  ---....T..  .....---.....  ....C...CA  T...A.TG..  ..A.....G.  ..........
alpha 1              .T....AA..  ....A.....  ..........  TA.-...CAC-  ---A...T..  .....---.....  ...AA....A  C.....A...  ...C......  ........
alpha 2              ..........  ...TA.....  ..AG......  T..-...CAC-  ---....T..  .....---.....  ..C.A....A  ....A.A...  ..........  ...T..T..T
alpha 2 dimer 1      ......T.C.  ...T....C.  ..........  ...-....T--  ---.......  .....---.....  .....C...-  ..........  ..........  ..........
alpha 2 dimer 2      ......A...  ..TT......  ..........  ...-...C.--  --T.....A.  .....---T....  ..A-CG...-  C.........  .T........  ....T...T.T
pseudo alpha         .C.....A..  ...TTC...C  ..........  .A.-A.C.---  --T.......  .....---.....  G....GT..-  ....A..GGC  .TA...T...  ..C.....C..
3 prime alpha        --------------------..  ..T......T  .A.-......-  ----...TC.  .....---....T  ....C....A  ......A...  .T........  ......
RANDOM               ...-......  ....A.....  ..T...A...  T..-...CAC-  ---A...T..  .....---.....  .....G...-  ...T......  ..A.......  ..........
PROTHROMBIN 1        ...G...AA.  ....T....C  .T....A...  T..-CTG.---  ---A...T..  .....---.....  .....G...A  ...T.....C  .T........  ..T...T...
PROTHROMBIN 2        .........A  ....A..A..  ..........  T..-...C.  27 nts      .....---.....  .A...G...-  ....A.....  ..A...A.G.  ..C.......
PROTHROMBIN 3        .......A..  ...T......  ..........  .A.-....AT-  ---....T..  .....---.....  .....G.---  ...T......  ..A.......  ..........
PROTHROMBIN 4        .......A..  ..........  ..........  T..-...A.--  --G.......  .....CAA.....  ....C..GG-  ...T......  ..T..T..G  ...G......
pJP53                ..A...T..-  ...T....C.  ..........  .A.C......-  ---....... T..-..----.T.A  .....G...-  ...A......  ..........  ..........
Globin Region        ......G...  ...T....T.  ....C.....  A.. ...CAT      ....T..  .....---.....  .-----C..-  ..........  .T....A..G  .....CG...
```

Figure 1. (continued)

C., unpublished results; Tsukada *et al.*, 1982; Baralle *et al.*, 1980). Interestingly, the truncation sites are nearly identical for two of the three partial *Alu*'s.

There is good agreement between the genomic and renatured DNA consensus sequences in Fig. 1. The two consensus sequences differ in 15 out of 296 sites. Half of these differences involve sites where the sequence is highly variable, so that the assignments are within the uncertainty of the average. At other sites where there is a difference, the base reported for the Blur (i.e., S1-treated) consensus is usually present at an appreciable frequency in the genomic clones. Since there is no significant difference between the consensus sequences derived by these two different approaches, we will only briefly review those sequence features that are relevant to the issues raised here. The original literature and subsequent reviews contain more detailed sequence analysis (Deininger *et al.*, 1981; Schmid and Jelinek, 1982; Jelinek and Schmid, 1982).

The average divergence of individual cloned members of the *Alu* family from the consensus is about 14% (Fig. 1). Large deletions or substitutions are rare and most sequence differences are point mutations (usually transitions) from the consensus sequence. The positions of these mutations are noticeably clustered at certain sites, whereas other sites are relatively well conserved. The divergence at positions 120–130, 180–190, 220–230, and 247–257 is only about 5%, as shown in Fig. 1. Jelinek *et al.* (1980) highlighted positions 180–190 as resembling the SV40 replication origin/T-antigen binding site and reported that it is well conserved in the evolution of divergent species, human and hamster. The highly conserved nine-nucleotide palindrome at positions 85–93 and 251–259 occupies a structurally interesting region within the repeat, which is described below.

Individual cloned members of the *Alu* family share a precisely defined 5′ end, which is designated position +1 in Fig. 1. The length and sequence of the A-rich 3′ end are variable. For graphical reasons many of these A-rich 3′ ends have been arbitrarily truncated in the representation depicted in Fig. 1 and in their fully extended forms consist of long tandem arrays of A-rich oligonucleotides (Jelinek and Schmid, 1982; Schmid and Jelinek, 1982).

The presence of the 3′ A-rich region suggests that these sequences may be inserted repeats. Most *Alu* family members are flanked by recognizable short direct repeats (see Fig. 2). The complex transcriptional activity of *Alu* family members is discussed in detail in Sections 4.3.1 and 6.2. However, for our present purposes two points are immediately relevant. First, *Alu* family members are actively transcribed *in vitro* by RNA polymerase III (Duncan *et al.*, 1979; Elder *et al.*, 1981; Fuhrman *et al.*, 1981; Shen and Maniatis, 1982). Second, the start site of this *in vitro*

```
5' DIRECT REPEAT ALU start site....(ALU).....poly a........DIRECT REPEAT 3'
                 *                                          *
  AAACAAGCAGGAGA CGCT   INSULIN             AAACAAGCAGGAGG GGCT
                  *  *                                    *  *
AAGATTCACTTGTTTAGA GGCT   G gamma        AAGATTCACTTGTTTAG- GCCT
                   *                                       *
           TTCTT GGCT   delta A/gamma                TTCTT -GCT
                 *                                      *
   AAGACCTTATCCT Ggct   delta globin          AAGACCTTATC-T Gtta
     *  **       *                              *  **       *
   AAACAGTTGCGGG GGcT   pseudo alpha          AAACACTT--GGG GGaT
      *     *                                      *    *
 AATAAAACTAAAAAT Cgac   alpha 2            AATAAA-CTAAAA-T Ctat
      *   *                                         *   *
    AAAGTGATGGTTT ggcc   alpha 2 dimer          AAAGTCATGATTT tttgggggg
                  *                                           *
        TAAAAATT GGCc   3 prime alpha               TAAAAATT AGCa
              *     *                                     *     *
     AATCAAGCCAGT CGCT   RANDOM                   AATCAAGCTAGT C-CT
    ***                                           ***
 AAAAAGCAGGAGGCA Ggtc   PROTHROMBIN            AAAAAGTGTGAGGCA Gccc
        **                                            **
AGAAGTAAAGGACCCAT ggcc   PROTHROMBIN         AGAAGTAAACAACCCAT ctca
          **                                           **
 AAGACCAATACCAGG ggcc   PROTHROMBIN            AAGACCAACCCCAGG aatg
       *       *                                     *       *
AAAATGTTTAGATAAG ggct   pJP53                AAAA-GTTTAGATAAA acaa
                  *                                            *
  AAAAGTGTGATTC TCGC   truncated alu            AAAAGTGTGATTC T-GC
         *                                                 *
  CATCAAATTCCTT GAgg   Globin Region            CATCAAATTCCCT GAcc
         *                                                 *
   AATCACTGGGTA GGcg   ACTH a                    AATCACTGGGCA GGtt
       *     *                                          *    *
       AAGAAACTG ggct   ACTH b                        AACAAACAG aatg
            *   *                                          *   *
  AAAAGACCGGGCT GgCC   ACTH c                   AAAAGACCGGGTT GtCC
       *     *                                          *    *
       CACACTGA gacc   ACTH d                         CATACTGT tagt
   *           *                                   *      *
  AATGAATATTGAG ggcc   ACTH f                   AAAGAATATTGAA ataa
```

Figure 2. Evolution of direct repeats flanking human *Alu* members. The direct repeats are written as capital letters. The initial four nucleotides of each *Alu* family member (e.g., GGCT for insulin) are offset from the remaining flanking direct repeat by a space. Asterisks mark differences, hyphens mark deletions, and lowercase letters designate the end of the duplicated unit. The individual *Alu's* are referenced in Fig. 1. The dinucleotide and trinucleotide sequences TG and CA(C) in the direct repeats are underlined.

transcript closely approximates the start site of the *Alu* family consensus depicted in Fig. 1 (see Duncan *et al.*, 1981). It must, however, be noted that no *discrete in vivo* polymerase III transcript of the human *Alu* family has yet been detected (Weiner, 1980; Jelinek and Schmid, 1982; Schmid and Jelinek, 1982). In summary, *Alu* family members usually share a precise 5′ end corresponding to a transcription start site, have variable A-rich 3′ ends, and are flanked by short direct repeats. This fulfills the conditions that identify an inserted RNA pseudogene. No discrete RNA molecule is known to be coded for by a 300-bp *Alu* family member. In this limited sense *Alu* family members fail to fulfill one condition in our previous description of small RNA pseudogenes. We conclude that *Alu* family members, like many small RNA pseudogenes, are inserted repeats (Van Ardsell *et al.*, 1981; Jagadeeswaran *et al.*, 1981).

4.2. Direct Repeats Flanking *Alu* Family Members

4.2.1. The Origin of Flanking Direct Repeats

The flanking short direct repeats are proposed to be duplicated upon insertion of a repeat into an unoccupied genomic site (Bell *et al.*, 1980; Van Ardsell *et al.*, 1981; Jagadeeswaran *et al.*, 1981). The duplication occurs by way of a staggered break in the genomic DNA to reveal cohesive ends, which are subsequently subjected to repair replication. In five separate cases we can compare the unoccupied genomic insertion site and the direct repeats that result from the insertion of an *Alu*-like family member. These comparisons are shown in Fig. 3. In agreement with the insertion models of Van Ardsell *et al.* (1981) and Jagadeeswaran *et al.* (1981), these direct repeats are clearly derived from preexisting genomic sequences.

4.2.2. Structure of Flanking Direct Repeats

The direct repeats of different *Alu* family members share a number of features (Fig. 2). We define the direct repeats as those sequences that immediately flank and may include the 5′ terminus of the *Alu* family member (Fig. 1) and also occur following the 3′ poly A-rich tail of the *Alu* family. As discussed below, some investigators choose to exclude the beginning of the *Alu* family member from the direct repeat, so that their definition of the direct repeat is different than that used here. In most cases, the 3′ end of the direct repeat includes homology with the extreme 5′ end of the *Alu* family. The clearest example is the *Alu* family member that flanks the human insulin gene (Bell *et al.*, 1980). The 3′ end of this short direct repeat GGCT (Fig. 2) is also the 5′ end of the insulin *Alu* family (Fig. 1). In other cases, e.g., the δ-globin *Alu*, the homology only includes the first nucleotide of the *Alu* family member. Especially interesting examples of this type include the direct repeats flanking the α2 *Alu*, the random *Alu*, and the truncated *Alu*. In each of these cases the 5′ end of the *Alu* sequence is different than the consensus as shown in Figs. 1 and 2, yet appears to be part of the direct repeat. The direct repeats of α2 dimer *Alu* and the third prothrombin *Alu* may or may not include the beginning of the *Alu* family member, depending on how they

Comparison Source	Pre-existing Site	5' Direct Repeat	3' Direct Repeat
Rat tubulin gene and pseudo gene	AAGACCAAGCGTACC	AAGACCAAGCGTACC	AAGACCAAGCGTACC
Region flanking duplicate human α globins	AAA*ACAT*CACTTT	AAA*TCAT*GACTTT	AAA*CCAT*CACTTT
monkey α satellite consensus	aaG*GCYGTTCTTGTGg	aaG*ACTGTTCTTGTGg	aaG*ACTGTTCTTGTGg
mouse ribosomol spacer	AAAAGAAACCTCCC	AAAAGAAACCTCCC	AAAAGAAACCTCCC
rat prolactin gene	AAAAAATGCGCTACTCT	AAAAAATGCGCTAATCT	AAAAAATGCGCTAATCT

Figure 3. Comparison of flanking direct repeats and the ancestral genomic target site. The direct repeats of an *Alu*-like sequence inserted into a rat α-tubulin pseudogene is compared to the same site in the functional gene (Lemishka and Sharp, 1982). The direct repeats flanking a dimeric *Alu* pair 5′ to the human α2-globin gene is compared to the homologous site flanking the α1-globin gene (Hess *et al.*, 1983). The direct repeats of an *Alu* inserted into the African green monkey α satellite is compared to the consensus α-satellite sequence (Grimaldi *et al.*, 1982). This last direct repeat is represented in a slightly different form than that by the original authors to reemphasize two points made in the text: The ancestral 5′ dinucleotide "aa" may be duplicated upon insertion or may merely resemble the 3′ end of the poly A tail of the *Alu*. The underlined dinucleotide Gg is homologous to the 5′ end of this *Alu* family member. The direct repeats of an *Alu*-like sequence inserted into the mouse ribosomal spacer is compared to an unoccupied spacer (Kominami *et al.*, 1983a). The polymorphic flanking regions of rat prolactin gene variants that do and do not contain an *Alu*-like element are compared (Schuler *et al.*, 1983).

are represented. Direct repeats that clearly do not duplicate one or more bases at the beginning of the *Alu* repeat are the second prothrombin example, clone pJP53, and three ACTH examples (Fig. 2). It is easy to imagine that the insertion of a repetitive sequence at a staggered chromosomal break is assisted by one or more steps involving homologous recombination.

A second feature of the flanking direct repeats is that their 5′ ends always include an A-rich run, as indicated in Fig. 2. Since most *Alu* family members end in an A-rich 3′ end, which naturally precedes the subsequent

3′ direct repeat, this causes some ambiguity in identifying the short direct repeat. If the 5′ flanking direct repeat happened to be preceded by an A-rich region, it could be mistakenly identified as part of the overall direct repeat. According to this idea, it may be more accurate to represent the first prothrombin *Alu* direct repeat as . . .GAGGCA, which is perfectly repeated about this particular *Alu* family member (Fig. 2). As another example of this ambiguity, notice that the direct repeats of the α-satellite *Alu* family member (Fig. 3) are *ancestrally* preceded by the dinucleotide AA. Does AA flanking the 3′ direct repeat result from polyadenylation or duplication of the target site? Since this is occasionally a matter of subjective judgment, we generally assume that any additional homology involving A-rich regions is part of the duplication. Whether or not this assumption is correct, we can conclude that direct repeats either begin with or are preceded by an A-rich region (Fig. 2).

A third interesting characteristic of the direct repeats is that many (15 out of 20) contain the nucleotide sequences CA(C) or TG. These di- or trinucleotide sequences are underlined in Fig. 2. They are located 7–12 nucleotides away from the 5′ end of the *Alu* repeat. As mentioned above, there are two different ways to identify and assign the direct repeats for the *Alu* family repeats. If we assume that the A-rich regions are not part of, but instead precede, the direct repeats, then many of these CA(C) or TG sequences would correspond to the 5′ ends of the direct repeats. It is most interesting to note here that several eukaryotic transposons—*copia* in *Drosophila* (Levis *et al.*, 1980), Ty 1 of yeast (Gafner and Philippsen, 1980; Farabaugh and Fink, 1980), and retrovirus chicken spleen necrosis virus (Shimotono et al., 1980), consist of the sequence TG(T) at their 5′ termini and CA at their 3′ termini.

4.2.3. Evolution of Short Direct Repeats

The 5′ and 3′ flanking direct repeats are inexact duplicates. Fukumaki *et al.* (1983) reason that these mutations must be introduced either during or following the duplication of the direct repeats. The number of base differences between the 5′ and 3′ direct repeats might provide a clock to estimate the average age of *Alu* family members. From a compilation of nine *Alu* family repeats Fukumaki *et al.* (1983) estimate the average divergence between the 5′ and 3′ flanking direct repeats to be 5% The compilation of 20 pairs of direct repeats analyzed in Fig. 2 suggests a 12% divergence (34 differences/287 bases) between the 5′ and 3′ repeats. This substantial difference in estimated values for the divergence (5% vs. 12%) primarily results from a different representation of direct repeats by Fukumaki *et al.* and by the analysis of Fig. 2. They truncate the 3′ end of the repeat to exclude homology of the *Alu* family. For example, they exclude GGCT for the insulin *Alu* family member, in contrast to the representation in Fig. 2. As a second contrasting example, they represent the direct repeat of the γ *Alu* as ending in TAG, which introduces an extra base (the mismatched A in Fig. 2) between the *Alu* family member and the 5′ direct repeat. Instead, the same direct repeat can be extended to include the mismatched tetranucleotides GGCT and GCCT, which are homologous to the 5′ end of the *Alu* family. Since this is again a matter of judgment, it would be just as reasonable to end the direct repeat at the first G of the *Alu* family and avoid one more mismatch. By extending the direct repeat further in the 3′ direction than Fukumaki *et al.* (1983), we arrive at a higher level of mismatching than they. For the reasons given above, we are confident that the direct repeat should include some homology to the 5′ terminal sequence of the *Alu*; how much is debatable.

A somewhat different issue is to use the divergence of the 5′ and 3′ direct repeats as a measure of the time elapsed since insertion of the *Alu* family member. For this purpose, Fukumaki *et al.*'s analysis may be preferable to our own. Insertion of the *Alu* member may be error-prone and could, for example, result in single-base deletions and insertions at the junction between the direct repeat and the *Alu* family member. It is noteworthy that this is the case in many of the examples of Fig. 2. Since such concomitant mutations are unrelated to the subsequent mutational time of the direct repeat, we would expect the value for the divergence estimated here to be somewhat greater than the value estimated by Fukumaki *et al.* (1983).

Since at present we cannot decide the exact value for this divergence, we will employ the range of values suggested by the alternate analyses: 8 ± 4% per pair of direct repeats or 4 ± 2% per sequence.

Individual *Alu* members differ by 14% from the consensus (Fig. 1). Assuming that the repeats diverge at the same rate as the flanking direct repeats, then individual *Alu* family members would differ from the consensus by about 10 ± 2% (14% − 4 ± 2%) at the time of their insertion. This initial sequence diversity must result from a very large number of genes that can code for the insertion of new *Alu*'s (Fukumaki *et al.*, 1983). As discussed below, nearly every *Alu* member has an internal RNA polymerase III promoter, so that each may be competent to code for subsequent family members (Van Ardsell *et al.*, 1983; Jagadeeswaren *et al.*, 1983).

Based on the divergence of the short direct repeats, we argued above that individual *Alu* family members have resided in the genome for a time sufficient to accumulate 4 ± 2% average sequence divergence. Assuming the present population of *Alu* repeats is at steady state, then individual members would accumulate an average divergence of 8 ± 4% prior to their removal from the genome. Eight percent divergence happens to approximate the divergence of human and monkey single-copy DNAs (Kohne *et al.*, 1972; Hoyer *et al.*, 1972). Without reviewing the arithmetic and its underlying assumptions, the implication of this comparison is that roughly half of the 500,000 ancestral primate repeats have turned over since the separation of the human of monkey lineages. Such a turnover rate is certain to have significant evolutionary effects and we believe, as discussed in Section 6, that some of these effects have been observed.

4.3. Nonhuman *Alu*-like Sequences

4.3.1. Rodent *Alu*-like Families

The *Alu* family is not restricted to the human genome, and related 300-bp-long repeats have been observed in Old World monkeys, New World monkeys, and prosimians (Grimaldi and Singer, 1982; Daniels *et al.*, 1983; Daniels and Deininger, 1983). A closely related family of sequences has also been observed in Chinese hamster (called type 1 *Alu* equivalent) and in mouse (called the B1 family) (Haynes *et al.*, 1981; Krayev *et al.*, 1980). As described for the human *Alu* family, members of these rodent families begin with precisely defined 5′ ends, terminate in an A-rich 3′ end of variable length and sequence, and are usually flanked by short direct repeats. There is a major structural difference between the rodent and human *Alu* family repeats. The rodent repeats are only 130 bp in length, considerably shorter than the 300-bp human sequence. The human *Alu* family consists of a head-to-tail dimer of 130-bp monomer units as shown in Fig. 4 (Deininger *et al.*, 1981; Schmid and Jelinek, 1982). Thirty bp of extra DNA in the right monomer unit completes the 300-bp sequence (Fig. 4). The data do not distinguish between an insertion of the extra 30 bp of DNA in the right monomer or a deletion of 30 bp from the left monomer. For purposes of discussion we shall nominally call it an insertion. The type 1 *Alu* equivalent in hamster and the nearly identical B1 mouse family are monomeric *Alu* sequences, which, allowing for a 30-bp insertion, best match the right human monomer (Fig. 4). The nine-nucleotide palindromic sequence immediately following the human insertion and preceding the CHO/mouse insertion is evolutionarily well conserved (see position 85 in Fig. 4). (see also Jelinek and Schmid, 1982). We previously noted that this same palindrome is well conserved among the individual members of the human *Alu* family (positions 251–259, Fig. 1).

A second type of short interspersed repeat has been observed in rat,

```
                          1                                                   51
HUMAN monomer 1           --GGCcGGGCGt-GGTGGCTCACgCCTGTAATCCCAGC-ACTTTGGGAGGCCGAG
                          ***  *             *  *               *   **       *
HUMAN monomer 2           TTAGCTGGGcGT-GGTGGcgCATGCCTGTAATCCCAGCTACTC-GGGAGGCTGAG
                          134                                                186
CHO TYPE 1                ----CCAGGCATTGGTGGyACACACCTTTAGTCCCAGC-ACTCAGG-AGGCAGAC
                          #### ##   # #     ##  ##   #  #       #    #  #    #

                          52                             84
HUMAN monomer 1           GCGGGtgGATCACCTGAGGTCAGGAGTTCGAGA----------------------
                            *  * *   * *   *** *    ** **
HUMAN monomer 2           GCAGGAGAATCGCTTGAACCcgGGAGG-C-gGAGGTTGCAGTGAGCCgAGATTGn
                          187                           215
CHO TYPE 1                GCAGGAGGATCACTT--------GAGTTCnAGAG---------------------
                                 #   # #          ## ##

                                     85
HUMAN monomer 1           -----------CCAGCCTGG-----------------------------------
HUMAN monomer 2           GCCACTGCACTCCAGCCTGG-----------------------------------
CHO TYPE 1                -----------CCAGCCTGGTCTACCAnAGTTCCTGAGTTCAAG--CCAGGCTA
                                        + +   +   +  ++  +   +        +++
fragment type 2 sequence         xxxxCCTGACTGCTCTTCCGAAGGTCC-GAGTTCAATTCCCAGxxxx

                          94                                    133
HUMAN monomer 1           CCAACATGGTGAAACCCCGTCTCTACTAAAAATACAAAAA
                          ***   ** *  *  *
HUMAN monomer 2           GtgACAGAGcgAGACTCCGTCTC ( A RICH TAIL )
                          260                 282
CHO TYPE 1                -TA-CAGAG--AAACCCTGTCT  ( A RICH TAIL )
                          # ##     ## #  # #
```

Figure 4. Comparison of the human *Alu* and rodent equivalent sequences. The genomic consensus of the human *Alu* family (Fig. 1) can be represented as a heat-to-tail dimer with a 33-nt insertion at position 217. Asterisks mark differences between the human monomers. The CHO type 1 *Alu* (Haynes *et al.*, 1981) is a monomeric sequence that contains a 32-nt insertion relative to either human monomer. Differences between the CHO type 1 *Alu* and the second human monomer are designated by the symbol #. A fragment of the rodent B2/type 2 *Alu*-like consensus sequence is closely related to this CHO type 1 32-nt insertion (Krayev *et al.*, 1982). Differences between the CHO type 1 *Alu* and B2-type 2 rodent sequence are designated by a plus sign. The underlined regions designated promoter-like elements as discussed in the text.

Chinese hamster (called type 2 *Alu* equivalent), and mouse (called B2) (Page *et al.*, 1981; Haynes *et al.*, 1981; Haynes and Jelinek, 1981; Krayev *et al.*, 1982). The type 2 (B2) rodent families are not homologous to the human *Alu* family (except for a promoter region discussed below), but they do share significantly homology with the type 1 (B1) rodent families. The region of maximum homology between the type 2 (B2) and type 1 (B1) sequences includes the region previously described as the rodent insertion sequence (Fig. 4).

The human *Alu* dimer is transcribed *in vitro* by RNA polymerase III by virtue of a promoter active in the left monomer unit (Elder *et al.*, 1981; Duncan *et al.*, 1981; Fuhrman *et al.*, 1981). Interestingly, the corresponding promoter region in the nearly homologous right monomer unit is inactive. Elder *et al.* (1981) identify the sequence GAGTTCGAGACC, which spans position 84 in Fig. 4, as part of the RNA polymerase III internal promoter. This promoter element abuts and includes part of the well-conserved nine-nucleotide palindrome described above (Fig. 4). The homologous element in the right monomer (about position 215) is interrupted by the 30-nucleotide insertion sequence and has two single-base length mutations (near position 214) compared to the left monomer promoter element. Any of these differences might contribute to the inactivation of the right promoter element.

A member of the type 1 rodent family was found to be transcriptionally inactive by RNA polymerase III *in vitro*, whereas the type 2 sequences are actively transcribed (Haynes and Jelinek, 1981). [Since another type 1 sequence is actively transcribed *in vivo* (Young *et al.*, 1982), it is not certain whether all type 1 sequences are inactive *in vitro*.] Further, the type 2 *Alu* equivalent or a closely related homologue in hamster has been shown to code to a small RNA *in vivo* (Haynes and Jelinek, 1981). To our knowledge the rodent polymerase III promoter sequences have not yet been mapped. Sequences homologous to the human *Alu* promoter element are present in both the type 2 and type 1 rodent sequence (Fig. 4). This element, GAGTTCAATCC, maps at position 57 in the B2/type 2 rodent equivalent and at both positions 62 and 102 in the type 1 sequence (both underlined in Fig. 4). Since the promoter has not yet been mapped in type 1 (B1) *Alu*'s, we do not know which of these elements are or are not active. There is also partial homology between the sequences that immediately precede the human promoter (near position 74, Fig. 4) and this putatitive type 2 promoter.

The highly conserved nine-nucleotide palindrome (position 85, Fig. 4), appears to be both a hot spot for insertions into the ancestral *Alu* family sequence as well as a part of the *Pol*III internal promoter. It is noteworthy that human 7S RNA is a sequence homologue to a monomeric *Alu* family member. Included in the 7S RNA is an extra sequence of 140 bases inserted at precisely the same position as the 30-bp insertion into the second human monomer (Ullu *et al.*, 1982). The nine-nucleotide palindrome is perfectly conserved in the 7S sequence at the boundary of this 140-base insertion.

4.3.2. Nonhuman Primate *Alu* Families

Dimeric *Alu* repeats resembling the human sequence have not been detected in rodent DNAs, and rodent monomeric structures have not been detected in human DNA. There must have been a complete turnover of *Alu* family members since the divergence of rodent and human from their common ancestor as well as a mechanism that homogenizes the *Alu* family members within a given species. Individual cloned members of Old World and New World monkey *Alu* families resemble the 300-bp human consensus sequence as closely as individual members of the human *Alu* family (Grimaldi *et al.*, 1982; Daniels *et al.*, 1983). In contrast, galago (prosimian primate) DNA contains two distinct types of *Alu* family members (Daniels and Deininger, 1983). One of these types (called type 1) closely resembles the human consensus sequence (Fig. 1), except that there are clear species-specific differences between the human and prosimian sequences (Daniels *et al.*, 1983). These species-specific differences are recognizable as positions at which all available galago sequences differ from all human sequences. This is again evidence for complete turnover and homogenization of *Alu* family members subsequent in this case to the divergence of human and galago from their common ancestor. A second type of *Alu* family, called type 2, is also present in galago. This type 2 galago *Alu* family is closely related to the previous described type 1 galago sequence throughout the right monomer unit, so that the species-specific differences between the human and type 1 galago sequences are shared by the type 2 galago family (Daniels and Deininger, 1983). The type 1 and type 2 galago sequences are distinguished by a number of mutational differences in the left monomer unit. Of particular importance, the 5′ ends of the type 2 sequences are about 10 bp shorter than 5′ ends of the type 1 sequences.

Since both *Alu* family types in galago share species-specific differences relative to human, these two types probably diverged from a common ancestral sequence, which amplified following the divergence of human and galago. The distinct 5′ ends of these two types identifies them as the products of two distinct sets of transcriptional units.

4.3.3. Homogenization of the *Alu* Family

There must have been a complete conversion of the presumably ancestral human monomeric sequences into dimeric sequences following the separation of the human and rodent lineages. The ancestral human dimeric sequences must have been homogenized at least once more to account for the species-specific differences between the prosimian and anthropoid dimeric sequences. These homogenizations could result from either a correction of existing members of the sequence or a replacement of the entire ensemble with new members. Based on the available data, we cannot decide which of these is the more important mechanism, but we do know that there has been ample time for replacement to occur. Further, the two types of galago *Alu* repeats are clearly the products of two separate sets of transcription units.

As reviewed above, the 5′ and 3′ direct repeats flanking *Alu* members are 8 ± 4% divergent. For comparison, galago and human single-copy sequences are at least 28% divergent (Deininger and Schmid, 1979). Assuming as before that single-copy DNA and the flanking direct repeats diverge at the same rate, there is sufficient time for complete replacement of the *Alu* repeats in each line following the divergence of human and galago.

This calculation does not predict that *Alu* repeats are necessarily homogenized within a species, but does indicate that several complete turnovers of the entire family could have occurred following the separation of species that have recognizably different *Alu* families. The question of how the family is homogenized is compounded when we recall our previous conclusion that a very large bank of *Alu* repeats codes for subsequent members of the family. We imagine that any of the following possibilities could explain the homogenization of *Alu* repeats. There might be a very strong biological selection for particular sequences. Alternatively, there could be a strong biochemical selection for certain sequence features that need not confer a selective advantage on the host but could confer an advantage on the numerical success of particular sequence variants. In other words, *Alu* repeats could be selfish DNA (Doolittle and Sapienza, 1980; Orgel and Crick, 1980). There might be a sequence correction process that functions on members of multigene families and coincidently corrects the *Alu* repeat family. Dover (1982) describes the sequence correction of multigene families. Until more is known about the function or lack of function of *Alu* repeats this issue of what drives their homogenization will remain undecided.

4.3.4. Alu-like Sequences in Other Species

Base sequence studies identify an *Alu*-related sequence in cow (Watanabe *et al.*, 1982) and *Alu*-like sequences are probably present in all placental mammals. Stumph *et al.* (1981) report a family of interspersed repeats in chicken that resembles the human *Alu* family in some features. However, mammalian *Alu* equivalents resemble certain low-molecular weight RNAs, including especially 7S RNA (Weiner, 1980; Ullu *et al.*, 1982). The base sequence of 7S RNA is known to be well conserved in chicken. Since the chicken *Alu*-like family does not resemble the 7S RNA sequence, it is possible that chicken DNA contains other interspersed repeats more closely related to the human *Alu* family.

Hybridization studies using a type 2 rat *Alu* as a probe show that *Alu*-like sequences are present in amphibians (*Xenopus*), fish (herring), sea urchin, *Drosophila*, and some, but not all, plants (Blin *et al.*, 1983). Whether these sequence homologies correspond to 7S-like genes or interspersed *Alu*-like repeats is not known.

5. Non-*Alu*-Family Interspersed Repeats

5.1. *Kpn/Bam*H1 Repeats in Primate and Rodent

5.1.1. Equivalence of the Primate and Rodent Versions

Evidently, *Alu* family members are dispersed by the same mechanism by which many small RNA pseudogenes are inserted into the genome. *Alu* and small RNA pseudogenes may, however, be selected examples that need not be typical of other interspersed repeats. The following examples suggest that most interspersed repeats result from a similar insertion mechanism.

Digestion of human and several nonhuman primate DNAs with the restriction enzyme *Kpn* reveals a number of high-molecular weight bands (~2–4 kb), which collectively are called the *Kpn* family (Shafit-Zagardo *et al.*, 1982). A member of this same family has been identified as a 6.4-kb repeat flanking the β-globin gene cluster (Adams *et al.*, 1980). Base sequence comparisons demonstrate that a family of mouse repeats, which has most often been called the *Bam*H1 family, is closely related to the primate *Kpn* family (Singer *et al.*, 1983).

Numerous groups have reported studies on restriction fragments that are now known to be derived from the exceptionally long repeats contained in either the primate *Kpn* or mouse *Bam*H1 families. Because individual investigators were usually unaware that they were working on different fragments of the same families of sequences, the literature is confused by the plethora of repeat sequence "family" names. We recommend the summary papers of Fanning (1983) and Singer *et al.* (1983) as reviews of the original literature as well as adequate cross-references for the many aliases adopted for the rodent *Bam*H1 and primate *Kpn* families.

5.1.2. Length Variants of the *Bam*H1/*Kpn* Family

An especially interesting feature of the mouse *Bam*H1 family is that members of this family share a common 3′ end but are often truncated on their 5′ ends (Wilson and Storb, 1983; Fanning, 1983). For example, discrete repeated sequences (called R elements) are found to be interspersed among mouse immunoglobulin genes (Gebhard *et al.*, 1982). The 500-bp-long R elements have an A-rich 3′ end and are surrounded by short flanking direct repeats. According to our previous discussion, these features identify R elements as insertion units. Some R elements are the A-rich 3′ ends of longer (~1 kb) discrete repeats, which are also flanked by short direct repeats (Wilson and Storb, 1983). A progression of longer variants of this family up to ~7 kb in length results from sequences with common 3′ ends and less extensive 5′ truncations (Fanning, 1983; Voliva *et al.*, 1983). Since these variants are flanked by short direct repeats and end in an A-rich 3′ end, they are also probably inserted by an RNA-mediated process (Kole *et al.*, 1983). We suggest that the 5′ truncation of these sequences might result from an incomplete reverse transcription of the RNA intermediate. Perhaps the secondary structure or variations in processing of the RNA template causes the 5′ truncations. It is interesting to compare these 5′ truncations to the previously reviewed examples of 3′-truncated U3 RNA pseudogenes, which are thought to result from internal priming of the RNA intermediate (Bernstein *et al.*, 1983).

Like the mouse *Bam*H1 family, the primate *Kpn* also consists of a progression of 5′ truncations (Thayer and Singer, 1983; Sun *et al.*, 1984). One member of the *Kpn* family consists of a scrambling of normal *Kpn* homologues and an insertion of heterologous DNA (Lerman *et al.*, 1983). How typical this member is of other family members is not known.

5.1.3. Genomic Abundance of the Human Kpn and *Alu* Families

The repetition frequency of the *Kpn*/*Bam*H1 family is complicated by the variable 5′ truncation of individual members; not all parts of the sequence are represented at the same frequency. The extreme 3′ end of the (human) sequence is present in 50,000 copies per haploid genome (Schmid *et al.*, 1983). In agreement with this value, Kole *et al.* (1983) estimate that restriction fragments that contain *Kpn* family members are present in 14,000–40,000 copies in the human genome. For technical reasons, their estimate can be taken as a lower limit. Length variants of members of this family as described above range in size from 600 to 6400 bp. If we guess the average length of members to be 2000 bp (in fact many known examples approximate this size), then the *Kpn* family comprises ~4% of the human genome ($50{,}000 \times 2000$ bp/2.5×10^9 bp). Kole *et al.* (1983) estimate about ~6% of the genome to be composed of the *Kpn* family. Since the human *Alu* family comprises about 7% of the human genome (Rinehart *et al.*, 1981), these two families of interspersed repeats (*Kpn* and *Alu*) account for about 11% of the human genome. Our best estimates based on renaturation data suggest that at least 20% of the human genome, including noninterspersed satellite DNA, is repetitive (Deininger and Schmid, 1975). In this case two families of sequences, *Alu* and *Kpn*, account for a substantial mass fraction of total repetitive DNA and a correspondingly larger fraction of the interspersed repeats within the human genome. Presumably the rodent equivalent families account for a similar fraction of their genome. By the criteria of either mass (*Kpn* and *Alu*) or number (*Alu* alone), most of the interspersed repeats in the primate and probably rodent genomes are inserted repeats.

5.2. Other Families of Inserted Repeats

Other, less abundant interspersed repeats in human DNA are also inserted repeats. The previously discussed small RNA pseudogenes that led to the recognition of the insertion mechanism certainly belong to this class. Randomly selected low-abundance interspersed repeats have also been identified as inserted repeats (Sun *et al.*, 1984). There is, however, at least one class of interspersed repeats that is not the result of an RNA insertion product.

5.3. Interspersed Tandem Repeats

5.3.1. "Poly CA"

Long runs of the dinucleotide CA are present in regions flanking known human genes (Miesfeld *et al.*, 1981; Nordheim and Rich, 1983). There are about 50,000 copies of this sequence in human DNA and most amazingly it is found in the same relative abundance in all eukaryotes examined (Hamada *et al.*, 1982). This is the single proven example of a repetitive DNA that transcends all divisions in eukaryotic evolution. The conservation of this sequence might be the result of stringest biological selection. As one possibility, "poly CA" can adopt the Z form of duplex DNA and it has been proposed in molecular detail that the B to Z transition regulates transcription (Rich, 1983). It is also possible that the strict evolutionary conservation of poly CA is an inevitable consequence of similar biochemical pathways in all eukaryotes and does not necessarily imply that this sequence has a selected function. In particular, we propose below that poly CA and other tandem repeats are generated *in situ* by a tandem amplification of existing sequences, perhaps by slippage during DNA replication.

5.3.2. Tandem Repetitions of Short Oligonucleotides

Runs of randemly repeated oligonucleotides have been observed in the flanking and intervening regions of a number of genes. As one example of such observations, the first intron of the human pseudo ζ gene consists of 39 tandem copies of variants to the sequence ACAGTGGGGAGGGG, whereas the second intron includes 52 tandem copies of variant forms of the sequence CGGGG (Proudfoot *et al.*, 1982). The first and second intron of the ζ gene are each occupied by corresponding tandem repeats (Proudfoot *et al.*, 1982). The genomic repetition of these simple sequences has not yet been examined in detail, and strictly speaking they may not be repetitive. However, simple sequence DNA, usually polypyrimidines (or equivalently polypurines), have been observed in regions flanking a number of known genes (Maroteaux *et al.*, 1983; Kominami *et al.*, 1983b; Heilig, *et al.*, 1982; Sawada *et al.*, 1983; Bell *et al.*, 1982; Dodgson and Engel, 1983). Further polypyrimidine tracts are reported to be interspersed repetitive sequences in all eukaryotes tested (Birnboim *et al.*, 1979).

For our purposes, the best-studied examples of interspersed tandem repeats involve the avian ovalbumin gene family. Tandem repeats of the oligopyrimidine CCTT occupy a 300-bp region flanking the ovalbumin Y gene (Heilig *et al.*, 1982). The first intron of the ovalbumin X gene contains 36 tandem copies of the pentanucleotide TCTCC (Maroteaux *et al.*, 1983). Yet another polypyrimidine tract consists of 27 tandem copies of CTTCC and flanks the chicken ovotransferin gene. In analogy to this chicken gene, the pheasant ovotransferin gene is also flanked by 55 tandem copies of another polypyrimidine variant, TTTCC. Blot hybridization studies show each of these variants to be repetitive and to be present in divergent avians (Maroteaux *et al.*, 1983).

5.3.3. Possible Origin of Interspersed Tandem Repeats

Neither poly CA nor the other examples of interspersed tandem repeats cited above are flanked by any special sequences, including, in particular, recognizable short direct repeats. Presumably, interspersed tandem repeats are generated by a different mechanism than the one that produces the inserted repeats.

We have observed cases in which the sequence homology of closely related genes breaks down in regions consisting of short runs of tandemly repeated oligonucleotides (Sawada *et al.*, 1983). As one example, the 5′ flanking regions of the human adult $\alpha 2$- and $\alpha 1$-globin genes are mostly homologous sequences (Lauer *et al.*, 1980). This homology probably results in part from the ancestral duplication of these related genes and more so from recent gene correction events. One 200-bp region of nonhomology flanks the $\alpha 1$ gene but is absent from the otherwise homologous region flanking the $\alpha 2$ gene. This sequence is either an insertion 5′ to $\alpha 1$ or a deletion 5′ to $\alpha 2$. This 200-bp nonhomology initiates with nine imperfect tandem copies of the pyrimidine CCT (Sawada *et al.*, 1983; Hess *et al.*, 1983). Also contained within this nonhomology are about 20 nucleotides, which consist of imperfect tandem repeats of alternating CA residues. It is tempting to view this sequence as an incipient run of poly CA. We imagine that slippage during the replication of simple sequence DNA would tend to expand the size of such a tract. According to this view, the nonhomology present in the region flanking the $\alpha 1$ gene is a recent expansion of sequence that in part results from the misreplication of simple sequences. While this interpretation remains to be tested, it is plausible to imagine that tandemly repeated oligonucleotides at other chromosomal sites are generated by a similar mechanism. It is especially noteworthy that regions containing tandem interspersed repeats are often polymorphic (Kominami *et al.*, 1983b; Bell *et al.*, 1982; B. S. Chapman, personal communication).

6. Biological Functions and Effects

6.1. Similarity of Inserted Repeats and Pseudogenes

The evolution of interspersed repeats would be better understood if we knew the functions of these sequences. The similarities between the inserted repeats and processed RNA pseudogenes compels the suggestion that this class of interspersed repeats is essentially functionless. Conceivably, a backlog of pseudogenes provides the organism some flexibility in evolving new genes, but we do not regard such potential as a function. Accepting the possibility that inserted repeats are functionless units, this opens the question of why certain pseudogenes are numerically successful whereas others are not. The success of the *Alu* family has been attributed to its internal RNA polymerase III promoter, the idea being that any one member of the family that is inserted might code for subsequent members of the family (Bernstein *et al.*, 1983). However, the same argument should apply to inserted pseudogenes for 5S RNA and tRNA, which are not present in high-copy number. The type of promoter cannot be the deciding factor, since the next most abundant inserted repeat family in human DNA, the *Kpn* family, almost certainly results from RNA polymerase II transcription products (Kole *et al.*, 1983; Sun *et al.*, 1984). Other, more abundantly expressed mRNAs have not spawned processed pseudogenes in numbers even remotely corresponding to the size of the *Kpn* family. Perhaps *Alu*, *Kpn*, and other numerically successful repeat families were actively transcribed at developmental times that happen to be ideal for insertion of new DNA into the genome. Alternatively, numerically successful repeats might have a biological function.

The close analogy between inserted repeats (e.g., *Alu* and *Kpn*) and processed RNA pseudogenes does not necessarily require that these repeats be functionless. Their function may be related to their genomic dispersion, and perhaps the processed RNA pseudogenes are cases in which functional mRNAs were accidentally shanghaied by the required dispersion mechanism. It is therefore worthwhile to speculate on the functions of such repeats.

6.2. Possible Functions of *Alu*-like Sequences

Since we do not know the biological function of any family of inserted repeats, we will restrict our discussion to the most thoroughly studied example, the *Alu* family and its rodent equivalents. Base sequence comparisons between the *Alu* family and other nucleic acids suggest a number of biological functions for the *Alu* family (Jelinek *et al.*, 1980). It is probably fair to state that none of these possible functions have been decisively tested, although some are now thought to be less likely.

The biological function of a closely related sequence, 7S RNA, involves an association with the signal recognition particle (Walter and Blobel, 1982). It should be emphasized that *Alu* family members do not code for 7S RNA, although they are closely related sequences (Li *et al.*, 1982; Ullu, 1982; Ullu *et al.*, 1982). This suggests two possibilities: either the function of *Alu* is fulfilled by discrete low-molecular weight RNAs that resemble 7S RNA or its function involves the interaction between *Alu* transcripts and specific recognition proteins. *Alu* is abundantly transcribed as part of longer hnRNA molecules, but these are mostly read-through rather than discrete transcripts (Schmid and Jelinek, 1982; Jelinek and Schmid, 1982). Although (as previously described) individual *Alu* family members are actively transcribed into low-molecular weight RNAs *in vitro*, the corresponding *in vivo* transcripts have been difficult to detect (Weiner, 1980). In one case a cloned *Alu* family member (mouse B1) is abundantly transcribed *in vivo* as a low-molecular weight RNA (Young *et al.*, 1982). However, this essentially demonstrates that the polymerase III promoter on unrepressed *Alu* family members can function *in vivo* and does not necessarily imply that they are normally transcriptionally active. One interpretation of this result is that *in vivo* transcription of *Alu* is strongly repressed. Allan and Paul (1984) report a short length transcript corresponding to a human Alu repeat upstream from the ϵ-globin gene. The transcription of this repeat is linked to the expression of the adjacent gene. The rodent type II *Alu* family equivalent (also called B2) appears to code for a discrete *in vivo* transcript (Haynes and Jelinek, 1981). However, as reviewed above, there are substantial sequence differences between the rodent type II and human *Alu* families and they are probably functionally distinct classes (Haynes and Jelinek, 1981; Krayev *et al.*, 1982).

There is the second possibility cited above, that the function of *Alu* involves the specific interaction of *Alu* transcripts with recognition proteins. These need not be specific transcripts, but could instead be the abundant hnRNA readthrough transcripts of interspersed *Alu* repeats. Because the complex formed between the cytoplasmic 7S RNA and the signal recognition particle is involved in the transport of proteins across the cell membrane, we are led to suggest that a specific complex might form between the abundant *Alu* transcripts in hnRNA and as yet unidentified nuclear protein(s). On this basis it has been reasonably suggested that *Alu* sequences might in some way identify unprocessed nuclear RNA (Sharp, 1983). Supporting this possibility is the observation that *Alu* family members are not inserted into otherwise functional coding regions, but are abundantly represented in intergenic regions and have been reported in intervening sequences (references in Fig. 1). With the single exception of a processed pseudogene, they have not been reported in the coding region of genes (Lemishka and Sharp, 1982).

An 82-nucleotide repetitive sequence is found in poly A brain RNA (Sutcliffe *et al.*, 1982). This repetitive element is homologous to both type 2 (B2) sequences and a 160-nucleotide RNA that is unique to brain RNA (Sutcliffe *et al.*, 1982; Schuler *et al.*, 1983). It is suggested that this sequence could serve an identification function.

6.3. Effects Attributable to the Presence and Insertion of *Alu*

Whether or not interspersed repeats are functional, their broad genomic distribution and high copy number must inevitably result in profound genetic effects. For perspective on the possibility of such effects it is worth remembering that highly repeated sequences are simply not tolerated in either prokaryotic or lower eukaryotic genomes, but are instead rapidly removed by homologous recombination. The highly repeated sequences that we now know to be present in mammalian genomes are probably not completely stable, but are also subject to the deletions and rearrangements we would expect in analogy to the case of lower organisms.

In three known examples, deletions involving *Alu* repeats have resulted in phenotypic changes. The hereditary persistence of fetal globin is attributed to a deletion of the β- and δ-globin genes, which occurs by a recombination within an *Alu* repeat (Jagadeeswaran *et al.*, 1982). This particular *Alu* family member is one arm of a pair of inverted *Alu* repeats that map 5′ to the normal δ- and β-globin gene pair. A different disorder, δ^0-β^0-thalassemia, is found to result from a similar deletion involving the other arm of this same pair of inverted *Alu* repeats (Ottolenghi and Giglioni, 1982). One form of α-thalassemia involves a deletion that occurs within and is bounded by an *Alu* repeat (Orkin and Michelson, 1980, cited in Jagadeeswaran *et al.*, 1982). The precise recombinational role of *Alu* in each of these deletions is not yet known. Presumably, as other gene families are subjected to the same scrutiny as the hemoglobins, additional examples of such deletions will be observed.

These deletions might answer a question raised above. If *Alu*'s are inserted at a half-time comparable to the divergence time of human and monkey, then how are they removed? One plausible explanation is that they are removed by homologous recombination between their flanking direct repeats. In this event those *Alu*'s that happened to lose their flanking direct repeats would not be removed and would preferentially accumulate within the genome. There are a few known examples of *Alu*'s that have lost their direct repeats. Unfortunately, there is not sufficient evidence to decide if their abundance is increasing. Alternatively, *Alu*'s might be removed by recombination between neighboring or near-neighboring members. Perhaps the globin deletions described above testify to such events. In this case the continual turnover of repeats would incidently prune the single-copy DNA content of the genome. Since such deletions would preferentially occur in regions of the genome that are *Alu*-rich, this mechanism would, as desired, tend to preferentially reduce the repetitive DNA content of the genome. Saunders and colleagues report that a region of the genome especially rich in *Alu* is genetically unstable (Calabretta *et al.*, 1982; Robberson *et al.*, 1983; Shmookler-Reis *et al.*, 1983). More recent studies questioned these claims at least in part (R. Shoomkler-Reis and S. Goldstein, personal communication).

Hess *et al.* (1983) has found that the insertion of *Alu* family repeats has induced large deletion/insertion in the human α-globin gene duplication units. Rearranged sequences flank the inserted *Alu* repeats and the 3′ end of one *Alu* repeat has been removed by a deletion. It is interesting to note that such evolution events are similar to the IS element-generated deletions in bacteria [for a review see Kleckner (1977)].

The presence of repetitive DNAs must have other effects in addition to promoting deletions. Barsh *et al.* (1983) find that the human growth hormone gene family probably results from unequal crossing over between *Alu* repeats that flanked the ancestral genes.

In many of the arguments presented here we accept the transposition of inserted repeats on an evolutionary time scale as a demonstrated fact. However, in the case of bacterial transposons or transposons in yeast and *Drosophila*, transposition can be demonstrated in contemporary populations. For higher eukaryotes it will be very difficult to demonstrate the transposition of an individual member of a hyperrepetitive sequence family. In the absence of a phenotypic selection, the other family members present a tremendous background in any proposed experimental test of the transposition. In spite of this obstacle, there are two examples of intraspecies polymorphisms resulting from the insertion of an *Alu*-like family member (Kominami *et al.*, 1983a; Schuler *et al.*, 1983). We may conclude that inserted repeats are mobile in existing populations.

6.4. Functions and Effects of Interspersed Tandem Repeats

Physical chemists have long recognized that simple nucleic acid sequences have a number of anomalous properties as compared to complex sequences. The facility with which alternating pyrimidine–purine sequences, such as poly CA, undergo the B to Z conformational change merely illustrates this general point (Rich, 1983). We can expect other simple sequence DNAs to exhibit novel structures and interactions that might be biologically significant. There is an especially detailed and attractive proposal that the B to Z structural transition could regulate transcription (Rich, 1983). In effect, this proposes a biological function for the sequence alternating poly CA, which is the most repetitive naturally occurring alternating pyrimidine–purine sequence. There is approximately one run of poly CA per "transcriptional domain," so that numerically it is sufficiently abundant to fulfill a transcriptional function (Nordheim and Rich, 1983; Hamada *et al.*, 1982).

The function of the interspersed tandem repeats also awaits a decisive test. In the absence of proven function, we once more recognize effects attributable to such sequences. Base sequence studies of the human γ^G- and γ^A-globin genes reveal a poly CA tract in the second intron of each of these genes (Slightom *et al.*, 1980). In one individual the poly CA tract defines the 3′ terminus of a gene conversion event between the γ^G and γ^A loci. It is suggested that such simple sequences, perhaps because of their unusual secondary structures, might provide specialized sites for recombination or sequence conversion.

Restriction mapping suggests a gene conversion event extending into the first intron of the human embryonic α-like ζ-globin gene and the pseudo ζ gene (Chapman *et al.*, 1985). Confirmation of this event as well as the precise definition of its 3′ end point required DNA base sequence comparisons. However, it is remarkable that the first introns of the pair of ζ- globin genes, like the second introns of the pair of γ genes described above, contains a tandem repetitive sequence. It has been found that each pair of genes is also subject to frequent correction.

7. Summary

The existence of repetitive sequences in eukaryotic DNAs has been known for 15 years and it is disappointing that the function of these sequences remains unknown. Perhaps, as Orgel and Crick (1980) suggest, a search for such a function is a sterile exercise.

During the last 5 years remarkable progress has been made in describing the structures, molecular origins, evolution, and genetic effects of interspersed repeats. In retrospect, it was premature to propose biological functions for these sequences in the absence of these more recent findings. It remains to be seen whether it is now timely to propose specific functions. What has been established by these more recent studies is that a number of genetic effects are associated with the presence and turnover of interspersed repeats. It is certainly not surprising that a family of repeats, such as *Alu* or *Kpn*, that comprises 5% of an organism's DNA and is broadly distributed throughout the genome would have important genetic effects. Nonetheless, it is satisfying that we can now document some of the ways in which repetitive families have led to genetic changes and introduced sequence diversity into a genome.

ACKNOWLEDGMENTS. The research programs of C.-K. J. S. and C.W.S. are supported by U.S. Public Health Service grants AM29800 and GM21346. We especially wish to thank Gloria Marquez for her skillful preparation of this manuscript. We also thank Dr. Catherine Houck and Eric Paulson for providing us with unpublished results, and Eric Paulson for his incisive analysis of the *Pol*III promoter question.

References

Adams, J. W., Kaufman, R. E., Kretschner, P. J., Harrison, M., and Nienhuis, A. W., 1980, A family of long repeated DNA sequences, *Nucleic Acids Res.* **9**:6113–6128.

Allan, M., and Paul, J., 1984, Transcription *in vivo* of an Alu family member upstream from the human ϵ-globin gene, *Nucleic Acids Res.* **12**:1193–1200.

Baralle, F. E., Shoulders, C. D., Goodbourn, S., Jeffreys, A., and Proudfoot, N. J., 1980, The 5′ flanking region of human ε-globin gene, *Nucleic Acids Res.* **8**:4393–4404.

Barsh, G. S., Seeburg, P. H., and Gelinas, R. E., 1983, The human growth hormone gene family: Structure and evolution of the chromosomal locus, *Nucleic Acids Res.* **11**:3939–3959.

Bell, G. I., Pictet, R., and Rutter, W. J., 1980, Analysis of the regions flanking the human insulin gene and sequence of an *Alu* family member, *Nucleic Acids Res.* **8**:4091–4109.

Bell, G. I., Selby, M. J., and Rutter, W. J., 1982, The highly polymorphic region near the human insulin gene is composed of simple tandemly repeating sequences, *Nature* **295**:31–35.

Bernstein, L. B., Mount, S. M., and Weiner, A. M., 1983, Pseudogenes for human small nuclear RNA U3 appear to arise by self primed reverse transcripts of the RNA into new chromosomal sites, *Cell* **32**:461–472.

Birmboim, H. C., Sederoff, R. R., and Paterson, M. C., 1979, Distribution of poly-pyrimidine · polypurine segments in DNA from diverse organisms, *Eur. J. Biochem.* **98**:301–307.

Blin, N., Weber, T., and Alonso, A., 1983, Cross reaction of snRNA and an *Alu* I-like sequence from rat with DNAs from different eucaryotic species, *Nucleic Acids Res.* **11**:1375–1389.

Britten, R. J., and Davidson, E. H., 1971, Repetitive and nonrepetitive DNA sequences and a speculation on the origins of evolutionary novelty, *Q. Rev. Biol.* **46**:111–133.

Britten, R. J., and Kohne, D. E., 1968, Repeated sequences in DNA, *Science* **161**:529–540.

Calabretta, B., Robberson, D. L., Barrera-Saldana, H. A., Lambrou, T. P., and Saunders, G. F., 1982, Genome instability in a region of human DNA enriched in *Alu* repeat sequences, *Nature* **296**:219–225.

Chapman, B. S., Vincent, K. A., and Wilson, A. C., 1985, Length polymorphisms, gene correction and recombination in the zeta globin region of humans, *Genetics* (submitted).

Daniels, G. R., and Deininger, P. L., 1983, A second major class of *Alu* family repeated DNA sequences in a primate genome, *Nucleic Acids Res.* **11**:7595–7610.

Daniels, G. R., Fox, M., Lowensteiner, D., Schmid, C., and Deininger, P. L., 1983, Species specific homogeneity of the primate *Alu* family of repeated DNA sequence, *Nucleic Acids Res.* **11**:7579–7593.

Davidson, E. H., and Britten, R. J., 1973, Organization, transcription and regulation in the animal genome, *Q. Rev. Biol.* **48**:565–613.

Davidson, E. H., Galau, G. A., Angerer, R. C., and Britten, R. J., 1975, Comparative aspects of DNA organization in metazoa, *Q. Rev. Biol.* **51**:253–259.

Deininger, P. L., and Schmid, C. W., 1976, An electron microscope study of the DNA sequence organization of the human genome, *J. Mol. Biol.* **106**:773–790.

Deininger, P. L., and Schmid, C. W., 1979, A study of the evolution of repeated DNA sequences in primates and the existence of a new class of repetitive sequences in primates, *J. Mol. Biol.* **127**:437–460.

Deininger, P. L., Jolly, D. J., Rubin, C. M., Friedmann, T., and Schmid, C. W., 1981, Base sequence studies of 300 nucleotide renatured repeated human DNA clones, *J. Mol. Biol.* **151**:17–33.

Denison, R. A., and Weiner, A. M., 1982, Human U1 RNA pseudogenes may be generated by both DNA and RNA mediated mechanisms, *Mol. Cell Biol.* **2**:815–282.

Dodgson, J. B., and Engel, J. D., 1983, The nucleotide sequence of the adult chicken α-globin genes, *J. Biol. Chem.* **258**:4623–4629.

Doolittle, W. F., and Sapienza, C., 1980, Selfish genes, the phenotype paradigm and genome evolution, *Nature* **284**:601–603.

Dover, G., 1982, Molecular drive: A cohesive mode of species evolution, *Nature* **111–117.**

Duncan, C. H., Biro, P. A., Choudary, P. V., Elder, J. T., Wang, R. R. C., Forget, B. G., deRiel, J. K., and Weissman, S. M., 1979, RNA polymerase III transcription units are interspersed among human non-α-globin genes, *Proc. Natl. Acad. Sci. USA* **76:**5095–5099.

Duncan, C. H., Jagadeeswaran, P., Wang, R. R. C., and Weissman, S. M., 1981, Structural analysis of templates and RNA polymerase III transcripts of *Alu* family sequences interspersed among human β-like globin genes, *Gene* **13:**185–196.

Elder, J. T., Pan, J., Duncan, C. H., and Weissman, S. M., 1981, Transcriptional analysis of interspersed repetitive polymerase III transcription units in human DNA, *Nucleic Acids Res.* **9:**1171–1189.

Fanning, T. G., 1983, Size and structure of the highly repetitive *Bam*HI element in mice, *Nucleic Acids Res.* **11:**5073–5091.

Farabaugh, P. J., and Fink, G. R., 1980, Insertion of the eukaryote transposable element T1 creates a 5 base pair duplication, *Nature* **286:**352–356.

Friezner Degen, S. J., MacGillivray, R. T. A., and Davie, E. W., 1983, Characterization of the complementary DNA and gene coding for human prothrombin, *Biochemistry* **9:**2087–2098.

Fuhrman, S. A., Deininger, P. L., LaPorte, P., Friedmann, T., and Geiduschek, E. P., 1981, Analysis of transcription of the human *Alu* family ubiquitous repeating element by eukaryotic RNA polymerase III, *Nucleic Acid Res.* **9:**6439–6456.

Fukumaki, Y., Collins, F., Kole, R., Stoeckert, C. J., Jagadeeswaran, P., Duncan, C. H., and Weissman, S. M., 1983, Sequence of human repetitive DNA, *Cold Spring Harbor Symp. Quant. Biol.* **XLVII:**1079–1081.

Gafner, J., and Philippsen, P., 1980, Common features of transposition—A yeast transposon also generates duplications of the target sequence, *Nature* **286:**414–418.

Gebhard, W., Meitinger, T., Höchtl, J., and Zachau, H. G., 1982, A new family of interspersed repetitive DNA sequences in the mouse genome, *J. Mol. Biol.* **157:**453–471.

Grimaldi, G., and Singer, M. F., 1982, A monkey *Alu* sequence is flanked by 13-base-pair direct repeats of an interrupted α satellite DNA sequence, *Proc. Natl. Acad. Sci. USA* **79:**1497–1500.

Hamada, H., Petrino, M. G., and Kakunga, T., 1982, A novel repeated element with Z-DNA-forming potential is widely found in evoluntarily diverse eukaryotic genomes, *Proc. Natl. Acad. Sci. USA* **79:**6465–6469.

Hammarström, K., Westin, G., and Pettersson, U., 1982, A pseudogene for human U4 RNA with a remarkable structure, *EMBO J.* **1:**737–739.

Hayashi, K., 1981, Organization of sequences related to U6 RNA in the human genome, *Nucleic Acids Res.* **9:**3379–3388.

Haynes, S. R., and Jelinek, W. R., 1981, Low molecular weight RNAs are transcribed *in vitro* and *in vivo* by RNA polymerase III from *Alu* type dispersed repeats in Chinese hamster DNA, *Proc. Natl. Acad. Sci. USA* **78:**6130–6134.

Haynes, S. R., Toomey, T. P., Leinwand, L., and Jelinek, W. R., 1981, The Chinese hamster *Alu*-equivalent sequence: A conserved highly repetitious interspersed DNA sequence in mammals has a structure suggestive of a transposable element, *Mol. Cell Biol.* **1:**573–583.

Heilig, R., Muraskowsky, R., and Mandel, J.-L., 1982, The ovalbumin gene family—The 5′ end region of the X and Y genes, *J. Mol. Biol.* **1956:**1–19.

Hess, J. F., Fox, G. M., Schmid, C., and Shen, C.-K. J., 1983, Molecular evolution of the human adult α-like globin gene region: Insertion and deletion of *Alu* family repeats and non-*Alu* DNA sequences, *Proc. Natl. Acad. Sci. USA* **80:**5970–5974.

Hollis, G. F., Hieter, P. A., McBride, O. W., Swan, D., and Leder, P., 1982, Processed genes: A dispersed human immunoglobulin gene bearing evidence of RNA-type processing, *Nature* **296**:321–325.

Houck, C. M., Rinehart, F. P., and Schmid, C. W., 1979, A ubiquitous family of repeated DNA sequences in the human genome, *J. Mol. Biol.* **132**:289–306.

Hoyer, B. H., van de Velde, N. W., Goodman, M., and Roberts, R. B., 1972, Examination of hominoid evolution by DNA sequence homology, *J. Hum. Evol.* **1**:645–649.

Hudspeth, M. E. S., Timberlake, W. E., and Goldberg, R. B., 1977, DNA sequence organization in the water mold *Achlya*, *Proc. Natl. Acad. Sci. USA* **74**:4332–4336.

Jagadeeswaran, P., Forget, B. G., and Weissman, S. M., 1981, Short interspersed repetitive DNA elements in eukaryotes: Transposable DNA elements generated by reverse transcription of RNA *pol* III transcripts, *Cell* **26**:141–142.

Jagadeeswaran, P., Tuan, D., Forget, B. G., and Weissman, S. M., 1982, A gene deletion ending in the midpoint of a repetitive DNA sequence in one form of hereditary persistence of fetal globin, *Nature* **296**:770–771.

Jelinek, W. R., and Schmid, C. W., 1982, Repetitive sequences in eukaryotic DNA and their expression, *Annu. Rev. Biochem.* **51**:813–844.

Jelinek, W. R., Toomey, T. P., Leinwand, L., Duncan, C., Biro, P. A., Choudary, P. V., Weissman, C. M., Rubin, C. M., Houck, C. M., Deininger, P. L., and Schmid, C. W., 1980, Ubiquitous interspersed repeated sequences in mammalian genomes, *Proc. Natl. Acad. Sci. USA* **77**:1398–1402.

Karin, M., and Richards, R. I., 1982, Human metallothionein genes—Primary structure of the metallothionein-II gene and a related processed gene, *Nature* **299**:797–802.

Kleckner, N., 1977, Transposable elements in procaryotes, *Cell* **11**:11–28.

Kohne, D. E., Chiscon, J. A., and Hoyer, B. H., 1972, Evolution of primate DNA sequences, *J. Hum. Evol.* **1**:627–644.

Kole, L. B., Haynes, S. R., and Jelinek, W. R., 1983, Discrete and heterogeneous high molecular weight RNAs complementary to a long dispersed repeat family (a possible transposon) of human DNA, *J. Mol. Biol.* **165**:257–286.

Kominami, R., Muramatsu, M., and Moriwaki, K., 1983a, A mouse type 2 *Alu* sequence (M2) is mobile in the genome, *Nature* **301**:87–89.

Kominami, R., Urano, Y., Mishima, Y., Muramatsu, M., Moriwaki, K., and Yoshikura, H., 1983b, Novel repetitive sequence families showing size and frequency polymorphism in the genome of mice, *J. Mol. Biol.* **165**:209–228.

Krayev, A. J., Kramerov, D. A., Skryabin, K. G., Ryskov, A. P., Bayev, A. S., and Georgiev, G. P., 1980, The nucleotide sequence of the most ubiquitous repetitive DNA sequence B1 complementary to the most abundant class of mouse fold back RNA, *Nucleic Acids Res.* **8**:1201–1215.

Krayev, A. S., Markusheva, T. V., Kramerov, D. A., Ryskov, A. P., Skryabin, K. G., Bayev, A. A., and Georgiev, G. P., 1982, Ubiquitous transposon-like repeats B1 and B2 of the mouse genome B2 sequencing, *Nucleic Acids Res.* **10**:7461–7475.

Kurnit, D. M., and Maio, J. J., 1973, Subnuclear redistribution of DNA species in confluent and growing mammalian cells, *Chromosoma* **42**:23–36.

Lauer, J., Shen, C.-K. J., and Maniatis, T., 1980, The chromosomal arrangement of human α-like globin genes: Sequence homology and α-globin gene deletions, *Cell* **20**:119–130.

Lemishka, I. R., and Sharp, P. A., 1982, The sequences of an expressed rat α-tubulin gene and a pseudogene with an inserted repetitive element, *Nature* **300**:330–335.

Lerman, M. I., Thayer, R. E., and Singer, M. F., 1983, *Kpn* I family of long interspersed repeated DNA sequences in primates: Polymorphism of family members and evidence for transcription, *Proc. Natl. Acad. Sci. USA* **80**:3966–3970.

Levis, R., Dunsmuir, P., and Rubin, G. M., 1980, Terminal repeats of the *Drosophila* transposable element copia: Nucleotide sequence and genomic organization, *Cell* **21**:581–588.

Li, W.-Y., Reddy, R., Henning, D., Epstein, P., and Busch, H., 1982, Nucleotide sequence of 7S RNA (homology to *Alu* DNA and LA 4.5S RNA), *J. Biol. Chem.* **257**:5136–5142.

Manning, J. E., Schmid, C. W., and Davidson, N., 1975, Interspersion of repetitive and non repetitive DNA sequences in the *D. melanogaster* genome, *Cell* **4**:141–155.

Manuelidis, L., 1976, Repeating restriction fragments of human DNA, *Nucleic Acids Res.* **3**:3063–3076.

Maroteaux, L., Helig, R., Dupret, D., and Mandel, J. L., 1983, Repetitive satellite-like sequences are present within and upstream from 3 avian protein-coding genes, *Nucleic Acids Res.* **11**:1227–1245.

Miesfeld, R., Krystal, M., and Arnheim, N., 1981, A member of a new repeated family which is conserved throughout eucaryotic evolution is found between the human δ and β globin genes, *Nucleic Acids Res.* **9**:5931–5447.

Nishioka, Y., Leder, A., and Leder, P., 1980, Unusual α-globin-like gene that has cleanly lost both globin intervening sequences, *Proc. Natl. Acad. Sci. USA* **77**:2806–2809.

Nordheim, A., and Rich, A., 1983, The sequence $(dC\text{–}dA)_n \cdot (dG\text{–}dT)_n$ forms left handed Z-DNA in negatively supercoiled plasmids, *Proc. Natl. Acad. Sci. USA* **80**:1821–1825.

Orgel, L. E., and Crick, F. H. C., 1980, Selfish DNA: The ultimate parasite, *Nature* **284**:604–607.

Orkin, S. H., and Michelson, A., 1980, Partial deletion of the α globin structural gene in human α-thalassaemia, *Nature* **286**:538–540.

Ottolenghi, S., and Giglioni, B., 1982, The deletion in a type of δ^0 β^0 thalassaemia begins in an inverted *Alu* I repeat, *Nature* **300**:770–771.

Page, G. S., Smith, S., and Goodman, H. M., 1981, DNA sequence of the rat growth hormone gene: Location of the 5′ terminus of the growth hormone mRNA and identification of an internal transposon-like element, *Nucleic Acids Res.* **9**:2087–2104.

Proudfoot, N. J., Gill, A., and Maniatis, T., 1982, The structure of the human zeta-globin gene and a closely linked, nearly identical pseudogene, *Cell* **31**:553–563.

Rich, A., 1983, Right handed and left handed DNA: Conformational information in genetic material, *Cold Spring Harbor Symp. Quant. Biol.* **XLVII**:1–12.

Rice, N., 1971, Thermal stability of reassociated repeated DNA from rodents, *Carnegie Inst. Yearb.* **69**:472–479.

Rinehart, F. P., Ritch, T. G., Deininger, P. L., and Schmid, C. W., 1981, Renaturation rate studies of a single family of interspersed repeated sequences in human deoxyribonucleic acid, *Biochemistry* **20**:3003–3010.

Robberson, D. L., Calabretta, B., Barrera-Saldana, H. A., Somasundaram, T., Lambrou, T., Stubblefield, E., Yen, N., Brock, D. L., Peterson, C. A., and Saunders, G. F., 1983, Genome rearrangements and extra chromosomal circular DNAs in human cells, in: *Perspective on Genes and the Molecular Biology of Cancer* (D. L. Robberson and G. F. Saunders, eds.), Raven Press, New York, pp 51–80.

Sawada, I., Beal, M. P., Shen, C.-K. J., Chapman, B., Wilson, A. C., and Schmid, C., 1983, Intergenic DNA sequences flanking the pseudo alpha globin genes of human and chimpanzee, *Nucleic Acids Res.* **11**:8087–8101.

Scarpulla, R. C., and Wu, R., 1983, Non allelic members of the cytochrome *c* multigene family of rat may arise through different messenger RNAs, *Cell* **32**:473–482.

Schmid, C. W., and Deininger, P. L., 1975, Sequence organization of the human genome, *Cell* **6**:345–358.

Schmid, C. W., and Jelinek, W. R., 1982, The *Alu* family of dispersed repetitive sequences, *Science* **216**:1065–1070.

Schmid, C. W., Fox, G. M., Dowds, B., Lowensteiner, D., Paulson, K. E., Shen, C.-K. J., and Leinwand, L., 1983, Families of repeated human DNA sequences and their arrangements, in: *Perspectives on Genes and the Molecular Biology of Cancer* (D. L. Robberson and G. F. Saunders, eds.), Raven Press, New York, pp. 35–41.

Schuler, L. A., Weber, J. L., and Gorski, J., 1983, Polymorphism near the rat prolactin gene caused by insertion of an *Alu*-like element, *Nature* **305:**159–160.

Seeburg, P. H., 1982, The human growth hormone gene family: Nucleotide sequences show recent divergence and predict a new polypeptide hormone, *DNA* **1**(3):239–249.

Shafit-Zagardo, B., Maio, J. J., and Brown, F. L., 1982, *Kpn* I families of long, interspersed repetitive DNAs in human and other primate genomes, *Nucleic Acids Res.* **10:**3175–3193.

Sharp, P. A., 1983, Conversion of RNA to DNA in mammals *Alu*-like elements and pseudogenes, *Nature* **301:**471–472.

Shen, C.-K. J., and Maniatis, T., 1982, The organization, structure and *in vitro* transcription of *Alu* family RNA polymerase III transcription units in the human α-like globin gene cluster, *J. Mol. Appl. Genet.* **1:**343–360.

Shen, S.-H., and Smithies, O., 1982, Human globin $\psi\beta_2$ is not a globin related sequence, *Nucleic Acids Res.* **10:**7809–7818.

Shimotono, K., Mizutani, S., and Temin, H., 1980, Sequence of retrovirus provirus resembles that of bacterial transposable elements, *Nature* **285:**550–554.

Shmookler Reis, R. J., Lumpkin, C. K., McGill, J. R., Riabonol, K. T., and Goldstein, S., 1983, Extra chromosomal circular copies of an 'inter *Alu*' unstable sequence in human DNA are amplified during *in vitro* and *in vivo* ageing, *Nature* **301:**394–398.

Singer, M. F., Thayer, R. E., Grimaldi, G., Lerman, M. I., and Fanning, T. G., 1983, Homology between the *Kpn* I primate and *Bam*H1 (MIF 1) rodent families of long interspersed repeated sequences, *Nucleic Acids Res.* **11:**5739–5745.

Slightom, J. L., Blechl, A. E., and Smithies, O., 1980, Human fetal G γ- and A γ-globin genes: Complete nucleotide sequences suggest that DNA can be exchanged between these duplicated genes, *Cell* **21:**627–638.

Stumph, W. E., Kristo, P., Tsai, M.-J., and O'Malley, B. W., 1981, A chicken middle-repetitive DNA sequence which shares homology with mammalian ubiquitous repeats, *Nucleic Acids Res.* **9:**5383–5397.

Sun, L., Paylson, K. E., Schmid, C. W., Kady, K. L., and Leinwand, L. 1984, Non-Alu family interspersed repeats in human DNA and their transcriptional activity, *Nucleic Acids Res.* **12:**2669–2690.

Sutcliffe, J. G., Milner, R. J., Bloom, F. E., and Lerner, R. A., 1982, Common 82-nucleotide sequence unique to brain RNA, *Proc. Natl. Acad. Sci. USA* **79:**4942–4946.

Thayer, R. E., and Singer, M. F., 1983, Interruption of an α-satellite array by a short member of the *Kpn* I family of interspersed highly repeated monkey DNA sequences, *Mol. Cell Biol.*, **3:**967–973.

Tsukada, T., Watanabe, Y., Nakai, Y., Nakanishi, S., and Numas, 1982, Repetitive DNA sequences in the human corticotropin β lipotropin gene region: *Alu* family members, *Nucleic Acids Res.* **10:**1471–1479.

Ullu, E., Murphy, S., and Melli, M., 1982, Human 7S RNA consists of a 140 nucleotide middle repetitive sequence inserted in an *Alu* sequence, *Cell* **29:**195–202.

Van Ardsell, S. W., Denison, R. A., Bernstein, L. B., and Weiner, A. M., 1981, Direct repeats flank three small nuclear RNA pseudogenes in the human genome, *Cell* **26:**11–17.

Vanin, E. F., Goldberg, G. I., Tucker, P. W., and Smithies, O., 1980, A mouse α globin related pseudogene lacking intervening sequences, *Nature* **286:**222–226.

Voliva, C. F., Jahn, C. L., Comer, M. B., Hutchison, C. A., and Edgell, M. H., 1983, The

L1Md long interspersed repeat family in the mouse: Almost all examples are truncated at one end, *Nucleic Acids Res.* **11**:8847–8859.

Walter, P., and Blobel, G., 1982, Signal recognition particle contains a 7S RNA essential for protein translocation across the endoplasmic reticulum, *Nature* **299**:691–698.

Watanabe, Y., Tsukaga, T., Notake, M., Nakanishi, S., and Numa, S., 1982, Structural analysis of repetitive DNA sequences in bovine corticotropin–β-lipotropin precursor gene region, *Nucleic Acids Res.* **10**:1459–1469.

Weiner, A. M., 1980, An abundant cytoplasmic 7S RNA is complementary to the dominant interspersed middle repetitive DNA sequence family in the human genome, *Cell* **22**:207–218.

Wilde, C. D., Crowther, C. E., Cripe, T. P., Lee, G.-S., and Cowan, N. J., 1982, Evidence that a human β-tubulin pseudogene is derived from its corresponding mRNA, *Nature* **297**:83–84.

Wilson, R., and Storb, U., 1983, Association of two different repetitive DNA elements near immunoglobulin light chain genes, *Nucleic Acids Res.* **6**:1803–1819.

Young, P. R., Scott, R. W., Hamer, D. H., and Tilghman, S. M., 1982, Construction and expression *in vivo* of an internally deleted mouse α fetoprotein gene: Presence of a transcribed *Alu*-like repeat within the first intervening sequences, *Nucleic Acids Res.* **10**:3099–3116.

CHAPTER 6

The Contributions of Retroviruses to the Study of Mammalian Evolution

RAOUL E. BENVENISTE

> The viruses, instead of being single-minded agents of disease and death, now begin to look more like mobile genes. Evolution is still an infinitely long and tedious biologic game, with only the winners staying at the table, but the rules are beginning to look more flexible. We live in a dancing matrix of viruses; they dart, rather like bees, from organism to organism, from plant to insect to mammal to me and back again, and into the sea, tugging along pieces of this genome, strings of genes from that, transplanting grafts of DNA, passing around heredity as though at a great party. They may be a mechanism for keeping new, mutant kinds of DNA in the widest circulation among us. If this is true, the odd virus disease, on which we must focus so much of our attention in medicine, may be looked on as an accident, something dropped.
>
> (Lewis Thomas, 1974, p. 5)

1. Introduction

Until recently, the taxonomic classification of the approximately 4000 living species of mammals (Nowak and Paradiso, 1983) was based primarily on their morphological and behavioral characteristics. During the past two decades, there has been an increasing emphasis on various molecular parameters as a basis for assessing similarities. These methods have included primarily a comparison of proteins, either amino acid sequence (Romero-Herrera *et al.*, 1973), immunologic relatedness (Baba *et al.*, 1975; Cronin and Sarich, 1975; Goodman, 1975; Wilson *et al.*, 1977),

RAOUL E. BENVENISTE • Laboratory of Viral Carcinogenesis, National Cancer Institute, Frederick, Maryland 21701.

or homologous isozyme loci (Avise and Aquadro, 1982); or a comparison of nucleic acid homologies by hybridization of single-copy genomic DNA (Kohne, 1970; Kohne *et al.*, 1972; Benveniste and Todaro, 1976; Brownell, 1983; Sibley and Ahlquist, 1984) or by comparison of restriction enzyme digests of mitochondrial or cellular DNA (Ferris *et al.*, 1981; Adams and Rothman, 1982). In addition, the karyotypes of mammalian species, studied by chromosomal banding techniques, have also been used as a method of establishing phylogenetic relationships (Dutrillaux, 1979; Seuanez, 1982; Wurster-Hill and Centerwall, 1982), although these studies are often complicated by the extreme heterogeneity of chromosome number and organization, even among closely related species. The biochemical and cytogenetic analysis of interspecific somatic cell hybrids (Ruddle and Creagan, 1975; O'Brien *et al.*, 1982, 1985) using either *in situ* hybridization techniques (Harper *et al.*, 1981) or the analysis of electrophoretic migrations of various isozymes in extracts of cells (O'Brien, 1984) has also yielded information on mammalian relationships. Although the emphasis of many recent publications in these areas has been to point out minor differences in phylogenetic classifications (perhaps because of the increased resolving power of the techniques employed), what emerges is a large degree of consensus among the various methods.

This chapter will deal primarily with the use of retroviruses as tools in the study of mammalian evolution. It may be presumptuous for a virologist to delve into the study of evolution, but retroviruses share some properties that render them uniquely suited for these studies. These RNA-containing viruses possess structural genes which are part of an 8-kb-long genome terminated at each end by a 1-kb direct redundancy. The viruses can form a DNA copy of their viral RNA and can then integrate into chromosomal DNA. Infectious retroviruses (Table I) are present only in infected animals as a few copies of proviral DNA per cell, and in that sense fit into the usual definition of a virus. Most of these retroviruses are pathogenic, often producing neoplasia, and may represent infection by viruses originating from still another group of animals.

Endogenous retroviruses, on the other hand (Table II), are present

Table I
Properties of Infectious Retroviruses

1. Present only in the cellular DNA of infected animals
2. Present as only a few (1–2) copies of proviral DNA per cell
3. Horizontally transmitted among members of a species
4. Generally infectious for cells of the same species

Table II
Characteristics of Endogenous Retroviruses

1.	Present in the cellular DNA of all somatic and germ cells of all animals of the species of origin
2.	Vertically transmitted from parent to offspring; may be used as an evolutionary index of genetic distance
3.	Present as multigene families—10–100 copies per haploid genome; these multiple copies seem to evolve more rapidly than other cellular genes; useful as an indicator of taxonomic relationships among closely related species
4.	Generally not infectious for cells of the same species, but may replicate readily in cells of heterologous species
5.	Retroviruses have been transmitted between vertebrate species that are only remotely related phylogenetically, with subsequent incorporation into the germ line; they may thus serve to transfer genes between sympatric species

in the cellular DNA of all somatic and germ cells of all animals of a species, and are thus vertically transmitted from one generation to the next like any other set of cellular genes. In that regard, it is difficult to think of these genes as viruses, since they are subject to the same pressures of selection and mutation that govern the inheritance of other cellular elements. Their ability to replicate occasionally as viruses in a heterologous cell means that these genes can be readily packaged and purified away from the bulk of the cellular DNA; they thus can also be used to generate nucleic acid probes that provide a fresh approach to the study of evolution. These types of studies have facilitated the investigation of the origin and modes of transmission of retroviruses. These integrated endogenous virogenes have remained stable enough to make it possible to detect related nucleic acid sequences among mammalian species that diverged millions of years ago. Studies with the baboon type C virus have also revealed that it is possible to distinguish those Old World monkey and apes whose present habitat is Africa from those that have evolved in Eurasia; these studies have suggested that the recent evolution of macaques, langur monkeys, orangutans, gibbon apes, and man has taken place outside Africa.

Endogenous mammalian viruses have been shown to be present as a multigene family, with approximately ten to several hundred copies of virogenes per haploid genome (Benveniste and Todaro, 1974c). These multiple copies seem to evolve more rapidly than the bulk of the unique sequence cellular DNA and are therefore useful as an indicator of taxonomic relationships among closely related species.

Finally, I also review evidence that retroviruses have in the past been transferred between vertebrate species that are only remotely related phylogenetically. Viral genes from one group of animals can give rise to infectious particles that can integrate into the DNA of another species, be incorporated into the germ line, and then be transmitted as cellular genes. There have now been several documented examples of retrovirus transfer between vertebrate species, including transfers from ancestors of primates to ancestors of carnivores, from rodents to carnivores, from rodents to primates, from rodents to artiodactyls, from primates to primates, and from primates to birds. By examining the donor and recipient species for nucleic acid sequences related to the transferred retrovirus, it is possible to establish the evolutionary date of the virus transmission. The time of virus transfer can then be used as an accurate marker for species divergence that is independent of the fossil record or of any assumptions about the relatedness of species. These findings have permitted studies of the rate of DNA evolution in different mammalian orders.

2. Isolation of Primate Retroviruses

2.1. Introduction

Although retroviruses had been obtained from avian and murine species, it was not until it was realized that viruses could be induced from cells in culture (Lowy *et al.*, 1971), and that many mammalian cell lines spontaneously released retroviruses (Lieber *et al.*, 1973), that an attempt was made to systematically isolate these viruses from man's closest relatives, the primates. The first endogenous primate virus was isolated from baboons after cocultivation of normal tissues with a variety of heterologous (human, canine, feline, bat, mink) cell lines (Benveniste *et al.*, 1974b).

Table III lists the frequency of isolation of endogenous primate viruses. It is evident that among Old World monkeys, the different species of baboons (*Papio cynocephalus, P. anubis, P. papio, P. hamadryas*) as well as the closely related genus *Theropithecus* are unusual in their high propensity to release endogenous retroviruses (77 and 57%, respectively). These isolates are all morphologically and biochemically typical of mammalian type C viruses and are closely related to each other by host range, viral neutralization and interference, and immunologic and nucleic acid hybridization criteria. Among the New World primates, the squirrel monkey also possesses a retrovirus that is readily released (82% frequency of isolation) from normal tissues and cell lines in culture. This virus was initially described by Heberling *et al.* (1977) and belongs to the type D morphological group of retroviruses.

The other endogenous primate viruses isolated from macaques (Todaro *et al.*, 1978a), colobus (Sherwin and Todaro, 1979), langur (Benveniste and Todaro, 1977), and owl monkeys (Todaro *et al.*, 1978b) have been isolated at a much lower frequency; the individual isolates were only

Table III
Frequency of Isolation of Endogenous Primate Retroviruses

Primate	Number of tissues tested[a]	Number of isolates	Percent positive
Old World Monkeys			
Cercopithecinae			
Papio (baboon)	22	17	77
Theropithecus (gelada)	7	4	57
Cercocebus (mangabey)	10	0	0
Erythrocebus (patas)	7	0	0
Cercopithecus (vervet)	16	0	0
Macaca (macaque)	65	1	2
Colobinae			
Colobus (colobus)	18	1	6
Presbytis (langur)	9	1	11
New World Monkeys			
Saimiri (squirrel)	17	14	82
Lagothrix (woolly)	8	0	0
Callithrix (marmoset)	16	0	0
Ateles (spider)	12	0	0
Aotus (owl)	9	1	11

[a]Tissues from the indicated genera were cocultivated with heterologous mammalian cells known to support the growth of primate retroviruses (Benveniste *et al.*, 1974b). No endogenous ape retroviruses have been obtained; the infectious viruses isolated from gibbon, macaque, and man are not listed. Data taken in part from Benveniste and Todaro (1978).

obtained after 5–7 months of cocultivation. The failure to isolate endogenous viruses from the other primate species by these methods may reflect the fact that the production of nondefective exogenous viruses is rare among certain groups of primates, or it may reflect the limitations of these *in vitro* cocultivation techniques, which often depend on a fortuitous choice of the correct host mammalian cell line. For example, the isolate from *P. papio* has only been shown to replicate in a rhesus monkey cell line, and if that cell line had not been included in the original cocultivation experiments, this isolate would not have been obtained.

2.2. Comparison of Primate Viral Isolates

The various primate retroviruses have been tested for their relatedness at the nucleic acid sequence level; Table IV summarizes the data. These isolates are distinct from each other, except for the macaque and colobus type C viruses, which share some protein sequence homology (Oroszlan *et al.*, 1981b), and the various baboon and gelada isolates, which

Table IV
Nucleic Acid Homology between the Genomes of Endogenous Primate Retroviruses

Source of viral RNA[a]		^{3}H-DNA transcripts (percent hybridization)					
		Baboon	Langur	Macaque	Colobus	Squirrel monkey	Owl monkey
Baboon	*P. cynocephalus*	100	1	2	1	4	2
	P. papio	82	1	2	2	5	1
Gelada	*T. gelada*	72	2	2	1	4	2
Langur	*P. obscurus*	2	100	2	2	2	2
Macaque	*M. arctoides*	2	2	100	21	3	3
	M. mulatta	2	2	96	24	2	2
Colobus	*C. polykomos*	2	2	29	100	2	2
Squirrel monkey	*S. sciureus*	1	3	1	2	100	1
Owl monkey	*A. trivirgatus*	2	1	3	2	2	100

[a]The endogenous retroviruses isolated from these species include the baboon viruses M7 (Benveniste *et al.*, 1974b) and PP-1-Lu (Todaro *et al.*, 1976), gelada virus TG-1-K (Todaro *et al.*, 1976), langur virus PO-1-Lu (Benveniste and Todaro, 1977), macaque virus MAC-1 (Todaro *et al.*, 1978a), *M. mulatta* viruses (Rabin *et al.*, 1979; Stromberg and Benveniste, 1983), colobus virus CPC-1 (Sherwin and Todaro, 1979), squirrel monkey virus (Heberling *et al.*, 1977), and owl monkey virus OMC-1 (Todaro *et al.*, 1978b).

are highly related to each other. These results have been extended, in a series of elegant experiments, to a comparison of some of these viruses by restriction endonuclease analysis (Cohen *et al.*, 1982). Although the data are not as quantitative as those obtained by liquid hybridization, the increased specificity permits an analysis of homologies between specific viral genes.

Table V lists all of the endogenous and infectious primate retroviruses that have been isolated. The infectious oncornaviruses include a type C isolate obtained from a woolly monkey fibrosarcoma (Theilen *et al.*, 1971), and a group of closely related type C viruses isolated from gibbons (Kawakami *et al.*, 1972; Todaro *et al.*, 1975) that have been shown to cause myelogenous leukemias and lymphosarcomas in these apes (Kawakami and Buckley, 1974). Infectious type D isolates include the Mason–Pfizer monkey virus (MPMV) from a rhesus monkey carcinoma (Chopra and Mason, 1970) and related isolates obtained from various species of macaques with simian acquired immunodeficiency syndrome (SAIDS) (Daniel *et al.*, 1984; Marx *et al.*, 1984; Stromberg *et al.*, 1984). The infectious human viruses include the HTLV-I and -II isolates, which have been obtained predominantly from patients with T-cell malignancies (Poiesz *et al.*, 1980), and isolates from patients with the acquired immune deficiency syndrome (AIDS): LAV (Barré-Sinoussi *et al.*, 1983) and HTLV-III (Popovic *et al.*, 1984). These lymphotropic viruses are classified in the Lentivirus class of retroviruses (Gonda *et al.*, 1985).

3. Phylogeny of Primates as Indicated by DNA–DNA Hybridization

Earlier studies with nucleic acid probes prepared from the baboon type C viruses showed that the cellular DNA of all baboon tissues and

Table V
Endogenous and Infectious Primate Retrovirus Isolates

	New world monkeys	Old world monkeys		Apes
		Cercopithecinae	Colobinae	
Endogenous				
Oncornavirus type C	Owl	Baboon, macaque	Colobus	—
Oncornavirus type D	Squirrel	—	Langur	—
Infectious				
Oncornavirus type C	Woolly	—	—	Gibbon,
Oncornavirus type D	—	Macaque	—	—
Lentivirus	—	Macaque	—	Man
Spumavirus	Squirrel,	Macaques		Chimpanzee
	Spider	African green	—	—

cell lines contained multiple copies of the virus, and that other Old World monkey and ape tissues also contained partially related viral sequences (Benveniste and Todaro, 1974a). In order to determine if the baboon virogenes were evolving in the same fashion as the rest of the primate cellular DNA, it became necessary to establish the relationships among primates as determined by DNA hybridization.

The extent of reassociation and the thermal stability of DNA hybrids derived from different species are a measure of their evolutionary relationships (Kohne *et al.*, 1972). Nonrepeated sequences comprise a large fraction of mammalian DNA; each sequence is believed to occur only once per haploid genome (Britten and Kohne, 1968). This DNA is therefore suitable for determining the extent of nucleotide diversity that has accumulated since two species diverged from their last common ancestor.

Table VI lists the final percent of hybridization and the thermal stability of hybrids obtained after human cellular DNA is hybridized under varying conditions to the DNA of various primate species. As is evident, the relative relationships between species remain the same regardless of the hybridization conditions employed. For example, the extent of nucleic acid sequence divergence between man and chimpanzee is approximately four times less than between man and the Old World monkeys. The percentage of hybridization is the value obtained after digestion of the hybrids with nuclease S1, an enzyme that specifically digests single-stranded (nonhybridized) nucleic acids (Ando, 1966). Some investigators have also used hydroxyapatite chromatography instead of nuclease S1 to measure the extent of hybridization (Kohne, 1970; Bonner and Todaro, 1980). The thermal stability of hybrids can also be used as an index of the degree of base pair mismatching between DNA strands. Mismatching results in the formation of hybrids that melt at a lower temperature; the effect of mismatched base pairs on thermal stability has been estimated at approximately 1°C per 1% altered pairs (Bonner *et al.*, 1973). As suggested by Kohne *et al.* (1972), a comparison of the same sequences can be accomplished by using the temperature at which 50% of the hybridized sequences are dissociated (T_m). The ΔT_m is the difference in T_m between the other DNA–DNA hybrids and the T_m of the homologous hybrid. ΔT_mR is the thermal stability difference for all nonrepeated DNA, and is the ΔT_m value corrected for the normalized percentage of hybridization to the labeled index DNA species; it corresponds to the $T_{50}H$ values used, for example, by Sibley and Ahlquist (1984). This correction is based on the assumption that the thermal stability values of DNA–DNA hybrids derive from the extent of nucleotide sequence divergence from common ancestral sequences.

Figure 1 shows the thermal melting curves obtained when radioactively labeled unique-sequence baboon cellular DNA is hybridized to the

Table VI
Relationship between Percent Hybridization and Thermal Stability under Various Hybridization Conditions

	Stringent conditions[b]			Moderately relaxed conditions[c]			
Species[a]	Percent hybridization	ΔT_m (°C)	ΔT_mR (°C)	Percent hybridization	ΔT_m (°C)	ΔT_mR (°C)	$T_{50}H$[d] (°C)
Apes							
Man (*H. sapiens*)	100	0.0	0.0	100	0.0	0.0	0.0
Chimpanzee (*P. troglodytes*)	90	1.4	2.3	93	1.3	1.8	1.8
Gibbon (*H. lar*)	64	3.4	6.4	82	3.6	5.0	5.2
Old World monkey							
Baboon (*P. hamadryas*)	56	4.7	9.0	72	5.5	7.7	7.7
New World monkey							
Capuchin (*Cebus* spp.)	33	8.2	15.1	56	9.3	13.0	NT
Carnivore							
Lion (*P. leo*)	4	NT	NT	9	NT	NT	NT

[a] Cellular DNA extracted from these species was hybridized to radioactively labelled human nonrepetitive cellular DNA using the conditions described

[b] Stringent hybridization conditions: hybridization at 65°C, 0.75 M NaCl; S1 nuclease digestion at 45°C, 0.15 M NaCl (Benveniste and Todaro, 1976). NT, Not tested.

[c] Moderately relaxed hybridization conditions: hybridization at 65°C, 0.75 M NaCl; S1 nuclease digestion at 37°C, 0.6 M NaCl (Benveniste *et al.*, 1978). NT, Not tested.

[d] Data from Sibley and Ahlquist (1984); hybridization at 60°C, 0.48 M sodium phosphate; hydroxyapatite detection of hybrids.

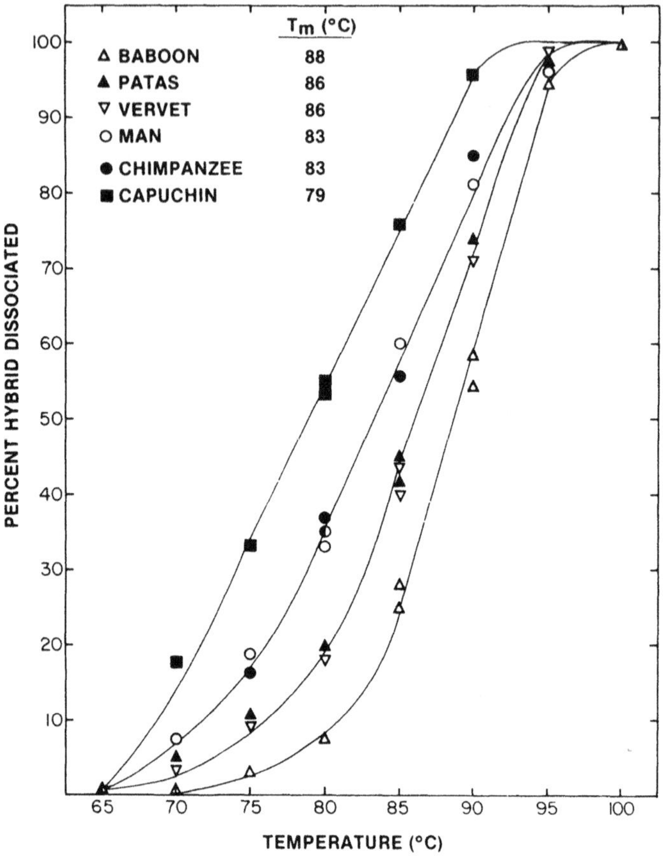

Figure 1. Thermal stability of hybrids formed between radioactively labeled baboon nonrepetitive cellular DNA and various primate DNAs. The T_m is the temperature at which 50% of the hybrids are dissociated.

cellular DNA of the other primate species listed. The T_m, or the temperature at which 50% of the hybrids have dissociated, decreases as a function of the phylogenetic distance of the species from the baboon.

The relationship between percent hybridization and $\Delta T_m R$ is shown in Fig. 2. The lower the percent hybridization detected between two species, the lower the thermal stability of their DNA hybrids. This curve, the slope of which varies depending on the hybridization and S1 nuclease conditions used, would be the same for any diverging set of nucleic acid sequences. The data shown include hybridizations obtained with both carnivore and primate cellular DNAs. It should not be possible to generate

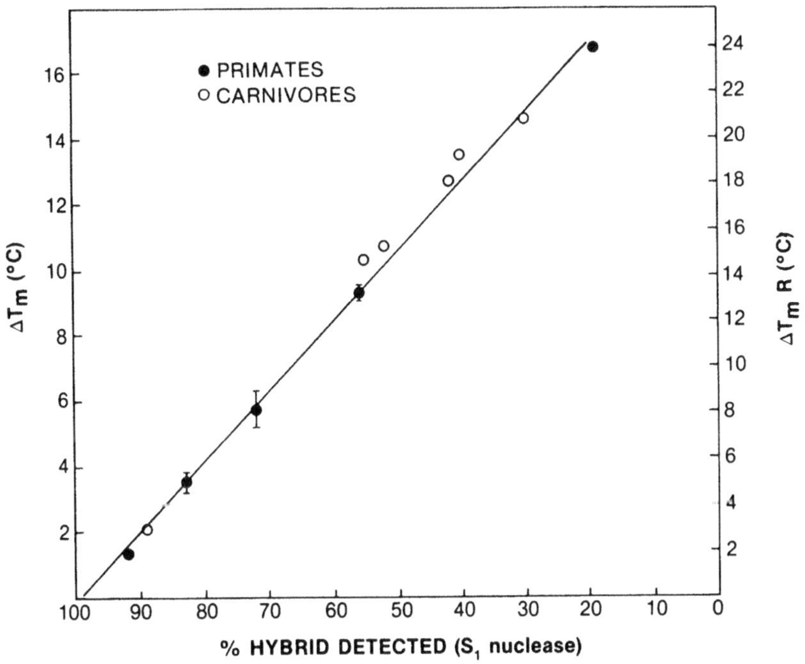

Figure 2. Relationship between percent hybrid detected with S1 nuclease and the thermal melting point difference (ΔT_m or ΔT_mR) of DNA–DNA hybrids. Shown are data obtained for (●) the hybridization of baboon nonrepetitive cellular DNA to the cellular DNA of various primates, (○) and hybridization of domestic cat nonrepetitive cellular DNA to the DNA of various carnivore species.

data points that lie in the region above the curve (no fully homologous hybrid could have a low thermal stability), but points lying significantly below the curve would occur if the sequences being examined contained large numbers of deletions or recombinations instead of point mutations.

The thermal stabilities obtained with various DNA hybrids were used to construct a phylogenetic tree of primate evolution. Table VII lists the data obtained when the cellular DNA from 23 genera of primates was hybridized to DNA from three species of Old World monkeys (baboons, rhesus, langur), a New World monkey (squirrel monkey), and man. Figure 3 summarizes the data obtained for the Old World monkeys. The Cercopithecinae most closely related to the baboon include, in relative order of proximity, the gelada, mandrill, macaques and mangabeys, and patas and guenon. The data suggest that macaques and mangabeys are equally distant from baboons. The cellular DNAs of the more primitive subfamily

Table VII
Thermal Stabilities for DNA-DNA Hybridization Comparisons among Various Primate Species

	^{3}H Unique sequence cellular DNA, $\Delta T_{m}R$ (°C)[a]				
Species	Baboon	Rhesus	Langur	Man	Squirrel monkey
Apes					
Man (*H. sapiens*)	7.6	7.7	7.7	0.0	12.7
Chimpanzee (*P. troglodytes*)	7.7	7.8	7.7	1.8	—
Gorilla (*G. gorilla*)	7.6	7.8	7.4	1.8	—
Orangutan (*P. pygmaeus*)	7.8	—	7.5	3.5	—
Siamang (*S. syndactylus*)	7.9	8.0	—	5.1	—
Gibbon (*H. lar*)	8.0	8.1	7.8	5.0	—
Old World monkeys					
Cercopithecinae					
Baboon (*P. anubis*)	0.0	2.1	—	7.7	—
(*P. cynocephalus*)	0.2	—	—	7.7	—
(*P. hamadryas*)	0.7	—	4.1	7.6	—
Gelada (*T. gelada*)	1.3	—	—	7.5	—
Mandrill (*M. sphinx*)	1.8	—	—	8.0	—
Macaque (*M. mulatta*)	2.1	0.0	4.2	7.9	—
(*M. arctoides*)	2.1	0.4	4.3	7.7	13.0
(*M. nemestrina*)	2.2	1.0	4.2	7.8	13.1
Mangabey (*C. atys*)	2.2	2.0	4.4	7.7	—
Guenon (*C. campbelli*)	3.2	2.8	4.0	7.6	—
Patas (*E. patas*)	3.1	3.0	—	7.7	—
Colobinae					
Colobus (*C. guereza*)	4.2	4.3	3.0	7.6	—
Langur (*P. obscurus*)	4.3	4.5	0.0	7.8	—
(*P. senex*)	4.3	—	1.4	—	—
New World monkeys					
Woolly (*Lagothrix* spp.)	13.0	13.5	—	13.0	9.2
Capuchin (*Cebus* spp.)	13.4	—	—	13.1	6.4
Howler (*Alouatta* spp.)	12.9	—	—	13.0	—
Spider (*A. paniscus*)	13.3	—	—	13.3	8.2
Squirrel (*S. sciureus*)	13.0	—	13.2	13.2	0.0
Owl (*A. trivirgatus*)	13.4	—	—	13.1	7.9
Prosimians					
Galago (*G. senegalensis*)	24.5	—	—	24.5	24.0
Slow loris (*N. coucang*)	—	—	—	23.7	—

[a] The thermal stabilities listed are from various experiments (5–20) involving multiple individuals of the species listed. Some of these data have been previously published (Benveniste and Todaro, 1976, 1977). Dashed Lines indicate hybridizations not tested. The homologous DNA–DNA hybrids melted at temperatures that ranged from 88.0 to 94.0°C; unique sequence DNA from baboon, rhesus, langur, man, and squirrel monkey were used as the radioactively labeled index species probe. The standard deviations for these values are approximately ±0.4°C for $\Delta T_{m}R$ values less than 5°C, and ±0.7°C for $\Delta T_{m}R$ values greater than 5°C.

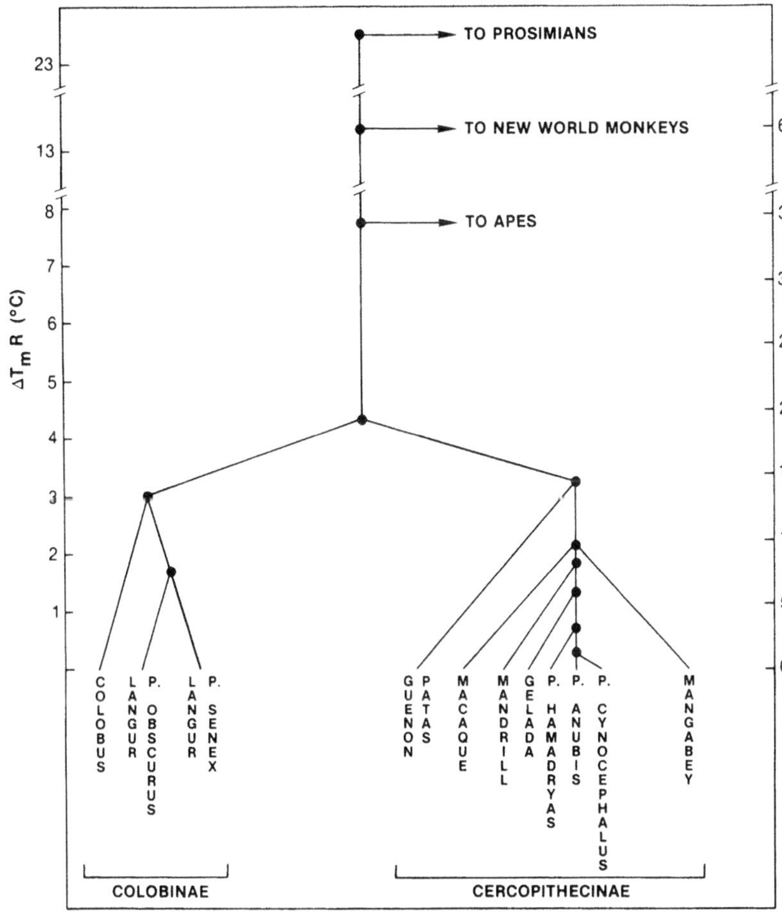

Figure 3. Phylogenetic tree depicting relationships among the cellular DNAs of Old World monkeys as determined by DNA hybridization. Baboon, macaque, langur, and colobus cellular DNAs were hybridized to the various species shown. Some of these data are listed in Table VII. The time of divergence of prosimians can only be estimated (see Section 3).

Colobinae show a lower thermal stability to baboon and rhesus DNA than does the cellular DNA of any of the monkeys of the subfamily Cercopithecinae. When radioactively labeled langur DNA is used as the index species, it is evident that the two species of langurs (*Presbytis obscurus* and *P. senex*) are closely related to each other, followed by *Colobus* and then the various members of the other Old World monkey subfamily, the Cercopithecinae. Ape cellular DNA has a ΔT_mR of 7.4–8.0°C, New World monkeys a $\Delta T_m R$ of 12.9–13.4°C, and prosimians a $\Delta T_m R$ of 24.0°C relative to the DNA of the Old World monkey species tested. The branching points for these groups of primates are also shown in Fig. 3. Although there is some disagreement about the exact times of divergence of the various primate species, we have used the time scale suggested by Pilbeam (1982), which places the divergence of the orangutan from man at approximately 16 million years ago (mya). A comparison of that time scale and these hybridization studies suggests that a $\Delta T_m R$ of 1°C corresponds to approximately 4.4 million years of evolution. This value is in excellent agreement with that obtained by Sibley and Ahlquist (1983, 1984) using hydroxyapatite to measure avian and primate cellular DNA reltionships. The relationship between $\Delta T_m R$ and time may not be linear at values greater than 15–20°C; the dates of prosimian divergence are therefore only approximate. In any case, even if the actual dates chosen for the divergence of the various primate species are shown to be wrong by future fossil or geological findings, the relative relationships among the various primate species would still remain the same.

Similar studies with the cellular DNA of six genera of apes and man are summarized in Fig. 4. To generate this phylogenetic tree, radioactively labeled unique sequence DNAs from chimpanzee, gorilla, man, and gibbon were hybridized to cellular DNAs extracted from tissues of several representatives of each species. As is evident, man is quite closely related to the chimpanzee and gorilla; the human–chimpanzee and human–gorilla cellular DNA hybrids dissociate with a $\Delta T_m R$ of approximately 1.8°C lower than the dissociation temperature obtained for the reannealing of human DNA. Our data do not reveal any two of these three species as being closer to each other than to the third (Benveniste and Todaro, 1976); they seem to have diverged from a common ancestor approximately 8 mya. Our next closest relatives are the orangutan, then the gibbon and siamang. The gibbon and siamang are as distantly related to each other ($\Delta T_m R$ of 2.1°C) as chimpanzee and gorilla are to man. This nucleic acid comparison of ape cellular DNA supports the classification proposed on an immunologic basis by Goodman and Moore (1971) that places man, chimpanzee, and gorilla in one subfamily (Homininae), since they clearly had a common ancestor more recently than any of these three species had with the orangutan or with the gibbon.

The recent divergence of chimpanzee, gorilla, and man appears by our data to be a trichotomy (Fig. 4). The statistical error in these liquid hybridization studies (± 0.3 to ± 0.5°C out of a total of 1.8°C $\Delta T_m R$ measured) suggests that one could not detect two of the species as being phylogenetically closer if they had diverged less than 22% of the total time since the oldest divergence. Recently, Sibley and Ahlquist (1984),

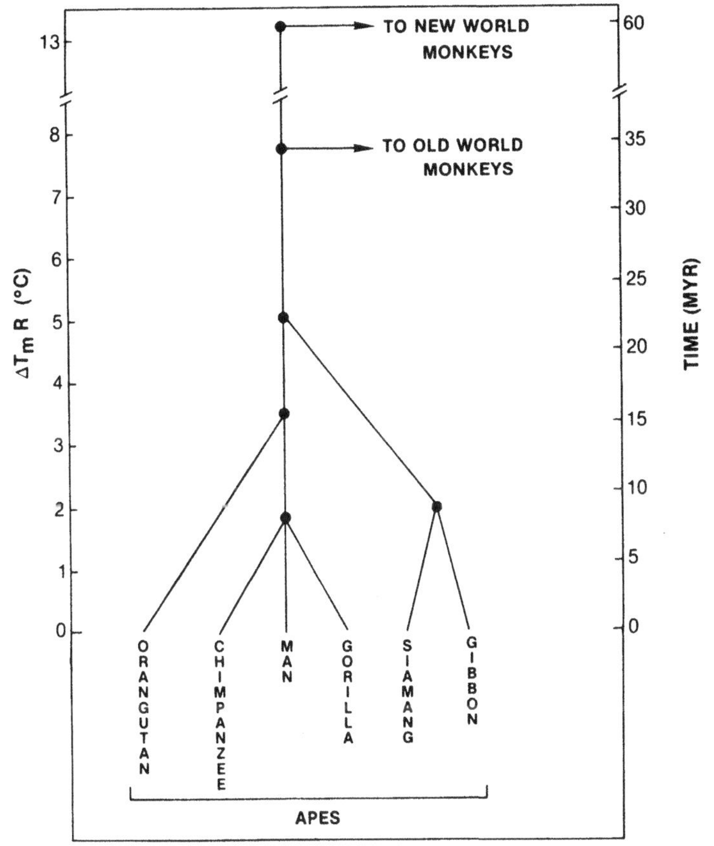

Figure 4. Phylogenetic tree depicting relationships among the cellular DNAs of the apes as determined by DNA hybridization. Data primarily from Benveniste and Todaro (1976).

using similar liquid hybridization techniques, have suggested that the chimpanzee and man had a common ancestor 22% (1.8°C $T_{50}H$ out of 2.3°C) more recently than the gorilla divergence.

Since the rate of evolution of the mitochondrial genome is approximately ten times faster than that of single-copy nuclear DNA (Brown *et al.*, 1979), it has also been used for the analysis of closely related species such as the hominoids. Mitochondrial DNA also differs from nuclear DNA in that it is present in multiple copies per cell and is maternally inherited. Recently, Ferris *et al.* (1983) and Powell (1983) have shown that mitochondrial DNA can cross species barriers, so that closely related but distinct species can share a common mitochondrial genome. This may render this DNA less reliable as an indicator of phylogenetic relationships.

The DNA sequence of a cloned 0.9-kb mitochondrial fragment (Brown *et al.*, 1982) and a restriction endonuclease cleavage analysis (Ferris *et al.*, 1981) support the view that humans are more closely related to the chimpanzee and the gorilla than to the orangutan. However, alternative phylogenies that do not resolve the chimpanzee–gorilla–human trichotomy could not be excluded. Templeton (1983) has developed an algorithm that attempts to deal with the nonrandomness in the distribution of restriction sites in DNA (Adams and Rothman, 1982) and with the assumptions that shared restriction sites reflect common evolutionary origins and are not the result of convergent evolution. He concluded that the data of Ferris *et al.* (1981) could be resolved by his algorithm at the 50% level of significance to show that chimpanzee and gorilla are more closely related to each other, and had diverged from each other after the hominid–African ape split. However, Nei *et al.* (1985) have developed statistical methods for computing the standard errors of branching points of evolutionary trees and have reconstructed an evolutionary tree of human and several ape species using four types of molecular data. Although the DNA sequence data suggested that the chimpanzee is the closest and that the gorilla is the next closest to the human species, the differences between the branching points for the human and the chimpanzee species and that for the gorilla species and the human-chimpanzee group was not statistically significant. The fact that the details of hominid phylogeny thus depend on the choice of the algorithm used to analyze the data may suggest that in fact the chimpanzee, gorilla, and human ancestors diverged at approximately the same time.

All of the molecular data, including those shown here, reveal that the orangutan is more distantly related to man than are the gorilla and the chimpanzee. However, after reviewing selected morphological characteristics, Schwartz (1984) concluded that man and orangutan had a common ancestor not shared with any African apes. The difficulty with these types of studies, however, is that the choice of the morphological parameters are often arbitrary. In addition, there need not be a correlation between molecular changes and certain morphological parameters (King and Wilson, 1975; Cheny *et al.*, 1978).

Figure 5 examines the relationships among the cellular DNAs of New World monkeys, using squirrel monkey as the index species. The five genera examined are approximately equidistant from squirrel monkeys, although the squirrel monkey (*Saimiri*) appears closer to the capuchin monkey (*Cebus*). Since the $\Delta T_m R$ values measured between the New World primate genera were approximately 6–9°C, these primates are as distantly related to the squirrel monkey as the Old World monkeys are to apes and man ($\Delta T_m R$ = 8°C). These primate relationships correlate

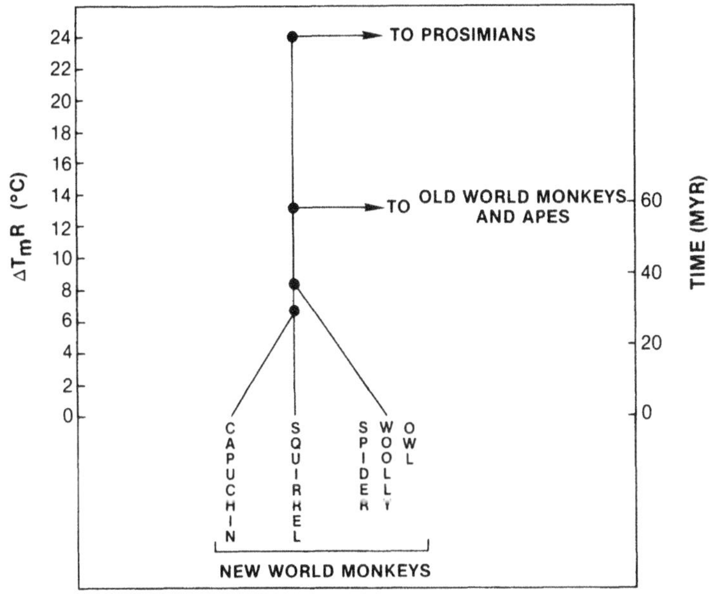

Figure 5. Phylogenetic tree depicting relationships among the cellular DNAs of New World monkeys. The radioactively labeled nonrepetitive cellular DNA extracted from a squirrel monkey lung cell line was hybridized to the DNA of the various species shown. The actual ΔT_mR values are shown in Table VII. The time of divergence of prosimians can only be estimated (see Section 3).

well with those found by most other investigators who have used DNA reassociation techniques (Kohne *et al.*, 1972; Hoyer *et al.*, 1972; Sibley and Ahlquist, 1984) or protein data (Goodman and Moore, 1971; Goodman, 1975; King and Wilson, 1975; Cronin and Sarich, 1975; Sarich and Cronin, 1976; Bruce and Ayala, 1979) to classify the primates (see, for example, Table VIII), as well as with the phylogenetic relationships established on the basis of anatomical considerations and the fossil record (Jolly, 1970; Simons, 1970, 1976).

3.1. Generation Time Effect on the Average Rate of DNA Evolution

Some investigators (Kohne, 1970; Kohne *et al.*, 1972) have proposed that the generation length of a species influences the rate of accumulation of base pair mutations in DNA. This conclusion was reached because the cellular DNA difference between presumably recently diverged rodent species such as mouse and rat was much greater than expected. The thermal stability measurements for the mouse–rat hybrids (ΔT_m, ΔT_mR,

Table VIII
Comparison of Albumin Immunologic Distance and DNA Hybridization Measurements among the Primates

Groups compared[a]	Albumin[b]	DNA[c]
Homo–Pan	0.12	0.14
Homo–Pongo	0.25	0.27
Hominoidea–Cercopithecoidea	0.58	0.59
Catarrhini–Platyrrhini	1.0	1.0
Anthropoidea–Prosimii	2.1	1.8

[a] The relative molecular differences between these primate groups was determined relative to a catarrhine–platyrrhine value of 1.0.
[b] The albumin data were obtained from Sarich and Cronin (1976, p. 155).
[c] The DNA data were derived by dividing the thermal stability values ($\Delta T_{\mathrm{m}}\mathrm{R}$) shown in Table VII by the corresponding catarrhine–platyrrhine value ($\Delta T_{\mathrm{m}}\mathrm{R}$ = 13.1°C).

T_{50}H, or Δ mode) have ranged from approximately 10.5 to 20°C (Kohne, 1970; Benveniste and Todaro, 1975b; Benveniste *et al.*, 1977; Brownell, 1983). This would correspond to a divergence time (based on a $\Delta T_{\mathrm{m}}\mathrm{R}$ of 1°C corresponding to 4.4 million years) of 46–88 mya, which was thought to be too ancient a divergence. Sarich and Wilson (1973) used a relative rate test in which an outside mammalian reference order (carnivores) was compared immunologically to the hominoid (longer generation time) and Old World monkey (shorter generation time) lineages. They found no evidence of a rate change, thus supporting the "molecular clock" model of evolutionary rates (Fitch, 1976).

Table IX represents a relative rate test in primates using DNA hybridization techniques. The reference species in one study was rhesus, an Old World monkey. DNA hybrids formed between rhesus cellular DNA and the cellular DNA of two apes with different generation times (man and gibbon) melted at approximately the same temperature. Similarly, when the index species was a New World monkey, both human and rhesus DNA formed hybrids that again melted at approximately the same temperature. Although the data do show the species with longer generation time as forming slightly more stable hybrids, the differences are not significant. Thus, to the extent that these experiments are valid relative rate tests (and assuming that present-day generation lengths provide an estimate of the generation time of the ancestors of the species), the data do not show an effect of generation length on the average rate of primate

Table IX
Effect of Generation Length of Primates on the Average Rate of DNA Evolution[a]

Index species	Species tested	$\Delta T_m R$ (°C)
Rhesus	Man	7.7 (±0.4)
Rhesus	Gibbon	8.0 (±0.6)
Owl monkey	Man	12.7 (±0.5)
Owl Monkey	Rhesus	13.0 (±0.6)

[a] ^{3}H-labeled nonrepetitive cellular DNA from rhesus and owl monkey cell lines was hybridized to the species listed. The decrease in thermal stability ($\Delta T_m R$) of the hybrids formed is listed, along with the standard deviation.

DNA evolution. Sibley and Ahlquist (1983) came to the same conclusion in a relative rate test involving DNA comparisons among birds with different generation lengths. However, the key parameter in generation time may be the number of cell replications per year in the germ line instead of gestation lengths. At this level the difference among hominoids and catarrhines may be very small and may account for the apparent "success" of the DNA hybridization rate tests. The rates of molecular evolution may also not always be constant. Goodman *et al.* (1983) have reported that the rate of evolution among hominoid globin genes has been decelerating. There is also evidence that the rate of nucleotide substitutions in rodents may be faster than in man (Li *et al.*, 1985), and Bonner *et al.*, (1980) have reported that the evolution of cellular DNA sequences has been retarded in Malagasy primates.

3.2. Thermal Stability Measurements of Mammalian Species Capable of Producing Viable Hybrids

Hybrid animals have been obtained by breeding different species of baboons as well as *Papio* and *Theropithecus*, and *Papio* and *Macaca*. The *Papio–Theropithecus* hybrids are apparently fertile, while the *Papio–Macaca* hybrids are reported to be sterile (Markarjan *et al.*, 1974). Myers and Shafer (1979) have also reported hybrids between gibbon and siamang apes. Table X lists the thermal stability differences between the cellular DNA of the parents of various primate hybrids, as well as for the domestic cat and leopard cat, which can produce viable hybrids in which the females are usually fertile (Centerwall, 1974). The baboon–gelada divergence, where the offspring are fertile ($\Delta T_m R$ = 1.3°C), is thus more recent than the baboon–macaque divergence, where the offspring are

Table X
Thermal Stability Differences between the Nonrepetitive Cellular DNA of Various Mammalian Species That Can Mate to Produce Viable Hybrids

Parent species[a]	$\Delta T_m R$ (°C)[b]
Domestic cat × leopard cat (*Felis catus* × *Felis bengalensis*)	2.2 (±0.2)
Mangabey × macaque (*C. atys* × *M. mulatta*)	2.1 (±0.3)
Baboon × macaque (*P. hamadryas* × *M. mulatta*)	2.2 (±0.2)
Baboon × gelada (*P. hamadryas* × *T. gelada*)	1.3 (±0.2)
Langur × langur (*P. senex* × *P. entellus*)	1.7 (±0.3)
Gibbon × siamang (*H. moloch* × *S. syndactylus*)	2.0 (±0.2)

[a] Mammalian hybrid data are taken from Gray (1971), Markarjan *et al.* (1974), and Myers and Shafer (1979).
[b] The decrease in thermal stability ($\Delta T_m R$) and the standard deviation between the cellular DNA's of the various pairs of animals are listed.

sterile (2.2°C). The data suggest that a $\Delta T_m R$ difference of approximately 2°C (corresponding to about 2% base sequence divergence) for the reassociation of hybrids may represent the upper limit in carnivores and primates for genetic compatibility leading to the production of viable offspring. It will be interesting to extend these types of measurements to other mammalian orders such as rodents, as well as to other vertebrates.

Wilson *et al.* (1974) have compared the serum albumins of 31 pairs of mammals capable of producing viable hybrids. The albumins of these pairs differed by approximately 3 "immunologic distance units" (range 0–10), corresponding to a 0.6% amino acid sequence difference. Given the degeneracy of the genetic code, this value is in good agreement with the data shown in Table X.

4. Evolution of Primate Retrovirus Genes

4.1. Evolution of Baboon Type C Viral Genes

Previous studies had shown that nucleic acid sequences related to the baboon type C virus can be detected in the cellular DNA of all Old

World monkey species (Benveniste and Todaro, 1974a). Since the evolutionary relationships among Old World monkey and ape species had been established by DNA hybridization techniques (see Fig. 3 and 4), we decided to examine these primates for the extent of homology of their virogenes to the type C virus isolated from baboons.

Figure 6 lists the final percent homology obtained after hybridizing a DNA probe prepared from the baboon type C virus to primate DNA. These data are superimposed on the phylogenetic tree of Old World monkey and ape evolution shown in Fig. 3 and 4. The data show that sequences related to the baboon virus can be detected in all Old World monkey and ape species; no homology can be detected in New World monkey DNA (Benveniste and Todaro, 1976). The presence of virogene sequences in

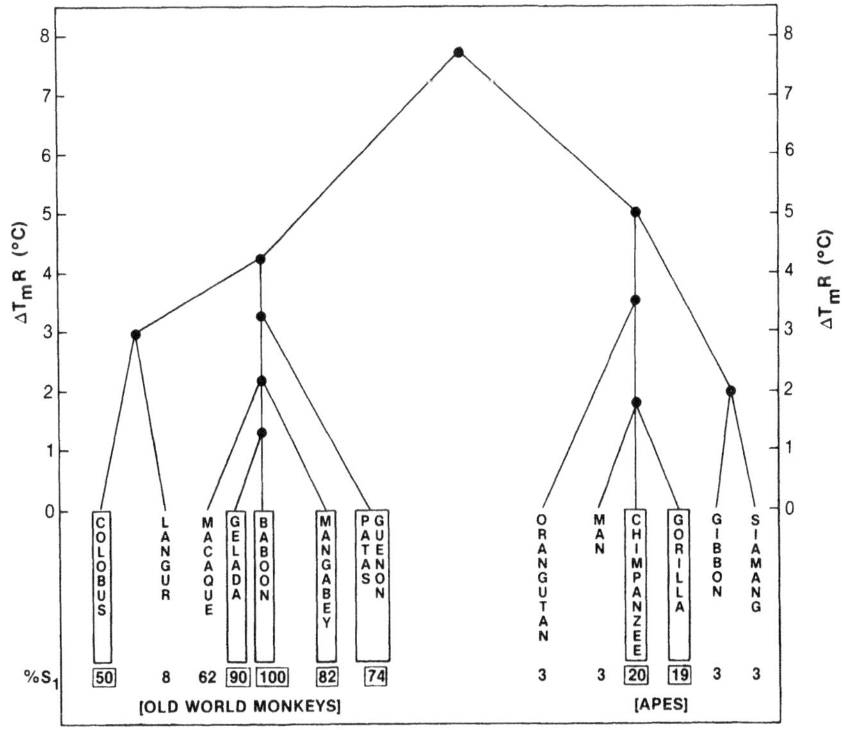

Figure 6. Phylogenetic tree showing relationships among the cellular DNAs of Old World monkeys and apes; the $\Delta T_m R$ of the DNA hybrids is shown along the ordinate. A ^{3}H-labeled DNA probe prepared from the baboon type C virus grown in bat cells was hybridized to the cellular DNA of the various species. The final percent hybridization after digestions of the hybrids with nuclease S1 is shown along the abscissa. The species enclosed by boxes represent primates whose habitat is Africa; the other species inhabit Eurasia.

ape DNA suggests that this virus has been present in the primate stock as a stable cellular element ever since the divergence of Old World monkeys and apes approximately 30 mya.

A striking finding from Fig. 6, however, is that the extent of nucleic acid homology for these type C virogene sequences does not exactly parallel the evolutionary relationships based on the nonrepetitive cellular DNA. For example, even though the macaques and mangabeys are approximately equally distant phylogenetically from the baboon ($\Delta T_{m}R$ = 2.2°C; see also Table VII), the virogene sequences of these animals are quite different, hybridizing 62 and 82% to baboon viral DNA, respectively. The data obtained with the Colobinae are even more striking. Colobus and langur are equidistant relative to baboon cellular DNA ($\Delta T_{m}R$ = 4.3°C), but the virogenes of these two species hybridize 50 and 8% to the baboon viral probe, respectively. Finally, although the cellular DNA of all six genera of apes shown in Fig. 6 are phylogenetically equidistant from the baboon (cellular DNA $\Delta T_{m}R$ = 7.8°C), two of the apes, chimpanzee and gorilla, contain virogene sequences that are much more closely related to baboon virogene DNA than are those of the other apes (orangutan, gibbon, siamang) or man. In fact, in contrast to the data obtained with the baboon cellular DNA probe, chimpanzee and gorilla cellular DNA hybridize more to the baboon virogene probe than does the DNA of the langur, an Old World monkey.

4.2. Effect of Geographic Habitat of a Species on the Rate of Baboon Virogene Evolution

The Old World monkeys and apes inhabit both Africa and Eurasia. In Fig. 6, those primates whose natural habitat is Africa are enclosed in a box; the other primates inhabit Eurasia. The results show that the extent of nucleic acid dissimilarity between the primate virogenes depends on two factors: (1) the phylogenetic distance of the species from *Papio*; and (2) the geographic habitat of that species.

Figure 7 plots the thermal stability ($\Delta T_{m}R$) of primate cellular DNA hybrids obtained after hybridization to either the baboon nonrepetitive cellular DNA or to the baboon viral DNA probe. The lower curve is a plot of the nonrepetitive cellular DNA data. When these same primate DNAs are hybridized to the baboon viral probe, the other two curves are obtained. All of the African primates have viral gene sequences whose thermal stability falls on a straight line as expected based on their evolutionary distance from the baboon. Similarly, the data obtained with the Asian primates also fall on a straight line; however, their viral sequences show correspondingly less homology than their African counterparts. Thus,

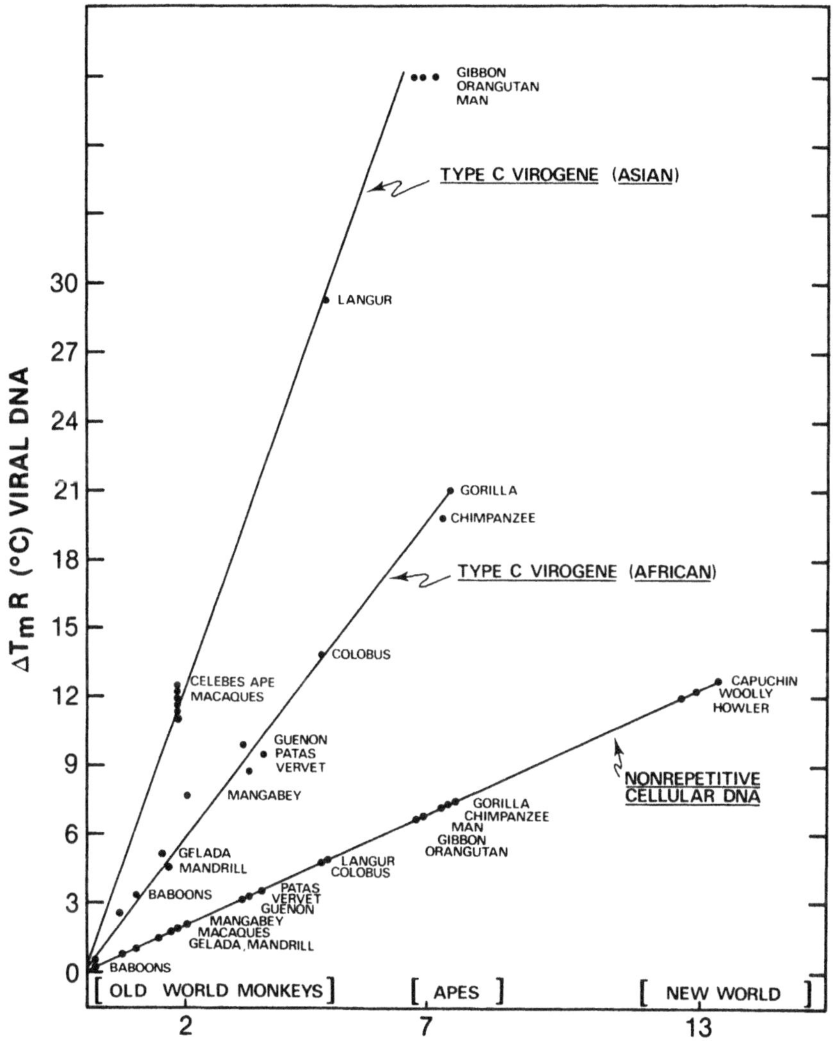

Figure 7. Relationship among primate DNAs as determined by homology to baboon nonrepetitive cellular DNA and to the baboon type C viral DNA probe. The genetic distance from the baboon (*Papio cynocephalus*) of various Asian and African primate species as determined by the thermal stability (ΔT_mR) of hybrids formed between their cellular DNA and baboon *cellular* DNA is plotted along the abscissa (lower curve). The thermal stability of hybrids as determined with the baboon *viral* probe is plotted along the ordinate (African and Asian virogene curves).

with the endogenous primate type C viral genes (although not with the overall cellular DNA), we are able to discriminate quite clearly between African and Asian primates. Among the apes, the gibbon and orangutan stand out as Asian, and gorilla and chimpanzee as African. The data obtained with the baboon virogenes suggest that after diverging from the chimpanzee and gorilla, man's ancestors were developing in Asia. These studies with the baboon viral probe have also been extended to show that even African prosimian (*Galago*) cellular DNA contains related virogene sequences (Bonner and Todaro, 1980).

Of all the primates, man and his ancestors have been the most successful at adapting to various habitats, and have migrated all over the world; their geographic origin is thus more obscure than that of the other primates. The earliest undoubted hominid fossils include *Australopithecus afarensis* ("Lucy"), discovered by Donald Johanson and his colleagues at Hadar in Ethiopia, and dated at 3.5 mya. *Homo erectus* also seems to have originated in Africa 1.5 mya. *Homo sapiens* first appeared 100,000 years ago and has spread rapidly all over the world. Hominid evolution in Africa before 3.5 mya is obscure because, for most of the early Pliocene, the African hominid branch is virtually without fossil representation. *Ramapithecus* from India has been dated at 7 mya and its morphological jaw and tooth resemblance to *Australopithecus* is either dismissed (as an example of parallel or convergent evolution) or emphasized, depending on the investigator's beliefs in hominoid evolution (Pilbeam, 1983). Because of man's close morphological resemblance to the African apes, and because the majority of the recent paleontological discoveries have been made in Africa, many investigators have proposed an African origin for man's ancestors. The results presented here suggest a significant portion of man's recent evolution occurred outside Africa. The time of divergence of man from chimpanzee and gorilla has been estimated at between 4 and 20 mya; our data indicate a divergence time of approximately 8 mya. Thus, for virtually all of the Pliocene, man's direct ancestors were evolving outside Africa. The nucleic acid changes we are measuring are the result of millions of years of evolution. The data, therefore, do not preclude a more recent migration, perhaps in the last 2–3 million years, of hominids into Africa. The more recent migrations by man all over the world are also events too recent for the baboon viral probe to detect.

Why should the baboon virus be able to distinguish the geographic habitat of Old World primates? Table XI lists some of the possible factors that might be involved. As was shown in Table III, baboons and their close relatives are unusual among Old World monkeys in their ability to release infectious retroviruses frequently. In addition, an endogeneous domestic cat virus has been postulated to have originated from an en-

Table XI
Possible Bases for Geographic Effect on Rates of Baboon Virogene Evolution

1. Baboons and their close relatives are unusual among Old World monkeys in their ability to release infectious retroviruses frequently (see Table III)
2. Ancestors of the baboon also transmitted a virus to ancestors of the domestic cat; this virus is now part of normal cat genome (see Section 6.1)
3. The five species of baboons range throughout most of Africa
4. Baboon virus replicates well in cells of Asian primate origin (macaque, gibbon), but does not infect African primate cell lines (mangabey, vervet)
5. African primates may have had to maintain a close nucleic acid sequence homology in order to restrict the growth of this potentially pathogenic virus
6. Asian primates, not exposed to baboons, have not had to conserve their virogene sequences

dogenous Old World monkey type C virus horizontally transmitted to an ancestor of the domestic cat (Benveniste and Todaro, 1974b). This suggests the possibility that baboons and their ancestors, which ranged throughout Africa (Napier and Napier, 1967), may have been releasing infectious virus particles for several million years. The viruses they released, while perhaps harmless to baboons (endogenous viruses do not replicate in the cells of the species of origin), may have had a significant effect on animals coming in contact with them. The other African primates may well have had to conserve a greater degree of nucleic acid sequence homology relative to the baboon type C virus in order to restrict the replication of this potentially pathogenic virus in their cells. Evidence that this may have been the case comes from the fact that cell lines derived from African primate species do not support the growth of the baboon type C virus in culture, whereas cells of Asian primate origin allow virus replication.

Asian primates, not exposed to the baboon virus because of their geographic isolation, would be under less selective pressure to maintain cellular nucleic acid sequences similar to those present in the baboon type C viruses; their virogenes, then, would be free to accumulate point mutations at an apparently faster rate relative to the African primate virogene sequences. The ancestors of man, which are postulated to have evolved in a geographic area (Eurasia) where they would not have been in contact with baboons, would be expected to be susceptible to infection by the baboon type C virus. Human cell cultures can in fact be easily infected by the various baboon viral isolates (Benveniste *et al.*, 1974b).

The DNA hybridization data obtained with the baboon virogenes has thus had an unexpected anthropological twist, since they reveal primate and human geographic origins and habitats. These genes, if they provide even a slight selective advantage to the host by offering protection against disease, would be strongly selected for during the course of evolution. Other cellular genes have been described that also have a geographic component to their evolutionary selection. For example, the genes that code for sickle-cell anemia (hemoglobin S) are very common throughout sub-Saharan Africa and are absent in other human populations. These genes are strongly selected for in the African environment because humans that are heterozygous for hemoglobin S are more resistant to malarial infection. The genes for hemoglobin S can thus be used as a reliable indicator of the recent African ancestry of some modern men. In an analogous fashion, the baboon virogenes seem to be reliable indicators of the non-African habitat, millions of years ago, of macaque, langur, gibbon, orangutan, and man.

The multiple retroviral genomes present in primate DNA have also been examined by restriction enzyme analysis of cloned viral fragments. Birkenmeier *et al.* (1982) showed that colobus DNA contains multiple families of viral sequences that are only partially related to each other. Liquid hybridization data had shown that colobus DNA contains multiple copies of nucleic acid sequences related to the baboon, langur, and macaque endogenous retroviruses, all of which are distinct (Table IV). Bonner *et al.* (1982a) cloned a viral fragment from chimpanzee DNA that was related to one of the colobus families of retroviruses (but not to the baboon type C virus related family). A qualitative analysis of restriction patterns revealed that closely related viral sequences were found in closely related species, and different fragments in more distantly related species, as would be expected for a long-term germ-line transmission of these sequences. Under moderately stringent conditions, no hybridization was detected to human, gibbon, or orangutan DNA, whereas chimpanzee and gorilla DNA yielded multiple related fragments. The authors concluded that these sequences were therefore absent or had been deleted in those three Asian primate genomes. However, since gibbon diverged before orangutan, and both species diverged before the human–chimpanzee–gorilla trichotomy, a total absence of viral sequences in gibbon, orangutan, and man would necessitate three independent deletion events. Since the multiple viral copies are known to be dispersed at various chromosomal sites throughout the genome, it is difficult to envision how they all were deleted in just these three Asian species. Moreover, hybridization under low stringency conditions has since revealed the presence of multiple viral sequences in the cellular DNA of these Asian primates.

Endogenous retroviral sequences have now been cloned from human cellular DNA by Martin *et al.* (1981), Bonner *et al.* (1982b), Noda *et al.* (1982), Callahan *et al.* (1982), Repaske *et al.* (1983), and O'Connell *et al.* (1984). One such clone, ERV-1 (Bonner *et al.*, 1982b), has nucleic acid sequences related to the polymerase region of the baboon endogenous virus and is present in approximately 12 copies in human DNA. The presence of identical cellular flanking sequences in chimpanzee and human DNA suggests that this virus has been in its present chromosomal location prior to the divergence of man and the chimpanzee. Another cloned human retrovirus, ERV-3, is located on a different chromosome than ERV-1, and is a full-length but defective human provirus, which was isolated by hybridization at low stringency with a chimpanzee viral clone (O'Connell *et al.*, 1984). ERV-3 and ERV-1 are partially related to each other. These studies have confirmed that viruses distantly related to the baboon endogenous virus *are* present in human DNA, and that the extent of this homology is much less than that found in gorilla or chimpanzee DNA. The determination of the molecular basis for the differences in the Asian and African baboon-related virogenes will require a comparison among all the hominoids of nucleic acid sequences related to these viruses.

4.3. Evolution of Langur Type D Viral Genes

Langurs belong to the other subfamily of Old World monkeys, the Colobinae or leaf-eating primates. A type D virus has been isolated (Table III) from normal langur (*Presbytis obscurus*) fibroblasts. This virus is endogenous, since there are multiple copies integrated in the DNA of all langurs (Benveniste and Todaro, 1977), and sequences related to the langur virogene can be detected in both subfamilies of Old World monkeys. Since the estimated time of divergence between these two subfamilies is about 20 mya (Fig. 3), this then represents a minimum time that type D viral genes have been present in the primate lineage. The highest degree of homology is obtained with primates that are phylogenetically closest to the langur, such as other langur species and *Colobus*.

A correlation between the phylogenetic proximity of a species to the langur and the extent of hybridization to the langur viral probe is plotted in Fig. 8B. Members of the genus *Presbytis* range throughout Southeast Asia, while the genus *Colobus* is restricted to Africa. Since both genera are primarily tree-dwellers, the desert between Africa and Asia has constituted an impenetrable barrier, preventing contact between these species for several million years (Napier, 1970). Nevertheless, both have retained viral gene sequences in their DNA. Unlike the data obtained with the baboon type C virogenes (Figs. 7 and 8A), no environmental effect is evident between species with African or non-African habitats. A comparison of the slopes of the lines in Figs. 8A and 8B shows that the rate of evolution of the langur viral sequences parallels the rate seen with the baboon viral sequences among Asian primates. This comparison thus

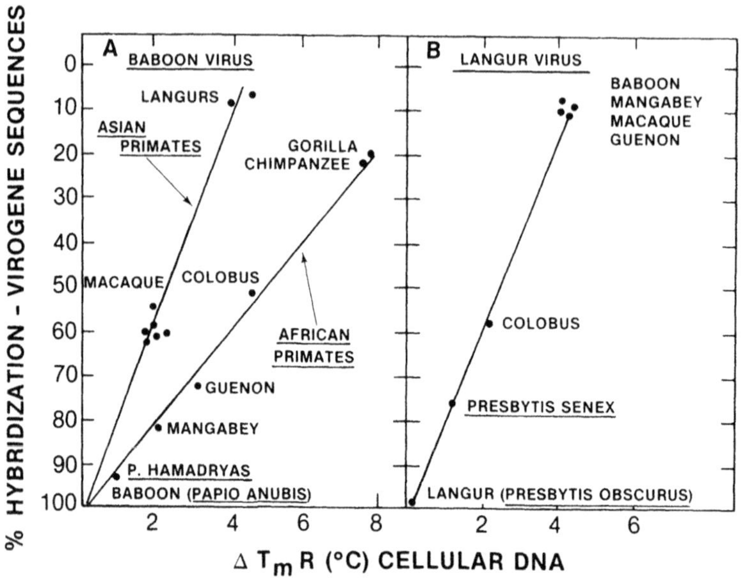

Figure 8. Evolution of langur and baboon virogenes. (A) The data shown in Fig. 7, except that the homology of the primate cellular DNAs to the baboon viral DNA is shown as percent hybridization instead of ΔT_mR. (B) The relationship among primate cell DNAs as determined by homology to langur *cellular* DNA (abscissa) and to the langur *viral* DNA probe (ordinate). The rate of langur virogene evolution appears to be the same as the Asian component of the baboon virogene data.

reveals that what is unusual about the baboon virogenes is their abnormally *slow* rate of evolution among African primates.

The langur virus was obtained as a rare, one-time isolate, and grows very poorly in cell culture. The lack of an environmental effect may therefore reflect the low infectivity and lack of pathogenicity of this class of viruses. The other Asian primates would not have to conserve their langur virogene sequences in order to restrict the growth of this class of virus, because they would seldom be encountered as an infectious agent.

4.4. Evolution of Colobus, Macaque, Squirrel Monkey, and Owl Monkey Virogenes

As listed in Table III, endogenous retroviruses have also been obtained from colobus monkeys, macaques (both *Macaca arctoides* and *M. mulatta*), and squirrel and owl monkeys. In a fashion analogous to that described for the baboon and langur viruses, the evolution of these viruses among primate species has been examined.

Nucleic acid sequences related to the endogenous colobus and macaque viruses have been found in the cellular DNA of all Old World monkey and ape species (Todaro *et al.*, 1978a; Sherwin and Todaro, 1979). These viruses are related to each other (Table IV); they are thus members of the same retrovirus class isolated from two different Old World species (see Table V). These viruses are not related to the baboon type C or langur type D viruses. Hybridization data obtained with the colobus and macaque isolates also show that there may be a geographic component to the evolution of these viruses (Todaro *et al.*, 1978a; Tainsky, 1981), but this finding has not been investigated further.

The type D viruses isolated from squirrel monkeys (Heberling *et al.*, 1977) and the type C virus isolated from owl monkeys (Todaro *et al.*, 1978b) are both endogenous retroviruses, since the cellular DNA of all normal squirrel monkey or owl monkey tissues contains multiple copies of their respective virogenes. The other New World primate genera do not contain nucleic acid sequences related to these viruses (Benveniste and Todaro, 1978); it is therefore not possible to conclude how long these sequences have been present in the New World monkey germ line. However, the lack of detectable hybridization to the cellular DNA of these other New World primates does not necessarily imply that they contain no related viruses. As is evident from Fig. 5, the various genera of New World monkeys are evolutionarily as far apart from each other ($\Delta T_m R$ = 6–9°C) as, for example, apes are from an Old World monkey such as the langur. With the langur viral probe, which shows no geographic effect on its rate of evolution, it is not possible to detect related viral sequences in ape DNA using these techniques (Fig. 8B) (Benveniste and Todaro, 1977). Thus, if the squirrel and owl monkey virogenes are evolving at the same rate as the langur virogenes, and if there is no South American habitat effect on the rate of evolution of these New World primate virogenes, then related viral sequences would not be detected in other New World primates. The squirrel monkey and owl monkey retroviruses are therefore of limited use in deciphering the phylogeny of New World primates. However, the squirrel monkey retrovirus has played a fascinating role in New World carnivore evolution (see Section 6.5).

In summary, then, the results presented in this section support the conclusion that primate virogenes have evolved as the species have evolved, with virogenes from more closely related genera and families showing more sequence homology than those from more distantly related taxons. That such species as baboon and langur, which have diverged genetically and have been geographically separated for several million years, still retain related virogene sequences, and the consistently observed hybridization to ape DNA with the baboon viral probe, demonstrate that this virogene information has been conserved in the primate stock during the course of evolution for at least 30–40 million years.

Furthermore, it has become clear that primates possess, as a normal component of their cellular DNA, multiple copies of several unrelated classes of retroviruses. For example, one can show that colobus cellular DNA contains 50 to several hundred copies of virogene sequences related to the endogenous colobus type C virus, to the endogenous baboon type C virus, and to the langur type D virus. There may also be other, as yet undiscovered virogenes in this primate species. Assuming 3×10^9 total nucleotide bases per haploid genome in primate DNA (Kohne, 1970), these 200–600 primate virogene copies [$(1.5–5) \times 10^6$ bp] represent approximately 0.1% of the total primate genome. The sets of retroviral genes are evolutionarily conserved to different extents, thus suggesting that different evolutionary pressures independently affect the rate of divergence of each class of viral genes. In any case, the preservation of this much DNA suggests that these genes may provide functions with a selective advantage to the species possessing them.

5. Endogenous Mammalian Retroviruses

Table XII lists the various mammalian species from which endogenous retroviruses have been isolated. The endogenous primate viruses have already been described. A large number of retroviruses have also been obtained from rodents—the majority of isolates have been obtained from Old World rodents belonging to the Myomorpha suborder, including mice, rats, and hamsters. The Asian mouse species *Mus caroli* and *Mus cervicolor* and related rodents have yielded endogenous retroviruses belonging to four different classes, which are distinct from each other both immunologically and by nucleic acid hybridization criteria (Lieber *et al.*, 1975; Benveniste *et al.*, 1977; Callahan and Todaro, 1978). When the cellular DNA of myomorphs is examined for DNA sequences related to the four classes of murine retroviruses, it is evident that their rate of evolution is not uniform (Callahan and Todaro, 1978; Callahan *et al.*, 1979). One class evolves at the same rate as the nonrepetitive cellular DNA, while the other classes evolve at much faster rates and are therefore useful in establishing phylogenetic relationships between closely related *Mus* species. The extensive genetic data available on the laboratory mouse (*Mus musculus*) and its close relatives make these animals especially

Table XII
Mammalian Species from Which Endogenous Retroviruses Have Been Isolated

	Virus[a]
Rodents	
Chinese hamster (*Cricetulus griseus*)	CCL 14.1
Syrian hamster (*Mesocricetus auratus*)	Y22
Mouse (*Mus musculus*)	MMTV, BALB/c
Asian mice (*M. cervicolor*)	M432, cerv CI, cerv CII
(*M. caroli*)	caro CI
(*M. cookii*)	cookii CI
(Other related species)	
Rat (*Rattus norvegicus*)	VNRK, RT21C
Guinea pig (*Cavia porcellus*)	GPV
Tree mouse (*Vandeleuria oleracea*)	Vand C-I
Agouti (*Dasyprocta punctata*)	DPC-1
Carnivores	
Domestic cat (*Felis catus*)	RD-114
Mink (*Mustela vison*)	
Primates	
Baboon (*Papio cynocephalus*)	M7, M28
(*P. anubis*)	BAB-8-K
(*P. papio*)	PP-1-Lu
(*P. hamadryas*)	BILN
Gelada (*Theropithecus gelada*)	TG-1-K
Macaque (*Macaca mulatta*)	MMC-1
(*M. arctoides*)	MAC-1
Colobus (*Colobus polykomos*)	CPC-1
Langur (*Presbytis obscurus*)	PO-1-Lu
Owl monkey (*Aotus trivirgatus*)	OMC-1
Squirrel monkey (*Saimiri sciureus*)	SMRV, M534
Artiodactyls	
Pig (*Sus scrofa*)	PK-15
Mule deer (*Odocoileus hemionus*)	

[a] Some species, such as rodents, contain multiple distinct retroviruses; not all are listed. In addition, a retrovirus has also been isolated from tree shrew (*Tupaia belangeri*) (Flügel *et al.*, 1978), but there is no definitive evidence that it is endogenous (i.e., sequences in related species). Numerous nonmammalian retroviruses have been described; these are listed in the text (Section 5).

attractive for studies on the evolution and function of these viral gene sequences.

In the rodent suborder Caviomorpha, endogenous type B retroviruses have been identified in guinea pig cells (Nadel *et al.*, 1967), but they have not yet been transmitted *in vitro* to other cell lines. A type C virus has also been isolated from the New World rodent agouti (*Dasyprocta punctata*) by cocultivation with human cells (Sherwin *et al.*, 1979). This virus is present in multiple copies in agouti cellular DNA and related sequences can be detected in the DNA of another New World caviomorph, the guinea pig.

Among the carnivores, an endogenous type C virus has been isolated from the domestic cat, *Felis catus* (McAllister *et al.*, 1972), and from mink, *Mustela vison* (Sherr *et al.*, 1978). The cat virus (RD-114) is present in multiple copies in the cellular DNA of the domestic cat and five other closely related *Felis* species. We have previously shown that this virus was acquired by ancestors of the domestic cat from ancestors of baboons several million years ago (see Section 6.1). Sequences related to the mink virus can be found in the cellular DNA of only a few closely related mustelids, such as weasels and ermines; the other carnivores tested are too distantly related to the mink to permit the detection of related virogenes with the hybridization conditions employed.

Endogenous retroviruses have also been isolated from pigs (see Section 6.4) and from deer (*Odocoileus hemionus*). The latter type C isolate is present in all cells of the species of origin and related virogene sequences are present in other genera of the family Cervidae, such as deer, caribou, elk, and muntjac, and in the more distantly related family Bovidae (sheep) (Tronick *et al.*, 1977). The data show that the deer virus has been genetically transmitted in artiodactyls for millions of years, in a fashion that parallels the evolution of the nonrepetitive cellular DNA in this group of animals.

Numerous retroviruses have also been isolated from other vertebrate classes, including chickens (Vogt and Friis, 1971), reptiles (Zeigel and Clark, 1969; Andersen *et al.*, 1979), and fish (Papas *et al.*, 1976). The viruses from reptiles and fish have been poorly characterized and insufficient evidence (such as sequences in related species) is available to classify them as endogenous retroviruses.

Not included in Table XII are the *infectious* mammalian retrovirus isolates, which include bovine leukemia virus from domestic cattle (Miller *et al.*, 1969), visna virus from sheep (Sigurdsson *et al.*, 1957), caprine arthritis-encephalitis virus (CAEV) from goats (Cork *et al.*, 1974), equine infectious anemia virus (EIAV) from horses (Stein *et al.*, 1942), primate foamy viruses [reviewed by Hooks and Gibbs (1975)], feline leukemia virus from domestic cats (Hardy *et al.*, 1973), gibbon leukemia virus (Kawakami *et al.*, 1972), Mason–Pfizer monkey virus (Chopra and Mason, 1970), the various isolates from macaques with an immunodeficiency syndrome (Daniel *et al.*, 1984 and 1985; Marx *et al.*, 1984; Stromberg *et al.*, 1984), and various human isolates from patients with cutaneous T-cell neoplasias (Poiesz *et al.*, 1980) and the acquired immune deficiency syndrome (AIDS) (Barré-Sinoussi *et al.*, 1983; Popovic *et al.*, 1984). These infectious mammalian viruses, most of which cause disease, can only be found in the cells of infected animals.

6. Interspecies Transmission and Germ Line Integration of Mammalian Retroviruses

Millions of years before man began moving genes between species in the laboratory, retroviruses were transferring their genes between eukaryotes. It has been demonstrated that retroviruses have been transferred, under natural conditions, between vertebrate species that are only remotely related phylogenetically, and to be subsequently inherited as cellular genes. The first example showed that the baboon type C viruses, which are genetically transmitted in primates, had been transferred to ancestors of the domestic cat in the Mediterranean Basin area several million years ago (Benveniste and Todaro, 1974b). These viruses were incorporated into the germ line of cats, where they are present in multiple copies per haploid genome, and are inherited in the same Mendelian fashion as other cellular genes (Benveniste and Todaro, 1975a).

Several additional examples of interspecies gene transmission have subsequently been described. The other retrovirus of cats, feline leukemia virus, was also acquired from ancestors of the Old World rat (Benveniste *et al.*, 1975). Domestic pigs and their feral relatives, the bush pig and wart hog, have also acquired their genetically transmitted viruses from rodent endogenous viruses. Hybridization data suggest that the virus-related sequences in the pig were acquired from members of the family Muridae after the mouse had diverged from the rat but before speciation of surviving mice had occurred (Benveniste and Todaro, 1975b). In addition, retroviral genes have also been shown to have been transferred between a New World primate (squirrel monkey) and a New World carnivore (ancestor of the skunk) (Benveniste *et al.*, 1978). All of these transferred viral genes are now an integral part of the new host species genomes and are present as moderately repetitive genes.

There are also two well-documented examples where retroviruses have been transferred between species that cohabit the same environment, but integration into the germ line has not occurred. One case involves the transfer of endogenous viruses from certain Southeast Asian rodents (*Mus* spp.) to gibbons (Benveniste and Todaro, 1973, Lieber *et al.*, 1975). Another example is the transfer of endogenous retroviruses from langurs to macaques (Benveniste and Todaro, 1977). In both these cases, where integration into germ cells has not occurred, the viruses are pathogenic and cause leukemias, lymphosarcomas, and immunosuppressive diseases.

6.1. Transfer of Virus from Baboon Ancestor to Domestic Cat Ancestor

Sequences partially homologous to the baboon type C virus can be detected in the cellular DNA of Old World monkeys and apes. These virogenes have therefore evolved for at least 30 million years among anthropoid primates in a fashion that correlates closely with the taxonomic relatedness of the species. Domestic cat DNA also contains type C virogenes, which can give rise to the RD-114 group of cat viruses (McAllister *et al.*, 1972). By several criteria, including nucleic acid sequence homology (Benveniste *et al.*, 1974a) and immunologic relatedness of the viral proteins (Sherr *et al.*, 1975), these viruses are related to each other. The unexpected relatedness of the endogenous cat and the endogenous baboon virus suggested that they had a common origin.

The origin of the RD-114 group of cat viruses was determined by examining other felids, which consist of approximately 36 species, for the presence of RD-114 viral sequences. Table XIII lists the data obtained when the cellular DNA of 20 felids was tested for relatedness to the index species, the domestic cat. As is evident, the modern cats are a very closely related family. The greatest degree of divergence, corresponding to approximately $\Delta T_m R = 3.0°C$ is found between the DNA of the domestic cat and cellular DNA from the larger cats, such as the lion and jaguar. This degree of divergence is less than that observed, among the primates, between man and orangutan, and corresponds to about 15 million years of evolutionary divergence. The data indicate that the other carnivore species tested (mink, dog, hyena) are more distantly related. The phylogeny of the carnivores has been examined by DNA-DNA hybridization and will not be discussed further here (Benveniste and Wilson, manuscript in preparation). Cat cellular DNA hybridizes to a low extent (10%) to primate cellular DNA, but this level is significantly above that obtained with rodent DNA (5%) and may reflect a distant homology between primate and carnivore orders (Romero-Herrera *et al.*, 1973).

When a DNA probe prepared from the endogenous cat virus RD-114 was hybridized to the cellular DNA of these same cat species, the results shown in Table XIII were obtained (see also Benveniste and Todaro, 1974b). Even though all of the cats are very closely related, sequences related to the endogenous cat virus RD-114 are present in only five other species of Felidae. The data also show that sequences related to RD-114 can be found in the cellular DNA of all Old World monkeys as well as in chimpanzee and gorilla DNA. The fact that RD-114 sequences can be found in all anthropoid primates but are present in just six cat species suggests that RD-114 is a primate virus, acquired by cats recently in their evolution. The cats that contain RD-114 sequences are from the Mediterranean Basin area, while those from Southeast Asia and the New World lack these sequences. The larger African cats (lion and cheetah) also do not contain RD-114 sequences in their cellular DNA. Therefore, the successful transmission of gene sequences was subsequent to the major radiation of the Felidae and was limited to one geographic region.

The time of viral transmission can be pinpointed more exactly. As shown in Table XIII, the final extent of hybridization of the RD-114 virus to primate DNA is not uniform. Since the rate of baboon virogene evolution (to which RD-114 is related) is influenced by the geographic habitat of a species (see Section 4.2), it is important to examine only those primates whose habitat is Africa. The data reveal that various baboon species and the gelada possess the most sequence homology to the RD-114 virus (32% hybridization). Mandrill, mangabey, and patas form another node at 23% hybridization, colobus DNA hybridizes 20%, and the African apes 12% These data allow us to conclude that the time of virus transfer occurred prior to the speciation of the baboons and gelada, but shortly after these primates had diverged from macaques, mangabeys, and the mandrill. Since RD-114 was transferred to cats not long after the macaques had diverged from baboons, the Asian–African effect on the rate of virogene evolution is not evident between mangabeys and macaques when assayed with this "escaped baboon virus." However, the Asian–African effect is pronounced when more distantly related primates such as the colobus and langur are compared.

Figure 9 shows a phylogenetic tree of Old World monkey and felid evolution indicating the postulated time of virus transfer. The six species of cats that have diverged since acquiring the virus (domestic cat, European wildcat, sand cat, African wildcat, jungle cat, and black-footed cat) all contain sequences related to RD-114. A comparison of the thermal stability of cellular DNA hybrids among the Felidae and the Cercopithecinae suggests that the time of virus transfer occurred approximately 6 mya.

Table XIII
Nucleic Acid Homology and Thermal Stability between Domestic Cat Unique Sequence Cellular DNA and RD-114 Viral DNA and the DNA from Various Species

Species	Domestic cat cellular DNA[a]		RD-114 viral DNA	
	Percent hybridization	$\Delta T_m R$ (°C)	Percent hybridization	$\Delta T_m R$ (°C)
Carnivores				
Felidae				
Domestic cat (*F. catus*)	100	0.0	100	0.0
European wildcat (*F. sylvestris*)	98	0.3	97	0.6
Sand cat (*F. margarita*)	96	1.0	93	1.8
Jungle cat (*F. chaus*)	95	1.1	92	2.1
African wildcat (*F. lybica*)	96	1.1	90	2.2
Black-footed cat (*F. nigripes*)	93	1.4	89	3.0
Caracal (*F. caracal*)	93	2.2	2	—
Serval (*F. serval*)	92	2.2	2	—
African golden (*F. aurata*)	91	2.2	2	—
Leopard cat (*F. bengalensis*)	90	2.2	1	—
Margay (*F. wiedii*)	90	2.5	3	—
Geoffroy (*F. geoffroyi*)	89	2.8	2	—
Ocelot (*F. pardalis*)	90	2.9	1	—
Bobcat (*F. rufa*)	91	2.5	1	—
Bengal tiger (*P. tigris*)	88	3.2	1	—
African leopard (*L. pardus*)	90	2.8	3	—
Lion (*P. leo*)	89	2.8	2	—
Snow leopard (*U. uncia*)	87	3.2	2	—
Jaguar (*L. onca*)	89	2.9	2	—
Cheetah (*A. jubatus*)	89	2.9	1	—
Other carnivores				
Mink (*M. vison*)	30	20.5	2	—
Dog (*C. familiaris*)	31	20.5	2	—

Hyena (*C. crocuta*)	56	14.5	1	—
Primates				
Old World monkeys				
Baboon (*P. cynocephalus*)	9	—	33	20.4
(*P. papio*	—	—	33	—
(*P. hamadryas*)	—	—	31	21.0
Gelada (*T. gelada*)	—	—	32	20.5
Mandrill (*M. sphinx*)	—	—	25	—
Mangabey (*C. atys*)	—	—	23	23.5
Macaque (*M. mulatta*)	—	—	22	—
Patas (*E. patas*)	8	—	24	—
Colobus (*C. polykomos*)	—	—	20	24.2
Langur (*P. obscurus*)	—	—	5	—
Apes				
Chimpanzee (*P. troglodytes*)	10	—	12	—
Gorilla (*G. gorilla*)	—	—	12	
Gibbon (*H. lar*)	—	—	2	—
Man (*H. sapiens*)	—	—	2	—
New World monkeys				
Capuchin (*Cebus* spp.)	—	—	2	—
Howler (*Alouatta* spp.)	11	—	2	—
Other mammals				
Mouse (*M. musculus*)	5	—	2	—
Cow (*B. taurus*)	9	—	1	—
Fish				
Salmon (*S. salar*)	0.5	—	0	—

[a] Domestic cat cellular DNA or RD-114 viral DNA was hybridized to the cellular DNA of the species listed. The results shown are the average of multiple experiments, involving several representatives of each species. Dashes denote hybrids not tested.

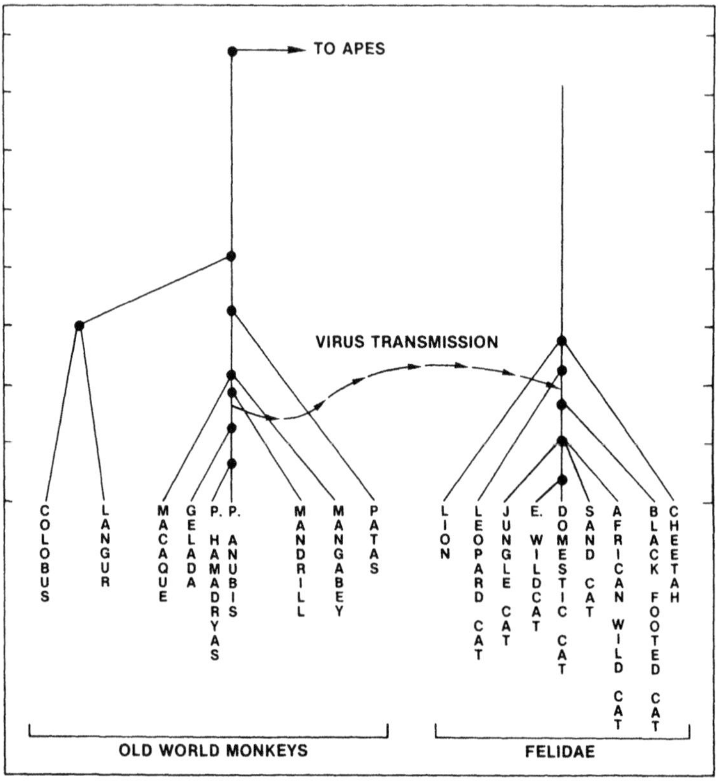

Figure 9. Transfer of a retrovirus from an ancestor of baboons and geladas to an ancestor of the domestic cat. The phylogenetic trees depict the relationship among Old World monkey cellular DNA (see Fig. 3) and Felidae DNA (see Table XIII). The six species of Felidae that diverged after the time of virus transfer several million years ago all contain endogenous virogene sequences related to the acquired RD-114 virus.

The mechanism by which the virus was acquired by cats is unknown. As previously discussed, baboons and their close reltives are unusual among Old World monkeys in their frequency of spontaneous release of type C viruses (Table III). The ancestors of the domestic cat, perhaps reaching Africa for the first time 5–10 mya, came in contact with the ancestors of baboons. Since cats are carnivores, the virus might have been acquired as part of their food chain. In any case, there must have been a selective pressure that favored the integration of the virus in the germ line of domestic cats. Perhaps those cats that were able to integrate the baboon virus in their DNA and suppress its expression were protected from infection by the potentially pathogenic baboon virus. Domestic cat cells in culture today do not support the growth of the baboon virus, whereas the cells of the leopard cat, a closely related species that never acquired the baboon virus, readily support the growth of this class of viruses. In any event, this acquired baboon virus is now present as 10–50 multiple copies in the cellular DNA of six cat species, and is inherited, in strict Mendelian fashion, like any other set of cellular genes (Benveniste and Todaro, 1975a).

6.2. Use of Interspecies Retrovirus Transmission as a Marker for Species Divergence

The derivation of phylogenetic relationships by DNA hybridization techniques assumes that the accumulation of nucleotide substitutions in nonrepetitive DNA proceeds at a constant average rate for the group of animals whose relationship is being examined. However, to convert the DNA hybridization data into absolute time requires the calibration of species divergences against an external dating source, which can be either the fossil record or geological events such as the separation of land masses. Both of these methods of dating divergences are inexact, the former because of its incomplete and fragmented nature, and the latter because of uncertainties in dates or the effects of rafting, for example, between adjacent continents.

The transmission of retroviruses between species can be used as a marker for the evolutionary divergence of species that is independent of any assumptions about the fossil record. In the case of the baboon to cat viral transfer, one can examine the evolution of any set of genes in those species of cats that have diverged since the RD-114 virus was acquired and compare their rate of evolution to that obtained in the species of primates that have diverged since transmitting the virus. Table XIV shows the data obtained by examining the evolution of the overall nonrepetitive cellular DNA in those cat and primate species that have evolved since viral transfer. It is evident that cellular DNA has evolved at approximately the same rate in the Felidae and the Cercopithecinae during the past several million years. One immediate application of this observation is that the DNA hybridization relationships among carnivores and primates can be equated to each other in terms of absolute times of divergence. It has not escaped our attention that many cat species that are now classified as belonging to the same genus (*Felis*) in fact show as much divergence at the nucleotide level as do several primate genera. The constant genetic distance measured between the parents of cat and primate hybrids (see Table X) argues further for the validity of a molecular classification of mammalian species.

Table XIV
Retroviruses as Markers for Evolutionary Divergence of Species[a]

	$\Delta T_m R$ (°C)
Old World monkey species that have diverged since transfer of RD-114 virus from ancestor of baboons to ancestors of domestic cats:	
Baboon–gelada	1.3
Felidae species that have diverged since acquiring RD-114 virus from ancestors of baboons:	
Domestic cat–black-footed cat	1.4

[a] Figure 9 lists these pairs of species in relation to the transmission and acquisition of viral sequences. The $\Delta T_m R$ is the thermal stability difference obtained after hybridizing the cellular DNA from each listed pair of animals.

6.3. Interspecies Transmission from Ancestors of Rodents to Cats

Several additional examples of virus transmission between species have subsequently been described. Besides the RD-114 cat viruses that were transmitted from primates to cats, another distinct class of type C retroviruses (feline leukemia virus, FeLV) was acquired by cats and is now present in their germ line. Studies showed that FeLV-related gene sequences are found in the same six felids that contain RD-114 viral genes. More distant *Felis* species lack genes related to FeLV, while the cellular DNA of rodents, and in particular that of rats, contains related virogene sequences. The absence of FeLV sequences in most Felidae and the presence of sequences in rodents suggests that FeLV-related genes were introduced into the *Felis* lineage after trans-species infection by a type C virus of rodent origin (Benveniste *et al.*, 1975). It is interesting that the same six cat species that contain sequences related to RD-114 virus also contain FeLV-related genes, whereas other felids lack both sets of sequences. Both groups of viral genes appear, therefore, to have been introduced into the cat lineage at the same time in the Mediterranean Basin area from different groups of animals (rodents and primates).

All members of the family Felidae, however, have been shown to contain viral gene sequences related to the Old World monkey virus MAC-1 (Bonner and Todaro, 1979). MAC-1 transcripts hybridize to the cellular DNA of felids that have RD-114- and FeLV-related sequences as well as to those cats lacking these sequences, such as lions and jaguars. The MAC-1 viral transcripts are thus detecting viral-related sequences in Felidae that correspond to viral genes not yet isolated. Since these sequences are found in all Felidae, the time of transmission must have been greater than 10 mya, but since MAC-1-related sequences have been present in primates for at least 30 million years, it is not possible to determine if carnivores acquired the sequences from ancestral primates or *vice versa*.

6.4. Interspecies Transmission from Rodent Ancestors to Ancestors of the Pig

Another example of trans-species infection of retroviruses, with subsequent integration into the germ line of the recipient species, is that of an endogenous virus acquired by an ancestor of the domestic pig from an ancestor of the mouse (Benveniste and Todaro, 1975b). Pig cell cultures produce type C viruses (Todaro *et al.*, 1974) that are present in multiple copies in the DNA of all pig tissues and cells, and are therefore genetically transmitted. However, when the cellular DNA of other artiodactyls is examined, related viral sequences are found only in very close feral relatives such as the bush pig and wart hog. Other ungulates, such as a close relative of the pig, the South American peccary, do not have detectable virogene sequences (Table XV). Extensive homology is detected, how-

Table XV
Evidence for Transfer of Retroviruses from Rodents to Pigs

	Percent hybridization[b]			
	Mouse DNA		Pig DNA	
Species	Cellular	Viral	Cellular	Viral
Rodents				
Mouse (*M. musculus*)	100	100	3	23
(*M. caroli*)	61	13	4	26
(*M. cervicolor*)	55	10	4	25
Rat (*R. norvegicus*)	21	2	4	13
Hamster (*C. griseus*)	11	1	4	5
Guinea pig (*C. porcellus*)	3	2	4	1
Artiodactyls				
Domestic pig (*S. scrofa*)	4	1	100	100
Wild boar (*S. scrofa*)	4	1	100	100
Bush pig (*P. porcus*)	5	2	92	86
Wart hog (*P. aethiopicus*)	4	2	89	71
Peccary (*T. tajacu*)	3	2	57	2
Hippopotamus (*H. amphibius*)	4	1	18	2
Cow (*B. taurus*)	4	2	12	2

[a]Data from Benveniste and Todaro (1975b).

ever, in normal mouse cellular DNA as well as in rat and hamster cellular DNA.

Based on the extent of hybridization of the pig type C viral DNA probes to rodent cellular DNA, it has been concluded that the type C virogenes were introduce into the Suidae lineage by trans-species infection from a member of the family Muridae after the mouse had separated from the rat, but before the different species of mice had diverged from one another. This transfer is shown schematically in Fig. 10.

The evidence suggests that endogenous retroviruses have been genetically transmitted in rodents for millions of years [reviewed by Callahan and Todaro (1978)]. However, rodents have diverged considerably from one another, so that murine virogene sequences cannot be detected even in a member of the same family, such as the rat. As shown in Table XV, there is no detectable hybridization to rat cellular DNA (2%) with the murine viral probe. However, with the pig type C virus, related sequences can be detected not only in all mouse species examined, but also in rats and hamsters. The similarities and differences between the primate to cat viral gene transfer described in Section 6.1 and the mouse to pig transfer discussed here are summarized in Table XVI. An intriguing possibility to explain the reason the contemporary mouse viruses do not hybridize to pig DNA, but the pig viruses hybridize to rat and hamster DNA, is that

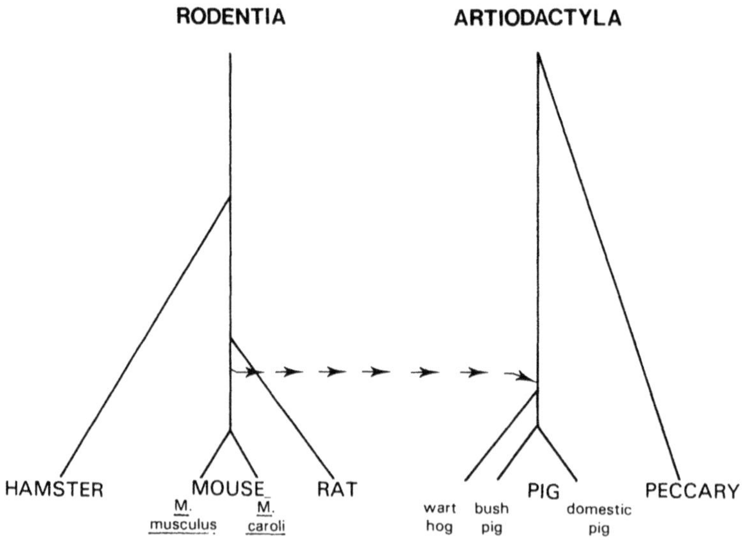

Figure 10. Transmission of murine virogenes to the porcine germ line. The approximate divergences of the various species are shown.

Table XVI
Comparison of Transmission of Retroviruses

Similarities	
Primate → Cat	Mouse → Pig
1. Primate DNA has sequences related to endogenous cat virus	1. Rodent DNA has sequences related to endogenous pig virus
2. Primate and cat viral proteins closely related	2. Mouse and pig viral proteins closely related
3. Wild relatives of the cat have related sequences	3. Wild relatives of the pig have related sequences
Differences	
Primate → Cat	Mouse → Pig
1. Contemporary primate viruses hybridize to cat cellular DNA	1. Contemporary mouse viruses *do not* hybridize to pig cellular DNA
2. Virus has moved from longer to shorter lived species	2. Virus has moved from shorter to longer lived species

in being transferred from a shorter lived animal (the rodent) to a longer lived one (the pig) the rate of evolution of the transferred murine virogene has decreased. If that is the case, the present-day porcine type C viral genes would most closely resemble the viral genes as they were several million years ago in the rodent lineage. The pig viruses may thus represent "fossilized" mouse viruses.

Brownell (1983) compared immunological distance measurements and DNA hybridization values in rodents, and suggested that this mammalian order may evolve more rapidly than primates at the DNA sequence level. The results with the pig virogenes also suggest that rodent virogene evolution has proceeded at a faster rate.

6.5. Interspecies Transmission from New World Monkeys to New World Carnivores

In a striking parallel to the transfer of retroviruses that occurred between an Old World monkey (an ancestor of the baboon) and an Old World carnivore (an ancestor of the domestic cat), evidence has also been presented for the transfer of a retrovirus between a New World primate (squirrel monkey) and a New World carnivore (skunk). Gene sequences homologous to the endogenous type D squirrel monkey viral genes are found not only in the cellular DNA of all squirrel monkey tissues and cells (see Section 4.4), but also in the DNA of normal tissues of the skunk

(*Mephitis mephitis* and *Spilogale putorius*) (Benveniste *et al.*, 1978). This was the first example of interspecies transmission between New World species and the first involving endogenous type D retroviruses. Sequences related to the squirrel monkey virus cannot be detected in other carnivore species, including the supposedly closely related mustelids, the mink and weasel. Recent studies of carnivore evolution have shed new light on these results, since skunks appear to be members of a separate carnivore family and are only distantly related to the minks and weasels (Benveniste and Wilson, in preparation). The exact direction of transfer (primate to skunk or skunk to primate) cannot be established with certainty because of the lack of new World species tested that are closely related to either the squirrel monkey or the skunk. However, since squirrel monkeys are unusual among New World monkeys in their ability to release infectious viruses (Table III), the direction of transfer was probably from primate to carnivore.

6.6. Interspecies Transmission from Rodent Ancestors to Ancestors of Minks and Weasels

An endogenous, genetically transmitted virus of the mink (*Mustela vison*) has been isolated by cocultivation with carnivore cell lines (Sherr *et al.*, 1978). Related nucleic acid sequences can be detected in all mink tissues as well as in the closely related species weasel and ermine; other carnivores do not contain related nucleic acid sequences. However, a low level of homology can be detected to rodent cellular DNA, and furthermore, the virus is antigenically related to rodent viruses. The possibility of interspecies transmission from rodents to mink has been proposed, but the exact rodent species of origin has not been defined.

7. Interspecies Transmission of Infectious Retroviruses

The examples of virus transmission summarized above demonstrate that viral genes from one group of animals can release infectious particles that not only can infect animals of another species, but can also be incorporated into the germ line and be transmitted as normal cellular genes. In the five examples known, the viruses are no longer pathogenic for the recipient species. All of the viral transmissions occured millions of years ago, as determined by appropriate evolutionary studies.

Several examples of transmission between species where the virus has not yet integrated into the germ cells of the recipient species have also been documented. In all of these cases, the viruses are pathogenic, usually causing neoplastic diseases in various mammalian species, including man. All of the transfers are believed to be recent, perhaps contemporary, events, and they therefore serve as a useful model for the evolutionarily more ancient examples.

7.1. Origin of Gibbon Leukemia Virus

Infectious type C viruses have been isolated from a fibrosarcoma of a woolly monkey house pet (SSAV) (Theilen *et al.*, 1971) and from gibbons with lymphoma from vivarium colonies in Thailand (GALV) (Kawakami *et al.*, 1972). These viruses spread from animal to animal under natural conditions and induce tumors when inoculated into other primates (Wolfe *et al.*, 1971). They are related to one another by several immunologic criteria (Sherr *et al.*, 1975; Oroszlan *et al.*, 1977) and by viral interference, and also contain related RNA genomes (Todaro *et al.*, 1975). Since nucleic acid sequences homologous to this class of viruses cannot be detected in the cellular DNA of normal woolly monkeys, gibbons, or other primates (Benveniste *et al.*, 1974a; Scolnick *et al.*, 1974), these viruses are horizontally transmitted from animal to animal, rather than vertically transmitted as are the endogenous type C viruses of primate origin. Epidemiologic studies have also suggested that the woolly monkey acquired its virus from a gibbon housed in an adjacent cage, so the single isolate from woolly monkey (a New World primate) is in reality the same virus as the one more frequently isolated from gibbons.

Hybridization studies with the infectious primate viruses SSAV and GALV have shown that they share a significant degree of nucleic acid sequence homology with the endogenous type C viruses of the laboratory mouse, *Mus musculus* (Benveniste and Todaro, 1973). Normal mouse cellular DNA also contains multiple virogene copies related to the gibbon infectious virus. The proteins of the gibbon and mouse viruses also shared unique interspecies determinants (Sherr *et al.*, 1975). These unexpected findings suggested the possibility that the infectious viruses of the GALV–SSAV group were derived from endogenous mouse viruses or from a type C virus of a rodent closely related to the mouse.

The ease with which infectious type C viruses can be isolated from an Asian primate, the gibbon, and their partial homology to mouse cellular DNA suggested that an Asian species of *Mus* might have an endogenous virus more closely related to these primate viruses. Indeed, endogenous viruses isolated from cells of the Asian species *Mus caroli* and *Mus cervicolor* were shown to be related antigenically to the gibbon and woolly monkey type C virus and to be distinctly different from the previously studied type C viruses of *Mus musculus* (Lieber *et al.*, 1975; Benveniste *et al.*, 1977). These results led to the conclusion that this group of horizontally transmitted, infectious primate retroviruses originated by trans-species infection of gibbons from an endogenous type C virus from *Mus caroli, Mus cervicolor*, or another, closely related sympatric species. This trans-species infection, which is depicted in Fig. 11, appears to be a relatively recent event, with the viruses not yet being incorporated into the genomes of the recipient primate species. After infection of the woolly monkey, that retrovirus has also acquired a cellular oncogene (v-*sis*) that codes for a platelet-derived growth factor (Doolittle *et al.*, 1983.)

7.2. Origin of Infectious Macaque Viruses

Type D retroviruses have been isolated from a rhesus monkey mammary carcinoma (MPMV) (Chopra and Mason, 1970) and from several species of macaques with fibrosarcoma and an immunodeficiency disease called simian AIDS (Daniel *et al.*, 1984; Marx *et al.*, 1984; Stromberg *et al.*, 1984). These viruses are related to each other, both immunologically and by nucleic acid hybridization criteria. The viruses of this class are not endogenous to macaques (Benveniste and Todaro, 1977; Stomberg *et al.*, 1984), since viral sequences related to them are found only in infected animals.

Since MPMV shared antigenic and nucleic acid similarities with the endogenous langur virus, we examined primate cellular DNA for homology to MPMV. The data showed that sequences related to MPMV are present in the cellular DNA of all Old World monkeys, with the highest extent being present in the cellular DNA of members of the Colobinae subfamily. The Colobinae consist of six genera: *Presbytis, Rhinopithecus, Pygathrix, Nasalis, Simias* (all from Southeast Asia), and *Colobus* (from Africa). Since the highest final extent of hybridization is obtained with

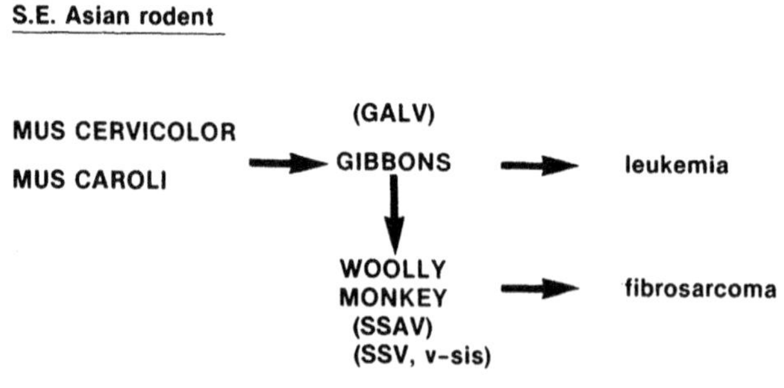

Figure 11. Epidemiology of infectious gibbon leukemia virus.

langur DNA, MPMV and the SAIDS group of infectious macaque viruses appear to have been derived from one of the Asian genera and perhaps from langurs themselves.

We have therefore suggested (Benveniste and Todaro, 1977) that an endogenous type D virus from a langur (or a close relative) has become infectious for rhesus monkeys. The transmissions may have occurred under natural conditions, since the range and habitats of langurs and rhesus monkeys overlap throughout much of Southeast Asia. The finding that MPMV and the SAIDS group of viruses are only partially homologous to langur cell DNA may be attributable to a more rapid evolution of these horizontally transmitted viruses as compared to the genetically inherited viral gene sequences. This would be analogous to the partial relatedness seen between the infectious and endogenous type C viruses of chickens (Wright and Neiman, 1974), mice (Benveniste and Todaro, 1973), and cats (Baluda and Roy-Burman, 1973).

7.3. Mammalian Retroviruses of Unknown Origin

The known examples of transmission of retroviruses between mammalian species are shown schematically in Fig. 12. Not all of these virus transmissions have led to incorporation of the virogenes into the germ line of the recipient species, as summarized in Table XVII. A retrovirus called bovine leukemia virus (BLV) has been isolated from leukemic cattle

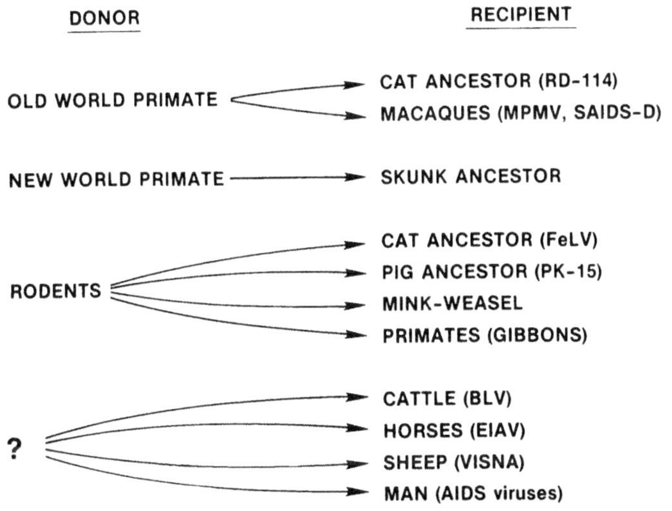

Figure 12. Transmission of retroviruses between species.

Table XVII
Transmission of Retrovirus Genes between Mammalian Species

Donor	Recipient	Genetically transmitted in recipient	Disease
Primate: Old World monkey	Carnivore: cat ancestor (RD-114)	Yes	—
Primate: Old World monkey	Primate: macaque (MPMV)	No	Immunosuppression
	(SAIDS-D)	No	Fibromatosis
Primate: New World monkey	Carnivore: skunk ancestor	Yes	—
Rodent: mouse ancestor	Artiodactyl: pig ancestor (PK-15)	Yes	—
Rodent: rat ancestor	Carnivore: cat ancestor (FeLV)	No[a]	Leukemia
Rodent: unknown ancestor	Carnivore: mink–weasel ancestor	Yes	—
Rodent: Asian *Mus* species	Primate: gibbon (GaLV)	No	Lymphosarcoma
Unknown	Cows (BLV)	No	Leukemia
Unknown	Sheep (visna)	No	Neurologic
Unknown	Goats (CAEV)	No	Neurologic, arthritis
Unknown	Horses (EIAV)	No	Anemia
Unknown	Primates (Foamy)	No	—
Unknown	Man (HTLV I, II)	No	T-cell malignancies
	(HTLV III, LAV)	No	AIDS

[a] The genetically transmitted feline leukemia virus (FeLV) has not been obtained from cat cell cultures. The infectious FeLV, which is partially related to endogenous FeLV (Baluda and Roy-Burman, 1973), is responsible for most cases of hematopoietic neoplasia in cats.

(Miller *et al.*, 1969). Sequences related to the virus are found only in the DNA of tissues from leukemic cattle with the adult form of the disease, and not in the cellular DNA of normal cattle (Callahan *et al.*, 1976). BLV is therefore an infectious virus of cattle that causes lymphosarcomas or leukemia in that species. However, the cellular origin of this virus is unknown, since related nucleic acid sequences have not been detected in any other mammalian species. Other retroviruses that cause disease in their mammalian hosts and have no known species of origin include visna virus of sheep, CAEV virus of goats, and EIAV virus in horses (Table XVII). The identification of the donor species might permit control of the vectors for virus spread and lead to prevention of these diseases.

Infectious retroviruses have also been isolated from man. A class of viruses called HTLV-I and -II have been isolated from patients with T-cell malignancies (Poiesz *et al.*, 1980); and lymphadenopathy-associated viruses (LAV) (Barré-Sinoussi *et al.*, 1983) and HTLV-III (Popovic *et al.*, 1984) have been isolated from lymphocytes of patients with the acquired immune deficiency syndrome (AIDS). These viruses are detected only in the cells of infected patients. The origin of these viruses, which are pathogenic in man, is unknown; the identification of a donor species would have immediate public health consequences. It is also possible that some of the viruses of unknown origin shown in Table XVII may turn out to have a common evolutionary origin.

7.4. Interspecies Transmission of Nonmammalian Retroviruses

The viruses listed in Table XVII summarize our knowledge about the origin and modes of transmission of mammalian retroviruses. This chapter has not dealt with retroviruses of other vertebrate orders, but there is also evidence for transmission of retroviruses between nonmammalian species. Frisby *et al.* (1979) have shown that nucleic acid sequences related to the endogenous chicken virus RAV-O have not diverged in parallel with the normal host genes during the evolution of the Phasianidae. Although RAV-O sequences are endogenous in all domestic chickens (*Gallus gallus*), it appears from a study of the cellular DNA in related species that the RAV-O genome has been introduced recently into the germ line of *Gallus gallus*, following speciation but before domestication.

A retrovirus isolated from a Russell's viper tumor was shown by molecular hybridization to be an endogenous virus of that reptilian species (Andersen *et al.*, 1979) and to be immunologically related to certain Old World monkey primate viruses. These data suggest an evolutionary relatedness or possibly another example of virus transmission, but other reptilian species have not been examined for nucleic acid sequence homology in order to identify the potential direction of transfer.

The northern pike (*Esox lucius*) has a 21% incidence of lymphosarcomas, which is an unusually high frequency for a malignancy in feral vertebrates. A retrovirus has been described in the cells from the tumor tissue (Papas *et al.*, 1976), but evolutionary studies have not been performed to establish whether this is an endogenous or infectious virus of the pike. Epizootiologic evidence, however, suggests that the disease is transmitted horizontally by contact during spawning.

The reticuloendotheliosis virus family, which includes spleen necrosis virus, duck infectious anemia virus, chick syncytial virus, and reticuloendotheliosis virus (REV) [for review see Purchase *et al.* (1973)], causes a variety of diseases in several avian species and appears to be spread horizontally. Studies have established that these viruses are antigenically distinct from the other chicken viruses and closely resemble mammalian viruses (Hunter *et al.*, 1978; Barbacid *et al.*, 1979). An analysis of the sequence of viral proteins (Oroszlan *et al.*, 1981a) shows that the REV genome is more closely related to the macaque group of primate retroviruses (MAC-1, MMC-1, CPC-1; see Table XII). The current hypothesis, therefore, is that REV was derived by rescue of a normal avian cell gene by a primate retrovirus horizontally transmitted to these avian species.

8. Conclusion

The evidence that retroviruses are genetically transmitted and that they have been conserved during the course of evolution in several distinct vertebrate lineages suggests that they may provide some functions advantageous to the species carrying them. One system that might be advantageous to the host is exemplified by the baboon type C viral genes. These genetically transmitted viruses may serve to protect African primates against more virulent related viruses present in sympatric species. Thus, one of the ways that an animal has of protecting itself against a potentially harmful virus is to integrate it, and as a consequence acquire a certain measure of resistance to repeated infection by the same or closely related viruses. Most vertebrate cells are resistant to infection by their own endogenous viruses as a result of a variety of mechanisms operating at the cell surface receptor and postpenetration stages of the infective cycle. This resistance to self-infection may indicate that the host needs to limit the infectious spread of endogenous viruses in its somatic cells.

It has now been well documented that retroviruses can transfer their genes between distantly related species, with subsequent integration into the germ line. These viruses are the only ones that have been shown to transfer genes among the germ lines of different mammalian species under natural conditions. They are uniquely suited for this role because they must integrate into cellular DNA in order to replicate, but they do not usually kill the cells that they infect. They are also able to extricate defined host genes [reviewed by Bishop (1981)]. This natural retrovirus transmission into the germ line of a recipient species has been duplicated by Jaenisch (1976), who has shown that exogenous viruses can also be experimentally manipulated into the germ line of mice and subsequently be inherited as cellular genes.

Recent cloning and DNA sequencing techniques have shown a high incidence of structural gene mobility within prokaryotic and eukaryotic species due to transposons in bacteria, and Ty 1 and *copia* elements in yeast and in *Drosophila* (Camerow *et al.*, 1979; Calos and Miller, 1980). When integrated, retroviruses form a segment of DNA flanked by 0.6-kb direct repeats and resemble the structure of the moderately repetitive transposable elements Ty 1 and *copia* (Temin, 1980). While moving from cell to cell, retroviruses may carry with them other host cell genes and thus serve to maintain a species in contact with its ecologic as well as genetic neighbors. Whether this has been a major force in evolution is unknown.

Busslinger *et al.* (1982) have identified a set of histone genes in the sea urchin that has not evolved at the same average rate as the rest of the unique sequence cellular DNA. They postulate a recent horizontal transfer of the histone gene clusters from a distantly related sea urchin species. Although there is as yet no evidence for retroviruses in invertebrates, a viral gene transfer among sea urchin species remains an intriguing possibility.

There is also a parallel in bacteria for the acquisition of genes from sympatric species. The actinomycetes are a group of bacteria that produce antibiotics, presumably to compete effectively for nutrients with other soil microorganisms. They contain various enzymes that can acetylate, adenylylate, and phosphorylate these antibiotics to render them inactive during their synthesis. Gram-negative bacteria that inhabit the same soil samples have been shown to possess, perhaps as a defense mechanism, the same antibiotic-inactivating enzymes that are present in the actinomycetes (Benveniste and Davies, 1973). The mechanism of gene transfer in this case is unknown, but may involve several vectors as intermediates.

Classical Darwinian theory states that the reproductive barrier between species prevents the transfer of genetic material. When a species becomes sufficiently distinct from its neighbors so that it can no longer exchange genes, its ability to change is limited to rearranging, mutating, and duplicating its own genes. Retroviruses may lower the barriers that normally isolate chromosomal genes, as plasmids and other vector systems do in bacteria (Reanney and Ackermann, 1982). The movement of genes between species by retroviruses may thus be part of a process by which genetically isolated species maintain contact with one another. This movement, as reviewed here, has been occurring in vertebrates over millions of years and may provide some of the genetic plasticity that is being increasingly revealed among eukaryotes.

ACKNOWLEDGMENTS. I am grateful to Elise Brownell, Morris Goodman, David Pilbeam, Vincent Sarich, Charles Sibley, and Allan Wilson for their helpful review of this manuscript.

References

Adams, J., and Rothman, E. D., 1982, Estimation of phylogenetic relationships from DNA restriction patterns and selection of endonuclease cleavage sites, *Proc. Natl. Acad. Sci. USA* **79:**3560–3564.

Andersen, P. R., Barbacid, M., Tronick, S. R., Clark, H. F., and Aaronson, S. A., 1979, Evolutionary relatedness of viper and primate endogenous retroviruses, *Science* **204:**318–321.

Ando, T., 1966, A nuclease specific for heat-denatured DNA isolated from a product of *Aspergillus oryzae*, *Biochem. Biophys. Acta* **114:**158–168.

Avise, J. C., and Aquadro, C. F., 1982, A comparative summary of genetic distances in the vertebrates: Problems and correlations, in: *Evolutionary Biology*, Vol. 15 (M. K. Hecht, and B. Wallace, and G. T. Prance, eds.), Plenum Press, New York, pp. 151–185.

Baba, M. L., Goodman, M., Dene, H., and Moore, G. W., 1975, Origins of the Ceboidea viewed from an immunological perspective, *J. Hum. Evol.* **4:**89–102.

Baluda, M. A., and Roy-Burman, P., 1973, Partial characterisation of RD114 virus by DNA–RNA hybridization studies, *Nature New Biol.* **244:**59–62.

Barbacid, M., Hunter, E., and Aaronson, S. A., 1979, Avian reticuloendotheliosis viruses: Evolutionary linkage with mammalian type C retroviruses, *J. Virol.* **30:**508–514.

Barré-Sinoussi, I. F., Chermann, J. C., Rey, F., Nugeyre, M. T., Chamaret, S., Gruest, J., Dauguet, C., Axler-Blin, C., Vézinet-Brun, F., Rouzioux, C., Rozenbaum, W., and Montagnier, L., 1983, Isolation of a T-lymphocyte retrovirus from a patient at risk for acquired immune deficiency syndrome (AIDS), *Science* **220:**868–870.

Benveniste, R. E., and Davies, J., 1973, Aminoglycoside antibiotic-inactivating enzymes in actinomycetes similar to those present in clinical isolates of antibiotic-resistant bacteria, *Proc. Natl. Acad. Sci. USA* **70:**2276–2280.

Benveniste, R. E., and Todaro, G. J., 1973, Homology between type C viruses of various species as determined by molecular hybridization, *Proc. Natl. Acad. Sci. USA* **70:**3316–3320.

Benveniste, R. E., and Todaro, G. J., 1974a, Evolution of type C viral genes. I. Nucleic acid from baboon type C virus as a measure of divergence among primate species, *Proc. Natl. Acad. Sci. USA* **71**:4513–4518.

Benveniste, R. E., and Todaro, G. J., 1974b, Evolution of type-C viral genes. Inheritance of exogenously acquired viral genes, *Nature* **252**:456–459.

Benveniste, R. E., and Todaro, G. J., 1974c, Multiple divergent copies of endogenous type C virogenes in mammalian cells, *Nature* **252**:170–173.

Benveniste, R. E., and Todaro, G. J., 1975a, Segregation of RD-114 and FeLV-related sequences in crosses between domestic cat and leopard cat, *Nature* **257**:506–508.

Benveniste, R. E., and Todaro, G. J., 1975b, Evolution of type C viral genes. Preservation of ancestral murine type C viral sequences in pig cellular DNA, *Proc. Natl. Acad. Sci. USA* **72**:4090–4094.

Benveniste, R. E., and Todaro, G. J., 1976, Evolution of type C viral genes: Evidence for an Asian origin of man, *Nature* **261**:101–108.

Benveniste, R. E., and Todaro, G. J., 1977, Evolution of primate oncornaviruses: An endogenous virus from langurs (*Presbytis* spp.) with related virogene sequences in other Old World monkeys, *Proc. Natl. Acad. Sci. USA* **74**:4557–4561.

Benveniste, R. E., and Todaro, G. J., 1978, Approaches to the isolation of RNA tumor viruses from primates, *Arthritis Rheum.* **21**:S2–S16

Benveniste, R. E., Heinemann, R., Wilson, G. L., Callahan, R., and Todaro, G. J., 1974a, Detection of baboon type C viral sequences in various primate tissues by molecular hybridization, *J. Virol.* **14**:56–67.

Benveniste, R. E., Lieber, M. M., Livingston, D. M., Sherr, C. J., Todaro, G. J., and Kalter, S. S., 1974b, Infectious C-type virus isolated from a baboon placenta, *Nature* **248**:17–20.

Benveniste, R. E., Sherr, C. J., and Todaro, G. J., 1975, Evolution of type C viral genes: Origin of feline leukemia virus, *Science* **190**:886–888.

Benveniste, R. E., Callahan, R., Sherr, C. J., Chapman, V., and Todaro, G. J., 1977, Two distinct endogenous type C viruses isolated from the Asian rodent *Mus cervicolor*; Conservation of virogene sequences in related rodent species, *J. Virol.* **21**:849–862.

Benveniste, R. E., Wilson, G. L., and Todaro, G. J., 1978, Skunks have gene sequences in their cellular DNA related to squirrel monkey retroviruses: Transmission between species of a New World primate endogenous type D retrovirus, *Biochem. Biophys. Res. Commun.* **81**:1363–1369.

Birkenmeier, E. H., Bonner, T. I., Reynolds, K., Searfoss, G. H., and Todaro, G. J., 1982, Colobus type C virus: Molecular cloning of unintegrated viral DNA and characterization of the endogenous viral genomes of colobus, *J. Virol.* **41**:842–854.

Bishop, J. M., 1981, Enemies within: The genesis of retrovirus oncogenes, *Cell* **23**:55–6.

Bonner, T. I., and Todaro, G. J., 1979, Carnivores have sequences in their cellular DNA distantly related to the primate endogenous virus, MAC-1, *Virology* **94**:224–227.

Bonner, T. I., and Todaro, G. J., 1980, The evolution of baboon endogenous type C virus: Related sequences in the DNA of distant species, *Virology* **103**:217–227.

Bonner, T. I., Brenner, D. J., Neufeld, B. R., and Britten, R. J., 1973, Reduction in rate of DNA reassociation by sequence divergence, *J. Mol. Biol.* **81**:123–135.

Bonner, T. I., Heinemann, R., and Todaro, G. J., 1980, Evolution of DNA sequences has been retarded in Malagasy primates, *Nature* **286**:420–423.

Bonner, T. I., Birkenmeier, E. D., Gonda, M. A., Mark, G. E., Searfoss, G. E., and Todaro, G. J., 1982a, Molecular cloning of a family of retroviral sequences found in chimpanzee but not human DNA, *J. Virol.* **43**:914–924.

Bonner, T. I., O'Connell, C., and Cohen, M., 1982b, Cloned endogenous retroviral sequences from human DNA, *Proc. Natl. Acad. Sci. USA* **79**:4709–4713.

Britten, R. J., and Kohne, D. E., 1968, Repeated sequences in DNA, *Science* **161**:529–540.

Brown, W. M., George, M., Jr., and Wilson, A. C., 1979, Rapid evolution of animal mitochondrial DNA, *Proc. Natl. Acad. Sci. USA* **76**:1967–1971.

Brown, W. M., Prager, E. M., Wang, A., and Wilson, A. C., 1982, Mitochondrial DNA sequences of primates: Tempo and mode of evolution, *J. Mol. Evol.* **18**:225–239.

Brownell, E., 1983, DNA/DNA hybridization studies of muroid rodents: Symmetry and rates of molecular evolution, *Evolution* **37**:1034–1051.

Bruce, E. J., and Ayala, F. J., 1979, Phylogenetic relationships between man and the apes: Electrophoretic evidence, *Evolution* **33**:1040–1056.

Busslinger, M., Rusconi, S., and Birnstiel, M. L., 1982, An unusual evolutionary behaviour of a sea urchin histone gene cluster, *EMBO J.* **1**:27–33.

Callahan, R., and Todaro, G. J., 1978, Four major endogenous retrovirus classes each genetically transmitted in various species of *Mus*, in: *Workshop on the Origins of Inbred Mice* (H. Morse, ed.), Academic Press, New York, pp. 689–713.

Callahan, R., Lieber, M. M., Todaro, G. J., Graves, D. C., and Ferrer, J. F., 1976, Bovine leukemia virus genes in the DNA of leukemic cattle, *Science* **192**:1005–1007.

Callahan, R., Meade, C., and Todaro G. J., 1979, Isolation of an endogenous type C virus related to the infectious primate type C viruses from the asian rodent *Vandeleuria oleracea*, *J. Virol.* **30**:124–131.

Callahan, R., Drohan, W., Tronick, S., and Schlom, J., 1982, Detection and cloning of human DNA sequences related to the mouse mammary tumor virus genome, *Proc. Natl. Acad. Sci. USA* **79**:5503–5507.

Calos, M. P., and Miller, J. H., 1980, Transposable elements, *Cell* **20**:579–595.

Camerow, J. R., Loh, E. Y., and Davis, R. W., 1979, Evidence for transposition of dispersed repetitive DNA families in yeast, *Cell* **16**:739–751.

Centerwall, W. R., 1974, The truth about hybrids, *Cats Mag.* **31**:8–9.

Cheny, L. M., Case, S. M., and Wilson, A. C., 1978, Frog perspective on the morphological difference between humans and chimpanzees, *Science* **200**:209–211.

Chopra, H. C., and Mason, M. M., 1970, A new virus in a spontaneous mammary tumor of a rhesus monkey, *Cancer Res.* **30**:2081–2086.

Cohen, M., Rice, N., Stephens, R., and O'Connell, C., 1982, DNA sequence relationship of the baboon endogenous virus genome to the genomes of other type C and type D retroviruses, *J. Virol.* **41**:801–812.

Cork, L. C., Hadlow, W. J., Gorham, J. R., Piper, R. C., and Crawford, T. B., 1974, Infectious leukoencephalomyelitis of goats, *J. Infect. Dis.* **129**:134–141.

Cronin, J. E., and Sarich, V. M., 1975, Molecular systematics of the New World monkeys, *J. Hum. Evol.* **4**:357–375.

Daniel, M. D., King, N. W., Letvin, N. L., Hunt, R. D., Sehgal, P. K., and Desrosiers, R. C., 1984, A new type D retrovirus isolated from macaques with an immunodeficiency syndrome, *Science* **223**:602–605.

Daniel. M. D., Letvin, N. L., King, N. W., Kannagi, M., Sehgal, P. K., Hunt, R. D., Kanki, P. J., Essex, M., and Desrosiers, R. C., 1985, Isolation of T-cell tropic HTLV-III-like retrovirus from macaques, *Science* **228**:1201–1204.

Doolittle, R. F., Hunkapiller, M. W., Hood, L. E., Devare, S. G., Robbins, K. C., Aaronson, S. A., and Antoniades, H. N., 1983, Simian sarcoma virus *onc* gene, v-*sis*, is derived from the gene (or genes) encoding a platelet-derived growth factor, *Science* **221**:275–277.

Dutrillaux, B., 1979, Chromosomal evolution in primates: Tentative phylogeny from *Microcebus murinus* (prosimian) to man, *Hum. Genet.* **48**:251–314.

Ferris, S. D., Wilson, A. C., and Brown, W. M., 1981, Evolutionary tree for apes and humans based on cleavage maps of mitochondrial DNA, *Proc. Natl. Acad. Sci. USA* **78**:2432–2436.

Ferris, S. D., Sage, R. D., Huang, C.-M., Nielsen, J. T., Ritte, U., and Wilson, A. C., 1983, Flow of mitochondrial DNA across a species boundary, *Proc. Natl. Acad. Sci. USA* **80**:2290–2294.

Fitch, W. M., 1976, Molecular evolutionary clocks, in: *Molecular Evolution* (F. J. Ayala, ed.), Sinauer, Sunderland, Massachusetts, pp. 160–178.

Flügel, R. M., Zentgraf, H., Munk, K., and Darai, G., 1978, Activation of an endogeneous retrovirus from *Tupaia* (tree shrew), *Nature* **271**:543–545.

Frisby, D. P., Weiss, R. A., Roussel, M., and Stehelin, D., 1979, The distribution of endogenous chicken retrovirus sequences in the DNA of galliform birds does not coincide with avian phylogenetic relationships, *Cell* **17**:623–634.

Goodman, M., 1975, Protein sequence and immunological specificity: Their role in phylogenetic studies of primates, in: *Phylogeny of the Primates* (W. P. Luckett and F. S. Szalay, eds.), Plenum Press, New York, pp. 219–248.

Goodman, M., and Moore, G. W., 1971, Immunodiffusion systematics of the primates. I. The Catarrhini, *Syst. Zool.* **20**:19–62.

Goodman, M., Braunitzer, G., Stangl, A., and Schrank, B., 1983, Evidence on human origins from haemoglobins of African apes, *Nature* **303**:546–548.

Gonda, M. A., Wong-Staal, F., Gallo, R. C., Clements, J. E., Narayan, O., and Gilden, R. V., 1985, Sequence homology and morphologic similarity of HTLV-III and visna virus, a pathogenic lentivirus, *Science* **227**:173–177.

Gray, A. P., 1971, *Mammalian Hybrids*, Commonwealth Agricultural Bureau, Farnham Royal, England.

Hardy, W. D., Jr., Old, L. J., Hess, P. W., Essex, M., and Cotter, S. N., 1973, Horizontal transmission of feline leukemia virus, *Nature* **244**:266–269.

Harper, M. E., Ullrich, A., and Sanders, G. F., 1981, Localization of the human insulin gene to the distal end of the short arm of chromosome 11, *Proc. Natl. Acad. Sci. USA* **78**:4458–4460.

Heberling, R. L., Barker, S. T., Kalter, S. S., Smith, G. C., and Helmke, R., 1977, Oncornavirus: Isolation from a squirrel monkey (*Saimiri sciureus*) lung culture, *Science* **195**:289–292.

Hooks, J. J., and Gibbs, C. J., Jr., 1975, The foamy viruses, *Bacteriol. Rev.* **39**:169–185.

Hoyer, B. H., van de Velde, N. W., Goodman, M., and Roberts, R. B., 1972, Examination of hominoid evolution by DNA sequence homology, *J. Hum. Evol.* **1**:645–649.

Hunter, E., Brown, A. S., and Bennett, J. C., 1978, Amino-terminal amino acid sequence of the major structural polypeptides of avian retroviruses: Sequence homology between reticuloendotheliosis virus p30 and p30s of mammalian retroviruses, *Proc. Natl. Acad. Sci. USA* **75**:2708–2712.

Jaenisch, R., 1976, Germ line integration and Mendelian transmission of the exogenous Moloney leukemia virus, *Proc. Natl. Acad. Sci. USA* **73**:1260–1264.

Jolly, C. J., 1970, The large African monkeys as an adaptive array, in: *Old World Monkeys, Evolution, Systematics, and Behavior* (J. R. Napier and P. H. Napier, eds.), Academic Press, London, pp. 139–174.

Kawakami, T. G., and Buckley, P. M., 1974, Antigenic studies in gibbon type-C viruses, *Transplant. Proc.* **6**:193–196.

Kawakami, T. G., Huff, S. D., Buckley, P. M., Dungworth, D. L., Snyder, S. P., and Gilden, R. V., 1972, C-type virus associated with gibbon lymphosarcoma, *Nature New Biol.* **235**:170–171.

King, M., and Wilson, A. C., 1975, Evolution at two levels in humans and chimpanzees, *Science* **188:**107–116.

Kohne, D. E., 1970, Evolution of higher-organism DNA, *Q. Rev. Biophys.* **33:**327–375.

Kohne, D. E., Chiscon, J. A., and Hoyer, B. H., 1972, Evolution of primate DNA sequences, *J. Hum. Evol.* **1:**627–644.

Li, W.-H., Luo, C.-C., and Wu, C.-I, 1985, Evolution of DNA sequences, in: *Molecular Evolutionary Genetics* (R. MacIntyre, ed.), Plenum Press, New York, pp. 1–94.

Lieber, M. M., Benveniste, R. E., Livingston, D. M., and Todaro, G. J., 1973, Mammalian cells in culture frequently release type C viruses, *Science* **182:**56–59.

Lieber, M. M., Sherr, C. J., Todaro, G. J., Benveniste, R. E., Callahan, R., and Coon, H. G. 1975, Isolation from the asian mouse *Mus caroli* of an endogenous type C virus related to infectious primate type C viruses, *Proc. Natl. Acad. Sci. USA* **72:**2315–2319.

Lowy, D. R., Rowe, W. P., Teich, N., and Hartley, J. W., 1971, Murine leukemia virus: High frequency activation *in vitro* by 5-iododeoxyuridine and 5-bromodeoxyuridine, *Science* **174:**155–157.

Markarjan, D. S., Isakov, E. P., and Kondakov, G. I., 1974, Intergeneric hybrids of the lower (42-chromosome) monkey species of the Sukhumi monkey colony, *J. Hum. Evol.* **3:**247–255.

Martin, M. A., Bryan, T., Rasheed, S., and Khan, A. S., 1981, Identification and cloning of endogenous retroviral sequences present in human DNA, *Proc. Natl. Acad. Sci. USA* **78:**4892–4896.

Marx, P. A., Maul, D. H., Osborn, K. G., Lerche, N. W., Moody, P., Lowenstine, L. J., Henrickson, R. V., Arthur, L. O., Gilden, R. V., Gravell, M., London, W. T., Sever, J. L., Levy, J. A., Munn, R. J., and Gardner, M. B., 1984, Simian AIDS: Isolation of a type D retrovirus and transmission of the disease, *Science* **223:**1083–1086.

McAllister, R. M., Nicholson, M. O., Gardner, M. B., Rongey, R. W., Rasheid, S., Sarma, P. S., Huebner, R. J., Hatanaka, M., Oroszlan, S., Gilden, R. V., Kabigting, A., and Vernon, L., 1972, C-type virus released from cultured human rhabdomyosarcoma cells, *Nature* **235:**3–6.

Miller, J. M., Miller, L. D., Olson, C., and Gillette, K. G., 1969, Virus-like particles in phytohemagglutinin-stimulated lymphocyte cultures with reference to bovine lymphosarcoma, *J. Natl. Cancer Inst.* **43:**1297–1305.

Myers, R. H., and Shafer, D. A., 1979, Hybrid ape offspring of a mating of gibbon and siamang, *Science* **205:**308–310.

Nadel, E., Banfield, W., Burstein, S., and Tousimis, A. J., 1967, Virus particles associated with strain 2 guinea pig leukemia (L2C/N-B), *J. Natl. Cancer Inst.* **38:**979–981.

Napier, J. R., 1970, Paleoecology and catarrhine evolution, in: *Old World Monkeys, Evolution, Systematics, and Behavior* (J. R. Napier, and P. H. Napier, eds.), Academic Press, New York, pp. 53–96.

Napier, J. R., and Napier, P. H., 1967, *A Handbook of Living Primates*, Academic Press, New York, p. 247.

Nei, M., Stephens, J. C., and Saitou, N., 1985, Methods for computing the standard errors of branching points in an evolutionary tree and their application to molecular data from humans and apes, *Mol. Biol. Evol* **2**(1)**:**66–85.

Noda, M., Kurihara, M., and Takano, T., 1982, Retrovirus-related sequences in human DNA: Detection and cloning of sequences which hybridize with the long terminal repeat of baboon endogenous virus, *Nucl. Acids Res.* **10:**2865–2878.

Nowak, R. M., and Paradiso, J. L., 1983, *Walker's Mammals of the World*. 4th ed., Johns Hopkins University Press, Baltimore.

O'Brien, S. J. (ed.), 1984, *Genetic Maps*, Vol 3, Cold Spring Harbor Press, New York, pp. 1–581.

O'Brien, S. J., Simonson, J. M., Eichelberger, M. A., 1982, Genetic analysis of hybrid cells using isozyme markers as monitors of chromosome segregation, in: *Techniques in Somatic Cell Genetics* (J. W. Shay, ed.), Plenum Press, New York, pp. 513–524.

O'Brien, S. J., Seuanez, H. N., and Womack, J. E., 1985, On the evolution of genome organization in mammals, in: *Molecular Evolutionary Genetics* (R. MacIntyre, ed.), Plenum Press, New York, pp. 519–589.

O'Connell, C., O'Brien, S., Nash, W. G., and Cohen, M., 1984, ERV3, a full-length human endogenous provirus: Chromosomal localization and evolutionary relationships, *Virology* **138**:225–235.

Oroszlan, S., Copeland, T. D., Smythers, G., Summers, M. R., and Gilden, R. V., 1977, Comparative primary structure analysis of the p30 protein of woolly monkey and gibbon type C viruses, *Virology* **77**:413–417.

Oroszlan, S., Barbacid, M., Copeland, T. D., Aaronson, S. A., and Gilden, R. V., 1981a, Chemical and immunological characterization of the major structural protein (p28) of MMC-1, a rhesus monkey endogenous type C virus: Homology with the major structural protein of the avian reticuloendotheliosis virus, *J. Virol.* **39**:845–854.

Oroszlan, S., Copeland, T. D., Gilden, R. V., and Todaro, G. J., 1981b, Structural homology of the major internal proteins of endogenous type C viruses to two distantly related species of Old World monkeys: Macaca arctoides and Colobus polykomos, *Virology* **115**:262–271.

Papas, T. S., Dahlberg, J. E., and Sonstegard, R. S., 1976, Type C virus in lymphosarcoma in northern pike (*Esox lucius*), *Nature* **261**:506–508.

Pilbeam, D., 1982, New hominoid skull material from the miocene of Pakistan, *Nature* **295**:232–234.

Pilbeam, D., 1983, Hominoid evolution and hominid origins, in: *Recent Advances in the Evolution of the Primates* (C. Chaga, ed.), Pontificae Academiae Scientiarum Scripta Varia 50, Vatican, Rome, pp. 43–61.

Poiesz, B. J., Ruscetti, F. W., Gazdar, A. F., Bunn, P. A., Minna, J. D., and Gallo, R. C., 1980, Detection and isolation of type C retrovirus particles from fresh and cultured lymphocytes of a patient with cutaneous T-cell lymphoma, *Proc. Natl. Acad. Sci. USA* **77**:7415–7419.

Popovic, M., Sarngadharan, M. G., Read, E., and Gallo, R. C., 1984, Detection, isolation, and continuous production of cytopathic retroviruses (HTLV-III) from patients with AIDS and pre-AIDS, *Science* **224**:497–500.

Powell, J. R., 1983, Interspecific cytoplasmic gene flow in the absence of nuclear gene flow: Evidence from *Drosophila*, *Proc. Natl. Acad. Sci. USA* **80**:492–495.

Purchase, H. G., Ludford, C., Nazerian, K., and Cox, H. W., 1973, A new group of oncogenic viruses: Reticuloendotheliosis, chick syncytial, duck infectious anemia, and spleen necrosis viruses, *J. Natl. Cancer Inst.* **51**:489–499.

Rabin, H., Benton, C. V., Tainsky, M. A., Rice, N. R., and Gilden, R. V., 1979, Isolation and characterization of an endogenous virus from Rhesus monkeys, *Science* **204**:841–842.

Reanney, D. C., and Ackerman, H.-W., 1982, Comparative biology and evolution of bacteriophages, in: *Advances in Virus Research* (M. A. Laufer, F. B. Bang, K. Maramorosch, and K. M. Smith eds.), Academic Press, New York, pp. 205–280.

Repaske, R., O'Neill, R. R., Steele, P. E., and Martin, M. A., 1983, Characterization and partial nucleotide sequence of endogenous type C retrovirus segments in human chromosomal DNA, *Proc. Natl. Acad. Sci. USA* **80**:678–682.

Romero-Herrera, A. E., Lehmann, H., Joysey, K. A., and Friday, A. E., 1973, Molecular evolution of myoglobin and the fossil record: A phylogenetic synthesis, *Nature* **246**:389–395.

Ruddle, F. H., and Creagan, R. P., 1975, Parasexual approaches to the genetics of man, *Annu. Rev. Genet.* **9**:407–486.

Sarich, V. M., and Cronin, J. E., 1976, Molecular systematics of the primates, in: *Molecular Anthropology* (M. Goodman and R. E. Tashian, eds.), Plenum Press, New York, pp. 141–170.

Sarich, V., and Wilson, A. C., 1973, Generation time and genomic evolution in primates, *Science* **179**:1144–1147.

Schwartz, J. H., 1984, The evolutionary relationships of man and orangutans, *Nature* **308**:501–505.

Scolnick, E. M., Parks, W., Kawakami, T., Kohne, D., Okabe, H., Gilden, R., and Hatanaka, M., 1974, Primate and murine type-C viral nucleic acid association kinetics: Analysis of model systems and natural tissues, *J. Virol.* **13**:363–369.

Seuanez, H. N., 1982, Chromosome banding and primate phylogeny, in: *Advanced Views in Primate Biology* (A. B. Chiarelli and R. S. Corruccini, eds.), Springer, Berlin, pp. 224–235.

Sherr, C. J., Fedele, L. A., Benveniste, R. E., and Todaro, G. J., 1975, Interspecies antigenic determinants of the reverse transcriptase and p30 proteins of mammalian type C viruses, *J. Virol.* **15**:1440–1448.

Sherr, C. J., Benveniste, R. E., and Todaro, G. J., 1978, Endogenous mink (*Mustela vison*) type C virus isolated from sarcoma virus-transformed mink cells, *J. Virol.* **25**:738–749.

Sherwin, S. A., and Todaro, G. J., 1979, A new endogenous primate type C virus isolated from the Old World monkey *Colobus polykomos*, *Proc. Natl. Acad. Sci. USA* **76**:5043–5045.

Sherwin, S. A., Bonner, T. I., Heine, U., and Todaro, G. J., 1979, The isolation of an endogenous retrovirus from the New World rodent *Dasyprocta punctata* (Agouti), *Virology* **94**:409–416.

Sibley, C. G., and Ahlquist, J. E., 1983, Phylogeny and classification of birds based on the data of DNA–DNA hybridization, in: *Current Ornithology*, Vol. 1 (R. F. Johnston, ed.), Plenum Press, New York, pp. 245–292.

Sibley, C. G., and Ahlquist, J. E., 1984, The phylogeny of the hominoid primates, as indicated by DNA–DNA hybridization, *J. Mol. Evol.* **20**:2–15.

Sigurdsson, B., Palsson, P. A., and Grimsson, H., 1957, Visna, a demyelinating transmissible disease of sheep, *J. Neuropathol. Exp. Neurol.* **16**:389–403.

Simons, E. L., 1970, The deployment and history of Old World monkeys (Cercopithecidae, primates), in: *Old World Monkeys, Evolution, Systematics, and Behavior* (J. R. Napier and P. H. Napier, eds.) Academic Press, London, pp. 97–129.

Simons, E. L., 1976, The fossil record of primate phylogeny, in: *Molecular Anthropology* (M. Goodman and R. E. Tashian, eds.), Plenum Press, New York, pp. 35–62.

Stein, C. D., Lotze, C. J., and Mott, L. O., 1942, Transmission of equine infectious anemia by the stablefly, *Stomoxys calcitrans*, the horsefly, *Tabanus sulcifrons* (Macquart) and by the injection of minute amounts of virus, *Am. J. Vet. Res.* **3**:183–193.

Stromberg, K., and Benveniste, R., 1983, Efficient isolation of endogenous rhesus retrovirus from trophoblast, *Virology* **128**:518–523.

Stromberg, K., Benveniste, R. E., Arthur, L. O., Rabin, H., Giddens, W. E., Ochs, H. D., Morton, W. R., and Tsai, C.-C., 1984, Characterization of exogenous type D retrovirus from a fibroma of a macaque with simian AIDS and fibromatosis, *Science* **224**:289–292.

Tainsky, M. A., 1981, Analysis of the virogenes related to the rhesus monkey endogenous type C retrovirus in monkeys and apes, *J. Virol.* **37**:922–930.

Temin, H. M., 1980, Origin of retrovirus from cellular moveable genetic elements, *Cell* **21**:599–600.

Templeton, A. R., 1983, Phylogenetic inference from restriction endonuclease cleavage site maps with particular reference to the evolution of humans and the apes, *Evolution* **37**(2):221–224.

Theilen, G. H., Gould, D., Fowler, M., and Dungworth, D. C., 1971, C-type virus in tumor tissue of a woolly monkey (Lagothrix spp.) with fibrosarcoma, *J. Natl. Cancer Inst.* **47**:881–899.

Thomas, L., 1974, *The Lives of a Cell*, Viking Press, New York.

Todaro, G. J., Benveniste, R. E., Lieber, M. M., and Sherr, C. J., 1974, Characterization of a type C virus released from the porcine cell line PK(15), *Virology* **58**:65–74.

Todaro, G. J., Lieber, M. M., Benveniste, R. E., Sherr, C. J., Gibbs, C. J., Jr., and Gajdusek, D. C., 1975, Infectious primate type C viruses: Three isolates belonging to a new subgroup from the brains of normal gibbons, *Virology* **67**:335–343.

Todaro, G. J., Sherr, C. J., and Benveniste, R. E., 1976, Baboons and their close relatives are unusual among primates in their ability to release nondefective endogenous type C viruses, *Virology* **72**:278–282.

Todaro, G. J., Benveniste, R. E., Sherwin, S. A., and Sherr, C. J., 1978a, MAC-1, A new genetically transmitted type C virus of primates: "Low frequency" activation from stumptail monkey cell cultures, *Cell* **13**:775–782.

Todaro, G. J., Sherr, C. J., Sen, A., King, N., Daniel, M. D., and Fleckenstein, B., 1978b, Endogenous New World primate type C viruses isolated from an owl monkey (*Aotus trivirgatus*) kidney cell line, *Proc. Natl. Acad. Sci. USA* **75**:1004–1008.

Tronick, S. R., Golub, M. M., Stephenson, J. R., and Aaronson, S. A., 1977, Distribution and expression in mammals of genes related to an endogenous type C RNA virus of *Odocoileus hemionus*, *J. Virol.* **23**:1–9.

Vogt, P. K., and Friis, R. R., 1971, An avian leukosis virus related to RSV(0): Properties and evidence for helper activity, *Virology* **43**:223–230.

Wilson, A. C., Maxson, L. R., and Sarich, V. M., 1974, Two types of molecular evolution. Evidence from studies of interspecific hybridization, *Proc. Natl. Acad. Sci. USA* **71**:2843–2847.

Wilson, A. C., Carlson, S. S., and White, T. J., 1977, Biochemical evolution, *Annu. Rev. Biochem.* **46**:573–639.

Wolfe, L. G., Deinhardt, F., Theilen, G. H., Rubin, H., Kawakami, T., and Bustad, L. K., 1971, Induction of tumors in marmoset monkeys by simian sarcoma virus type 1 (*Lagothrix*): A preliminary report, *J. Natl. Cancer Inst.* **47**:1115–1120.

Wright, S. E., and Neiman, P. E., 1974, Base-sequence relationships between avian ribonucleic acid endogenous and sarcoma viruses assayed by competitive ribonucleic acid–deoxyribonucleic acid hybridization, *Biochemistry* **13**:1549–1554.

Wurster-Hill, D. H., and Centerwall, W. R., 1982, The interrelationships of chromosome banding patterns in canids, mustelids, hyena and felids, *Cytogenet. Cell Genet.* **34**:178–192.

Zeigel, R. F., and Clark, H. F., 1969, Electron microscopic observations on a "C"-type virus in cell cultures from a tumor-bearing viper, *J. Natl. Cancer Inst.* **43**:1097–1102.

CHAPTER 7

Evolution of Ribosomal DNA

SUSAN A. GERBI

1. Origin of Genes for the Translational Apparatus

Darwinian evolution requires selection operating on the phenotype to favor which changes in the genotype are to be maintained and perpetuated. Since proteins are generally responsible for the phenotype, most of Darwinian evolution as we think of it today could not occur until the establishment of protein synthesis. It is primarily the translational apparatus that allows expression of the genotype at the level of the phenotype. In this chapter I review ideas relating to the evolution of the genes coding for a major constituent of the translational apparatus, namely the genes for ribosomal RNA (rRNA). Since the products of these genes are RNAs and not proteins, this is one case where evolutionary selection may act at the RNA level.

The terms "higher" and "lower" organisms are carefully avoided in this review [see Dobell (1911) for a delightful critique of these terms], but occasional lapses must be pardonned for simplicity of expression.

1.1. An RNA Basis for the Primitive Translational Machinery

The present-day protein synthetic apparatus contains a multitude of components: the small ribosomal subunit consists of a ribosomal RNA molecule 16S–18S in size and complexed with over 20 ribosomal proteins; the large ribosomal subunit has over 30 ribosomal proteins plus 5S RNA,

SUSAN A. GERBI • Division of Biology and Medicine, Brown University, Providence, Rhode Island 02912. This chapter is dedicated to my father, Claudio Gerbi, M.D., on the joyous occasion of his 77th birthday.

23S–28S rRNA, and in the case of eukaryotes also a 5.8S RNA. Several protein factors are needed in addition for the initiation and elongation processes of protein synthesis.

Many people supposed that the numerous ribosomal proteins and protein factors are the major ingredients of importance for the protein synthesis mechanism; their reasoning led to the notion that the only function of rRNA was to be a scaffold on which the ribosomal proteins were placed. Although there are indeed specific binding regions on the rRNA for the early binding ribosomal proteins, it is now becoming clear that rRNA plays other important roles in the ribosome, too. In fact, an appealing hypothesis is that rRNA itself is responsible for the basic translational mechanism. The points in favor of this hypothesis are summarized below; for a fuller account the reader should see the thought-provoking source of these arguments (Woese, 1980):

1. It is likely that only a few molecular components were used for a primitive translational machinery; with time, more components could be added to refine the mechanism into a more efficient process.
2. Since the process of translation is used for the biosynthesis of proteins, one would have to devise an alternate way to make proteins if the protein endproduct of the translational process was also used as the basis of this process. This is rather like the paradox, "Which came first, the chicken or the egg?" From this point of view it is easier to imagine that the biosynthesis of RNA evolutionarily preceded the biosynthesis of proteins, and that the first translational apparatus had only RNA components.
3. Translation can take place even in the absence of some proteins thought to play an important role in this process (Pestka, 1968; Garilova and Spirin, 1974).
4. There is precedent for RNA interactions playing a vital role in translation. One example of this is the codon–anticodon interaction by which the nucleic acid information carried by messenger RNA (mRNA) is decoded by tRNA and translated into the appropriate amino acid sequence. A second example of the importance of RNA interactions for protein synthesis is the association of the 5′ end of mRNA with the Shine and Dalgarno CCUCC sequence at the 3′ end of 16S rRNA for initiation of translation in a large number of prokaryotes [reviewed by Steitz (1980)].

If the early translational apparatus only consisted of RNA, how did it work? It has been proposed that an allosteric change in the anticodon arm of tRNA could result in ratchet-like movement relative to the mRNA

(Woese, 1970). The notion of allosteric transitions in RNA also appears in the recent hypothesis that an RNA switch mechanism is at the heart of the translational process [reviewed by Brimacombe *et al.* (1983)]. If a stretch of RNA "switched" its base pairing from one RNA pairing partner to another, a conformational change could be induced. Such a conformational change could be the driving force that translocates tRNA from the A to the P site of the ribosome. A series of such RNA switches could cycle the conformation of the ribosome back to its ground state. Such a chain of switches has been formulated for *E. coli* 16S rRNA (Thompson and Hearst, 1983b). RNA pairing interactions important for switching could be between two stretches both within rRNA, or it could be between a region of rRNA and a small RNA pairing partner (e.g, 5S RNA, tRNA, 5.8S RNA).

Three evolutionary lines of descent led to the eubacteria, archaebacteria, and eukaryotes (Woese and Fox, 1977; Woese *et al.*, 1978). Recently a fourth line of descent leading to the sulfur-dependent bacteria, eocytes, has been proposed (Lake *et al.*, 1984). Since both three and four Kingdoms are consistant with phylogeny, the arguments degenerate to the criteria to be used to determine Kingdom classification. Whether there are three or four Kingdoms is not a resolved issue of debate; for simplicity in this chapter I will refer to four Kingdoms—which the reader can take with a grain of salt. The ribosomes of these four kingdoms share striking similarities at both the ultrastructural (Lake *et al.*, 1982, 1984; Henderson *et al.*, 1984) and molecular levels. Thus, the basic translational mechanism must have been in place before these four kingdoms diverged. If RNA switches are at the heart of this mechanism, then one might anticipate that these switches would generally be conserved even between kingdoms. Interestingly, two switches that have been proposed for *E. coli* 23S rRNA (Glotz *et al.*, 1981) can also be drawn on paper for 28S rRNA from the eukaryote *Xenopus laevis* (Clark *et al.*, 1984). One of these switches involves alternative rRNA–rRNA pairing at a candidate region for the A site for tRNA binding. In the second conserved switch, an rRNA–rRNA base-paired stem opens up to allow one side of the stem to pair with $tRNA^{Met}_{init}$ (P site) and the other side of the stem to pair with 5S RNA (Fig. 1). A dynamic model to depict how the second switch works is being built by Dr. Christian Zwieb (personal communication). It will be exciting to see if experimental data can be garnered that support the validity and importance of rRNA switches.

Several people have reservations about the likelihood that switches exist that involve helical unwrapping, since topological constraints might not permit rotations needed for unwrapping the helix. Moreover, a large

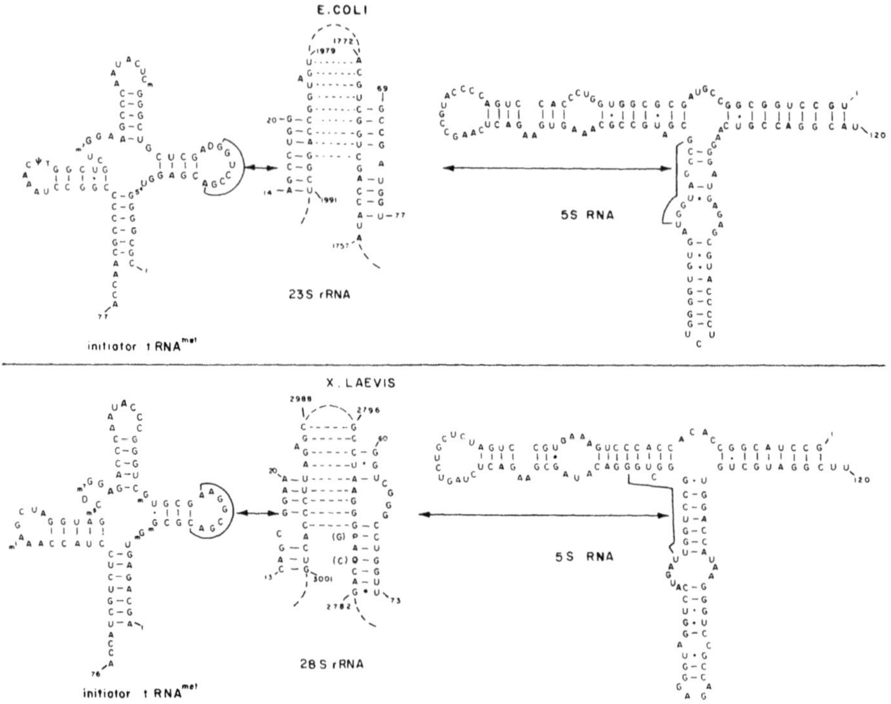

Figure 1. A possible RNA "switch." rRNA can either pair with itself to form a hairpin stem, or else the stem could open up and one side could pair with initiator $tRNA^{Met}$ and the other side of the open rRNA stem could pair with 5S RNA. This switch was first proposed for *E. coli* (Glotz *et al.*, 1981) as shown in the upper panel, but also can be drawn for *Xenopus laevis* as in the lower panel.

amount of energy might be needed to open up the hydrogen bonds of a base-paired stem; however, this second point may be moot if helical unwrapping is progressive and a new base pair is formed immediately after breakage of the old base pair.

It is also possible, and to some more plausible, that RNA switches that could exist do not involve helical unwrappings, but instead involve changes in coaxial stacking. Adjacent helices with no single-stranded nucleotides between them can be coaxial; often a noncanonical AG base pair is found at the base of the helix and might be used to cause minor bends in coaxial structure [for a fuller discussion see Woese *et al.* (1983)]. A conformational change could be effected by two coaxial helices simply unstacking (or *vice versa*), or by one of the helices, having unstacked, now entering into an alternate coaxial stacking with a third helix (Fig. 2).

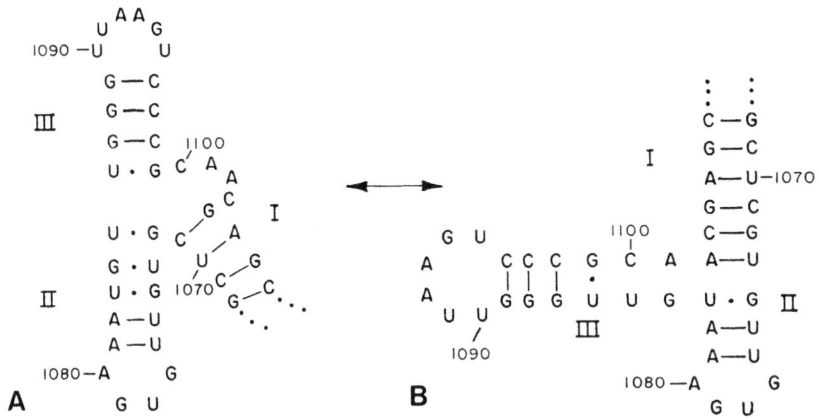

Figure 2. A possible coaxial stacking "switch." The region shown for *E. coli* 16S rRNA could be in two coaxial stacking conformations with either (A) stems II and III aligned or (B) stems I and II aligned. [Adapted from Woese *et al.* (1983).]

1.2. Was There a tRNA Ancestor to rRNA?

The argument has been presented in the preceding section that the antecedent to the present-day ribosome consisted only of RNA. I postulated that this evolutionary origin may still be manifested as the RNA switches proposed to be at the heart of the mechanism of protein synthesis. If we go back to still earlier evolutionary times, we can inquire into the origins of rRNA and the translation mechanism itself. Woese (1980) has speculated that tRNA could carry out the four functions of translation: amino acid attachment, codon recognition, peptidyl transfer, and mRNA movement. In other words, tRNA–mRNA interactions might predate ribosomes. An earlier version of tRNA may have been smaller than it is today; for example, it may have consisted only of the anticodon arm, which could carry out the four functions of translation listed above (Woese, 1980). A duplication in the primitive tRNA gene that coded for only the anticodon arm could have led to the isomorphic common arm of tRNA. Bloch *et al.* (1983) have speculated that sequence homologies between tRNA and rRNA suggest a common evolutionary origin for these two RNA species. They suggest that recurring patterns of alignment of tRNAs with rRNA can be found at a nine-base spacing (Bloch *et al.*, 1985; Nazarea *et al.*, 1985). Bloch *et al.* (1985) hypothesize that nine-base duplications gave rise to tRNA and to rRNA. If tRNA and rRNA share a common origin, it is clear that rRNA could have taken over some of the translation functions initially carried out only by tRNA (for example, perhaps peptidyl transfer and mRNA movement). In so doing, rRNA may have retained the use of RNA allosteric changes utilized previously by tRNA to carry out these functions. Therefore, RNA switches not only may be crucial for carrying out translation function, but may have even preceded the appearance of a primitive ribosome composed of rRNA.

1.3. Ribosomal Proteins May Be a Later Addition to the Ribosome

In the discussion above I summarized hypotheses that a shorter tRNA could carry out translation using RNA interactions; for example, translocation could result from ratchet-like allosteric changes in tRNA (Woese, 1970). Duplication events may have given rise to longer tRNAs, and also the rRNAs that took over some translation functions previously carried out by the primitive tRNA. At some later point in evolution, ribosomal proteins may have become complexed with rRNA to increase the efficiency of the ribosome. In studying rRNA structure and function today we should expect to find some regions in rRNA that are important for RNA–RNA interactions that may be critical for the translation process, and other regions of rRNA that are specifically recognized as binding sites for ribosomal proteins.

It has been possible to reconstitute ribosomal subunits of *E. coli* from its isolated RNA and protein components. In the 30S ribosomal subunit, up to seven proteins are implicated as strong binding proteins needed for the first step of 30S reconstitution [S4, (S7), S8, (S13), S15, S17, and (S20) (Nomura and Held, 1974)]. Similarly, there are 8–9 strong binding proteins [L1, (L2), L3, L4, L9, L11, L20, L23, and L24] needed for the first step of reconstitution of the 50S subunit (Röhl and Nierhaus, 1982). One of these proteins, L24, binds to the 5′ end of 23S rRNA and may be the initiator protein for normal 50S subunit biogenesis (Nowotny and Nierhaus, 1982), but L24 is not absolutely obligatory for 50S subunit assembly (Dabbs, 1982; Nowotny *et al.*, 1985).

There is coordinate regulation in *E. coli* of the synthesis of ribosomal proteins and rRNA. A translational feedback mechanism operates so that synthesis of ribosomal proteins is turned off when there is no more naked rRNA with which it can complex. Specific ribosomal proteins (generally those that bind early in reconstitution) have been identified that regulate the translation of several different ribosomal proteins on the same polycistronic mRNA [reviewed by Nomura *et al.* (1984)]. It has been proposed that there may be similarity in the binding sites for the regulatory ribosomal proteins on both the rRNA and on its own polycistronic mRNA, and that the rRNA and mRNA sites compete for regulatory ribosomal proteins—which will bind preferentially to the rRNA (Nomura *et al.*, 1984). When no naked rRHA is available for binding (i.e., the ribosomal proteins are in excess above what is needed for ribosome biogenesis), then the regulatory ribosomal protein may bind to the 5′ end of its polycistronic mRNA to shut off further translation of all the ribosomal proteins on that polycistronic unit (Nomura *et al.*, 1980). Deletion and mutagenesis studies of the L11 leader mRNA have defined the target structure for regulatory binding of ribosomal protein L1 on the polycistronic L11–L1 unit (Baughman and Nomura, 1983, 1984). The rRNA binding site for L1 is evolutionarily conserved, since we found that *E. coli* ribosomal protein L1 can bind to *Dictyostelium* 26S rRNA, and sequence analysis showed that primary and secondary structure features of the L1 binding site are conserved among rRNAs from *E. coli, Dictyostelium*, and *Xenopus* (Gourse *et al.*, 1981). However, *E. coli* ribosomal protein L1 is the only one found so far that can cross-react with eukaryotic rRNA.

It would be desirable for eukaryotes also to have a mechanism to regulate coordinately the synthesis of rRNA and ribosomal proteins, and, indeed, several cases can be found. For example, in *Xenopus* embryos a complete set of newly synthesized ribosomal protein mRNAs appear first at gastrulation (Pierandrei-Amaldi *et al.*, 1982) in parallel with the appearance of rRNA (Brown and Littna, 1964). Fusion of CHO and mouse 3T3 cells has shown that the feedback loop for coordinate synthesis of ribosomal proteins with rRNA seems to be species independent (Wejksnora and Warner, 1981). A few ribosomal proteins seem to escape this coordinate control and can be found prior to the onset of rRNA synthesis in *Xenopus* embryos (Pierandrei-Amaldi *et al.*, 1982), *Drosophila* embryos (Santon and Pellegrini, 1980), regenerating rat liver (Faliks and Meyuhas, 1982), and yeast (Gorenstein and Warner, 1976). For a fuller discussion of coordinate regulation of rRNA and ribosomal protein synthesis, the reader is referred to a review by Warner *et al.* (1980).

In order to ascertain if coordinate regulation may be controlled at the translational level in eukaryotes, studies have been helpful in which rRNA synthesis is abnormally depressed. It has been known for some time that when rRNA synthesis is inhibited in mouse L cells, ribosomal protein synthesis continues (Craig and Perry, 1971), but apparently these ribosomal proteins do not accumulate (Craig, 1971). A similar observation of ribosomal protein turnover has been made in HeLa cells (Warner, 1977). Finally, anucleolate mutants of *Xenopus*, which lack rRNA genes, only accumulate four ribosomal proteins (Hallberg and Brown, 1969). These anucleolate mutants synthesize ribosomal protein mRNAs (Pierandrei-Amaldi *et al.*, 1982) which are somewhat reduced in stability and are translated into ribosomal proteins that are rapidly degraded (Pierandrei-Amaldi *et al.*, 1985b).

In addition to protein turnover, other posttranscriptional regulatory steps besides translational control may also be utilized by eukaryotes. For example, when the gene for *Xenopus* ribosomal protein L1 is injected into *Xenopus* oocytes, it is transcribed efficiently, but two of its nine introns are not removed by splicing; this is in contrast to injected ribosomal protein L14 genes, which are transcribed and processed normally (Bozzoni *et al.*, 1984). Similarly, most but not all ribosomal protein genes in yeast contain introns (Rosbash *et al.*, 1981; Fried *et al.*, 1981; Bollen *et al.*, 1982; Teem *et al.*, 1984); the temperature-sensitive yeast mutant rna2 causes an accumulation of precursor mRNA for many ribosomal proteins (Rosbash *et al.*, 1981; Fried *et al.*, 1981). However, some ribosomal proteins (S2, L8, L16) are produced normally in the yeast rna mutants (Warner and Gorenstein, 1977); since L16 is one of the few yeast ribosomal protein genes that lacks an intron (Rosbash *et al.*, 1981; Bollen *et al.*, 1982), this suggests that splicing may be one level of control for coordinate regulation of rRNA and ribosomal proteins.

Also, when yeast cells are transformed with a cloned ribosomal protein gene, the transcription rate for that ribosomal protein mRNA increased in proportion to the increased gene dosage, but no major increase in ribosomal protein synthesis was seen (Pearson *et al.*, 1982). When an intron-lacking ribosomal protein gene is constructed and transformed into yeast, now a concomitant increase in ribosomal protein synthesis is seen (Warner *et al.*, 1985). This is a direct demonstration that posttranscriptional regulation occurs at the splicing step. It has been proposed that a ribosomal protein may bind to part of the intron in its own mRNA to inhibit splicing in yeast (Planta and Mager, 1985). Thus antogenous regulation at the translational level in *E. coli* would seem to be shifted to the splicing level in eukaryotes.

Still another posttranscriptional step for regulation could be at the level of mobilization of ribosomal protein mRNA from mRNP to polysomes, as suggested by Geyer *et al.* (1982) for growth-stimulated mouse fibroblasts.

It is unlikely that eukaryotes share with prokaryotes the same mechanism of autogenous regulation of translation of ribosomal protein mRNA, especially as no translational blockage was seen for *in vitro* translation of mRNA by its own ribosomal protein from *Xenopus* (Pierandrei-Amaldi *et al.*, 1985a). However, it appears that eukaryotes have added other posttranscriptional controls to improve the efficiency of control, including control at the levels of mRNA processing (splicing), mRNA export to the cytoplasm, utilization of mRNA on polysomes, and ribosomal protein turnover.

It is not surprising that posttranscriptional control of ribosomal protein mRNA differs in eukaryotes, since certain differences exist as compared to the bacterial system. First of all, the eukaryotic ribosomal protein genes studied so far are monocistronic, unlike many of the *E. coli* operons; therefore each ribosomal protein in eukaryotes must be individually regulated. In yeast there is a single copy of each of the five ribosomal protein genes studied (Fried *et al.*, 1981). On the other hand, in *Xenopus* (Bozzoni *et al.*, 1981b) there are 2–5 copies of the several ribosomal protein genes that have been investigated; it is not clear if these multiple copies are all functional. In mouse it is known that there is a single functional copy or only a few copies for each ribosomal protein gene, but there may be over a dozen apparently nonfunctional "processed" copies (i.e., lacking introns) for ribosomal protein L32 and probably others as well (Dudov and Perry, 1984).

A second difference between prokaryotes and eukaryotes is that transcription and translation are not coupled in the latter. Eukaryotes have a double nuclear membrane, which spatially segregates transcription from translation. It is difficult to imagine how rRNA and the ribosomal protein mature mRNA could simultaneously compete for binding the ribosomal protein, since the former is found as precursor rRNA in the nucleus, where it assembles with ribosomal proteins prior to export to the cytoplasm, and the translated mRNA for the ribosomal protein is found in the cytoplasm on polysomes. Clearly it is easier to understand a competition for binding ribosomal protein between rRNA and the unspliced precursor mRNA coding for that ribosomal proteins. In this case both rRNA and precursor mRNA are located in the nucleus.

The situation for coordinate regulation of ribosomal protein and rRNA synthesis is likely to be even more complex for organelles, since in that case there is still more compartmentalization of the presumed interacting components. Most of the ribosomal proteins are encoded by the nuclear genome and synthesized in the cytoplasm; for a coordinate control system they must be able to sense the level of organellar rRNA being synthesized in the organelle. Usually one or a few organellar ribosomal proteins are encoded by the organellar genome; they conceivably might play an important regulatory role. Examples of the latter are the *var* 1 gene on yeast mitochondrial DNA (Hudspeth *et al.*, 1982), ribosomal protein s5, which might be encoded within the rDNA intron of *Neurospora crassa* mitochondria (Burke and RajBhandary, 1982), and a few ribosomal protein genes encoded by the chloroplast genome of *Chlamydomonas reinhardtii* (Schmidt *et al.*, 1983, 1984) and *Euglena gracilis* (Montandon and Stutz, 1984). However this does not hold true for mammalian mitochondria where all the mitochondrial ribosomal proteins are encoded by the nuclear genome, as demonstrated by total inhibition of their synthesis by cycloheximide, and no inhibition by chloramphenicol (T. W. O'Brien, personal communication).

1.4. Coordinate Regulation of Synthesis of RNA Components of the Ribosome

In order to function at maximum efficiency the cell must coordinate the synthesis of all the components needed for protein synthesis. For biogenesis of the ribosome, not only must the synthesis of ribosomal proteins be coordinated with the synthesis of 16S–18S and 23S–28S rRNAs, but also other molecules involved in translation. An intriguing feedback loop has recently been proposed to regulate the quantity of IF3 relative to ribosomes in *E. coli*; specifically, it is hypothesized that IF3 mRNA might bind to the ribosome in the absence of IF3 protein (Gold *et al.*, 1984).

Synthesis of smaller RNA components of the ribosome must also be coordinately regulated. As will be discussed in the next section, a small RNA (e.g., 5.8S RNA) may be transcribed as a part of the rRNA precursor, thereby ensuring a 1:1 molar ratio of the small RNA to the large rRNAs. In the case of 5S RNA, it is found as part of the rRNA precursor in prokaryotes, but in eukaryotes it may be found instead at other loci. When the 5S RNA genes are physically unlinked from the rDNA, is there a way to regulate transcription of these two gene families to ensure molar equality of their products? One mechanism for this is to have about the same number of repeated copies of 5S RNA genes and rRNA genes; this is the case in *Drosophila melanogaster* [about 160 copies of 5S RNA genes and a comparable number of rRNA genes per haploid genome (Ritossa and Spiegelman, 1965; Vermeulen and Atwood, 1965; Tartof and Perry, 1970; Quincey, 1971)], *Drosophila hydei* [320 5S RNA genes per haploid genome (Renkawitz-Pohl, 1978) as compared to 250 rDNA genes (Hennig and Meer, 1971; Schäfer and Kunz, 1976)], and in sea urchins [about 175 copies of 5S RNA genes and rDNA per haploid genome (Patterson and Stafford, 1971; Nijhawan and Marzluff, 1979)]. If the transcription rate is roughly equivalent for these two gene families, then equimolar amounts of 5S RNA and rRNA should be synthesized. There are other cases, however, where the number of gene copies for 5S RNA differs from that for rDNA. For example, humans have 2000 copies of 5S RNA genes per haploid genome (Hatlen and Attardi, 1971), but under 200 copies of rDNA (Schmickel, 1973; Gaubatz and Cutler, 1975; Henderson *et al.*, 1976; B. D. Young *et al.*, 1976). In another example, the amphibian *No-*

tophthalmus viridescens has 300,000 copies of 5S RNA genes in a haploid genome (Pukkila, 1975), yet only 5800 copies of rDNA (Brown and Dawid, 1968). In such cases either all the 5S RNA genes are not transcribed in the soma as described below, or else perhaps the one RNA species that is made in excess relative to the other RNA may simply be degraded to ensure equimolar amounts of 5S RNA and rRNA.

Some instances suggest a lack of tight coordination of 5S RNA and rRNA synthesis. In sea urchins, 5S RNA is made as early as the 32-cell stage of embryonic development and its accumulation is not coordinate with that of rRNA (O'Melia, 1979). Similarly, in *Xenopus* embryos 5S RNA synthesis occurs at blastula stage, but rRNA synthesis does not begin until gastrulation (L. Miller, 1974). The additional finding that anucleolate mutants of *Xenopus*, which lack rDNA altogether, can still synthesize 5S RNA at a normal rate (L. Miller, 1973) demonstrates that 5S RNA transcription can occur independent of rDNA control.

Even though 5S RNA synthesis may not be obligatorily under rDNA control, the dual 5S RNA gene system first described in *Xenopus* (Brown and Sugimoto, 1973) suggests that under normal conditions the organism tries to coordinately regulate 5S RNA and rRNA synthesis. *Xenopus laevis* has about 400 somatic-type 5S RNA genes (Peterson *et al.*, 1980), comparable to the rDNA copy number of 450 (Brown and Weber, 1968; Birnstiel *et al.*, 1968). On the other hand, there are 20,000 oocyte-type 5S RNA genes (Brown and Sugimoto, 1973; Peterson *et al.*, 1980), which may be needed to provide enough 5S RNA to couple with the rRNA produced by the 1000-fold amplified copies of rDNA found in the oocyte's multiple nucleoli (Brown and Dawid, 1968; Gall, 1968; Perkowska *et al.*, 1968). The somatic 5S RNA genes are located on one chromosome within a cluster of oocyte 5S RNA genes (Harper *et al.*, 1983); the oocyte 5S RNA genes are additionally found on the telomeres of all chromosomes of *Xenopus* (Pardue *et al.*, 1973). Both types of 5S RNA are synthesized in oocytes, but the oocyte-type 5S RNA genes are inactive in somatic cells. The dual 5S RNA gene system is an elegant biological solution to maintain equimolar amounts of 5S RNA and rRNA in different development stages; this dual 5S RNA gene system predates the evolution of amphibians, since it is also found in a fish (Denis and Wegnez, 1977).

Specific regulation of the 5S RNA dual gene family resides at an internal control region (residues 50–55 to 80–83) within the 5S RNA gene (Bogenhagen *et al.*, 1980; Sakonju *et al.*, 1980, 1981). A 40,000-dalton protein, transcription factor IIIA (TF IIIA), binds to the noncoding (RNA-like) strand of this internal control region (Sakonju and Brown, 1982), and this permits initiation of transcription. Competition experiments showed that somatic 5S RNA genes bind TF IIIA four times tighter than oocyte

5S RNA genes (Wormington *et al.*, 1981), and this can be explained by base differences at positions 53 and 55 between the two types of genes (Sakonju and Brown, 1982). This fact alone, however, is not sufficient to explain the differential usage of the dual 5S gene system. It has been postulated that a stable transcriptional complex (involving transcription factors TF III A, B, and C) prevents inactivation of the 5S RNA gene by histone H1 (Brown, 1984; Schlissel and Brown, 1984); once histone H1 binds to a transcriptionally inactive copy of the 5S RNA gene, this gene cannot be reactivated by subsequent addition of transcription factors and RNA polymerase III (Bogenhagen *et al.*, 1982; Gottesfeld and Bloomer, 1982).

A different but related transcription factor for 5S RNA synthesis is found in somatic cells of *Xenopus* (Honda and Roeder, 1980; Pelham *et al.*, 1981). Transcription factors are important not only for regulation of gene transcription, but also may play a role in utilization of the newly synthesized 5S RNA, since TF IIIA complexes with oocyte 5S RNA to form a 7S RNP particle which is stored in the previtellogenic oocyte (Picard and Wegnez, 1979; Pelham and Brown, 1980; Honda and Roeder, 1980). The TF IIIA binding region on 5S RNA is similar but not indentical to the binding region on the RNA-like strand of the 5S RNA gene (Wool, 1985).

2. Basic Organization of rDNA

It seems likely that the genes for the RNAs of the ribosome were initially clustered in one transcription unit; this organization can still be found in prokaryotes. During evolution, parts of the coding region were moved away from this cluster to other locations, as exemplified by the 5S RNA gene. Separation of physical linkage posed the need for a regulatory mechanism to maintain equimolar production of the RNA components of the ribosome (see preceding section).

2.1. Organization of rDNA in Bacteria

Escherichia coli, a representative eubacterium, has seven rDNA units per haploid chromosome, as determined by ..ybridization to Southern blots (Kiss *et al.*, 1977). All of these seven clusters have been cloned on plasmids and/or transducing phages. Figure 3 summarizes studies [reviewed by Morgan (1982)] showing that the gene order is 5′-16S rRNAtRNA-23S rRNA-5S RNA-(tRNA)-3′. The spacer tRNA in four of the rDNA units is $tRNA_2^{Glu}$ and in three of the rDNA units is $tRNA_1^{Ile}$–$tRNA_{1B}^{Ala}$. The distal tRNAs are more variable and are absent in

Precursor RNA

5′ ——————————————————→ 3′

-16S rRNA-tRNA$^{(1)}$ -23S rRNA-5S RNA-tRNA$^{(2)}$

rDNA operon	$^{(1)}$Spacer tRNA	$^{(2)}$Distal tRNA
rrnA	tRNA$_1^{Ile}$–tRNA$_{1B}^{Ala}$	None?
rrnB	tRNA$_2^{Glu}$	None
rrnC	tRNA$_2^{Glu}$	tRNA^{Asp}–tRNA^{Tryp}
rrnD	tRNA$_1^{Ile}$–tRNA$_{1B}^{Ala}$	tRNA$_1^{Thr}$
rrnE	tRNA$_2^{Glu}$	None?
rrnG	tRNA$_2^{Glu}$	None?
rrnH	tRNA$_1^{Ile}$–tRNA$_{1B}^{Ala}$	tRNA$_1^{Asp}$

Figure 3. rDNA organization in *E. coli*. The arrangement of the coding regions in the seven rDNA operons of *E. coli* is shown. [Data summarized by Morgan (1982).]

some cases; two promoters are found upstream of the 16S rRNA coding region and transcription proceeds through the end of the cluster to generate a 30S RNA precursor [reviewed by Morgan (1982)]. Like many of the *E. coli* rDNA genes, the spacer tRNAs in *Bacillus subtilis* are tRNA^{Ile} and tRNA^{Ala} (Loughney *et al.*, 1982).

The situation differs somewhat in the archaebacteria. For example, *Halobacterium halobium* also shares the gene order 5′-16S rRNA-23S rRNA-5S RNA-3′, but this operon is only present in a single copy in the genome (Hofman *et al.*, 1979). A common rRNA precursor has not been found in *Halobacterium*, and so, unlike *E. coli*, the possibility remains open that each of the RNA coding regions in this cluster may have its own promoter (Hofman *et al.*, 1979).

This possible separation between RNA coding regions seems to be taken to an even greater extreme in the archaebacterium *Thermoplasma acidophilum*. In this species there is a single gene for 16S, 23S, and 5S RNAs, but these genes are quite far from one another (≥7.5 kb between 16S rRNA and 23S rRNA genes; ≥6 kb between 23S rRNA and 5S RNA genes; ≥1.5 kb between 16S rRNA and 5S RNA genes (Tu and Zillig, 1982)]. Since a common RNA precursor also has not been found in this case, it again may be that each RNA coding region has its own promoter (Tu and Zillig, 1982).

In the archaebacterium *Methanococcus vannielii* there are four rDNA operons and a single unlinked 5S RNA gene (Jarsch *et al.*, 1983). Sequencing of spacer between the 16S and 23S rRNA coding regions demonstrated the presence of $tRNA^{Ala}$, but no tRNA was found in a second operon of this species (Jarsch and Böck, 1983). A five-nucleotide direct repeat that flanks the spacer $tRNA^{Ala}$ might have been responsible for its excision from the operon, that lacks a spacer tRNA (Jarsch and Böck, 1983).

From the examples above it would seem that the 16S, 23S, and 5S RNA genes once were linked in a common operon, as in *E. coli*, but later became separated from each other and possibly obtained a promoter for each of the separated coding regions. Since two of the archaebacteria mentioned above only have a single copy of each rDNA coding region and this is also true for *Mycoplasma capricolum* (Amikan *et al.*, 1982), it is likely that the ancestor for *E. coli* only had a single rDNA gene copy, which later became duplicated.

2.2. Organization of rDNA in Mitochondria

The mitochondria of most species seem to have lost the 5S RNA gene altogether (Borst and Grivell, 1971; Lizardi and Luck, 1971; O'Brien and Matthews, 1976; Buetow and Wood, 1978), and it is intriguing to question if this happened as a result of initial separation of the once present 5S RNA gene from the rest of the rDNA. This observation also raises the serious question of the function, if any, of 5S RNA in the ribosome, although it has been speculated that an insertion in domain V of the large rRNA of yeast mitochondria may be a substitute for 5S RNA (Thurlow *et al.*, 1984). This 5S-like insertion cannot be found in the same location in human mitochondrial rRNA, but it has been postulated that a 23-nucleotide remnant of the 5S RNA gene may remain at the 3′ end of the large rRNA (Nierlich, 1982). The exception to mitochondrial 5S RNA loss is found in the mitochondria of flowering plants, which maintain a 5S RNA gene Leaver and Harmey, 1976; Cunningham and Gray, 1977). In maize and wheat mitochondria the 5S RNA gene is close to the 18S rRNA coding region but at least 16 kb away from the 28S rRNA coding region (Bonen and Gray, 1980; Bonen *et al.*, 1980; Stern *et al.*, 1982).

The size and organization for the remaining rDNA coding regions vary depending on the species, as can be seen in Fig. 4 [reviewed by Grant and Lambowitz (1982)], but in almost all cases there is only a single copy per genome of the small- and large-subunit rRNA coding regions.

Mammalian mitochondria have a large precursor RNA, which is a transcript of almost all of the H strand of their circular genome; processing

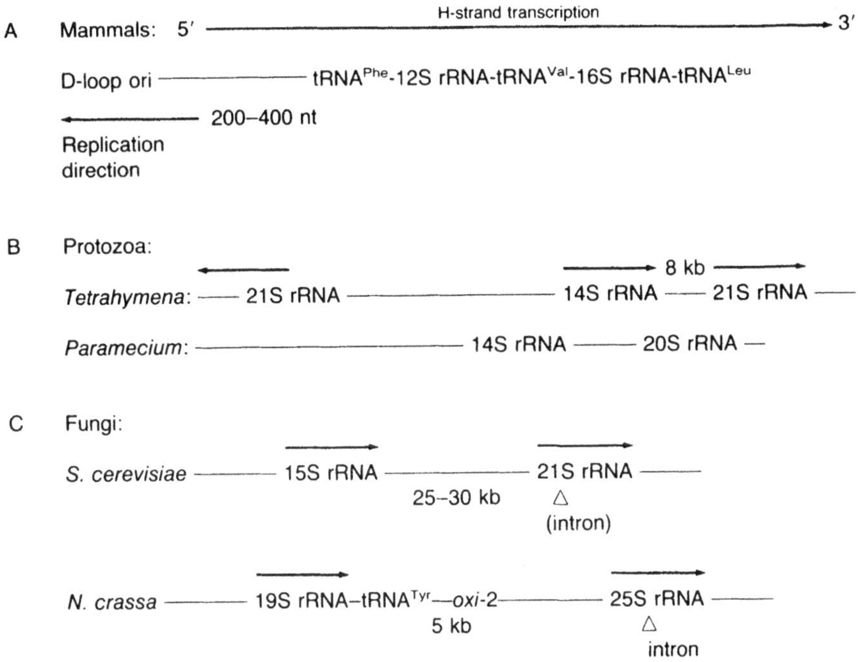

Figure 4. rDNA organization in mitochondria. The arrangement of coding regions in mitochondrial rDNA from (A) mammals, (B) protozoa, and (C) fungi. [Data summarized by Grant and Lambowitz (1982).] Recently *Drosophila* mitochondrial rDNA has been found to be the same organization as in mammals, but without the leading $tRNA^{phe}$ (Clary and Wolstenholme, 1985).

of this precursor releases the 12S rRNA (small subunit) and 16S rRNA (large subunit) rRNA (Aloni and Attardi, 1971a, b; Amalric *et al.*, 1978). Since tRNA coding regions on the H strand tightly abut the rRNA coding regions (Crews and Attardi, 1980; Eperon *et al.*, 1980; Kobayashi *et al.*, 1980; van Etten *et al.*, 1980), it may be that they act as secondary structure processing signals as well (Battey and Clayton, 1980; Ojala *et al.*, 1980, 1981). The rRNAs of mammalian mitochondria are among the smallest known, and raise questions for rRNA structure and function that will be addressed later in this review. The interspersion of tRNAs with rRNAs is similar to the bacterial situation, and this organization may reflect a common origin.

The rDNA organization in mitochondria of *Tetrahymena pyriformis* (Goldbach *et al.*, 1978) and *Paramecium aurelia* (Cummings *et al.*, 1980) is shown in Fig. 4. In *Tetrahymena* mitochondria there is a double copy of the 21S rRNA coding region, found just within the 3.4-kb inverted repeats at the ends of this linear genome. It would be of interest to know how this organelle maintains molar equivalence between 14S and 21S rRNAs, if at all, despite the 2:1 ratio of 21S to 14S rRNA coding regions.

In yeast the 15S and 21S rRNA coding regions of the mitochondria are separated by about 25–30 kb, which is nearly half of this circular genome (Faye *et al.*, 1975; Sanders *et al.*, 1975). The distance between the two rRNA coding regions is not as great in *Neurospora crassa* or other fungal mitochondrial DNAs [reviewed by Grant and Lambowitz (1982)]. Each rRNA coding region may have its own promoter in mitochondria of both yeast and *Neurospora* (Grimm and Lambowitz, 1979; Green *et al.*, 1981; reviewed by Grant and Lambowitz, 1982).

By comparing mitochondrial and bacterial rDNA organization, one could guess that they shared a common ancestor that had a single gene copy per genome of closely spaced small- and large-subunit rRNA coding regions interspersed with tRNA genes. rDNA sequence and structure comparisons support the notion that mitochondria arose from endosymbiotic bacteria; the lack of a common root for fungal and mammalian mitochondria suggests independent invasions by different bacteria (Küntzel and Köchel, 1981). As the spacing became larger, each rRNA coding region seems to have obtained its own promoter. All mitochondria except those of higher plants lost the 5S RNA gene, and deletions within the rRNA coding regions generally reduced their size relative to bacteria. The large rRNA gene of some fungal mitochondria obtained an intron (Bos *et al.*, 1978, 1980; de Vries *et al.*, 1979; Hahn *et al.*, 1979; Heckman and RajBhandary, 1979; Lazarus *et al.*, 1980; Green *et al.*, 1981), which is a characteristic of some nuclear genomes. Perhaps this reflects interchange between DNA of the mitochondria and its host cell's nuclear genome, as will be discussed below.

2.3. Organization of rDNA in Chloroplasts

The organization of rDNA in chloroplast genomes has been reviewed recently (Palmer and Thompson, 1982; Kössel *et al.*, 1983), and is summarized in Fig. 5. An inverted repeat of about 22–25 kb is found in the chloroplast genomes of spinach, mung bean, petunia, cucumber, corn, mustard, *Oenothera*, tobacco, wheat, daffodil, and nastursium, and one rDNA operon is found in each of these inverted repeats. The chloroplast genome of *Chlamydomonas* is somewhat larger than the previous examples, and its inverted repeats are a bit further distant from one another. Since the inverted repeats are even found in *Chlamydomonas*, this supports the idea that these repeats are evolutionarily quite old. From re-

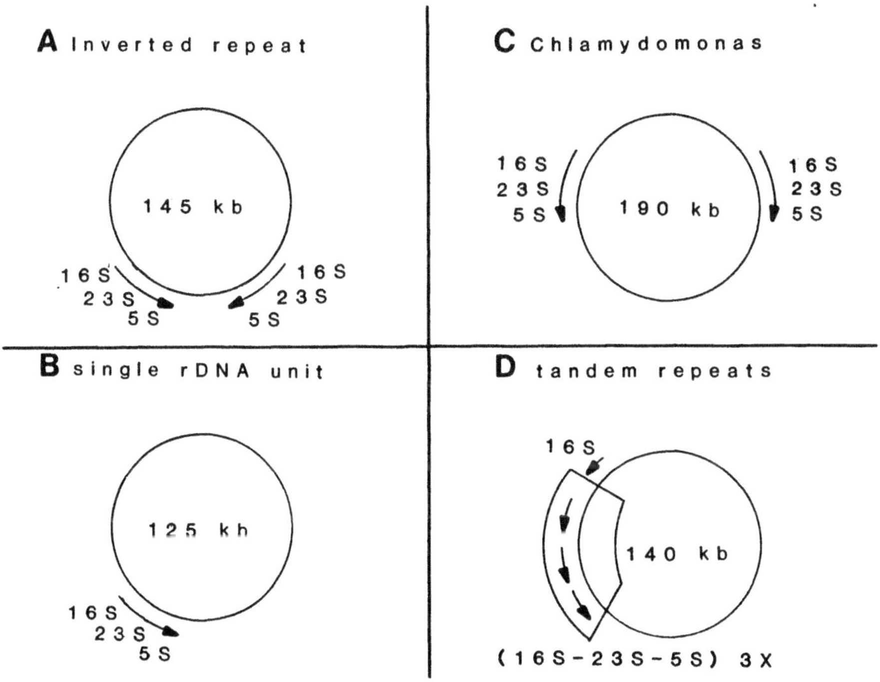

Figure 5. rDNA organization in chloroplasts. (A) The inverted repeat arrangement of rDNA found in spinach, mung bean, petunia, cucumber, corn, mustard, and *Oenothera*. (B) The single rDNA unit found in legumes such as pea and *Vicia*. (C) The rDNA inverted repeat of *Chlamydomonas*. (D) The three complete direct repeats of rDNA plus an additional 16S direct repeat as found in *Euglena gracilis* strain Z. [Data reviewed by Palmer and Thompson (1982) and Kössel *et al.* (1983).]

striction map alignments it would apppear that one of the entire repeat units has been lost in the chloroplast genomes of broad bean (Koller and Delius, 1980) and pea (Kolodner and Tewari, 1979; Chu *et al.*, 1981; Palmer and Thompson, 1981). Thus, in these two cases there is only a single rDNA operon. It seems that the presence of the inverted repeat stabilizes the chloroplast genome, since there is increased genome rearrangement in broad bean, which lacks it (Palmer and Thompson, 1982). Finally, the rDNA organization in the chloroplast genome of *Euglena gracilis* strain Z is rather unusual; in this case one of the tandem repeat units seems to have been lost, and the remaining unit has gone through tandem triplication to yield three tandem direct repeats of the rDNA operon (P. W. Gray and Hallick, 1978; Jenni and Stutz, 1978; Rawson *et al.*, 1978). In addition, a fourth copy of the 16S rRNA gene is found nearby, but outside the complete direct repeat region (Jenni and Stutz, 1979).

The organization of coding regions within the chloroplast rDNA is similar to that of *E. coli*, with the gene order running 16S–23S–5S RNA. Moreover, the intercistronic spacer between 16S and 23S rDNA almost universally codes for $tRNA^{Ile}$ and $tRNA^{Ala}$ (although only $tRNA^{Ala}$ has been reported in *Euglena* and only $tRNA_2^{Ile}$ has been reported in spinach) [reviewed by Kössel *et al.* (1983)]. These two intercistronic tRNAs are the same ones found in three of the *E. coli* rDNA operons (Fig. 3) and two of the *B. subtilis* rDNA operons. $tRNA^{Val}$ is also found 5′ to the 16S rRNA coding region, but is not found in the primary rRNA transcript [reviewed by Kössel *et al.* (1983)]. Like *E. coli*, a promoter for the rRNA precursor in chloroplasts is probably located 5′ to the 16S rRNA gene, and a precursor rRNA for 16S + 23S has been found in chloroplasts (Hartley and Ellis, 1973). However, unlike *E. coli*, the chloroplast rRNA precursor seems to terminate 3′ to the 23S rRNA gene (Kössel *et al.*, 1983). 5S RNA has not been found as part of the chloroplast rRNA precursor (Hartley, 1979; Kössel *et al.*, 1982), suggesting that it has its own promoter. In some nuclear genomes discussed below the coding region of the 5S RNA gene may move to the other DNA strand or spatially separate itself from the rDNA locus, and perhaps the acquisition of its own promoter is the first obligatory step in the evolution of separation of the 5S RNA gene from the rDNA operon.

Additional processing steps may occur in some chloroplast rRNA precursors, as compared to *E. coli*, to break pieces off from the 23S rRNA molecule. Primary sequence and secondary structure comparisons demonstrate that the 4.5S RNA found in the chloroplast ribosomes of higher plants is comparable to the 3′ end of *E. coli* 23S rRNA (Branlant *et al.*, 1981; Edwards and Kössel, 1981; Edwards *et al.*, 1981; MacKay, 1981; Clark and Gerbi, 1982). In the chloroplast genome of *Chlamydomonas* the gene order is 16S–7S–3S–23S–5S RNA, and the 7S and 3S RNA molecules may be equivalent to the 5′ end of 23S rRNA of *E. coli* (Rochaix, 1981). I will return to the subject of DNA insertions into rDNA genes in section 3.2 of this review.

The similarities between the rDNA organization in the chloroplast genome and that of *E. coli* support the endosymbiont hypothesis for the origin of organelles such as chloroplasts and mitochondria (Margulis, 1981). A corollary of this model is that some of the genes for components of the chloroplast ribosome were retained in the chloroplast genome (e.g., RNA components of the chloroplast ribosome), but other genes, such as those for some of the ribosomal proteins, were transferred to the host nucleus. There is a present-day example of gene transfer from the genome of a plant endosymbiont to the nuclear genome of the host: in crown gall tumors, part of the Ti plasmid from *Agrobacterium tumifaciens* is incorporated into the plant's nuclear genome (Schell and Van Montagu, 1980).

To take the endosymbiont model one step further, not only might gene transfer have occurred between the organelle and the host nucleus, but DNA transfer from the host nucleus into the organellar genome may also be possible. The presence of an intron in the 23S rRNA gene of *Chlamydomonas* chloroplasts (Rochaix and Malnoe, 1978) could be seen as a eukaryotic trait that has been transferred to the organelle. Introns are also found in the intercistronic spacer tRNAs of *Zea mays* chloroplast DNA (Koch *et al.*, 1981). None of these observations, however, prove the endosymbiont model for the origin of chloroplasts, and the opposing cluster-clone model [reviewed by Bogorad (1982)] is also plausible. In the latter model, membrane enclosure separated the nuclear and organellar genomes into different cellular compartments.

Regardless of the mode by which organelles arose, there is recent evidence to support the transfer of genetic information between chloroplasts and mitochondria, and between these organelles and the nucleus. Recently it was reported that the 16S rRNA and associated spacer $tRNA^{Ile}$ and $tRNA^{Val}$ coding regions of maize chloroplast DNA are also present in maize mitochondrial DNA (Stern and Lonsdale, 1982). The apparent mobility of DNA from chloroplasts to mitochondria was given the term "promiscuous DNA" (Ellis, 1982). It is now clear that regions spanning almost all of the chloroplast genome may be promiscuous, but the strongest homologies between chloroplast and mitochondrial genomes are still found within the chloroplast inverted repeat region (Stern and Palmer, 1984). Since the promiscuous DNA fragments are conserved in sequence and arrangement in chloroplast genomes but have a more varied organization in the mitochondrial genomes of different species, it is likely that the DNA was transferred from chloroplasts to mitochondria and not *vice versa* (Stern and Palmer, 1984). There is also recent evidence for the evolutionary transfer of organellar DNA to the nuclear genome: mitochondrial DNA sequences can be found by molecular hybridization in nuclear DNA (Jacobs *et al.*, 1982; van den Boogaart *et al.*, 1982; Gellissen *et al.*, 1983; Kemble *et al.*, 1983; Farrelly and Butow, 1983; Lewin, 1983) and chloroplast DNA sequences can be found in the nucleus (Scott and Timmis, 1983). An exciting recent finding was that during senescence of the fungus *Podospora anserina* there is excision and amplification of mitochondrial sequences, which are then transposed and integrated into the chromosomes of the nucleus (Wright and Cummings, 1983).

Three possible mechanisms for promiscuous DNA transfer have been suggested (Ellis, 1982; Stern and Palmer, 1984):

1. Lysis of the organelle would liberate DNA, which could be taken up by the other organelle.
2. Organelles could fuse and their genomes recombine [for examples

of chloroplast–mitochondrion fusion see Wildman *et al.* (1962) and Crotty and Ledbetter (1973)].

3. A vector such as a transposon could move DNA from one organelle to another [see Leaver and Gray (1982) for a discussion of transposon-like DNA in maize mitochondria].

The observations above on promiscuous DNA are consistent with but do not prove the idea that mitochondria and chloroplasts arose as endosymbionts (compare the similarity of their rDNA organization with that of bacteria), and subsequent gene transfer may have moved some generic loci coding for ribosome components (e.g., ribosomal protein genes) to the nucleus. See Palmer (this volume, Chapter 3), for a thorough discussion of the endosymbiotic origins of organellar genomes.

2.4. Organization of rDNA in Eukaryotic Nuclei

The rDNA found in the nuclear genome of eukaryotes generally consists of tandem repeated units with the structure shown in Fig. 6 [reviewed by Long and Dawid (1980b)]. The rRNA operon starts with an external transcribed spacer (ETS) and transcribes a large rRNA precursor; subsequent processing yields the mature rRNA species, which are 17S–18S, 5.8S, and 25S–28S in size [reviewed by Perry (1976)]. The transcripts of two internal transcribed spacers (ITS 1 and ITS 2) are discarded during rRNA maturation. A so-called nontranscribed spacer (NTS) is found between each pair of rRNA operons.

The length of the rDNA unit may vary considerably among different eukaryotic species, ranging from 6.8 kb in the protozoan *Stylonychnia* (Lipps and Steinbruck, 1982) and 7.5 kb in the protozoa *Euplotes* and

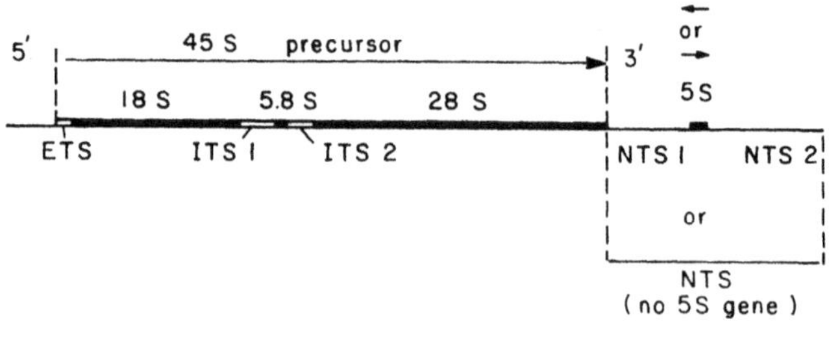

Figure 6. rDNA organization in eukaryotic nuclei. One typical rDNA unit is shown; ETS, external transcribed spacer; ITS, internal transcribed spacer; NTS, the so-called nontranscribed spacer. Depending on the species, the 5S RNA coding region may be found in either orientation within the NTS or else outside the rDNA cluster altogether (see text).

Oxytricha (Swanton *et al.*, 1982) to 42 kb in the slime mold *Dictyostelium* (Cockburn *et al.*, 1976; Maizels, 1976) and 44 kb in the rat (Stumph *et al.*, 1979). A major factor for this difference is the variable length of the NTS, which will be discussed more fully in section 5.1 of this review.

The number of copies of tandemly repeated rDNA units also varies among species, ranging from a single chromosomal copy in the micronucleus of *Tetrahymena* (Yao and Gall, 1977) to a few hundred copies in most metazoan animals and up to 6400 copies in wheat (Appels *et al.*, 1980). The rDNA copy number may also vary in one species during its development due to rDNA underreplication during polytenization (Hennig and Meer, 1971; Spear and Gall, 1973), the disproportionate replication of *Drosophila* rDNA compensation [reviewed by Kunz *et al.* (1982)], rDNA amplification in oocytes of many animals (Brown and Dawid, 1968; Gall, 1968) [reviewed by Tobler (1975)], or amplification in the polyploid macronuclei of protozoa [reviewed by Blackburn (1982) and by Yao (1982)]. The amplified rDNA in oocytes and protozoan macronuclei is extrachromosomal. The mechanisms of magnification and reduction in *Drosophila* also act to vary rDNA copy number between generations [reviewed by Kunz *et al.* (1982)]. A full discussion of the control of rDNA copy number is outside the scope of this review, but suffice it to say that control of gene copy number appears to be a mode of regulation added on in the evolution of eukaryotes, and is not known to occur in bacteria or organelles.

In most eukaryotes the 5S RNA genes are found at another location(s) on the nuclear genome, but in some fungi and protozoa the 5S RNA genes are within the rDNA locus, as shown in Table I. In those cases where the 5S RNA gene is part of the rDNA, the former may be transcribed

Table I
Organisms Where the 5S RNA Gene Is Part of rDNA

		Reference
Ascomycetes	*Saccharomyces cerevisiae*	Rubin and Sulston (1973)
	Saccharomyces carlsbergensis	Bell *et al.* (1977)
	Saccharomyces rosei	Verbeet *et al.* (1983)
	Torulopsis utilis	Tabata (1980)
	Kluyveromyces lactis	Verbeet *et al.* (1984a)
	Hansenula wingei	Verbeet *et al.* (1984a)
Oomycete	*Achyla ambixexualis*	Rozek and Timberlake (1979)
Zygomycete	*Mucor racemosus*	Cihlar and Sypherd (1980)
Basidiomycete	*Coprinus cinereus*	Cassidy *et al.* (1984)
Slime mold	*Dictyostelium discoideum*	Maizels (1976)
Protozoa	*Euglena gracilis*	Curtis and Rawson (1981)

from the same strand and thus in the same direction as the rRNA, such as in *Coprinus cinereus* (Cassidy *et al.*, 1984), or it may be transcribed from the opposite strand and thus in the opposite direction, such as in *Saccharomyces cerevisiae* (Maxam *et al.*, 1977; Valenzuela *et al.*, 1977; Kramer et al., 1978).

The 5S RNA genes are not found in the rDNA cluster universally in all fungi, but are located elsewhere in the genome in the ascomycetes *Neurospora crassa* (Free *et al.*, 1979; Selker *et al.*, 1981), *Schizosaccharomyces pombe* (Mao *et al.*, 1982), and *Aspergillus nidulans* (Bartnik *et al.*, 1981; Borsuk *et al.*, 1982; Lockington *et al.*, 1982). Regardless of the location of 5S RNA, it is produced by RNA polymerase III transcription, whereas the rRNA precursor is produced by RNA polymerase I transcription [reviewed by Roeder (1976)]. The acquisition of separate polymerases for 5S RNA and rRNA, coupled with the introduction of a separate promoter for the 5S RNA genes, may have been the first step in evolution for separation of these two gene families. Having gained its own promoter, the 5S RNA gene could then move to either the opposite DNA strand than that coding for rRNA (as in *S. cerevisiae*), or leave the rDNA cluster altogether. It is easiest to imagine that loss of the 5S RNA gene from the rDNA cluster occurred when there was only a single rDNA operon/genome, and was subsequently followed by tandem amplification of the rDNA operon now bereft of the 5S RNA gene, as well as tandem amplification of the 5S RNA genes in their new location away from rDNA. However, data is lacking to prove the mechanism by which the 5S RNA gene moved from one strand of DNA to the other or to another chromosomal locus.

The mechanism by which a tandem array of rDNA units was generated in the nuclear genome is unknown. Models such as unequal crossing-over (Smith, 1973) or excision of an rDNA copy coupled with rolling circle replication and subsequent reintegration have been proposed to explain the tandem duplication of rDNA. In any event, it is of interest to ask what the structure is of the rDNA copies at the two extreme ends of the rDNA tandem array. For example, are they complete or incomplete rDNA units? This question has been addressed by Zamb and Petes (1982), who have analyzed the boundary of the tandem array of 100 rDNA copies on chromosome XII of yeast. They observed that the centromere–distal boundary (J2) of the rDNA cluster has a complete coding region and that the fusion to a non-rDNA single-copy sequence occurred within the NTS (Fig. 7). Observations such as these will have to be compatible with any viable model for the generation of tandem repeats of rDNA.

A few years ago it was discovered that a solitary genetic element

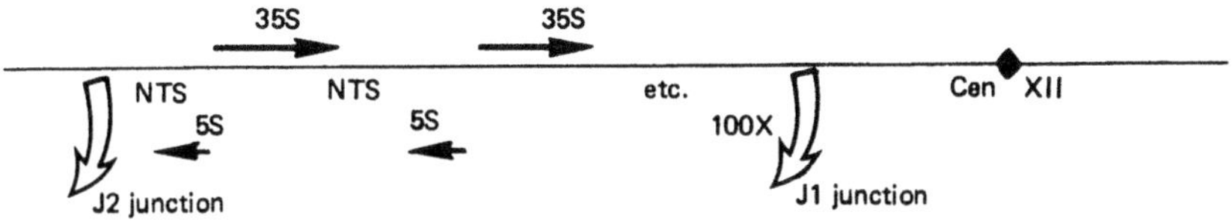

Figure 7. Boundaries of the rDNA cluster in yeast. The 100 copies of rDNA are found on chromosome XII of *Saccharomyces cerevisiae*. The centromere proximal (J1) and centromere distal (J2) boundaries of the rDNA cluster are indicated; the J2 junction occurs within the NTS of the last rDNA unit. [Data from Zamb and Petes (1982).]

may exist dispersed away from its tandem multigene family; the element, which has "lost its family," is termed an "orphon" (Childs *et al.*, 1981). The evidence for orphons is most detailed for histone genes, but also exists for rDNA of *Drosophila* and yeast. How distant the rDNA orphons are from the rest of the rDNA tandem cluster, or whether the orphons are on other chromosomes from the rDNA cluster, remains to be discovered. A description of the DNA that surrounds an rDNA orphon would be helpful in determining the mechanism by which the orphon moved away from its family.

3. Evolutionary Conservation in rRNA Structure

Certain areas within rRNA show remarkable conservation in primary and/or secondary structure among diverse species. Such conservation argues for a common ancestor linking the four kingdoms together with a prototype ribosome. Comparison of conserved sequences in rRNA has been used in a molecular taxonomy approach to build a global phylogeny (M. W. Gray *et al.*, 1984). Highly conserved regions within rRNA presumably are crucial for the function and/or structure of rRNA; mutations within the rDNA for these regions would be lethal to the organism and not perpetuated during evolution. Functions that correlate with some of the regions of conserved sequence are summarized below. Early information on the structure and function of eukaryotic ribosomes has been summarized previously (Wool, 1979). Although conserved regions are deduced to be functionally important, the converse of this does not apply. It is false to assume that variable regions of sequence are not important; it could be that as the sequence of a variable region evolved, the component with which it reacts coevolved with it. One example of this is the compensatory base changes that maintain secondary structure, and will be reviewed in Section 3.2.

3.1. Primary Structure of rRNA

It has been known for some time from heterologous hybridization experiments that at least portions of rDNA sequence are highly conserved; for example, *Xenopus* rRNA can cross-react with DNA from numerous species (Sinclair and Brown, 1971). Competition hybridization showed that this cross-reaction was due to a conserved subset of sequences within rDNA (Birnstiel and Grunstein, 1972; Gerbi, 1976). The conserved subset was mapped to defined restriction fragments within the 18S and 28S coding regions of rDNA by Southern blot hybridization (Gourse and Gerbi, 1980a; Cox and Thompson, 1980). These data showed that conserved regions are scattered throughout 18S and 28S at distinct locations. Now that the complete rDNA sequences are available for many examples (summarized in Tables II and III), it is possible to align sequences to determine regions of conservation at the nucleotide level [for example, see Olsen *et al.* (1983) for 16S–18S and Ware *et al.* (1983) for 23S–28S alignments].

Many of the conserved regions include sites that are methylated in the rRNA. Only a relatively small number of base methylations occur in prokaryotes such as *E. coli.* In eukaryotes such as yeast and *Xenopus*, a large number of 2′-O ribose methylations occur; this methylation occurs early in the life of the rRNA precursor. By contrast, eukaryotes have a fairly small amount of base methylation, and this modification occurs later in rRNA maturation. Most of the methylated oligonucleotides in eukaryotes have been conserved during evolution (Khan *et al.*, 1978). The location of rRNA methylation has been positioned on the restriction enzyme map of *Xenopus* rDNA (Brand and Gerbi, 1979; Maden and Reeder, 1979; Maden, 1980), and has been determined exactly on the nucleotide sequence of *Xenopus* 18S (Salim and Maden, 1981) and yeast 26S (Veldman *et al.*, 1981a). The correlation of rRNA methylation with evolutionary conservation also applies to mitochondrial rRNAs (Baer and Dubin, 1981). Unfortunately, the functional significance of rRNA methylation is not yet clear.

One methylation site that has been correlated with a function are the two highly conserved *N*-(6)-dimethyl A's found in the apical loop of the last base-paired stem at the 3′ end of 16S-18S rRNAs. Immunoelectron microscopy has positioned the dimethyl A's of *E. coli* at the surface of the 30S ribosomal subunit, at the base of the 30S platform (Politz and Glitz, 1977). Antibodies against dimethyl A inhibit IF-3 binding (Thammana and Cantor, 1978), thereby suggesting a role in initiation of translation for this region. Both eIF-2 and eIF-3 have been found cross-linked to 18S rRNA (Westermann *et al.*, 1980; Nygard and Westermann, 1982). Finally, kasugamycin is a prokaryotic antibiotic that inhibits mRNA-

Table II
Primary Sequences of Small-Subunit rRNA

	Reference
Eubacteria	
Escherichia coli 16S	Brosius *et al.* (1978), Carbon *et al.* (1979)
Proteus vulgaris 16S	Carbon *et al.* (1981)
Anacystis nidulans 16S	Tomioka and Sugiura (1983)
Bacillus brevis 16S	Kop *et al.* (1984a)
Mycoplasma capricolum 16S	Iwami *et al.* (1984)
Archaebacteria	
Halobacterium volcanii 16S	Gupta *et al.* (1983)
Halococcus morrhua 16S	Leffers and Garrett (1984)
Eukaryotes (cytoplasmic)	
Zea mays 17S	Messing *et al.* 1984
Oryza sativa 17S	Takaiwa *et al.* 1984
Saccharomyces cerevisiae 18S	Rubtsov *et al.* (1980), Mankin *et al.* (1981a), Hogan *et al.* (1984a)
Dictyostelium discoideum 17S	McCarroll *et al.* (1983)
Caenorhabditis elegans 18S	R. Ellis (personal communication)
Artemia salina 18S	Nelles *et al.* (1984)
Xenopus laevis 18S	Salim and Maden (1981), Atmadja *et al.* (1984)
Rat 18S	Torczynski *et al.* (1983), Chan *et al.* (1984)
Mouse 18S	Raynal *et al.* (1984)
Rabbit 18S	Connaughton *et al.* (1984b)
Human 18S	Torczynski *et al.* (1985)
Mitochondria	
Zea mays mitochondrion 18S	Chao *et al.* (1984)
Triticum aestivum mitochondrion 18S	Spencer *et al.* (1984)
Trypanosoma brucei 9S	Eperon *et al.* (1983)
Leishmania tarentolae mitochondrion 9S	de la Cruz *et al.* (1985a)
Crithidia fasciculata mitochondion 9S	Sloof *et al.* (1985)
Yeast mitochondrion 15S	Sor and Fukuhara (1980), Li *et al.* (1982)
Aspergillus nidulans mitochondrion 15S	Köchel and Küntzel (1981)
Paramecium primaurelia mitochondrion 15S	Seilhammer *et al.* (1984)
Paramecium tetraurelia mitochondrion 15S	Seilhammer *et al.* (1984)
Drosophila yakuba mitochondrion	Clary and Wolstenholme (1985)
Human mitochondrion 12S	Eperon *et al.* (1980)
Mouse mitochondrion 12S	Van Etten *et al.* (1980)
Rat mitochondrion 12S	Kobayashi *et al.* (1981)
Bovine mitochondrion 12S	Anderson *et al.* (1982)
Chloroplasts	
Zea mays chloroplast 16S	Schwarz and Kössel (1980)
Euglena gracilis chloroplast 16S	Graf *et al.* (1982)
Chlamydomonas reinhardii chloroplast 16S	Dron *et al.* (1982)
Nicotiana tabacum chloroplast 16S	Tohdoh and Sugiura (1982)

Table III
Primary Sequences of Large-Subunit rRNA

	Reference
Eubacteria	
Escherichia coli 23S	Brosius *et al.* (1980)
Anacystis nidulans 23S	Kumano *et al.* (1983), Douglas and Doolittle (1984)
Bacillus stearothermophilus 23S	Kop *et al.* (1984b)
Eukaryotes (cytoplasmic)	
Saccharomyces cerevisiae 26S	Georgiev *et al.* (1981)
Saccharomyces carlsbergensis 26S	Veldman *et al.* (1981a)
Physarum polycephalum 26S	Otsuka *et al.* (1983)
Caenorhabditis elegans 28S	R. Ellis (personal communication)
Xenopus laevis 28S	Ware *et al.* (1983)
Rat 28S	Chan *et al.* (1983)
	Hadjiolov *et al.* (1984)
Mouse 28S	Hassouna *et al.* (1984)
Human 28S	Gonzalez *et al.* (1985)
Mitochondria	
Zea mays mitochondrion 23S	Dale *et al.* (1984)
Trypanosoma brucei mitochondrion 12S	Eperon *et al.* (1983)
Crithidia fasciculata mitochondria 12S	Sloof *et al.* (1985)
Leishmania tarentolae mitochondrion 12S	de la Cruz *et al.* (1985b)
Paramecium primaurelia mitochondrion 20S	Seilhammer and Cummings (1981)
Aspergillus nidulans mitochondrion 21S	Netzker *et al.* (1982)
Yeast mitochondrion 21S	Sor and Fukuhara (1983)
Drosophila yakuba mitochondrion	Clary and Wolstenholme (1985)
Human mitochondrion 16S	Eperon *et al.* (1980)
Mouse mitochondrion 16S	Van Etten *et al.* (1980)
Rat mitochondrion 16S	Saccone *et al.* (1981)
Bovine mitochondrion 16S	Anderson *et al.* (1982)
Chloroplasts	
Zea mays chloroplast 23S	Edwards and Kössel (1981)
Nicotiana tabacum chloroplast 23S	Takaiwa and Sugiura (1982)

dependent tRNA binding; the alteration in kasugamycin-resistant mutants is a lack of the dimethyl group (Helser *et al.*, 1971).

Other data also implicate an initiation role for the 3′ end of 16S–18S rRNA. Shine and Dalgarno (1974) were the first to note that the sequence CUCC found next to the 3′ end of 16S rRNA has a complementary region just upstream of the initiation codon in mRNA. Almost but not all prokaryotic mRNAs have the Shine–Dalgarno sequence [reviewed by Kozak (1983)], which is on the average seven nucleotides 5′ to the AUG codon (Stormo *et al.*, 1982). Direct proof of the reality of the Shine–Dalgarno 16S rRNA–mRNA interaction came from the studies of Steitz and Jakes (1975), who found mRNA complexed with the 49-nucleotide oligomer clipped from the 3′ end of 16S rRNA by colicin. The Shine–Dalgarno sequence is found in the 16S rRNAs of eubacteria, archaebacteria, and chloroplasts, but is missing in rRNA of the cytoplasm of eukaryotes (Hagenbüchle *et al.*, 1978) and mitochondria. The latter led Kozak to propose a "scanning" model for initiation of translation in eukaryotes, whereby the ribosome attaches to the 5′ end of eukaryotic mRNA and slides along until it reaches the first AUG signal for initiation (Kozak, 1978, 1980), but there are a few eukaryotic mRNAs that do not obey this rule (Kozak, 1983). Although eukaryotic 18S rRNA lacks the Shine–Dalgarno sequence, psoralen cross-linking has shown that mRNA may associate nonetheless with the 3′ region of 18S rRNA (Nakashima *et al.*, 1980). An additional 18S rRNA–mRNA interaction in eukaryotes has been proposed by Thompson and Hearst (1983b): the hypermodified base amψ found one-third of the way in from the 3′ end of 18S rRNA may be brought near the 3′ end of 18S by long-range base pairing, which would allow it to make potential specific interactions with the m^7G cap found at the 5′ end of eukaryotic mRNAs.

Slightly interior to the 3′ end of 16S–18S rRNAs, another known function can be correlated with a conserved sequence. The 17-mer 5′...GUACACACC*GCCCGUC...3′ is found in the 3′ region of the small-subunit rRNAs of archaebacteria, eubacteria, eukaryotes, chloroplasts, and mitochondria. The C* represents C_{1400} in *E. coli* 16S rRNA and the corresponding C_{1626} in yeast 18S rRNA, which have both been cross-linked to *E. coli* $tRNA^{Val}$ (Ofengand *et al.*, 1982; Prince *et al.*, 1982; Ehresmann and Ofengand, 1984), demonstrating that this region of rRNA is part of the P site. Recently a crosslink has also been found for the 5′-anticodon base of tRNA when it is in the A site and the same C_{1400} residue of *E. coli* 16S rRNA (Ofengand *et al.*, 1985). Additional evidence that this area is part of the decoding site is the observation that a mutation in yeast mitochondrial 15S rRNA nine nucleotides to the 3′ side of the $tRNA^{Val}$ cross-linking site renders it resistant to paromomycin, an antibiotic that increases translational errors (Li *et al.*, 1982; Martin *et al.*, 1982); a comparable paromomycin resistance mutation has recently been mapped in the nuclear rDNA of *Tetrahymena* (E. H. Blackburn, personal communication). The cross-linked C of the P site has been localized by affinity immunoelectron microscopy to deep within the cleft between the head and the large protrusion of the 30S subunit (Keren-Zur *et al.*, 1979; Gornicki *et al.*, 1983). It is available for interactions, since this area can be kethoxylated (Herr *et al.*, 1979) and digested by RNAse (Stiegler *et al.*, 1981b; Vassilenko *et al.*, 1981). This region is part of the interface between the two ribosomal subunits (Herr *et al.*, 1979; Vassilenko *et al.*, 1981). Since initiator tRNA directly enters the P site, this area may also play a role in initiation of translation. Tertiary folding detected by psoralen cross-linking of residues 919–923 with 1530–1534 of *E. coli* 16S rRNA may bring the Shine-Dalgarno sequence into close proximity with the C_{1400} tRNA site (Hui and Cantor, 1985); the conformation of this region of 16S rRNA differs in active and inactive forms of the ribosome (Noller *et al.*, 1985). Recently it has been proposed that the TψC-loop of elongator tRNAs may bind to residues 1496–1501 of *E. coli* 16S rRNA (Sprinzl and Helk, 1985).

Although many other conserved stretches occur in 16S–18S rRNA, the only functional correlations known are for the areas discussed above, which are implicated for tRNA association and for mRNA binding and initiation. Similarly, only a few of the conserved sequence tracts in 23S–28S rRNA have functional correlates, and these are summarized by Clark *et al.* (1984). In domain III of 23S rRNA, an area of some sequence conservation correlates with the GTPase center. The antibiotic thiostrepton blocks EF-G association to the ribosome and thus stops EF-G-dependent GTPase activity and inhibits translocation (Thompson *et al.*, 1982). *Streptomyces azureus* produces thiostrepton and is itself resistant to the antibiotic due to a methylation of an A residue in the presumed GTPase center (Thompson *et al.*, 1982). Eukaryotes such as *Xenopus* have a G instead of A at this position, which may explain the decreased sensitivity of eukaryotes to thiostrepton. Finally, EF-G has been cross-linked to this proposed GTPase center (Sköld, 1983).

A conserved region in domain VI of 23S–28S rRNA most likely contains the peptidyl transferase center. This region includes the sites of mutation rendering mitochondrial ribosomes resistant to erythromycin (Sor and Fukuhara, 1982) and chloramphenicol (Dujon, 1980; Kearsay and Craig, 1981; Sor and Fukuhara, 1982). Recent studies using photoaffinity labels on tRNA have implicated nucleotides 2445–2668, 2451 plus 2452, and 2484 plus 2485 of this area of *E. coli* 23S rRNA as sites for tRNA association in the peptidyl transferase center (C. C. Hall, J. Smith, and B. S. Cooperman, personal communication; A. Barta, G. Steiner, and E. Kuechler, personal communication; Barta *et al.*, 1984). It has been postulated that the CCA end of tRNA pairs with the evolutionarily conserved bases UGG (nucleotides 807–809 in *E. coli*) in domain II of 23S–28S rRNA (Barta *et al.*, 1984; Noller *et al.*, 1985), necessitating this region of domain II to be in close proximity to the peptidyl transferase center of VI. In support of this, crosslinks have been found between domains II and VI (Stiege *et al.*, 1983), and also a synthetic oligonucleotide to residues 803-811 inhibits tRNA binding in *E. coli* (Hill, 1985). It has also been deduced that a crosslink to a derivative of puromycin is also close to the peptidyl transferase center of domain VI (Branlant *et al.*, 1981).

Other regions of sequence conservation in rRNA may play a critical role in translation by virtue of RNA–RNA interactions, as discussed in the discussion on "switches" in Section 1.1. Finally, some regions of rRNA sequence conservation doubtless play a scaffold role for association with ribosomal proteins; mechanisms of protein–RNA recognition in ribosome biogenesis have recently been reviewed (Garrett *et al.*, 1984b). A striking example of this is *E. coli* ribosomal protein L1, which can even bind to a conserved region in *Dictyostelium* 26S rRNA (Gourse *et al.*, 1981).

3.2. Secondary Structure of rRNA

The preceding section showed that functional constraints may preserve certain sequences within rRNA; presumably an alteration of these conserved sequences would impair rRNA function. Other regions within rRNA do not seem to have the same requirement for preservation of the actual nucleotide sequence; however, structural constraints may still be imposed upon some of these other regions. Specifically, if it is important to retain a base-paired stem as part of the rRNA secondary structure, then mutation of a base on one side of this stem may favor a compensatory mutation for its complementary base pairing partner. An example of this is shown in Fig. 8, where mutation of G to A favors the compensatory mutation of C to U to retain base pairing. The mismatched bulge of AC may not significantly detract from the stability of the stem, so the first mutation of G to A can be considered effectively neutral in terms of selection. However, the compensatory mutation of C to U will be slightly favored by selection. This is essentially a two-step mutation process, where positive selection acts on the second step.

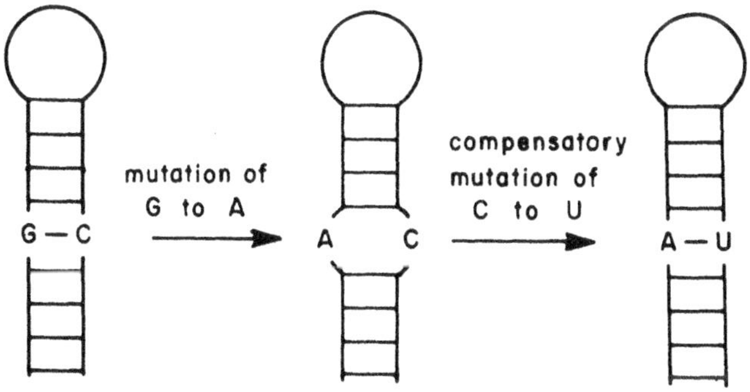

Figure 8. Compensatory base change. When G mutates to A it causes a bulge in the paired stem, but a compensatory mutation of C to U on the opposite side of the stem allows hydrogen bonding to be regained.

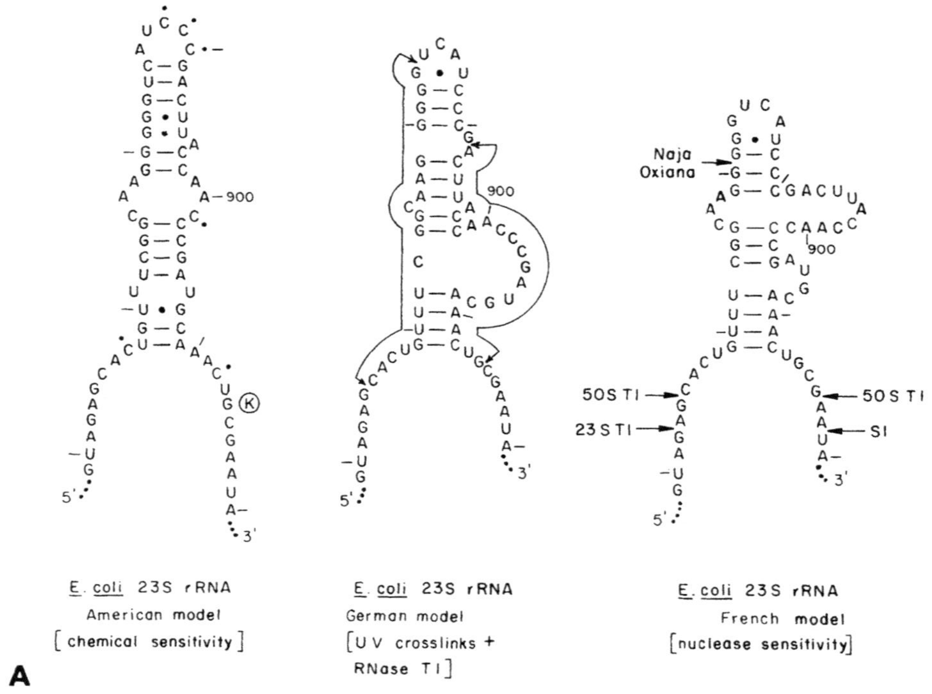

Figure 9A. Methods for deduction of rRNA secondary structure. Three experimental methods for the analysis of rRNA secondary structure: (1) sensitivity of single-stranded regions to chemicals such as kethoxal (Ⓚ) or bisulfite (•) as used by Noller and Woese; (2) UV cross-linking followed by isolation of RNAse T1 paired oligomers as used by Brimacombe; and (3) sensitivity to nucleases specific for single-stranded (RNAse T1) or double-stranded (*Naja oxiana* nuclease) regions as used on 50S ribosomal subunits or purified 23S rRNA by the Ebel group.

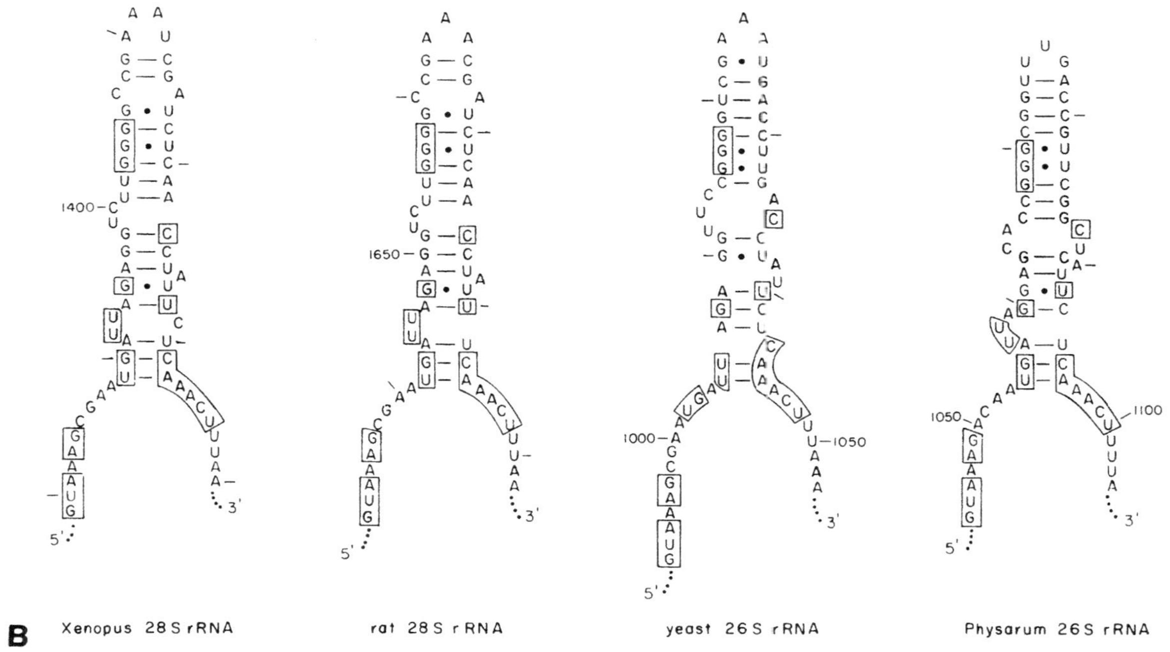

Figure 9B. The secondary structure deduced for the comparable region in four eukaryotes. The American model for this region is the only one compatible with all lines of experimental data plus the four eukaryotic sequences. Residues enclosed by a box are evolutionarily conserved between *E. coli* and eukaryotes.

Compensatory base changes have been taken as evolutionary "proof" for the reality of specific stems in the secondary structure of rRNA. Three groups have been instrumental in working out the secondary structure of *E. coli* 16S and 23S rRNAs; they are the American group of the Noller and Woese laboratories, the German group of the Brimacombe laboratory, and the French group of Ebel's coworkers. As a starting point for deducing rRNA secondary structure, the following types of experimental evidence were used: (1) sensitivity to chemicals such as kethoxal modification of G residues or bisulfite modification of C residues, thereby indicating that the sensitive bases are in single-stranded regions; (2) UV-induced cross-links of paired regions followed by RNAse digestion to resolve the cross-linked oligomers on two-dimensional gels; (3) sensitivity to single-strand- or double-strand-specific nucleases. The upper portion of Fig. 9 illustrates the structures arrived at for a stem in *E. coli* 23S rRNA using each of these three experimental approaches. Unfortunately in this example, although such is not always the case, three different structures for the same stem were derived; in this example only the American model can fit all three lines of data for this stem. The bottom portion of Fig. 9 presents evolutionary support for the American model for this particular stem. Strictly speaking, two or more compensatory base changes should be present in the same stem in two or more organisms as evolutionary "proof" for its existence (Woese *et al.*, 1983). A growing body of data support the occurrence of noncanonical base pairs in rRNA such as U·G. It has been suggested that the occurrence of an A·G pair at the base of a stem may enhance bending of the stem to bring it into coaxial alignment (Woese *et al.*, 1983). A bulged base, especially a bulged A, with no opposing pairing partner, may play a role for recognition by a protein for binding (Peattie *et al.*, 1981; Garrett *et al.*, 1984b); modification or deletion of a bulged A in *E. coli* 5S RNA reduces but does not totally abolish the binding of ribosomal protein L18 (Garrett *et al.*, 1985). Long-range base-pairing interactions fold the rRNA into domains; some of these long-range interactions have been supported by psoralen cross-linking and electron microscopy (Wollenzein *et al.*, 1979; Wollenzein and Cantor, 1982; Klein *et al.*, 1983, 1984; Thompson and Hearst, 1983a). A recent review by Noller (1984) discusses the various methods used to determine rRNA secondary structure.

Evolutionary comparison to *E. coli* rRNA has allowed the deduction of rRNA secondary structure from organisms where generally no direct experimental proof exists; these comparisons also permit refinement of the existing models for *E. coli* rRNA. Tables IV and V are a summary of the cases for which complete secondary structure models have been proposed. It will be helpful to have experimental proof for some of the cases besides *E. coli* rRNA, and this is starting to appear for limited parts of

Table IV
Secondary Structure Models for Small-Subunit rRNA

Ribosome source	Reference
Eubacterium	
Escherichia coli 16S	Woese *et al.* (1980), Glotz and Brimacombe (1980), Brimacombe (1980), Noller and Woese (1981), Stiegler *et al.* (1981a,b), Zwieb *et al.* (1981), Brimacombe *et al.* (1983), Ebel *et al.* (1983), Maly and Brimacombe (1983), Woese *et al.* (1983), Noller (1984)
Proteus vulgaris 16S	Carbon *et al.* (1981)
Archaebacteria	
Halobacterium volcanii 16S	Gupta *et al.* (1983), Woese *et al.* (1983)
Eukaryotes	
Yeast 18S	Mankin *et al.* (1981b), Stiegler *et al.* (1981a), Zwieb *et al.* (1981), Brimacombe *et al.* (1983), Woese *et al.* (1983), Hogan *et al.* (1984a)
Dictyostelium discoideum 17S	Olsen *et al.* (1983)
Artemia salina 18S	Nelles *et al.* (1984)
Xenopus laevis 18S	Zwieb *et al.* (1981), Stiegler *et al.* (1981a), Atmadja *et al.* (1984)
Rat 18S	Chan *et al.* (1984)
Chloroplasts	
Zea mays chloroplast 16S	Zwieb *et al.* (1981), Stiegler *et al*, (1981a)
Euglena gracilis chloroplast 16S	Graf *et al.* (1982)
Mitochondria	
Zea mays mitochondrion 18S	Chao *et al.* (1984)
Triticum aestivum mitochondrion 18S	Spencer *et al.* (1984)
Aspergillus nidulans mitochondrion 15S	Küntzel and Köchel (1981)
Drosophila yakuba mitochondrion	Clary and Wolstenholme (1985)
Yeast mitochondrion 15S	Stiegler *et al.* (1981a)
Mouse mitochondrion 12S	Zwieb *et al.* (1981)
Human mitochondrion 12S	Mankin and Kopylov (1981), Zwieb *et al.* (1981), Stiegler *et al.* (1981a), Brimacombe *et al.* (1983)
Leishmania tarentolae mitochondrion 9S	de la Cruz *et al.* (1985)
trypanosome mitochondrion 9S	Sloof *et al.* (9185)

Table V
Secondary Structure Models for Large-Subunit rRNA

Ribosome source	Reference
Eubacteria	
Escherichia coli 23S	Branlant *et al.* (1981), Glotz *et al.* (1981), Noller *et al.* (1981), Brimacombe *et al.* (1983), Ebel *et al.* (1983), Maly and Brimacombe (1983), Noller (1984)
Anacystis nidulans 23S	Douglas and Doolittle (1984)
Eukaryotes	
Yeast 26S	Veldman *et al.* (1981a), Hogan *et al.* (1984b)
Xenopus laevis 28S	Clark *et al.* (1984)
Rat 28S	Hadjiolov *et al.* (1984)
Mouse 28S	Michot *et al.* (1984)
Chloroplast	
Zea mays chloroplast 23S	Branlant *et al.* (1981), Glotz *et al.* (1981)
Mitochondria	
Mouse mitochondrion 16S	Branlant *et al.* (1981), Glotz *et al.* (1981)
Drosophilia yakuba mitochondrion	Clary and Wolstenholme (1985)
Human mitochondrion 16S	Branlant *et al.* (1981), Glotz *et al.* (1981)
Leishmania tarentolae mitochondrion 12S	de la Cruz *et al.* (1985b)
trypanosome mitochondrion 12S	Sloof *et al.* (1985)

Bacillus stearothermophilus 16S (Douthwaite *et al.*, 1983), yeast 18S (Mankin *et al.*, 1981a; Douthwaite *et al.*, 1983; Morgan *et al.*, 1984a), *Xenopus* 18S (Cox and Kelly, 1982; Atmadja *et al.*, 1984), rabbit 18S (Connaughton *et al.*, 1984a), *Bacillus stearothermophilus* 23S (Garrett *et al.*, 1984a), yeast 26S (Hogan *et al.*, 1984b), and *Xenopus* 28S (Cox and Kelly, 1982).

Despite the enormous diversity of ribosome sources listed in Tables IV and V, there is nonetheless an obvious core structure common to all rRNAs of small or large subunits (e.g., Ebel *et al.*, 1983). This common secondary structure was guessed at a decade ago based upon a common melting curve component (Cox *et al.*, 1973). The common core structure suggests that the architecture of the primitive ribosome was laid down before separation of the four kingdoms. With subsequent evolution, some changes appeared within the secondary structures of rRNAs from different species, such as extension or shortening of stems, or slight slippage of stems. The most dramatic change is the increase in size in eukaryotic ribosomes due to apparent insertion of blocks of sequence. We have called such insertions "expansion segments" (Clark *et al.*, 1984), since they expand the eukaryotic structure relative to *E. coli* rRNA. The location of expansion segments within regions of rRNA of little sequence conservation supports the idea that the expansion segments are tolerated in rRNA because they do not disrupt any region of preexisting functional importance. A secondary structure model for *Xenopus laevis* 18S rRNA is shown in Figs. 10 and 11 and for *Xenopus laevis* 28S rRNA in Figs.

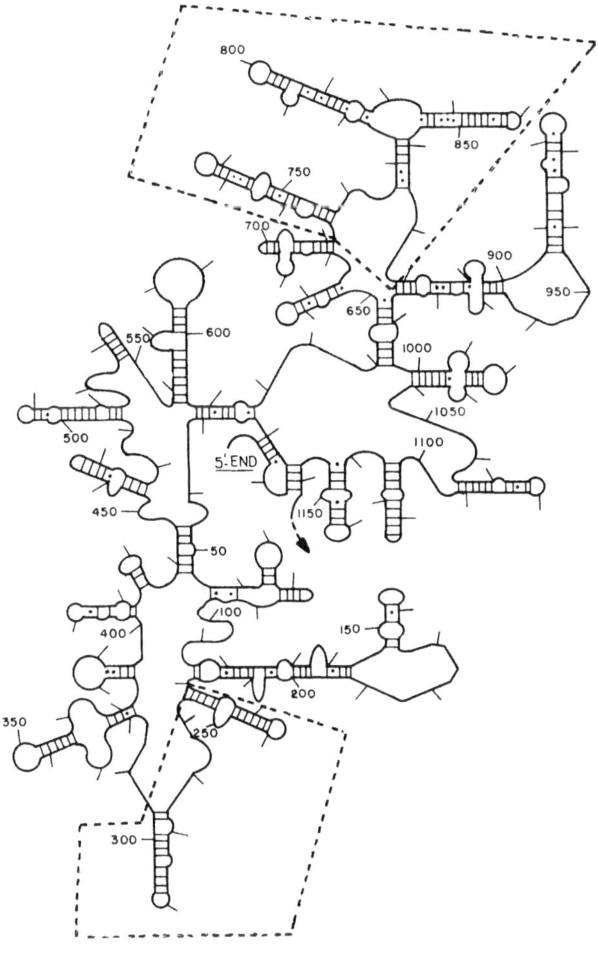

Figure 10. Secondary structure for the 5′ half of *Xenopus laevis* 18S rRNA. G · U base pairs are shown by a dot rather than a bar. Eukaryotic-specific expansion segments are indicated by dashed boxes, using the sequence alignment from Olsen *et al.* (1983). [Secondary structure from Atmadja *et al.* (1984).]

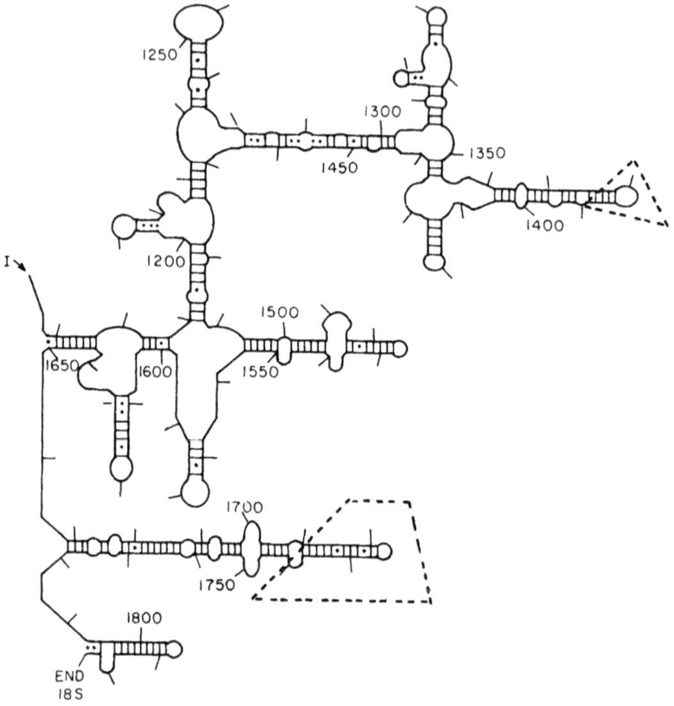

Figure 11. Secondary structure for the 3′ half of *Xenopus laevis* 18S rRNA. Details are the same as for Fig. 10.

12–18. The positions of expansion segments and other noteworthy features are indicated in these figures.

Expansion segments are generally found in the same locations in the rRNAs from different eukaryotes, but their length and sequence can be highly variable between species. This variability makes it difficult to guess accurately their secondary structure by evolutionary comparisons, and experimental data will be needed for a final judgment. Moreover, the general variability of sequence length of each expansion segment suggests that they may play no functional role in eukaryotic ribosomes; alternatively, one would have to argue for coevolution of the molecules with which they might interact. An exception to the variability in expansion segment sequence is a stretch of 25 bases found in expansion segment 2 (shown by a bracket in Fig. 13) and conserved in all eukaryotes, suggesting that this location may play a role specific to eukarotic ribosomes (Ware *et al.*, 1983; Clark *et al.*, 1984).

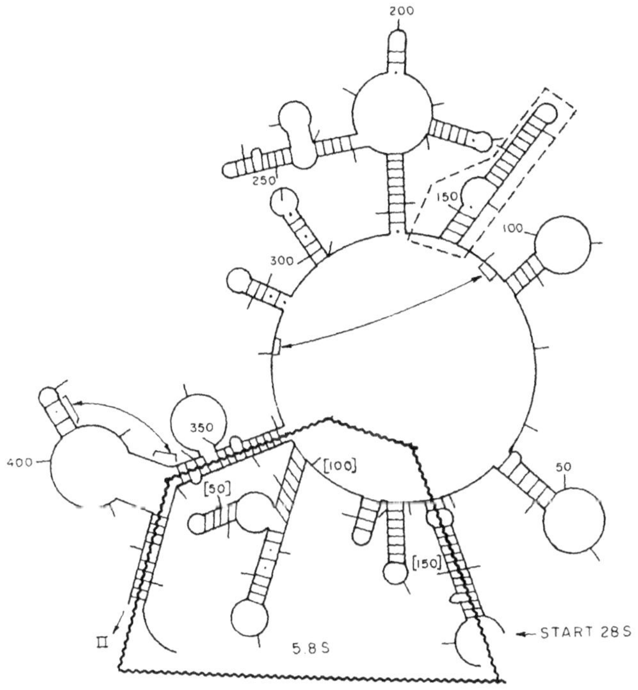

Figure 12. Secondary structure model for domain I of *Xenopus laevis* 28S rRNA. G · U base pairs are shown by a dot rather than a bar. The expansion segments are shown by dashed boxes; other equally plausible structures may be drawn for them, since no evolutionary or experimental data exist for these regions. Alternative base pairing partners are shown by arrows. 5.8S RNA is enclosed by a wiggly line, and its residue numbers are in brackets. [Model from Clark *et al.* (1984).]

In some species the RNA transcript of an expansion segment region may be removed by processing cuts; this further supports the idea that not only are expansion segments without function, but also that they may be entirely dispensable. An example of the removal of an expansion segment transcript is the processing cut in insects and many lower eukaryotes, which removes most of expansion segment 5 (Fig. 15) of domain IV and subdivides 28S rRNA into 28Sα and 28Sβ halves (Ware *et al.*, 1985). Expansion segment 5 interrupts the binding site for yeast ribosomal protein L25 (Planta and Mager, 1985), yet is apparently tolerated because yeast does not remove this expansion segment by processing. However yeast ribosomal protein L25 binds to *E. coli* 23S rRNA (which lacks expansion segments) better than it does to its own homologons yeast 26S rRNA (Planta and Mager, 1985). Therefore there is probably a slight

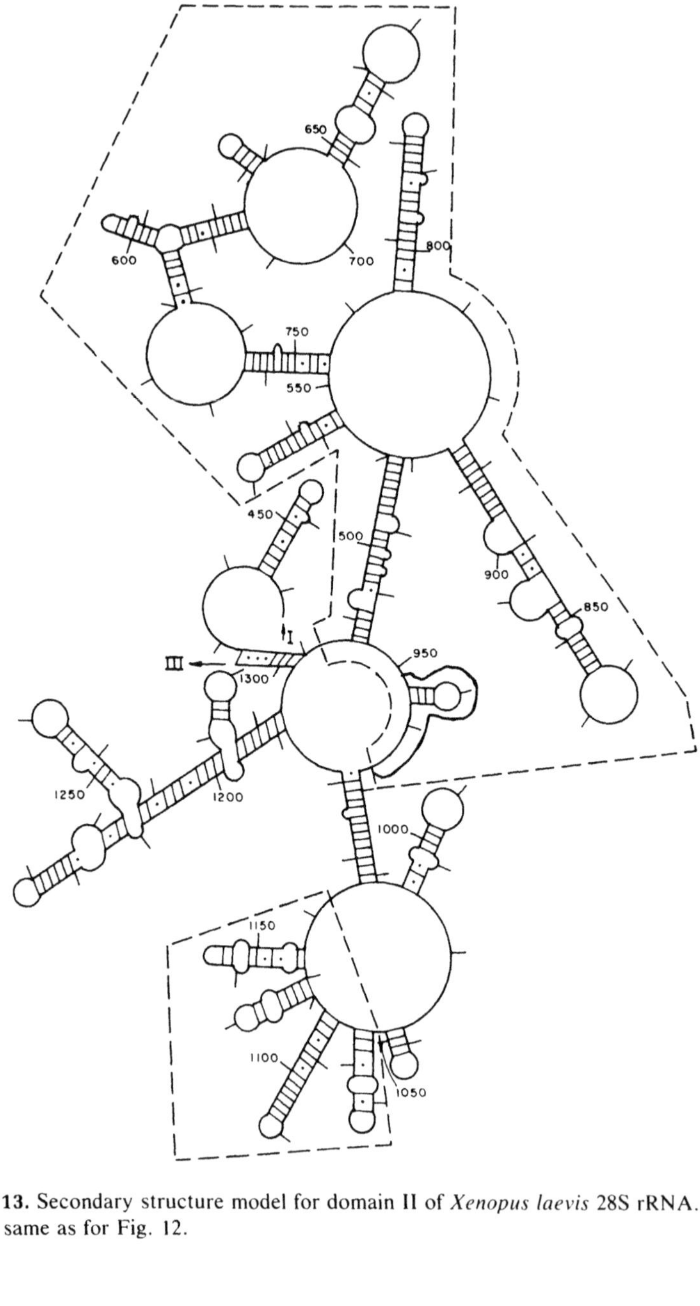

Figure 13. Secondary structure model for domain II of *Xenopus laevis* 28S rRNA. Details are the same as for Fig. 12.

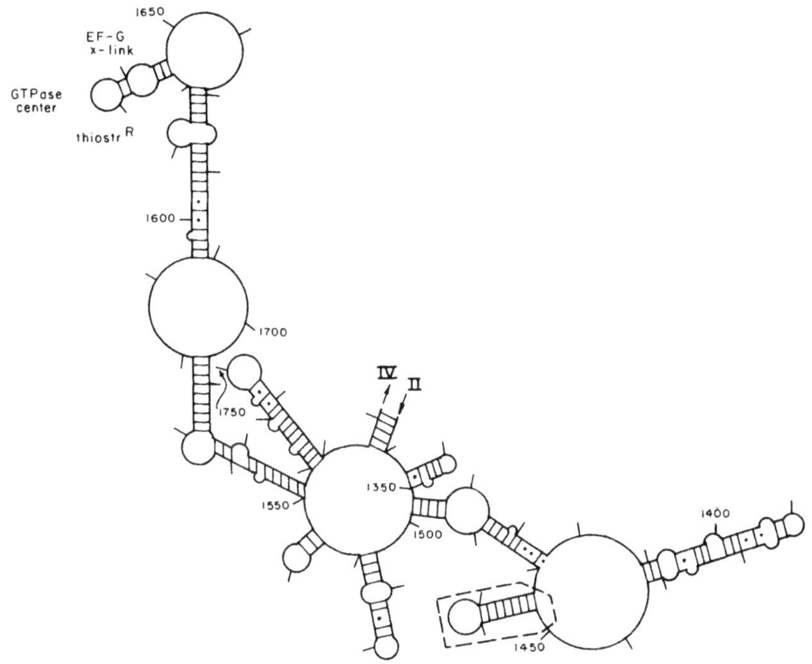

Figure 14. Secondary structure model for domain III of *Xenopus laevis* 28S rRNA. Details are the same as for Fig. 12. The GTPase center is indicated (see text).

negative selection against expansion segment 5 in yeast. Presumably when expansion segment 5 grows larger in other organisms the negative selection now becomes sufficiently great that a new processing event must be invoked to remove it, thereby giving rise to 28S α and 28S β rRNA.

Similarly, the ninth expansion segment is cleaved out during processing of higher plant chloroplast 23S rRNA. The RNA 3′ to this expansion segment is chloroplast 4.5S RNA (Fig. 18), and corresponds to the 3′ end of *E. coli* 23S rRNA (Edwards *et al.*, 1981; Machatt *et al.*, 1981; Mackay, 1981; Clark and Gerbi, 1982).

In contrast to the expanded structure of eukaryotic rRNA, the rRNA of mammalian mitochondria appears to be decreased in size relative to *E. coli* rRNA, due to several "amputations" of blocks of sequence. Sometimes, but not always, these amputations coincide with the positions at which eukaryotic expansion segments are found inserted into the core structure (e.g., Mankin and Kopylov, 1981). One of these amputations removes the region in domain VI that in *E. coli* binds the 5S RNA, L5,

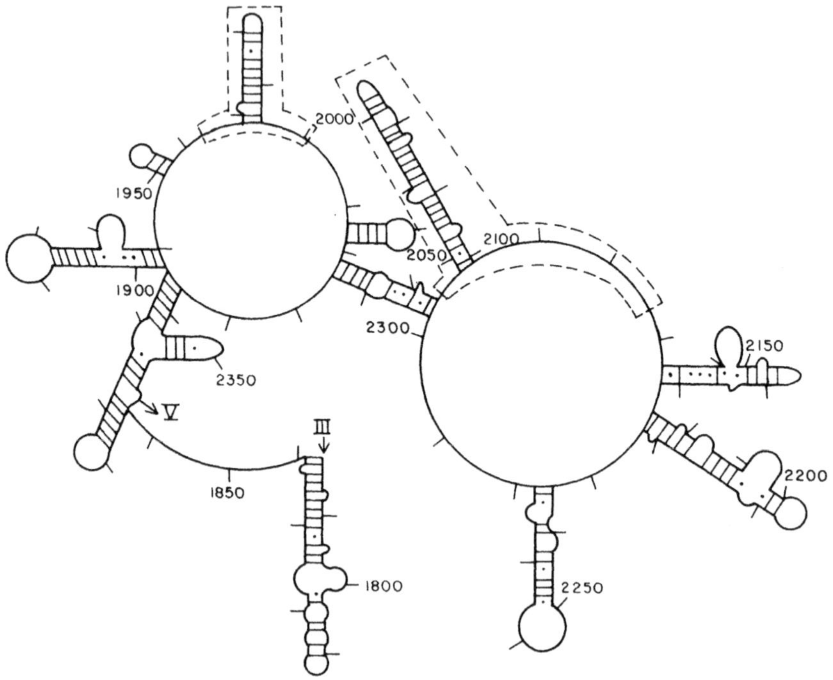

Figure 15. Secondary structure model for domain IV of *Xenopus laevis* 28S rRNA. Details are the same as for Fig. 12. The expansion segment centered at position 1980 is removed during "gap" processing in some species other than *Xenopus*. (Ware *et al.*, 1985)

L18, L25 complex, but this may not be surprising, since mitochondria seem to lack 5S RNA (see Section 2.2).

Chloroplast rRNA exhibits both small deletions and small insertions of sequence tracts relative to *E. coli* rRNA (Kössel *et al.*, 1983). The insertions are all extra stem + loop structures. The mechanism for insertion or deletion in the cases described above is unknown, but one could imagine that insertions are remnants of transposable elements that "hopped" into the rDNA, and that deletions resulted from the transposable element "hopping" out and taking adjacent DNA along with it. It is, however, impossible to judge the directionality of events. For example, one cannot rule out the possibility that the ancestral rDNA was as long as the longest eukaryotic rDNA and that with time it progressively shrunk to the size found in *E. coli* and even smaller to the size of mitochondrial rDNA. Another formal possibility is that the expansion segments represent diverged duplications of the rRNA sequence itself, rather than insertions

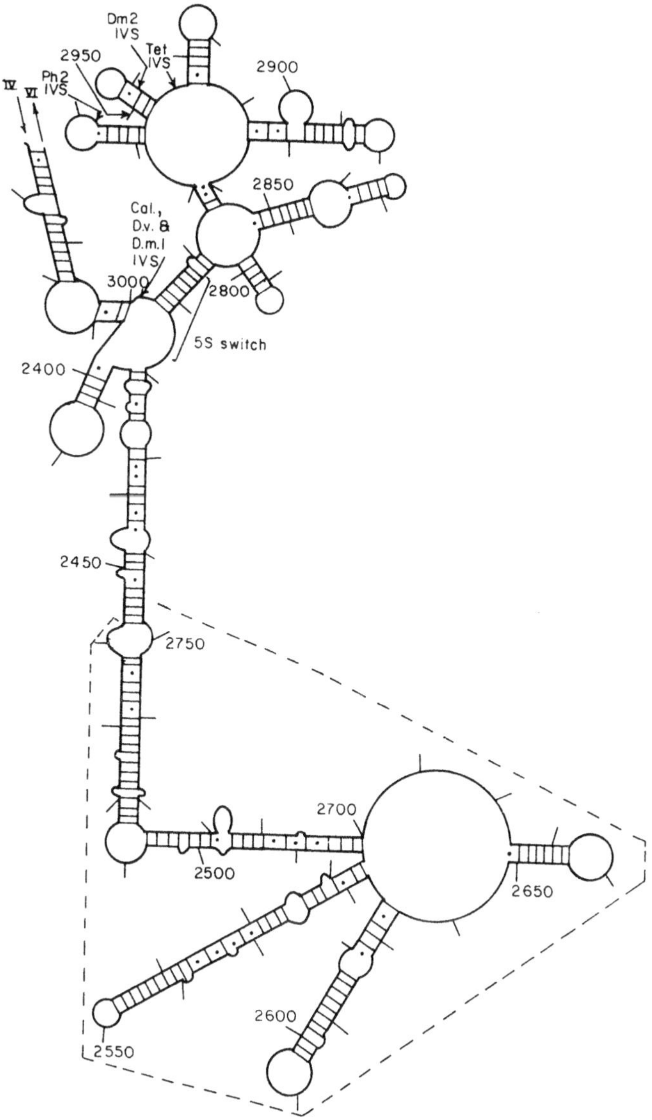

Figure 16. Secondary structure model for domain V of *Xenopus laevis* 28S rRNA. Details are the same as for Fig. 12. Sites of intervening sequences (IVS) in *Physarum* (Ph), *Drosophila melanogaster* (Dm), *Drosophila virilis* (Dv), and *Calliphora* (Cal) are indicated. The potential 5S RNA switch drawn in Fig. 1 is shown here.

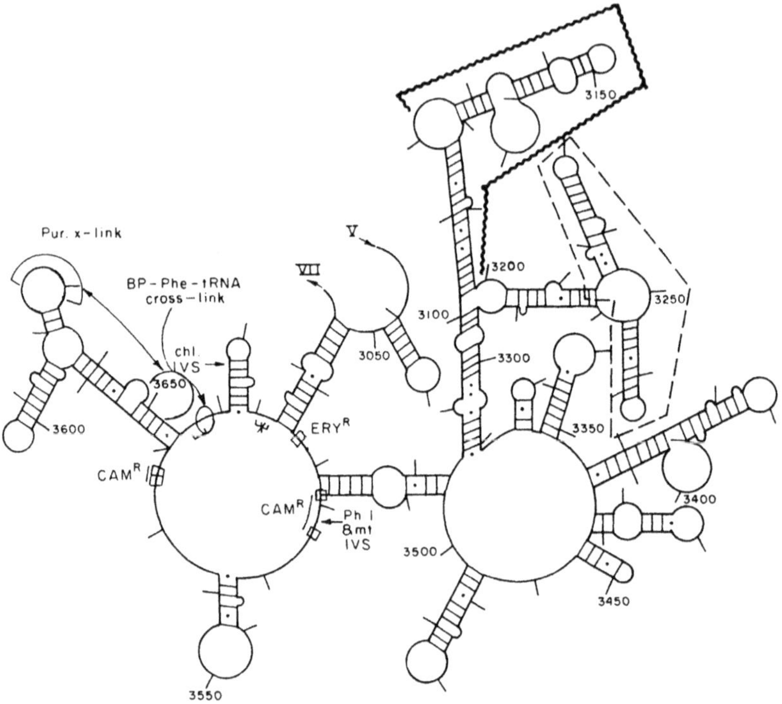

Figure 17. Secondary structure model for domain VI of *Xenopus laevis* 28S rRNA. Details are the same as for Fig. 12. The sites of intervening sequence (IVS) for chloroplasts (chl), mitochondria (mt) and *Physarum* (Ph) are shown. The peptidyl transferase center includes sites for resistance to chloramphenicol (CAM) and erythromycin (ERY) and has been cross-linked to a derivative of $tRNA^{Phe}$. The region that corresponds to the L1 ribosomal protein-binding site is enclosed by a wiggly line.

of foreign DNA. For ease of discussion, and to favor my own bias, I will nonetheless refer to insertions as well as deletions as possible events in rDNA evolution.

I suggested above that expansion segments are relics of DNA insertion into nonconserved areas that may be nonfunctional regions, and that these inserted pieces are tolerated in the mature rRNA. What happens if DNA is inserted into an evolutionarily conserved sequence of probable function? One may imagine that either such an event would be lethal and not perpetuated during the course of evolution, the inserted copy may not be transcribed if other noninsertion copies of rDNA exist, or the system might cope by removal of the insertion transcript during RNA processing with subsequent ligation. The latter corresponds to intervening

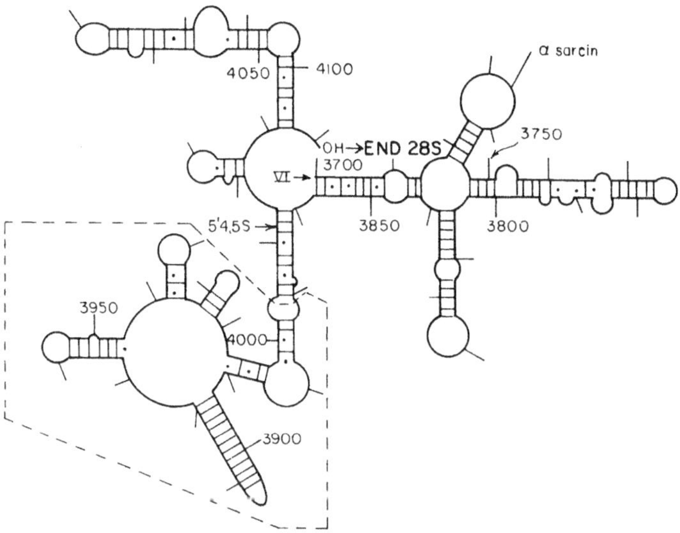

Figure 18. Secondary structure model for domain VII of *Xenopus laevis* 28S rRNA. Details are the same as for Fig. 12. THe α-sarcin cleavage site is shown. The 5′ end of 4.5S RNA, which is found in higher plant chloroplasts, is indicated; the expansion segment enclosed by the dashed line is removed during 4.5S RNA processing.

sequences ("introns"), which are found interrupting highly conserved sequences of rDNA [summarized by Gerbi *et al.* (1982)]. For example, 87 bases in a row with no mismatch are conserved between *Xenopus* 28S rDNA (which lacks an intron) and the region surrounding the intron in *Tetrahymena* rDNA (Gourse and Gerbi, 1980b). Insertion sequences are also found in the 26S rRNA region of some of the rDNA units of *Drosophila*; however, the rDNA copies with an insertion are generally not transcribed [reviewed by Beckingham (1982)]. So far intervening sequences have been described in invertebrates and unicellular eukaryotes; they also exist in mitochondria and chloroplast rDNA. Although no intervening sequences have been found in *Xenopus* rDNA, the corresponding position for intervening sequence occurrence in other rDNA systems has been marked in the secondary structure figures (Figs. 16 and 17). There is no obvious primary or secondary structural feature associated with the sites of intron placement. It is curious that they are clustered in domains V and VI of 28S rDNA. Evidence has recently appeared that a DNA insertion is also found in 18S rDNA of *Plasmodium lophurae* (Unnasch and Wirth, 1983).

The origin of the internal transcribed spacer 2 (ITS 2) may be similar to introns or expansion segments. By primary and secondary structure comparisons, it has been deduced that 5.8S RNA is homologous to the 5′ end of *E. coli* 23S rRNA (Nazar, 1980; Jacq, 1981; Clark and Gerbi, 1982). The ITS 2 appears to be an insertion that separated 5.8S RNA from the main body of the 28S rRNA gene. Like introns, the ITS 2 transcript is removed during processing, but unlike introns, this is not followed by covalent ligation. Ligation may not be needed in this case, since both ends of 5.8S RNA are held by hydrogen bonds to 28S rRNA (Pace *et al.*, 1977; Nazar and Sitz, 1980), as shown in Fig. 12.

To sum up, there is a core secondary structure of rRNA common to all ribosomes, but nucleic acid may have been added or deleted from this basic plan. The insertions may either persist in the mature rRNA species (e.g., expansion segments) or they may be removed by processing (e.g., introns and ITS 2).

4. Processing of rRNA

The small and large ribosomal subunit rRNAs are cotranscribed as one long precursor in both eubacteria and eukaryotes. In this section I review similarities and differences between these two kingdoms for the processing events that ultimately yield the mature rRNA species. The reader should be aware that most or all ribosomal proteins are already associated with rRNA before the cleavage events described below.

4.1. rRNA Processing in Eubacteria

In the presence of chloramphenicol, RNAse III$^-$ mutant cells of *E. coli* were found to accumulate a 30S precursor of rRNA; this 30S species could be cleaved *in vitro* by purified RNAse III to yield rRNA molecules a bit larger than the mature rRNAs (Dunn and Studier, 1973; Nikolaev *et al.*, 1973, 1974, 1975; Schlessinger *et al.*, 1974; Ginsburg and Steitz, 1975). As shown in Fig. 19, the 30S rRNA precursor appears to fold into a stem whose apical loop is the entirety of 16S rRNA and another stem whose apical loop has most of the 23S rRNA (seven nucleotides at the 5· end and eight nucleotides at the 3′ end of 23S rRNA form part of the stem). This long-range folding has probably been seen by electron microscopy (Wu and Davidson, 1975). RNAse III clips within the stem, with a 2-bp stagger in the cut, to produce the 17S precursor of 16S rRNA (R. A. Young and Steitz, 1978) and a molecule slightly larger than the mature 23S rRNA (Bram *et al.*, 1980).

Additional enzymatic cleavages are needed to form the mature rRNAs.

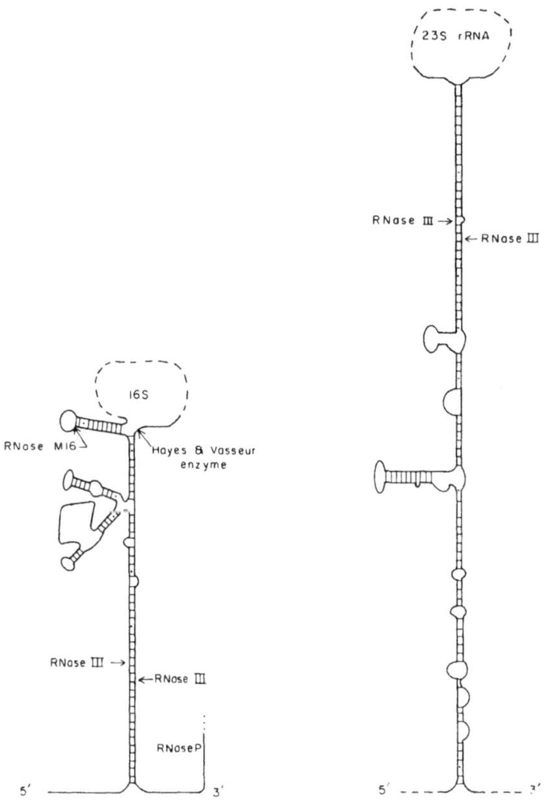

Figure 19. Processing of *E. coli* rRNA. Cleavage sites are shown on the 30S rRNA precursor that generate (left) 16S rRNA and (right) 23S rRNA. [Data are from R. A. Young and Steitz (1978) and Bram *et al.* (1980).]

RNAse M16 generates the 5′ end of mature 16S rRNA (Dahlberg *et al.*, 1978), and another enzyme clips at the 3′ end of mature 16S rRNA (Hayes and Vasseur, 1976).

Other enzymatic events must be employed to unlink 5S RNA and the intercistronic spacer tRNAs from the rRNA precursor. For example, RNAse P generates the 5′ end of $tRNA_I^{Ile}$ (McClain, 1977).

4.2. rRNA Processing in Eukaryotes

4.2.1. Processing Cuts Common to All Eukaryotes

The events in eukaryotic rRNA processing differ a fair bit from those described above. As in eubacteria, a single rRNA precursor is formed; it is 40S–45S in size in most eukaryotes. The spacers abutting 18S and 28S rRNA in *Xenopus laevis* (Hall and Maden, 1980; Maden *et al.*, 1982b) and mouse (Goldman *et al.*, 1983) cannot form long stems such as in *E. coli* (R. A. Young and Steitz, 1978; Bram *et al.*, 1980). However, stems of lower stability than those of *E. coli* are possible in yeast (Veldman *et al.*, 1981b). In *E. coli* the first rRNA processing event is cleavage within the spacers by RNAse III; in contrast, in eukaryotes the initial cuts are at the 5′ ends of 18S and 5.8S rRNAs (Bowman *et al.*, 1983). In other words, the initial processing cleavages in eukaryotic rRNA may be analogous to the final cleavages in prokaryotic rRNA.

The order of processing cuts in prokaryotic rRNA seems to have only one sequence. Eukaryotes differ, in that a variability has been described in mammalian rRNA for the order of processing cuts (Winicov, 1976; Dudov *et al.*, 1978; Bowman *et al.*, 1981, 1983), as shown in Fig. 20. The first cut may generate either the 5′ end of 18S rRNA or the 5′ end of 5.8S RNA. Removal of the ETS portion of the precursor may result in a small nucleoplasmic RNA, which has been characterized (Reichel *et al.*, 1982). The next cut generates the 3′ end of 18S rRNA. This order of cuts differs between mouse L cells and HeLa cells, and in a temperature-sensitive mutant of BHK cells the cutting order may follow either the mouse L-cell or HeLa-cell pattern, depending on the temperature (Winicov, 1976).

There is heterogeneity in the 5′ end of 5.8S RNA from yeast (Veldman *et al.*, 1980a), *Xenopus laevis* (Boseley *et al.*, 1978; Ford and Mathieson, 1978), HeLa (Khan and Maden, 1977), and mouse and rat (Bowman *et al.*, 1983; Smith *et al.*, 1984) due to inclusion of a few additional upstream bases. The separation of 5.8S from 28S rRNA still needs to be further studied, but in yeast (Veldman *et al.*, 1981b), rat (Reddy *et al.*, 1983), and mouse (Bowman *et al.*, 1983) an intermediate can be found of 5.8S RNA with a part of the ITS 2 at its 3′ end. Conserved features in the ITS 2 have recently been summarized, and show that vertebrate ITS 2 sequences fall into one class and lower eukaryote ITS 2 sequences fall into another class (Tague and Gerbi, 1984). It may be that the processing mechanism to separate 5.8S from 28S RNA differs between these two categories (Tague and Gerbi, 1984). During evolution a DNA insertion event may have separated the 5.8S RNA gene from the main body of 28S RNA in the eukaryotic lineage (Section 3.2). This has led to an analogy being drawn between ITS 2 and mRNA introns; since U1 small nuclear RNA (snRNA) is required for mRNA splicing (Lerner *et al.*, 1980; Rogers and Wall, 1980), a similar role to aid ITS 2 processing has been hypothesized for U3 snRNA, which is found in the nucleolus (Bachellerie *et al.*, 1983; Crouch *et al.*, 1983; Tague and Gerbi, 1984).

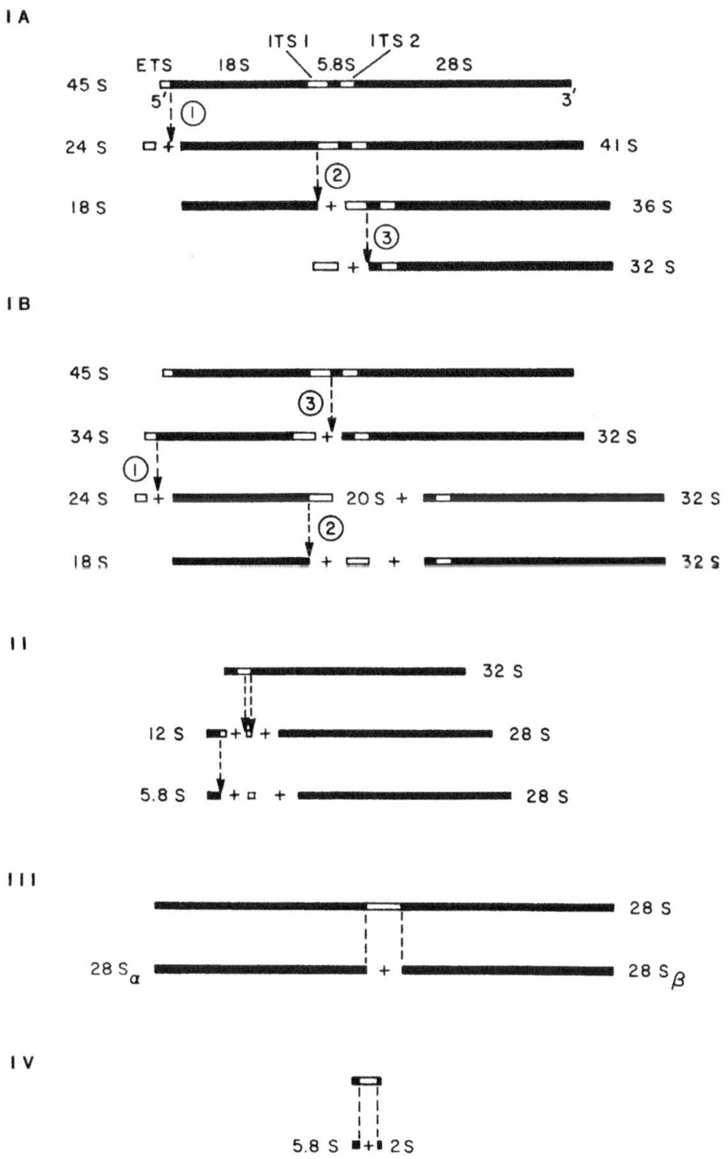

Figure 20. Processing scheme for maturation of eukaryotic rRNA. The white boxes represent transcribed spacers and the black boxes are the mature RNA regions. Rapid initial processing may trim the rRNA precursor to the mature 3′ terminus. (I) The separation of 18S RNA from 32S RNA; shown is the order of cleavage events for (A) mouse L cells and BHK 21ts422E cells at 38.5°C, and (B) HeLa cells and BHK 21ts422E cells at 33.5°C. (II) The processing of 32S rRNA to yield 5.8S RNA and 28S rRNA. (III) "Gap" processing found in some species, which divides 28S rRNA into α and β halves. (IV) The separation of 2S from 5.8S RNA found in some dipteran insects. [Data taken from Winicov (1976), Long and Dawid (1980a), and Bowman *et al.* (1983).]

4.2.2. Additional rRNA Processing Events in Some Eukaryotic Systems

4.2.2a. Excision of Some Expansion Segments. In the basic plan for processing rRNA in all eukaryotes described above, the first set of events separates 18S rRNA from 32S rRNA, although the mechanism is not parallel with that used by *E. coli* to separate 16S from 23S rRNA. The second set of events in eukaryotic rRNA processing sever 5.8S RNA from 28S rRNA; this has no parallel in *E. coli*, since 5.8S RNA is homologous with the 5′ end of 23S rRNA. We have previously argued that the general order of processing events follows the order of their evolutionary emergence (Clark and Gerbi, 1982), making this an example of Haeckel's law, "ontogeny recapitulates phylogeny." The most complex rRNA maturation known occurs in insects, where additional processing events occur beyond those described above common to all eukaryotes. The sequence of these events has been studied in depth in *Drosophila melanogaster* by Long and Dawid (1980a).

The first of these additional processing events involves cleavage of 26S rRNA into two halves, 26Sα and 26Sβ. In Section 3.2 I described the recent finding in the fly *Sciara coprophila* that this particular processing event removes part of expansion segment 5 from 26S rRNA (Ware *et al.*, 1985). The second of these additional processing events breaks off a 2S RNA fragment from the 3′ end of 5.8S RNA in the insects *Drosophila* (Pavlakis *et al.*, 1979) and *Sciara* (Jordan *et al.*, 1980). The material removed in this case may also be likened to an expansion segment that has been inserted in the 5.8S gene of insects. It is likely that 4.5S RNA processing in higher plant chloroplasts, which may remove expansion segment 9, also occurs as a late event (Hartley, 1979). The occurrence of these processing events late in the rRNA maturation pathway suggests that the expansion segments were inserted during evolution of the eukaryotic lineage rather than deleted, and their addition has presumably not altered the secondary and tertiary structure of rRNA, which may be recognized for the earlier processing events.

4.2.2b. Intron Excision. The second eukaryotic gene to be cloned was *D. melanogaster* rDNA, which contained a 5-kb insertion sequence (Glover *et al.*, 1975). Since that time insertion sequences have been described in the rDNA of many different insects [reviewed by Beckingham (1982)]. In all these cases the insect genome contains rDNA units both with and without insertions; in the former category there may be more than one type of insertion present [summarized in Table III of Beckingham (1982)]. Presumably there are a sufficient number of insertion-minus rDNA gene copies so that it is not necessary to transcribe the insertion-plus rDNA, which current molecular results suggest is generally silent transcriptionally [*D. melanogaster* type 1 insertions (Long and Dawid, 1979a; and Jolly and Thomas, 1980); *D. melanogaster* type 2 insertions (Long *et al.*, 1980; and Kidd and Glover, 1981)]. A different conclusion was reached in the electron microscopic study of Chooi (1979) of very young embryos and nurse cells, but technical features of her experiment may explain the discrepancy with the biochemical data of others [see Beckingham (1982) for a fuller discussion of this discrepancy]. The electron microscopic study of Glätzer (1979) on spermatocyte nuclei agreed with the biochemical conclusion above that *Drosophila* rDNA that contains insertions is transcriptionally inactive. It would be interesting to know the signals recognized in these rDNA units that render them transcriptionally silent. Since more than 200 bases upstream of the gene are identical for *Drosophila* rDNA units with or without insertions (Long *et al.*, 1981; Simeone *et al.*, 1985), it may be that transcriptional regulation lies at the level of chromatin organization rather than at the sequence level.

Two categories of rDNA intervening sequences seem to exist [summarized in Table III of Gerbi *et al.* (1982)]. *Drosophila* rDNA insertions are representative of the first category, which are chromosomal, transcriptionally silent, only are in a portion of the rDNA copies of the genome, and have a structure of direct repeats at their boundaries reminiscent of transposable elements. The second category are the rDNA introns found in *Tetrahymena, Physarum*, mitochondria, and chloroplasts. In this case all of the rDNA copies contain an intron and by necessity are transcriptionally active; rDNA of this category is extrachromosomal or organellar, and direct repeats are not found at the intron boundaries.

The transcript of the rDNA intron of this second category is able to self-excise, as shown by Cech and coworkers for *Tetrahymena*. The rDNA intron of *Tetrahymena* is excised as a linear molecule (Carin *et al.*, 1980; Zaug and Cech, 1980, 1982), which is then cyclized (Grabowski *et al.*, 1981; Zaug *et al.*, 1983). An exciting finding is the observation that splicing of the *Tetrahymena* intron can occur *in vitro* in a protein-independent manner (Kruger *et al.*, 1982), and no external energy source is needed. The steps of the splicing reaction are shown in Fig. 21 (Zaug *et al.*, 1983). A guanosine cofactor is covalently linked to the 5′ end of the intron to initiate the excision reaction (Cech *et al.*, 1981; Bass and Cech, 1984). Both for splicing of the rRNA and cyclization of the excised intron, the cleavage of a phosphodiester bond provides the energy for the subsequent ligation (Zaug *et al.*, 1983). For cyclization of the intron there is coincident cleavage to remove 15 nucleotides from the 5′ end of the linear form and ligation to form a circle, with no net change in the number

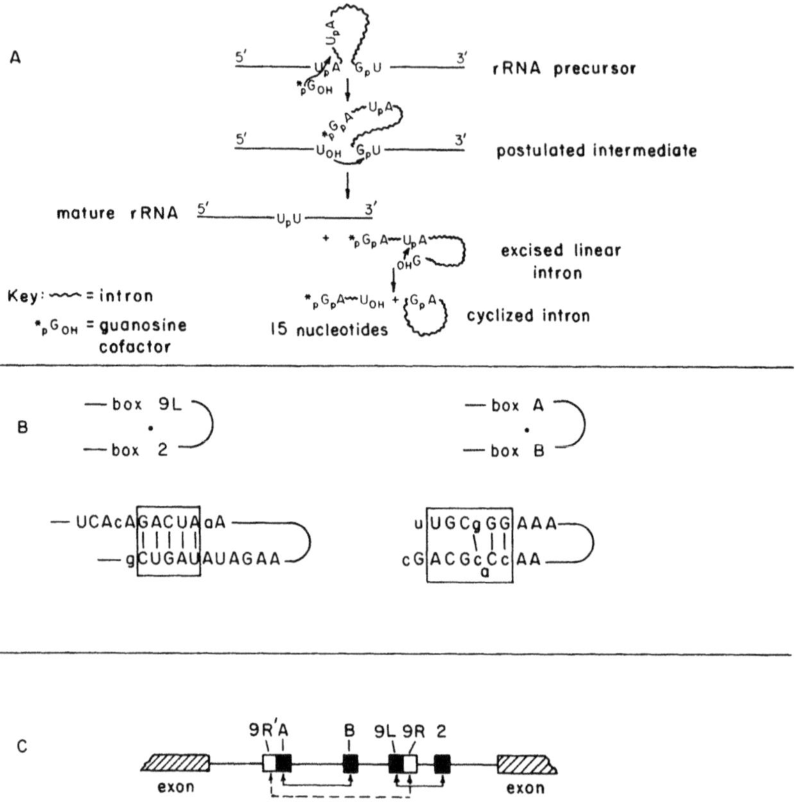

Figure 21. Intron processing. (A) The concerted phosphoester transfer mechanism for splicing out the *Tetrahymena* rRNA intron (Zaug *et al.*, 1983). (B) The conserved box pairs between mitochondrial introns and the *Tetrahymena* rRNA intron. A lowercase letter is a *Tetrahymena* nucleotide that differs from the mitochondrial consensus sequence. A third box pair is conserved in structure but not sequence: box 9R (downstream from box 9L) and box 9R′ (near the 5′ exon–intron boundary. (C) The linear arrangement of intron boxes. See text for further details. [Data summarized by Cech *et al.* (1983).]

of phosphodiester bonds (Zaug *et al.*, 1983). It might be that the excised intron is cyclized to be recognized for degradation and to prevent the splicing reaction from running backward. Some sequences of the Tetrahymena rDNA intron which are dispensable or important for its self-splicing have recently been defined in certain gene constructs (Price *et al.*, 1985).

In the splicing of the *Tetrahymena* rDNA intron the rRNA itself is acting as a biological catalyst: the specific structure of the RNA permits it to lower the activation energy for cleavage–ligation reactions at precise sites, the RNA has a specific binding site for the guanosine cofactor, and the splicing reaction occurs at a higher rate than a nonenzymatic reaction (Bass and Cech, 1984). Unlike most enzyme catalysts, the RNA catalyst cannot be reused for additional reactions. It may be that RNA catalysts ("ribozymes") preceded enzymatic catalysts in their evolutionary appearance. As discussed in Section 1.1, rRNA itself may be a major catalytic component for protein synthesis, if changes in its conformation drive the reactions of translation. Recently it has been demonstrated that the RNA component of RNAse P is the catalytic subunit for tRNA processing (Guerrier-Takada *et al.*, 1983; Guerrier-Takada and Altman, 1984); in this case the RNA catalyst does not undergo any net change and the enzyme can catalyze additional reactions.

The splicing of fungal mitochondrial rRNA and some mitochondrial mRNAs also seems to involve RNA catalysis [reviewed by Cech (1983)]. Self-splicing of yeast mitochondrial ribosomal and messenger RNA precursors have recently been reported (van der Horst and Tabak, 1985) However, proteins are also utilized in mitochondrial splicing: intron-encoded maturases are needed for cytochrome *b* splicing (Lazowska *et al.*, 1980), and nuclear mutations affect mitochondrial mRNA and rRNA splicing (Mannella *et al.*, 1979; Dieckmann *et al.*, 1982). It might be that these proteins are needed to help fold the RNA precursor into a suitable structure for RNA-catalyzed splicing. Indeed, the presence of *cis*-acting splicing mutants has demonstrated the importance of mitochondrial RNA structure for splicing (Anziano *et al.*, 1982; DeLaSalle *et al.*, 1982; Netter *et al.*, 1982; Weiss-Brummer *et al.*, 1982). These *cis*-acting mutations reside within conserved sequences and structures that are found in both mitochondrial introns and the *Tetrahymena* rDNA intron (Burke and RajBhandary, 1982; Cech *et al.*, 1983). These conserved "boxes" are shown in Fig. 21, and may be important to hold the intron in a specific secondary structure needed for splicing (Michel *et al.*, 1982; Waring *et al.*, 1982, 1983; Davies *et al.*, 1982; Cech *et al.*, 1983; Michel and Dujon, 1983).

There may be a common evolutionary path for the appearance of introns with conserved "boxes" in both the nuclear gene of *Tetrahymena* rDNA and in fungal mitochondrial genes. For example, a transposable element may have entered genes in both these organellar compartments; alternatively, an intron may have arisen in one of these organelles and then been transferred to the other organelle, as we saw possible in Section 2.3 (Cech *et al.*, 1983). At any rate, although the intron may have originated as an insertional event, its processing fate differs from that described in Section 4.2.2a for expansion segments and spacers, in that in this case the intron transcript is removed as an initial event in rRNA processing (Cech and Rio, 1979; Din *et al.*, 1979) rather than occurring as a late processing step. This may reflect a basic difference in the mechanism for removal of the inserted material in processing: the self-catalysis of the *Tetrahymena* intron may occur spontaneously due to the RNA structure, whereas the other processing events for spacers and some expansion segments might require other components.

5. Nontranscribed Spacers

Unlike the remarkable evolutionary conservation in sequence within regions of the rRNA transcript, the rDNA spacers show great diversity among different species. Such variability raises the question of whether the spacers carry out any functions, and this question will be discussed below.

5.1. Length Variability and Sequence Repetition in the NTS

Over a decade ago Brown *et al.* (1972) used molecular hybridization to show that the 18S and 28S rRNA coding regions are highly homologous between *Xenopus laevis* and *Xenopus borealis*, unlike the dissimilarity between their NTS sequences, and this was further borne out by other interspecies experiments on *Xenopus* (Forsheit *et al.*, 1974; Wellauer *et al.*, 1976a; Botchan *et al.*, 1977; Boseley *et al.*, 1979; Bach *et al.*, 1981b), as well as intergenic comparisons with *Xenopus* (Gourse and Gerbi, 1980a). On the other hand, molecular hybridization comparisons between closely related species of *Drosophila* showed homology between their NTS sequences (Tartof, 1979; Rae *et al.*, 1981), and the NTS homology dropped off as the taxonomic distance between the species being compared increased (Rae *et al.*, 1981).

The NTS structure, which has been studied in many organisms by restriction enzyme digestion and sequencing, often shows internal repetition. In *Xenopus laevis* the repeating units are 60 or 81 bp, which begin 260 bp upstream from the rRNA initiation site and extend about 1.5–5 kb further upstream (Boseley *et al.*, 1979; Sollner-Webb and Reeder, 1979; Moss *et al.*, 1980). Similarly, the mouse NTS has a 135-bp repeat, which begins about 170 bp upstream of the initiation site for transcription and extends 650–2700 bp upstream (Arnheim and Kuehn, 1979). Internal repetition in the NTS is also found in various *Drosophila* species, where 250-bp repeats begin about 50–80 bp upstream of the initiation site (*D. melanogaster*: Long and Dawid (1979b), Long *et al.* (1981); *D. virilis*: Rae *et al.* (1981); *D. hydei*: Renkawitz-Pohl *et al.* (1980)]. A variable number of repeat units within the NTS results in different overall lengths of the entire rDNA unit both between organisms and even in one individual [for example, in *Xenopus* (Morrow *et al.*, 1974; Wellauer *et al.*, 1974, 1976a; Botchan *et al.*, 1977), Chinese hamster (Stambrook, 1978), and mouse and man (Arnheim and Southern, 1977)]. In some cases, such as yeast, no length heterogeneity is obvious for the NTS (Meyerink and Retèl, 1976; Nath and Bollon, 1977) and sequence analysis shows that the larger repeated stretches described above do not occur, although shorter direct repeats of 4–16 nucleotides are found in variable quantities in the NTS 1 (Skryabin *et al.*, 1984). Spacer length heterogeneity must be reckoned with in hypotheses of rDNA evolution, and I will discuss this further in Section 6.

5.2. Functional Regions within the NTS

The variability in NTS sequence and length suggests that the NTS may lack a functional role; however, some sites with specific functions have been identified within the NTS and I will now examine these to see if they share any evolutionarily conserved structural features.

5.2.1. Promoter for Transcription

Initial experiments in eukaryotes located the promoter at +450 (Bach *et al.*, 1981a; Grummt, 1981b) or at +650 (Mishima *et al.*, 1980; Urano *et al.*, 1980) within the ETS; these positions were incorrectly located too far downstream due to the vagaries of S1 nuclease and other considerations described by K. G. Miller and Sollner-Webb (1982). It appears that the 40S–45S rRNA precursor is the primary transcript, since 10–30% of these molecules are polyphosphorylated at their 5′ ends (Reeder *et al.*, 1977; Sollner-Webb and Reeder, 1979; Bach *et al.*, 1981a). The difficulties in S1 nuclease mapping of the 5′ end of the 40S–45S rRNA have been overcome by primer extension experiments, as well as by oocyte injection or cell-free transcription systems, which allow genetic engineering to refine the boundaries of the promoter. Table VI presents sequences determined for the rDNA promoter region of various eukaryotes; structural features of RNA polymerase I promoters have been reviewed recently (Reeder, 1984; Sommerville, 1984). No homology is apparent between widely diverse species, which agrees with the demonstrated species specificity of factors for rDNA transcription (Grummt *et al.*, 1982; Kohorn and Rae, 1982a; Learned and Tjian, 1982; Mishima *et al.*, 1982a, b; Miesfeld and Arnheim, 1984). One of these species-specific factors that has

Table VI
Sequences of the rDNA Promoter Region[a]

	−40	−30	−20	−10	+1
Human	TTTCGCTCCG	AGTCGGCAAT	TTTGGGCCGC	CGGGTTATAT	GCTGACACGC
Rat	TCATCTTTGC	TATCTGT_CC	TATTGTACT_	GGAGATATAT	GCTGACACGC
Mouse	TGATCTTTTC	TATCTGTTCC	TATTGGACCT	GGAGATAGGT	ACTGACACGC
X. laevis	CTCCATGCTA	CGCTTTTTTG	GCATGTGCGG	GCAGGAAGGT	AGGGGAAGAC
X. clivii	CTCTAATCTA	CGCGTTTTAG	GCATGTGCCG	ACAGGAAGGT	AGGGAGAGAA
X. borealis	GGGGCCTCCC	CTCCCCCGGG	TACTGCTCCG	GCAGGAAGGT	AGGGACTGAG
D. melanogaster	AAATACCCGC	TTTGAGGACA	GCGGGTTCAA	AAACTACTAT	AGGTAGGCAG
P. polycephalum	CACCAGCTAT	GCTTCTTAAA	AAGAAACCCA	AGATACATAT	AGGGGGGGTA
D. discoideum	CCATTAGCCA	TTGAGCTCGG	TGACTACCCC	AAATACATAT	ACAAGAAGAG
T. pyriformis	AAAAAAAAAA	AAAGTATCAG	GGGGGTAAAA	ATGCATATTT	AAGAAGGGAA
S. rosei	_GTTTATAG_	TTCAA_TATT	TTTGGG_CAG	AAGGAACTTC	ATGCGAAAGC
K. lactis	_GANCATCGG	GTGA___GAAA	AATAGGCCGG	GAGGTACTTC	ATGCGAAAGC
S. cerevisiae	_GTACAAGT_	GTGAG_GAAA	AGTAGT_TGG	GAGGTACTTC	ATGCGAAAGC
H. wingei	TGTTTAAGGC	TAAAACACAT	TTCAGTGCCA	AAGGAACAGT	ATGCGAAGGC

[a] Sequences are for the noncoding strand of rDNA (= RNA-like strand). Here +1 is the first nucleotide of the 40S–45S precursor. Alignments are from Financsek *et al.* (1982), Verbeet *et al.* (1984a), and Skinner *et al.* (1984) using sequences determined from the following: human (Miesfeld and Arnheim, 1982; Learned and Tjian, 1982; Financsek *et al.* 1982), rat (Financsek *et al.*, 1982), mouse (Urano *et al.*, 1980; Mishima *et al.*, 1981; K. Miller and Sollner-Webb, 1981; Grummt, 1981b), *Xenopus laevis* (Sollner-Webb and Reeder, 1979; Boseley *et al.*, 1979; Moss *et al.*, 1980; Maden *et al.*, 1982b), *Xenopus clivii* (Bach *et al.*, 1981b), *Xenopus borealis* (Bach *et al.*, 1981b; Furlong *et al.*, 1983), *Drosophila melanogaster* (Long *et al.*, 1981), *Physarum polycephalum* (Blum *et al.*, 1983), *Dictyostelium discoideum* (Hoshikawa *et al.*, 1983), *Tetrahymena pyriformis* (Saiga *et al.*, 1982), *Saccharomyces rosei* (Verbeet *et al.*, 1983), *Kluyveromyces lactis* (Verbeet *et al.*, 1984b), *Saccharomyces cerevisiae* (Bayev *et al.*, 1980; Klemenz and Geiduschek, 1980) [identical with *Saccharomyces carlsbergensis* (Veldman, 1982)], *Hansenula wingei* (Verbeet *et al.*, 1984b). Regions of major homology are boxed. The bold-face G-16 and G-7 of mouse is of known importance by *in vitro* mutagenesis (Skinner *et al.*, 1984; Kishimoto *et al.*, 1985). Recently the promoter regions of *Tetrahymena thermophila* and *Glaucoma chattoni* have also been sequenced (Challoner *et al.*, 1985).

been isolated has been shown to bind to the promoter region of rDNA (Miesfeld and Arnheim, 1984). Deletion mutants demonstrate that −50 to +20 is required for transcription initiation factor binding to rDNA of *Acanthamoeba* (Iida *et al.*, 1985). The species specificity of RNA polymerase I transcription coupled with the lack of general sequence conservation in the promoter region is in marked contrast to conserved promoter sequences and lack of species specificity for RNA polymerase II or III transcription. Perhaps since several types of genes are transcribed by the latter two enzymes, there are more constraints to prevent divergence. RNA polymerase I, on the other hand, only transcribes rDNA, suggesting that changes in the rDNA promoter sequence may be coordinated with changes in the polymerase I factor(s), thereby giving rise to species specificity.

Closely related species do show some conserved sequences among their rDNA promoters, notably from −9 onward downstream into the ETS. This has been suggestive of at least one important region within the promoter region. Oocyte injections of rDNA and *in vitro* transcription of rDNA have provided further insight into other important stretches within the promoter. For example, deletion constructions of *Xenopus* rDNA that were injected into *Xenopus* oocytes and examined as Miller spreads in the electron microscope showed that −320 to +115 was the minimum region needed for transcription; this truncated promoter was sufficient to load RNA polymerase I at its characteristic high density of about every 100 bp along the rDNA gene (Bakken *et al.*, 1982; Morgan *et al.*, 1983). Other deletion studies of *Xenopus* rDNA narrowed down the promoter to −140 through +6 (Moss, 1982; Sollner-Webb *et al.*, 1983). Similar deletion studies in *Drosophila* indicated that the major promoter activity resided within −240 to −30, but that only −34 to +30 was necessary to support low levels of initiation (Kohorn and Rae, 1982b); further research showed that the major promoter of *Drosophila* rDNA resides at −43 to −27 (Kohorn and Rae, 1983a), with an additional major component at +1 to +4 (Kohorn and Rae, 1983b). Studies by Grummt (1981b, 1982) in mouse demonstrated a decrease in initiation once the boundary of the deletion moved between −39 and −34, and when the boundary of the deletion moved from −34 to −13 initiation was abolished. In accordance with these observations for mouse rDNA, Yamamoto *et al.* (1984) saw some initiation with only −22 to −2, but the larger region −40 to +52 was needed for full activity. As depicted in part by the boxed regions in Table VI, this larger region of mouse rDNA includes conserved sequences at −38 to −33, −20 to −16, and −1 to +18 held in common by rat, mouse, and human. As might be predicted, deletion studies for human rDNA give similar boundaries for the promoter region as was true for mouse (Learned *et al.*, 1983). Recently, *in vitro* mutagenesis within the region of −35 to −14 of mouse rDNA has yielded the result that a transition of the conserved G to A at position −16 caused a decrease in transcription by 95% (Skinner *et al.*, 1984) and bisulfite modification of G at position −7 eliminated promoter activity entirely (Kishimoto *et al.*, 1985).

All of the studies concur that only a relatively short distance upstream of the ETS is needed to promote transcription of rRNA, and important sites within this promoter region may reside within sequences conserved between closely related species.

Studies on *Xenopus* rDNA have shown that 2–8 duplicated promoters can be found in the NTS (Boseley *et al.*, 1979; Moss and Birnstiel, 1979; Sollner-Webb and Reeder, 1979; Moss *et al.*, 1980). These duplicated promoters are located at about 1-kb intervals from each other and from the gene promoter found adjacent to the ETS (Fig. 22); each duplicated promoter region contains a *Bam* H1 restriction site, thereby defining the "*Bam* islands" in the sea of other NTS sequences. Separating each *Bam* island with a duplicated promoter are 60- or 81-bp repeats. A 42-bp stretch of sequence found in the gene promoter at −73 to −114 can also be found within each 60/81-bp repeat as well as within each duplicated promoter. The NTS duplications have been invoked to explain the enhancer-like properties of the NTS in DNA transcription [for a review of enhancers see Khoury and Gruss (1983)].

For example, when two rDNA genes of varying length are coinjected into *Xenopus* oocytes, the rDNA copy with the longer NTS is transcriptionally dominant over a spacerless rDNA copy (Moss, 1983a), and this is due to the increased number of 60/81-bp repeats found in the longer

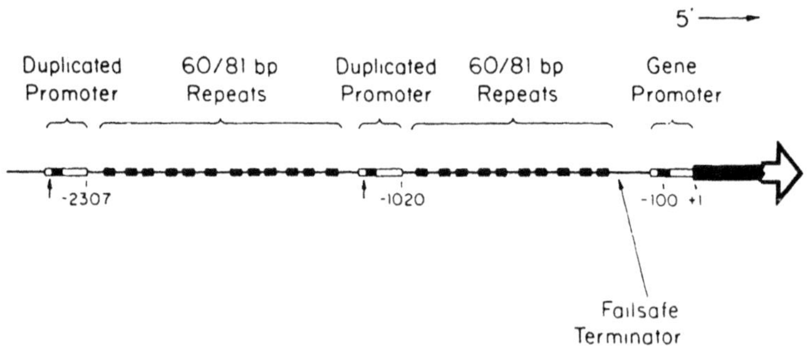

Figure 22. Organization of the NTS of *Xenopus laevis* rDNA. The white boxes represent the gene promoter and duplicated spacer promoters; the black boxes represent the conserved 42-bp region found in the promoters and in the 60/81-bp repeats. [Adapted from Busby and Reeder (1983).]

NTS (Reeder *et al.*, 1983). The enhancer-like influence of the 60/81-bp repeats can be transmitted through at least 4.3 kb of plasmid DNA to a distant rDNA promoter, and the enhancer effect is found regardless of orientation and upstream or downstream location of the 60/81-bp repeats relative to the rDNA promoter in an injected oocyte assay (Labhart and Reeder, 1984). However, a polarity effect is observed when rDNA is injected into developing embryos, such that no enhanced transcription can be found if the NTS is inverted (Busby and Reeder, 1983).

Based on biochemical evidence, Moss (1983a) has proposed that the duplicated spacer promoter elements in the *Bam* islands bind RNA polymerase I and drive it by spacer transcription to the major rDNA gene promoter; the question of the validity of this model has recently been discussed (Murtif and Rae, 1983; Moss, 1983b). Evidence does in fact exist demonstrating a low level of NTS transcription (hence the term "nontranscribed" spacer is a bit of a misnomer and the terms "apparent" spacer or "intergenic" spacer used by others on occasion may be more appropriate). For example, Miller spreads for electron microscopy have shown "prelude transcripts" in the NTS (Scheer *et al.*, 1973, 1977; Franke *et al.*, 1976). Depending on the frog being examined, prelude transcripts may be in none, or, most generally, 2–5%, or occasionally up to 70%, of the spacers of the rDNA units (Trendelenburg, 1981). The variable and differential usage of spacer promoters for prelude transcripts cannot be accounted for by DNA sequence (Morgan *et al.*, 1984). The low frequency of prelude transcript occurrence has been taken as evidence against the model of Moss (Murtif and Rae, 1983), but if spacer transcripts have a short half-life, they may not be visualized often (Moss, 1983b). The observation that FUdR slows down RNA processing and degradation and increases the frequency of prelude transcripts seen in the electron microscope (Rungger *et al.*, 1978) supports the latter explanation. Biochemical estimates of prelude transcripts also suggest a very low occurrence of only 1% of the rRNA precursor in *Xenopus* or *Drosophila* (Sollner-Webb and Schultz, cited by Sollner-Webb *et al.*, 1982; Kohorn and Rae, 1982a), but other data yield a frequency for prelude transcripts of 10% of the rRNA precursor (Moss, 1983b). Again, however, these measurements reflect the steady-state frequency of prelude transcripts and not whether each rDNA unit has rapidly turning over prelude transcripts. A small nuclear RNA transcribed by RNA polymerase I in mammals initially had been likened to the prelude transcripts of the *Xenopus* NTS (Reichel *et al.*, 1982), but map corrections for the rDNA promoter now seem to suggest that the ETS may code for this small RNA.

A somewhat different model has been proposed by Reeder *et al.* (1983) in which the 42 bp within the 60/81-bp repeats of the NTS and homologous sections of the spacer promoters is hypothesized to bind a factor needed to activate the gene promoter. This model affords an explanation for nucleolar dominance of transcription of *Xenopus laevis* rDNA over *Xenopus borealis* rDNA, since the former has at least 24 and the latter has only 4 of the 42-bp repeats in the NTS (Bach *et al.*, 1981b; LaVolpe *et al.*, 1982). Gene constructs in *Xenopus laevis* and *Xenopus borealis* have further supported this model for nucleolar dominance (Reeder and Roan, 1984). However, this model is not a general explanation for all cases of nucleolar dominance, since the dominance of mouse over human rDNA transcription in cell hybrids appears to be due to inactivation or loss of a species-specific factor needed to activate human rDNA (Miesfeld and Arnheim, 1984; Miesfeld *et al.*, 1984; Onishi *et al.*, 1984). Finally, the control of nucleolar dominance may be still more complicated, at least in the case of *Drosophila*, where a genetic locus that controls this phenomenon maps outside the nucleolus organizer region (Durica and Krider, 1978).

Additional evidence from chromatin studies points to the potential importance of the spacer promoters of rDNA. Specifically, a DNAse I hypersensitive site has been located more than 1 kb upstream from the gene promoter, in the first *Bam* island of the NTS (LaVolpe *et al.*, 1983). The position and sequence of this hypersensitive site are similar for *Xenopus laevis* and *Xenopus borealis*, and this hypersensitive site is absent when *X. borealis* rDNA is unexpressed. In other words, the first spacer promoter is available for interaction with external components (e.g., RNA polymerase I, a species-specific factor, etc.) when its rDNA unit will be transcribed.

If spacer promoters are truly important for enhancement of rDNA transcription, one might expect to find them in a wide variety of species. Most of the discussion above has centered on the spacer duplications of the sequence adjacent to that coding for the 5′ end of the *Xenopus* rRNA precursor (Boseley *et al.*, 1979; Sollner-Webb and Reeder, 1979; Moss *et al.*, 1980; Bach *et al.*, 1981b). A similar duplication of the promoter region occurs in *Drosophila* (Kohorn and Rae, 1982a; Coen and Dover, 1982; Simeone *et al.*, 1982; J. R. Miller *et al.*, 1983), and spacer transcipts have been reported for *Drosophila* (Murtif and Rae, 1985). The duplications are not as obvious in yeast, since they have diverged in sequence, but a counterpart of the rDNA initiation region may be found at −46 to −18, −17 to +9, +10 to +32, and +33 to +55 (Verbeet *et al.*, 1984a). Recently the region −2230 to −2420 of the yeast NTS has been shown to have enhancer-like activity for rRNA transcription (Elion and Warner, 1984; Warner *et al.*, 1985). Even *E. coli* has duplicated rDNA promoters: P1 (the major promoter *in vivo*) and P2 are separated by 120 bp, and P3 and P4 are found at more than 1 kb upstream (Boros *et al.*, 1983a,b).

It is known that control of transcription resides at the level of the individual rDNA unit, since during developmental activation of the rDNA cluster some units may be fully active while others are entirely inactive in transcription (Scheer *et al.*, 1976). It would be interesting to know if this differential activation of rDNA units parallels the variation in the NTS repeats according to the models discussed above.

5.2.2. Termination Signal of Transcription

The signals for termination of transcription appear to be highly conserved in evolution, both among different RNA polymerase gene families and also among widely divergent species. This signal includes a region of dyad symmetry (with the potential to form a hairpin stem) followed by a cluster of T residues on the RNA-like (noncoding) DNA strand. Such a signal for termination is even recognized by *E. coli* RNA polymerase for rho-independent termination (Adhya and Gottesman, 1978), and is found in the rDNA genes of *E. coli*, which have been studied at the sequence level (R. A. Young, 1979; Duester and Holmes, 1980; Sekiya *et al.*, 1980; Brosius *et al.*, 1981).

In eukaryotes such as *Xenopus* the dyad symmetry followed by a cluster of T's causes RNA polymerase III termination of 5S RNA transcription (Korn and Brown, 1978; Bogenhagen and Brown, 1981). A naturally occurring 5S RNA gene variant with three instead of four T residues reads through the usual termination site and stops at another cluster of T's downstream (Brown *et al.*, 1979). Gene fusion experiments have shown that when the 3′ end of the 5S RNA gene is replaced by the 3′ end of the rDNA gene, RNA polymerase III stops transcription at the rDNA termination site, thereby indicating the universality of the termination signal (Sollner-Webb and Bogenhagen, 1981).

Table VII summarizes the termination sequences found for a variety of eukaryotic species, although it should be noted that in some cases such as yeast the dyad symmetry follows rather than precedes the T cluster (Veldman *et al.*, 1980b). Initially the *S. cerevisiae* termination site was hypothesized to be 220 bp further downstream (Valenzuela *et al.*, 1977), but this region is partially deleted in *S. carlsbergensis* and is apparently incorrect (Veldman *et al.*, 1980b). There also has been some confusion over the *Xenopus* rDNA termination site, which was originally stated to be $-CCAAGC_{OH}$ (Sollner-Webb and Reeder, 1979), but S1 nuclease may have artificially nibbled off the end of the gene and the true termination site may therefore be further downstream (K. Miller and Sollner-Webb, 1981). Oocyte injection experiments have shown that the *Xenopus* rDNA

Table VII

Sequences of the Termination Region of rDNA[a]

	−30	−20	−10	+10	+20	+30
Mouse	TGAAAGTCAG	CCCTCGACAC	AAGGGTTTGT	CTCTGCGGGC	TTTCCCGTGG	CACGCCCGCT CGC
			28S			45S
Rat	TGAAAGTCAG	CCCTCGACAC	AAGGGTTTGT			
			28S			
X. laevis rDNA gene	TGAAAGTCAT	CCCTTGAGCC	AAGCTTTTGT	CGGAAGGAGC	AGGCCGGAAG	GGCGCCCCCG CCG
			28S			
X. laevis NTS termination	−250	−240	−230	−220	−210	
	TCCCCGGGCC	CTTTGGCGCC	CGTTTTTTCG	CAAAGTGCGG	C	
D. melanogaster	CTGAAGCTTA	TCCTTTGCTT	GATGATTCGA	TATAAAATAA	ATGGTTGCCA	AACAGCTCGT
			28S			

N. crassa	CTATAAGCCC	TCCTTCGGCA	TTAGATTTCA			
			↑ 26S			
P. polycephalum	CTGAGCCAGT	GTTTGCGTTT	TTTTGTCGGC	GCCTGGGGAG	GCCTAATAAT	GCCCCGTCCC
			↑ 26S			
T. thermophila	CTGAGATTCA	GCCCGTCTCC	TTAGATTTAT	CTCATCTCCC	TTTATTTTTT	ACTTCTGCT
			↑ 26S		35S	
S. carlsbergensis	CTGAGATTAA	GCCTTTGTTG	TCTGATTTGT	TTTTTATTTC	TTTCTAAGTG	GGTACTGGCA
			↑ 26S	37S		
E. coli	CGGUACTAAT	GAACCGTGAG	GCTTAACCTT			
			↑ 23S			

[a]Sequences given are for the noncoding (RNA-like) strand of DNA. References: mouse: Kominami *et al.* (1982); rat: Chan *et al.* (1983) and Hadjiolov *et al.* (1984); *Xenopus laevis*: Sollner-Webb and Reeder (1979); *Drosophila melanogaster*: Mandel and Dawid (1981); *Neurospora crassa*: Kelly and Cox (1981); *Physarum polycephalum*: Otsuka *et al.* (1983); *Tetrahymena thermophila*: Din *et al.* (1982); *Saccharomyces carlsbergensis*: Veldman *et al.* (1980b); *Escherichia coli*: Brosius *et al.* (1980). Recently the termination region of *Glaucoma chattoni* has also been sequenced (Challoner *et al.*, 1985).

termination signal is not beyond the cluster of T residues (Bakken *et al.*, 1982); furthermore, *Xenopus* rDNA transcription does not terminate properly if the cluster of four T's is shortened to two T's (Bakken *et al.*, 1982).

Although it is clear that the dyad symmetry and cluster of T's are generally signals for termination, the actual site for occurrence of termination could be further downstream. In fact, 3′ processing of the rRNA precursor occurs in yeast (Veldman *et al.*, 1980a), *Tetrahymena* (Din *et al.*, 1982), and mouse (Kominami *et al.*, 1982) to remove seven, 14, or 30 nucleotides, respectively, relative to the mature 26S–28S rRNA 3′ terminus. Recently it has been discovered that over 330 nucleotides beyond the 3′ end of 28S are transcribed as part of the mouse rRNA precursor (Grummt *et al.*, 1985), which is probably rapidly processed to the 30 nucleotide extension at the 3′ end and finally to the mature 28S 3′ terminus. In *Xenopus* the 3′ end of the 40S rRNA precursor matches that of 28S rRNA (Sollner-Webb and Reeder, 1979), but one could argue that 3′ processing has been so rapid that the true 3′ end of the precursor has not been detected.

A similar termination signal of a short palindrome followed by an oligo-T tract is found at −225 of the *Xenopus* NTS, and acts as a "fail-safe" terminator to end prelude piece transcription before reaching the true gene promoter for rRNA precursor transcription (Sollner-Webb and Reeder, 1979; Morgan *et al.*, 1983). Miller spread observations with the electron microscope demonstrated that the ribonucleoprotein fibril gradient of the prelude piece ends before the 40S rRNA precursor fibril gradient begins; generally two transcript-free RNA polymerase molecules are seen spanning the distance between the two transcription gradients (Trendelenburg, 1982). From this data one could therefore argue that prelude piece transcription has helped to load polymerase I molecules in an array before the true gene promoter, as hypothesized by Moss (1983a). A factor has been isolated from *Tetrahymena* needed for termination (Leer *et al.*, 1979), and it will be interesting to discover if such a termination factor also plays a role for prelude piece termination.

5.2.3. Origin of Replication of the rDNA Unit

An autonomously replicating sequence (ARS) has been found within the NTS 2 of yeast near the promoter region for transcription (Szostak and Wu, 1979; Larionov and Shubochkina, 1982; Skryabin *et al.*, 1984), and this location coincides with the position of small replication bubbles observed in synchronized yeast cells (Saffer and Miller, 1983; L. P. Saffer and O. L. Miller, Jr., personal communication). A replication origin has also been attributed to the NTS of *Lytechinus* (Botchan and Dayton, 1982) and *Drosophila* (McKnight and Miller, 1975; McKnight *et al.*, 1977). Surprisingly, the origin of replication has been placed within the 28S rRNA coding region of *Xenopus* (Benbow *et al.*, 1978; Hines and Benbow, 1982), which is in contrast to the studies above and another study on *Xenopus* rDNA (Bozzoni *et al.*, 1981a). Even if all rDNA units contained an origin of replication, they need not all be activated and replicons longer than the rDNA unit length could exist.

The origin of replication for extrachromosomal rDNA has been located to roughly midway along the NTS in *Tetrahymena pyriformis* (Truett and Gall, 1977) and to 600 bp from the NTS center in *Tetrahymena thermophila* (Cech and Brehn, 1981). This region of *T. thermophila* rDNA can also autonomously replicate in yeast (Kiss *et al.*, 1981; Amin and Pearlman, 1985), but whether the ARS element is identical to the origin utilized by *Tetrahymena* is unknown. The NTS of extrachromosomal copies of rDNA of *Physarum* also seems to contain an origin (Vogt and Braun, 1977). It is yet to be discovered if the origins mapped to the NTS regions of the extrachromosomal rDNA of the protozoa listed above are also recognized as origins for chromosomal replication; they might well differ for extrachromosomal rDNA amplification and chromosomal duplication.

6. Evolutionary Forces Acting on rDNA

6.1. The Riddle of Concerted Evolution

An evolutionary riddle was posed by Brown and his co-workers after their comparison of the rDNAs of *Xenopus laevis* and *Xenopus borealis* (originally mistaken to be *Xenopus mulleri*): the sequence of the tandem repeats of rDNA was basically the same in all repeated units of rDNA within one species, but when the comparison was drawn between these two species it was observed that the NTS sequences differed, although the coding region was about the same (Brown *et al.*, 1972). This phenomenon of intraspecific homogeneity coupled with interspecific differences for sequences of a tandemly repeated gene family has been termed horizontal, coincidental, or concerted evolution, and has recently been reviewed (Dover, 1982; Dover *et al.*, 1982; Arnheim, 1983; Dover and Flavell, 1984).

Concerted evolution could arise from either (1) a sudden correction mechanism; or (2) a gradual correction mechanism (Brown and Sugimoto, 1973). Some of the sudden correction mechanisms hypothesized to exist are saltatory expansion, expansion–contraction, and the "master–slave" model (Callan and Lloyd, 1960; Callan, 1967; Thomas, 1970; Buongiono-Nardelli *et al.*, 1972). After rectification by a sudden correction mechanism, all tandemly repeated units in the organism would be identical. By contrast, a gradual correction mechanism (such as unequal crossing-over) permits some differences between adjacent tandem copies. These two opposing classes of mechanisms were first experimentally distinguished for the 5S RNA repeated gene family of *Xenopus laevis* (Carroll and Brown, 1976); their observation that adjacent repeat units could be heterogeneous in length ruled out a sudden correction mechanism and favored a gradual mechanism. We now know that this conclusion is also valid for rDNA tandem arrays, where the individual repeat units may differ in length due to variations attributable to the NTS and in some cases also attributable to intervening sequences.

6.2. Mechanisms of Molecular Drive

What are the gradual rectification mechanisms that act on mutations arising in a tandemly repeated gene family? These mechanisms must act to homogenize the repeated copies of a gene family, which may be in tandem arrays in not only just one locus, but sometimes in several loci. Dover has proposed that "molecular drive," which includes the mechanisms of unequal crossing-over, gene conversion, and transposition, is the dynamic process acting upon the constant turnover in DNA sequence (Dover, 1982; Dover and Flavell, 1984). Molecular drive spreads variants through the tandem population and fixes these changes within all individuals of the species. We will now review the details of how molecular drive is theorized to work.

6.2.1. Unequal Crossing-over

Unequal crossing-over is one of the mechanisms of the molecular drive process. Computer and mathematical simulation has shown that repeated rounds of unequal crossing-over between two members of a repeated gene family, one of which has a mutation and the other which lacks this mutation, can homogenize the gene cluster so that all tandem members either have or lack the mutation (Smith, 1973, 1976; Perelson and Bell, 1977; Koch, 1978; Ohta, 1980). Experimental data have been obtained that support the occurrence of unequal crossovers in the rDNA gene family of yeast (Petes, 1980; Szostak and Wu, 1980). It is surprising that unequal crossing-over has not yet randomized the placement of intervening sequence containing copies of rDNA in *Drosophila hydei* and *Calliphora*; instead, they are found clustered together within the tandem array (Renkawitz-Pohl *et al.*, 1981; Beckingham, 1981).

Other data also support unequal crossovers in rDNA clusters on nonhomologous chromosomes in eukaryotes (Arnheim *et al.*, 1980; Appels and Dvořák, 1982; Coen *et al.*, 1982a; Mikus and Petes, 1982; Coen and Dover, 1983). However, it seems likely that the rate of homogenization between *Drosophila* X and Y chromosomes is less than the rate at which the X and Y chromosomes are each homogenized separately (Coen and Dover, 1983): in *Drosophila melanogaster* the X and Y chromosomes differ with regard to their 18S rDNA sequences (Yagura *et al.*, 1979), X-specific type 1 insertions (Tartof and Dawid, 1976; Wellauer *et al.*, 1978), and NTS variants (Endow and Glover, 1979; Coen *et al.*, 1982b). Similarly, rDNA length variants in the rDNA of mouse are specific for one homologue versus another (Arnheim *et al.*, 1982). In contrast, the rDNA copies are indistinguishable on the five homologues to which they map in humans (Arnheim *et al.*, 1980; Krystal *et al.*, 1981); these five homologues associate together in one common nucleolus. Taken together, the above data suggest that the rate of nonhomologous exchange is lower in flies and mice than in man.

Meiotic recombination seems to be inhibited in the rDNA of yeast 70- to 100-fold (Petes and Botstein, 1977) and also in *Coprinus* (Cassidy *et al.*, 1984) relative to the higher rates of meiotic recombination observable for genetic loci of unique genes. A suppression of crossing-over between genes in tandem arrays has been discussed for other cases as well (Fogel *et al.*, 1983). On the other hand, an increase in both inter- and intrachromosomal mitotic recombination in yeast can be localized to a hotspot ("HOT 1") in the NTS near the 5S RNA gene (Keil and Roeder, 1984). "HOT 1" only works when placed in one orientation, and its lack of function when placed opposite heterologous DNA suggests that it is used as a pairing site between homologous sequences.

It could also be suggested that the internal subrepeats found in the NTS of some rDNAs are selectively advantageous, as they could function to promote unequal exchange to increase homogeneity of the rDNA cluster; moreover, a consequence of unequal crossovers is the generation of rDNA length heterogeneity due usually to variable members of the NTS internal subrepeats [reviewed by Federoff (1979)]. The general stability of rDNA length variants in *Drosophila* has been used to estimate that the maximum rate of unequal exchange in rDNA is low (Coen *et al.*, 1982b, 1984; Barrett, 1984). Another consequence of unequal exchange is that the rDNA copy number of one individual may differ a bit from the rDNA copy number of another individual, as has been found, for example, in humans (Warburton *et al.*, 1976).

6.2.2. Gene Conversion

Gene conversion is a second, major component of the molecular drive process. Nagylaki (1984) has presented a mathematical argument for its importance in the evolution of multigene families. Holliday (1964) has proposed that gene conversion results from mismatch repair within a heteroduplex DNA region formed during recombination, but other models for how gene conversion is effected are also possible (Szostak *et al.*, 1983). Gene conversion is nonreciprocal, such that if one parental chromosome has a mutant form of the gene and the other parental chromosome has the nonmutated gene, then either the daughter chromosomes will *both* have the mutant gene or *both* have the nonmutant gene (Radding, 1978). The extent of DNA involved in a gene conversion event may be up to several kb (Jackson and Fink, 1981; Klar *et al.*, 1981; Klein and Petes, 1981), and gene conversion can occur between homologous (Edelman and Gally, 1970; Finnerty, 1976; Fogel *et al.*, 1978; Fincham *et al.*, 1979; Klein and Petes, 1981; Lamb and Helmi, 1982) or between nonhomologous chromosomes (Scherer and Davis, 1980; Ernst *et al.*, 1981; Munz and Leopold, 1981; Mikus and Petes, 1982). Gene conversion may be either random with respect to whether the daughter products carry the mutant or nonmutant gene, or biased if one gene form is preferentially kept over the other form. Finally, gene conversion events have been shown to be theoretically capable of producing concerted evolution (Birky and Skavaril, 1976; Ohta, 1977; Nagylaki and Petes, 1982; Ohta and Dover, 1983, 1985).

6.2.3. DNA Transposition

Like gene conversion, DNA transposition may also be duplicative and result in an increase in the number of copies of the gene being transposed (or duplicated). DNA transposition could theoretically move a rDNA copy to a new chromosomal location. Perhaps this is an explanation for the occurrence of orphons (Childs *et al.*, 1981). With expansion of the transposed rDNA, a new cluster of tandem rDNA units will be found in this additional chromosomal locus. In organisms such as *Xenopus* or *Drosophila*, there is only one chromosomal locus for rDNA, but in other organisms, such as humans, several rDNA loci have been mapped in the genome.

6.2.4. Fixation of Changes

These mechanisms of molecular drive serve continuously to homogenize the constant turnover in a repeated gene family. The changes in the repeated family spread rapidly through the population, so that all organisms of the species carry the changes. In other words, the population evolves cohesively because the randomization of chromosomes via the sexual process moves at a much faster rate than the rates of unequal exchange, gene conversion, or transposition (Dover, 1982). If there is a bias to favor a directionality for gene conversion or transposition, this could accelerate somewhat the rate of fixation of the changes (Dover, 1982). It has been estimated that if there are 200 repeated rDNA copies and the rate of unequal exchange is 3×10^{-4} per generation, then it will take 10^7 generations to spread one variant to all copies of the rDNA cluster (Dover *et al.*, 1982); this calculation disregards any effects that other molecular drive processes or selection might have on the homogenization rate. Similarly, the rates of gene conversion and unequal crossovers in yeast have been calculated to be less than the rate of fixation of mutation (Petes, 1980; Szostak and Wu, 1980; Klein and Petes, 1981).

6.3. Selection Superimposed on Molecular Drive

Since the initial mutation events are essentially random, but can be rapidly transmitted to the population as a whole, one can predict that the rDNA of one species will evolve away from the rDNA of another species. How then can one explain the different rates of evolution within one rDNA copy? For example, when the rDNA is compared among *Drosophila melanogaster* subgroups, the NTS is found to vary most greatly, next the transcribed spacers, and finally the rRNA-coding regions, which are virtually identical (Coen *et al.*, 1982a; Coen *et al.*, 1982b; Strachan *et al.*, 1982). Sequence comparisons between the rDNA of *Xenopus laevis* and *Xenopus borealis* revealed sequence divergence interspersed with conserved regions in the transcribed spacers (Furlong and Maden, 1983; Furlong *et al.*, 1983), unlike the rare microheterogeneities found in the rRNA-coding region (Furlong and Maden, 1983). The differing amounts of variation within one rDNA unit in the population reflects the same phenomenon at the level of the individual organism. Comparative sequencing of clones of different rDNA repeats from *Xenopus laevis* found no sequence microheterogeneities in the 18S rRNA-coding region (Maden *et al.*, 1982a), although such microheterogeneities have been found in the transcribed spacers of *Xenopus* rDNA (Stewart *et al.*, 1983). Of course, the major differences in Xenopus rDNA units of one organism reside in the NTS and give rise to length variants (Wellauer *et al.*, 1976b). If the processes of molecular drive were the only ones acting on rDNA, then the degree of variation should be approximately uniform throughout the entire rDNA repeat. Since this is not the case, it is likely that selection pressures are superimposed upon the resulting products of molecular drive. It is unknown what percentage of the rDNA repeated copies must carry a mutation before it can cause a discernible difference upon which selection can act. Negative or neutral selection occurs more frequently than positive selection (Kimura, 1983). It is obvious that the importance of selection increases as the percentage of rDNA copies that carry a mutation increases in the cluster (Ohta and Dover, 1985).

In this chapter I have reviewed data on the function of various regions within the rDNA unit, so that it is now possible to assess selective pressures on these regions. Some regions of the rRNA-coding sequence are conserved from *E. coli* through mammals, suggesting that variants in these vitally important regions have been selected against negatively. Other parts of the coding region are important at the secondary structure level, and a mutation on one side of the stem may have minimal negative selection impact, but a compensatory mutation by its pairing partner on the other side of the stem would have slight selective advantages. It is outside the scope of this review to discuss whether any mutations really can be selectively neutral (Crow, 1981; Kimura, 1983). For example, some hypervariable regions within cytochrome *c* that were originally cited by some to support the neutralist theory were later found to be important for plastic structural requirements; one such case is a hypervariable region in cytochrome *c* that correlates with the coevolving oxidase that reacts with it (Ferguson-Miller *et al.*, 1979). In the same vein, it may be that there are no truly neutral mutations in the rRNA-coding region. Probably an accurate measure of the neutral or nearly neutral mutation rate can best be assessed within introns or parts of the NTS.

Finally, we have seen that mutations in the promoter region of the NTS seem to be coupled to coevolution of species-specific RNA transcription factors. If the initial spread of a mutation in a promoter is slow through the rDNA cluster, it allows time for transcription factor genes to coevolve (Dover and Flavell, 1984; Wilson *et al.*, 1984). Coevolution may also occur for changes in the rRNA-coding region, such as areas that intereact directly with ribosomal proteins. If more than one other component reacts with a region of rDNA, then the constraints increase to permit change, since coevolution in several genes for these multiple components would be much more difficult.

Our increasing understanding of the structure and function of regions within rDNA provides a model system to test the dynamic processes leading to evolutionary change.

ACKNOWLEDGMENTS. I am grateful to the numerous authors who sent me reprints and preprints of their recent work. I thank Lynn Rothschild and David Brautigan for stimulating discussions on evolution, William Yue, Jurka Jerzy and Jim McIlwain for their help with the references, Beth Destasio for proofreading, and Missy Stebbins-Boaz for grading final exams so I could complete this review. Carol King is to be highly commended for the monumental job of typing this manuscript, and I appreciate the helpful assistance from Jean Waage. This work was supported by PHS-GM20261.

References

Adhya, S., and Gottesman, M., 1978, Control of transcription termination, *Annu. Rev. Biochem.* **47**:967–996.

Aloni, Y., and Attardi, G., 1971a, Expression of the mitochondrial genome in HeLa cells. II. Evidence for complete transcription of mitochondrial DNA, *J. Mol. Biol.* **55**:251–270.

Aloni, Y., and Attardi, G., 1971b, Symmetrical *in vivo* transcription of mitochondrial DNA in HeLa cells, *Proc. Natl. Acad. Sci. USA* **68**:1757–1761.

Amalric, F., Merkel, C., Gelfand, R., and Attardi, G., 1978, Fractionation of mitochondrial RNA from HeLa cells by high-resolution electrophoresis under strongly denaturing conditions, *J. Mol. Biol.* **118**:1–25.

Amikam, D., Razin, S., and Glaser, G., 1982, Ribosomal RNA genes in *Mycoplasma*, *Nucleic Acids Res.* **10**:4215–4222.

Amin, A. A., and Pearlman, R. E., 1985, Autonomously replicating sequences from the nontranscribed spacers of *Tetrahymena thermophila* ribosomal DNA, *Nucleic Acids Res.* **13**:2647–2659.

Anderson, S., DeBruijn, M. H. L., Coulson, A. R., Eperon, I. C., Sanger, F., and Young, I. G., 1982, Complete sequence of bovine mitochondrial DNA. Conserved features of the mammalian genome, *J. Mol. Biol.* **156**:683–717.

Anziano, P. Q., Hanson, D. K., Mahler, H. R., and Perlman, P. S., 1982, Functional domains in introns: *trans*-acting and *cis*-acting regions of intron 4 of the *cob* gene, *Cell* **30**:925–932.

Appels, R., and Dvořák, J., 1982, The wheat ribosomal DNA spacer region: Its structure and variation in populations and among species, *Theor. Appl. Genet* **63**:337–348.

Appels, R., Gerlach, W. L., Dennis, E. S., Swift, H., and Peacock, W. J., 1980, Molecular and chomosomal organization of DNA sequences coding for the ribosomal RNAs in cereals, *Chromosoma* **78**:293–311.

Arnheim, N., 1983, Concerted evolution of multigene families, in: *Evolution of Genes and Proteins* (M. Nei and R. K. Koehn, eds.), Sinauer, Sunderland, Massachusetts, pp. 38–61.

Arnheim, N., and Kuehn, M., 1979, The genetic behaviour of a cloned mouse ribosomal DNA segment mimics mouse ribosomal gene evolution, *J. Mol. Biol.* **134**:743–765.

Arnheim, N., and Southern, E. M., 1977, Heterogeneity of the ribosomal genes in mice and men, *Cell* **11**:363–370.

Arnheim, N., Krystal, M., Schmickel, R., Wilson, G., Ryder, O., and Zimmer, E., 1980, Molecular evidence for genetic exchanges among ribosomal genes on nonhomologous chromosomes in man and ape, *Proc. Natl. Acad. Sci. USA* **77**:7323–7327.

Arnheim, N., Treco, D., Taylor, B., and Eicher, E., 1982, Distribution of ribosomal gene length variants among mouse chromosomes, *Proc. Natl. Acad. Sci. USA* **79**:4677–4680.

Atmadja, J., Brimacombe, R., and Maden, B. E. H., 1984, *Xenopus laevis* 18S ribosomal RNA: Experimental determination of secondary structure elements and locations of methyl groups in the secondary structure model, *Nucleic Acids Res.* **12**:2649–2667.

Bach, R., Grummt, I., and Allet, B., 1981a, The nucleotide sequence of the initiation region of the ribosomal transcription unit from mouse, *Nucleic Acids Res.* **9**:1559–1569.

Bach, R., Allet, B., and Crippa, M., 1981b, Sequence organization of the spacer in the ribosomal genes of *Xenopus clivii* and *Xenopus borealis*, *Nucleic Acids Res.* **9**:5311–5330.

Bachellerie, J. P., Michot, B., and Raynal, F., 1983, Recognition signals for mouse pre-rRNA processing. A potential role for U3 nucleolar RNA, *Mol. Biol. Rep.* **9**:79–86.

Baer, R. J., and Dubin, D. T., 1981, Methylated regions of hamster mitochondrial ribosomal RNA: Structural and functional correlates, *Nucleic Acids Res.* **9**:323–337.

Bakken, A., Morgan, G., Sollner-Webb, B., Roan, J., Busby, S., and Reeder, R. H., 1982, Mapping of transcription initiation and termination signals on *Xenopus laevis* ribosomal DNA, *Proc. Natl. Acad. Sci. USA* **79**:56–60.

Barrett, J. A., 1984, DNA variation and evolution, *Nature* **309**:285.

Barta, A., Steiner, G., Brosius, J., Noller, H. F., and Kuechler, E., 1984, Identification of a site on 23S ribosomal RNA located at the peptidyl transferase center, *Proc. Natl. Acad. Sci. USA* **81**:3607–3611.

Bartnik, E., Strugala, K., and Stepien, P. P., 1981, Cloning and analysis of recombinant plasmids containing genes for *Aspergillus nidulans* 5S rRNA, *Curr. Genet.* **4**:173–176.

Bass, B. L., and Cech, T. R., 1984, Specific interaction between the self-splicing RNA of *Tetrahymena* and its guanosine substrate: Implications for biological catalysis by RNA, *Nature* **308**:820–826.

Battey, J., and Clayton, D. A., 1980, The transcription map of human mitochondrial DNA implicates transfer RNA excision as a major processing event, *J. Biol. Chem.* **255**:11599–11606.

Baughman, G., and Nomura, M., 1983, Localization of the target site for translational regulation of the L11 operon and direct evidence for translational coupling in *Escherichia coli*, *Cell* **34**:979–988.

Baughman, G., and Nomura, M., 1984, Translational regulation of the L11 ribosomal protein operon of *Escherichia coli*: Analysis of the mRNA target site using oligonucleotide-directed mutagenesis, *Proc. Natl. Acad. Sci. USA* **81**:5389–5393.

Bayev, A. A., Georgiev, O. I., Hadjiolov, A. A., Kermekchiev, M. B., Nikolaev, N., Skryabin, K. G., and Zakharyev, V. M., 1980, The structure of the yeast ribosomal RNA genes, 2. The nucleotide sequence of the initiation site for ribosomal RNA transcription, *Nucleic Acids Res.* **8**:4919–4926.

Beckingham, K., 1981, The ribosomal DNA of *Calliphora erythrocephala*: The cistron classes of total genomic DNA, *J. Mol. Biol.* **149**:141–169.

Beckingham, K., 1982, Insect rDNA, in: *The Cell Nucleus: rDNA*, Vol. X, Part A (H. Busch and L. Rothblum, eds.), Academic Press, New York, pp. 205–269.

Bell, G. I., DeGennaro, L. J., Gelfand, D. H., Bishop, R. J., Valenzuela, P., and Rutter, W. J., 1977, Ribosomal RNA genes of *Saccharomyces cerevisiae*. I. Physical map of the repeating unit and location of the regions coding for 5S, 5.8S, 18S and 25S rRNAs, *J. Biol. Chem.* **252**:8118–8125.

Benbow, R. M., Krauss, M. R., and Reeder, R., 1978, DNA synthesis in a multi-enzyme system from *Xenopus laevis* eggs, *Cell* **13**:307–318.

Birky, C. W., and Skavaril, R. V., 1976, Maintenance of genetic homogeneity in systems with multiple genomes, *Genet. Res.* **27**:249–265.

Birnstiel, M. L., and Grunstein, M., 1972, The ribosomal cistrons of eukaryotes—A model system for the study of evolution of serially repeated genes, *FEBS Symp.* **23**:349–365.

Birnstiel, M. L., Spiers, J., Purdom, I., Jones, K., and Loening, U. E., 1968, Properties and composition of the isolated ribosomal DNA satellite of *Xenopus laevis*, *Nature* **219**:454–463.

Blackburn, E. H., 1982, Characterization and species differences of rDNA: protozoans, in: *The Cell Nucleus: rDNA*, Vol. X, Part A (H. Busch and L. Rothblum, eds.), Academic Press, New York pp. 145–170.

Bloch, D. P., McArthur, B., Widdowson, R., Spector, D., Guimaraes, R. C., and Smith, J., 1983, tRNA–rRNA sequence homologies: Evidence for a common evolutionary origin? *J. Mol. Evol.* **19**:420–428.

Bloch, D. P., McArthur, B., and Mirrop, S., 1985, tRNA–rRNA sequence homologies: Evidence for an ancient modular format shared by tRNAs and rRNAs, *Biosystems* **17**:209–225.

Blum, B., Seebeck, T., Braun, R., Ferris, P., and Vogt, V., 1983, Localization and DNA sequence around the initiation site of ribosomal RNA transcription in *Physarum polycephalum Nucleic Acids Res,* **11**:8519–8533.

Bogenhagen, D. F., and Brown, D. D. 1981, Nucleotide sequences in *Xenopus* 5S DNA required for transcription termination, *Cell* **24**:261–270.

Bogenhagen, D. F., Sakonju, S., and Brown, D. D., 1980, A control region in the center of the 5S RNA gene directs specific initiation of transcription. II. The 3′ border of the region, *Cell* **19**:27–35.

Bogenhagen, D. F., Wormington, W. M., and Brown, D. D., 1982, Stable transcription complexes of *Xenopus* 5S RNA genes: A means to maintain the differentiated state, *Cell* **28**:413–421.

Bogorad, L., 1982, Regulation of intracellular gene flow in the evolution of eukaryotic genomes, in: *On The Origins of Chloroplasts* (J. A. Schiff, ed.), Elsevier-North Holland, New York, pp. 277–295.

Bollen, G. H. P. M., Molenaar, C. M. T., Cohen, L. H., van Raamsdonk-Duin, M. M. C., Mager, W. H., and Planta, R. J., 1982, Ribosomal protein genes of yeast contain intervening sequences, *Gene* **18**:29–37.

Bonen, L., and Gray, M. W., 1980, Organization and expression of the mitochondrial genome of plants I. The genes for wheat mitochondrial ribosomal and transfer RNA: Evidence for an unusual arrangement, *Nucleic Acids Res.* **8**:319–335.

Bonen, L., Huh, T. Y., and Gray, M. W., 1980, Can partial methylation explain the complex fragment patterns observed when plant mitochondrial DNA is cleaved with restriction endonucleases?, *FEBS Lett.* **111**:340–346.

Boros, I., Csordás-Tóth, E., Kiss, A., Kiss, I., Tŏrŏk, I., Udvardy, A., Udvardy, K., and Venetianer, P., 1983a, Identification of two new promoters probably involved in the transcription of a ribosomal RNA gene of *Escherichia coli, Biochim. Biphys. Acta* **739**:173–180.

Boros, I., Kiss, A., Sain, B., Somlyai, G., and Venetianer, P., 1983b, Cloning of the promoters of an *Escherichia coli* rRNA gene. New experimental system to study the regulation of rRNA transcription, *Gene* **22**:191–201.

Borst, P., and Grivell, L. A., 1971, Mitochondrial ribosomes, *FEBS Lett.* **13**:73–88.

Borsuk, P. A., Nagiec, M. M., Stepien, P. P., and Bartnick, E., 1982, Organization of the ribosomal RNA gene cluster in *Aspergillus nidulans, Gene* **17**:147–152.

Bos, J. L., Heyting, C., Borst, P., Arnberg, A. C., and van Bruggen, E. F. J., 1978, An insert in the single gene for the large ribosomal RNA in yeast mitochondrial DNA, *Nature* **275**:336–338.

Bos, J. L., Osinga, K. A., Van der Horst, G., Hecht, N. B., Tabak, H. F., Van Ommen, G.-J. B., and Borst, P., 1980, Splice point sequence and transcripts of the intervening sequence in the mitochondrial 21S ribosomal RNA gene of yeast, *Cell* **20**:207–214.

Boseley, P. G., Tuyns, A., and Birnstiel, M. L., 1978, Mapping of *Xenopus laevis* 5.8S rDNA by restriction and DNA sequencing, *Nucleic Acids Res.* **5**:1121–1137.

Boseley, P., Moss, T., Mächler, M., Portmann, R., and Birnsteil, M., 1979, Sequence organization of the spacer DNA in a ribosomal gene unit of *X. laevis, Cell* **17**:19–31.

Botchan, P. M., and Dayton, A. I., 1982, A specific replication origin in the chromosomal rDNA of *Lytechinus variegatus, Nature* **299**:453–456.

Botchan, P., Reeder, R., and Dawid, I. B., 1977, Restriction analysis of the nontranscribed spacers of *Xenopus laevis* ribosomal DNA, *Cell* **11**:599–607.

Bowman, L. H., Rabin, B., and Schlessinger, D., 1981, Multiple ribosomal RNA cleavage pathways in mammalian cells, *Nucleic Acids Res.* **9**:4951–4966.

Bowman, L. H., Goldman, W. E., Goldberg, G. I., Hebert, M. B., and Schlessinger, D., 1983, Location of the initial cleavage sites in mouse pre-rRNA, *Mol. Cell. Biol.* **3**:1501–1510.

Bozzoni, I., Baldari, C. T., Amaldi, F., and Buongiorno-Nardelli, M., 1981a, Replication of ribosomal DNA in *Xenopus laevis*, *Eur. J. Biochem.* **118**:585–590.

Bozzoni, I., Beccari, E., Luo, Z. X., Amaldi, F., Pierandrei-Amaldi, P., and Campioni, N., 1981b, *Xenopus laevis* ribosomal protein genes: Isolation of recombinant cDNA clones and study of the genomic organization, *Nucleic Acids Res.* **9**:1069–1086.

Bozzoni, I., Fragapane, P., Annesi, F., Pierandrei-Amaldi, P., Amaldi, F., and Beccari, E., 1984, Expression of two *Xenopus laevis* ribosomal protein genes in injected frog oocytes. A specific splicing block interferes with the L1 RNA maturation, *J. Mol. Biol.* **180**:987–1005.

Bram, R. J., Young, R. A., and Steitz, J. A., 1980, The ribonuclease III site flanking 23S sequences in the 30S ribosomal precursor RNA of *E. coli*, *Cell* **19**:393–401.

Brand, R. C., and Gerbi, S. A., 1979, Fine structure of ribosomal RNA. II. Distribution of methylated sequences within *Xenopus laevis* rRNA, *Nucleic Acids Res.* **7**:1497–1511.

Branlant, C., Krol, A., Machatt, M. A., Pouyet, J., Ebel, J. P., Edwards, K., and Kössel, H., 1981, Primary and secondary structures of *E. coli* MRE 23S rRNA. Comparison with models of secondary structure for maize chloroplast 23S rRNA and for large portions of mouse and human 16S mitochondrial rRNAs, *Nucleic Acids Res.* **9**:4303–4324.

Brimacombe, R., 1980, Secondary structure homology between *Escherichia coli* 16S and *Saccharomyces cerevisiae* 18S ribosomal RNA, *Biochem. Int.* **1**:162–171.

Brimacombe, R., Maly, P., and Zwieb, C., 1983, The structure of ribosomal RNA and its organization relative to ribosomal protein, *Prog. Nucleic Acid Res. Mol. Biol.* **28**:1–48.

Brosius, J., Palmer, M. L., Kennedy, P. J., and Noller, H. F., 1978, Complete nucleotide sequence of a 16S ribosomal RNA gene from *Escherichia coli*, *Proc. Natl. Acad. Sci. USA* **75**:4801–4805.

Brosius, J., Dull, T. J., and Noller, H. F., 1980, Complete nucleotide sequence of a 23S ribosomal RNA gene from *Escherichia coli*, *Proc. Natl. Acad. Sci. USA* **77**:201–204.

Brosius, J., Dull, T. J., Sleeter, D. D., and Noller, H. F., 1981, Gene organization and primary structure of a ribosomal RNA operon from *Escherichia coli*, *J. Mol. Biol.* **148**:107–127.

Brown, D. D., 1984, The role of stable complexes that repress and activate eucaryotic genes, *Cell* **37**:359–365.

Brown, D. D., and Dawid, I. B., 1968, Specific gene amplification in oocytes, *Science* **160**:272–280.

Brown, D. D., and Littna, E., 1964, RNA synthesis during development of *Xenopus laevis*, the South African clawed toad, *J. Mol. Biol.* **8**:669–687.

Brown, D. D., and Sugimoto, K., 1973, The structure and evolution of ribosomal and 5S DNAs in *Xenopus laevis* and *Xenopus mulleri*, *Cold Spring Harbor Symp. Quant. Biol.* **38**:501–505.

Brown, D. D., and Weber, C. S., 1968, Gene linkage by RNA–DNA hybridization. I. Unique DNA sequences homologous to 4S RNA, 5S RNA and ribosomal RNA, *J. Mol. Biol.* **34**:661–680.

Brown, D. D., Wensink, P. C., and Jordan, E., 1972, A comparison of the ribosomal DNAs of *Xenopus laevis* and *Xenopus mulleri*: The evolution of tandem genes, *J. Mol. Biol.* **63**:57–73.

Brown, D. D., Birkenmeier, E., Bogenhagen, D., Korn, L. J., Jordan, E., Petersen, R., and Sakonju, S., 1979, The dual 5S RNA gene system in *Xenopus*, *Carnegie Inst. Yearb.* **78**:71–84.

Buetow, D. E., and Wood, W. M., 1978, The mitochondrial translation system, *Subcell. Biochem.* **5**:1–85.

Buongiorno-Nardelli, M., Amaldi, F., and Lava-Sanchez, P. A., 1972, Amplification as a rectification mechanism for the redundant rDNA genes, *Nature New Biol.* **238**:134–137.

Burke, J. M., and RajBhandary, U. L., 1982, Intron within the large rRNA gene of *N. crassa* mitochondria: A long open reading frame and a consensus sequence possibly important in splicing, *Cell* **31**:509–520.

Busby, S. J., and Reeder, R. H., 1983, Spacer sequences regulate transcription of ribosomal gene plasmids injected into *Xenopus* embryos, *Cell* **34**:989–996.

Callan, H. G., 1967, The organization of genetic units in chromosomes, *J. Cell Sci.* **2**:1–7.

Callan, H. G., and Lloyd, L., 1960, Lampbrush chromosomes of crested newts *Triturus cristatus* (Laurenti), *Phil. Trans. R. Soc. Lond. B.* **243**:135–219.

Carbon, P., Ehresmann, C., Ehresmann, B., and Ebel, J. P., 1979, The complete nucleotide sequence of the ribosomal 16S RNA from *Escherichia coli*, *Eur. J. Biochem.* **100**:399–410.

Carbon, P., Ebel, J. P., and Ehresmann, C., 1981, The sequence of the ribosomal 16S RNA from *Escherichia coli*, *Nucleic Acids Res.* **9**:2325–2333.

Carin, M., Jensen, B. F., Jentsch, K. D., Leer, J. C., Nielson, O. I., and Westergaard, O., 1980, *In vitro* splicing of the ribosomal RNA precursor in isolated nucleoli of *Tetrahymena*, *Nucleic Acids Res.* **8**:5551–5566.

Carroll, D., and Brown, D. D., 1976, Adjacent repeating units of *Xenopus laevis* 5S DNA can be heterogeneous in length, *Cell* **7**:477–486.

Cassidy, J. R., Moore, D., Lu, B. C., and Pukkila, P. J., 1984, Unusual organization and lack of recombination in the ribosomal RNA genes of *Coprinus cinereus*, *Curr. Genet.* **8**:607–613.

Cech, T. R., 1983, RNA splicing: Three themes with variations, *Cell* **34**:713–716.

Cech, T. R., and Brehn, S. L., 1981, Replication of the extrachromosomal ribosomal RNA genes of *Tetrahymena thermophila*, *Nucleic Acids Res.* **9**:3531–3543.

Cech, T. R., and Rio, D. C., 1979, Localization of transcribed regions on extrachromosomal ribosomal RNA genes of *Tetrahymena thermophila* by R-loop mapping, *Proc. Natl. Acad. Sci. USA* **76**:5051–5055.

Cech, T. R., Zaug, A. J., and Grabowski, P. J., 1981, *In vitro* splicing of the ribosomal RNA precursor of *Tetrahymena*: Involvement of a guanosine nucleotide in the excision of the intervening sequence, *Cell* **27**:487–496.

Cech, T. R., Tanner, N. K., Tinoco, I., Weir, B. R., Zuker, M., and Perlman, P. S., 1983, Secondary structure of the *Tetrahymena* ribosomal RNA intervening sequence: Structural homology with fungal mitochondrial intervening sequences, *Proc. Natl. Acad. Sci. USA* **80**:3903–3907.

Challoner, P. B., Amin, A. A., Pearlman, R. E., and Blackburn, E. H., 1985, Conserved arrangements of repeated DNA sequences in nontranscribed spacers of ciliate ribosomal RNA genes: Evidence for molecular coevolution, *Nucleic Acids Res.* **13**:2661–2680.

Chan, Y.-L., Olvera, J., and Wool, I. G., 1983, The structure of rat 28S ribosomal ribonucleic acid inferred from the sequence of nucleotides in a gene, *Nucleic Acids Res.* **11**:7819–7831.

Chan, Y.-L., Gutell, R., Noller, H. F., and Wool, I. G., 1984, The nucleotide sequence of a rat 18S ribosomal ribonucleic acid gene and a proposal for the secondary structure of 18S ribosomal ribonucleic acid, *J. Biol. Chem.* **259**:224–230.

Chao, S., Sederoff, R., and Levings, C. S., 1984, Nucleotide sequence and evolution of the 18S ribosomal RNA gene in maize mitochondria, *Nucleic Acids Res.* **12**:6629–6644.

Childs, G., Maxson, R., Cohn, R. H., and Kedes, L., 1981, Orphons: Dispersed genetic elements derived from tandem repetitive genes of eukaryotes, *Cell* **23**:651–655.

Chooi, W. Y., 1979, The occurrence of long transcription units among the X and Y ribosomal genes of *Drosophila melanogaster*: Transcription of insertion sequences, *Chromosoma* **74:**57–74.

Chu, N. M., Oishi, K. K., and Tewari, K. K., 1981, Physical mapping of the pea chloroplast DNA and localization of the ribosomal RNA genes, *Plasmid* **6:**279–292.

Cihlar, R. L., and Sypherd, P. S., 1980, The organization of the ribosomal RNA genes in the fungus *Mucor racemosus*, *Nucleic Acids Res.* **8:**793–804.

Clark, C. G., and Gerbi, S. A., 1982, Ribosomal RNA evolution by fragmentation of the 23S progenitor: Maturation pathway parallels evolutionary emergence, *J. Mol. Evol.* **18:**329–336.

Clark, C. G., Tague, B. W., Ware, V. C., and Gerbi, S. A., 1984, *Xenopus laevis* 28S ribosomal RNA: A secondary structure model and its evolutionary and functional implications, *Nucleic Acids Res.* **12:**6197–6220.

Clary, D. O., and Wolstenholme, D. R., 1985, The ribosomal RNA genes of *Drosophila* mitochondrial DNA, *Nucleic Acids Res.* **13:**4029–4045.

Cockburn, A. F., Newkirk, M. J., and Firtel, R. A., 1976, Organization of the ribosomal RNA genes of *Dictyostelium discoideum*: Mapping of the nontranscribed spacer regions, *Cell* **9:**605–613.

Coen, E. S., and Dover, G. A., 1982, Multiple Pol I initiation sequences in rDNA spacers of *Drosophila melanogaster*, *Nucleic Acids Res.* **10:**7017–7026.

Coen, E. S., and Dover, G. A., 1983, Unequal exchanges and the coevolution of X and Y rDNA arrays in *Drosophila melanogaster*, *Cell* **33:**849–855.

Coen, E. S., Strachan, T., and Dover, G. A., 1982a, The dynamics of concerted evolution in the ribosomal and histone gene families in the *Drosophila melanogaster* species subgroup, *J. Mol. Biol.* **158:**17–35.

Coen, E. S., Thoday, J. M., and Dover, G. A., 1982b, The rate of turnover of structural variants in the ribosomal gene family of *Drosophila melanogaster*, *Nature* **295:**564–568.

Coen, E. S., Thoday, J. M., and Dover, G. A., 1984, Reply to Barrett, *Nature* **309:**286.

Connaughton, J. F., Kumar, A., and Lockard, R. E., 1984a, Nucleotide sequence and structure determination of rabbit 18S ribosomal RNA, in: *Eukaryotic Gene Expression* (A. Kumar, ed.), Plenum Press, New York, pp. 203–221.

Connaughton, J. F., Rairkar, A., Lockard, R. E., and Kumar, A., 1984b, Primary structure of rabbit 18S ribosomal RNA, determined by direct RNA sequence analysis *Nucleic Acids Res.* **12:**4731–4745.

Cox, R. A., and Kelly, J. M., 1982, Structural aspects of eukaryotic ribosomes, *Biochem. Soc. Symp.* **47:**11–48.

Cox, R. A., and Thompson, R. D., 1980, Distribution of sequences common to the 25–28S ribonucleic acid genes of *Xenopus laevis* and *Neurospora crassa*, *Biochem. J.* **187:**75–90.

Cox, R. A., Hüvös, P., and Godwin, E. A., 1973, Unusual features of the structure of the major RNA component of the larger subribosomal particle of the rabbit reticulocyte, *Israel J. Chem.* **11:**407–422.

Craig, N. C., 1971, On the regulation of the synthesis of ribosomal proteins in L-cells, *J. Mol. Biol.* **55:**129–134.

Craig, N., and Perry, R. P., 1971, Persistant cytoplasmic synthesis of ribosomal proteins during the selective inhibition of ribosomal RNA synthesis, *Nature New Biol.* **229:**75–80.

Crews, S., and Attardi, G., 1980, The sequences of the small ribosomal RNA gene and the phenylalanine tRNA gene are joined end to end in human mitochondrial DNA, *Cell* **19:**775–784.

Crotty, W. J., and Ledbetter, M. C., 1973, Membrane continuities involving chloroplasts and other organelles in plant cells, *Science* **182:**839–841.

Crouch, R. J., Kanaya, S., and Earl, P. L., 1983, A model for the involvement of the small nucleolar RNA (U3) in processing eukaryotic ribosomal RNA, *Mol. Biol. Rep.* **9:**75–78.

Crow, J. F., 1981, The neutralist–selectionist controversy: An overview, in: *Population and Biological Aspects of Human Mutations* (E. B. Hook, ed.), Academic Press, New York, pp. 3–14.

Cummings, D. J., Maki, R. A., Conlon, P. J., and Laping, J., 1980, Anatomy of mitochondrial DNA from *Paramecium aurelia, Mol. Gen. Genet.* **178:**499–510.

Cunningham, R. S., and Gray, M. W., 1977, Isolation and characterization of ^{32}P-labeled mitochondrial and cytosol ribosomal RNA from germinating wheat embryos, *Biochim. Biophys. Acta* **475:**476–491.

Curtis, S. E., and Rawson, J. R. Y., 1981, Characterization of the nuclear ribosomal DNA of *Euglena gracilis, Gene* **15:**237–247.

Dabbs, E. R., 1982, A spontaneous mutant of *Escherichia coli* with protein L24 lacking from the ribosome, *Mol. Gen. Genet.* **187:**453–458.

Dahlberg, A. E., Tokimatsu, H., Zahalak, M., Reynolds, F., Calvert, P. C., Rabson, A. B., Lund, E., and Dahlberg, J. E., 1978, Processing of the 5′ end of *Escherichia coli* 16S ribosomal RNA, *Proc. Natl. Acad. Sci. USA* **75:**3598–3602.

Dale, R. M. K., Mendu, N., Ginsburg, H., and Kridl, J. C., 1984, Sequence analysis of the maize mitochondrial 26S rRNA gene and flanking regions, *Plasmid* **11:**141–150.

Davies, R. W., Waring, R. B., Ray, J. A., Brown, T. A., and Scazzocchio, C., 1982, Making ends meet: A model for RNA splicing in fungal mitochondria, *Nature* **30:**719–724.

de la Cruz, V. F., Lake, J. A., Simpson, A. M., and Simpson, L., 1985a, A minimal ribosomal RNA: Sequence and secondary structure of the 9S Kinetoplast ribosomal RNA from *Leishmania tarentolae, Proc. Nat. Acad. Sci.* **82:**1401–1405.

de la Cruz, V. F., Simpson, A. M., Lake, J. A., and Simpson, L., 1985b, Primary sequence and partial secondary structure of the 12S kinetoplast (mitochondrial) ribosomal RNA from *Leishmania tarentolae*: Conservation of peptidyl transferase structural elements, *Nucleic Acids Res.* **13:**2337–2356.

DeLaSalle, H., Jacq, C., and Slonimski, P. P., 1982, Critical sequences within mitochondrial introns: Pleiotropic mRNA maturase and *cis*-dominant signals of the box intron controlling reductase and oxidase, *Cell* **28:**721–732.

Denis, H., and Wegnez, M., 1977, Biochemical research on oogenesis, oocytes and liver cells of the teleost fish *Tinca tinca* contain different kinds of 5S RNA, *Dev. Biol.* **59:**228–236.

De Vries, M., de Jonge, J. C., Bakker, H., Meurs, H., and Kroon, A., 1979, The anatomy of the tRNA–rRNA region of the *Neurospora crassa* mitochondrial DNA, *Nucleic Acids Res.* **6:**1791–1803.

Dieckmann, C. L., Pape, L. K., and Tzagoloff, A., 1982 Identification and cloning of a yeast nuclear gene (CBP1) involved in expression of mitochondrial cytochrome b, *Proc. Nat. Acad. Sci. USA* **79:**1805–1809.

Din, N., Engberg, J., Kaffenberger, W., and Eckert, W. A., 1979, The intervening sequence in the 26S rRNA coding region of *T. thermophila* is transcribed within the largest stable precursor for rRNA, *Cell* **18:**525–532.

Din, N., Engberg, J., and Gall, J. G., 1982, The nucleotide sequence at the transcription termination site of the ribosomal RNA gene in *Tetrahymena thermophila, Nucleic Acids Res.* **10:**1503–1513.

Dobell, C. C., 1911, The principles of protistology, *Arch. Protistenk.* **23:**269–310.

Douglas, S. E., and Doolittle, W. F., 1984, Complete nucleotide sequence of the 23S rDNA gene of the cyanobacterium *Anacystis nidulans, Nucleic Acids Res.*, **12:**3373–3386.

Douthwaite, S., Christensen, A., and Garrett, R. A., 1983, Higher order structure in the 3′ minor domain of small subunit ribosomal RNAs from a gram negative bacterium, a gram positive bacterium, and a eukaryote, *J. Mol. Biol.* **169:**249–279.

Dover, G., 1982, Molecular drive: A cohesive mode of species evolution, *Nature* **299:**111–117.

Dover, G. A., and Flavell, R. B., 1984, Molecular co-evolution: rDNA divergence and the maintenance of function, *Cell* **38**:622–623.

Dover, G., Brown, S., Coen, E., Dallas, J., Strachan, T., and Trick, M., 1982, The dynamics of genome evolution and species differentiation, in: *Genome Evolution* (G. A. Dover and R. B. Flavell, eds.), Academic Press, London, pp. 343–372.

Dron, M., Rahire, M., and Rochaix, J., 1982, Sequence of the chloroplast 16S rRNA gene and its surrounding regions of *Chlamydomonas reinhardtii, Nucleic Acids Res.* **10**:7609–7620.

Dudov, K. P., and Perry, R. P., 1984, The gene family encoding the mouse ribosomal protein L32 contains a uniquely expressed intron-containing gene and an unmutated processed gene, *Cell* **37**:457–468.

Dudov, K. P., Dabeva, M. D., Hadjiolov, A. A., and Todorov, B. N., 1978, Processing and migration of ribosomal ribonucleic acids in the nucleolus and nucleoplasm of rat liver nuclei, *Biochem. J.* **171**:375–383.

Duester, G. L., and Holmes, W. M., 1980, The distal end of the ribosomal RNA operon rrnD of *Escherichia coli* contains a $tRNA^{thr}$ gene, two 5S rRNA genes and a transcription terminator, *Nucleic Acids Res.* **8**:3793–3807.

Dujon, B., 1980, Sequence of the intron and flanking exons of the mitochondrial 21S rRNA gene of yeast strains having different alleles at the ω and *rib-1* loci, *Cell* **20**:185–197.

Dunn, J. J., and Studier, F. W., 1973, T7 early RNAs and *Escherichia coli* ribosomal RNAs are cut from large precursor RNAs *in vivo* by ribonuclease III, *Proc. Natl. Acad. Sci. USA* **70**:3296–3300.

Durica, D. S., and Krider, H. M., 1978, Studies on the ribosomal cistrons in interspecific *Drosophila* hybrids. II. Heterochromatic regions mediating nucleolar dominance, *Genetics* **89**:37–64.

Ebel, J. P., Branlant, C., Carbon, P., Ehresmann, B., Ehresmann, C., Krol, A., and Stiegler, P., 1983, Sequence and secondary structure conservation in ribosomal RNAs in the course of evolution, in *Nucleic Acids: The Vectors of Life* (B. Pullman and J. Jortner, eds.), D. Reidel, Durdrecht, pp. 387–401.

Edelman, G. M., and Gally, J. A., 1970, Arrangement and evolution of eukaryotic genes, in: *Neurosciences: Second Study Program* (F. O. Schmitt, ed.), Rockefeller University Press, New York, pp. 962–972.

Edwards, K., and Kössel, H., 1981, The rRNA operon from *Zea mays* chloroplasts: Nucleotide sequence of 23S rDNA and its homology with *E. coli* 23S rDNA, *Nucleic Acids Res.* **9**:2853–2869.

Edwards, K., Bedbrook, J., Dyer, T. A., and Kössel, H., 1981, 4.5S rRNA from *Zea mays* chloroplasts shows structural homology with the 3′ end of prokaryotic 23S rRNA, *Biochem Int.* **2**:533–538.

Ehresmann, C., and Ofengand, J., 1984, Two-dimensional gel electrophoresis technique for determination of the cross-linked nucleotides in cleavable covalent RNA–RNA complexes. Application to *Escherichia coli* and *Bacillus subtilis* acetylvalyl-tRNA covalently linked to *E. coli* 16S and yeast 18S ribosomal RNA, *Biochemistry* **23**:438–445.

Elion, E. A., and Warner, J. R., 1984, The major promoter element of rRNA transcription in yeast lies 2 kb upstream, *Cell* **39**:663–673.

Ellis, J., 1982, Promiscuous DNA—Chloroplast genes inside plant mitochondria, *Nature* **299**:678–679.

Endow, S. A., and Glover, D. M., 1979, Differential replication of ribosomal gene repeats in polytene nuclei of *Drosophila, Cell* **17**:597–605.

Eperon, I. C., Anderson, S., and Nierlich, D. P., 1980, Distinctive sequence of human mitochondrial ribosomal RNA genes, *Nature* **286**:460–467.

Eperon, I. C., Janssen, J. W. G., Hoeijmakers, J. H. J., and Borst, P., 1983, The major transcripts of the kinetoplast DNA of *Trypanosoma brucei* are very small ribosomal RNAs, *Nucleic Acids Res.* **11:**105–125.

Ernst, J. F., Stewart, J. W., and Sherman, F. 1981, The cycl-11 mutation in yeast reverts by recombination with a non-allelic gene: Composite genes determining the iso-cytochromes *c*, *Proc. Nat. Acad. Sci. USA* **78:**6334–6338.

Faliks, D., and Meyuhas, O., 1982, Coordinate regulation of ribosomal protein mRNA in regenerating rat liver. Study with the corresponding mouse cloned cDNAs, *Nucleic Acid Res.* **10:**789–801.

Farrelly, F., and Butow, R. A., 1983, Rearranged mitochondrial genes in the yeast nuclear genome, *Nature* **301:**296–301.

Faye, G., Kujawa, C., Dujon, B., Bolotin-Fukuhara, M., Wolf, K., Fukuhara, H. and Slonimski, P. P., 1975, Localization of the gene coding for the mitochondrial 16S ribosomal RNA using rho$^-$ mutants of *Saccharomyces cerevisiae*, *J. Mol. Biol.* **99:**203–217.

Federoff, N. V., 1979, On spacers, *Cell* **16:**697–710.

Ferguson-Miller, S., Brautigan, D. L., and Margoliash, E., 1979, The electron transfer function of cytochrome C, in: *The Porphyrins*, Vol. VII, *Biochemistry*, Part B (D. Dolphin, ed.), Academic Press, New York, pp. 149–240.

Financsek, J., Mizumoto, K., and Muramatsu, K., 1982, Human ribosomal RNA gene: Nucleotide sequence of the transcription initiation region and comparison of three mammalian genes, *Proc. Natl. Acad. Sci. USA* **79:**3092–3096.

Fincham, J. R. S., Day, P. R., and Radford, A., 1979, *Fungal Genetics*, 4th ed., Blackwell, Oxford.

Finnerty, V., 1976, Gene conversion in *Drosophila*, in: *The Genetics and Biology of Drosophila*, Vol. 1a (M. Ashburner and E. Novitski, eds.), Academic Press, London, pp. 331–347.

Fogel, S. R., Mortimer, K., and Tavares, F., 1978, Meiotic gene conversion: A signal of the basic recombination event in yeast, *Cold Spring Harbor Symp. Quant. Biol.* **43:**1325–1341.

Fogel, S., Welch, J. W., Cathala, G., and Karin, M., 1983, Gene amplification in yeast: *cuP1* copy number regulates copper resistance, *Curr. Genet.* **7:**347–355.

Ford, P. Y., and Mathieson, T., 1978, The nucleotide sequences of 5.8S ribosomal RNA from *Xenopus laevis* and *Xenopus borealis*, *Eur. J. Biochem.* **87:**199–214.

Forsheit, A. B., Davidson, N., and Brown, D. D., 1974, An electron microscope heteroduplex study of the ribosomal DNAs of *Xenopus laevis* and *Xenopus mulleri*, *J. Mol. Biol.* **90:**301–314.

Franke, W., Scheer, U., Spring, H., Trendelenburg, M., and Krohne, G., 1976, Morphology of transcriptional units of rDNA. Evidence for transcription in apparent spacer intercepts and cleavages in the elongating nascent RNA, *Exp. Cell Res.* **100:**233–244.

Free, S. J., Rice, P. W., and Metzenberg, R. L., 1979, Arrangement of the genes coding for ribosomal ribonucleic acids in *Neurospora crassa*, *J. Bacteriol.* **137:**1219–1226.

Fried, H. M., Pearson, N. J., Kim, C. H., and Warner, J. R., 1981, The genes for fifteen ribosomal proteins of *Saccharomyces cerevisiae*, *J. Biol. Chem.* **256:**10176–10183.

Furlong, J. C., and Maden, B. E. H., 1983, Patterns of major divergence between the internal transcribed spacers of ribosomal DNA in *Xenopus borealis* and *Xenopus laevis*, and of minimal divergence within ribosomal coding regions, *EMBO J.* **2:**443–448.

Furlong, J. C., Forbes, J., Robertson, M., and Maden, B. E. H., 1983, The external transcribed spacer and preceding region of *Xenopus borealis* rDNA: Comparison with the corresponding region of *Xenopus laevis* rDNA, *Nucleic Acids Res.* **11:**8183–8196.

Gall, J. G., 1968, Differential synthesis of the genes for ribosomal RNA during amphibian oogenesis, *Proc. Natl. Acad. Sci. USA* **60:**553–560.

Garilova, L. P., and Spirin, A. S., 1974, Interaction of SH-reagents with the ribosomal 30S subparticle and 'non-enzymatic' translocation, *FEBS Lett.* **39:**13–16.

Garrett, R. A., Christensen A., and Douthwaite, S., 1984a, Higher order structure in the 3′-terminal domain VI of the 23S ribosoml RNA from *Escherichia coli* and *Bacillus stearothermophilus*, *J. Mol. Biol.* **179:**689–712.

Garrett, R. A., Vester, B., Leffers, H., Sørensen, P. M., Kjems, J., Olesen, S. O., Christensen, A., Christiansen, J., and Douthwaite, S., 1984b, Mechanisms of protein–RNA recognition and assembly in ribosomes, in: *The Alfred Benzon Symposium*, No. 19 (B. Clark, and U. Pederson, eds.), Munksgaard, Copenhagen, pp. 71–89.

Garrett, R. A., Christiansen, J., Douthwaite, S., and Christensen, A., 1985, L18 and 5S RNA. in: *Structure, Function and Genetics of Ribosomes* (B. Hardesty and G. Kramer, eds.), Springer-Verlag (in press).

Gaubatz, J., and Cutler, R. G., 1975, Hybridization of ribosomal RNA labeled to high specific radioactivity with dimethyl sulfate, *Biochemistry* **14:**760–764.

Gellissen, G., Bradfield, J. Y., White, B. N., and Wyatt, G. R., 1983, Mitochondrial DNA sequences in the nuclear genome of a locust, *Nature* **301:**631–634.

Georgiev, O. I., Nikolaev, N., Hadjiolov, A. A., Skryabin, K. G., Zakharyev, V. M., and Bayev, A. A., 1981, The structure of the yeast ribosomal RNA genes. 4. Complete sequence of the 25S rRNA gene from *Saccharomyces cerevisiae, Nucleic Acids Res.* **9:**6953–6958.

Gerbi, S. A., 1976, Fine structure of ribosomal RNA I. Conservation of homologous regions within ribosomal RNA of eukaryotes, *J. Mol. Biol.* **106:**791–816.

Gerbi, S. A., Gourse, R. L., and Clark, C. G., 1982, Conserved regions within ribosomal DNA: Locations and some possible functions, in: *The Cell Nucleus*, Vol. X (H. Busch and L. Rothblum, eds.), Academic Press, New York, pp. 351–386.

Geyer, P. K., Meyuhas, O., Perry, R. P., and Johnson, L. F., 1982, Regulation of ribosomal protein messenger RNA content and translation in growth-stimualted mouse fibroblasts, *Mol. Cell. Biol.* **2:**685–693.

Ginsburg, D., and Steitz, J. A., 1975, The 30S ribosomal precursor RNA from *E. coli, J. Biol. Chem.* **250:**5647–5654.

Glätzer, K. H., 1979, Lengths of transcribed rDNA repeating units in spermatocytes of *Drosophila hydei*: Only genes without an intervening sequence are expressed, *Chromosoma* **75:**161–175.

Glotz, C., and Brimacombe, R., 1980, An experimentally-derived model for the secondary structure of the 16S RNA from *Escherichia coli, Nucleic Acids Res.* **8:**2377–2395.

Glotz, C., Zwieb, C., Brimacombe, R., Edwards, K., and Kössel, H., 1981, Secondary structure of the large subunit ribosomal RNA from *Escherichia coli, Zea mays* chloroplast, and human and mouse mitochondrial ribosomes, *Nucleic Acids Res.* **9:**3287–3306.

Glover, D. M., White, R. L., Finnegan, D. J., and Hogness, D. S., 1975, Characterization of six cloned DNAs from *Drosophila melanogaster*, including one that contains the genes for rRNA, *Cell* **5:**149–157.

Gold, L., Stormo, G., and Saunders, R., 1984, *Escherichia coli* translational initiation factor IF3: A unique case of translational regulation, *Proc. Nat. Acad. Sci. USA* **81:**7061–7065.

Goldbach, R. W., Borst, P., Bollen-DeBoer, J. E., and Van Bruggan, E. F. F., 1978, The organization of ribosomal RNA genes in the mitochondrial DNA of *Tetrahymena pyriformis* strain ST, *Biochim. Biophys. Acta* **521:**169–186.

Goldman, W. E., Goldberg, G., Bowman, L. H., Steinmetz, D., and Schlessinger, D., 1983, Mouse rDNA: Sequences and evolutionary analysis of spacer and mature RNA regions, *Mol. Cell. Biol.* **3:**1488–1500.

Gonzalez, I. L., Gorski, J. L., Campen, T. J., Dorney, D. J., Erickson, J. M., Sylvester, J. E., and Schmickel, R. D., 1985, Variation among human 28S ribosomal RNA genes (submitted).

Gorenstein, C., and Warner, J. R., 1976, Coordinate regulation of the synthesis of eukaryotic ribosomal proteins, *Proc. Natl. Acad. Sci. USA* **73**:1547–1551.

Gornicki, P., Nurse, K., and Ofengand, J., 1983, Involvement of 16–18S rRNA in the ribosomal decoding site, *Fed. Proc.* **42**:2186.

Gottesfeld, J., and Bloomer, L. S., 1982, Assembly of transcriptionally active 5S RNA gene chromatin *in vitro, Cell* **28**:781–791.

Gourse, R. L., and Gerbi, S. A., 1980a, Fine structure of ribosomal RNA III. Location of evolutionarily conserved regions within ribosomal DNA, *J. Mol. Biol.* **140**:321–339.

Gourse, R. L., and Gerbi, S. A., 1980b, Fine structure of ribosomal RNA. IV. Extraordinary evolutionary conservation in sequences that flank introns in rDNA, *Nucleic Acids Res.* **8**:3623–3637.

Gourse, R. L., Thurlow, D. L., Gerbi, S. A., and Zimmermann, R. A., 1981, Specific binding of a prokaryotic ribosomal protein to a eukaryotic ribosomal RNA: Implications for evolution and autoregulation, *Proc. Natl. Acad. Sci. USA* **78**:2722–2726.

Grabowski, P. J., Zaug, A. J., and Cech, T. R., 1981, The intervening sequence of the ribosomal RNA precursor is converted to a circular RNA in isolated nuclei of *Tetrahymena, Cell* **23**:467–476.

Graf, L., Roux, E., Stutz, E., and Kössel, H., 1982, Nucleotide sequence of a *Euglena gracilis* chloroplast gene coding for the 16S rRNA: Homologies to *E. coli* and *Zea mays* chloroplast 16S rRNA, *Nucleic Acids Res.* **10**:6369–6381.

Grant, D. M., and Lambowitz, A. M., 1982, Mitochondrial ribosomal RNA genes, in: *The Cell Nucleus: rDNA*, Vol. X, Part A (H. Busch and L. Rothblum, eds.), Academic Press, New York, pp. 387–408.

Gray, M. W., Sankoff, D., and Cedergren, R. J., 1984, On the evolutionary descent of organisms and organelles: A global phylogeny based on a highly conserved structural core in small subunit RNA, *Nucleic Acids Res.* **12**:5837–5852.

Gray, P. W., and Hallick, R. B., 1978, Physical mapping of the *Euglena gracilis* chloroplast DNA and ribosomal RNA gene region, *Biochemistry* **17**:284–289.

Green, M. R., Grimm, M. F., Goewert, R. R., Collins, R. A., Cole, M. D., Lambowitz, A. M., Heckman, J. E., Yin, S., and RajBhandary, U. L., 1981, Transcripts and processing patterns for the ribosomal RNA and transfer RNA region of *Neurospora crassa* mitochondrial DNA, *J. Biol. Chem.* **256**:2027–2034.

Grimm, M. F., and Lambowitz, A. M., 1979, The 32S RNA of *Neurospora crassa* mitochondria is not a precursor of the mitochondrial ribosomal RNAs, *J. Mol. Biol.* **134**:667–672.

Grummt, I., 1981a, Specific transcription of mouse ribosomal DNA in a cell free system that mimics control *in vivo, Proc. Natl. Acad. Sci. USA* **78**:727–731.

Grummt, I., 1981b, Mapping of a mouse ribosomal DNA promoter by *in vitro* transcription, *Nucleic Acids Res.* **9**:6093–6102.

Grummt, I., 1982, Nucleotide sequence requirements for specific initiation of transcription by RNA polymerase I, *Proc. Natl. Acad. Sci. USA* **79**:6908–6911.

Grummt, I., Roth, E., and Paule, M., 1982, Ribosomal RNA transcription *in vitro* is species specific, *Nature* **296**:173–174.

Grummt, I., Sorbaz, H., Hofmann, A., and Roth, E., 1985, Spacer sequences downstream of the 28S RNA coding region are part of the mouse rDNA transcription unit, *Nucleic Acids Res.*, **13**:2293–2304.

Guerrier-Takada, C., and Altman, S., 1984, Catalytic activity of an RNA molecule prepared by transcription *in vitro, Science* **223**:285–286.

Guerrier-Takada, C., Gardiner, K., Marsh, T., Pace, N., and Altman, S., 1983, The RNA moiety of ribonuclease P is the catalytic subunit of the enzyme, *Cell* **35**:849–857.

Gupta, R., Lanter, J. M., and Woese, C. R., 1983, Sequence of the 16S ribosomal RNA from *Halobacterium volcanii*, an archaebacterium, *Science* **221**:656–659.

Hadjiolov, A. A., Georgiev, O. I., Nosikov, V. V., and Yavachev, L. P., 1984, Primary and secondary structure of rat 28S ribosomal RNA, *Nucleic Acids Res.* **12**:3677–3693.

Hagenbüchle, O., Santer, M., Steitz, J. A., and Mans, R. J., 1978, Conservation of the primary structure at the 3′ end of 18S rRNA from eukaryotic cells, *Cell* **13**:551–563.

Hahn, U., Lazarus, C. M., Lunsdorf, H., and Küntzel, H., 1979, Split gene for mitochondrial 24S ribosomal RNA of *Neurospora crassa*, *Cell* **17**:191–200.

Hall, L. M. C., and Maden, B. E. H., 1980, Nucleotide sequence through the 18S–28S intergene region of a vertebrate ribosomal transcription unit, *Nucleic Acids Res.* **8**:5993–6005.

Hallberg, R. L., and Brown, D. D., 1969, Coordinated synthesis of some ribosomal proteins and ribosomal RNA in embryos of *Xenopus laevis*, *J. Mol. Biol.* **46**:393–411.

Harper, M. E., Price, J., and Korn, L. J., 1983, Chromosomal mapping of *Xenopus* 5S genes: Somatic-type versus oocyte-type, *Nucleic Acids Res.* **11**:2313–2323.

Hartley, M. R., 1979, The synthesis and origin of chloroplast low-molecular-weight ribosomal ribonucleic acid in spinach, *Eur. J. Biochem.* **96**:311–320.

Hartley, M. R., and Ellis, R. J., 1973, Ribonucleic acid synthesis in chloroplasts, *Biochem. J.* **134**:249–262.

Hassouna, N., Michot, B., and Bachellerie, J. P., 1984, The complete nucleotide sequence of mouse 28S rRNA gene. Implications for the process of size increase of the large subunit rRNA in higher eukaryotes, *Nucleic Acids Res.* **12**:3563–3583.

Hatlen, L., and Attardi, G., 1971, Proportion of the HeLa cell genome complementary to transfer RNA and 5S RNA, *J. Mol. Biol.* **56**:535–553.

Hayes, F., and Vasseur, M., 1976, Processing of the 17S *Escherichia coli* precursor RNA in the 27S pre-ribosomal particle, *Eur. J. Biochem.* **61**:433–442.

Heckman, J. E., and RajBhandary, U. L., 1979, Organization of tRNA and rRNA genes in *N. crassa* mitochondria: Intervening sequence in the large rRNA gene and strand distribution of the RNA genes, *Cell* **17**:583–595.

Helser, T. L., Davies, J. E., and Dahlberg, J. E., 1971, Change in methylation of 16S ribosomal RNA associated with mutation to kasugamycin resistance in *Escherichia coli*, *Nature New Biol.* **233**:12–14.

Henderson, A. S., Atwood, K. C., and Warburton, D., 1976, Chromosomal distribution of rDNA in *Pan paniscus*, *Gorilla gorilla beringei*, and *Symphalangus syndactylus*: Comparison to related primates, *Chromosoma* **59**:147–155.

Henderson, E., Oakes, M., Clark, M. W., Lake, J. A., Matheson, A. T., and Zillig, W., 1984, A new ribosome structure, *Science* **225**:510–512.

Hennig, W., and Meer, B., 1971, Reduced polyteny of ribosomal RNA cistrons in giant chromosomes of *Drosophila hydei*, *Nature New Biol.* **233**:70–72.

Herr, W., Champman, N. M., and Noller, H. F., 1979, Mechanism of ribosomal subunit association: Discrimination of specific sites in 16S RNA essential for association activity, *J. Mol. Biol.* **130**:433–449.

Hill, W., 1985, Probing ribosomal structure and function using short cDNA oligomers, in: *Structure, Function, and Genetics of Ribosomes* (B. Hardesty and G. Kramer, eds.), Springer-Verlag (in press).

Hines, P. J., and Benbow, R. M., 1982, Initiation of replication at specific origins in DNA molecules microinjected into unfertilized eggs of the frog *Xenopus laevis*, *Cell* **30**:459–468.

Hofman, J. D., Lau, R. H., and Doolittle, W. F., 1979, The number, physical organization and transcription of ribosomal RNA cistrons in an archaebacterium: *Halobacterium halobium*, *Nucleic Acids Res.* 7:1321–1333.

Hogan, J. J., Gutell, R. R., and Noller, H. F., 1984a, Probing the conformation of 18S rRNA in yeast 40S ribosomal subunits with Kethoxal, *Biochemistry* **23**:3322–3329.

Hogan, J. J., Gutell, R. R., and Noller, H. F., 1984b, Probing the conformation of 26S rRNA in Yeast 60S ribosomal subunits with kethoxal, *Biochemistry* **23**:3330–3335.

Holliday, R., 1964, A mechanism for gene conversion in fungi, *Genet. Res.* **5**:282–304.

Honda, B. M., and Roeder, R. G., 1980, Association of a 5S gene transcription factor with 5S RNA and altered levels of the factor during cell differentiation, *Cell* **22**:119–126.

Hoshikawa, Y., Iida, Y., and Iwabuchi, M., 1983, Nucleotide sequence of the transcriptional initiation region of *Dictyostetium discoideum* rRNA gene and comparison of the initiation regions of three lower eukaryotes' genes, *Nucleic Acids Res.* **11**:1725–1734.

Hudspeth, M. E. S., Ainley, W. M., Shumard, D. S., Butow, R. A., and Grossman, L. I., 1982, Location and structure of the *var1* gene on yeast mitochondrial DNA: nucleotide sequence of the 40.0 allele, *Cell* **30**:617–626.

Hui, C.-F., and Cantor, C. R., 1985, Mapping the location of psoralen cross-links on RNA by mung bean nuclease sensitivity of RNA-DNA hybrids, *Proc. Nat. Acad. Sci.* **82**:1381–1385.

Iida, C. T., Kownin, P., and Paule, M. R., 1985, Ribosomal RNA transcription: Proteins and DNA sequences involved in preinitiation complex formation, *Proc. Nat. Acad. Sci.* **82**:1668–1672.

Iwami, M., Muto, A., Yamao, F., and Osawa, S., 1984, Nucleotide sequence of the rrnB 16S ribosomal RNA gene from *Mycoplasma capricolum*, *Mol. Gen. Genet.* **196**:317–322.

Jackson, J. A., and Fink, G., 1981, Gene conversion between duplicated genetic elements in yeast, *Nature* **292**:306–307.

Jacobs, H. T., Posakony, J. W., Grula, J. W., Roberts, J. W., Xin, J.-H., Britten, R. J., and Davidson, E. H., 1982, Mitochondrial DNA sequences in the nuclear genome of *Strongylocentrotus purpuratus*, *J. Mol. Biol.* **165**:609–632.

Jacq, B., 1981, Sequence homologies between eukaryotic 5.8S rRNA and the 5′ end of prokaryotic 23S rRNA: Evidences for a common evolutionary origin, *Nucleic Acids Res.* **9**:2913–2932.

Jarsch, M., and Böck, A., 1983, DNA sequence of the 16S rRNA/23S rRNA intercistronic spacer of two rDNA operons of the archaebacterium *Methanococcus vannielii*, *Nucleic Acids Res.* **11**:7537–7544.

Jarsch, M., Altenbuchner, J., and Böck, A., 1983, Physical organization of the genes for ribosomal RNA in *Methanococcus vannielii*, *Mol. Gen. Genet.* **189**:41–47.

Jenni, B., and Stutz, E., 1978, Physical mapping of the ribosomal RNA gene region of *Euglena gracilis* chloroplast DNA, *Eur. J. Biochem.* **88**:127–134.

Jenni, B., and Stutz, E., 1979, Analysis of *Euglena gracilis* chloroplast DNA, mapping of a sequence complementary to 16S rRNA outside of the rRNA gene sets, *FEBS Lett.* **102**:95–99.

Jolly, D. J., and Thomas, C. A., 1980, Nuclear RNA transcripts from *Drosophila melanogaster* ribosomal RNA genes containing introns, *Nucleic Acids Res.* **8**:67–84.

Jordan, B. R., Latil-Damotte, M., and Jourdan, R., 1980, Coding and spacer sequences in the 5.8S–2S region of *Sciara coprophila* ribosomal DNA, *Nucleic Acids Res.* **8**:3565–3573.

Kearsay, S. E., and Craig, I. W., 1981, Altered ribosomal RNA genes in mitochondria from mammalian cells with chloramphenicol resistance, *Nature* **290**:607–608.

Keil, R. L., and Roeder, G. S., 1984, *Cis*-acting recombination-stimulating activity in a fragment of the ribosomal DNA of *Saccharomyces cerevisiae*, *Cell* **39**:377–386.

Kelly, J. M., and Cox, R. A., 1981, The nucleotide sequence at the 3′-end of *Neurospora crassa* 25S-rRNA and the location of a 5.8S-rRNA binding site, *Nucleic Acids Res.* **9**:1111–1121.

Kemble, R. J., Mans, R. J., Gabay-Laughnan, S., and Laughnan, J. R., 1983, Sequences homologous to episomal mitochondrial DNAs in the maize nuclear genome, *Nature* **304**:744–747.

Keren-Zur, M., Boublik, M., and Ofengand, J., 1979, Localization of the decoding region on the 30S *Escherichia coli* ribosomal subunit by affinity immunoelectron microscopy, *Proc. Natl. Acad. Sci. USA* **76**:1054–1058.

Khan, M. S. N., and Maden, B. E. H., 1977, Nucleotide sequence relationships between vertebrate 5.8S ribosomal RNAs, *Nucleic Acids Res.* **4**:2495–2505.

Khan, M. S. N., Salim, M., and Maden, B. E. H., 1978, Extensive homologies between the methylated nucleotide sequences in several vertebrate ribosomal ribonucleic acids, *Biochem. J.* **169**:531–542.

Khoury, G., and Gruss, P., 1983, Enhancer elements, *Cell* **33**:313–314.

Kidd, S. J., and Glover, D. M., 1981, *Drosophila melanogaster* ribosomal DNA containing type II insertions is variably transcribed in different strains and tissues, *J. Mol. Biol.* **151**:645–662.

Kimura, M., 1983, The neutral theory of molecular evolution, in: *Evolution of Genes and Proteins* (M. Nei and R. K. Koehn, eds.), Sinauer, Sunderland, Massachusetts, pp. 208–233.

Kishimoto, T., Nagamine, M., Sasaki, T., Takakusa, N., Miwa, T., Kominami, R. and Muramatsu, M., 1985, Presence of a limited number of essential nucleotides in the promoter region of mouse ribosomal RNA gene, *Nucleic Acids Res.* **13**:3515–3532.

Kiss, A., Sain, B., and Venetianer, P., 1977, The number of rRNA genes in *Escherichia coli, FEBS Lett.* **79**:77–79.

Kiss, G. B., Amin, A. A., and Pearlman, R. E., 1981, Two separate regions of the extra-chromosomal ribosomal deoxyribonucleic acid of *Tetrahymena thermophila* enable autonomous replication of plasmids in *Saccharomyces cerevisiae, Mol. Cell. Biol.* **1**:535–543.

Klar, A. J. S., Stathern, J. N., and Hicks, J. B., 1981, A position-effect control for gene transposition: State of expression of yeast mating-type genes affects their ability to switch, *Cell* **25**:517–524.

Klein, H. L., and Petes, T., 1981, Intrachromosomal gene conversion in yeast: A new type of genetic exchange, *Nature* **289**:144–148.

Klein, B. K., King, T. C., and Schlessinger, D., 1983, Structure of partially denatured *Escherichia coli* 23S ribosomal RNA determined by electron microscopy, *J. Mol. Biol.* **168**;809–830.

Klein, B. K., Forman, P., Shiomi, Y., and Schlessinger, D., 1984, Electron microscopy of secondary structure in partially denatured *Escherichia coli* 16S-rRNA and 30S subunits, *Biochemistry* **23**:3927–3933.

Klemenz, R., and Geiduschek, E. P., 1980, The 5′ terminus of the precursor ribosomal RNA of *Saccharomyces cerevisiae, Nucleic Acids Res.* **8**:2679–2689.

Kobayashi, M., Yaginuma, K., Seki, T., and Koike, K., 1980, Nucleotide sequences of the cloned *Eco*A fragment of rat mitochondrial DNA, in: *The Organization and Expression of the Mitochondrial Genome* (A. M. Kroon and C. Saccone, eds.), Elsevier, Amsterdam, pp. 221–229.

Kobayashi, M., Seki, T., Yaginuma, K., and Koike, K., 1981, Nucleotide sequences of small ribosomal RNA and adjacent transfer RNA genes in rat mitochondrial DNA, *Gene* **16**:297–307.

Koch, A. L., 1978, Selection and recombination in populations containing tandem multiplicate genes, *Genetics* **88**:s53–s54.

Koch, W., Edwards, K., and Kössel, H., 1981, Sequencing of the 16S–23S spacer in a ribosomal RNA operon of *Zea mays* chloroplast DNA reveals two split tRNA genes, *Cell* **25**:203–213.

Köchel, H. G., and Küntzel, H., 1981, Nucleotide sequence of the *Aspergillus nidulans* mitochondrial gene coding for the small ribosomal subunit RNA: Homology to *E. coli* 16S rRNA, *Nucleic Acids Res*. **9**:5689–5696.

Kohorn, B. D., and Rae, P. M. M., 1982a, Accurate transcription of truncated ribosomal DNA templates in a *Drosophila* cell-free system, *Proc. Nat. Acad. Sci. USA* **79**:1501–1505.

Kohorn, B. D., and Rae, P. M. M., 1982b, Nontranscribed spacer sequences promote *in vitro* transcription of *Drosophila* ribosomal DNA, *Nucleic Acids Res*. **10**:6879–6886.

Kohorn, B. D., and Rae, P. M. M., 1983a, Localization of DNA sequences promoting RNA polymerase I activity in *Drosophila, Proc. Natl. Acad. Sci. USA* **80**:3265–3268.

Kohorn, B. D., and Rae, P. M. M., 1983b, A component of *Drosophila* RNA polymerase I promoter lies within the rRNA transcription unit, *Nature* **304**:179–181.

Koller, B., and Delius, H., 1980, *Vicia faba* chloroplast DNA has only one set of ribosomal RNA genes as shown by partial denaturation mapping and R-loop analysis, *Mol. Gen. Genet*. **178**:261–269.

Kolodner, R., and Tewari, K. K., 1979, Inverted repeats in chloroplast DNA from higher plants, *Proc. Natl. Acad. Sci. USA* **76**:41–45.

Kominami, R., Mishima, Y., Urano, Y., Sakai, M., and Muramatsu, M., 1982, Cloning and determination of the transcription termination site of ribosomal RNA gene of the mouse, *Nucleic Acids Res*. **10**:1963–1979.

Kop, J., Kopylov, A. M., Magrum, L., Siegel, R., Gupta, R., Woese, C. R. and Noller, H. F., 1984a, Probing the structure of 16S ribosomal RNA from *Bacillus brevis, J. Biol. Chem*. **259**:15287–15293.

Kop, J., Wheaton, V., Gupta, R., Woese, C. R., and Noller, H. F., 1984b, Complete nucleotide sequence of a 23S ribosomal RNA gene from *Bacillus stearothermophilus, DNA* **3**:347–357.

Korn, L. M., and Brown, D. D., 1978, Nucleotide sequence of *Xenopus borealis* oocyte 5S DNA: Comparison of sequences that flank several related eukaryotic genes, *Cell* **15**:1145–1156.

Kössel, H., Edwards, K., Koch, W., Langridge, P., Schiefermayr, E., Schwarz, Zs., Strittmatter, G. and Zenke, G., 1982, Structural and functional analysis of an rRNA operon and its flanking tRNA genes from *Zea mays* chloroplasts, *Nucleic Acids Res. Symp. Ser*. **11**:117–120.

Kössel, H., Edwards, K., Fritzsche, E., Koch, W., and Schwarz, Zs., 1983, Phylogenetic significance of nucleotide sequence analysis, in: *Protein and Nucleic Acids in Plant Systematics* (U. Jensen and D. E. Fairbrothers, eds.), Springer-Verlag, Berlin, pp. 36–57.

Kozak, M., 1978, How do eukaryotic ribosomes select initiation regions in mRNA?, *Cell* **15**:1109–1123.

Kozak, M., 1980, Evaluation of the "scanning model" for initiation of protein synthesis in eukaryotes, *Cell* **22**:7–8.

Kozak, M., 1983, Comparison of initiation of protein synthesis in procaryotes, eucaryotes, and organelles, *Microbiol. Rev*. **47**:1–45.

Kramer, R., Phillippsen, P., and Davis, R. W., 1978, Divergent transcription in the yeast ribosomal RNA coding region as shown by hybridization to separated strands and sequence analysis of cloned DNA, *J. Mol. Biol*. **123**:405–416.

Kruger, K., Grabowski, P. J., Zaug, A. J., Sands, J., Gottschling, D. E., and Cech, T. R., 1982, Self-splicing RNA: Autoexcision and autocyclization of the ribosomal RNA intervening sequence of *Tetrahymena, Cell* **31**:147–157.

Krystal, M., D'Eustachio, P., Ruddle, F. H., and Arnheim, N., 1981, Human nucleolus organizers on non-homologous chromosomes can share the same ribosomal gene variants, *Proc. Natl. Acad. Sci. USA* **78**:5744–5748.

Kumano, M., Tomioka, N., and Sugiura, M., 1983, The complete nucleotide sequence of a 23S rRNA gene from a blue-green alga, *Anacystis nidulans, Gene* **24**:219–225.

Küntzel, H., and Köchel, H. G., 1981, Evolution of rRNA and origin of mitochondria, *Nature* **293**:751–755.

Kunz, W., Grimm, C., and Franz, G., 1982, Amplification and synthesis of rDNA: *Drosophila* in: *The Cell Nucleus: rDNA*, Vol. XII, Part C (H. Busch and L. Rothblum, eds.), Academic Press, New York, pp. 155–184.

Labhart, P., and Reeder, R. H., 1984, Enhancer-like properties of the 60/81 bp elements in the ribosomal gene spacer of *Xenopus laevis, Cell* **37**:285–289.

Lake, J. A., Henderson, E., Clark, M. W., and Matheson, A. T., 1982, Mapping evolution with ribosome structure: Intralineage constancy and interlineage variation, *Proc. Natl. Acad. Sci. USA* **79**:5948–5952.

Lake, J. A., Henderson, E., Oakes, M., and Clark, M. W., 1984, Eocytes: A new ribosome structure indicates a kingdom with a close relationship to eukaryotes, *Proc. Natl. Acad. Sci. USA* **81**:3786–3790.

Lamb, B. C., and Helmi, S., 1982, The extent to which gene conversion can change allele frequencies in populations, *Genet Res.* **39**:199–217.

Larionov, V. L., and Shubochkina, E. A., 1982, Construction of hybrid plasmids containing yeast replicator, *Molek. Biol.* **16**:948–955.

LaVolpe, A., Taggart, M., Macleod, D., and Bird, A., 1982, Coupled demethylation of sites in a conserved sequence of *Xenopus* ribosomal DNA, *Cold Spring Harbor Symp. Quant. Biol.* **47**:585–592.

LaVolpe, A., Taggart, M., McStay, B., and Bird, A., 1983, DNAase I-hypersensitive sites at promoter-like sequences in the spacer of *Xenopus laevis* and *Xenopus borealis* ribosomal DNA, *Nucleic Acids Res.* **11**:5361–5380.

Lazarus, C. M., Lünsdorf, H., Hahn, U., Stepien, P. P., and Küntzel, H., 1980, Physical map of *Aspergillus nidulans* mitochondrial genes coding for ribosomal RNA: An intervening sequence in the large rRNA cistron, *Mol. Gen. Genet.* **177**:389–397.

Lazowska, J., Jacq, C., and Slonimski, P. P., 1980, Sequence of introns and flanking exons in wild-type and box3 mutants of cytochrome *b* reveals an interlaced splicing protein coded by an intron, *Cell* **22**:333–348.

Learned, R. M., and Tjian, 1982, *In vitro* transcription of human ribosomal RNA genes by RNA polymerase I, *J. Mol. Appl. Genet.* **1**:575–584.

Learned, R. M., Smale, S. T., Haltiner, M. M., and Tjian, R., 1983, Regulation of human ribosomal RNA transcription, *Proc. Natl. Acad. Sci. USA* **80**:3558–3562.

Leaver, C. J., and Gray, M. W., 1982, Mitochondrial genome organization and expression in higher plants, *Annu. Rev. Plant Physiol.* **33**:373–402.

Leaver, C. J., and Harmey, M. A., 1976, Higher-plant mitochondrial ribosomes contain a 5S ribosomal ribonucleic acid component, *Biochem. J.* **157**:275–277.

Leer, J. C., Tiryaki, D., and Westergaard, O., 1979, Termination of transcription in nucleoli isolated from *Tetrahymena, Proc. Natl. Acad. Sci. USA*, **76**:5563–5566.

Leffers, H., and Garrett, R. A., 1984, The nucleotide sequence of the 16S ribosomal RNA gene of the archaebacterium *Halococcus morrhua, EMBO J.* **3**:1613–1619.

Lerner, M. R., Boyle, J. A., Mount, S. M., Wolin, S. L., and Steitz, J. A., 1980, Are snRNPs involved in splicing?, *Nature* **283**:220–224.

Lewin, R., 1983, Promiscuous DNA leaps all barriers, *Science* **219**:478–479.

Li, M., Tzagoloff, A., Underbrink-Lyon, K., and Martin, N. C., 1982, Identification of the paromomycin-resistance mutation in the 15S rRNA gene of yeast mitochondria, *J. Biol. Chem.* **257**:5921–5928.

Lipps, H. J., and Steinbruck, G., 1982, Free genes for rRNAs in the macronuclear genome of the ciliate *Stylonychia mytilis*, *Chromosoma* **69**:21–26.

Lizardi, P. M., and Luck, D. J. L., 1971, Absence of 5S RNA component in the mitochondrial ribosomes of *Neurospora crassa*, *Nature New Biol.* **229**: 140–142.

Lockington, R. A., Taylor, G. G., Winther, M., Scazzocchio, C., and Davies, R. W., 1982, A physical map of the ribosomal DNA repeat unti of *Aspergillus nidulans*, *Gene* **20**:135–137.

Long, E. O., and Dawid, I. B., 1979a, Expression of ribosomal DNA insertions in *Drosophila melanogaster*, *Cell* **18**:1185–1196.

Long, E. O., and Dawid, I. B., 1979b, Restriction analysis of spacers in ribosomal DNA of *Drosophila melanogaster*, *Nucleic Acids Res.* **7**:205–215.

Long, E. O., and Dawid, I. B., 1980a, Alternative pathways in the processing of ribosomal RNA precursor in *Drosophila melanogaster*, *J. Mol. Biol.* **138**:873–878.

Long, E. O., and Dawid, I. B., 1980b, Repeated genes in eukaryotes, *Annu. Rev. Biochem.* **49**:727–764.

Long, E. O., Rebbert, M. L., and Dawid, I. B., 1980, Structure and expression of ribosomal RNA genes of *Drosophila melanogaster* interrupted by type-2 insertions, *Cold Spring Harbor Symp. Quant. Biol.* **45**(part 2):667–672.

Long, E. O., Rebbert, M. L., and Dawid, I. B., 1981, Nucleotide sequence of the initiation site for ribosomal RNA transcription in *Drosophila melanogaster:* Comparison of genes with and without insertions, *Proc. Natl. Acad. Sci.* **78**:1513–1517.

Loughney, K., Lund, E., and Dahlberg, J. E., 1982, tRNA genes are found between the 16S and 23S rRNA genes in *Bacillus subtilis*, *Nucleic Acids Res.* **10**:1607–1624.

Machatt, M. A., Ebel, J. P., and Branlant, C., 1981, The 3′-terminal region of bacterial 23S ribosomal RNA: Structure and homology with the 3′-terminal region of eukaryotic 28S rRNA and with chloroplast 4.5S rRNA, *Nucleic Acids Res.* **9**:1533–1549.

MacKay, R. M., 1981, The origin of plant chloroplast 4.5S ribosomal RNA, *FEBS Lett.* **123**:17–18.

Maden, B. E. H., 1980, Methylation map of *Xenopus laevis* ribosomal RNA, *Nature* **288**:293–296.

Maden, B. E. H., and Reeder, R. H., 1979, Partial mapping of methylated sequences in *Xenopus laevis* ribosomal RNA by preparative hybridization to cloned fragments of ribosomal DNA, *Nucleic Acids Res.* **6**:817–830.

Maden, B. E. H., Forbes, J. M., Stewart, M. A., and Eason, R., 1982a, 18S coding sequences in amplified ribosomal DNA from *Xenopus laevis* oocytes are highly homogeneous, unmethylated, and lack major open reading frames, *EMBO J.* **1**:597–601.

Maden, B. E. H., Moss, M., and Salim, M., 1982b, Nucleotide sequence of an external transcribed spacer in *Xenopus laevis* rDNA: Sequences flanking the 5′ and 3′ ends of 18S rRNA are non-complementary, *Nucleic Acids Res.* **10**:2387–2398.

Maizels, N., 1976, *Dictyostelium* 17S, 25S and 5S rDNAs lie within a 38,000 base pair repeated unit, *Cell* **9**:431–438.

Maly, R., and Brimacombe, R., 1982, Refined secondary structure models for the 16S and 23S ribosomal RNA of *Escherichia coli*, *Nucleic Acids Res.* **11**:7263–7286.

Mandel, R. K., and Dawid, I. B., 1981, The nucleotide sequence at the transcription termination site of ribosomal RNA in *Drosophila melanogaster*, *Nucleic Acids Res.* **9**:1801–1811.

Mankin, A. S., and Kopylov, A. M., 1981, A secondary structure model for mitochondrial 12S rRNA: An example of economy in rRNA structure, *Biochem. Int.* **3**:587–593.

Mankin, A. S., Kopylov, A. M., and Bogdanov, A. A., 1981a, Modification of 18S rRNA in the 40S ribosomal subunit of yeast with dimethyl sulfate, *FEBS Lett.* **134**:11–14.

Mankin, A. C., Kopylov, A. M., Rubtsov, P. M., and Skryabin, K. G., 1981b, 18S rRNA of eucaryotic ribosomes: The model of the secondary structure, *Dokl. Akad. Nauk USSR* **256**:1006–1010.

Mannella, C. A., Collins, R. A., Green, M. R., and Lambowitz, A. M., 1979, Defective splicing of mitochondrial rRNA in cytochrome-deficient nuclear mutants of *Neurospora crassa*, *Proc. Natl. Acad. Sci. USA* **76**:2635–2639.

Mao, J., Appel, B., Schaack, J., Sharp, S., Yamada, H., and Söll, D., 1982, The 5S RNA genes of *Schizosaccharomyces pombe*, *Nucleic Acids Res.* **10**:487–500.

Margulis, L., 1981, *Symbiosis in Cell Evolution; Life and Its Environment on the Early Earth*, Freeman, San Francisco.

Martin, R., Bordonne, R., and Dirheimer, G., 1982, The paromomycin region in the yeast mitochondrial genome, in: *Proceedings of the Special FEBS Meeting on Cell Function and Differentiation*, Vol. 102B, Alan R. Liss, New York, pp. 355–365.

Maxam, A. M., Tizard, R., Skryabin, K. G., and Gilbert, W., 1977, Promoter region for yeast 5S ribosomal RNA, *Nature* **267**:643–645.

McCarroll, R., Olsen, G. J., Stahl, Y. D., Woese, C. R., and Sogin, M. L., 1983, Nucleotide sequence of the *Dictyostelium discoideum* small-subunit ribosomal ribonucleic acid inferred from the gene sequence: Evolutionary implications, *Biochemistry* **22**:5858–5868.

McClain, W. H., 1977, Seven terminal steps in a biosynthetic pathway leading from DNA to transfer RNA, *Acc. Chem. Res.* **10**:418–425.

McKnight, S. L., and Miller, O. L., Jr., 1975, Transcription and replication during early blastoderm formation in *Drosophila melanogaster*, *J. Cell. Biol.* **67**:276a.

McKnight, S. L., Bustin, M., and Miller, O. L., 1977, Electron microscopic analysis of chromosome metabolism in the *Drosophila melanogaster* embryo, *Cold Spring Harbor Symp. Quant. Biol.* **42**:741–754.

Messing, J., Carlson, J., Hagen, G., Rubenstein, I. R., and Oleson, A., 1984, Cloning and sequencing of the ribosomal RNA genes in maize: The 17S region, *DNA* **3**:31–40.

Meyerink, J. H., and Retèl, J., 1976, Topographical analysis of yeast ribosomal DNA by cleavage with restriction endonucleases. II. The physical map of *Eco* RI fragments, *Nucleic Acids Res.* **3**:2697–2707.

Michel, F., and Dujon, B., 1983, Conservation of RNA secondary structures in two intron families including mitochondrial-, chloroplast- and nuclear-encoded members, *EMBO J.* **2**:33–38.

Michel, F., Jacquier, A., and Dujon, B., 1982, Comparison of fungal mitochondrial introns reveals extensive homologies in RNA secondary structure, *Biochimie* **64**:867–881.

Michot, B., Hassouna, N., and Bachellerie, J.-P., 1984, Secondary structure of mouse 28S rRNA and general model for the folding of the large rRNA in eukaryotes, *Nucleic Acids Res.* **12**:4259–4279.

Miesfeld, R., and Arnheim, N., 1982, Identification of the *in vivo* and *in vitro* origin of transcription in human rDNA, *Nucleic Acids Res.* **10**:3933–3949.

Miesfeld, R., and Arhnheim, N., 1984, Species-specific rDNA transcription is due to promoter-specific binding factors, *Mol. Cell Biol.* **4**:221–227.

Miesfeld, R., Sollner-Webb, B., Croce, C., and Arnheim, N., 1984, The absence of a human-specific rDNA transcription factor leads to nucleolar dominance in mouse>human hybrid cells, *Mol. Cell Biol.* **4**:1306–1312.

Mikus, M. D., and Petes, T. D., 1982, Recombination between genes located on nonhomologous chromosomes in the yeast *Saccharomyces cerevisiae*, *Genetics* **101**:369–404.

Miller, J. R., Hayward, D. C., and Glover, D. M., 1983, Transcription of the 'non-transcribed' spacer of *Drosophila melanogaster* rDNA, *Nucleic Acids Res.* **11**:11–19.

Miller, K., and Sollner-Webb, B., 1981, Transcription of mouse ribosomal RNA genes by RNA polymerase I: *In vitro* and *in vivo* initiation and processing sites, *Cell* **27**:165–174.

Miller, K. G., and Sollner-Webb, B., 1982, Transcription of mouse ribosomal RNA genes, in: *The Cell Nucleus: rDNA*, Vol. XII, Part C (H. Busch and L. Rothblum, eds.), Academic Press, New York, pp. 69–100.

Miller, L., 1973, Control of 5S RNA synthesis during early development of anucleolate and partial nucleolate mutants of *Xenopus laevis*, *J. Cell Biol.* **59**:624–632.

Miller, L., 1974, Metabolism of 5S RNA in the absence of ribosome production, *Cell* **3**:275–281.

Mishima, Y., Kominami, R., Honjo, T., and Muramatsu, M., 1980, Cloning and determination of a putative promoter region of a mouse ribosomal deoxyribonucleic acid fragment, *Biochemistry* **19**:3780–3786.

Mishima, Y., Yamamoto, O., Kominami, R., and Muramatsu, M., 1981, *In vitro* transcription of a cloned mouse ribosomal RNA gene, *Nucleic Acids Res.* **9**:6773–6785.

Mishima, Y., Financzek, I., Kominami, R., and Muramatsu, M., 1982a, Fractionation and reconstitution of factors required for accurate transcription of mammalian ribosomal RNA genes: Identification of a species-dependent factor, *Nucleic Acids Res.* **10**:3933–3949.

Mishima, Y., Finansek, I., Kominami, R., and Muramatsu, M., 1982b, Fractionation and reconstitution of factors required for accurate transcription of mammalian ribosomal RNA genes: Identification of a species dependent initiation factor, *Nucleic Acids Res.* **10**:6659–6670.

Montandon, P. E. and Stutz, E., 1984, The genes for the ribosomal proteins S12 and S7 are clustered with the gene for the EF-Tu protein on the chloroplast genome of *Euglena gracilis*, *Nucleic Acids Res.* **12**:2851–2859.

Morgan, E. A., 1982, Ribosomal RNA genes in *Escherichia coli*, in: *The Cell Nucleus: rDNA*, Vol. X, Part A (H. Busch and L. Rothblum, eds.), Academic Press, New York, pp. 1–29.

Morgan, G. T., Bakken, A. H., and Reeder, R. H., 1983, Transcription in cloned spacers of *Xenopus laevis* rDNA, *Proc. Natl. Acad. Sci. USA* **80**:6490–6494.

Morgan, G. T., Roan, J. G., Bakken, A. H., and Reeder, R. H., 1984, Variations in transcriptional activity of rDNA spacer promoters, *Nucleic Acids Res.* **12**:6043–6052.

Morrow, J. F., Cohen, S. N., Chang, A. C. Y., Boyer, H. W., Goodman, H. M., and Hellings, R. B., 1974, Replication and transcription of eukaryotic DNA in *Escherichia coli*, *Proc. Natl. Acad. Sci. USA* **71**:1743–1747.

Moss, T., 1982, Transcription of cloned *Xenopus laevis* ribosomal DNA microinjected into *Xenopus* oocytes, and the identification of an RNA polymerase I promoter, *Cell* **30**:835–842.

Moss, T., 1983a, A transcriptional function for the repetitive ribosomal spacer in *Xenopus laevis*, *Nature* **302**:223–228.

Moss, T., 1983b, A transcriptional function for repetitive ribosomal spacers in *Xenopus*? Reply, *Nature* **304**:562.

Moss, T., and Birnstiel, M. L., 1979, The putative promoter of a *Xenopus laevis* ribosomal gene is reduplicated, *Nucleic Acids Res.* **6**:3733–3743.

Moss, T., Boseley, P. G., and Birnstiel, M. L., 1980, More ribosomal spacer sequences from *Xenopus laevis*, *Nucleic Acids Res.* **8**:467–485.

Munz, P., and Leopold, U., 1981, Heterologous recombination between redundant tRNA genes in *Schizosaccharomyces pombe*, in: *Molecular Genetics of Yeast* (D. von Wettstein, A. Stenderup, M. Kielland-Brandt, and J. Friis, eds.), Munksgaard, Copenhagen, pp. 264–275.

Murtif, V. L., and Rae, P. M. M., 1983, A transcriptional function for repetitive ribosomal spacers in *Xenopus?*, *Nature* **304**:561–562.

Murtif, V. L., and Rae, P. M. M., 1985, *In vivo* transcription of rDNA Spacers in *Drosophila*, *Nucleic Acids Res.* **13**:3221–3239.

Nagylaki, T., 1984, Evolution of multigene families under interchromosomal gene conversion, *Proc. Natl. Acad. Sci. USA* **81**:3796–3800.

Nagylaki, T., and Petes, T. D., 1982, Intrachromosomal gene conversion and the maintenance of sequence homogeneity among repeated genes, *Genetics* **100**:315–337.

Nakashima, K., Darzynkiewicz, E., and Shatkin, A. J., 1980, Proximity of mRNA 5′-region and 18S rRNA in eukaryotic initiation complexes, *Nature* **286**:226–230.

Nath, K., and Bollon, A. P., 1977, Organization of the yeast ribosomal RNA gene cluster via cloning and restriction analysis, *J. Biol. Chem.* **252**:6562–6471.

Nazar, R. N., 1980, A 5.8S rRNA-like sequence in prokaryotic 23S rRNA, *FEBS Lett* **119**:212–214.

Nazar, R. N., and Sitz, T. O., 1980, Role of the 5′-terminal sequence in the RNA binding site of yeast 5.8S rRNA, *FEBS Lett* **115**:71–76.

Nazarea, A., Bloch, D. P., and Semrau, A., 1985, Detection of a principle modular format in tRNAs and rRNAs: Second order spectral analysis, *Proc. Nat. Acad. Sci.* (in press).

Nelles, L., Fang, B-L., Volckaert, G., Vandenberghe, A., and De Wachter, R., 1984, Nucleotide sequence of a crustacean 18S ribosomal RNA gene and secondary structure of eukaryotic small subunit ribosomal RNAs, *Nucleic Acids Res.* **12**:8749–8768.

Netter, P., Jacq, C., Carighani, G., and Slonimski, P. P., 1982, Critical sequences within mitochondrial introns: *Cis*-dominant mutations of the "cytochrome-*b*-like" intron of the oxidase gene, *Cell* **28**:733–738.

Netzker, R., Köchel, H., Basak, N., and Küntzel, H., 1982, Nucleotide sequence of *Aspergillus nidulans* mitochondrial genes coding for ATPase subunit 6, cytochrome oxidase subunit 3, seven unidentified proteins, four tRNAs and L-rRNA, *Nucleic Acids Res.* **10**:4783–4801.

Nierlich, D. P., 1982, Fragmentary 5S rRNA gene in the human mitochondrial genome, *Mol. Cell Biol.* **2**:207–209.

Nijhawan, P., and Marzluff, W. F., 1979, Metabolism of low molecular weight ribonucleic acid in early sea urchin embryos, *Biochemistry* **18**:1353–1360.

Nikolaev, N., Silengo, L., and Schlessinger, D., 1973, Synthesis of a large precursor to ribosomal RNA in a mutant of *Escherichia coli*, *Proc. Nat. Acad. Sci. USA* **70**:3361–3365.

Nikolaev, N., Schlessinger, D., and Wellauer, P. K., 1974, 30S pre-ribosomal RNA of *Escherichia coli* and products of cleavage by ribonuclease III: Length and molecular weight, *J. Mol. Biol.* **86**:741–747.

Nikolaev, N., Birenbaum, M., and Schlessinger, D., 1975, 30S pre-ribosomal RNA of *Escherichia coli*: Primary and secondary processing. *Biochim. Biophys. Acta* **395**:478–489.

Noller, H. F., 1984, Structure of ribosomal RNA, *Annu. Rev. Biochem.* **53**:119–162.

Noller, H. F., and Woese, C. R., 1981, Secondary structure of 16S ribosomal RNA. *Science* **212**:403–411.

Noller, H. F., Kop, J., Wheaton, V., Brosius, J., Gutell, R. R., Kopylov, A. M., Dohme, F., Herr, W., Stahl, D. A., Gupta, R., and Woese, C. R., 1981, Secondary structure model for 23S ribosomal RNA, *Nucleic Acids Res.* **9**:6167–6189.

Noller, H. F., Asire, M., Douthwaite, S., Gutell, R., Moazed, D., Normanly, J., Stern, S., Triman, K., Turner, S., Van Stolk, B., Weiser, B., Wheaton, V., and Woese, C. R., 1985, Probing the structure and function of ribosomal RNA by chemical modification, *in vitro* genetics and computer analysis, in: *Structure, Function, and Genetics of Ribosomes* (B. Hardesty and G. Kramer, eds.), Springer-Verlag (in press).

Nomura, M., and Held, W. A., 1974, Reconstitution of ribosomes: Studies of ribosome structure, function and assembly, in: *Ribosomes* (M. Nomura, A. Tissières, and P. Lengyel, eds.), Cold Spring Harbor Laboratory, Cold Spring Harbor, New York, pp. 193–223.

Nomura, M., Yates, J. L., Dean, D., and Post, L., 1980, Feedback regulation of ribosomal protein gene expression in *Escherichia coli*: Structural homology between ribosomal RNA and ribosomal protein mRNA, *Proc. Natl. Acad. Sci. USA* **77**:7084–7088.

Nomura, M., Gourse, R., and Baughman, G., 1984, Regulation of the synthesis of ribosomes and ribosomal components, *Annu. Rev. Biochem.* **53**:75–117.

Nowotny, V., and Nierhaus, K. H., 1982, Initiator proteins for the assembly of the 50S subunit from *Escherichia coli* ribosomes, *Proc. Natl. Acad. Sci. USA* **79**:7238–7242.

Nowotny, V., Herold, M., and Nierhaus, K. H., 1985, Initiation phase of the *in vitro* assembly of *E. coli* ribosomes, in: *Structure, Function, and Genetics of Ribosomes* (B. Hardesty and G. Kramer, eds.), Springer-Verlag (in press).

Nygard, O., and Westermann, P., 1982, Specific interaction of one subunit of eukaryotic initiation factor eIF-3 with 18S ribosomal RNA within the binary complex, eIF-3 small ribosomal subunit, as shown by cross-linking experiments, *Nucleic Acids Res.* **10**:1327–1334.

O'Brien, T. W., and Matthews, D. E., 1976, Mitochondrial ribosomes, *Handb. Genet.* **5**:535–580.

Ofengand, J., Gornicki, P., Chakraburtty, K., and Nurse, K., 1982, Functional conservation near the 3′ end of eukaryotic small subunit RNA: Photochemical crosslinking of P site-bound acetylvalyl-tRNA to 18S RNA of yeast ribosomes, *Proc. Natl. Acad. Sci. USA* **79**:2817–2821.

Ofengand, J., Ciesiolka, J., Denman, R., and Nurse, K., 1985, Structure of the tRNA-ribosome complex as revealed by photoaffinity labelling, in: *Structure, Function, and Genetics of Ribosomes* (B. Hardesty and G. Kramer, eds.), Springer-Verlag (in press).

Ohta, T., 1977, On the gene conversion model as a mechanism for maintenance of homogeneity in systems with multiple genomes, *Genet. Res.* **30**:89–91.

Ohta, T., 1980, *Evolution and Variation of Multigene Families*, Springer-Verlag, Berlin.

Ohta, T., and Dover, G. A., 1983, Population genetics of multigene families that are dispersed into two or more chromosomes, *Proc. Natl. Acad. Sci. USA* **80**:4079–4083.

Ohta, T., and Dover, G. A., 1985, The cohesive genetics of molecular drive and its interaction with natural selection (submitted).

Ojala, D., Merkel, C., Gelfand, R., and Attardi, G., 1980, The tRNA genes punctuate the reading of genetic information in human mitochondrial DNA, *Cell* **22**:393–403.

Ojala, D., Montoya, J., and Attardi, G., 1981, tRNA punctuation model of RNA processing in human mitochondria, *Nature* **290**:470–474.

Olsen, G. J., McCarroll, R., and Sogin, M. L., 1983, Secondary structure of the *Dictyostelium discoideum* small subunit ribosomal RNA, *Nucleic Acids Res.* **11**:8037–8049.

O'Melia, A. F., 1979, The synthesis of 5S RNA and its regulation during early sea urchin development, *Dev. Growth Differ.* **21**:99–108.

Onishi, T., Berglund, C., and Reeder, R. H., 1984, On the mechanism of nucleolar dominance in mouse–human somatic cell hybrids, *Proc. Natl. Acad. Sci. USA* **81**:484–487.

Otsuka, T., Nomiyama, H., Yoshida, H., Kukita, T., Kuhara, S., and Sakaki, Y., 1983, Complete nucleotide sequence of the 26S rRNA gene of *Physarum polycephalum*: Its significance in gene evolution, *Proc. Natl. Acad. Sci. USA* **80**:3163–3167.

Pace, N. R., Walker, T. A., and Schroeder, E., 1977, Structure of the 5.8S RNA component of the 5.8S–28S ribosomal RNA junction complex, *Biochemistry* **16**:5321–5328.

Palmer, J. D., and Thompson, W. F., 1981, Rearrangements in the chloroplast genomes of mung bean and pea, *Proc. Natl. Acad. Sci. USA* **78**:5533–5537.

Palmer, J. D., and Thompson, W. F., 1982, Chloroplast DNA rearrangements are more frequent when a large inverted repeat sequence is lost, *Cell* **29**:537–550.

Pardue, M. L., Brown, D. D., and Birnstiel, M. L., 1973, Location of the genes for 5S ribosomal RNA in *Xenopus laevis*, *Chromosoma* **42**:191–203.

Patterson, J. B., and Stafford, D. W., 1971, Characterization of sea urchin ribosomal satellite deoxyribonucleic acid, *Biochemistry* **10**:2775–2779.

Pavlakis, G. N., Jordan, B. R., Wurst, R. M., and Vournakis, J. N., 1979, Sequence and secondary structure of *Drosophila melanogaster* 5.8S and 2S rRNAs and of the processing site between them, *Nucleic Acids Res.* **7**:2213–2238.

Pearson, N. J., Fried, H. M., and Warner, J. R., 1982, Yeast use translational control to compensate for extra copies of a ribosomal protein gene, *Cell* **29**:347–355.

Peattie, D. A., Douthwaite, S., Garrett, R. A., and Noller, H. F., 1981, A "bulged" double helix in a RNA–protein contact site, *Proc. Natl. Acad. Sci. USA* **78**:7331–7335.

Pelham, H. R. B., and Brown, D. D., 1980, A specific transcription factor that can bind either the 5S RNA gene or 5S RNA, *Proc. Natl. Acad. Sci. USA* **77**:4170–4174.

Pelham, H. R. B., Wormington, W. M., and Brown, D. D., 1981, Related 5S RNA transcription factors in *Xenopus* oocytes and somatic cells, *Proc. Natl. Acad. Sci. USA* **78**:1760–1764.

Perelson, A. S., and Bell, G. I., 1977, Mathematical models for the evolution of multigene families by unequal crossing over, *Nature* **265**:304–310.

Perkowska, E., MacGregor, H. C., and Birnstiel, M. L., 1968, Gene amplification in the oocyte nucleus of mutant and wild-type *Xenopus laevis*, *Nature* **217**:649–650.

Perry, R. P., 1976, Processing of RNA, *Annu. Rev. Biochem.* **45**:605–629.

Pestka, S., 1968, Studies on the formation of transfer ribonucleic acid–ribosome complexes, *J. Biol. Chem.* **243**:2810–2820.

Peterson, R. C., Doering, J. L., and Brown, D. D., 1980, Characterization of two *Xenopus* somatic 5S DNAs and one minor oocyte-specific 5S DNA, *Cell* **20**:131–141.

Petes, T. D., 1980, Unequal meiotic recombination within tandem arrays of yeast ribosomal DNA genes, *Cell* **19**:765–774.

Petes, T. D., and Botstein, D., 1977, Simple Mendelian inheritance of the reiterated DNA of yeast, *Proc. Natl. Acad. Sci. USA* **74**:5091–5095.

Picard, B., and Wegnez, M., 1979, Isolation of a 7S particle from *Xenopus laevis* oocytes: A 5S RNA–protein complex, *Proc. Nat. Acad. Sci. USA* **76**:241–245.

Pierandrei-Amaldi, P., Campioni, N., Beccari, E., Bozzoni, I., and Amaldi, F., 1982, Expression of ribosomal-protein genes in *Xenopus laevis* development, *Cell* **30**:163–171.

Pierandrei-Amaldi, P., Campioni, N., Gallinari, P., Beccari, E., Bozzoni, I., and Amaldi, F., 1985a, Ribosomal protein synthesis is not autogenously regulated at the translational level in *Xenopus laevis*, *Dev. Biol.* **107**:281–289.

Pierandrei-Amaldi, P., Beccari, E., Bozzoni, I., and Amaldi, F., 1985b, Ribosomal protein production in normal and anucleolate *Xenopus laevis* embryos: Regulation at the post-transcriptional and translational levels, *Cell* (in press).

Planta, R. J., and Mager, W. H., 1985, Structure and expression of ribosomal protein genes in yeast, in: *Structure, Function,* and *Genetics* of *Ribosomes* (B. Hardesty and G. Kramer, eds.), Springer-Verlag (in press).

Politz, S. M., and Glitz, D. G., 1977, Ribosome structure: Localization of N^6,N^6-dimethyladenosine by electron microscopy of a ribosome–antibody complex, *Proc. Natl. Acad. Sci. USA* **74**:1468–1472.

Price, J. V., Kieft, G. L., Kent, J. R., Sievers, E. L., and Cech, T. R., 1985, Sequence requirements for self-splicing of the *Tetrahymena thermophila* pre-ribosomal RNA, *Nucleic Acids Res.* **13**:1871–1889.

Prince, J. B., Taylor, B. H., Thurlow, D. L., Ofengand, J., and Zimmermann, R. A., 1982,

Covalent crosslinking of tRNA$_1^{Val}$ to 16S RNA at the ribosomal P site: Identification of crosslinked residues, *Proc. Natl. Acad. Sci. USA* **79**:5450–5454.

Pukkila, P. J., 1975, Identification of the lampbrush chromosome loops which transcribe 5S ribosomal RNA in *Notophthalmus (Triturus) viridescens, Chromosoma* **53**:71–89.

Quincey, R. V., 1971, The number and location of genes for 5S ribonucleic acid within the genome of *Drosophila melanogaster, Biochem. J.* **123**:227–233.

Radding, C. M., 1978, Genetic recombination: Strand transfer and mismatch repair, *Annu. Rev. Biochem.* **47**:847–880.

Rae, P. M. M., Barnett, T., and Murtif, V. L., 1981, Nontranscribed spacers in *Drosophila* ribosomal DNA, *Chromosoma* **82**:637–655.

Rawson, J. R. Y., Kushner, S. R., Vapnek, D., Alton, N. K., and Boema, C. L., 1978, Chloroplast ribosomal RNA genes in *Euglena gracilis* exist as three clustered tandem repeats, *Gene* **3**:191–209.

Raynal, F. Michot, B., and Bachellerie, J-P., 1984, Complete nucleotide sequence of mouse 18S rRNA gene: Comparison with other available homologs, *FEBS Lett.* **167**:263–268.

Reddy, R., Rothblum, L. I., Subrahmanyam, C. S., Liu, M. H., Henning, D., Cassidy, B., and Busch, H., 1983, The nucleotide sequence of 8S RNA bound to preribosomal RNA of Novikoff hepatoma, *J. Biol. Chem.* **258**:584–589.

Reeder, R. H., 1984, Enhancers and ribosomal gene spacers, *Cell* **38**:349–351.

Reeder, R. H., and Roan, J. G., 1984, The mechanism of nucleolar dominance in *Xenopus* hybrids, *Cell* **38**:39–44.

Reeder, R. H., Sollner-Webb, B., and Wahn, H. L., 1977, Sites of transcription initiation *in viro* on *Xenopus laevis* ribosomal DNA, *Proc. Natl. Acad. Sci. USA* **74**:5402–5406.

Reeder, R. H., Roan, J. G., and Dunaway, M., 1983, Spacer regulation of *Xenopus* ribosomal gene transcription: Competition in oocytes, *Cell* **35**:449–456.

Reichel, R., Monstein, H.-J., Jansen, H.-W., Philipson, L., and Benecke, B.-J., 1982, Small nuclear RNAs are encoded in the nontranscribed region of ribosomal spacer DNA, *Proc. Nat. Acad. Sci. USA* **79**:3106–3110.

Renkawitz-Pohl, R., 1978, Number of the repetitive euchromatic 5S RNA genes in polyploid tissues of *Drosophila hydei, Chromosoma* **66**:249–258.

Renkawitz-Pohl, R., Glätzer, K. H., and Kunz, W., 1980, Characterization of cloned ribosomal DNA from *Drosophila hydei, Nucleic Acids Res.* **8**:4593–4611.

Renkawitz-Pohl, R., Glätzer, K. H., and Kunz, W., 1981, Ribosomal RNA genes with an intervening sequence are clustered within the X chromosomal ribosomal DNA of *Drosophila hydei, J. Mol. Biol.* **148**:95–101.

Ritossa, F. M., and Spiegelman, S., 1965, Localization of DNA complementary to ribosomal RNA in the nucleolus organizer region of *Drosophila melanogaster, Proc. Natl. Acad. Sci. USA* **53**:737–745.

Rochaix, J. D., 1981, Organization, function and expression of the chloroplast DNA of *Chlamydomonas reinhardtii, Experientia* **37**:323–332.

Rochaix, J. D. and Malnoe, P., 1978, Anatomy of the chloroplast ribosomal DNA of *Chlamydomonas reinhardtii, Cell* **15**:661–670.

Roeder, R., 1976, Eukaryotic RNA polymerases, in: *RNA Polymerase* (R. Losick and M. Chamberlin, eds.), Cold Spring Harbor Laboratory, Cold Spring Harbor, New York, pp. 285–330.

Rogers, J., and Wall, R., 1980, A mechanism for RNA splicing, *Proc. Natl. Acad. Sci. USA* **77**:1877–1879.

Röhl, R., and Nierhaus, K. H., 1982, Assembly map of the large subunit (50S) of *Escherichia coli* ribosomes, *Proc. Natl. Acad. Sci. USA* **79**:729–733.

Rosbash, M., Harris, P. K. W., Woolford, J. L., and Teem, J. L., 1981, The effect of temperature-sensitive RNA mutants on the transcription products from cloned ribosomal protein genes of yeast, *Cell* **24:**679–686.

Rozek, C. E., and Timberlake, W. E., 1979, Restriction endonuclease mapping by crossed contact hybridization: The ribosomal RNA genes of *Achyla ambisexualis*, *Nucleic Acids Res.* **7:**1567–1578.

Rubin, G. M., and Sulston, J. E., 1973, Physical linkage of the 5S cistrons to the 18S and 28S ribosomal RNA cistrons in *Saccharomyces cerevisiae*, *J. Mol. Biol.* **79:**521–530.

Rubstov, P. M., Musakhanov, M. M., Zakharyev, V. M., Krayev, A. S., Skryabin, K. G., and Bayev, A. A., 1980, The structure of the yeast ribosomal RNA genes. I. The complete nucleotide sequence of the 18S ribosomal RNA gene from *Saccharomyces cerevisiae*, *Nucleic Acids Res.* **8:**5779–5794.

Rungger, D., Crippa, M., Trendelenburg, M. F., Scheer, U., and Franke, W. W., 1978, Visualization of rDNA spacer transcription in *Xenopus* oocytes treated with fluoruridine, *Exp. Cell Res.* **116:**481–486.

Saccone, C., Cantatore, P., Gadaleta, G., Gallerani, R., Lanave, C., Pepe, G., and Kroon, A. M., 1981, The nucleotide sequence of the large ribosomal RNA gene and the adjacent tRNA genes from rat mitochondria, *Nucleic Acids Res.* **9:**4139–4148.

Saffer, L. D., and Miller, O. L., Jr., 1983, Electron microscopic study of replication and its interaction with transcription in synchronized yeast cells, *J. Cell Biol.* **97:**111a.

Saiga, H., Mizumoto, K., Matsui, T., and Higashinakagawa, T., 1982, Determination of the transcription initiation site of *Tetrahymena pyriformis* rDNA using *in vitro* capping of 35S pre-rRNA, *Nucleic Acids Res.* **10:**4223–4236.

Sakonju, S., and Brown, D. D., 1982, Contact points between a positive transcription factor and the *Xenopus* 5S RNA gene, *Cell* **31:**395–405.

Sakonju, S., Bogenhagen, D. F., and Brown, D. D., 1980, A control region in the center of the 5S RNA gene directs specific initiation of transcription. I. The 5′ border of the region, *Cell* **19:**13–25.

Sakonju, S., Brown, D. D., Engelke, D., Ng, S.-Y., Shastry, B. S., and Roeder, R. G., 1981, The binding of a transcription factor to deletion mutants of a 5S ribosomal RNA gene, *Cell* **23:**665–669.

Salim, M., and Maden, B. E. H., 1981, Nucleotide sequence of *Xenopus laevis* 18S ribosomal RNA inferred from gene sequence, *Nature* **291:**205–208.

Sanders, J. P. M., Heyting, C., and Borst, P., 1975, The organization of genes in yeast mitochondrial DNA. I. The genes for large and small ribosomal RNA are far apart, *Biochem. Biophys. Res. Commun.* **65:**699–707.

Santon, J. B., and Pellegrini, M., 1980, Expression of ribosomal proteins during *Drosophila* early development, *Proc. Natl. Acad. Sci. USA* **77:**5649–5653.

Schäfer, U., and Kunz, W., 1976, Ribosomal DNA content and bobbed phenotype in *Drosophila hydei*, *Heredity* **37:**351–355.

Scheer, U., Trendelenburg, M. F., and Franke, W. W., 1973, Transcription of ribosomal RNA cistrons. Correlation of morphological and biochemical data, *Exp. Cell Res.* **80:**175–190.

Scheer, U., Trendelenburg, M. F., and Franke, W. W. 1976, Regulation of transcription of genes of ribosomal RNA during amphibian oogenesis. A biochemical and morphological study, *J. Cell Biol.* **69:**465–489.

Scheer, U., Trendelenburg, M., and Franke, W., 1977, Lengths and patterns of transcriptional units in the amplified nucleoli of oocytes of *Xenopus laevis*, *Chromosoma* **60:**147–167.

Schell, J., and Van Montagu, M., 1980, The Ti-plasmids of *Agrobacterium tumefaciens* in their role in crown gall formation, in: *Genome Organization and Expression in Plants* (C. J. Leaver, ed.), Plenum Press, New York, pp. 453–470.

Scherer, S., and Davis, R. W., 1980, Recombination of dispersed repeated DNA sequences in yeast, *Science* **209:**1380–1384.

Schlessinger, D., Ono, M., Nikolaev, N., and Silengo, L., 1974. Accumulation of 30S preribosomal ribonucleic acid in an *Escherichia coli* mutant treated with chloramphenicol, *Biochemistry* **13:**4268–4271.

Schlissel, M. S., and Brown, D. D., 1984, The transcriptional regulation of *Xenopus* 5S RNA genes in chromatin: The roles of active stable transcription complexes and histone H1, *Cell* **37:**903–913.

Schmickel, R. D., 1973, Quantitation of human ribosomal DNA. Hybridization of human DNA with ribosomal RNA for quantitation and fractionation, *Pediatr. Res.* **7:**5–12.

Schmidt, R. J., Richardson, C. B., Gillham, N. W., and Boynton, J. E., 1983, Sites of synthesis of chloroplast ribosomal proteins in *Chlamydomonas, J. Cell Biol.* **96:**1451–1463.

Schmidt, R. J., Myers, A. M., Gillham, N. W., and Boynton, J. E., 1984, Immunological similarities between specific chloroplast ribosomal proteins from *Chlamydomonas reinhardtii* and ribosomal proteins from *E. coli, Mol. Biol. Evol.* **1:**317–334.

Schwarz, Z., and Kössel, H., 1980, The primary structure of 16S rDNA from *Zea mays* chloroplast is homologous to *E. coli* 16S rRNA, *Nature* **283:**739–742.

Scott, N. S., and Timmis, J. N., 1983, Sequence homology between spinach nuclear and chloroplast genomes, *Nature* **305:**65–67.

Seilhammer, J. J., and Cummings, D. J., 1981, Structure and sequence of the mitochondrial 20S rRNA and tRNA tyr gene of *Paramecium primaurelia, Nucleic Acids Res.* **9:**6391–6406.

Seilhammer, J. J., Olsen, G. J., and Cummings, D. G., 1984, *Paramecium* mitochondrial genes: 1. Small subunit rRNA, *J. Biol. Chem.*, **259:**5167–5172.

Sekiya, T., Mori, M., Takahashi, N., and Nishimura, S., 1980, Sequence of the distal $tRNA_{I}^{Asp}$ gene and the transcription termination signal in the *Escherichia coli* ribosomal RNA operon rrnF (or G), *Nucleic Acids Res.* **8:**3809–3827.

Selker, E. U., Yanofsky, C., Driftmier, K., Metzenberg, R. L., Alzner-DeWeerd, B., and RajBhandary, U. L., 1981, Dispersed 5S RNA genes in *N. crassa:* Structure, expression and evolution, *Cell* **24:**819–828.

Shine, J., and Dalgarno, L., 1974, The 3′-terminal sequence of *E. coli* 16S ribosomal RNA: Complementarity to nonsense triplets and ribosome binding sites, *Proc. Natl. Acad. Sci. USA* **71:**1342–1346.

Simeone, A., de Falco, A., Macino, G., and Boncinelli, E., 1982, Sequence organization of the ribosomal spacer of *D. melanogaster, Nucleic Acids Res.* **10:**8263–8272.

Simeone, A., La Volpe, A., and Boncinelli, E., 1985, Nucleotide sequence of a complete ribosomal spacer of *D. melanogaster, Nucleic Acids Res.* **13:**1089–1101.

Sinclair, J., and Brown, D. D., 1971, Retention of common nucleotide sequences in the ribosomal deoxyribonucleic acid of eukaryotes and some of their physical characteristics, *Biochemistry* **10:**2761–2769.

Skinner, J. A., Öhrlein, A., and Grummt, I., 1984, *In vitro* mutagenesis and transcriptional analysis of a mouse ribosomal promoter element, *Proc. Natl. Acad. Sci. USA* **81:**2137–2141.

Sköld, S.-E., 1983, Chemical crosslinking of elongation factor G to the 23S RNA in 70S ribosomes from *Escherichia coli, Nucleic Acids Res.* **11:**4923–4932.

Skryabin, K. G., Eldarov, M. A., Larionov, V. L., Bayev, A. A., Klootwijk, J., deRegt, V. C. H. F., Veldman, G. M., Planta, R. J., Georgiev, O. I., and Hadjiolov, A. A., 1984, Structure and function of the nontranscribed spacer regions of yeast rDNA, *Nucleic Acids Res.* **12:**2955–2968.

Sloof, P., Van den Burg, J., Voogd, A., Benne, R., Agostinelli, M., Borst, P., Gutell, R., and Noller, H., 1985, Further characterization of the extremely small mitochondrial ribosomal RNAs from trypanosomes: A detailed comparison of the 9S and 12S RNAs from *Crithidia fasciculata* and *Trypanosoma brucei* with rRNAs from other organisms, *Nucleic Acids Res.* **13**:4171–4190.

Smith, G. P., 1973, Unequal crossover and the evolution of multigene families, *Cold Spring Harbor Symp. Quant. Biol.* **38**:507–513.

Smith, G. P., 1976, Evolution of repeated DNA sequences by unequal cross overs, *Science* **191**:528–534.

Smith, S. D., Banerjee, N., and Sitz, T. O., 1984, Gene heterogeneity: A basis for alternative 5.8S rRNA processing, *Biochemistry* **23**:3648–3652.

Sollner-Webb, B. and Bogenhagen, D., 1981, A termination signal for RNA polymerase I stops transcription by RNA polymerase III, *Plasmid* **5**:229.

Sollner-Webb, B., and Reeder, R., 1979, The nucleotide sequence of the initiation and termination sites for ribosomal RNA transcription in *Xenopus laevis, Cell* **18**:485–499.

Sollner-Webb, B., Wilkinson, J. K., and Miller, K. G., 1982, Transcription of *Xenopus* ribosomal RNA genes, in: *The Cell Nucleus: rDNA*, Vol. XII, Part C (H. Busch and L. Rothblum, eds.), Academic Press, New York, pp. 31–67.

Sollner-Webb, B., Wilkinson, J. A. K., Roan, J., and Reeder, R. H., 1983, Nested control regions promote *Xenopus* ribosomal RNA synthesis by RNA polymerase I, *Cell* **35**:199–206.

Sommerville, J., 1984, RNA polymerase I promoters and transcription factors, *Nature* **310**:189–190.

Sor, F., and Fukuhara, H., 1980, Sequence nucleotidique du gene de l'ARN ribosomique 15S mitochondrial de la levure, *C. R. Acad. Sci. Paris D* **291**:933–936.

Sor, F., and Fukuhara, H., 1982, Identification of two erythromycin resistance mutations in the mitochondrial gene coding for the large ribosomal RNA in yeast, *Nucleic Acids Res.* **10**:6571–6577.

Sor, F., and Fukuhara, H., 1983, Complete DNA sequence coding for the large ribosomal RNA of yeast mitochondria, *Nucleic Acids Res.* **11**:339–348.

Spear, B. B., and Gall, J. G., 1973, Independent control of ribosomal gene replication in polytene chromosomes of *Drosophila melanogaster, Proc. Natl. Acad. Sci. USA* **70**:1359–1363.

Spencer, D. F., Schnare, M. N., and Gray, M. W., 1984, Pronounced structural similarities between the small subunit ribosomal RNA genes of wheat mitochondria and *Escherichia coli, Proc. Nat. Acad. Sci.* **81**:493–497.

Sprinzl, M. and Helk, B., 1985, Interaction of the TψC-loop of tRNA with the conserved $C_pG_pmU_pA_pA_pC_p$ sequence of 16S rRNA, in: *Structure, Function, and Genetics of Ribosomes* (B. Hardesty and G. Kramer, eds.), Springer-Verlag (in press).

Stambrook, P. J., 1978, Heterogeneity in Chinese hamster ribosomal DNA, *Chromosoma* **65**:153–159.

Steitz, J. A., 1980, RNA–RNA interactions during polypeptide chain initiation, in: *Ribosomes: Structure, Function and Genetics* (G. Chambliss, G. R. Craven, J. Davies, K. Davis, L., Kahan, and M. Nomura, eds.), University Park Press, Baltimore, pp. 479–495.

Steitz, J. A., and Jakes, K., 1975, How ribosomes select initiator regions in mRNA: Base pair formation between the 3′ terminus of 16S rRNA and the mRNA during initiation of protein synthesis in *Escherichia coli, Proc. Natl. Acad. Sci. USA* **72**:4734–4738.

Stern, D. B., and Lonsdale, D. M., 1982, Chloroplast genomes of maize have a 12-kilobase DNA sequence in common, *Nature* **299**:698–702.

Stern, D. B., and Palmer, J. D., 1984, Extensive and widespread homologies between mitochondrial DNA and chloroplast DNA in plants, *Proc. Natl. Acad. Sci. USA* **81**:1946–1950.

Stern, D. B., Dyer, T. A., and Lonsdale, D. M., 1982, Organization of the mitochondrial ribosomal RNA genes of maize, *Nucleic Acids Res.* **10:**3333–3340.

Stewart, M. A., Hall, L. M. C., and Maden, B. E. H., 1983, Multiple heterogeneities in the transcribed spacers of ribosomal DNA from *Xenopus laevis, Nucleic Acids Res.* **11:**629–646.

Stiege, W., Glotz, C., and Brimacombe, R., 1983, Localisation of a series of intra-RNA cross-links in the secondary and tertiary structure of 23S RNA, induced by ultraviolet irradiation of *Escherchia coli* 50S ribosomal subunits, *Nucleic Acids Res.* **11:**1687–1706.

Stiegler, P., Carbon, P., Ebel, J. P., and Ehresmann, C., 1981a, A general secondary-structure model for procaryotic and eucaryotic RNAs of the small ribosomal subunits, *Eur. J. Biochem.* **120:**487–495.

Stiegler, P., Carbon, P., Zuker, M., Ebel, J.-P., and Ehresmann, C., 1981b, Structural organization of the 16S ribosomal RNA from *E. coli.* Topography and secondary structure, *Nucleic Acids Res.* **9:**2153–2172.

Stormo, G., Schneider, T., and Gold, L., 1982, Characterization of translational initiation sites in *E. coli, Nucleic Acids Res.* **10:**2971–2996.

Strachan, T., Coen, E. S., Webb, D. A., and Dover, G. A., 1982, Modes and rates of change of complex DNA families of *Drosophila, J. Mol. Biol.* **158:**37–54.

Stumph, W. E., Wu, J. R., and Bonner, J., 1979, Determination of the size of rat ribosomal DNA repeating units by electron microscopy, *Biochemistry* **18:**2864–2871.

Swanton, M. T., McCarroll, R. M., and Spear, B. B., 1982, The organization of macronuclear rDNA molecules of four hypotrichous ciliated protozoans, *Chromosoma* **85:**1–9.

Szostak, J. W., and Wu, R., 1979, Insertion of a genetic marker into the ribosomal DNA of yeast, *Plasmid* **2:**536–554.

Szostak, J. W., and Wu, R., 1980, Unequal crossing over in the ribosomal DNA of *Saccharomyces cerevisiae, Nature* **284:**426–430.

Szostak, J. W., Orr-Weaver, T. L., Rothstein, R. J., and Stahl, F. W., 1983, The double-strand-break repair model for recombination, *Cell* **33:**25–35.

Tabata, S., 1980, Structure of the 5S ribosomal RNA gene and its adjacent regions in *Torulopsis utilis, Eur. J. Biochem.* **110:**107–114.

Tague, B. W., and Gerbi, S. A., 1984, Processing of the large rRNA precursor: Two proposed categories of RNA–RNA interactions in eukaryotes, *J. Mol. Evol.* **20:**362–367.

Takaiwa, F., and Sugiura, M., 1982, The complete nucleotide sequence of a 23S rRNA gene from tobacco chloroplasts, *Eur. J. Biochem.* **124:**13–19.

Takaiwa, F., Oono, K., and Sugiura, M., 1984, The complete nucleotide sequence of a rice 17S rRNA gene, *Nucleic Acids Res.* **12:**5441–5448.

Tartof, K. D., 1979, Evolution of transcribed and spacer sequences in the ribosomal RNA genes of *Drosophila, Cell* **17:**607–614.

Tartof, K. D., and Dawid, I. B., 1976, Similarities and differences in the structure of X and Y chromosome rRNA genes in *Drosophila, Nature* **263:**27–30.

Tartof, K. D., and Perry, R. P., 1970, The 5S RNA genes of *Drosophila melanogaster, J. Mol. Biol.* **51:**171–183.

Teem, J. L., Abovich, N., Kanfer, N. F., Schwindinger, W. F., Warner, J. R., Levy, A., Woolford, J., Leer, R. J., van Raamsdonk-Duin, M. M. C., Mager, W. H., Planta, R. J., Schultz, L., Friesen, J. D., Fried, H., and Rosbash, M., 1984, A comparison of yeast ribosomal protein gene DNA sequences, *Nucleic Acids Res.* **12:**8295–8312.

Thammana, P., and Cantor, C., 1978, Studies on ribosome structure and interactions near the $m^6_2Am^6_2A$ sequence, *Nucleic Acids Res.* **5:**805–823.

Thomas, C. A., 1970, The theory of the master gene, in: *The Neurosciences: Second Study Program* (F. O. Schmitt, ed.), Rockefeller University Press, New York, p. 973.

Thompson, J., Schmidt, F., and Cundliffe, E., 1982, Site of action of a ribosomal RNA methylase conferring resistance to thiostrepton, *J. Biol. Chem.* **257**:7915–7917.

Thompson, J. F., and Hearst, J. E., 1983a, Structure of *E. coli* 16S RNA elucidated by psoralen crosslinking, *Cell* **32**:1355–1365.

Thompson, J. F., and Hearst, J. E., 1983b, Structure–function relations in *E. coli* 16S RNA, *Cell* **33**:19–24.

Thurlow, D. L., Mason, T. L., and Zimmermann, R. A., 1984, 5S RNA-like structures in large ribosomal subunit RNAs of fungal mitochondria (Hypothesis), *FEBS Lett.* **173**:277–282.

Tobler, H., 1975, Occurrence and developmental significance of gene amplification, in: *Biochemistry of Animal Development*, Vol. 3 (R. Weber, ed.), Academic Press, New York, pp. 91–143.

Tohdoh, N., and Sugiura, M., 1982, The complete nucleotide sequence of a 16S ribosomal RNA gene from tobacco chloroplasts, *Gene* **17**:213–218.

Tomioka, N., and Sugiura, M., 1983, The complete nucleotide sequence of a 16S ribosomal RNA gene from a blue-green alga, *Anacystis nidulans*, *Mol. Gen. Genet.* **191**:46–50.

Torczynski, R., Bollon, A. P., and Fuke, M., 1983, The complete nucleotide sequence of the rat 18S ribosomal RNA gene and comparison with the respective yeast and frog genes, *Nucleic Acids Res.* **11**:4879–4890.

Torczynski, R. M., Bollon, A. P., and Fuke, M., 1985. Isolation and nucleotide sequence of a human 18S ribosomal RNA gene (submitted).

Trendelenburg, M. F., 1981, Initiation of transcription at distinct promoter sites in spacer regions between pre-rRNA genes in oocytes of *Xenopus laevis*: An electron microscopic analysis, *Biol. Cell* **42**:1–12.

Trendelenburg, M. F., 1982, Chromatin structure of *Xenopus* rDNA transcription termination sites, *Chromosoma* **86**:703–715.

Truett, M. A., and Gall, J. G., 1977, The replication of ribosomal DNA in the macronucleus of *Tetrahymena*, *Chromosoma* **64**:295–303.

Tu, J., and Zillig, W., 1982, Organization of rRNA structural genes in the archaebacterium *Thermoplasma acidophilum*, *Nucleic Acids Res.* **10**:7231–7245.

Unnasch, T. R., and Wirth, D. F., 1983, The cloned rRNA gene of *P. lophurae*: A novel rDNA structure, *Nucleic Acids Res.* **11**:8461–8472.

Urano, Y., Kominami, R., Mishima, Y., and Muramatsu, M. 1980, The nucleotide sequence of the putative transcription initiation site of a cloned ribosomal RNA of the mouse, *Nucleic Acids Res.* **8**:6043–6058.

Valenzuela, P., Bell, G. I., Venegas, A., Sewell, E. T., Masiarz, F. R., DeGennaro, L. J., Weinberg, F., and Rutter, W. J., 1977, Ribosomal RNA genes of *Saccharomyces cerevisiae*, *J. Biol. Chem.* **252**:8126–8135.

Van den Boogaart, P., Samallo, J., and Agsteribbe, E., 1982, Similar genes for a mitochondrial ATPase subunit in the nuclear and mitochondrial genomes of *Neurospora crassa*, *Nature* **298**:187–189.

van der Horst, G. and Tabak, H. F., 1985, Self-splicing of yeast mitochondrial ribosomal and messenger RNA precursors, *Cell* **40**:759–766.

Van Etten, R. A., Walberg, M. W., and Clayton, D. A., 1980, Precise localization and nucleotide sequence of the mouse mitochondrial rRNA genes and three immediately adjacent novel tRNA genes, *Cell* **22**:157–170.

Vassilenko, S. K., Carbon, P., Ebel, J.-P., and Ehresmann, C., 1981, Topography of 16S RNA in 30S subunits and 70S ribosomes accessibility to cobra venom ribonuclease, *J. Mol. Biol.* **152**:699–721.

Veldman, G. M., 1982, Regulatory and Structural Sequences of the Yeast Ribosomal RNA Operon, Ph.D. Thesis, Vrije Universiteit, Amsterdam.

Veldman, G. M., Brand, R. C., Klootwijk, J., and Planta, R. J., 1980a, Some characteristics of processing sites in ribosomal precursor RNA of yeast, *Nucleic Acids Res.* **8**:2907–2920.

Veldman, G. M., Klootwijk, J., de Jonge, P., Leer, R. J., and Planta, R. J., 1980b, The transcription termination site of the ribosomal RNA operon in yeast, *Nucleic Acids Res.* **8**:5179–5192.

Veldman, G. M., Klootwijk, J., de Regt, V. C. H. F., Planta, R. J., Branlant, C., Krol, A., and Ebel, J. P., 1981a, The primary and secondary structure of yeast 26S rRNA, *Nucleic Acids Res.* **9**:6935–6952.

Veldman, G. M., Klootwijk, J., Heerikhuizan, H., and Planta, R. J., 1981b, The nucleotide sequence of the intergenic region between 5.8S and 26S rRNA genes of the yeast ribosomal RNA operon. Possible implications for the interaction between 5.8S and 26S rRNA and the processing of the primary transcript, *Nucleic Acids Res.* **9**:4847–4862.

Verbeet, M. Ph., Klootwijk, J., Van Heerikhuizen, H., Fontijn, R. D., Vreugdenhil, E., and Planta, R. J., 1983, Molecular cloning of the rDNA of *Saccharomyces rosei* and comparison of its transcription initiation region with that of *Saccharomyces carlsbergensis*, *Gene* **23**:53–63.

Verbeet, M. Ph., Klootwijk, J., van Heerikhuizen, H., Fontijn, R. D., Vreugdenhil, E., and Planta, R. J., 1984a, A conserved sequence element is present around the transcription initiation site for RNA polymerase A in Saccharomycetoideae, *Nucleic Acids Res.* **12**:1137–1148.

Verbeet, M. Ph., van Heerikhuizen, H., Klootwijk, J., Fontijn, R. D., and Planta, R. J., 1984b, Evolution of yeast ribosomal DNA: Molecular cloning of the rDNA units of *Kluyveromyces lactis* and *Hansenula wingei* and their comparison with the rDNA units of other Saccharomycetoideae, *Mol. Gen. Genet.* **195**:116–125.

Vermeulen, C. W., and Atwood, K. C., 1965, The proportion of DNA complementary to ribosomal RNA in *Drosophila melanogaster, Biochem. Biophys. Res. Commun.* **19**:221–226.

Vogt, V. M., and Braun, R., 1977, The replication of ribosomal DNA in *Physarum polycephalum, Eur. J. Biochem.* **80**:557–566.

Warburton, D., Atwood, K. C., and Henderson, A. S., 1976, Variation in the number of genes for rRNA among human acrocentric chromosomes: Correlation with frequency of satellite association, *Cytogenet. Cell Genet.* **17**:221–230.

Ware, V. C., Tague, B. W., Clark, C. G., Gourse, R. L., Brand, R. C., and Gerbi, S. A., 1983, Sequence analysis of 28S ribosomal DNA from the amphibian *Xenopus laevis, Nucleic Acids Res.* **11**:7795–7817.

Ware, V. C., Renkawitz, R., and Gerbi, S. A., 1985, rRNA processing: Removal of only nineteen bases at the gap between 28Sα and 28Sβ rRNAs in *Sciara coprophila, Nucleic Acids Res.* **13**:3581–3597.

Waring, R. B., Davies, R. W., Scazzocchio, C., and Brown, T. A., 1982, Internal structure of a mitochondrial intron of *Aspergillus nidulans, Proc. Nat. Acad. Sci. USA* **79**:6332–6336.

Waring, R. B., Scazzocchio, C., Brown, T. A., and Davies, R. W., 1983, Close relationship between nuclear and mitochondrial introns—Implications for the mechanism of RNA splicing, *J. Mol. Biol.* **167**:595–605.

Warner, J. R., 1977, In the absence of ribosomal RNA synthesis, the ribosomal proteins of HeLa cells are synthesized normally and degraded rapidly, *J. Mol. Biol.* **115**:315–333.

Warner, J. R., and Gorenstein, C., 1977, The synthesis of eucaryotic ribosomal proteins *in vitro, Cell* **11**:201–212.

Warner, J. R., Tushinski, R. J., and Wejksnora, P. J., 1980, Coordination of RNA and proteins in eukaryotic ribosome production, in: *Ribosomes: Structure, Function and Genetics* (G. Chambliss, G. R. Craven, J. Davies, K. Davis, L. Kahan, and M. Nomura, eds.), University Park Press, Baltimore, pp. 889–902.

Warner, J. R., Elion, E. A., Kruse, C., and Johnson, S. P., 1985, Ribosomal RNA transcription and ribosomal protein function in yeast, in: *Structure, Function, and Genetics of Ribosomes* (B. Hardesty and G. Kramer, eds.), Springer-Verlag (in press).

Weiss-Brummer, B., Rodel, G., Schweyen, R. J., and Kaudewitz, F., 1982, Expression of the split gene *cob* in yeast: Evidence for a precursor of a "maturase" protein translated from intron 4 and preceding exons, *Cell* **29**:527–536.

Wejksnora, P. J., and Warner, J. R., 1981, Regulation of ribosomal RNA and proteins in mouse–hamster hybrid cells, *J. Biol. Chem.* **256**:9406–9413.

Wellauer, P. K., Reeder, R. H., Carroll, D., Brown, D. D., Deutsch, A., Higashinakagawa, T., and Dawid, I. B., 1974, Amplified ribosomal DNA from *Xenopus laevis* has heterogeneous spacer lengths, *Proc. Natl. Acad. Sci. USA* **71**:2823–2827.

Wellauer, P. K., Dawid, I. B., Brown, D. D., and Reeder, R., 1976a, The molecular basis for length heterogeneity in ribosomal DNA from *Xenopus laevis, J. Mol. Biol.* **105**:461–486.

Wellauer, P. K., Reeder, R. H., Dawid, I. B., and Brown, D. D., 1976b, The arrangement of length heterogeneity in repeating units of amplified and chromosomal ribosomal DNA from *Xenopus laevis, J. Mol. Biol.* **105**:487–505.

Wellauer, P. K., Dawid, I. B., and Tartof, K. D., 1978, X and Y chromosomal ribosomal DNA of *Drosophila:* Comparison of spacers and insertions, *Cell* **14**:269–278.

Westermann, P., Nygard, O., and Bielka, H., 1980, The α and γ subunits of initiation factor eIF-2 can be cross-linked to 18S ribosomal RNA within the quaternary initiation complex, eIF-2.Met-tRNA$_f$.GDCP.small ribosomal subunit, *Nucleic Acids Res.* **8**:3065–3071.

Wildman, S. G., Hongladarom, T., and Honda, S. I., 1962, Chloroplasts and mitochondria in living plant cells: Cinephotomicrographic studies, *Science* **138**:434–436.

Wilson, G. N., Knoller, M., Schmickel, R. D., and Szura, L. L., 1984, Individual and evolutionary variation of primate ribosomal DNA transcription initiation regions, *Mol. Biol. Evol.* **1**:221–237.

Winicov, I., 1976, Alternate temporal order in ribosomal RNA maturation, *J. Mol. Biol.* **100**:141–155.

Woese, C. R., 1970, Molecular mechanics of translation: A reciprocating ratchet mechanism, *Nature* **226**:817–820.

Woese, C. R., 1980, Just so stories and Rube Goldberg machines: Speculations on the origin of the protein synthetic machinery, in *Ribosomes: Structure, Function and Genetics* (G. Chambliss, G. R. Craven, J. Davies, K. Davis, L. Kahan, and M. Nomura, eds.), University Park Press, Baltimore, pp. 357–373.

Woese, C. R., and Fox, G. E., 1977, Phylogenetic structure of the prokaryotic domain: The primary kingdoms, *Proc. Natl. Acad. Sci. USA* **74**:5088–5090.

Woese, C. R., Magrum, L. J., and Fox, G. E., 1978, Archaebacteria, *J. Mol. Evol.* **11**:245–252.

Woese, C. R., Magrum, L. J., Gupta, R., Siegel, R. B., Stahl, D. A., Kop, J., Crawford, N., Brosius, J., Gutell, R., Hogan, J. J., and Noller, H. F., 1980, Secondary structure model for bacterial 16S ribosomal RNA: Phylogenetic, enzymatic and chemical evidence, *Nucleic Acids Res.* **8**:2275–2293.

Woese, C. R., Gutell, R., Gupta, R., and Noller, H. F., 1983, A detailed analysis of the higher order structure of 16S-like ribosomal ribonucleic acids, *Microbiol. Rev.* **47**:621–669.

Wollenzein, P. L., and Cantor, C. R., 1982, Gel electrophoretic technique for separating crosslinked RNAs. Application to improved electron microscopic analysis of psoralen crosslinked 16S ribosomal RNA, *J. Mol. Biol.* **159**:151–166.

Wollenzein, P. L., Hearst, J. E., Thammana, P., and Cantor, C. R., 1979, Base-pairing between distant regions of the *Escherichia coli* 16S ribosomal RNA in solution, *J. Mol. Biol.* **135**:255–269.

Wool, I., 1979, The structure and function of eukaryotic ribosomes, *Ann. Rev. Biochem.* **48**:719–754.

Wool, I. G., 1985, The use of a cytotoxic nuclease to analyze ribosome structure, in: *Structure, Function, and Genetics of Ribosomes* (B. Hardesty and G. Kramer, eds.), Springer-Verlag (in press).

Wormington, W. M., Bogenhagen, D. F., Jordan, E., and Brown, D. D., 1981, A quantitative assay for *Xenopus* 5S RNA gene transcription *in vitro, Cell* **24**:809–817.

Wright, R. M., and Cummings, D. J., 1983, Integration of mitochondrial gene sequences within the nuclear genome during senescence in a fungus, *Nature* **302**:86–88.

Wu, M., and Davidson, N., 1975, Use of gene 32 protein staining of single-strand polynucleotides for gene mapping by electron microscopy: Application to the ø80d$_3$ ilvsu$^+$7 system, *Proc. Natl. Acad. Sci. USA* **72**:4506–4510.

Yagura, T., Yagura, M., and Muramatsu, M., 1979, *Drosophila melanogaster* has different ribosomal RNA sequences on X and Y chromosomes, *J. Mol. Biol.* **133**:533–547.

Yamamoto, O., Takakusa, N., Mishima, Y., Kominami, R., and Muramatsu, M., 1984, Determination of the promoter region of mouse ribosomal RNA gene by an *in vitro* transcription system, *Proc. Natl. Acad. Sci. USA* **81**:299–303.

Yao, M.-C., 1982, Amplification of ribosomal RNA gene in *Tetrahymena,* in: *The Cell Nucleus: rDNA*, Vol. XII, Part C (H. Busch and L. Rothblum, eds.), Academic Press, New York, pp. 127–153.

Yao, M.-C. and Gall, J. G., 1977, A single integrated gene for ribosomal RNA in a eucaryote, *Tetrahymena pyriformis, Cell* **12**:121–132.

Young, B. D., Hell, A., and Birnie, G. D., 1976, A new estimate of human ribosomal gene number, *Biochim. Biophys. Acta* **454**:539–548.

Young, R. A., 1979, Transcription termination in the *Escherichia coli* ribosomal RNA operon rrnC*, *J. Biol. Chem.* **254**:12715–12731.

Young, R. A., and Steitz, J. A., 1978, Complementary sequences 1700 nucleotides apart form a ribonuclease III cleavage site in *Escherichia coli* ribosomal precursor RNA, *Proc. Natl. Acad. Sci. USA* **75**:3593–3597.

Zamb, T. J., and Petes, T. D., 1982, Analysis of the junction between ribosomal RNA genes and single-copy chromosomal sequences in the yeast *Saccharomyces cerevisiae, Cell* **28**:355–364.

Zaug, A. J., and Cech, T. R., 1980, *In vitro* splicing of the ribosomal RNA precursor in nuclei of *Tetrahymena, Cell* **19**:331–338.

Zaug, A. J., and Cech, T. R., 1982, The intervening sequence excised from the ribosomal RNA precursor of *Tetrahymena* contains a 5′-terminal guanosine residue not encoded by the DNA, *Nucleic Acids Res.* **10**:2823–2838.

Zaug, A. J., Grabowski, P. J., and Cech, T. R., 1983, Autocatalytic cyclization of an excised intervening sequence RNA is a cleavage–ligation reaction, *Nature* **301**:578–583.

Zwieb, C., Glotz, C., and Brimacombe, R., 1981, Secondary structure comparisons between small subunit ribosomal RNA molecules from six different species, *Nucleic Acids Res.* **9**:3621–3640.

CHAPTER 8

On the Evolution of Genome Organization in Mammals

STEPHEN J. O'BRIEN, HECTOR N. SEUANEZ, and JAMES E. WOMACK

"Evolution mama, don't you make a monkey out of me"
(Earthquake McGoon, San Francisco, March 1953)

1. Introduction

The range of morphological diversity in extant species of mammals is as great as in any present-day biological class. The functional differentiation, for example, between the Chiropterans (bat family) and the Cetaceans (whales and porpoises) is so extreme as to belie the idea of their taxonomic placement in the same class. Nonetheless, it is generally accepted that, although the single common ancestor of modern mammals may have existed as long as 250 million years ago (mya), the modern mammalian radiations have occurred, probably from several primitive, small, insectivore-like ancestors that existed during the Jurassic and Cretaceous periods, some 120 mya (Eisenberg, 1981). There are approximately 19 mammalian orders recognized today, some with many species (rodents have 1700) and some with only one (Tubulidentates, the aardvarks). In all, over 4000 living mammalian species have been described (Nowak and Paradiso,

STEPHEN J. O'BRIEN • Section of Genetics, National Cancer Institute, Frederick, Maryland 21701. *HECTOR N. SEUANEZ* • Department of Genetics, Universidade Federal do Rio de Janeiro, Rio de Janeiro, Brazil. *JAMES E. WOMACK* • Department of Veterinary Pathology, College of Veterinary Medicine, Texas A&M University, College Station, Texas 77843.

1983). Taxonomic classification of mammals has historically been largely in the realm of morphology and in the purview of morphologists.

More recently, a growing discipline of molecular evolution has emerged and it is this field that forms the basis of this volume. The conclusions of investigators in molecular evolution are often based upon applications of the principles of maximum parsimony to distance/similarity metrics derived from molecular measurements of related taxa. The commonly used procedures include: (1) immunological distances based upon serum albumins or transferrins (Wilson *et al.*, 1977); (2) DNA-DNA hybridization of single-copy genomic DNA (Benveniste and Todaro, 1976); (3) genetic distance using homologous isozyme loci (Avise and Aquadro, 1982); and (4) restriction endonuclease maps of mitochondrial DNA (Brown *et al.*, 1982). Phylogenetic topologies derived from molecular data depend on the evolutionary ticking of the "molecular clock" (Wilson *et al.*, 1977; Goodman and Tashian, 1976). The clock hypothesis requires that nucleotide substitution in genomic DNA was accumulated in a stochastic fashion over long periods of evolutionary time (Wilson *et al.*, 1977). By utilizing (index) "anchor species" outside a particular phylad, the fidelity of the molecular clock hypothesis can actually be tested. The generally (but not universally) found result is that molecules do accumulate mutations stochastically and that, at least within the linear range of each molecular metric, the molecular clock is a fairly accurate timekeeper.

Although there is active debate among students of morphology and molecular evolution as to the strengths and weaknesses of the respective approaches (Wilson *et al.*, 1977; Goodman and Tashian, 1976; Thorpe, 1982; Gribbon and Cherfas, 1982), it is often striking how concordant phylogenies derived by the two approaches can be. Although the phylogenetic placement of mammalian species is beyond the scope of this chapter, we would like to emphasize that several of the phylogenetic topologies of certain mammalian families presented here are consensus phylogenies generated from both morphological and molecular studies.

One very important concept that has developed from discussions of morphological adaptation on the one hand and continuous molecular variation on the other is the theory of punctuated equilibrium (Gould and Eldridge, 1977; Gould, 1980; Stanley, 1979). This theory addresses the long, troublesome observation that the fossil record of most taxa is neither constant nor gradual; rather, it is characterized by long periods of morphological invariance, "punctuated" by precipitous saltatory changes in the lines of several phylads. These observations prompted Gould and Eldridge (1977), Gould (1980), and Stanley (1979) to propose that evolution is generally not steady or gradual (especially in terms of morphology), but is really a "jerky" process with rapid speciation and differentiation in some groups while others remain unchanged for millions of years. The notion of punctuated evolutionary change in its extreme posture contrasts with the more traditional gradualist interpretation of Darwinism and the so-called modern synthesis (Stebbins and Ayala, 1981). Gradualism contends that evolutionary change is a steady and continuous process by which one species slowly evolves into the next, every million years or so. An apparent conflict of these two versions of evolutionary change had developed because fossil evidence suggests that organismal evolution is seldom time dependent, while molecular evolution proceeds in a clocklike manner. This paradox has led to the notion that genome evolution can be further partitioned into punctuated and gradual components and that these components might have different roles in evolution at the organismal and/or species level. The dichotomy, however, is not a serious problem, since several different evolutionary studies have demonstrated that morphological change is uncoupled from molecular change (Bush *et al.*, 1977; Wilson *et al.*, 1975, 1977; Wayne *et al.*, 1985). The simplest interpretation of these data would suggest that morphological variation has a large, punctuated component, while most molecular genetic variation is gradual and continuous.

The idea that speciation may occur with dramatic morphological changes has been raised before in the macroevolutionary theory of Richard Goldschmidt (1940). The theory of macroevolution is largely based on two biological tenets. First, the major evolutionary branches were not the result of gradual acquisition, but involved dramatic (and rapid, in evolutionary time) divergence from existing stock. Second, these changes were often accompanied (and possibly mediated) by large chromosomal changes (additions, deletions, rearrangements, etc). Numerous reviews of chromosomal evolution have addressed this issue (Todd, 1970; Bush *et al.*, 1977; Bickham and Baker, 1979; Giusto and Margulis, 1981; Seu-anez, 1984) and a number of possible interpretations of the chromosomal role of evolution have been suggested and debated. We shall not review these theories, propositions, or disagreements here; however, we will summarize the cumulative genetic and cytological data that do bear on these questions from several points of view. In general, the picture that emerges about the evolution of chromosome and genome organization in mammals is as follows.

Mammalian orders vary tremendously with respect to the conservation of chromosome presentation. Certain mammalian families show extreme conservation of linkage and chromosomal integrity. For example, within the cat family, Felidae, 90% of the chromosomes are cytologically identical among the 37 member species. A second carnivore family, the dogs, Canidae, is characterized by extremely variable chromosomal numbers and appearance. With certain dramatic exceptions (e.g., gibbon and owl monkey), the primates have conserved chromosomal morphologies, which have allowed the tracing of chromosome phylogenies as far back as the prosimians (Dutrillaux, 1979a). The rodents, like the dogs, gibbons, bears, and others, have a highly shuffled or rearranged karyotype, within the order, within genera, and even within the same species. After reviewing these observations, our inevitable conclusion is that mammalian chromosomal evolution is not continuous; rather, it is punctuated with massive complex rearrangements in certain phyla, but with highly conserved arrangements in others.

2. Procedures for Studying Genome Evolution

The understanding of evolutionary processes has been greatly facilitated by the development in the last two decades of cytological, cellular, and molecular procedures. The advent of chromosome banding (Q, R, G, and C) has permitted the unequivocal identification of mammalian chromosomes. The conventional G-, R-, and Q-banding procedures allow the recognition of some 300 subregions in the human karyotype. The same chromosomal subregions are stained by these techniques, although the pattern of chromosome staining is reversed between different banding procedures. The G-trypsin and quinacrine banding techniques have been most frequently used procedures for analysis of well-studied karyotypes, such as human and murine. The increasing application of chromosome banding methods to cytogenetic studies of mammalian chromosomes have made it possible to monitor more accurately the divergence of chromosome structure over tens of millions of years of mammalian evolution (Dutrillaux, 1979a; Seuanez, 1984). By using the pattern of banding as a guide, it is possible to identify regions of presumptive homology in the chromosomes of two species even when the overall morphology of the chromosomes is quite different. Comparisons of this kind have shown karyotypic similarities in the primate order (Turleau *et al.*, 1972), and have led to the proposition of tentative chromosome phylogenies from man to the prosimians (Dutrillaux, 1979a). More recent studies, using high-resolution banding techniques (which resolve up to 1000 bands in the human karyotype) (Francke and Oliver, 1980), have shown that extensive chromosome banding homology exists not only between closely related primates, such as man and the great apes (Yunis *et al.*, 1980; Yunis and Prakash, 1982), but also to a lesser extent between distantly related primates such as man and the woolly monkey (Dutrillaux and Couturier, 1981). Comparative cytological analysis of more than 70 extant primate species has permitted the reconstruction of numerous chromosome rearrangements, such as Robertsonian translocations, paracentric and pericentric inversions, fusions, and fissions, that presumably occurred during the 60–80 million years of primate evolution.

The emergence of genetic maps of homologous loci in a number of mammalian species provides a basis for the evolutionary field of comparative gene mapping (Roderick *et al.*, 1984; O'Brien, 1984a; Womack, 1982). The occurrence of homologous linkage groups in related species is generally interpreted to reflect the presence of that linkage arrangement in species that were ancestral to the modern species. The genetic maps have been derived primarily by sexual genetic analysis or by parasexual procedures using somatic cell hybrids. The early mouse map was largely a product of controlled multigene test crosses between inbred strains, while the human gene map has grown largely through the efforts of somatic cell genetics.

Interspecific somatic cell hybrids provide an important technology for gene mapping because it is possible to generate hybrids that preferentially segregate the chromosomes of one parental species (for example, human) in different combinations among the hybrid clones (Fig. 1). Concordant expression and loss of two genetic markers (antigens, isozymes, etc.) form the basis for identification of a syntenic group in the segregant parent. These groups are analogous to linkage groups derived from sexual genetic crosses (Ruddle *et al.*, 1971) and represent groups of loci that reside on individual chromosomes. The empirical definition of a syntenic group of multiple loci depends upon two important observations: (1) the concordant appearance of the markers in a hybrid panel; and (2) substantial discordancy with all the other markers followed in the same cross. In addition, concordant segregation of gene markers and specific chromosomes identified by banding techniques permits the assignment of the syntenic groups to individual chromosomes.

The methods and coordinate findings of gene mapping by somatic cell genetics have been thoroughly reviewed (Shows *et al.*, 1982; Ruddle and Creagan, 1975; O'Brien *et al.*, 1982; Shay, 1982; Womack, 1982) and will not be recapitulated here. However, we would like to mention that virtually all of the comparative gene mapping data discussed here have been derived from combined biochemical and cytogenetic analyses of selected panels of somatic cell hybrids (Fig. 1). Until very recently, the majority of the genes mapped in mammalian species were structural genes for cytoplasmic "household" enzymes or isozymes. The reason for this emphasis was the feasibility of recognizing and resolving homologous enzyme activities histochemically following gel electrophoresis of extracts of hybrid cells. With the development of molecular cloning, hundreds of

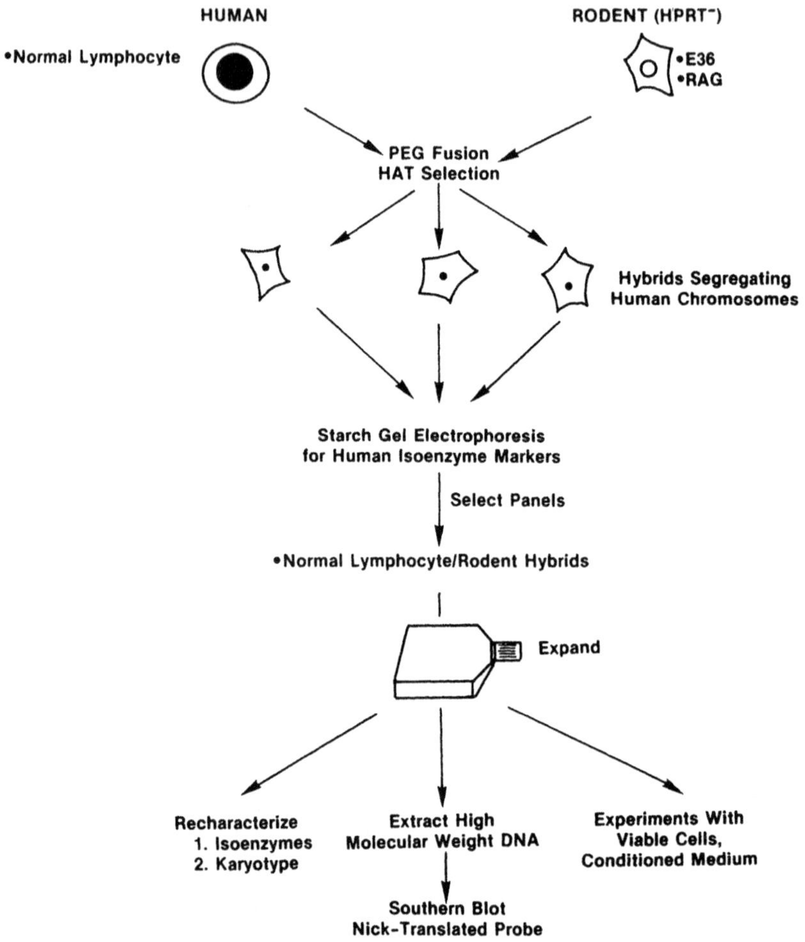

Figure 1. Procedures for preparation and analysis of human–rodent somatic cell hybrids. Hybrid clones are assayed for isozyme markers (O'Brien *et al.*, 1982) and for G-trypsin karyology (O'Brien and Nash, 1982). DNA is extracted and analyzed using a Southern (1975) transfer and specific, nick-translated probes (O'Brien *et al.*, 1983a,b). Experiments with viable cells using antiserum, cell receptors, etc., are assayed at the same passage as isozyme and karyology, because the hybrid's karyotype is often unstable.

cloned structural genes of virtually every type have become available for use in gene mapping as well. The molecular clones allow the genetic localization of virtually any gene or DNA segment regardless of whether the locus is expressed or functional. In general, the most commonly used procedures involved the "Southern" (1975) analysis of genomic DNA from somatic cell hybrid mapping panels (Fig. 1) (Shows *et al.*, 1982; D'Eustachio and Ruddle, 1983; Southern, 1975; O'Brien *et al.*, 1983a,b). It is also possible to radioactively label such probes by "nick-translation" and to hybridize *in situ* with metaphase chromosomal preparations (Harper *et al.*, 1981; Bonner *et al.*, 1984; Seigel *et al.*, 1984). *In situ* hybridization has been used for repetitive gene families, but more recently high-resolution techniques for mapping single-copy sequences have been developed. In combination with and in addition to cell hybrid analysis using cytogenetically characterized fragmented or rearranged human chromosomes, precise regional localization of several hundred human genes has been achieved (Shows *et al.*, 1984).

An important criterion for the efficacy of establishing homologous syntenic groups in different mammalian species is the cogent demonstration of homology of the included gene markers. The International Committee on Comparative Gene Mapping has prepared some recommendations for establishing homology of gene loci in related species (Pearson *et al.*, 1979). Because we wish to emphasize the importance of this demonstration in evolutionary considerations, we list those criteria for gene homology here: (1) DNA or amino acid sequence homology; (2) immunologic cross-reaction; (3) in the case of multimeric enzymes, formation of heteropolymeric molecules in interspecific somatic cell hybrids; (4) similar tissue distribution of isozymes; (5) similar developmental profiles; (6) similar pleiotropic phenotypes; (7) identical subcellular locations; (8) similar substrate specificity; and (9) similar response to specific inhibitors. A criterion that is not recommended because the logic is circular (but which cannot be ignored) is the chromosomal mapping of a putative gene homologue to a homologous syntenic group in a related species. The consideration of some or all of these criteria has been considered in the preparing of our tables that list homologous genes.

We review here the cumulative cytological and genetic data in several mammalian orders that have been studied in detail. The organization will begin with the human gene map, for which the mapping and cytological findings are the most extensive, and proceed to the great apes (Pongidae), the lesser apes (Hylobatidae), the Old World monkeys (Cercopithecidae), New World monkeys (Cebidae), and the mouse lemur (Lemuridae). From here the rodent data, most notably those of the house mouse (*Mus musculus*), will be compared to other rodents and to the primates. The carnivores will next be considered, with special reference to the Felidae, which has an extraordinary chromosomal conservation among the mammals. We shall also consider the cytological situation in certain specific mammalian families (Cervidae, Equidae) that tend to illustrate the abundant chromosome rearrangements characteristic of these taxa. The analysis will concentrate on cytological homologies and syntenic homologies largely of unique sequence structural genes. However, a discussion of certain multigene families will also be presented, since these bear in no small way on the character of linkage maps and have been suggested as operative in isolation mechanisms characteristic of biological speciation (Dover, 1982; Rose and Doolittle, 1983). Among these groups are the highly repetitive satellite DNA and ribosomal DNA families of the Pongidae primates, the oncogene family, which has, for nonevolutionary reasons, been intensively studied of late, and the families of endogenous retroviruses which invade mammalian genomes from without and have a fascinating evolutionary history of their own (see also Benveniste, this volume, Chapter 6).

3. The Human Gene Map: An Index Species

The human gene map has been developed over the past decade by a variety of genetics laboratories (McKusick and Ruddle, 1977; Ruddle *et al.*, 1971). This field of human gene mapping is updated every 2 years at the Human Gene Mapping Workshops, where *ad hoc* committees responsible for each of the 23 human chromosomes (plus the Y chromosome) tabulate and publish an updated version of the human gene map. The proceedings of the latest workshop, Human Gene Mapping-7, were published in 1984 (Human Gene Mapping-7, 1984) and summarized by McKusick (1984). Over 800 different loci have been assigned to the human map, many of them to specific regions or bands. In a review of the construction and details of the human gene map, Shows *et al.* (1982) list 48 different classes of genes that have been mapped. It is interesting that in the most recent mapping conference, over 100 of the loci were mapped with recombinant DNA techniques (Southern analysis of hybrids or *in situ* hybridization).

A G-trypsin-banded karyotype of man is presented in Fig. 2. We present in Fig. 3 an abridged version of the human gene map, which includes the high-resolution, standardized karyotype (850-band stage) and band identification number (ISCN, 1978). On the left of each chromosome are the regional positions in man of those loci whose homologous counterparts have been mapped in one or more additional mammalian species. On the right side of each chromosome are genes which are thought to participate in some way in neoplastic transformation. Most of these latter genes have been assigned using recombinant DNA techniques in the last few years. Many important human loci are not included in this figure, largely because comparative data in other species have not been reported.

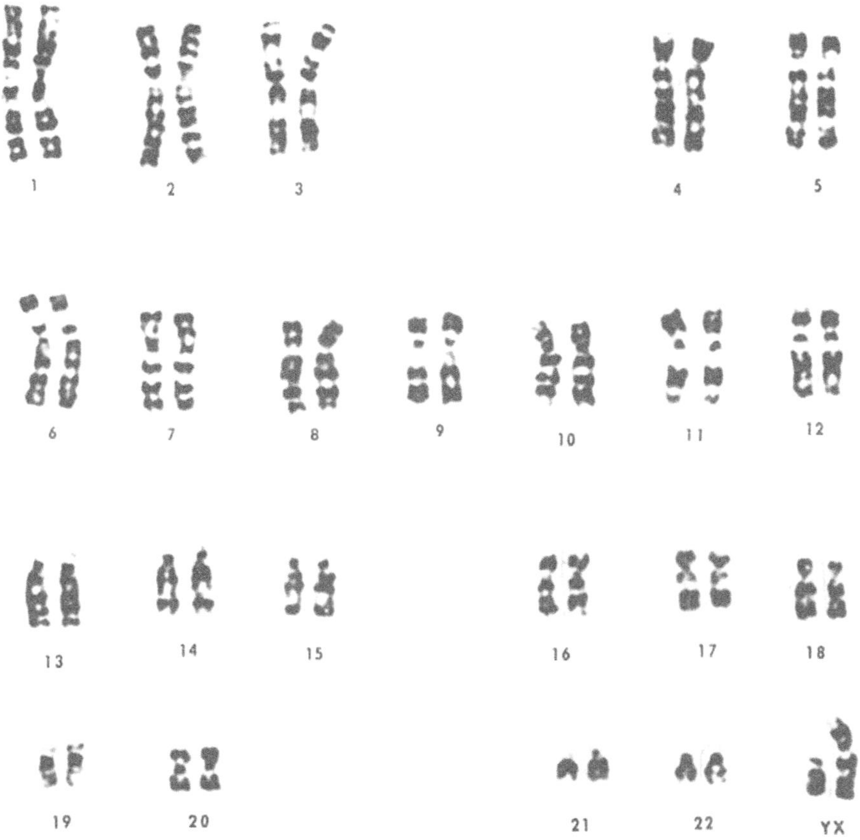

Figure 2. G-trypsin-banded human karyotype. (Prepared by W. G. Nash, National Cancer Institute.)

Since man is the most extensively mapped mammalian species, we shall use this human map as an index for primate and for mammalian comparisons.

4. Comparative Genetics and Cytogenetics of Primates

4.1. The Great Apes

Classical taxonomy had initially placed man, together with the great and lesser apes, within the superfamily Hominoidea, based on the gross morphological attributes of these species (Fig. 4). Morphological similarities between man and apes have been supported by biochemical studies

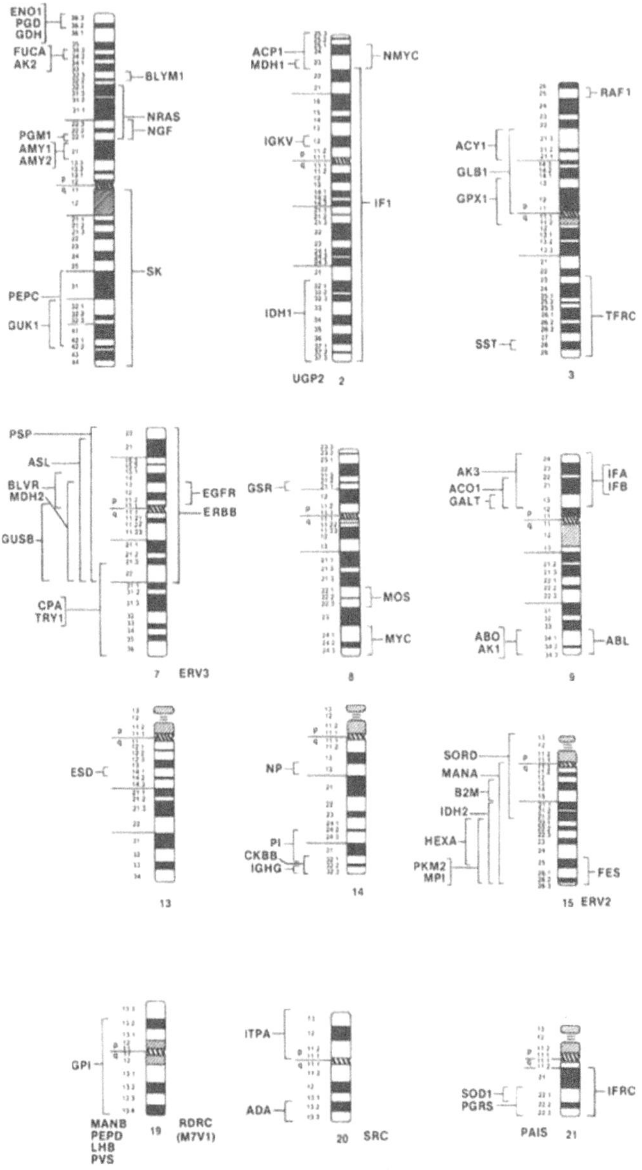

Figure 3. Regional mapping of the human genome derived from Human Gene Mapping-7 (1984) and summarized by McKusick (1984) and O'Brien (1984c). On the left of each chromosome is the regional position of any human genes whose homologous counterparts have also been mapped in other species discussed in this chapter. On the right of each chromosome are the regional positions of those human genes related to neoplastic transformation (on-

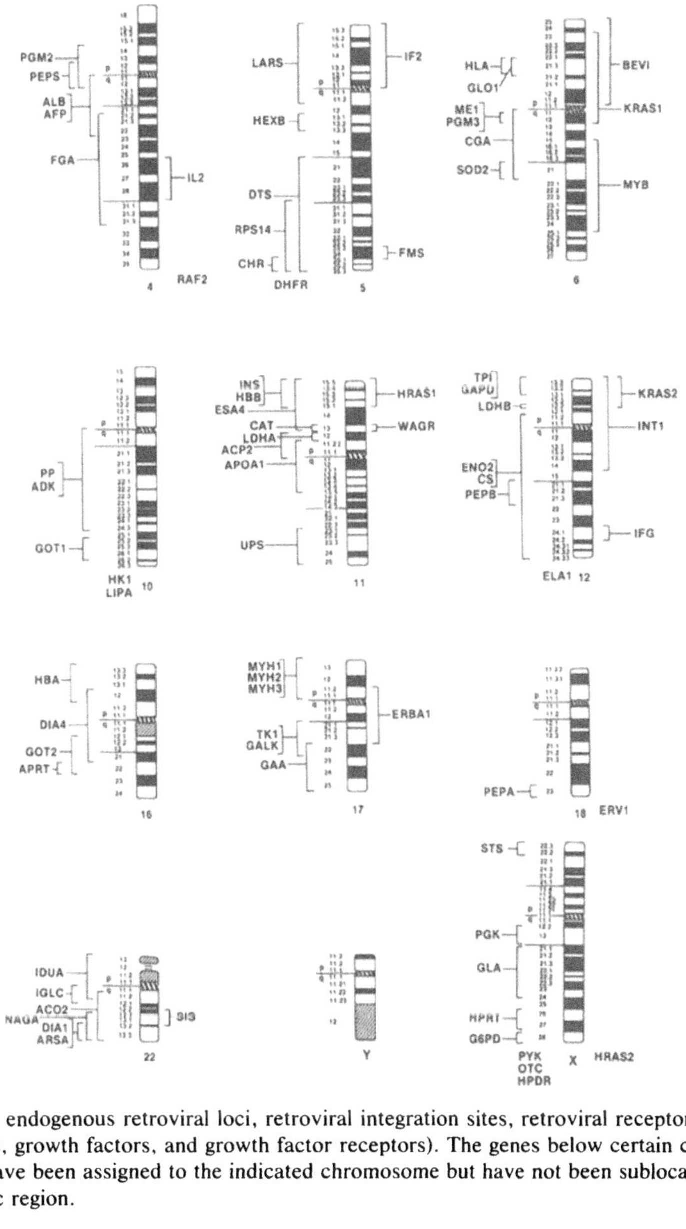

cogenes, endogenous retroviral loci, retroviral integration sites, retroviral receptors, lymphokines, growth factors, and growth factor receptors). The genes below certain chromosomes have been assigned to the indicated chromosome but have not been sublocalized to a specific region.

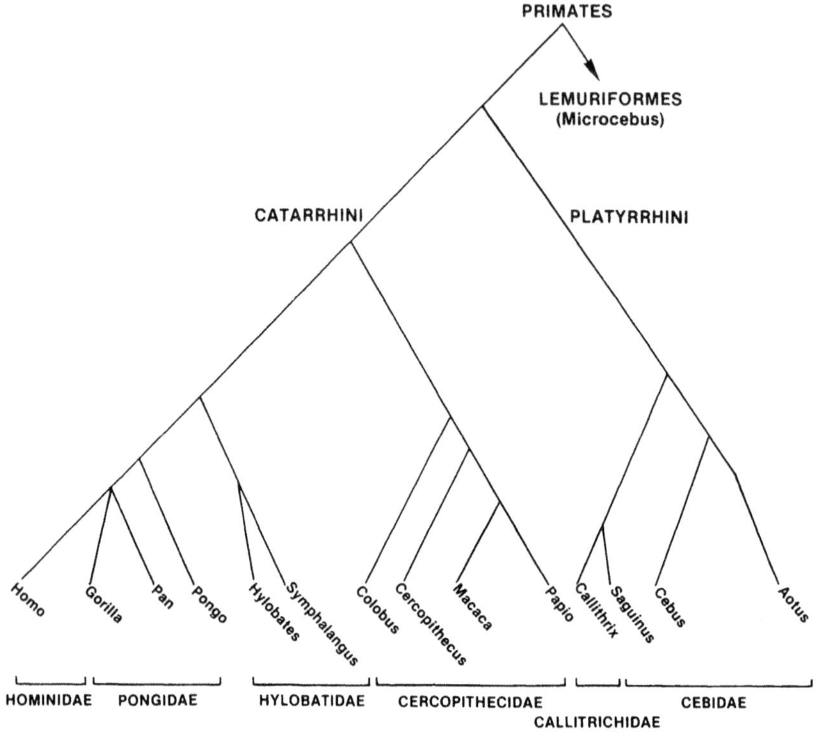

Figure 4. A consensus phylogeny of the primates based largely on morphological considerations (Eisenberg, 1981; Nowak and Paradiso, 1983); see also Benveniste and Todaro, 1976.)

of amino acid substitutions (Bruce and Ayala, 1978; Goodman,1975; King and Wilson, 1975). Moreover, molecular hybridization studies of single-copy genomic DNA (Benveniste and Todaro, 1976) as well as mitochondrial DNA (Ferris *et al.*, 1981a,b) have also indicated a very close phylogenetic relationship between these species, especially between man and the African apes. These findings led to the postulation that man and the great apes should be included in a single family (Hominidae). This taxonomic arrangement clearly contrasted with previous convention, based on gross morphometric characteristics, which split the great apes (Pongidae) from man and his extinct ancestors (Hominidae) within the superfamily Hominoidea (Simpson, 1945).

Man and chimpanzee (*Pan troglodytes*) were initially thought to share the same diploid chromosome number ($2n = 48$) (Painter, 1923; Yeager *et al.*, 1940). Later work, however, clearly showed that man had a diploid chromosome number of 46 (Tjio and Levan, 1956; Ford and Hamerton, 1956), whereas all great ape species had a diploid number of 48 chromosomes (Young *et al.*, 1960; Hamerton *et al.*, 1963; Chiarelli, 1961). Further studies of the nonhuman primates, gave partial evidence of karyotypic similarities between different species; a finding that established evolutionary conservation, at the cytological level, within the primate order [for reviews, see Chiarelli *et al.* (1979) and Seuanez (1984)]. A striking example of such similarity was the finding of a primary trisomy for a small acrocentric chromosome, similar to chromosome 21 in man, in a newborn chimpanzee with a pathological condition comparable to Down's syndrome in man (mongolism) (McClure *et al.*, 1969). Thus, a similar kind of chromosome aneuploidy was responsible, in two phylogenetically related species, for the appearance of the same clinical condition, apparently because these trisomies were unbalancing very similar genes in both man and the chimpanzee.

Comparative cytological analyses of human and great ape karyotypes have shown a striking conservation of chromosomal appearance. The banding homology is so conserved that almost every human chromosome has a recognizable counterpart in each of the great ape species (Paris Conference 1971, Supplement, 1975; Stockholm Conference, 1977). Recently, Yunis and Prakash (1982) have used high-resolution banding of the chromosomes of great apes and man to reconstruct the chromosomal rearrangements that have presumably occurred and become fixed during hominoid radiations. This report confirmed previous observations (Turleau *et al.*, 1972; Dutrillaux, 1975; D. A. Miller, 1977; Seuanez, 1979; Stanyon and Chiarelli,1982) that most of the postulated rearrangements were pericentric inversions, although other rearrangements (paracentric inversions, chromosome fusion, etc.) have also occurred. The conventionally established chromosomal homologue (based on cytological appearance) of each great ape is presented in Table I.

The cytogenetic rearrangements that have been postulated to occur were described using banding procedures that visualize euchromatin. Heterochromatic regions which are detected by brilliant Q fluorescence are present in man and great apes exclusive of the orangutan (*Pongo pygmaeus*), suggesting that this material appeared subsequent to divergence from the orangutan but prior to the man-African ape radiation. Constitutive heterochromatin (C-band) regions do not follow the same evolutionary pattern. Many chimp-gorilla homologous chromosomes show (telomeric) regions that are absent in man or the orangutan. This observation has been interpreted as evidence that the chimpanzee and the gorilla may be closer to one another than either is to man or the orangutan (Stanyon and Chiarelli, 1982). However, the transposition, addition, deletion, etc.,

Table I
Chromosomal Positions of Homologous Loci in Man and the Great Apes

Human chromosome	Marker	Chimpanzee chromosome	Gorilla chromosome	Orangutan chromosome
1p	*PGM1*	1	1	1q
1p	*PGD*	1	1	1q
1p	*ENO1*	1	1	1q
1p	*AK2*	—	—	1q
1p	*FUCA*	—	—	—
1q	*PEPC*	1	1	1
1q	*FH*	—	1	1p
2q	*IDH1*	12(13)	12(11)	12(11)
2p	*MDH1*	13(12)	11(12)	11(12)
2p	*ACP1*	13(12)	—	11(12)
3	*GPX1*	2	2	(2)
4p	*PGM2*	3	3	3
5	*HEXB*	(4)	(4)	4
6p	*GLO*	5	—	—
6p	*MHC*	5	5	5
6q	*PGM3*	5	5	5
6q	*SOD2*	5	5	5
6q	*ME1*	5	5	5
7q	*GUSB*	6	6	(10)
8p	*GSR*	7	7	6
9q	*AK1*	11	13	(13)
9p	*ACO1*	11	13	—
9p	*AK3*	11	—	—
10p	*GOT1*	8	8	7
11p	*LDHA*	9	9	8
11p	*ACP2*	9	—	—
12p	*GAPD*	10	10	9
12p	*TPI1*	10	10	9
12p	*LDHB*	10	10	9
12q	*PEPB*	10	10	9
13q	*ESD*	(14)	14	(14)
14q	*NP*	15	18	(15)

Table I
(*continued*)

Human chromosome	Marker	Chimpanzee chromosome	Gorilla chromosome	Orangutan chromosome
15q	*MPI*	16	—	(16)
15q	*PKM2*	16	15	—
15q	*HEXA*	16	15	—
17q	*TK*	19	19	19
17q	*GALK*	19	19	—
18q	*PEPA*	17	16	(17)
19	*GPI*	20	20	20
20p	*ITPA*	15(21)	18(21)	(21)
21q	*SOD1*	22	22	22
X	*G6PD*	X	X	X
X	*HPRT*	X	—	—
X	*GLA*	X	X	X
X	*PGK*	X	X	—

[a]The great ape cytological homologue (Stockholm Conference, 1977) and the syntenic homologues are the same in every case except human chromosomes 2p, 2q, and 20. In these cases and in those cases when no homologues have been mapped, the cytological homologue is in parentheses.

of either brilliant fluorescent or constitutive heterochromatin regions appear to be rather dynamic, and may be independent of euchromatic linkage within the Hominoidea. For this reason, these cytological characters should be considered cautiously as markers for chromosomal phylogeny.

Approximately 40 structural genes homologus to human loci have been mapped in each great ape species (Table I). With the exception of human chromosome 2 (see below), each of the testable human syntenic groups is conserved intact in the great apes. Further, the data on comparative gene assignment have been largely coincident with the observed morphological similarities, at the chromosomal level, between different species (Table I). In some cases, there is a very clear correspondence between morphological and syntenic similarities, as is the case of chromosome 1 in man and its homologous counterpart in the orangutan. The morphological homologue of human 1p in great apes is 1q, and *vice versa*, human 1q is homologous to great ape 1p. This is because chromosome 1 in the great apes lacks a secondary constriction region, which in human 1 has elongated one arm with respect to the other, making the arm ratio of human chromosome 1 opposite to that in the great apes. Regional (arm) assignments in human and orangutan chromosome 1 have shown a strict correspondence (Table I) to the predictions of comparative banding studies.

Although the majority of cytological homologies in the great apes have been largely confirmed by syntenic homologies, there are certain exceptions. Human chromosome 2 has no single counterpart in the great apes. Instead, there is an acrocentric (or subtelocentric) homologue pair for 2p, and another acrocentric (or subtelocentric) homologue pair for 2q (see Table I). A direct cytological comparison between species suggests that human chromosome 2 emerged as a result of a fusion between two nonhomologous ancestral chromosomes similar to these arm counterparts. This fusion event explains the difference in diploid number between man ($2n = 46$) and the great apes ($2n = 48$), and postulates that human chromosome 2 consists of an association of two primitive syntenic groups that are still separate from one another in the great apes. Surprisingly, when genes homologus to human chromosome 2p (*MDH1* and *ACP1*) and to chromosome 2q (IDH1) were assigned to great ape chromosomes, their position was exactly the opposite of the predictions on cytological grounds (Sun et al., 1978a,b). Thus, *MDH1* and *ACP1* were mapped to the great ape homologues of human 2q despite their regional position on human 2p (Fig. 3). Conversely, *IDH1* was mapped to the great ape homologues of 2p despite its location on human 2q. The lack of agreement of syntenic and banding homologies is not easily reconciled and could represent either a biological enigma or possibly a mistake either in the syntenic or cytological reports. Whatever the explanation, some caution seems warranted in making phylogenetic inferences from conflicting syntenic and cytological observations.

A second conflict of the cytological and syntenic conclusions was raised by Yunis and Prakash (1982), who hypothesized on cytological grounds that the gorilla has two chromosome pairs that were derived from a 5p–17q reciprocal translocation in man, one homologous to human 5q–17q, and another to human 5p–17p. This presumptive translocation would necessarily require the relocation of the thymidine kinase (*TK1*) locus, which is in human 17q (Fig. 3), to the short arm of gorilla chromosome 4, the tentative 5q–17q homologue of man. However, the *TK1* locus has been assigned to gorilla chromosome 19 (see Table I), a chromosome of similar size to human chromosome 17, although not precisely with similar banding patterns to its human counterpart.

A third discordance of the great ape-human comparison is seen with the inosine-triphosphatase (*ITPA*) locus. *ITPA* is located on human chromosome 20, but is not located on the identical cytological homologues in the chimpanzee and the gorilla (chromosome 21 in these two species).

Actually, the *ITPA* locus was found to be syntenic with the nucleoside phosphorylase (*NP*) locus, in chromosome 15 of the chimpanzee and 18 of the gorilla, respectively, which can thus be considered to be the syntenic homologues of human chromosomes 14 + 20. This suggests that a cytologically invisible transposition has occurred at some stage during phylogeny, although we still do not know when this event actually took place. This is because *NP* and *ITPA* show dissociation in the rhesus monkey and the baboon, while they are syntenic in the rabbit (Soulie and de Grouchy, 1982), and because *NP* and *ITPA* also show dissociation in the domestic cat (see below), the mouse, the Chinese hamster, and the dog (see below, Sections 5 and 6). Finally, two chromosomes in the gorilla (13 and 18) that are the syntenic homologues of human chromosomes 9 and 14 do not share any apparent morphological similarities between their human counterparts.

In spite of these few exceptions, it is striking how coincident the cytological banding patterns and syntenic homologies among the Hominoidea are. The general conclusion derived from both cytological and linkage analyses is that cytological variation between man and great apes is extremely conserved, with the odd rearrangement being the exception to the general rule of band-to-band coincidence throughout the genome of these species.

4.2. Comparative Mapping of Man and Other Primates

Chromosomal similarities between man and the nonhuman primates have now been extended along the catarrhine and platyrrhine lineages (Fig. 4) as well as to the prosimians (Dutrillaux, 1979a; Giusto and Margulis, 1981). These tentative homologies represent relict similarities, which have been evolutionarily conserved throughout the primate lineage, with certain striking exceptions, such as the gibbons and the owl monkeys, where chromosome rearrangements have been prominent. Comparative gene mapping has confirmed many of these morphological similarities (Table II), especially in the X chromosome. Human X-linked genes are invariably found to be X-linked in other primates and in other mammals, confirming Ohno's (1973) hypothesis that the mammalian X chromosome represents an obligate syntenic group held together throughout the mammals, probably as a consequence of X-inactivation.

Several human autosomal syntenic associations are generally conserved throughout the primates (Table II). The syntenic association of human 1p + 1q markers is retained in certain catarrhine monkeys (e.g., *Macaca mulatta*, rhesus), but not in others (the African green monkey, *Cercopithecus aethiops*). *Cercopithecus* contains a cytological homologue

Table II
Comparative Gene Mapping: Man, Old World and New World Monkeys, and the Mouse Lemur[a]

Human chromosome	Human loci	Gibbon chromosome	Rhesus chromosome	Baboon chromosome	African green monkey chromosome	Capuchin monkey chromosome	Owl monkey (*Aotus* K-VI) chromosome	Mouse lemur chromosome
1p	*PGM1*	5	1	1	1	15	12	3
1p	*PGD*	24	1	1	1	15	12	3
1p	*ENO1*	24	1	1	1	15	12	3
1p	*AK2*	—	—	—	—	—	12	—
1p	*FUCA*	3	—	—	1	—	—	3
1q	*PEPC*	—	—	1	6	—	—	U4
1q	*FH*	—	1	—	6	U2	6	—
1q	*GUK1*	5	1	—	—	U2	—	U5
2p	*MDH1*	—	15	—	—	4	2	4
2p	*ACP1*	19	—	—	—	4	—	4
2q	*IDH1*	—	9	12	—	14	16	—
2	*UGP2*	—	—	—	—	4	2	4
3	*GPX1*	4	3	—	5	—	—	1
3p	*ACY1*	—	—	—	—	18	—	—
4p	*PEPS*	—	—	5	—	—	—	—
4p	*PGM2*	—	6	5	—	2	14	—
5	*HEXB*	—	5	—	—	—	—	—
6p	*GLO*	—	—	4	—	—	9	6
6p	*HLA*	—	2	—	—	—	9	—

6q	*PGM3*	17	2	—	—	3	9	6
6q	*SOD2*	3	2	4	—	—	9	—
6q	*ME1*	—	—	4	—	3	9	6
7q	*GUSB*	—	2	3	—	1	—	—
7	*MDH2*	—	2	—	—	—	4	—
8p	*GSR*	—	8	—	10	—	—	—
9q	*AK1*	8	—	—	U1	—	—	10
9p	*ACO1*	8	—	—	U1	—	15	—
9p	*AK3*	—	—	—	—	12	—	10
10q	*GOT1*	3	—	—	—	—	—	15
11p	*LDHA*	15	11	14	12	16	19	5
11p	*ACP2*	15	11	—	—	16	—	—
12p	*GAPD*	—	12	—	13	—	—	7
12p	*TPI1*	U1	12	—	13	10	10	7
12p	*LDHB*	U1	12	11	13	10	10	7
12q	*PEPB*	—	12	—	13	10	—	7
12	*CS*	—	12	—	13	—	—	7
12	*ENO2*	U1	—	—	—	—	—	7
14q	*NP*	U2	7	7	U2	—	11	2
14q	*CKBB*	U2	—	7	U3	—	—	2
15q	*MPI*	6	7	7	U2	U1	11	2
15q	*PKM2*	6	7	7	U2	U1	11	2
15q	*HEXA*	6	7	—	—	—	—	2
15q	*IDH2*	—	—	7	—	—	—	—

(continued)

Table II
(continued)

Human chromosome	Human loci	Gibbon chromosome	Rhesus chromosome	Baboon chromosome	African green monkey chromosome	Capuchin monkey chromosome	Owl monkey (*Aotus* K-VI) chromosome	Mouse lemur chromosome
15q	*SORD*	6	—	7	—	—	11	2
15q	*B2M*	—	—	—	—	—	11	—
16p	*PGP*	—	—	—	—	1	—	—
19	*GPI*	—	19	—	25	8	25	U3
19	*PEPD*	—	—	—	—	—	—	U3
20q	*ADA*	—	—	10	—	—	—	—
20p	*ITPA*	—	13	10	—	—	—	—
21q	*SOD1*	—	—	3	—	9	—	—
22	*NAGA*	—	13	—	—	—	—	—
X	*G6PD*	—	X	—	X	—	X	X
X	*HPRT*	X	—	—	—	—	—	X
X	*GLA*	X	X	—	X	—	X	X
X	*PGK*	X	—	—	X	—	X	X

[a] Derived from Roderick *et al.* (1984).

pair to human 1p and another arm homologue to human 1q (Finaz *et al.*, 1977). A similar dissociation of 1p from 1q markers has been found in the platyrrhine New World monkeys (Creau-Goldberg *et al.*, 1981, 1984; Ma, 1983a,b, 1984: Ma *et al.*, 1982), as well as in the Lemuridae (mouse lemur, *Microcebus murinus*) (Cochet *et al.*, 1982). The human 2p + 2q association is unique to man, while the conservation of each 2p and 2q linkage group has been observed throughout the primates and these chromosome arms generally show recognizable counterparts. Human chromosome 6 syntenic associations are relatively conserved in the primates, mainly involving the genes for the major histocompatibility complex in these species. In the rhesus monkey (*Macaca mulatta*), however, an association of human 6 + 7 syntenic groups has been reported (Estop *et al.*, 1979; Garver *et al.*, 1980), although this association was not observed in the baboon (*Papio papio*) (Creau-Goldberg *et al.*, 1983), a species sharing an identical chromosome complement with the rhesus monkey (Finaz *et al.*, 1978). In the baboon, genes homologous to human 6 and 7 markers were assigned to different chromosomes.

The human 12 syntenic group is another example of evolutionary conservation, as is the human 14 + 15 association in the non-ape primates. Recall that in man and in the Pongidae, the syntenic groups on chromosomes 14 and 15 are dissociated (Table I). Since the 14 + 15 association is also present in the cat and the pig (see below), this syntenic association probably represents the ancestral condition in mammals.

The lesser apes (*Hylobatidae*) and the owl monkeys (*Aotus*) are good examples of catarrhine and platyrrhine groups, respectively, among which extensive karyotypic rearrangements have occurred. Chromosome studies in the lesser apes have shown a significant chromosome diversity within this phenotypically homogeneous group (Couturier *et al.*, 1982), even between species capable of cross-breeding and producing viable offspring, such as the gibbon (*Hylobates muelleri*) and the siamang (*Symphalangus syndactylus*) (Myers and Shafer, 1979). Few cytologically homologous chromosomes or chromosomal subgroups are apparent between man and gibbon (*Hylobates concolor*). Comparative gene assignment has confirmed only the morphological homologies of the X chromosomes between these two species (Turleau *et al.*, 1983). Two human syntenic groups, 1p and 6q, show dissociation in this gibbon species (Table II), contrary to the well-established evolutionary conservation of most of these groups in other primates. The *Aotus* groups display 11 distinct karyomorphic populations (Ma, 1984) and, as such, represent an extreme of rapid karyologic diversification. The syntenic groups present in other primates are by and large disrupted in *Aotus* (Table II). In other *Aotus* karyomorphic types (not shown in Table II), syntenic associations of human 1p + 19,

2p + 12, 11 + 12 groups have been observed, as well as dissociations of human 12q markers (Ma, 1984). These changes are in agreement with the proposed cytological chromosome homologies within the *Aotus* group, and seem to follow the position of proposed cytological rearrangements postulated by Ma *et al.* (1976).

A general conclusion from comparative mapping studies in man and the nonhuman primates is that many syntenic associations have been evolutionarily conserved and that such conservation can be traced to each of the primate suborders. A positive correlation seems to exist between morphological and syntenic changes at the chromosome level. This makes morphological attributes, such as banding patterns, useful indicators of syntenic homologies in the primate order, although in some cases, these two events may not be strictly coincident. At least two genera (*Aotus* and *Hylobates*) depart from the conservative chromosomal evolution seen in other primates and are characterized by highly rearranged karyotypes. The morphological rearrangements have been accompanied by apprecia-ble disruption of prototype human syntenic groups (Table II). These chro-mosomal rearrangements seem to be both recent and limited, since de-spite the numerous syntenic disruptions, several others (which span some distance in the human genome, see Fig. 3) have been retained intact (e.g., *NP-CKBB; MPI-PKM2-HEXA-SORD* in gibbon).

4.3. Primate Gene Families

4.3.1. Satellite DNA Sequences

The human genome, like that of other eukaryotes, contains a sub-stantial proportion of repetitive DNA sequences that are kinetically char-acterized by their fast reassociation rate with respect to single-copy DNA (Britten and Kohne, 1968). The isolation of native, long tandem DNA repeats has been achieved based upon differences in base composition between a highly repetitive fraction and the rest of the DNA. Four dif-ferent satellite fractions, designated I–IV, were initially characterized by their physical properties, including buoyant density in neutral CsCl or their melting profiles. More recently, Mitchell *et al.* (1979) further char-acterized these human satellite DNAs by their restriction endonuclease digestion patterns, and carried out cross-reassociation experiments to detect the degree of sequence homology between them. Satellites III and IV were shown to contain homologous sequences in spite of having been isolated as different isopycnic fractions. Moreover, cross-reassociation studies showed that 40% of satellite III was homologous to 10% of either satellite I or II, although satellites I and II did not cross-react with one another. Endonuclease digestion and Southern analyses showed that satellites II and III consisted of heterogeneous components that had been copurified and coisolated as single isopycnic fractions. Satellite DNA III from male individuals, for example, when digested with *Hae*III, showed (1) a high-molecular weight component; (2) a 3.4-kilobase (kb) repetitive fraction identical with male-specific DNA (Cooke, 1976; Bostock *et al.*, 1978); and (3) a restriction "ladder" formed by a multimeric series of a small monomeric fragment of 0.17 kb. A two-dimensional gel electrophoresis-hybridization transfer (Bostock *et al.*, 1978 and Mitchell *et al.*, 1979), using each component as a probe against all others, detected cross-hybridization only among ladder fragments. (See also Miklos, this volume, Chapter 4, for more information on primate satellite DNA).

Homologous sequences to the human satellite DNAs were also found in the genome of some nonhuman primates. *Hae*III restriction digests of primate DNA followed by filter hybridizations to a satellite III DNA probe showed that the *Hae*III ladder component of satellite III DNA had probably appeared in the catarrhine genome some 24 mya (Mitchell *et al.*, 1981). Conversely, satellite I DNA was found only in the hominoid genome, including the lesser apes, while satellite II sequences were found in man but not in the great apes. These findings suggested that the ladder component of satellite III DNA was older than the other two components of satellite III, and than satellites I and II DNAs. Since satellite III was found to cross-reassociate with both satellites I and II DNAs (Mitchell *et al.*, 1979), it is therefore probable that they have originated from satellite III DNA at different stages of the evolutionary divergence of the primates.

The chromosomal localization of human satellite DNA sequences has been achieved using *in situ* hybridization (Jones *et al.*, 1973a; 1974; Evans *et al.*, 1974a; Gosden *et al.*, 1975) and hybrid cell analyses (Schmeckpeper *et al.*, 1979; Beauchamp *et al.*, 1979). Interspecific *in situ* hybridizations, using human satellite probes to form molecular hybrids with chromosomal DNA of the great apes, have been useful for the detection of homologous sequences in the higher primates (Jones *et al.*, 1973b; Mitchell *et al.*, 1977; Gosden *et al.*, 1977). These studies (Table III) showed that homologous sequences to human satellites I–III could be detected by *in situ* hybridization in great ape chromosomes, except for satellite II sequences in the chimpanzee (*Pan troglodytes*). By identifying each primate chromosome with quinacrine staining before *in situ* hybridization, a detailed comparison was made between (1) the relative amount of hybridization among different chromosomes of a species and between different species; and (2) the chromosome region where hybridization was detected. The sites of hybridization were mainly the centromere region of many chromosomes, the Y chromosome, and the secondary constriction region of gorilla chro-

Table III
In Situ Hybridizations of Human Complementary cRNAs I–III Great Ape Chromosomes[a]

Man				Chimpanzee				Gorilla				Orangutan			
	cRNA				cRNA				cRNA				cRNA		
Chromosome	I	II	III	Chromosome	I	II	III	Chromosome	I	II	III	Chromosome	I	II	III
1	+	+	+	1	−	−	+	1	−	+	+	1	−	−	−
2p	−	−	−	12	−	−	−	12	+	+	+	12	+	+	+
2q	−	−	−	13	+	−	+	11	−	−	−	11	+	+	+
3	−	−	−	2	−	−	−	2	−	−	−	2	−	−	−
4	−	−	−	3	−	−	−	3	−	−	−	3	−	−	−
5	+	−	+	4	−	−	−	4	+	+	+	4	−	−	−
6	−	−	−	5	−	−	−	5	−	−	−	5	−	−	−
7	+	−	+	6	+	−	+	6	−	−	−	10	−	−	−
8	−	−	−	7	−	−	−	7	−	−	−	6	−	−	−
9	+	+	+	11	+	−	+	13	+	+	+	13	+	+	+
10	−	−	−	8	−	−	−	8	−	−	−	7	−	+	−
11	−	−	−	9	−	−	−	9	−	−	−	8	+	−	−
12	+	−	−	10	−	−	−	10	−	−	−	9	+	−	−
13	+	+	+	14	+	−	+	14	+	+	+	14	+	+	+
14	+	+	+	15	+	−	+	18	+	+	+	15	+	+	+
15	+	+	+	16	+	−	+	15	+	+	+	16	+	+	+
16	−	+	−	18	+	−	+	17	+	+	+	18	−	−	−
17	−	+	−	19	+	−	+	19	−	−	−	19	−	−	−
18	−	−	−	17	+	−	+	16	+	+	+	17	+	+	+
19	−	−	−	20	−	−	−	20	−	−	−	20	−	−	−
20	+	+	+	21	+	−	+	21	−	−	−	21	−	−	−
21	+	+	+	22	+	−	+	22	−	−	−	22	+	+	+
22	+	+	+	23	+	−	+	23	+	+	+	23	+	+	+
X	−	−	−	X	−	−	−	X	−	−	−	X	−	−	−
Y	+	+	+	Y	+	−	+	Y	+	+	+	Y	+	+	+

[a] From Mitchell *et al.* (1977) and Gosden *et al.* (1977). +, hybridization; −, no hybridization.

mosomes 17 and 18, which, like the secondary constriction regions of chromosomes 1, 9, and 16, corresponded to regions of constitutive heterochromatin (see Table III).

We know now from later work (Mitchell *et al.*, 1981) that satellite II DNA is present only in man, so that hybridization of gorilla and orangutan chromosomes to a satellite II probe probably represents a cross-reassociation with satellite III DNA (Mitchell *et al.*, 1979). However, since satellite III is also present in the chimpanzee, we cannot explain the lack of hybridization between chimpanzee chromosomes and the satellite II probe. Interspecific hybridizations with the satellite I probe must have resulted in some small amount of cross-hybridization to satellite III, but must have also detected most (90%) of satellite I DNA, which is different from (and does not cross-reassociate with) satellite III DNA. In fact, cross-hybridizations would probably tend to minimize differences in satellite DNA distributions between species rather than to enhance them, so that the actual differences of satellite DNA content between homologous chromosomes might, in fact, be underestimated.

The general conclusion from these studies is that satellite DNA amplification occurred independently in each lineage (Gosden *et al.*, 1977), because interspecific homologues do not always show the same kind (or amount) of hybridization (see Table III). Thus, the distribution of these repetitive sequences in the higher primates clearly contrasts with the evolutionary conservation of syntenic associations within the group, where homologous chromosomes generally contain homologous genes. Moreover, the evolution of human satellite DNAs in man and their related sequences in the higher primates has probably been independent of chromosome change, because the differences in satellite DNA distribution between homologue chromosomes could not be explained by rearrangements. This is contrary to the observations in the kangaroo rats, where the presence of satellite DNA is related to chromosome rearrangement, and, notably, to the evolutionary divergence of this group (Hatch *et al.*, 1976).

4.3.2. Y-Specific Repetitive DNA

Hybridization between male and female DNAs allowed the isolation of a small fraction of male DNA that did not cross-reassociate with female DNA (Kunkel *et al.*, 1976); this sequence has been localized in the long arm of the human Y chromosome (Kunkel *et al.*, 1977). *Hae*III digestion of male and female DNAs showed that males possessed two repetitive kilobase-long fragments, 3.4 and 2.1 kb, which accounted for 40 and 20% of the Y chromosome DNA, respectively (Cooke, 1976). The 3.4-kb sequence, which was also present in purified satellite III DNA (Bostock *et al.*, 1978), was found to cross-hybridized weakly with autosomal DNA (Cooke and McKay, 1978). This sequence was localized to the brilliant fluorescent block of the human Y chromosome (Bostock *et al.*, 1978) and a positive correlation was observed between the amount of the 3.4-kb sequence and the length of the brilliant fluorescent region of the human Y chromosome (McKay *et al.*, 1978). Conversely, the abundance of the 2.1-kb sequence was unrelated to variation in chromosome fluorescence, in spite of its location at the distal tip of the long arm of the human Y chromosome (Schmidtke and Schmid, 1980; Cooke *et al.*, 1982).

Endonuclease digestion of nonhuman primate DNA has so far failed to detect a male-specific DNA fragment; and *in situ* and filter hybridizations with a human, male-specific, 3.4-kb probe did not reveal a homologous sequence family in the genome of the nonhuman primates (Szabo *et al.*, 1979; Mitchell *et al.*, 1981). These findings suggested that the 3.4-kb, Y-specific sequence appeared in the human genome some time after the divergence of man from the common stock with the great apes. Cooke and McKay (1978) have postulated that this 3.4-kb sequence family in man was originally distributed in many chromosomes, but that the copies present on the modern Y chromosome have become more similar to each other than to those present in the rest of the genome. Autosomal sequences homologous to the human *Hae*III, 2.1-kb, male-specific DNA have been found in the chimpanzee and the gorilla, but these were not detected in their Y chromosomes (Cooke *et al.*, 1982). Apparently, the human copies of the 2.1-kb fragment present in the Y chromosome were recently amplified and kept homogeneous apart from those in the autosomes. In both gorilla and man, the autosomal sequences produce an orderly pattern of restriction enzyme fragments, which are consistent with the notion of a tandem array(s) of these sequences on the autosomes. Conversely, chimpanzee DNA produces a heterogeneous smear after restriction analysis, suggesting that the sequences may be either modified in those restriction sites that yield internal fragments or that the individual sequences may be chromosomally dispersed to such an extreme degree as to preclude resolution of individual segments.

4.3.3. Ribosomal Gene Evolution in the Primates

Multigene families coding for ribosomal RNA in eukaryotes comprise three different subunits of remarkable sequence conservation, which are designated 5S, 18S, and 28S. The 18S and 28S cistrons form part of a transcribed DNA segment approximately 13 kb pairs in length in man, which encodes the 45S precursor of the 18S and 28S rRHAs, and is present in approximately 200–300 copies in the human genome. In man and the great apes, the 5′ → 3′ orientation of this segment with respect to the origin of transcription is: (1) the external transcribed spacer; (2) the 18S structural gene; (3) the internal transcribed spacer; and (4) the 28S structural gene. Transcribed segments are interspersed by a nontranscribed spacer that, unlike the 18S and 28S cistrons, has accumulated substantial nucleotide base substitutions during the hominoid-ape radiation (Birnstiel and Grunstein, 1972; Nelkin *et al.*, 1980; Arnheim *et al.*, 1980; Cerbi, this volume, Chapter 7).

In man, the rDNA cistrons have been localized by *in situ* hybridization to the secondary constriction region of chromosomes 13, 14, 15, 21, and 22 (Henderson *et al.*, 1972: Evans *et al.*, 1974b); Table IV. The nucleolar organizer regions (NORs), which contain the rDNA genes, can be specifically stained by silver (Goodpasture and Bloom, 1975) when they are transcriptionally active. A comparison of the chromosome localization of the 18S–28S cistrons in the primates has shown no clear correlation between the regional distribution of these sequences and cytological or evolutionary syntenic homology of the regions that encode these cistrons, probably because chromosomal evolution and dispersion of rDNA genes are independent processes (Henderson *et al.*, 1977). For example, in man and chimpanzee, the ribosomal cistrons are distributed on five pairs of chromosomes, only two of which contain ribosomal genes

Table IV
Chromosome Distribution of 18S and 28S Ribosomal DNA Genes in Man and the Great Apes[a]

Man[b]	Chimpanzee[c]	Pygmy chimpanzee[c]	Gorilla[c]	Orangutan[d]
2p	12	12	12	**12**
2q	13	13	11	**11**
9	11	11	13	**13**
13	**14**	**14**	14	**14**
14	**15**	**15**	18	**15**
15	16	16	15	**16**
18	**17**	**17**	16	**17**
21	**22**	**22**	**22**	**22**
22	**23**	**23**	**23**	**23**

[a]Chromosomes containing rDNA genes in each species are boxed. Chromosomes of the great apes are cytological homologues to human chromosomes. (Stockholm Conference, 1977).
[b]From Henderson *et al.* (1972).
[c]From Henderson *et al.* (1974b, 1976a).
[d]From Gosden *et al.* (1978).

in the gorilla (Table IV). The lack of concordance of chromosome homology (or synteny) with the rDNA position is seen in other primates as well. In the cercopithecoid monkeys, the gibbon, and the siamang, rDNA sequences are restricted to one chromosome pair (Henderson *et al.*, 1974a, 1977; Warburton *et al.*, 1975; Tantravahi *et al.*, 1975; Henderson *et al.*, 1976a), which is characteristically evident as a "marker" chromosome pair in their chromosome complement (Chiarelli, 1971). A single chromosomal NOR is present in some platyrrhine species, such as the squirrel monkey (Lau and Arrighi, 1976), the owl monkey (C. K. Miller *et al.*, 1977), and the spider monkey (Henderson *et al.*, 1977), while in other platyrrhines, such as the marmoset and the tamarin (Bedart *et al.*, 1978), the saki (Henderson *et al.*, 1977), and the tufted capucin monkey (Freitas and Seuanez, 1982), the distribution is multichromosomal, as it is in two prosimians, the tree shrew and the lemur (Henderson *et al.*, 1977).

Molecular studies of the 18S–28S multigene families in the primates have shown a remarkable level of intraspecific sequence homogeneity, even in DNA segments of unknown function. This coincidental or "concerted" evolution (Arnheim *et al.*, 1980) is surprising because it seems as if 200–300 copies of the 18S–28S genes have become "homogenized" for mutational substitutions in each evolutionary step despite their multichromosomal distribution. Nelkin *et al.* (1980) has listed several explanations that might account for this event. First, tandem repeats of genes with related functions may be modified following a mutation in a master gene, according to the master-slave model. Second, rDNA genes could be replicated extrachromosomally, as in amphibians, to be later reinserted into chromosomes that would therefore contain identical copies. Third, crossovers between nonhomologous chromosomes could spread new mutations in rDNA genes. Fourth, the rDNA families are reduced to a single copy in germ cells (germ cell contraction) followed by extensive amplification in somatic tissues. Fifth, specific methylation of rDNA genes may produce the parallel modification of restriction patterns.

Arnheim *et al.* (1980) have argued that nonhomologous crossovers are the most likely explanation for the concerted evolution of these multigene families. By comparing the endonuclease cleavage maps of human and ape ribosomal genes, Arnheim *et al.* (1980) found some species-specific mutations in rDNA cistrons located on five different human chromosome pairs in man and the chimpanzee, two pairs in the gorilla, and nine pairs in the orangutan. Since independent mutations at multiple sites seemed highly unlikely, Arnheim *et al.* (1980) postulated that single mutations of this kind could have spread to the whole ribosomal multigene family by nonhomologous crossovers at nucleolar organizer regions. It may be important to note here that the rDNA data in mouse are somewhat different. In this species, length variations in the nontranscribed spacer were found to be restricted to each NOR, because variant sequences were not shared between nonhomologous murine chromosomes (Krystal *et al.*, 1981). The difference between human and mouse ribosomal genes was interpreted to be due to a lower rate of interchromosomal exchange in the mouse, as a consequence of the low number of NOR-bearing chromosomes in this species, and to the tight association of NORs to centromeres in murine chromosomes. In man, NORs are located in the satellite stalks of the D and G groups, and presumably are less tightly associated with centromere regions, around which crossover events may not occur. However, ribosomal multigene families in man and the great apes have been found to be in close association with satellite DNA sequences (Gosden *et al.*, 1978), the molecular variation of which is restricted to individual chromosome pairs in man, probably due to lack of nonhomologous crossovers (Beauchamp *et al.*, 1979; Cooke and Hindley, 1979).

The distribution of the 5S sequences in man and other primates shows a more stable evolutionary conservation. The 5S rDNA genes in man were localized by *in situ* hybridization to the terminal region of the long arm of chromosome 1 (Steffensen *et al.*, 1973; Atwood *et al.*, 1975). In the great apes, where the chromosome 1 homologues have a reversal of long and short arms, the 5S ribosomal cistrons are placed at the homologous position of the short arm of chromosome 1 (Henderson *et al.*, 1976b). In the baboon (*Papio cynocephalus*), the 5S genes were detected in the terminal region of the short arm of chromosome 1, a region that is also homologous to the distal region of human chromosome 1q (Warburton *et al.*, 1976).

5. Comparative Genetics and Cytogenetics of Rodents

5.1. Comparative Genetics within Rodentia

Of the 4000 extant mammalian species, approximately 1700 are rodents. Many of the 29 families, including Muridae (457 species), are relatively young, appearing in the lower Miocene. The large and diverse Cricetidae (567 species) appeared earlier, during the Oligocene. Compared to other vertebrates, rodent cytogenetics has received considerable attention. Most genera are karyotypically diverse relative to other mammalian orders, inter- and intraspecific variation being the rule rather than the exception (Bickham and Baker, 1979; Patton, 1970; Rogers *et al.*, 1984). The Muridae are characterized by both internal rearrangements and variation in chromosome number (Patton, 1970). On the other hand, the well-studied genus *Peromyscus* is invariant in chromosome number ($2n = 48$), but extensive variation in the form of pericentic inversions and presence of short-arm heterochromatin has been demonstrated (Rogers *et al.*, 1984). Despite the distance between the cricetid and murid genera (Fig. 5), G-banded chromosome homology has been suggested (Mascarello *et al.*, 1974). While the rodents offer bountiful opportunities to study the concordant evolution of karyotypes and linkage maps, comparative gene mapping in rodents still lags far behind the comparative cytogenetics.

The laboratory mouse, *Mus musculus,* has been the subject of many years of linkage analysis and its gene map consists of more than 650 genes distributed over each of its 19 autosomes and X and Y chromosomes (Roderick and Davisson, 1984). The mouse map is largely a product of controlled matings of variant phenotypes among inbred strains. The discovery and use of biochemically variant phenotypes permitted the recombination mapping of many gene loci whose homologues have now been mapped in other mammals, including humans. The use of somatic cell hybrids that segregate mouse chromosomes has resulted in a considerable growth of the mouse map, especially with loci that are invariant among inbred mouse strains (Minna *et al.*, 1975; Lalley *et al.*, 1978). The development of recombinant inbred strains (Bailey, 1971) and their use in gene mapping (Taylor, 1978) have added significantly to the mouse map.

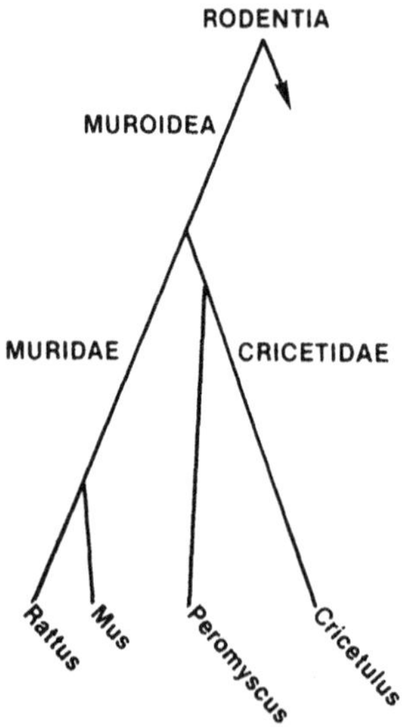

Figure 5. A consensus phylogeny of rodents (Eisenberg, 1981; Nowak and Paradiso, 1983).

The methods and history of mouse gene mapping are reviewed elsewhere (Eicher, 1981; Womack, 1982). The sophistication of its gene map makes *Mus musculus* the "standard" rodent for genetic comparisons with other mammalian orders and the focal point for comparison among rodents. An abridged mouse map that presents all the murine loci whose homologues have been mapped in other species is presented in Fig. 6.

The progress in constructing genetic maps of biochemical loci in other rodent species has not been extensive. Recent progress in gene mapping of rats (Cramer, 1984), Chinese hamsters (Stallings *et al.*, 1984), and deermice (Dawson, 1984), however, does allow the comparison of linkage relationships of several homologous gene loci (Table V) in four rodent genera from two families (Fig. 5). The Norway rat (*Rattus norvegicus*) has a diploid number of 44 chromosomes, and thus far its gene map contains 23 markers homologous to genes also assigned in the mouse (Table V). The two species show perfect concordance for the 11 rat linkage (syntenic) groups that can be compared. Furthermore, in one group (mouse chromosome 7), the gene order of *p-c-Hbb* is conserved in both species and recombination map distances are similar over the 25- to 30-cM span (Fig. 6). Additional homology of gene order and recombination frequency has been demonstrated for the esterase cluster and an erythrocyte antigen in rat Linkage Group V and mouse chromosome 8 (Womack and Sharp, 1976). Other notable conservations include mouse chromosome 17/rat chromosome 14 and mouse chromosome 4/rat chromosome 5.

The extent of linkage homology between rat and mouse raised the possibility that chromosomal banding homologies might exist between these murid genera. In an early study, Nesbitt (1974) unsuccessfully sought to identify the karyotypic homologue of the distal portion of mouse chromosome 7 after considering the *p-c-Hbb* linkage homology. More recently, Yoshida (1983a) has reported extensive cytological homology between mouse chromosome 7 and rat chromosome 7, both of which contain *Gpi-1*. Since *Gpi-1* is linked to *p-c-Hbb* in mice, it is tempting to speculate identity of rat Linkage Group I with chromosome 7 and broaden the borders of this rodent linkage conservation. This hypothesis is strengthened by the conservation of *Gpi-1-Hbb* in *Peromyscus* (Snyder, 1980). Yoshida (1983b) has also reported banding homology between mouse chromosome 4 and rat chromosome 5 while assigning rat genes homologous to mouse *Fuca* and *Gpd-1* to that chromosome.

The Chinese hamster (Cricetulus griseus) has a diploid number of 22 chromosomes and consequently should be expected to retain certain syntenic groups that are disrupted in *Mus musculus* ($2n = 40$). Despite the difference in chromosome number, however, there is extensive syntenic homology between these rodent species (Table V). There are also at least

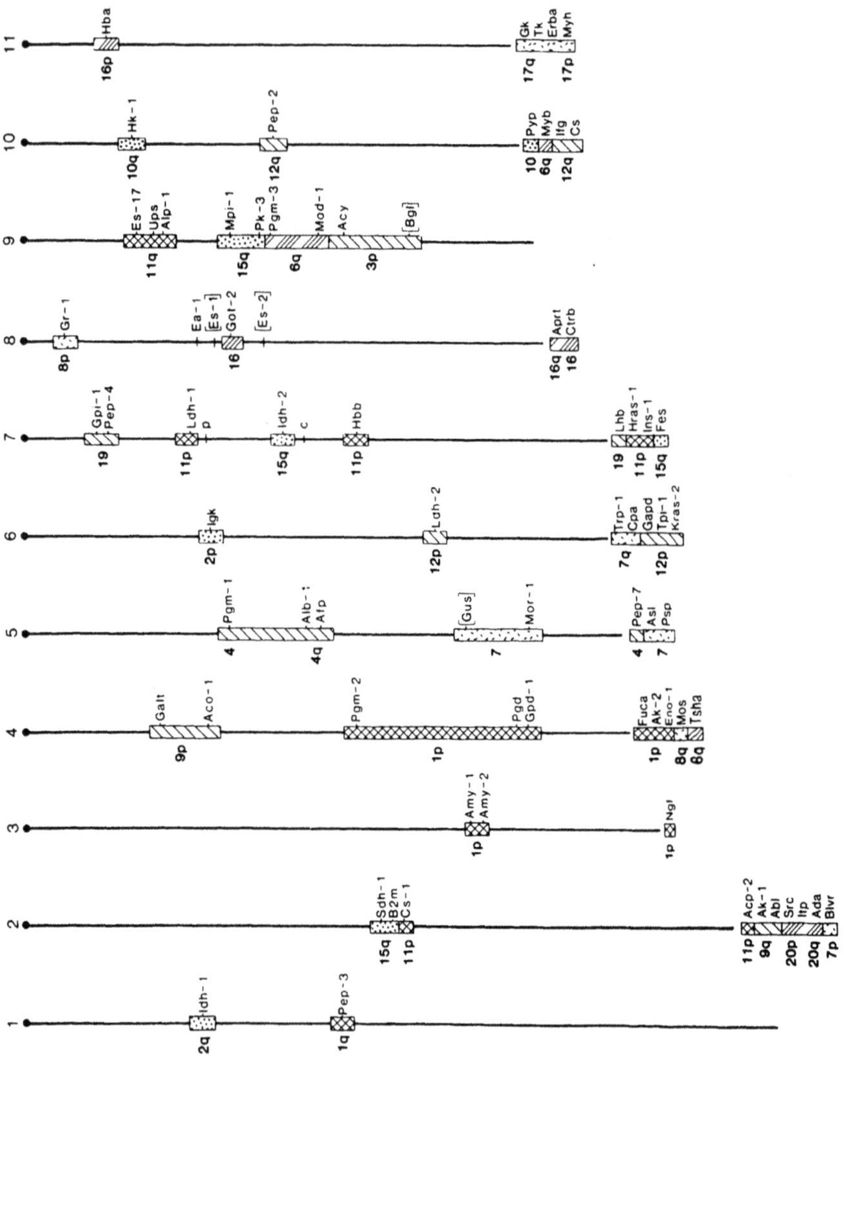
1
Idh-1
Pep-3
2q
1q
2
Sdh-1
B2m
Cs-1
15q
11p
Acp-2
Ak-1
Abl
Src
Itp
Ada
Blvr
11p
9q
20p
20q
7p
3
Amy-1
Amy-2
1p
Ngl
1p
4
Galt
Aco-1
9p
Pgm-2
Pgd
Gpd-1
1p
Fuca
Ak-2
Eno-1
Mos
Tsha
1p
8q
6q
5
Pgm-1
Alb-1
Afp
4
4q
[Gus]
Mor-1
7
Pep-7
Asl
Psp
4
7
6
Igk
2p
Ldh-2
12p
Trp-1
Cpa
Gapd
Tpi-1
Kras-2
7q
12p
7
Gpi-1
Pep-4
19
Ldh-1
p
11p
Idh-2
c
15q
Hbb
11p
Lhb
Hras-1
Ins-1
Fes
19
11p
15q
8
Gr-1
8p
Ea-1
[Es-1]
Got-2
[Es-2]
16
Aprt
Ctrb
16q
16
9
Es-17
Ups
Alp-1
11q
Mpi-1
Pk-3
Pgm-3
15q
Mod-1
6q
Acy
3p
[Bgl]
10
Hk-1
10q
Pep-2
12q
Pyp
Myb
Ifg
Cs
10
6q
12q
11
Hba
16p
Gk
Tk
Erba
Myh
17q
17p

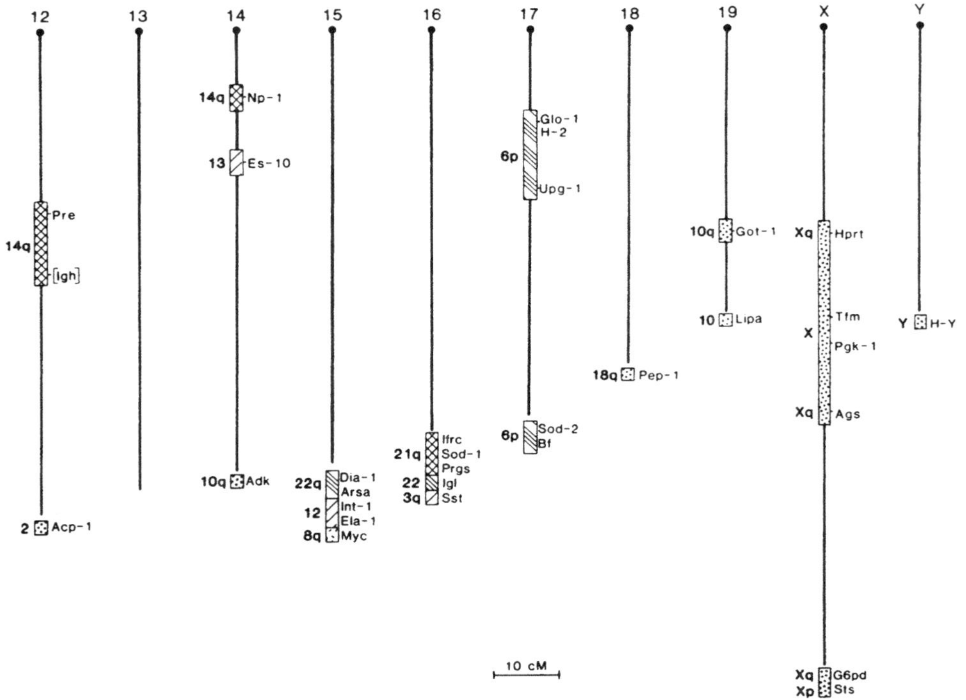

Figure 6. Biochemical genetic map of the house mouse (*Mus musculus*) demonstrating only those loci whose homologous counterparts have been mapped in other species discussed in this chapter (Roderick and Davisson, 1984; Womack, 1984). Numbers to left of mouse chromosomes indicate the chromosomal location of the gene in man (Fig. 3). Genes at the bottom of each chromosome have been assigned to that chromosome by somatic cell mapping, but their regional location is unknown. Conserved linkages in mice and man are indicated by hatched bars.

Table V
Homologous Gene Assignments in Mice and Other Rodents[a]

Gene	Mouse (*Mus musculus*)	Rat (*Rattus norvegicus*)	Chinese Hamster (*Cricetulus griseus*)	Deer mouse (*Peromyscus*)
Pep-3	1	—	5	—
Svp-1	2	IV	—	—
Ak-1	2	—	6	—
Ada	2	—	6	—
Itpa	2	—	6	—
Amy-1	3	—	—	VI
Adh-1	3	—	—	VI
b	4	II	—	II
Mup-1	4	II	—	—
Pgd-1	4	5	—	—
Gpd-1	4	5	—	—
Fuca	4	5	—	—
Galt	4	—	2q	—
Pgm-2	5	—	1	—
Pep-7	5	—	1	—
Alb-1	5	—	—	VI
Tpi-1	6	—	8	—
Idh-2	7	—	3q	—
Gpi-1	7	7	9	I
Ldh-1	7	7	—	—
Pep-4	7	7	9	—
Hbb	7	I	—	I
c	7	I	—	I
p	7	I	—	I
Aprt	8	—	3q	—
Gr-1	8	—	1	—
Es (family)	8	V	—	IV
Ea-1	8	V	—	—
Trf	9	—	—	V
Pgm-3	9	—	4	—
Mod-1	9	—	4	—
Mpi	9	2	4	—
Pk-3	9	—	4	—

Table V
(continued)

Gene	Mouse (*Mus musculus*)	Rat (*Rattus norvegicus*)	Chinese Hamster (*Cricetulus griseus*)	Deer mouse (*Peromyscus*)
Pep-2	10	—	1	VI
Tk-1	11	—	7q	—
Glk	11	—	7q	—
Igh	12	VIII	—	—
Acp-1	12	—	7	—
Adk	14	—	1	—
Es-10	14	—	1	—
Np-1	14	—	1	—
H-2	17	14 (IX)	—	—
Neu-1	17	14 (IX)	—	—
Glo-1	17	14 (IX)	—	
G6pd	X	X	X	—
Hprt	X	X	Xp	—
Ags	X	X	—	—
Pgk-1	X	X	X	—

[a] Gene symbols are from Roderick and Davisson (1984). Roman numerals are linkage groups as yet unassigned to specific chromosomes.

two disruptions in Chinese hamster chromosome linkages in mice (mouse chromosomes 7 and 8). As expected on the basis of chromosome number and size, hamster chromosome 1 contains genes from several (four) mouse chromosomes (5, 8, 10, and 14).

Conservation of linkage homology is also apparent between *Peromyscus* and *Mus musculus* (Table V) (Dawson *et al.*, 1983). An exception to this rule is Linkage Group VI of *Peromyscus*, which contains genes whose homologues are located on three different mouse chromosomes. The most extensively conserved linkage group present in these representative species of the orders Cricetidae and Muridae is in the *Gpi-1-p-c-Hbb* conservation (mouse chromosome 7). Unlike rats and mice, however, where additive recombination frequencies between *Gpi-1* and *Hbb* are 25–30%, the *Peromyscus* recombination frequency in this region is only 16% (Snyder, 1980). Since *p* and *c* are also 16 map units apart in deermice

(Robinson, 1964), they may not be bounded by *Gpi-1* and *Hbb* as expected from the mouse and rat maps. A four-point test cross to order the four homologous loci would be a useful way to resolve the disparate recombination distances. Unfortunately, such a cross has not been possible. Disruptions in gene order and/or in map distances, however, are not inconsistent with the cytogenetic observations of large numbers of inversions within the genus *Peromyscus* (Rogers *et al.*, 1984). Thus, while conserved blocks of genes persist throughout the four rodent genera, as a general rule, rodent linkages are not as conserved as those of primates.

5.2. Comparative Genetics between Rodents and Primates

By far the two most extensive mammalian gene maps are those of man and mouse. Of over 100 homologous genes that have been mapped in both species (Table VI), the comparative linkage arrangements allow a rather precise examination of the extent and character of chromosomal conservation between the rodent and primate orders, which have been separated by 100 million years. Table VI clearly demonstrates blocks of conserved linkage groups in these remotely related mammals. Twenty-nine groups of two or more genes are conserved, and several points regarding these conserved blocks are worthy of discussion.

Figures 3 and 6 provide a visual appreciation for the size of conserved autosomal segments in humans and mice. Since not all homologous genes are regionally assigned in humans or mapped by meiotic recombination in mice, this graphic is incomplete. It does, however, provide considerable insight into the upper and lower size limits of conserved autosomal segments. The smallest class of conserved segments is pairs of related genes, tightly linked and likely arising by tandem duplication of a single ancestral gene prior to divergence of the mammalian families. Such groups, exemplified by the *Amy-1* and *Amy-2* or the β-globin cluster, should be considered as a single locus for the analysis of chromosomal rearrangement. Several homologous segments are at least 10 cM in length, however, and the human chromosome 1p/mouse chromosome 4 homology is at least 25 cM in length on the mouse map. The corresponding human segment (Fig. 3) consists of the upper three-fourths of chromosome 1p. The mouse loci *Fuca, Ak-2,* and *Eno-1* have been assigned to mouse chromosome 4 using somatic cell hybrids and their position on the meiotic map is unknown. It is unlikely that they are in the region centromere-proximal to *Aco-1*, however, since *Aco-1* and *Galt* comprise a segment conserved in another human chromosome, 9p. In humans, *FUCA* and *AK2* are regionally assigned between *PGM1* and *PGD* (Fig. 3) and if gene order is conserved they will not add to the recombinant length of the conserved

Table VI
Homologous Gene Assignments in Mouse and Man[a]

Human chromosome	Human locus	Mouse locus	Mouse chromosome	Human chromosome	Human locus	Mouse locus	Mouse chromosome
1p	*PGD*	*Pgd*	4	12p	*GAPD*	*Gapd*	6
1p	*GDH*	*Gpd-1*	4	12p	*TPI1*	*Tpi-1*	6
1p	*ENO1*	*Eno-1*	4	12p	*KRAS2*	*Kras-2*	6
1p	*FUCA1*	*Fuca*	4	12p	*LDHB*	*Ldh-2*	6
1p	*AK2*	*Ak-2*	4	12q	*CS*	*Cs*	10
1p	*PGM1*	*Pgm-2*	4	12q	*PEPB*	*Pep-2*	10
1p	*AMY1*	*Amy-1*	3	12q	*IFG*	*Ifg*	10
1p	*AMY2*	*Amy-2*	3	12	*ELA1*	*Ela-1*	15
1p	*NGF*	*Ngf*	3	12	*INT-1*	*Int-1*	15
1q	*PEPC*	*Pep-3*	1				
				13	*ESD*	*Es-10*	14
				14q	*NP*	*Np*	14
				14q	*IGH*	*Igh*	12
2	*ACP1*	*Acp-1*	12	14q	*PI*	*Pre-1*	12
2p	*IGK*	*Igk*	6				
2q	*IDH1*	*Idh-1*	1	15	*SORD*	*Sdh-1*	2
				15q	*B2M*	*B2m*	2
3p	*ACY1*	*Acy*	9	15q	*MPI*	*Mpi*	9
3p	*GLB1*	*Bgl*	9	15q	*PKM2*	*Pk-3*	9
3q	*SST*	*Sst*	16	15q	*IDH2*	*Idh-2*	7
				15q	*FES*	*Fes*	7
4	*PGM2*	*Pgm-1*	5				
4	*PEPS*	*Pep-7*	5	16p	*HBA*	*Hba*	11
4q	*ALB*	*Alb-1*	5	16	*GOT2*	*Got-2*	8
4q	*AFP*	*Afp*	5	16	*CTRB*	*Ctrb*	8
				16q	*APRT*	*Aprt*	8

(continued)

Table VI
(continued)

Human chromosome	Human locus	Mouse locus	Mouse chromosome	Human chromosome	Human locus	Mouse locus	Mouse chromosome
6p	*HLA*	*H-2*	17				
6p	*GLO1*	*Glo-1*	17	17p	*MYH-1*	*Myh*	11
6p	*BF*	*Bf*	17	17p	*MYH-2*	*Myh*	11
6q	*SOD2*	*Sod-2*	17	17p	*MYH-3*	*Myh*	11
6q	*PGM3*	*Pgm-3*	9	17q	*TK1*	*Tk-1*	11
6q	*ME1*	*Mod-1*	9	17q	*GALK*	*Glk*	11
6q	*CGA*	*Tsha*	4	17	*ERBA1*	*Erba*	11
6q	*MYB*	*Myb*	10				
				18q	*PEPA*	*Pep-1*	18
7p	*BLVR*	*Blvr*	2				
7	*GUSB*	*Gus*	5	19	*GPI*	*Gpi-1*	7
7	*MDH2*	*Mor-1*	5	19	*PEPD*	*Pep-4*	7
7	*ASL*	*Asl*	5	19	*LHB*	*Lhb*	7
7	*PSP*	*Psp*	5				
7q	*TRV1*	*Trv-1*	6	20p	*ITPA*	*Itp*	2
7q	*CPA*	*Cpa*	6	20q	*ADA*	*Ada*	2
				20	*SRC*	*Src*	2
8p	*GSR*	*Gr-1*	8				
8q	*MOS*	*Mos*	4	21q	*SOD1*	*Sod-1*	16

8q	*MYC*	*Myc*	15	21q	*IFRC*	*Ifrc*	16
				21	*PRGS*	*Prgs*	16
9p	*ACO1*	*Aco-1*	4				
9p	*GALT*	*Galt*	4	22q	*ARSA*	*Arsa*	15
9q	*AK1*	*Ak-1*	2	22q	*DIA1*	*Dia-1*	15
9q	*ABL*	*Abl*	2	22q	*SIS*	*Sis*	15
				22	*IGLC*	*Igl*	16
10	*PP*	*Pyp*	10	Xq	*G6PD*	*G6pd*	X
10q	*HK1*	*Hk-1*	10	Xq	*HPRT*	*Hprt*	X
10	*LIPA*	*Lipa*	19	Xq	*GLA*	*Ags*	X
10q	*GOT1*	*Got1*	19	Xq	*PGK*	*Pgk-1*	X
10q	*ADK*	*Adk*	14	X	*TFM*	*Tfm*	X
				X	*PYK*	*Phk*	X
11p	*INS*	*Ins-1*	7	Xp	*MDD*	*Mdx*	X
11p	*HBB*	*Hbb*	7	Xp	*STS*	*Sts*	X
11p	*LDHA*	*Ldh-1*	7	X	*OTC*	*Spf*	X
11p	*HRAS1*	*Hras1*	7	X	*HPDR*	*Hyp*	X
11p	*ACP2*	*Acp-2*	2				
11p	*CAT*	*Cs-1*	2	Y	*HYA*	*H-Y*	Y
11q	*APOA1*	*Alp-1*	9				
11q	*UPS*	*Ups*	9				
11q	*ESA4*	*Es-17*	9				

[a]Gene symbols are from Shows *et al*. (1984) and Roderick and Davisson (1984). References for gene assignments of man are in Human Gene Mapping-7 (1984), those for mice in Womack (1984).

segment. Nadeau and Taylor (1984) have estimated the mean length of all conserved autosomal segments in mice and humans to be 8.1 ± 1.6 cM.

When the human and murine syntenies are compared *in toto*, two general trends become evident. First, considerable rearrangements have occurred during the divergence of man and mouse. Nadeau and Taylor (1984) have estimated that a minimum of 178 ± 39 different rearrangements have occurred since the two species diverged. The second observation is that, despite the number of required rearrangements, those syntenies that are conserved are today contiguous in both the human and in the murine genomes. For example, the lower end of mouse chromosome 4 matches human chromosome 1p and the centromeric end matches human 9p, as discussed above. The distal end of mouse 5 matches human 7, while the centromeric end matches human 4. The only clear disruption of this contiguous arrangement of conserved segments is on mouse chromosome 7, where *Idh-2* interrupts the *Ldh-1-Hbb* block. While the orientation of mouse *Pgm-3* and *Pk-3* has not been definitively established by recombination, mouse chromosome 9 is likely a tandem series of contiguous blocks of human chromosomes 11q (*Es-17, Ups, Alp-1*), 15q (*Mpi, Pk-3*), 6q (*Mod-1, Pgm-3*), and 3p (*Acy, Bgl*) proceeding from the centromeric to telomeric end. The hypothesis that conserved blocks are generally uninterrupted predicts the chromosomal location of chromosomally assigned but unmapped genes. For example, *Pep-7* likely maps to the centromeric half of mouse chromosome 5, while *As1* and *Psp* are likely in the telomeric half.

The mammalian X chromosome is highly, if not perfectly, conserved, as predicted by Ohno *et al.* (1964) as a consequence of evolution of the mammalian dosage compensation mechanism of X-inactivation. Mammalian X chromosomes vary, however, in their karyotypic appearance and the selective constraints applied by dosage compensation do not necessarily preclude internal rearrangements. Francke and Taggart (1980) studied the order of X-linked genes in humans and mice and found the gene order in mice (centromere-*Hprt-Ags*) to differ from the order of the homologues on the long arm of the human X (centromere-*GALA-HPRT*). These data predict that the karyotypic diversity of the X chromosome among mammalian groups may be coupled with rearrangements in the linear order of syntenically conserved X-linked genes.

6. Comparative Genetics and Cytogenetics of Carnivores

6.1. The Cat Family, Felidae

Carnivora is a morphologically variable order with several extremely specialized species (the high-speed cheetahs, the massive polar bear, the bamboo-eating giant panda). The Carnivora is an interesting order for this discussion because, as with the primates, certain families have a high degree of chromosomal conservation (e.g., the Felidae) while other families (e.g., Ursidae and Canidae) display a great deal of karyotypic diversity. Recently developed genetic and cytological data of the carnivores have been very informative in attempts to resolve the tempo of genomic evolution in mammals. A consensus phylogeny of Carnivora (Eisenberg, 1981; Nowak and Paradiso, 1983; Ewer, 1973) is presented in Fig. 7.

An emphasis on construction of a genetic map of the domestic cat has been a major focus in the laboratory by one of us (S.J.O.) for several years. By means of somatic cell genetics, a biochemical map of over 40 loci has been developed (O'Brien and Nash, 1982; O'Brien, 1984b). Thirty-six of these loci were homologous to genes previously mapped in man and in mouse (Table VII). Figure 8 presents a chromosome by chromosome comparison of 31 mapped feline genes and their homologous human counterparts. The genes are arranged in order of their location on human chromosomes (see also Fig. 3), since the subchromosomal locations of the cat genes are known in only a few cases.

A striking observation of this comparison between human and feline linkages is the extent of syntenic homology between the two species. Of the 17 groups compared in Fig. 8 (several with 2–4 loci), only three

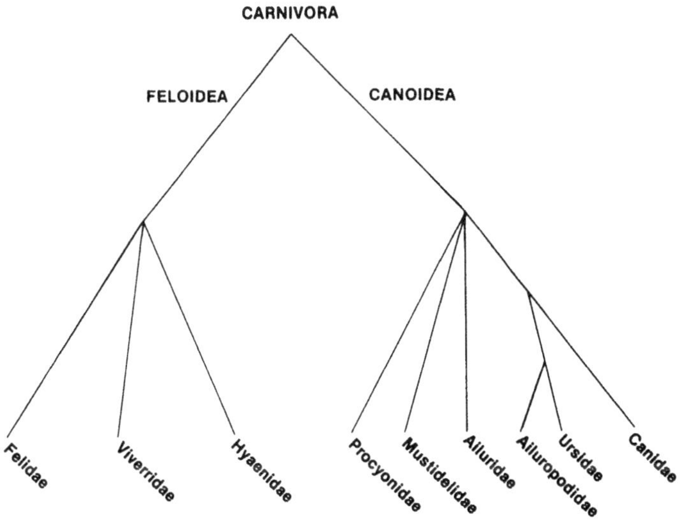

Figure 7. A consensus phylogeny of the carnivores based largely on morphological considerations (Eisenberg, 1981; Nowak and Paradiso, 1983; Ewer, 1973).

Table VII
Homologous Gene Assignments in Mammalian Species with Preliminary Gene Maps

Human chromosome	Human locus	Mouse locus	Cat (*Felis catus*)	Rabbit (*Oryctolagus cuniculus*)	Dog (*Canis familiaris*)	Cattle (*Bos taurus*)	Sheep (*Ovis ovis*)
1p	*PGD*	*Pgd*	C1	—	U1	U1	U1
1p	*ENO1*	*Eno-1*	—	U2	U1	U1	U1
1p	*PGM1*	*Pgm-2*	C1	U3	U2	U6	U1
1q	*PEPC*	*Pep-3*	—	—	U4	U13	—
1q	*GUK1*	*Guk-1*	—	15	U3	U19	—
2p	*ACP1*	*Acp-1*	A3	—	U5	—	—
2p	*MDH1*	*Mor-2*	A3	U5	U6	—	—
2q	*IDH1*	*Idh-1*	C1	—	U7	U17	—
3p	*ACY1*	*Acy*	—	9	U8	U12	—
3	*GPX1*	*Gpx*	—	9	U8	—	—
4	*PGM2*	*Pgm-1*	—	—	U9	U15	—
4	*IL2*	*Il-2*	B1	—	—	—	—
4	*PEPS*	*Pep-7*	B1	—	—	—	—
6p	*HLA*	*H-2*	—	—	—	—	U6
6p	*GLO1*	*Glo-1*	B2	—	—	—	—
6q	*SOD2*	*Sod-2*	B2	—	U10	U2	U8
6q	*PGM3*	*Pgm-3*	B2	—	U10	U2	—
6q	*ME1*	*Mod-1*	B2	—	U10	U2	—
7	*GUSB*	*Gus*	E3	—	—	—	—
7	*MDH2*	*Mor-1*	—	15	U11	U8	—
8p	*GSR*	*Gr-1*	C2	19	—	U14	—
9p	*AK3*	—	—	—	U12	—	—
9q	*AK1*	*Ak-1*	U5	—	—	U16	—
10	*PP*	*Pyp*	D4	—	U13	—	—
10q	*HK1*	*Hk-1*	D2	—	—	—	—
11p	*LDHA*	*Ldh-1*	D1	1	U14	U7	—
11p	*ACP2*	*Acp-2*	D1	1	U15	—	—
11p	*CAT*	*Cs-1*	—	—	—	U20	—
11q	*ESA4*	*Es-17*	—	—	U16	—	—
12p	*GAPD*	*Gapd*	B4	4	U17	U3	—
12p	*TPI1*	*Tpi-1*	B4	4	U17	U3	U2
12p	*LDHB*	*Ldh-2*	B4	4	U17	U3	U2
12q	*PEPB*	*Pep-2*	B4	17	—	U3	U2

Table VII
(continued)

Human chromosome	Human locus	Mouse locus	Cat (*Felis catus*)	Rabbit (*Oryctolagus cuniculus*)	Dog (*Canis familiaris*)	Cattle (*Bos taurus*)	Sheep (*Ovis ovis*)
13	*ESD*	*Es-10*	A1	—	U18	—	—
14q	*NP*	*Np*	B3	17	U19	U5	U3
15q	*HEXA*	—	B3	—	—	—	—
15q	*MPI*	*Mpi-1*	B3	—	U20	U4	U9
15q	*PKM2*	*Pk-3*	B3	—	U20	U5	U4
18q	*PEPA*	*Pep-1*	U4	—	—	—	—
19	*GPI*	*Gpi-1*	U2	U6	U22	U9	—
20p	*ITPA*	*Itp*	A3	17	U23	U11	—
20q	*ADA*	*Ada*	A3	—	—	U11	—
20	*SRC*	*Src*	—	—	—	—	—
21q	*SOD1*	*Sod-1*	C2	—	U24	U10	U10
21q	*IFRC*	*Ifrc*	—	—	—	U10	—
21	*PRGS*	*Prgs*	—	—	—	—	U10
21	*PAIS*	—	—	—	—	—	U10
22	*ACO2*	—	—	—	U25	—	—
22	*IDUA*	—	D4	—	—	—	—
Xq	*G6PD*	*G6pd*	X	X	X	X	X
Xq	*HPRT*	*Hprt*	X	—	X	X	X
Xq	*GLA*	*Ags*	X	—	X	X	X
Xq	*PGK*	*Pgk-1*	—	—	—	X	X

differences between the species are evident. First, in man, *IDH1* is linked to *MDH1* and *ACP1* on chromosome 2, while in cat, *IDH1* is linked to *PGM1* and *PGD* (chromosome C1). A discordance of *IDH1* and *MDH1-ACP1* is also observed in maps of all primates (Tables I and II), because human chromosome 2 is a recent derivation of two primitive acrocentric chromosomes that are separate in the great apes (see Section 4). The second exception is that *NP* is linked to *PKM2-MPI-HEXA* (chromosome B3) in cats, while in man *NP* is on a separate chromosome (chromosome 14) from *PKM2-MPI-HEXA* (chromosome 15) (see Fig. 3). The linkage association of *NP* with *PKM2-MPI-HEXA* is conserved in primates other

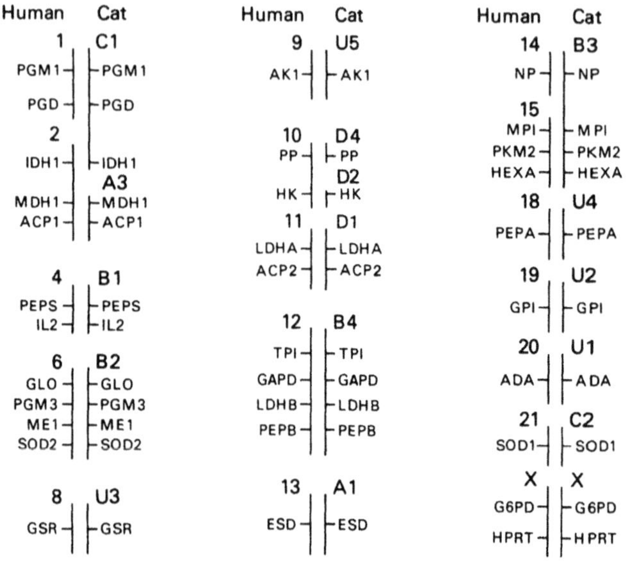

Figure 8. Linkage group associations comparing the cat and man. The gene order given is that determined in man (Fig. 3). Feline linkage groups U1–U5 have not yet been associated with a known cat chromosome. Gene names used are those of homologous human enzyme loci as recommended by the international system for human gene nomenclature (Shows *et al.*, 1984).

than the hominoids (Table II) and in pigs (Gellin *et al.*, 1981), indicating that dual chromosomes for these markers reflect a derived chromosomal characteristic of the Hominoidea. The third exception is the placement of *HK1* and *PP* on two chromosomes (D2 and D4, respectively) in the cat while these groups are linked in both mouse and man. The separation of these markers may be a recent Felidae development.

Several of the human-feline syntenies involve genes that are appreciably separated in the regional human map (Fig. 3). For example, homologous markers linked to human chromosomes 6 (*GLO1-ME1-PGM3-SOD2*) and 12 (*TPI-GAPD-LDHB-PEPB*) span the human centromere and a large portion of each arm (Fig. 3). In addition, several syntenic groups (homologous to human chromosomes 1p, 4, and X) include apparently large segments of chromosome arms if one makes the (admittedly unproven) assumption that the gene orders between the two species are the same. This extent of conservation among autosomal segments was not observed in the rodent-human analyses despite the presence of three times as many homologous murine loci to generate putative homologous segments. In fact, a similar attempt to compare linkage relationships between mouse and man using the same homologous loci (as are illustrated in Fig. 8) revealed that half of the cat-human homologous linkages are disrupted between rodents and man or cat (O'Brien and Nash, 1982).

In light of the similarities in linkage association observed between the cat and man, the extent of cytological homology between the syntenically homologous chromosomes in the two species was examined (Nash and O'Brien, 1982). G-banded chromosomes of each human type were compared with feline chromosomes thought to be homologous based on the linkage homologies shown in Fig. 8. The results of these analyses have been described in detail elsewhere (Nash and O'Brien, 1982) and will be summarized here. In general, four specific requirements for putative homology have been used, as listed in Table VIII. Essentially, only those band-for-band alignments that were apparent upon high-resolution banding and which include regionally mapped loci (in man and/or cat) were included. We have also taken advantage of the abundant cytological data which has been collected on primate cytogenetics (Dutrillaux, 1979a,b) and feline cytogenetics (Wurster-Hill and Gray, 1973; Wurster-Hill and Centerwall, 1982; Nash and O'Brien, 1982). We have already discussed the conservation of large portions of the human genome within the primate order. Then, one could identify those portions of the modern human genome that are "ancestral" and the parts that were recently rearranged. This was possible largely because a sizeable portion of the human genome has apparently been preserved intact over the approximately 80 million years of primate evolution.

For the comparison of feline and human chromosomes another serendipitous situation contributed to the feasibility of such an analysis. It turns out that the karyotype of the cat family is highly conserved (Wurster-Hill and Gray, 1973; Wurster-Hill and Centerwall, 1982). The domestic cat has 19 chromosomes, 16 of which are invariant in all 37 species of

Table VIII
Cat (Felidae) vs. Human (Primate):
Criteria for Probable Cytogenetic Homologies of Chromosomal Regions

1. Map placement of homologous genes to the same chromosome
2. Cytological correlation of G-banded thickness, spacing, and intensity, at both low-resolution banding (400-band stage) and high-resolution banding (800-band stage)
3. Placement of homologous genes, by regional gene mapping, in the region of cytological homology
4. Requirement that the putative region of homology be ancestral and evolutionarily conserved within the primate or carnivore orders[a]

[a] Based upon cytological conclusions of Wurster-Hill and Centerwall (1982), Dutrillaux (1979a,b), and Nash and O'Brien (1982).

the Felidae. Furthermore, of these 16 chromosomes, 15 are present in several other carnivore orders either intact or only slightly modified (Table IX). The domestic cat is characterized by a highly primitive karyotype, which, like the human karyotype, can be segmented into ancestral and rearranged portions. For the comparative analysis of human and feline syntenies, only the ancestral portions of chromosomes from each species were compared (Nash and O'Brien, 1982).

Six subchromosomal regions (homologous to human chromosomes 1p, 2p, 2q, 4q, 12, and X) were found to be conserved and homologous by all of the criteria listed in Table VIII. Presented in Fig. 9 are illustrations of the high-resolution banded chromosomes of cat, cheetah, and man, which were determined to be both syntenically and cytologically homologous. By length, they represent over 20% of the human genome. This is the most extensive group of syntenic and cytological homology known to exist between mammalian orders. In addition, another group of eight chromosomal regions may include homologous segments (as suggested by inclusion of homologous loci), but fail to adhere to one or more of the

Table IX
Karyology of Carnivora[a]

Family	Haploid chromosome number	Number of Felidae homologues[b]	Number of nonfelid carnivore homologues[c]	Number of family-specific homologues[d]
Felidae (*N* = 30 species)	18,19	16	0	2
Viverridae				
Palm civet	21	15	3	4
Procyonidae				
Racoon	22	12	4	4
Cacomistle	19	12	4	1
Canidae	18–36	0	0	ND
Ailuridae				
Red panda	18	7	7	4
Ailuropoda				
Giant panda	21	0	4	17

[a] These data are derived from the work of D. Wurster-Hill and her collaborators (Wurster-Hill and Gray, 1973, 1975; Wurster-Hill and Bush, 1980: Wurster-Hill and Centerwall, 1982). She has presented karyotypes of multiple species in several carnivore families, and the extent of recurrence of specific homologues in several families is indicated here. ND, Not determined.

[b] Feline homologue: chromosomes found in the Felidae family that are not rearranged.

[c] Nonfelid carnivore homologues: chromosome homologues found in several carnivore families, but not found in any felids.

[d] Family-specific: chromosomes found in several species of the indicated family, but not in any other carnivore family.

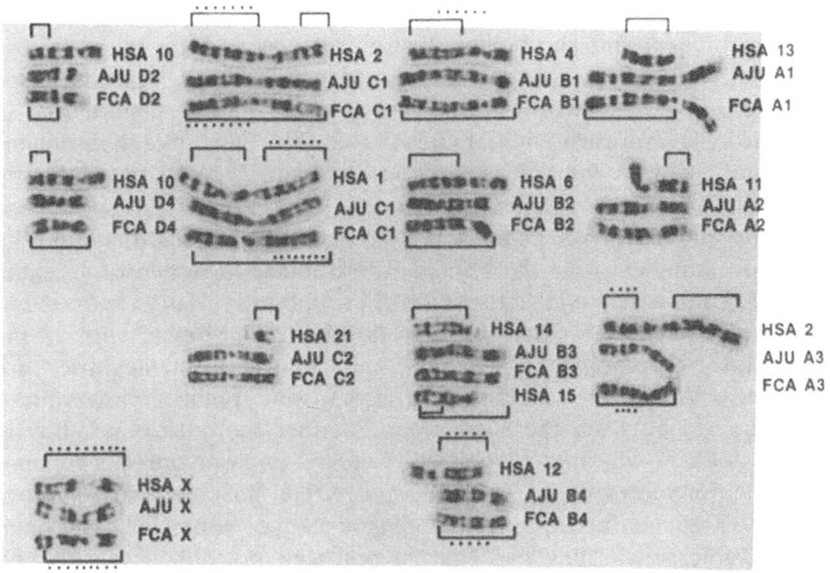

Figure 9. High-resolution G-banded comparison of the chromosomes of man (HSA), cheetah (AJU), and domestic cat (FCA), which contain syntenically and cytologically homologous regions (dotted) that fulfill each of the criteria in Table VIII. Open brackets delineate the regions of human (top) and feline (bottom) chromosomes thought to be ancestral (see text).

criteria listed in Table VIII. The simplest interpretation of these "possible homologies" is to postulate a series of chromosomal rearrangements that tend to obscure any syntenic homologies. Resolution of these chromosomes with culturing procedures that produce even higher levels of extension will clarify these regions with greater precision.

6.2. Other Carnivores

The carnivore order (Fig. 7) is generally divided into two suborders based upon the bone composition of the tympanic bullae (Eisenberg, 1981; Nowak and Paradiso, 1983; Ewer, 1973). The arctoid or canoid superfamily includes procyonids (racoon and coatimundi), dogs, bears, pandas, and mustelids, while the superfamily Feloidea contains the cats, hyenas, and viverrids (civets and mongooses). The chromosome characteristics of certain of these families have been summarized in Table IX. We have mentioned that only three of the 19 feline chromosomes vary within the Felidae. Further, the elegant cytogenetic analysis of Wurster-Hill and her colleagues has demonstrated the occurrence of a number of conserved chromosomes within the Felidae that have identical homologues in procyonids, viverrids, red pandas and hyenas. Also, certain "carnivore-specific" chromosomes which are not found in the Felidae nonetheless are present in several carnivore families (Table IX). Thus, like the primates, there are several evolutionary lines in carnivores which are highly conserved from a karyotypic perspective.

The rule of karyologic conservation is completely disrupted in other carnivore families. Like the Felidae, the Canidae have almost 40 extant species. Unlike the Felidae, the range of karyotypic variation in the canids is extreme (Table X). The domestic dog has a karyotype with 78 tiny acrocentric chromosomes, while the red fox has 36 metacentric chromosomes. Within the Canidae, the chromosome number varies almost continuously between these extremes. Neither the red fox nor the domestic dog contains any recognizable Felidae- or carnivore-specific chromosome homologues. Meera-Khan *et al.* (1984) have derived a syntenic map of 35 biochemical markers homologous to loci mapped in other mammals (Table VII). They find that several (but not all) of the linkages retained in the cat-human analysis are disrupted in the dogs. Although no mapping data are available, a similar level of cytological variability is evident in the Ursidae (bears), with certain species having large numbers of acrocentric chromosomes not homologous to carnivore-specific chromosomes (Wurster-Hill and Bush, 1980). Clearly, then, at least two carnivore radiations, Canidae and Ursidae, are karyotypically shuffled, while others including Felidae, Viverridae, and Procyonidae, are rather conserved. Furthermore, as stated in Section 6.1., some of these same primitive carnivore chromosomal segments were found intact in the primates as well.

Table X
Chromosome Numbers of Canidae (15 genera; 36 species)

Generic name	Common name	2*n*	Range
Vulpes vulpes	Red fox	36	North America, Eurasia, North Africa
Alopex lagopus	Arctic fox	48	Arctic
Fennecus zerda	Fennec fox	64	North Africa, Southwest Asia
Nycterentes procyonoides	Raccoon dog	42	East Asia
Urocyon clinereo-argenteus	Grey fox	66	North and South America
Speothos venaticus	Bush dog	74	South America
Chrysocyon brachyurus	Maned wolf	76	South America
Lycaon pictus	African hunting dog	78	Africa
Cuon aplinus	Dhole	78	Asia, Sumatra, Java
Canis familiaris	Dog	78	Worldwide

[a] For a fascinating discussion of canid chromosomal evolution, see Todd (1970).

7. Other Mammalian Families

Although representative species of most of the other mammalian orders have been karyotyped at least at the low-resolution stage, there has been only a beginning of comparative genetic analysis. The genetic maps of some species (rabbit, dog, cattle, sheep, pig) have been initiated (Table VII), but since only one or a few laboratories are actively working in these areas, progress has been slow. Nonetheless, the persistence of those few laboratories holds, we believe, much promise in the more thorough description of the processes of karyotypic evolution.

Two other mammalian orders which are characterized by genomic shuffling should be mentioned. Within the order Perissodactyla, the horse family, Equidae (Table XI), has evolved from a common ancestor in the last 4–5 million years. (For perspective, note that 4–5 mya is the most recent date suggested for the occurrence of the common ancestor of chimpanzee and man.) In fact, the equids are well known for their ability to produce interspecific hybrids. Nonetheless, the chromosome number in this family ranges from 32 in Hartman's mountain zebra to 66 in the wild precursor to domestic horses, the Mongolian wild horse (Ryder *et al.*, 1978). What is perhaps most striking is the ability to produce hybrids (albeit sterile) between such karyotypically diverse species as the donkey ($2n = 62$) and the mountain zebra ($2n = 32$).

A final and often quoted chromosomal shuffle is apparent in the Cervidae family of the order Artiodactyla (Table XII). The deer family, Cervidae, has 53 species and their karyotypes are extremely variable. The high extreme is the white-tailed deer, with a diploid number of 70. The opposite extreme is seen in the red or Indian muntjac (*Muntiacus muntjak*), which has a diploid number of six in females and seven in males due to an X-autosome translocation (Wurster and Benirschke, 1970) (Fig. 10).

Table XI
Chromosome Numbers of Equidae (one genus; seven species)[a]

Generic name	Common name	$2n$
E. przewalskii	Mongolian wild horse	66
E. caballus	Domestic horse	64
E. asinus	Ass, donkey	62
E. hemionus onager	Onager	56
E. grevyi	Grevy's zebra	46
E. burchelli boehmi	Grant's zebra	44
E. zebra hartmannae	Hartmann's mountain zebra	32

[a] For references, see Ryder *et al.* (1978).

Table XII
Chromosome Numbers in Cervidae (17 genera; 53 species)

Generic name	Common name	2*n*	Range
Muntiacus reevesi	Reeves muntjac	46	China
Muntiacus muntjak	Indian muntjac	6♀; 7♂	India
Odocoileus virginianus	White-tailed deer	70	North America

The closely related Chinese muntjac (*M. reevesi*) has a karyotype of 46 chromosomes. The observation that hybrid (although sterile) offspring were produced by crossing the two muntjac species illustrates the possible rapidity in evolutionary terms of the chromosomal reassortment that led to these modern groups. A comparative analysis of the two muntjac chromosomal complements present in the hybrid revealed that the huge Indian muntjac chromosomes are derived from a series of tandem and Robertsonian translocations of the chromosomes seen in the Chinese muntjac (Liming *et al.*, 1980). Eight of the latter species' chromosomes combined to produce chromosome 1 in the Indian muntjac.

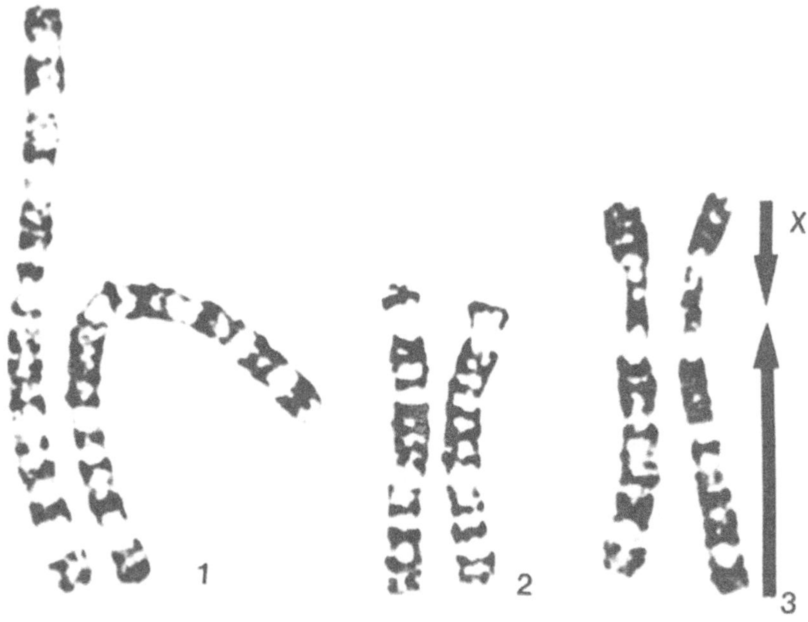

Figure 10. Karyotype of the red (Indian) muntjac, *Muntiacus muntjak*.

8. Endogenous Retroviruses

The elucidation of the existence and behavior of retroviruses has been among the most fascinating developments of mammalian molecular genetics. The RNA-containing virus was originally discovered in neoplastic tissues of chickens, mice, cats, primates, and a variety of other vertebrates (Stephenson, 1980; Weiss *et al.*, 1982). The different virus isolates all have a similar structure, containing three structural genes, *gag, pol,* and *env,* encoded in the approximately 8-kb-long genomes bounded at each end by a long (around 1.0-kb) terminal direct redundancy (LTR). The viruses often do not cause acute pathology of infected cells; rather, they form, by reverse transcription, a cDNA copy that can integrate into chromosomal DNA. The integrated provirus functions there as an autonomous genetic element, which replicates either during DNA synthesis at cell division or independently of host DNA as a proviral genome.

Since retroviruses were discovered in tumors as adventitious agents that were themselves horizontally transmitted, the discovery of endogenous retroviruses in the normal (noncancerous) cellular DNA was originally a surprise. However, multiple isolations of endogenous viruses from normal cell cultures treated with halogenated pyrimidines have been achieved. Further, the detection of cellular sequences homologous to the RNA genomes of tumor isolates confirmed the presence of these sequences in normal cell DNA (Aaronson *et al.*, 1969; Lowy *et al.*, 1971; Coffin, 1982). Endogenous retroviral sequences have been detected and isolated from a variety of vertebrate species (Table XIII; see also Benveniste, this volume, Chapter 6). The number of retrovirus copies that a particular species encodes in its genomic DNA varies from one copy to thousands. For example, it is estimated that 0.05% of the mouse genome is related to retroviruses (Callahan and Todaro, 1978). The evolutionary

Table XIII
Speices Known to Harbor Complete Endogenous Virogene Sequences in Cell DNA[a]

Chicken (RAV-O)	Baboon (M7, M28, PP-1)
Chinese hamster	Langur (PO-1-Lu, MPMV)
Syrian hamster	Squirrel monkey (SMRV, M534, M706)
House mouse (MuLV)	Owl monkey (OMC-1)
House cat (RD-114, FeLV)	Macaques (MAC-1)
Asian mouse	Colobus monkey (CPC-1)
Rat (RLV)	

[a] After Benveniste (1985).

history of endogenous retroviruses is a fascinating biological story, which is reviewed in detail by Benveniste (this volume, Chapter 6). Because of their ubiquity in mammalian genomes, however, we will present certain general observations characteristic of the gene family.

First, there are several different retroviral families in mammals, but virtually all studied species have shown some degree of retention of retroviral sequences. Some species (e.g., cat, mouse) harbor several distinct families. Although the number of copies per haploid genome varies from none to thousands, the general number of a given class is usually between 10 and 100 copies. The viral families evolve rapidly compared to unique sequence cellular DNA. For example, in Old World primates the rate of sequence divergence of the baboon endogenous virus family is 4–6 times faster than genomic DNA of the same species (Benveniste and Todaro, 1976). Rapid sequence variation is also evident within retroviral gene families of several species. Individual members of the RD-114 family in normal cat DNA vary greatly in genome length, restriction map, and sequence diversity, especially in the *env* gene (Reeves and O'Brien, 1984). A similar level of divergence was seen by Martin and hig coworkers (Repaske *et al.*, 1983; Martin *et al.*, 1981) in the murine leukemia virus-related family in human DNAs. In three species where chromosome localization of retroviral families has been achieved (cat, mouse, and man), the virogenes were dispersed to multiple chromosomes (Reeves *et al.*, 1985; Kozak, 1984; S. J. O'Brien and M. Martin, in preparation). Since *de novo* infections with retroviruses usually integrate at multiple chromosomal sites as well, the retroviral integration is thought not to be host sequence-specific (Weiss *et al.*, 1982).

Retroviral genomes share a number of biological parallels with prokaryotic and lower eukaryotic transposable elements (Temin, 1982). As such, they are capable of horizontal infection of the other individuals in a population or even different species. In fact, most of the recoverable endogenous viruses (Table XIII) are xenotropic in their host range. This means that they replicate better in a heterologous species than in the species in which they reside. Several cases of interspecies or even interordinal transmissions of endogenous viruses followed by fixation in the recipient species' genome have been reported (Benveniste, 1985; Frisby *et al.*, 1979; Benveniste and Todaro, 1974; Benveniste *et al.*, 1975).

For example, although cellular DNA from all cats tested showed 90% homology with domestic cat DNA, only four species [all members of the genus *Felis*, using the classification of Ewer (1973): *F. catus*, *F. sylvestris*, *F. chaus*, and *F. margarita*], contained DNA sequences homologous to the endogenous feline retrovirus RD-114 (Benveniste and Todaro, 1974). Ten other tested Felidae species, all members of genera other than *Felis*,

(*Prionailurus*, *Profelis*, *Lynx*, *Panthera*, and *Acinonyx*), were totally lacking in RD-114-related sequences (Benveniste and Todaro, 1974). Furthermore, an unexpected sequence homology was detected between RD-114 and the genomes of Old World monkeys (Cercopithecidae), which was subsequently shown to result from homology between baboon endogenous retrovirus (BaEV) and RD-114 (Benveniste and Todaro, 1974). Since BaEV is apparently a rather ancient primate virus (Benveniste and Todaro, 1976), these results indicate that endogenous feline RD-114 sequences in the members of the *Felis* genus were introduced to the germ line of a common ancestor of the modern *Felis* species at a point prior to the divergence of *Felis* but subsequent to the divergence of the Felidae.

With the apparent mobility that is conferred on retroviral families by their infectious nature, their ability to undergo genomic recombination with cellular mRNA becomes important. The presence of processed or intronless pseudogenes for a number of structural loci in mammals has been interpreted as reflecting a cycle of reverse transcription of a cellular mRNA mediated by a retrovirus (Leder *et al.*, 1981; Lueders, 1982; O'Brien *et al.*, 1983b). Retroviruses have been shown to recombine with another family of mammalian cellular genes, termed oncogenes, which confer on the viruses the ability to transform fibroblasts *in vitro* and to cause tumors *in vivo*. The transpositional nature of the retroviral families and their ability to capture and to move cellular genes add a new dimension to genome evolution in mammals, the ultimate significance of which is yet to be determined.

9. Comparative Genetics of Mammalian Oncogenes

Cellular transforming genes (designated c-*onc*) are a class of evolutionarily conserved vertebrate DNA segments and have been identified by two different approaches. One group of these cellular genes has been defined because of their close homology to the transforming genes of the acute transforming retroviruses (v-*onc*) (Bishop, 1983; Klein, 1981). These viruses induce oncogenic transformation of tissue culture cells and *in vivo* cause a variety of neoplasms with a short latency period. The high oncogenicity of the acute transforming retroviruses is largely due to their having transduced a cellular *onc* gene whose level of expression is now controlled by the retroviral long terminal repeat (LTR), which contains a strong promoter of RNA transcription. Acute transforming viruses have been isolated from tumors arising in widely divergent species, such as chickens, cats, rodents, and nonhuman primates (Bishop, 1983; Klein, 1981; Weiss *et al.*, 1982). Molecular analyses of these viruses has led to

Table XIV
Location of Human Genes That Participate in Neoplastic Transformation[a]

Gene family	Gene	Origin virus (species)	Human gene	Human chromosome position
Oncogenes				
src	*src*[b]	Rous sarcoma virus (chickens)	*SRC*	20
	fes/*fps*[b]	GA + ST sarcoma virus (cat), Fujinami sarcoma virus (chicken)	*FES*	15q25–q26
	abl[b]	Abelson leukemia virus (mouse)	*ABL*	9q34
	fms[b] (CSF)	McDonough sarcoma virus (cat)	*FMS*	5q34
	yes[b]	Yamaguchi sarcoma virus (chicken)	*YES*	18
	ros[b]	Rochester-2 sarcoma virus (chicken)	—	ND
	mos	Moloney sarcoma virus (mouse)	*MOS*	8q22
	erb B (EGFR)	Erythroblastosis virus (chicken)	*ERBB*	7pter–q22
	raf/*mht*/*mil*	3611 murine sarcoma virus (mouse), carcinoma virus MH2 (chicken)	*RAF1*	3p25
	raf	2611 MSV (mouse)	*RAF2*	4
	rel	Reticuloendotheliosis virus (turkey)	RELI	2p
ras	Ha-*ras*-1 (bas)	Harvey and Rasheed sarcoma virus (rat), Balb/c sarcoma virus (mouse)	*HRAS1*	11p15.1–p15.5
	Ha-*ras*-2	HSU (rat)	*HRAS2*	X
	Ki-*ras*-1	Kirsten sarcoma virus (rat)	*KRAS1*	6p23–q12
	Ki-*ras*	KiSV (rat)	*KRAS2*	12p12–pter
	N-*ras*	Human neuroblastoma	*NRAS*	1p22–p31
sis	*sis* (PDGF)	Simian sarcoma virus (Woolly monkey), Parodi–Irgens sarcoma virus (cat)	*SIS*	22q12.3–q13.1
myc	*myc*	Myelocytomatosis virus MC29 (chicken), myelocytomatosis virus OK10 (chicken), carcinoma virus MH2 (chicken)	*MYC*	8q24
	N-*myc*	Human neuroblastoma	*NMYC*	2p23–24

myb	*myb*	Avian myeloblastosis virus (chicken)	*MYB*	6q15–q24
ets	*ets*	Avian myeloblastosis/erythroblastosis virus E26 (chicken)	*ETS1*,*ETS2*	11q,21q
fos	*fos*	FBJ osteosarcoma virus (mouse)	*FOS*	14q21-q31
erb	*erb*	Erythroblastosis virus (chicken)	*ERBA1*	17p11–q21
ski	*ski*	SKV 770 (chicken)	SKV	1q12–qter
lym	B-*lym*	Human lymphoma	*BLYM1*	1p32
Endogenous retrovirus	*ERV1*	Normal human DNA	*ERV1*	18
Endogenous retrovirus	*ERV3*	Normal human DNA	*ERV2*	7
Growth factor	*NGF*	Normal human DNA	*NGF*	1p22
Growth factor	*TCGF* (*IL-2*)	Normal human DNA	*IL2*	4q26–q28
Growth factor	*PDGF* (*sis*)	Normal human DNA	*SIS*	22q12.3–q13.1
Growth factor receptor	*EGFR* (*erb B*)	Normal human fibroblasts	*EGFR*	7p11–p13
	IFRC	Normal human fibroblasts	*IFRC*	21q21–qter
Retroviral integration site	*BEVI*	Normal human fibroblasts	*BEVI*	6p
Retroviral integration site	*int-1*	Mouse tumor DNA	*INT1*	12q14–pter
Retroviral receptor	*RDRC/M7V1*	Normal human fibroblasts	*RDRC/M7V1*	19

[a] For references to chromosome position see O'Brien (1984c). ND, Not determined.

[b] Members of the *src* oncogene family whose protein products possess tyrosine-specific protein kinase activity.

the description and characterization of distinct transforming genes that have been transduced by retroviral genomes (Table XIV). Experiments in several systems have demonstrated that viral *onc* genes and some of their normal cellular homologues (when placed under the promotional control of a viral LTR) are capable of *in vitro* transformation of fibroblasts.

The second group of cellular transforming genes have been detected by the capacity of certain tumor cell genes to induce focal transformation of tissue culture cells. These transforming genes, which have been found in genomic DNA of cell lines derived from a variety of animal and human tumors (Cooper, 1982; Weinberg, 1982), represent activated forms of normal cellular genes. Control transfections using DNA from nontransformed cells and from normal tissues are generally negative in this transformation assay. Since both the retroviral and tumor cell transforming genes originated from normal cellular genes, it was not a total surprise when many of the oncogenes isolated by transfection of human tumor cells (*ras*, *myc*, N-*ras*) were shown to be identical or homologous to the retroviral oncogenes (Bishop, 1983; Cooper, 1982; Klein, 1981; Weinberg, 1982).

The normal cellular function(s) of cellular oncogenes is a slowly unfolding mystery. Several oncogene products exhibit tyrosine-specific protein kinase activity; however, several others do not. All of the oncogenes that express tyrosine kinase activity share a distant but significant DNA sequence homology, and can be grouped into a distantly related *src* family (Table XIV). Evolutionarily conserved homologues of the *src* family have been described in yeast and in *Drosophila*, suggesting that this group is an ancient primordial gene class (Lorincz and Reed, 1984; Hoffman *et al.*, 1983). At least four members of the *src* family (*mos*, *rel*, *raf*, and *erb B*) do not possess demonstrable tyrosine kinase activity (Table XIV). The functional dichotomy in the *src* family, in spite of conservation of sequence homology, has confused the interpretation of the kinase activity as the primary normal function of these oncogenes.

By means of computer-assisted DNA sequence alignment, a positive identification of the function of at least two oncogenes (*sis* and *erb B*) has been reported. The sequence of the woolly monkey virus *sis* gene precisely matched the sequence of human platelet-derived growth factor (PDGF) (Waterfield *et al.*, 1983; Doolittle *et al.*, 1983). Since this report, a number of biological studies has confirmed this conclusion (Robbins *et al.*, 1983; Deuel *et al.*, 1983). In the same manner, the oncogene *erb B* has been shown to be homologous to the cell surface receptor for epidermal growth factor (EGF) (Downward *et al.*, 1984). Since the EGF receptor is well known to contain and to utilize tyrosine-specific protein kinase activity (Cohen *et al.*, 1982), the absence of activity in the *erb B* viral protein can be reinterpreted. If the function of the *src* family of oncogenes is related to growth promotion mediated by growth factors and their receptors, the kinase activity probably is only one of several critical properties of these molecules. Thus, a variety of empirically testable models for the participation of a truncated EGF receptor gene in the alteration of normal growth control is being considered.

For a number of reasons, including their possible importance in the etiology of human cancer, the genetics of human oncogenes has recently experienced a high degree of research emphasis. All but a few of the human oncogene homologues have been mapped to a specific human chromosome, and in most cases these assignments have been confirmed in independent laboratories [for references see O'Brien (1984c)]. The present status of the human map of genes that participate in neoplasia are presented to the right of each chromosome in Fig. 3 (see also Table XIV). Included in Fig. 3 are proto-oncogene loci, growth factors, endogenous retroviral loci, retroviral integration sites, and cell surface receptors.

Because of the intense research activity and the resulting knowledge that has developed about the biology of oncogenes, this group represents a fertile class of genes in which to study genomic evolution. One class of oncogenes, the *ras* family, illustrates this point. The *ras* family was originally described as transforming genes captured by two isolates of sarcoma virus (Harvey and Kirsten) from murine leukemia virus (MuLV) infected rats. The *ras* gene product is a nucleotide (GTP)-binding protein (p21), which is localized in the cell nucleus (Shih *et al.*, 1980). Five distinct human c-*ras* genes have been molecularly cloned and characterized (Chang *et al.*, 1982; Marshall *et al.*, 1982; Taparowsky *et al.*, 1983; Hall *et al.*, 1983). Three of these (*HRAS1*, *KRAS2*, and *NRAS*) are functional and have been isolated on multiple occasions from human tumors, while the other two (*HRAS2* and *KRAS1*) are intronless pseudogenes, which fail to transform fibroblasts in reconstruction experiments, even when they have been ligated to strong LTR promoters (Chang *et al.*, 1982).

By conventional gene mapping, the five human *ras* genes were found to be dispersed to five distinct human chromosomes (Table XIV and Fig. 3). One of the pseudogenes, *HRAS2*, was found on the X chromosome. When the same specific *ras* probes were used to map the feline homologues of the *ras* gene family, they also dispersed to multiple chromosomes. Preliminary results have indicated that the feline versions of the functional genes were located on the predicted cat chromosomes based upon the homologies presented in Fig. 8 (E. Chang and S. J. O'Brien, in preparation). Further, the feline version of the *HRAS2* pseudogene was detected and this locus was found on the feline X chromosome. Thus, it appears that the transposition of the *HRAS2* pseudogene to the mammalian X chromosome preceded the divergence of the carnivores and the primates over 80 mya.

The development of homologous chromosomal segments between the species described in this review combined with the accumulated genetic and molecular information of the vertebrate oncogenes provide a good opportunity to look for transposition of cellular genes. The divergent radiations of the primate, rodent, and carnivore orders since the Jurassic period has provided over 100 million years of evolutionary opportunity for genic transposition. Fortuitously for such a consideration, the oncogenes are prime candidates for gene transposition because they are the best known targets for retroviral recombination. The best data for such an analysis are derived from the mouse.

At this writing, 12 oncogene homologues have been mapped in the mouse (Table XV). We have already stated that, despite the fact that the murine genome is appreciably shuffled compared to man, a number of short, genetically marked chromosomal segments have been conserved between the two species. Thus, it is possible to predict from the human mouse syntenies (Table VI) the expected conserved position of a murine oncogene homologue. When this was attempted, 8 of the 12 mouse homologues did map to a murine autosomal homologue predicted by the linkage homologies (Table XV). The four oncogenes that did not satisfy

Table XV
Oncogene Positions in Mouse and Man[a]

Locus	Human chromosome	Mouse homologue(s)	Mouse oncogene location	Conserved linkage
raf-1	3	9,16	6	−
myb	6q	9,4	10	−
myc	8q	8	15	−
mos	8q	8	4	−
abl	9q	2	2	+
Ha-*ras*	11p	7,2	7	+
Ki-*ras*	12p	6	6	+
int-1	12	6,10,15	15	+
fes	15	2,7,9	7	+
erb	17	11	11	+
src	20	2	2	+
sis	22	15,16	15	+

[a]References for oncogene mapping in mouse are Kozak (1984), Kozak *et al.*, (1983a,b, 1984), Goff *et al.* (1982), Crews *et al.* (1982), Swan *et al.* (1982) and Sakaguchi *et al.* (1984).

the predictions all mapped in regions of the human genome that were neither within nor very near a murine syntenic segment. In the absence of more markers in these regions to define more precisely the syntenic segments, our conclusion is that there is *no* evidence for functional molecular transposition of the oncogenes during the primate/rodent radiations. Thus, with the exception of pseudogene transpositions that have occurred with at least two oncogenes, *ras* and *raf* (O'Brien *et al.*, 1983b; Bonner *et al.*, 1984), the positions of murine and human oncogene homologues by and large have remained genomically inert, the only movement being consistent with chromosomal segmentation.

10. Conclusions

The cumulative data on comparative cytology, gene mapping, and gene family dispersal in mammals suggest several major generalizations about the character of genome organization. First, the recently developed procedures of chromosome banding (especially at the high-resolution stage) and comparative gene mapping have provided a powerful and lucid perception of the trends and events of genome evolution. Although there are certain disagreements between cytological and syntenic homologies, the most common experience has been the independent confirmation of chromosomal homologies. In several mammalian orders, most notably primates and carnivores, a precise cytological alignment of large numbers of the chromomeres is possible, suggesting a preservation of the integrity of chromosomal morphology since the emergence of these taxa. The syntenic relationships between homologous genes have convincingly reaffirmed the postulated cytological homologies of the great apes and, as such, established a standard for chromosomal homologies within and between mammalian families and, in some cases, between mammalian orders.

A second important observation is that chromosomal evolution in mammals is punctuated in a manner reminiscent of morphological variation. Thus, several phylogenetic lineages are rather conserved with a high degree of chromosomal homology, while other lines are characterized by abundant chromosomal rearrangements. The primates show levels of extreme conservation, with multiple chromosomal homologues being identified between such distant groups as man, New World monkeys, and mouse lemurs. In fact, the conservation of primate cytological integrity extends beyond primates to the carnivores, in which the genome of one species, the domestic cat, shows striking syntenic and cytological homologies with that of man. When several stringent criteria for cytogenetic homology were employed, over 20% of the human genome was aligned band for band with that of the cat. We have interpreted these concordant cytological and syntenic homologies as reflecting ancestral chromosomal arrangements that have been preserved intact within the modern groups that share them.

Conversely, several mammalian families are not at all conserved chromosomally and have chromosomal organizations that are dramatically rearranged or "shuffled" both between and within their respective groups. Within the primates, the gibbon family, Hylobatidae, and the New World monkey *Aotus* are particularly shuffled, with many diverse karyotypes between and within genera when compared to the human map. Several carnivore families (e.g., Canidae and Ursidae) are also karyotypically diverse when compared to conserved Felidae. The rodents are particularly diverse chromosomally and vary greatly between and even within certain species. It is important to note, however, that the "shuffled" rodent genome is not rearranged beyond resolution in a comparative sense. Thus, although cytological homologies between rodents and primates are not apparent, the comparison of the human and mouse syntenic maps does reveal the linkage of short regions of homologous loci between the two species. Thus, the human and mouse maps can be segmented into nearly 200 homologous chromosome segments, which vary in length from 1 to 25 cM. These segments contain groups of homologous loci that are contiguous in both species and, as such, illustrate the limits of cytogenetic divergence that occurred between the conserved human prototype arrangement and the shuffled rodent prototype.

Examination of the chromosomal dispersion of several mammalian gene families studied reveals that at least two levels of evolution have been operative. The present chromosomal locations of several of these families (satellite DNA, Y-chromosome repetitive DNA, and ribosomal DNA) do not fulfill the phylogenetic expectations predicted by chromosomal homologies. These results have been interpreted to reflect the end result of occasional gene amplification and subsequent nonspecific chromosomal transposition of gene family members prior to (or possibly as a part of) speciation. This sort of amplification has precedent in several modern biological species in somatic tissues (e.g., in tumors) or in the germ line (e.g., rDNA amplification of *bobbed* rDNA genes in *Drosophila*) [for review of gene amplification, see Schimke (1982)]. The retroviral gene families add another level of complexity to this discussion because their inherent ability to transpose and infect heterologous species provides an opportunity to introduce precipitously new gene families into new phylogenetic lineages. The role of gene amplification of single-copy genes, as well as of middle repetitive families of chromosomal DNA, is a fascinating area and is only beginning to be interpreted in evolutionary terms.

The second level of evolution of gene families in mammals is the one we are used to. As illustrated by the oncogenes (Section 8), the position of the majority of single-copy structural genes and apparently functional gene families has been transpositionally inert during the mammalian radiations. These linkage arrangements stay together in conserved lineages and sometimes become rearranged in shuffled groups. This, of course, does not preclude their transposition, by secondary mechanisms (e.g., retroviral recombination of spliced cDNA transcripts followed by dispersed chromosomal integration), which can result in intronless genes and often (or always?) pseudogenes. Despite these occasional exceptions, the tightly linked groups of genes have tended to remain together throughout the mammalian radiations. Whether these conserved arrangements are adaptive or simply frozen accidents (Ohno, 1974) is not a simple question to resolve at this stage. It may be sufficient to mention that spontaneous chromosome rearrangements in most species today tend to be deleterious and that there are several explanations for this, including: (1) disruption of functional genes; (2) disruption of functional linkage groups; and (3) abrogation of viable recombination in translocated chromosomes. It is difficult to apportion weight to the extent of damage of any specific chromosome rearrangement that has occurred in our evolutionary past. It is clear, however, that genomic reorganization is a dynamic process, which has influenced species isolation in the past and will continue to affect the patterns of mammalian evolution in the future.

ACKNOWLEDGMENTS. We are grateful to R. Wayne, E. Brownell, and P. Johnson for their critical review of this manuscript.

References

Aaronson, S. A., Hartley, J. W., and Todaro, G. J., 1969, Mouse leukemia virus: "Spontaneous" release by mouse embryo cells after long term *in vitro* cultivation, *Proc. Natl. Acad. Sci. USA* **64**:87–94.

Arnheim, N., Krystal, M., Schmickel, R., Wilson, G., Ryder, O., and Zimmer, E., 1980, Molecular evidence for genetic exchanges among ribosomal genes on nonhomologous chromosomes in man and apes, *Proc. Natl. Acad. Sci. USA* **77**:7323–7327.

Atwood, K. C., Yu, M. T., Johnson, L. D., and Henderson, A. S., 1975, The site of 5S RNA genes in human chromosome 1, *Cytogenet. Cell Genet.* **15**:50–54.

Avise, J. C., and Aquadro, C. F., 1982, A comparative summary of genetic distances in the vertebrates: Problems and correlations, in: *Evolutionary Biology*, Vol. 15 (M. K. Hecht, B. Wallace, and G. T. Prance eds.), Plenum Press, New York, pp. 151–246.

Bailey, D. W., 1971, Recombinant inbred strains: An aid to finding identity, linkage, and function of histocompatibility and other genes, *Transplantation* **11**:325–327.

Beauchamp, R. S., Mitchell, A. R., Buckland, R. A., and Bostock, C., 1979, Specific arrangements of human satellite III DNA sequences in human chromosomes, *Chromosoma* **71**:153–166.

Bedart, M. T., Ma, N. S., and Jones, T. C., 1978, Chromosome banding patterns and nucleolar organizing regions in three species of Callithricidae *(Saguinus oedipus, Saguinus fuscicollis* and *Callithrix jacchus), J. Med. Primatol.* **7**:82–97.

Benveniste, R., 1985, The contributions of retroviruses to the study of mammalian evolution, in: *Molecular Evolutionary Genetics.* (R. J. MacIntyre, ed.), Plenum Press, New York, pp. 359–410.

Benveniste, R. E., and Todaro, G. J., 1974, Evolution of C-type viral genes: Inheritance of exogenously acquired viral genes, *Nature* **252**:456–459.

Benveniste, R. E., and Todaro, G. J., 1976, Evolution of type C viral genes: Evidence for an Asian origin of man, *Nature* **261**:101–108.

Benveniste, R. E., Sherr, C. J., and Todaro, G. J., 1975, Evolution of type C viral genes: Origin of feline leukemia virus, *Science* **190**:886–888.

Bickham, J. W., and Baker, R. J., 1979, Canalization model of chromosomal evolution, *Bull. Carnegie Mus. Nat. Hist.* **13**:70–84.

Birnstiel, M. L., and Grunstein, M., 1972, The ribosomal cistrons of eukaryotes—A model system for the study of evolution of serially repeated genes, *FEBS Symp.* 349–365.

Bishop, J. M., 1983, Cellular oncogenes and retroviruses, *Annu. Rev. Biochem.* **52**:301–354.

Bonner, T., O'Brien, S. J., Nash, W. G., Rapp, U. R., Morton, C. C., and Leder, P, 1984, The human homologs of the *raf* (*mil*) oncogene are located on human chromosomes 3 and 4, *Science* **223**:71–74.

Bostock, C. J., Gosden, J. R., and Mitchell, A. R., 1978, Localization of a male-specific DNA fragment to a sub-region of the human Y chromosome, *Nature* **272**:324–328.

Britten, R. J., and Kohne, D. E., 1968, Repeated sequences in DNA, *Science* **161**:529–540.

Brown, W. M., Prager, E. M., Wang, A., and Wilson, A. C., 1982, Mitochondrial DNA sequences of primates: Tempo and mode of evolution, *J. Mol. Evol.* **18**:225–239.

Bruce, E. J., and Ayala, F. J., 1978, Humans and apes are genetically very similar, *Nature* **276**:264–265.

Bush, G. L., Case, S. M., Wilson, A. C., and Patton, J. L., 1977, Rapid speciation and chromosomal evolution in mammals, *Proc. Natl. Acad. Sci. USA* **74**:3942–3946.

Callahan, R., and Todaro, G. J., 1978, Four major endogenous retrovirus classes each genetically transmitted in various species of *Mus*, in: *Origins of Inbred Mice* (H. C. Morse, III, ed.), Academic Press, New York, pp. 689–713.

Chang, E. H., Gonda, M. A., Ellis, R. W., Scolnick, E. M., and Lowy, D. R., 1982, Human genome contains four genes homologous to transforming genes of Harvey and Kirsten murine sarcoma viruses, *Proc. Natl. Acad. Sci. USA* **79**:4848–4852.

Chiarelli, B., 1961, Chromosomes of the orangutan, *Nature* **192**:228–229.

Chiarelli, B., 1971, Comparative cytogenetics in primates and its relevance for human cytogenetics, in: *Comparative Genetics in Monkeys, Apes and Man* (B. Chiarelli, ed.), Academic Press, New York, pp. 276–304.

Chiarelli, B., Koen, A. L., and Ardito, G., 1979, *Comparative Karyology of the Primates,* Mouton, The Hague.

Cochet, C., Creau-Goldberg, N., Turleau, C., and de Grouchy, J., 1982, Gene mapping of *Microcebus murinus* (Lemuridae): A comparison with man and *Cebus capucinus* (Cebidae), *Cytogenet. Cell Genet.* **33**:213–221.

Coffin, J., 1982, Endogenous viruses, in: *Molecular Biology of Tumor Viruses: RNA Tumor Viruses,* 2nd ed. (R. Weiss, N. Teich, H. Varmus, and J. Coffin, eds.), Cold Spring Harbor Laboratory, Cold Spring Harbor, New York, pp. 1109–1204.

Cohen, S., Fava, R. A., and Sawyer, S. T., 1982, Purification and characterization of epidermal growth factor receptor/protein kinase from normal mouse liver, *Proc. Natl. Acad. Sci. USA* **79**:6237–6241.

Cooke, H. J., 1976, Repeated sequence specific to human males, *Nature* **262**:182–186.

Cooke, H. J., and Hindley, J., 1979, Cloning of human satellite III DNA: Different components are on different chromosomes, *Nucleic Acids Res.* **6**:3177–3197.

Cooke, H. J., and McKay, R. D. G., 1978, Evolution of a human Y chromosome-specific repeated sequence, *Cell* **13**:453–460.

Cooke, H. J., Schmidtke, J., and Gosden, J. R., 1982, Characterization of a human Y chromosome repeated sequence and related sequences in higher primates, *Chromosoma* **87**:491–502.

Cooper, G. M., 1982, Cellular transforming genes, *Science* **217**:801–806.

Couturier, J., Dutrillaux, B., Turleau, C., and de Grouchy, J., 1982, Comparaisons chromosomiques chez quartre espèces on sous-espèces de gibbons, *Ann. Genet.* **25**:5–10.

Cramer, D. V., 1984, Biochemical loci of the rat, in: *Genetic Maps*, Vol. 3, (S. J. O'Brien, ed.), Cold Spring Harbor Press, New York, pp. 370–373.

Creau-Goldberg, N., Cochet, C., Turleau, C., and de Grouchy, J., 1981, Comparative gene mapping of man and *Cebus capucinus*: A study of 23 enzymatic markers, *Cytogenet. Cell Genet.* **31**:228–239.

Creau-Goldberg, N., Turleau, C., Cochet, C., and de Grouchy, J., 1982, New gene assignments in the baboon and new chromosome homologies with man, *Ann. Genet.* **26**:75–78.

Creau-Goldberg, N., Cochet, C., Turleau, C., and de Grouchy, J., 1984, Gene map of the Capuchin monkey (*Cebus capucinus*), in: *Genetic Maps*, Vol. 3 (S. J. O'Brien, ed.), Cold Spring Harbor Press, New York, p. 414.

Crews, S., Barth, R., Hood, L., Prehn, J., and Calame, K., 1982, Mouse c-*myc* is located on chromosome 15 and translocated to chromosome 12 in plasmacytomas, *Science* **218**:1319–1321.

Dawson, W. D., 1984, Genetic map of *Peromyscus*, in: *Genetic Maps*, Vol. 3 (S. J. O'Brien, ed.), Cold Spring Harbor Press, New York, pp. 374–377.

Dawson, W. D., Huang, L. L., Felder, M. R., and Shaffer, J. B., 1983, Linkage relationships among eleven biochemical loci in *Peromyscus*, *Biochem. Genet.* **21**:1101–1114.

D'Eustachio, P., and Ruddle, F. H., 1983, Somatic cell genetics and gene families, *Science* **220**:919–924.

Deuel, T. F., Huang, J. S., Huang, S. S., Stroobant, P., and Waterfield, M. D., 1983, Expression of a platelet-derived growth factor-like protein in simian sarcoma virus transformed cells, *Science* **221**:1348–1350.

Doolittle, R. F., Hunkapiller, M. W., Hood, L. E., Devare, S. G., Robbins, K. C., Aaronson, S. A., and Antoniades, H. N., 1983, Simian sarcoma virus *onc* gene, v-*sis*, is derived from the gene (or genes) encoding a platelet-derived growth factor, *Science* **221**:275–277.

Dover, G., 1982, Molecular drive: A cohesive mode of species evolution, *Nature* **299**:111–117.

Downward, J., Yarden, Y., Mayes, E., Scrace, G., Totty, N., Stockwell, P., Ullrich, A., Schlessinger, J., and Waterfield, M. D., 1984, Close similarity of epidermal growth factor receptor and v-*erb-B* oncogene protein sequences, *Nature* **307**:521–527.

Dutrillaux, B., 1975, *Sur la Nature et l'origine des Chromosome Humaines*, L'Expansion Scientifique, Paris.

Dutrillaux, B., 1979a, Chromosomal evolution in primates: Tentative phylogeny from *Microcebus murinus* (prosimian) to man, *Hum. Genet.* **48**:251–314.

Dutrillaux, B., 1979b, Very large analogy of chromosome banding between *Cebus capucinus* (Platyrrhini) and man, *Cytogenet. Cell Genet.* **24**:84–94.

Dutrillaux, B., and Couturier, J., 1981, The ancestral karyotype of platyrrhine monkeys, *Cytogenet. Cell Genet.* **30**:232–242.

Eicher, E. M., 1981, Foundation for the future: Formal genetics of the mouse, in: *Mammalian Genetics and Cancer: The Jackson Laboratory Fiftieth Anniversary Symposium, Progress in Clinical and Biological Research,* Vol. 45 (E. S. Russell, ed.), Alan R. Liss, New York, pp. 7–49.

Eisenberg, J. F., 1981, *The Mammalian Radiations. An Analysis of Trends in Evolution, Adaptation and Behavior,* University of Chicago Press, Chicago.

Estop, A., Garver, J. J., Meera-Khan, P., and Pearson, P. L., 1979, Rhesus-human chromosome homologies via cytogenetic and gene mapping studies, Human Gene Mapping 5 (1979), *Cytogenet. Cell Genet.* **25:**150–151.

Evans, H. J., Gosden, J. R., Mitchell, A. R., and Buckland, R. A., 1974a, Localization of human satellite DNAs on the Y chromosome, *Nature* **251:**346–347.

Evans, H. J., Buckland, R. A., and Pardue, M. L., 1974b, Location of the genes coding for 18S and 28S ribosomal RNA in the human genome, *Chromosoma* **48:**405–426.

Ewer, R. F. (ed.), 1973, *The Carnivores,* Cornell University Press, Ithaca, New York.

Ferris, S. D., Brown, W. M., Davidson, W. S., and Wilson, A. C., 1981a, Extensive polymorphism in the mitochondrial DNA of apes, *Proc. Natl. Acad. Sci. USA* **78:**6319–6323.

Ferris, S. D., Wilson, A. C., and Brown, W. M., 1981b, Evolutionary tree for apes and humans based on cleavage maps of mitochondrial DNA, *Proc. Natl. Acad. Sci. USA* **78:**2432–2436.

Finaz, C., Van Cong, N., Cochet, C., Frezal, J., and de Grouchy, J., 1977, Histoire naturelle du chromosome 1 chez les primates, *Ann. Genet.* **20:**85–92.

Finaz, C., Cochet, C., and de Grouchy, J., 1978, Identité des caryotypes de *Papio papio* et *Macaca mulatta* en bandes R, G, C et Ag-NOR, *Ann. Genet.* **21:**149–151.

Ford, C. E., and Hamerton, J. L., 1956, The chromosomes of man, *Nature* **178:**1020–1023.

Francke, U., and Oliver, N., 1978, Quantitative analysis of high-resolution trypsin-Giemsa bands on human prometaphase chromosomes, *Hum. Genet.* **45:**137–165.

Francke, U., and Taggart, R. T., 1980, Comparative gene mapping: Order of loci on the X chromosome is different in mice and humans, *Proc. Natl. Acad. Sci. USA* **77:**3595–3599.

Freitas, L., and Seuanez, H. N., 1982, Chromosome heteromorphisms in *Cebus apella, J. Hum. Evol.* **10:**173–180.

Frisby, D. P., Weiss, R. A., Roussel, M., and Stehelin, D., 1979, The distribution of endogenous chicken retrovirus sequences in the DNA of galliform birds does not coincide with avian phylogenetic relationships, *Cell* **17:**623–634.

Garver, J. J., Estop, A. M., Meera-Khan, P., Balner, H., and Pearson, P. L., 1980, Evidence of similar organization of the chromosomes carrying the major histocompatibility complex in man and other primates, *Cytogenet. Cell Genet.* **27:**238–245.

Gellin, J., Echard, G., Benne, F., and Gillois, M., 1981, Pig gene mapping: *PKM2-MPI-NP* synteny, *Cytogenet. Cell Genet.* **30:**59–62.

Giusto, J. P., and Margulis, L., 1981, Karyotypic fission theory and the evolution of Old World monkeys and apes, *Biosystems* **13:**267–302.

Goff, S. P., D'Eustachio, P., Ruddle, F. H., and Baltimore, D., 1982, Chromosomal assignment of the endogenous proto-oncogene c-*abl, Science* **218:**1317–1319.

Goldschmidt, R. (ed.), 1940, *The Material Basis of Evolution,* Yale University Press, New Haven, Connecticut.

Goodman, M., 1975, Protein sequence and immunologic specificity: Their role in phylogenetic studies in the primates, in: *Phylogeny of the Primates* (W. P. Luckett and J. S. Szalay, eds.), Plenum Press, New York, pp. 219–248.

Goodman, M., and Tashian, R. E. (eds.), 1976, *Molecular Anthropology,* Plenum Press, New York.

Goodpasture, C., and Bloom, S. E., 1975, Visualization of nucleolar organizer regions in mammalian chromosomes using silver staining, *Chromosoma* **53:**37–50.

Gosden, J. R., Mitchell, A. R., Buckland, R. A., Clayton, R. P., and Evans, H. J., 1975, The location of four satellite DNAs on human chromosomes, *Exp. Cell Res.* **92:**148–158.

Gosden, J. R., Mitchell, A. R., Seuanez, H. N., and Gosden, C. M., 1977, The distribution of sequences complementary to human satellite DNAs I, II and IV in the chromosomes of chimpanzee (*Pan troglodytes*), gorilla (*Gorilla gorilla*), and the orangutan (*Pongo pygmaeus*), *Chromosoma* **63:**253–271.

Gosden, J. R., Lawrie, S., and Seuanez, H. N., 1978, Ribosomal and human-homologous repeated DNA distribution in the orangutan (*Pongo pygmaeus*). Comparison with distribution of these DNAs in the other species of the Hominidae, *Cytogenet. Cell Genet.* **21:**1–10.

Gould, S. J., 1980, Is a new and general theory of evolution emerging?, *Paleobiology* **6:**119–130.

Gould, S. J., and Eldridge, N., 1977, Punctuated equilibria: The tempo and mode of evolution reconsidered, *Paleobiology* **3:**115–151.

Gribbin, J., and Cherfas, J., 1982, *The Monkey Puzzle. Reshaping the Evolutionary Tree*, Pantheon, New York.

Hall, A., Marshall, C. J., Spurr, N. K., and Weiss, R. A., 1983, Identification of transforming gene in two human sarcoma cell lines as a new member of the *ras* gene family located on chromosome 1, *Nature* **303:**396–400.

Hamerton, J. L., Klinger, H. P., Mutton, D. E., and Langue, E. M., 1963, The somatic chromosomes of the Hominoidea, *Cytogenetics* **2:**240–263.

Harper, M. E., Ullrich, A., and Saunders, G. F., 1981, Localization of the human insulin gene to the distal end of the short arm of chromosome 11, *Proc. Natl. Acad. Sci. USA* **78:**4458–4460.

Hatch, F. T., Bodner, A. J., Mazrimas, J. A., and Moore, D. H., 1976, Satellite DNA and cytogenetic evolution: DNA quantity, satellite DNA and karyotypic variations in kangaroo rats (genus *Dipodomys*), *Chromosoma* **58:**155–168.

Henderson, A. S., Warburton, D., and Atwood, K. C., 1972, Location of ribosomal DNA in the human chromosome complement, *Proc. Natl. Acad. Sci. USA* **69:**3394–3398.

Henderson, A. S., Warburton, D., and Atwood, K. C., 1974a, Localization of rDNA in the chromosome complement of the rhesus (*Macaca mulatta*), *Chromosoma* **44:**367–370.

Henderson, A. S., Warburton, D., and Atwood, K. C., 1974b, Localizing of rDNA in the chimpanzee (*Pan troglodytes*) chromosome complement, *Chromosoma* **46:**435–441.

Henderson, A. S., Atwood, K. C., and Warburton, D., 1976a, Chromosomal distribution of rDNA in *Pan paniscus, Gorilla gorilla beringei* and *Symphalangus syndactylus:* Comparison to related primates, *Chromosoma* **59:**147–155.

Henderson, A. S., Atwood, K. C., Yu, M. T., and Warburton, D., 1976b, The site of 5S RNA genes in primates. I. The great apes, *Chromosoma* **56:**29–32.

Henderson, A. S., Warburton, D., Megraw-Ripley, S., and Atwood, K. C., 1977, The chromosomal location of rDNA in selected lower primates, *Cytogenet. Cell Genet.* **19:**281–302.

Hoffman, F. M., Fresco, L. D., Hoffman-Falk, H., and Shilo, B.-Z., 1983, Nucleotide sequences of the *Drosophila src* and *abl* homologs: Conservation and variability in the *src* family oncogenes, *Cell* **35:**393–401.

Human Gene Mapping-7, 1984, Los Angeles Conference 1983, *Cytogenet. Cell Genet.* **37:**1–666.

ISCN, 1978, An international system for human cytogenetic nomenclature, Report of the Standing Committee on Human Cytogenetic Nomenclature, *Cytogenet. Cell Genet.* **21:**309–404.

Jones, K. W., Prosser, J., Corneo, G., and Ginelli, E., 1973a, The chromosomal localization of human satellite III, *Chromosoma* **42:**445–451.

Jones, K. W., Prosser, J., Corneo, G., Ginelli, E., and Bobrow, M., 1973b, Satellite DNA, constitutive heterochromatin and human evolution, in: *Modern Aspects of Cytogenetics: Constitutive Heterochromatin in Man* (R. A. Pfeiffer, ed.), Shaltauer, New York, pp. 45–61.

Jones, K. W., Purdom, I. F., Prosser, J., and Corneo, G., 1974, The chromosomal location of human satellite DNA I, *Chromosoma* **49:**161–171.

King, M. C., and Wilson, A. C., 1975, Evolution at two levels in humans and chimpanzees, *Science* **188:**107–116.

Klein, G. (ed.), 1981, *Cell Derived Oncogenes,* Vol. I, Raven Press, New York.

Kozak, C. A., 1984, Retroviral and cancer associated loci of the mouse *Mus musculus,* in: *Genetic Maps,* Vol. 3 (S. J. O'Brien, ed.), Cold Spring Harbor Press, New York, pp. 356–359.

Kozak, C. A., Sears, J. F., and Hoggan, M. D., 1983a, Genetic mapping of the mouse oncogenes c-Ha-*ras*-1 and c-*fes* to chromosome 7, *J. Virol.* **47:**217–220.

Kozak, C. A., Sears, J. F., and Hoggan, M. D., 1983b, Genetic mapping of a mouse protooncogene c-*sis* to chromosome 15, *Science* **221:**867–869.

Kozak, C. A., Gunnell, M. A., and Rapp, U. R., 1984, A new oncogene, c-*raf,* is located on mouse chromosome 6, *J. Virol.* **49:**297–299.

Krystal, M., D'Eustachio, P., Ruddle, F. H., and Arnheim, N., 1981, Human nucleolus organizers on nonhomologous chromosomes can share the same ribosomal gene variants, *Proc. Natl. Acad. Sci. USA* **78:**5744–5748.

Kunkel, L. M., Smith, K. D., and Boyer, S. H., 1976, Human Y-chromosome-specific reiterated DNA, *Science* **191:**1189–1190.

Kunkel, L. M., Smith, K. D., Boyer, S. H., Borgaonkar, D. S., Wachtel, S. S., Miller, O. J., Breg, W. R., Jones, H. W., and Rary, J. M., 1977, Analysis of human Y-chromosome-specific reiterated DNA in chromosome variants, *Proc. Natl. Acad. Sci. USA* **74:**1245–1249.

Lalley, P. A., Minna, J. D., and Francke, U., 1978, Conservation of autosomal gene synteny groups in mice and man, *Nature* **274:**160–163.

Lalley, P. A., Dixon, D., and Held, M. B., 1979, Gene mapping studies in the baboon (*Papio papio*) using somatic cell hybrids, Human Gene Mapping 5 (1979), *Cytogenet. Cell Genet.* **25:**178.

Lau, Y. F., and Arrighi, F. E., 1976, Studies of the squirrel monkey, *Saimiri sciureus,* genome. I. Cytological characterization of chromosomal heterozygosity, *Cytogenet. Cell Genet.* **17:**51–60.

Leder, A., Swan, D., Ruddle, F. H., D'Eustachio, P., and Leder, P., 1981, Dispersion of α-like globin genes of the mouse to three different chromosomes, *Nature* **293:**196–200.

Liming, S., Yingying, Y., and Xingsheng, D., 1980, Comparative cytogenetic studies on the red muntjac, Chinese muntjac and their F_1 hybrids, *Cytogenet. Cell Genet.* **26:**22–27.

Lorincz, A. T., and Reed, S. I., 1984, Primary structure homology between the product of yeast cell division control gene CDC28 and vertebrate oncogenes, *Nature* **307:**183–185.

Lowy, D. R., Rowe, W. P., Teich, N., and Hartley, J. W., 1971, Murine leukemia virus: High frequency activation *in vitro* by 5-iododeoxyuridine and 5-bromodeoxyuridine, *Science* **174:**155–157.

Lueders, K., Leder, A., Leder, P., and Kuff, E., 1982, Association between a transposed α-globin pseudogene and retrovirus-like elements in the BALB/c mouse genome, *Nature* **295:**426–428.

Ma, N. S. F., 1983a, Gene map of the New World Bolivian owl monkey, *Aotus, J. Hered.* **74**:27–33.

Ma, N. S. F., 1983b, Comparative gene mapping of the owl monkey, *Aotus*, with karyotypes V ($2n = 46$) and VI ($2n = 49,50$), *Cytogenet. Cell Genet.* **35**:117–123.

Ma, N. S. F., 1984, Linkage and syntenic relationship in chromosomes of owl monkeys (with karyotypes I, III, V, VI, VII) homologous to man, in: *Genetic Maps*, Vol. 3 (S. J. O'Brien, ed.), Cold Spring Harbor Press, New York, pp. 410–413.

Ma, N. S. F., Jones, T. C., Miller, A. C., Morgan, L. M., and Adams, E. A., 1976, Chromosome polymorphism and banding patterns in the owl monkey (*Aotus*), *Lab. Anim. Sci.* **26**:1022–1036.

Ma, N. S. F., Simeone, T., McLean, J., and Parham, P., 1982, Chromosome localization and gene synteny of the major histocompatibility complex in the owl monkey, *Aotus, Immunogenetics* **15**:1–16.

Marshall, C. J., Hall, A., and Weiss, R. A., 1982, A transforming gene present in human sarcoma cell lines, *Nature* **299**:171–173.

Martin, M. A., Bryan, T., Rasheed, S., and Khan, A. S., 1981, Identification and cloning of endogenous retroviral sequences present in human DNA, *Proc. Natl. Acad. Sci. USA* **78**:4892–4896.

Mascarello, J. T., Stock, A. D., and Pathak, S., 1974, Conservatism in the arrangement of genetic material in rodents, *J. Mammal.* **55**:695–704.

McClure, H., Belden, K. H., and Pieper, W. A., 1969, Autosomal trisomy in a chimpanzee: Resemblance to Down's syndrome, *Science* **65**:1010–1012.

McKay, R. D. G., Bobrow, M., and Cooke, H. J., 1978, The identification of a repeated DNA sequence involved in the karyotype polymorphism of the human Y chromosome, *Cytogenet. Cell Genet.* **21**:19–32.

McKusick, V. A., 1984, The human gene map, in: *Genetic Maps*, Vol. 3 (S. J. O'Brien, ed.), Cold Spring Harbor Press, New York, pp. 417–441.

McKusick, V. A., and Ruddle, F. H., 1977, The status of the gene map of human chromosomes, *Science* **196**:390–405.

Meera-Khan, P., Brahe, C., and Wijnen, L. M. M., 1984, Biochemical loci of the dog, *Canis familiaris*, in: *Genetic Maps*, Vol. 3 (S. J. O'Brien, ed.), Cold Spring Harbor Press, New York, pp. 388–389.

Miller, C. K., Miller, D. A., Miller, O. J., Tantravahi, R., and Reese, R. T., 1977, Banded chromosomes of the owl monkey, *Aotus trivirgatus, Cytogenet. Cell Genet.* **19**:215–226.

Miller, D. A., 1977, Evolution of primate chromosomes, *Science* **198**:1116–1124.

Minna, J. D., Marshall, T. H., and Shaffer-Berman, P. V., 1975, Chinese hamster × mouse hybrid cells segregating mouse chromosomes and isozymes, *Somat. Cell Genet.* **1**:355–369.

Mitchell, A. R., Seuanez, H. N., Lawrie, S. S., Martin, D. E., and Gosden, J. R., 1977, The location of DNA homologous to human satellite III DNA in the chromosomes of the chimpanzee (*Pan troglodytes*), gorilla (*Gorilla gorilla*), and orangutan (*Pongo pygmaeus*), *Chromosoma* **61**:345–358.

Mitchell, A. R., Beauchamp, R. S., and Bostock, C. J., 1979, A study of sequence homologies in four satellite DNAs of man, *J. Mol. Biol.* **135**:127–149.

Mitchell, A. R., Gosden, J. R., and Ryder, O. A., 1981, Satellite DNA relationships in man and the primates, *Nucleic Acids Res.* **9**:3235–3249.

Myers, R. H., and Shafer, D. A., 1979, Hybrid ape offspring of a mating of gibbon and siamang, *Science* **205**:308–310.

Nadeau, J. H., and Taylor, B. A., 1984, Lengths of chromosomal segments conserved since the divergence of man and mouse, *Proc. Natl. Acad. Sci. USA* **81**:814–818.

Nash, W. G., and O'Brien, S. J., 1982, Conserved subregions of homologous G-banded chromosomes between orders in mammalian evolution: Carnivores and primates, *Proc. Natl. Acad. Sci. USA* **79**:6631–6635.

Nelkin, B., Strayer, D., and Vogelstein, B., 1980, Divergence of primate ribosomal RNA genes as assayed by restriction enzyme analysis, *Gene* **11**:89–96.

Nesbitt, M. N., 1974, Evolutionary relationships between rat and mouse chromosomes, *Chromosoma* **46**:217–224.

Nowak, R. M., and Paradiso, J. L., 1983, *Walker's Mammals of the World*, 4th ed., Johns Hopkins University Press, Baltimore.

O'Brien, S. J., (ed.), 1984a, *Genetic Maps*, Vol. 3, Cold Spring Harbor Press, New York.

O'Brien, S. J., 1984b, The genetic map of the domestic cat (*Felis catus*), in: *Genetic Maps*, Vol. 3 (S. J. O'Brien, ed.), Cold Spring Harbor Press, New York, pp. 385–387.

O'Brien, S. J., 1984c, The human map of oncogenes and retrovirus loci, in: *Genetic Maps*, Vol. 3 (S. J. O'Brien, ed.), Cold Spring Harbor Press, New York, pp. 451–456.

O'Brien, S. J., and Nash, W. G., 1982, Genetic mapping in mammals: Chromosome map of domestic cat, *Science* **216**:257–265.

O'Brien, S. J., Simonson, J. M., and Eichelberger, M. A., 1982, Genetic analysis of hybrid cells using isozyme markers as monitors of chromosome segregation, in: *Techniques in Somatic Cell Genetics* (J. W. Shay, ed.), Plenum Press, New York, pp. 513–524.

O'Brien, S. J., Bonner, T. I., Cohen, M., O'Connell, C., and Nash, W. G., 1983a, Mapping of an endogenous retroviral sequence to human chromosome 18, *Nature* **303**:74–77.

O'Brien, S. J., Nash, W. G., Goodwin, J. L., Lowy, D. R., and Chang, E. H., 1983b, Dispersion of the *ras* family of transforming genes to four different chromosomes in man, *Nature* **302**:839–842.

Ohno, S., 1973, Ancient linkage groups and frozen accidents, *Nature* **244**:259–262.

Ohno, S., Becak, W., and Becak, M. L., 1964, X-autosome ratio and the behavior pattern of individual X-chromosomes in placental mammals, *Chromosoma* **15**:14–16.

Painter, T. S., 1923, Studies in mammalian spermatogenesis. II. The spermatogenesis of man, *J. Exp. Zool.* **37**:291–336.

Paris Conference 1971, Supplement, 1975, Standardization in human cytogenetics, *Birth Defects* **8**(7):1–46.

Patton, J. L., 1970, Karyotypic variation following an elevational gradient in the pocket gopher, *Thonomys bottae grahamensis* Goldman, *Chromosoma* **31**:41–50.

Pearson, P. L., Roderick, T. H., Davisson, M. T., Garver, J. J., Warburton, D., Lalley, P.A., and O'Brien, S. J., 1979, Comparative mapping; report of the International Committee, *Cytogenet. Cell Genet.* **25**:82–95.

Reeves, R. H., and O'Brien, S. J., 1984, Molecular genetic characterization of the RD-114 family of endogenous feline retroviral sequences, *J. Virol.* **52**:164–171.

Reeves, R., Nash, W. G., and O'Brien, S. J., 1985, Genetic mapping of endogenous RD-114 retroviral sequences of the domestic cat, *J. Virol.* (in press).

Repaske, R., O'Neill, R. R., Steele, P. E., and Martin, M. A., 1983, Characterization and partial nucleotide sequence of endogenous type C retrovirus segments in human chromosomal DNA, *Proc. Natl. Acad. Sci. USA* **80**:678–682.

Robbins, K. C., Antoniades, H. N., Devare, S. G., Hunkapiller, M. W., and Aaronson, S. A., 1983, Structural and immunological similarities between simian sarcoma virus gene product(s) and human platelet-derived growth factor, *Nature* **305**:605–608.

Robinson, R., 1964, Linkage in *Peromyscus*, *Heredity* **19**:701–709.

Roderick, T. H. and Davisson, M. T., 1984, Linkage map of the mouse (*Mus musculus*), in: *Genetic Maps*, Vol. 3 (S. J. O'Brien, ed.), Cold Spring Harbor Press, New York, pp. 343–355.

Roderick, T. H., Lalley, P. A., Davisson, M. T., O'Brien, S. J., Womack, J. E., Creau-

Goldberg, N., Echard, G., and Moore, K. L., 1984, Comparative genetic mapping in mammals: Report of the International Committee, *Cytogenet. Cell Genet.* **37**:312–339.
Rogers, D. S., Greenbaum, I. F., Gunn, S. J., and Engstrom, M. D., 1984, Cytosystematic value of chromosomal inversion data in the genus *Peromyscus* (Rodentia : Cricetidae), *J. Mammal.*, **65**:457–465.
Rose, M. R., and Doolittle, W. F., 1983, Molecular biological mechanisms of speciation, *Science* **220**:157–162.
Ruddle, F. H., and Creagan, R. P., 1975, Parasexual approaches to the genetics of man, *Annu. Rev. Genet.* **9**:407–486.
Ruddle, F. H., and Nichols, E. A., 1971, Starch gel electrophoretic phenotypes of mouse-human somatic cell hybrids in mouse isozyme polymorphisms, *In Vitro* **7**:120–131.
Ruddle, F. H., Chapman, V. M., Ricciuti, F., Murnane, M., Klebe, R., and Meera-Khan, P., 1971, Linkage relationships of seventeen human gene loci as determined by man-mouse somatic cell hybrids, *Nature* **232**:69–73.
Ryder, O. A., Epel, N. C., and Benirschke, K., 1978, Chromosome banding studies of the Equidae, *Cytogenet. Cell Genet.* **20**:332–350.
Sakaguchi, A. Y., Lalley, P. A., Zabel, B. U., Ellis, R. W., Scolnick, E. M., and Naylor, S. C., 1984, Chromosomal assignments of four mouse cellular homologs of sarcoma and leukemia virus oncogenes, *Proc. Natl. Acad. Sci. USA* **81**:525–529.
Schimke, R. T., 1982, *Gene Amplification,* Cold Spring Harbor Laboratory, Cold Spring Harbor, New York.
Schmeckpeper, B. J., Smith, K. D., Dorman, B. P., Ruddle, F. H., and Talbot, C. C., Jr., 1979, Partial purification and characterization of DNA from the human X chromosome, *Proc. Natl. Acad. Sci. USA* **76**:6525–6528.
Schmidtke, J., and Schmid, M., 1980, Regional assignment of a 2.1-kb repetitive sequence to the distal part of the human Y heterochromatin, *Hum. Genet.* **55**:255–257.
Seigel, L. J., Harper, M. E., Wong-Staal, F., Gallo, R. C., Nash, W. G., and O'Brien, S. J., 1984, Gene for T-cell growth factor: Location on human chromosome 4q and feline chromosome B1, *Science* **223**:175–178.
Seuanez, H. N., 1979, *The Phylogeny of Human Chromosomes,* Springer-Vertag, Berlin.
Seuanez, H. N., 1984, Evolutionary aspects of human chromosomes, *Subcell. Biochem.* **10**:453–535.
Shay, J. W. (ed.), 1982, *Techniques in Somatic Cell Genetics,* Plenum Press, New York.
Shih, T. Y., Papageorge, A. G., Stokes, P. E., Weeks, M. O., and Scolnick, E. M., 1980, Guanine nucleotide-binding and autophosphorylating activities associated with p21^{src} protein of Harvey murine sarcoma virus, *Nature* **287**:686–691.
Shows, T. B., Sakaguchi, A. Y., and Naylor, S. L., 1982, Mapping the human genome, cloned genes, DNA polymorphisms and inherited disease, *Adv. Hum. Genet.* **12**:341–452.
Shows, T. B., McAlpine, P. J., and Miller, R. L., 1984, The 1983 catalogue of mapped human genetic markers and report of the Nomenclature Committee, *Cytogenet. Cell Genet.* **37**:340–397.
Simpson, G. G., 1945, The principles of classification and classification of the mammals, *Bull. Am. Mus. Nat. Hist.* **85**:1–350.
Snyder, L. R., 1980, Evolutionary conservation of linkage groups: Additional evidence from murid and cricetid rodents, *Biochem. Genet.* **18**:209–220.
Soulie, J., and de Grouchy, J., 1982, Of rabbit and man: Comparative gene mapping, *Hum. Genet.* **60**:172–175.
Southern, E. M., 1975, Detection of specific sequences among DNA fragments separated by gel electrophoresis, *J. Mol. Biol.* **98**:503–517.
Stallings, R. L., Adair, G. M., and Siciliano, M. J., 1984, Gene map of the Chinese hamster,

in: *Genetic Maps*, Vol. 3 (S. J. O'Brien, ed.), Cold Spring Harbor Press, New York, pp. 378–380.

Stanley, S. (ed.), 1979, *Macroevolution: Patterns and Processes*, Freeman, San Francisco.

Stanyon, R., and Chiarelli, B., 1982, Phylogeny of the Hominoidea: The chromosome evidence, *J. Hum. Evol.* **11**:493–504.

Stebbins, G. L., and Ayala, F. J., 1981, Is the new evolutionary synthesis necessary? *Science* **213**:967–971.

Steffensen, D. M., Prenski, W., and Dufy, P., 1974, Localization of the 5S ribosomal RNA genes in the human genome, Human Gene Mapping 1 (1973), *Birth Defects Orig. Artic. Ser.* **10**:153–154.

Stephenson, J. R. (ed.), 1980, *Molecular Biology of RNA Tumor Viruses*, Academic Press, New York.

Stockholm Conference, 1977, An International System for Human Cytogenetic Nomenclature, 1978 I.S.C.N., *Cytogenet. Cell Genet.* **21**:309–409.

Sun, N. C., Sun, C. R. Y., and Ho, T., 1978a, Chimpanzee chromosome 12 is homologous to human chromosome arm 2q, *Cytogenet. Cell Genet.* **22**:594–597.

Sun, N. C., Sun, C. R. Y., and Ho, T., 1978b, Chimpanzee chromosome 13 is homologous to human chromosome arm 2p, *Cytogenet. Cell Genet.* **22**:598–601.

Swan, D., Oskarsson, M., Keithley, D., Ruddle, F. H., D'Eustachio, P., and Vande Woude, G. F., 1982, Chromosomal localization of the Moloney sarcoma virus mouse cellular (c-*mos*) sequence, *J. Virol.* **44**:752–754.

Szabo, P., Kunkel, L., Hy, L. C., George, D., and Smith, K. K., 1979, Chromosomal distribution of DNA sequences derived from the human Y chromosome in human—higher primates, *Cytogenet. Cell Genet.* **25**:212–213.

Tantravahi, R., Dev, V. G., Firschein, I. L., Miller, D. A., and Miller, O. J., 1975, Karyotype of the gibbons, *Hylobates lar* and *H. moloch* inversion in chromosome No. 7, *Cytogenet. Cell Genet.* **15**:92–102.

Taparowsky, E., Shimizu, K., Goldfarb, M., and Wigler, M., 1983, Structure and activation of the human N-*ras* gene, *Cell* **34**:581–586.

Taylor, B. A., 1978, Recombinant inbred strains: Use in gene mapping, in: *Origins of Inbred Mice* (H. C. Morse, ed.), Academic Press, New York, pp. 423–438.

Temin, H. M., 1982, Function of the retrovirus long terminal repeat, *Cell* **28**:3–5.

Thorpe, J. P., 1982, The molecular clock hypothesis: Biochemical evolution, genetic differentiation and systematics, *Annu. Rev. Ecol. Syst.* **13**:139–168.

Tjio, H., and Levan, A., 1956, The chromosome number of man, *Hereditas* **42**:1–6.

Todd, N. B., 1970, Karyotypic fissioning and canid phylogeny, *J. Theor. Biol.* **26**:445–480.

Turleau, C., de Grouchy, J., and Klein, M., 1972, Phylogénie chromosomique de l'homme et des primates hominiens (*Pan troglodytes, Gorilla gorilla* et *Pongo pygmaeus*): Essai de reconstitution du caryotipe de l'ancêtre comun, *Ann. Genet.* **15**:225–240.

Turleau, C., Creau-Goldberg, N., Cochet, C., and de Grouchy, J., 1983, Gene mapping of the gibbon. Its position in primate evolution, *Hum. Genet.* **64**:65–72.

Warburton, D., Henderson, A. S., and Atwood, K. C., 1975, Localization of rDNA and Giemsa-banded chromosome complement of white-handed gibbon, *Hylobates lar, Chromosoma* **51**:35–40.

Warburton, D., Yu, M. T., Atwood, K. C., and Henderson, A. S., 1976, The location of 5S RNA genes on the chromosomes of primates, Human Gene Mapping 3 (1975), *Birth Defects Orig. Artic. Ser.* **12**:440–442.

Waterfield, M. D., Scrace, G. T., Whittle, N., Stroobant, P., Johnsson, A., Wasteson, A., Westermark, B., Heldin, C.-H., Huang, J. S., and Deuel, T. F., 1983, Platelet-derived growth factor is structurally related to the putative transforming protein p28^{sis} of simian sarcoma virus, *Nature* **304**:35–39.

Wayne, R., Festing, M., and O'Brien, S. J., 1985, Empirical demonstration that structural gene and morphometric variation are uncoupled (submitted).

Weinberg, R. A., 1982, Oncogenes of spontaneous and chemically induced tumors, *Adv. Cancer Res.* **36**:149–163.

Weiss, R., Teich, N., Varmus, H., and Coffin, J. (eds.), 1982, *RNA Tumor Viruses*, 2nd ed., Cold Spring Harbor Press, New York.

Wilson, A. C., Bush, G. L., Case, S. M., and King, M. C., 1975, Social structuring of mammalian populations and rate of chromosomal evolution, *Proc. Natl. Acad. Sci. USA* **12**:5061–5065.

Wilson, A. C., Carlson, S. S., and White, T. J., 1977, Biochemical evolution, *Annu. Rev. Biochem.* **46**:573–639.

Womack, J. E., 1982, Linkage of mammalian isozyme loci: A comparative approach, in: *Isozymes: Current Topics in Biological and Medical Research,* Vol. 6 (M. C. Rattazzi, J. G. Scandalios, and G. S. Whitt, eds.), Alan R. Liss, New York, pp. 207–246.

Womack, J. E., 1984, Biochemical loci of the mouse (*Mus musculus*), in: *Genetic Maps,* Vol. 3 (S. J. O'Brien, ed.), Cold Spring Harbor Press, New York, pp. 360–369.

Womack, J. E., and Sharp, M., 1976, Comparative autosomal linkage in mammals: Genetics of esterases in *Mus musculus* and *Rattus norvegicus*, *Genetics* **82**:665–675.

Wurster, D. H., and Benirschke, K., 1970, Indian muntjac, *Muntiacus muntjak:* A deer with a low diploid chromosome number, *Science* **168**:1364–1366.

Wurster-Hill, D. H., and Bush, M., 1980, The interrelationship of chromosome banding patterns in the giant panda (*Ailuropoda melanoleuca*), hybrid bear (*Ursus middendorfi* × *Thalarctos maritimus*), and other carnivores, *Cytogenet. Cell Genet.* **27**:147–154.

Wurster-Hill, D. H., and Centerwall, W. R., 1982, The interrelationships of chromosome banding patterns in canids, mustelids, hyena and felids, *Cytogenet. Cell Genet.* **34**:178–192.

Wurster-Hill, D. H., and Gray, C. W., 1973, Giemsa banding patterns in the chromosomes of twelve species of cats (Felidae), *Cytogenet. Cell Genet.* **12**:377–397.

Wurster-Hill, D. H., and Gray, C. W., 1975, The interrelationships of chromosome banding patterns in procyonids, viverrids, and felids, *Cytogenet. Cell Genet.* **15**:306–331.

Yeager, C. H., Painter, T. S., and Yerkes, R. M., 1940, The chromosomes of the chimpanzee, *Science* **91**:74–75.

Yoshida, M. C., 1983a, GPI, LDHA, and PEPD are syntenic and assigned to rat chromosome 7, *Cytogenet. Cell Genet.* **37**:613 (abstract).

Yoshida, M. C., 1983b, Assignment of *GDH* and *FUCA* to rat chromosome 5: Evidence for homologous linkage groups in rat, mouse, and man, *Cytogenet. Cell Genet.* **37**:613 (abstract).

Young, W. J., Merz, T., Ferguson-Smith, M. A., and Johnston, A. W., 1960, Chromosome number of the chimpanzee *(Pan troglodytes)*, *Science* **131**:1672–1673.

Yunis, J. J., and Prakash, O., 1982, The origin of man: A chromosomal pictorial legacy, *Science* **215**:1525–1529.

Yunis, J. J., Sawyer, J. R., and Dunham, K., 1980, The striking resemblance of high-resolution G-banded chromosomes of man and chimpanzee, *Science* **208**:1145–1148.

Index

GPSR Compliance

The European Union's (EU) General Product Safety Regulation (GPSR) is a set of rules that requires consumer products to be safe and our obligations to ensure this.

If you have any concerns about our products, you can contact us on ProductSafety@springernature.com

In case Publisher is established outside the EU, the EU authorized representative is:

Springer Nature Customer Service Center GmbH
Europaplatz 3
69115 Heidelberg, Germany

Batch number: 10837688

Printed by Printforce, the Netherlands